T0201195

Routes to Essential Medicines

A Workbook for Organic Synthesis

Peter J. Harrington
Better Pharma Processes, LLC
Louisville, CO, USA

WILEY

This edition first published 2022

Registered Office
John Wiley & Sons, Inc., 111 River Street, Hoboken, NJ 07030, USA

Editorial Office
111 River Street, Hoboken, NJ 07030, USA

For details of our global editorial offices, customer services, and more information about Wiley products visit us at www.wiley.com.

Wiley also publishes its books in a variety of electronic formats and by print-on-demand. Some content that appears in standard print versions of this book may not be available in other formats.

Library of Congress Cataloging-in-Publication Data

Names: Harrington, Peter J., author.
Title: Routes to essential medicines : a workbook for organic synthesis /
Peter J. Harrington, Better Pharma Processes, LLC, Louisville, CO, USA.
Description: First edition. | Hoboken, NJ : Wiley, 2022. | Includes index.
Identifiers: LCCN 2021043883 (print) | LCCN 2021043884 (ebook) | ISBN
9781119722861 (paperback) | ISBN 9781119722878 (adobe pdf) | ISBN
9781119722830 (epub)
Subjects: LCSH: Pharmaceutical chemistry. | Drugs–Synthesis.
Classification: LCC RS403 .H372 2022 (print) | LCC RS403 (ebook) | DDC
615.1/9–dc23/eng/20211004
LC record available at https://lccn.loc.gov/2021043883
LC ebook record available at https://lccn.loc.gov/2021043884

Cover Design: Wiley
Cover Image: Created by and courtesy of Chloe Schaub

Set in 9.5/12.5pt STIXTwoText by Straive, Pondicherry, India

SKY10031137_110321

Routes to Essential Medicines

This workbook is dedicated to everyone who made it their mission in life to discover and manufacture Essential Medicines and to:
Professor Louis S. Hegedus

Who inspired this workbook project with a lesson learned at Colorado State University:
An hour in the library will save you from two weeks of floundering around in the laboratory.

Contents

Introduction

While teaching undergraduate Organic Chemistry at University of Colorado, I was asked if I could teach a course in organic synthesis. I have worked as synthetic organic chemist and educator for my entire career (Princeton University, Colorado State University, SUNY Binghamton, Syntex, Roche, University of Denver, University of Colorado, and Better Pharma Processes, LLC) but I realized I was not prepared to teach the course. This workbook project began with that missed opportunity.

In my first Organic Chemistry class at Canisius College more than forty years ago, I recognized the power synthetic organic chemists have to create new medicines to improve human healthcare. With great power comes great responsibility. The synthetic organic chemistry community accepted this responsibility: the result is our World Health Organization (WHO) Model List of Essential Medicines.

Routes to Essential Medicines: A Workbook for Organic Synthesis highlights the synthetic organic chemistry in the manufacturing routes of nearly three hundred medicines on the World Health Organization (WHO) Model List of Essential Medicines (20th List from March 2017). The workbook includes all the medicines on the list for which synthetic organic chemistry plays an important role in the manufacturing process.

Routes to Essential Medicines: A Workbook for Organic Synthesis is intended for use by upper-level undergraduate students and graduate students participating in a course in organic synthesis or medicinal chemistry. Students using this workbook will become familiar with the structures and synthetic challenges associated with nearly three hundred essential medicines and gain an appreciation for the manufacture of specialty chemical starting materials. Students who use this workbook will develop a solid foundation for their academic and postacademic research: an extensive favorites list of key journal and information sites and a personal library of reagents, solvents, and conditions for many workhorse organic reactions. Classroom discussion and extended discussion time will provide valuable experience presenting route and reagent options and mechanisms.

For many students, this workbook will be a first encounter with the Essential Medicines on the WHO List. These medicines are the milestones in our progress in improving human healthcare in the last 100 years. The structural features of these medicines and the synthetic strategies used to create these medicines are the current state of the art and the foundation we will build upon to move human healthcare ahead for the next 100 years. How important are the Essential Medicines? In 2020 they are critically important. They are life-and-death important. I look out my office window on a world practicing social distancing to slow the spread of covid19. We wait for "the answer," a drug to speed recovery from the infection, better yet, a drug to stop the spread of the virus. An optimal drug would be a drug with known and tolerable side effects, a drug which can be produced and distributed quickly, an inexpensive drug. This description fits many of the Essential Medicines on the WHO List so it comes as no surprise that many of the medicines in this workbook are being evaluated as an answer for covid19.

Workbook Organization

The workbook is organized by INN (International Non-Proprietary Name) of the essential medicines in alphabetical order. The presentation for each target molecule begins with the structure of the target and the indications for the drug taken from the Model List. For example, for amiodarone:

From the List:

12. Cardiovascular Medicines
12.3 Antihypertensive Medicines

In the workbook:

Cardiovascular Medicines/Antihypertensive Medicines

In a group setting, each target presentation could begin with some discussion of indications, old and new. The breadth and depth of that discussion is tailored to match the time allotted and the talents and interests of the group. In many cases the discussion could be extended to include new analogs of an old drug.

One route is presented for each target molecule in most cases. The route usually has the shortest sequence from starting materials to target molecule, is adequately described in the literature, and is likely to be close to a manufacturing route still in use today.

At first glance at the target structure, previous knowledge and experience are used to create "back of the envelope" ideas for synthesis of the target. A **text box** highlights one idea from the list which was most influential in guiding the decision-making leading to the presented route. Students should add ideas to create their own list as they work through the synthesis.

The route is presented as a retrosynthetic analysis accompanied by a brief text **Discussion**. Experimental details, references and mechanisms are not provided so that students will first focus on the structures and key disconnections. Students are then tasked to search for, evaluate, and present the procedure details (reagent(s), solvent(s), temperature, reaction time, workup, percent yield) which are the keys to the success of each reaction and the overall route. A discussion of mechanisms could accompany or follow the discussion of procedure details.

The simple sentence structure and limited vocabulary used in the text **Discussion** are intended to facilitate understanding by non-English-speaking readers. The **Discussion** includes tested online search terms (names for reactions, intermediates, and starting materials). The name used for an intermediate or starting material is the name commonly encountered in an online search. Name reactions are highlighted in **bold**. Some questions or tasks are (embedded) in the **Discussion** to draw attention to multiple options for reagent(s) for a reaction, selectivity data, or critical separation procedures.

The schemes are usually presented left-to-right without the use of retrosynthetic arrows. Retrosynthetic arrows are used to avoid confusion in some schemes. Only carbon-containing reagents, intermediates, and products are shown. One structure in a scheme often has the positions relevant to the text Discussion numbered.

The discussion is often written so that each sentence describes one reaction. This allows the discussion to be read as a description of the retrosynthesis or the synthesis. To illustrate, for amiodarone:

Retrosynthesis: The phenol is iodinated in both *ortho* positions. The phenol is formed by demethylation of the ether. The ketone is formed by acylation of 2-butyl-2,3-benzofuran with 4-methoxybenzoyl chloride (**Friedel–Crafts Acylation**). 4-Methoxybenzoyl chloride is formed from *para*-anisic acid.

Synthesis: 4-Methoxybenzoyl chloride is formed from *para*-anisic acid. The ketone is formed by acylation of 2-butyl-2,3-benzofuran with 4-methoxybenzoyl chloride (**Friedel–Crafts Acylation**). The phenol is formed by demethylation of the ether. The phenol is iodinated in both *ortho* positions.

The retrosynthetic analysis ends with specialty chemical(s), petrochemical(s), or biochemical(s) starting materials. Specialty chemical and petrochemical starting materials are highlighted with a box to direct the reader to an Appendix. **Appendix A** contains the specialty chemicals used in this workbook. One retrosynthetic analysis is provided for each specialty chemical. It is important to emphasize that specialty chemicals are often manufactured by more than one route. The preferred route often changes with the implementation of new manufacturing technologies and with changes in availability and cost of petrochemical and biochemical raw materials. Each specialty chemical retrosynthetic analysis ends with petrochemical(s) or biochemical(s) available in bulk from many suppliers. **Appendix B** lists the petrochemicals extracted or produced from coal, crude oil, or natural gas that are used in this workbook. **Appendix C** lists the biochemicals that are used in this workbook.

In the optional **Extended Discussion**, students are tasked to propose and evaluate alternative routes, reactions, or reagents encountered in their online search work.

A course based on this workbook could be organized in many ways: medicines for pain, medicines containing fluorine, medicines made using a Diels–Alder Reaction, etc. Course development will be facilitated by the provided **Indices** (name reactions, starting materials from the chiral pool, diazonium salt reactions, epoxide ring-opening reactions, fluorine-containing target molecules, furans, imidazoles, nucleophilic aromatic substitution reactions, oxidations, peptides, photochemistry, purines, pyrazines, pyridines, pyrimidines, quinolines, symmetrical molecules, thiazoles, thioethers, triazoles.)

Appendices A, B, and C and a complete list of the **References** used in preparation of the workbook are available on the companion website.

www.wiley.com/go/Harrington/routes_essential_medicine

About the Companion Website

www.wiley.com/go/Harrington/routes_essential_medicine

A

Abacavir

Antiviral Medicines/Antiretrovirals/Nucleoside or Nucleotide Reverse Transcriptase Inhibitors

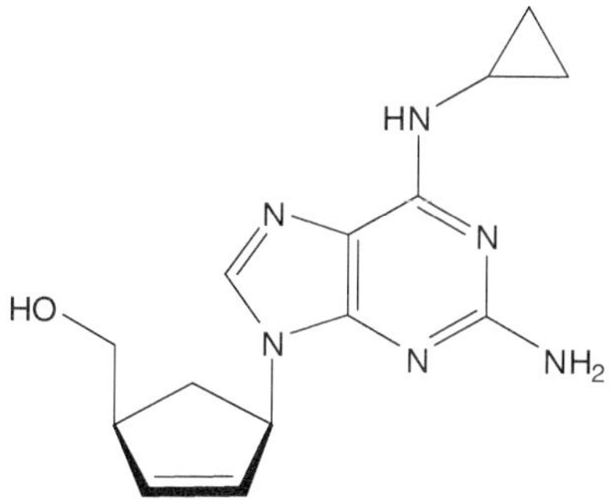

When two chiral carbons are separated by one or more atoms, disconnections often lead back to an intermediate with the two chiral carbons next to each other and formed in the same reaction.

Discussion. Cyclopropanamine is introduced by chloride displacement in the final step (the chloropurine partner is also used to make the AIDS drug carbovir.) The imidazole ring of the purine is formed from the formamide (**Traube Purine Synthesis**). A C—N bond is formed by displacement of chloride from the symmetrical dichloropyrimidine by the amine of (1*S*,4*R*)-4-amino-2-cyclopentene-1-methanol.

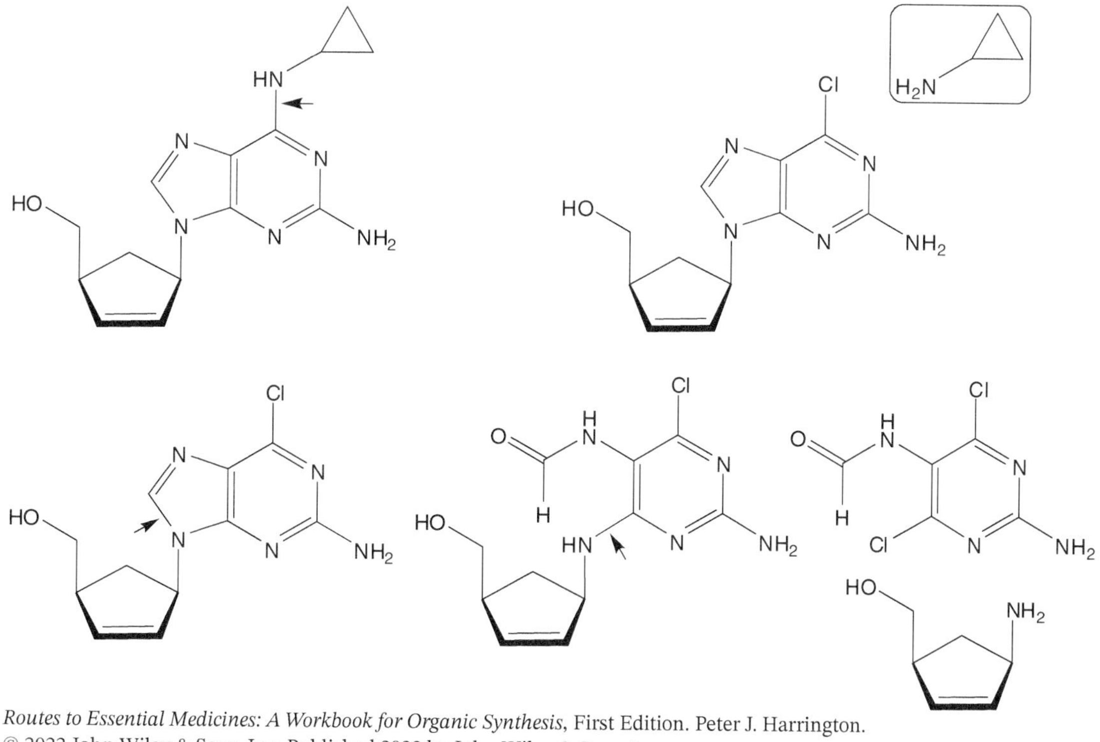

Routes to Essential Medicines: A Workbook for Organic Synthesis, First Edition. Peter J. Harrington.

Companion website: www.wiley.com/go/Harrington/routes_essential_medicine

The 2,5-diaminopyrimidine 5-formamide is formed from the 2,5-diaminopyrimidine and formic acid. 2,5-Diamino-4,6-dichloropyrimidine is formed from 2,5-diamino-4,6-dihydroxypyrimidine. 2,5-Diamino-4,6-dihydroxypyrimidine is formed by hydrolysis of the 5-acetamide. 5-Acetamido-2-amino-4,6-dihydroxypyrimidine ring is formed from guanidine and diethyl acetamidomalonate (**Pinner Pyrimidine Synthesis**).

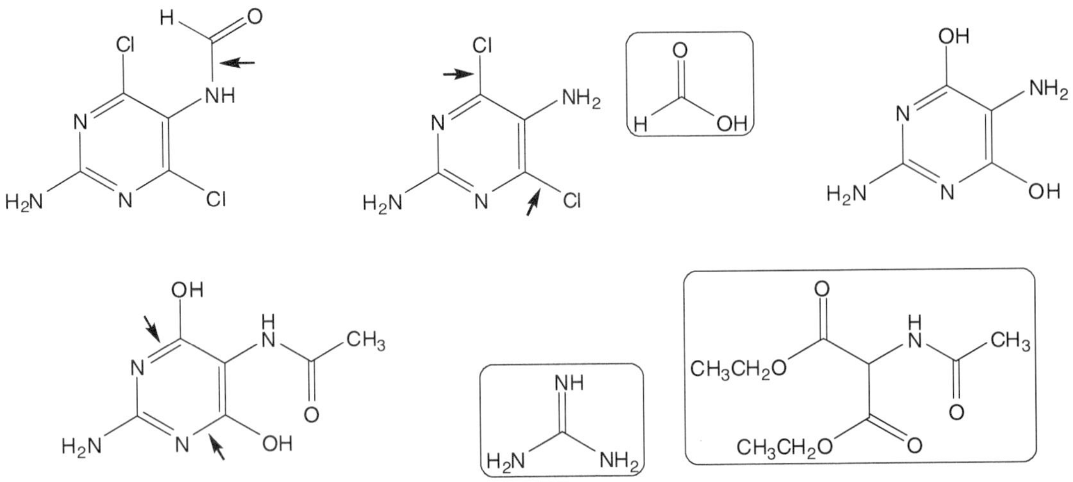

(1*S*,4*R*)-4-Amino-2-cyclopentene-1-methanol and the (1*R*,4*S*)-enantiomer are separated by resolution. The amino alcohols are formed by reduction of the amides (**Vince Lactam**). The amides are formed by displacement of methanesulfinate by hydroxide.

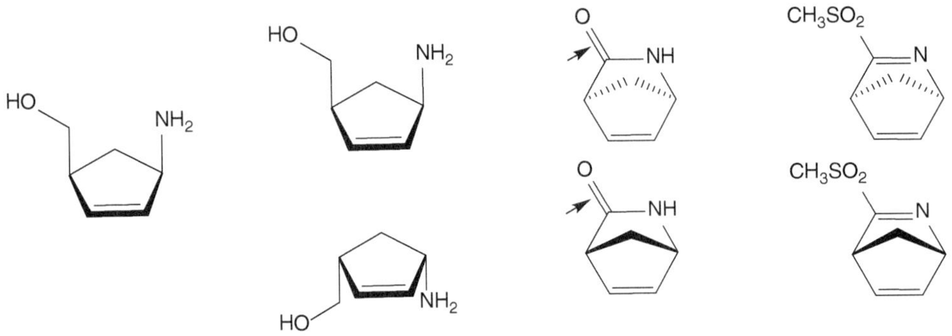

The 2-azabicyclo[2.2.1]hepta-2,5-dienes are formed by [4+2]-cycloaddition of cyclopentadiene with methanesulfonyl cyanide (**Diels–Alder Cycloaddition**). Methanesulfonyl cyanide is formed from sodium methanesulfinate and cyanogen chloride. Sodium methanesulfinate is formed by reduction of methanesulfonyl chloride.

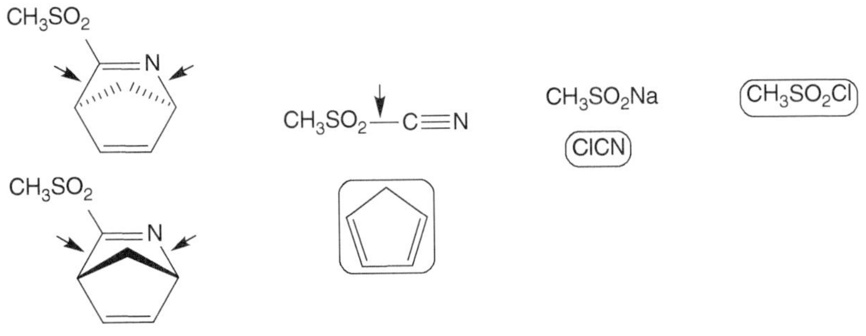

Extended Discussion

Draw the structures of the retrosynthetic analysis of one alternative route to abacavir using a disconnection of the C—N bond joining the purine ring to the cyclopentene ring. Include the structures of the retrosynthetic analysis of any organic starting material(s) from petrochemical or biochemical raw materials.

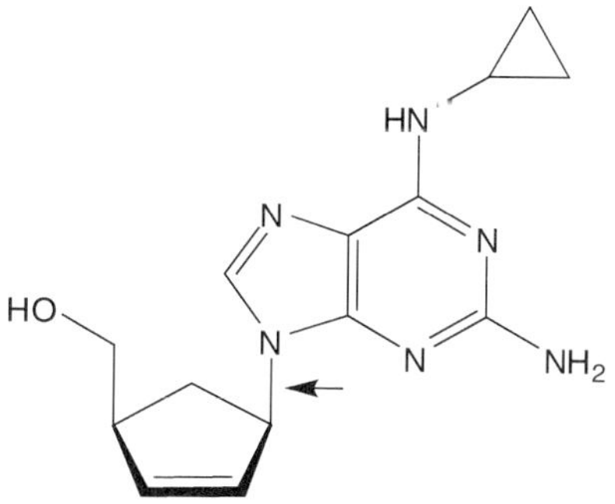

Acetazolamide

Ophthalmological Preparations/Miotics and Antiglaucoma Medicines

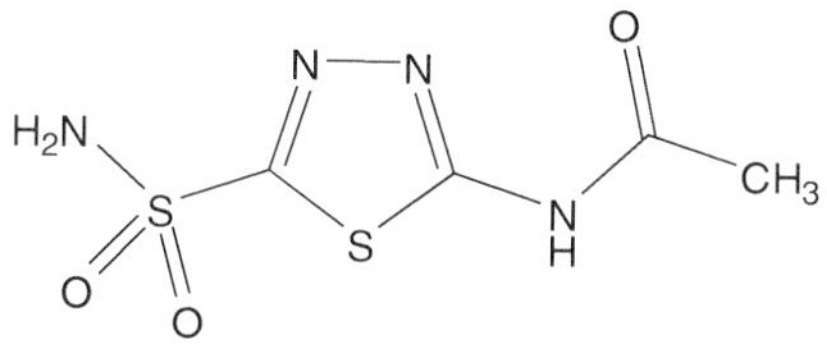

A sulfonyl chloride attached to an aromatic ring may be formed by oxidation of the thiol.

Discussion. Acetazolamide is formed in just three steps from 5-amino-1,3,4-thiadiazole-2-thiol. The sulfonamide is formed from the sulfonyl chloride in the final step. The sulfonyl chloride is formed by oxidation of the thiol. The amide is formed by acetylation of the amine using acetic anhydride.

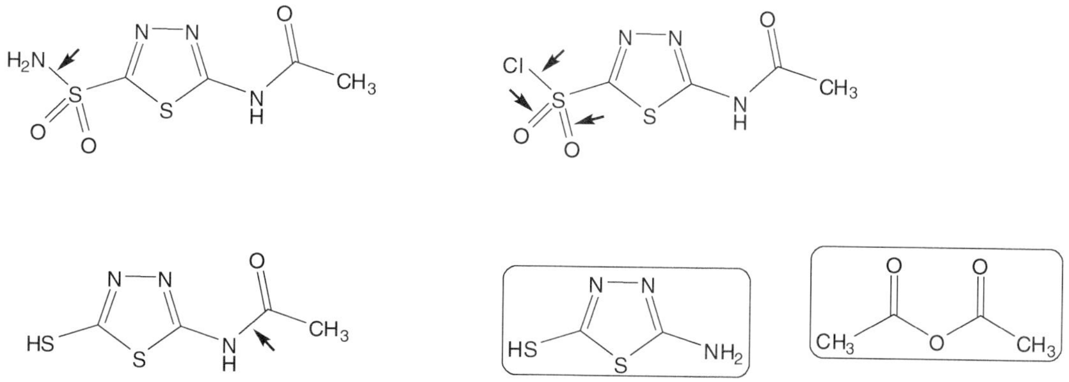

Extended Discussion

Draw the structures of the retrosynthetic analysis of one alternative route to the starting material 5-amino-1,3,4-thiadiazole-2-thiol. List the pros for the route presented and the alternative route and select one route as the preferred route.

Acetylcysteine

Antidotes/Specific

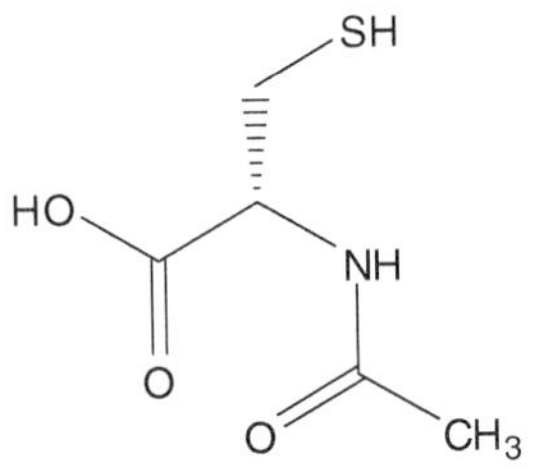

A chiral carbon in a single-enantiomer molecule is often delivered in a starting material.

Discussion. *N*-Acetyl-L-cysteine is formed by acetylation of L-cysteine with acetic anhydride. L-Cysteine is produced by fermentation.

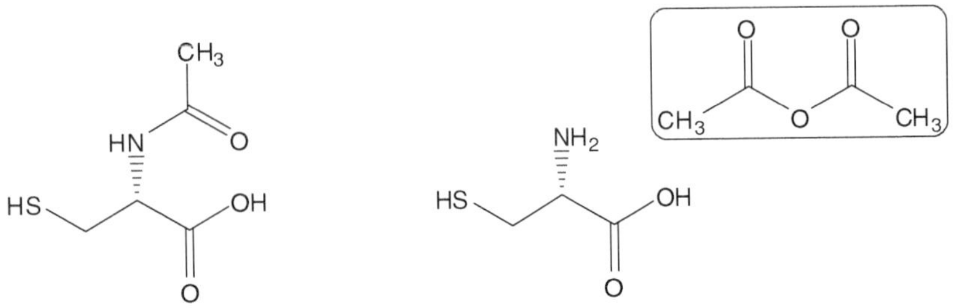

L-cysteine

Acetylsalicylic Acid

Medicines for Pain and Palliative Care/Non-Opioids and Non-Steroidal Anti-Inflammatory Medicines
Antimigraine Medicines/For Treatment of Acute Attack
Cardiovascular Medicines/Antithrombotic Medicines/Anti-Platelet Medicines
Medicines for Diseases of Joints/Juvenile Joint Diseases

salicylic acid
Dermatological Medicines/Medicines Affecting Skin Differentiation and Proliferation

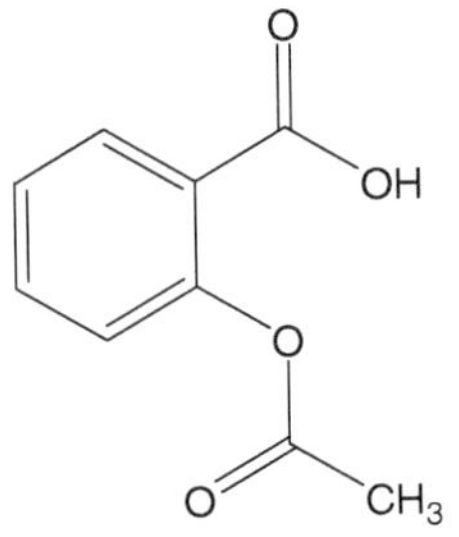

acetylsalicylic acid

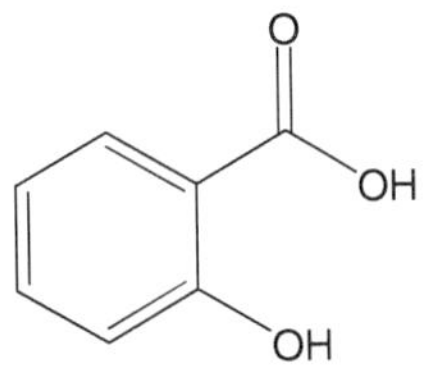

salicylic acid

Discussion. Acetylsalicylic acid (aspirin) is formed from another essential medicine, salicylic acid, and acetic anhydride. Salicylic acid is formed from phenol and carbon dioxide (**Kolbe–Schmitt Reaction**).

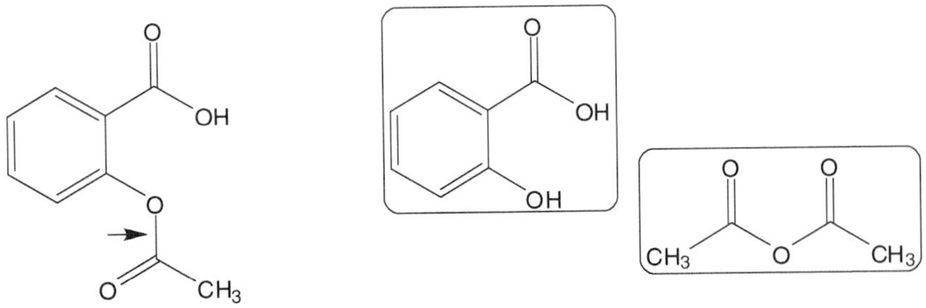

Acyclovir

Anti-Infective Medicines/Antiviral Medicines/Antiherpes Medicines
Ophthalmological Preparations/Anti-Infective Agents

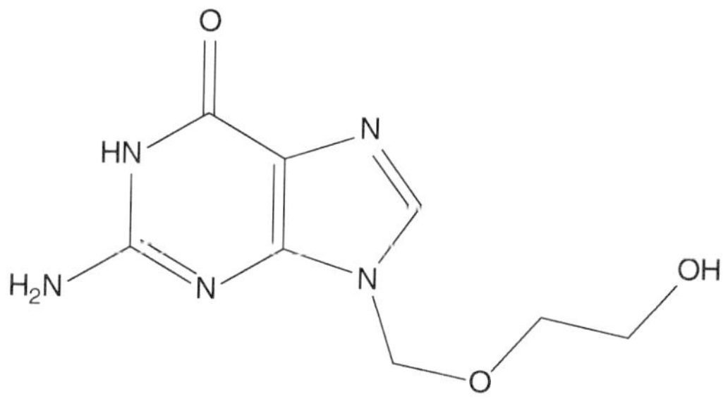

Guanine is often converted to an acylated or silylated derivative to increase solubility in organic solvents. These derivatives react with alkylating agents to form a mixture of N7-alkylated (kinetic) product and N9-alkylated (thermodynamic) product.

Discussion. The concepts and challenges common to the many routes to acyclovir are featured in a comparison of two preferred routes. In route A, the alcohol is released by *O*-desilylation in the final step. Acyclovir *O*-trimethylsilyl ether is formed by desilylation of persilyl acyclovir. A mixture of the N9-alkylated persilyl acyclovir and the N7-alkylated regioisomer is formed in situ by in the reaction of persilyl guanine with 1,3-dioxolane (What is the highest ratio of persilyl acyclovir to the N7-alkylated regioisomer? What reaction conditions are associated with the highest ratio? How is the N7-alkylated side product separated from the N9-alkylated product?). Persilyl guanine is a mixture of N7-TMS and N9-TMS regioisomers formed in situ by the reaction of guanine with excess hexamethyldisilazane (HMDS).

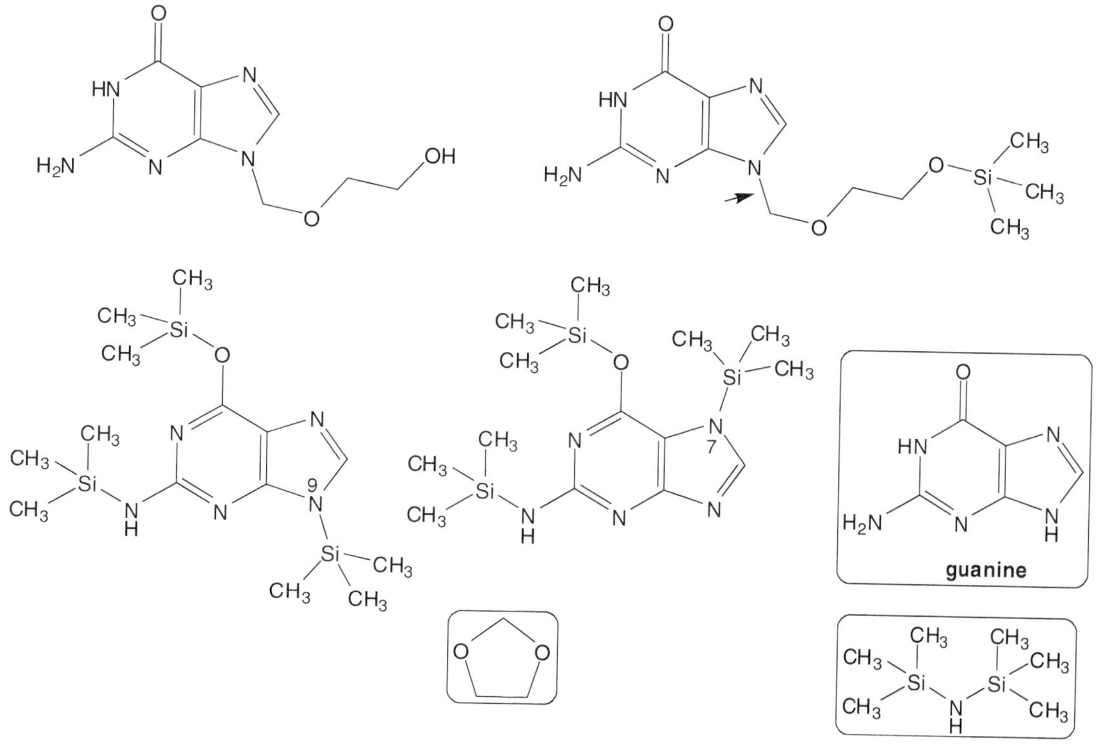

In route B, the alcohol and amino group are released by hydrolysis of the ester and amide in the final step. The alkylation of N2,9-diacetylguanine with 2-(acetoxyethoxy)methyl acetate affords a mixture of the N7- and N9-regioisomers. (Draw the structure of the N7-regioisomer. What is the highest N9:N7 ratio? What reaction conditions are associated with the highest ratio? How is the N7-alkylated side product separated from N9-alkylated product?) N2,9-Diacetylguanine is formed by reaction of guanine with acetic anhydride.

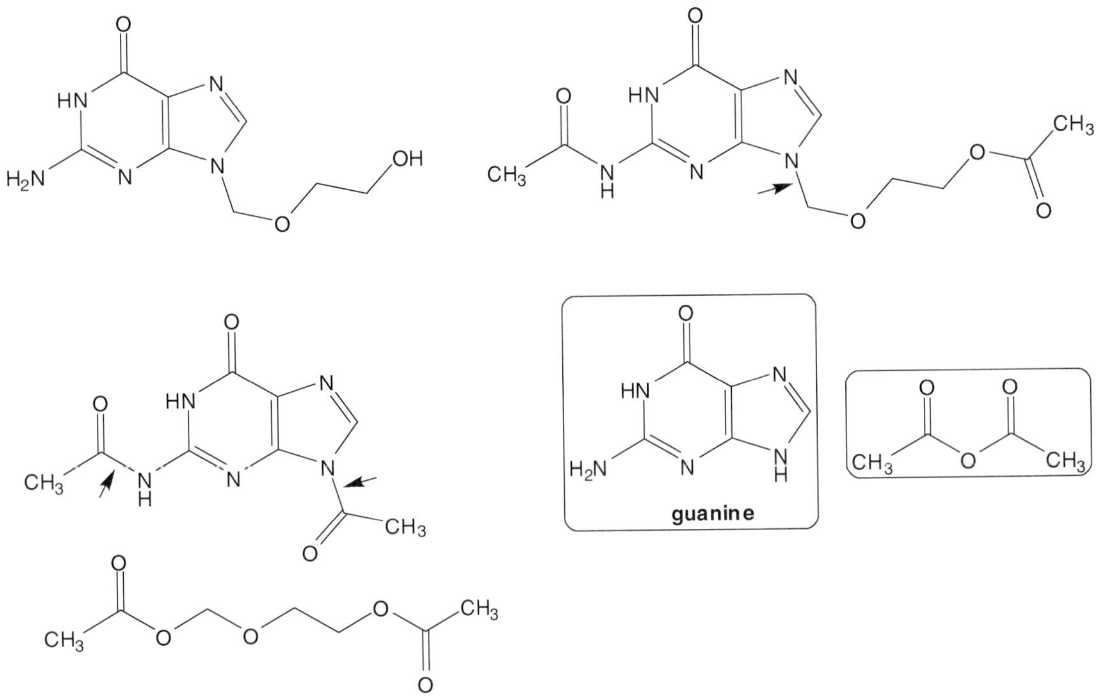

2-(Acetoxyethoxy)methyl acetate is formed from 1,3-dioxolane, acetic acid, and acetic anhydride.

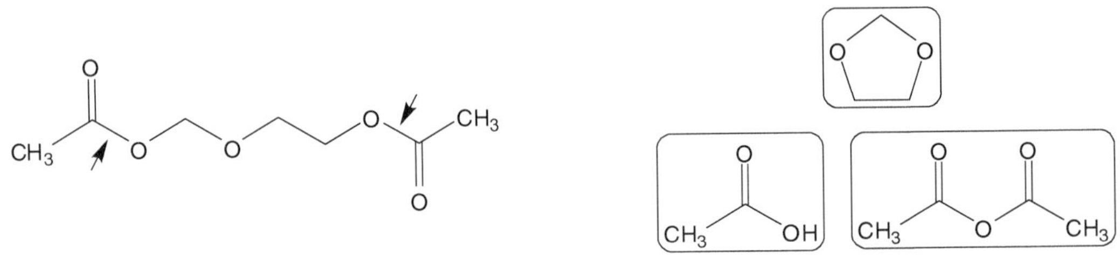

Extended Discussion

List the pros and cons for routes A and B and select one route as the preferred route.

Albendazole

Anti-Infective Medicines/Anthelmintics/Antifilarials

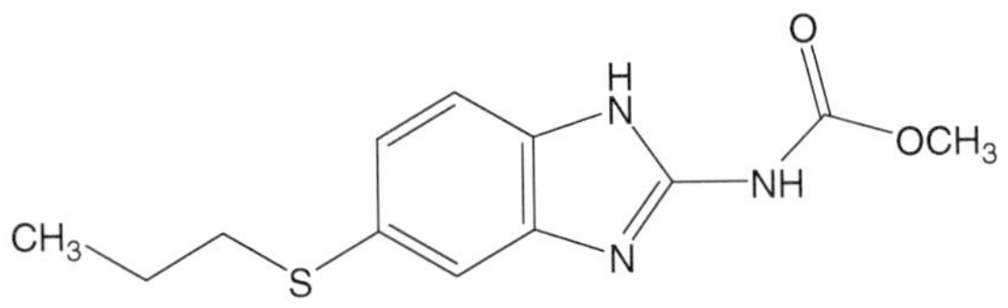

A benzimidazole is often formed from a 1,2-phenylenediamine.

Discussion. The benzimidazole is formed in the final step from the benzene-1,2-diamine and *N*-methoxycarbonylcyanamide. *N*-Methoxycarbonylcyanamide is formed from cyanamide and methyl chloroformate. 4-(Propylthio)benzene-1,2-diamine is formed by reduction of 2-nitro-4-propylthioaniline. A C—S bond is formed by displacement of chloride from 4-chloro-2-nitroacetanilide by sodium propanethiolate. The acetanilide is also hydrolyzed under the chloride displacement reaction conditions. 4-Chloro-2-nitroacetanilide is formed from 4-chloro-2-nitroaniline and acetic anhydride.

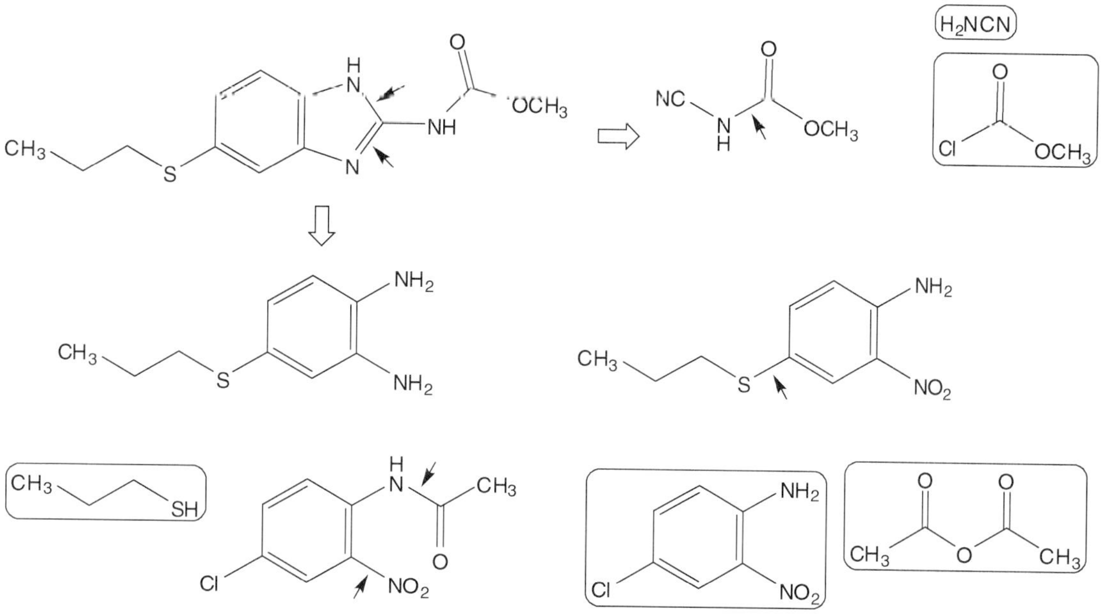

Extended Discussion

2-Nitro-4-propylthioaniline can also be manufactured from 1-chloro-2-nitrobenzene. Draw the structures of the retrosynthetic analysis of this route. List the pros and cons for both routes. Which route is preferred?

Allopurinol

Antineoplastics and Immunosuppressives/Cytotoxic and Adjuvant Medicines
Medicines for Diseases of Joints/Medicines Used to Treat Gout

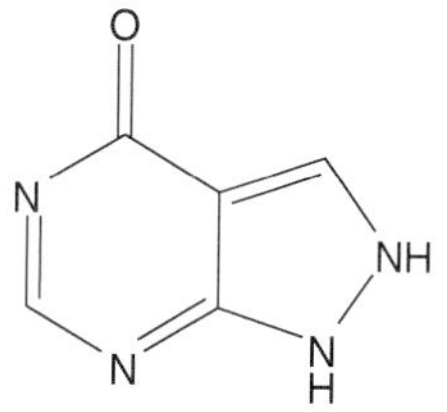

Hydrazine is often the source of the two nitrogen atoms in a pyrazole ring.

Discussion. The pyrimidine ring of the pyrazolo[3,4-d]pyrimidine is formed in the final step by reaction of 3-aminopyrazole-4-carboxamide with formamide. The pyrazole ring is formed from 2-cyano-3-morpholinoacrylamide and hydrazine. The enamine of 2-cyano-3-morpholinoacrylamide is formed from the enol ether by the displacement of ethanol by morpholine. The enol ether of 2-cyano-3-ethoxyacrylamide is formed by the reaction of 2-cyanoacetamide with triethyl orthoformate.

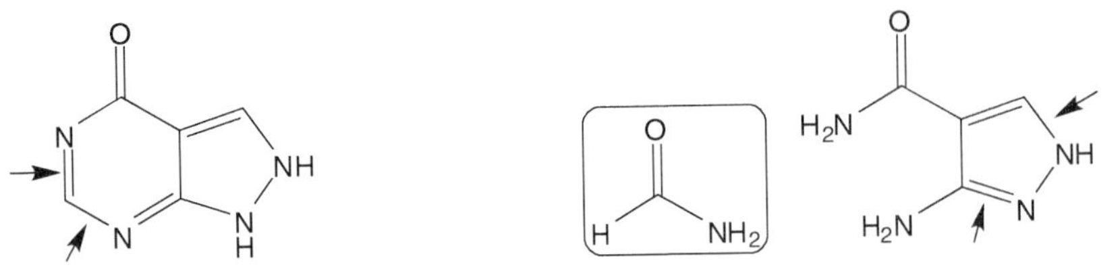

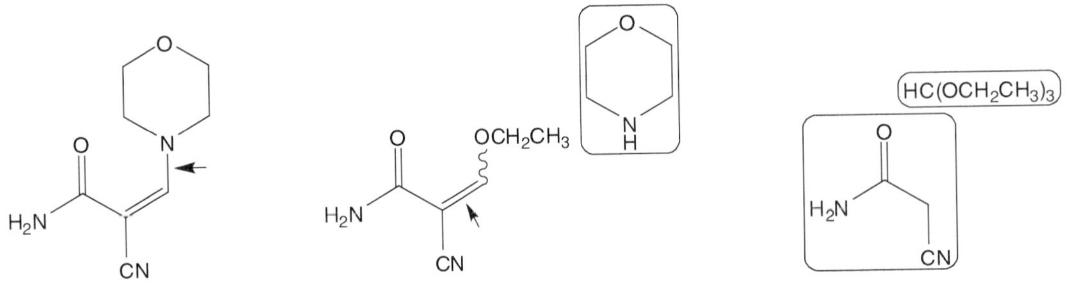

Extended Discussion

The pyrimidine ring is also formed by reaction of ethyl 3-aminopyrazole-4-carboxylate with formamide. Draw the structures of the retrosynthetic analysis of ethyl 3-aminopyrazole-4-carboxylate. List the pros and cons for both routes and select one route as the preferred route.

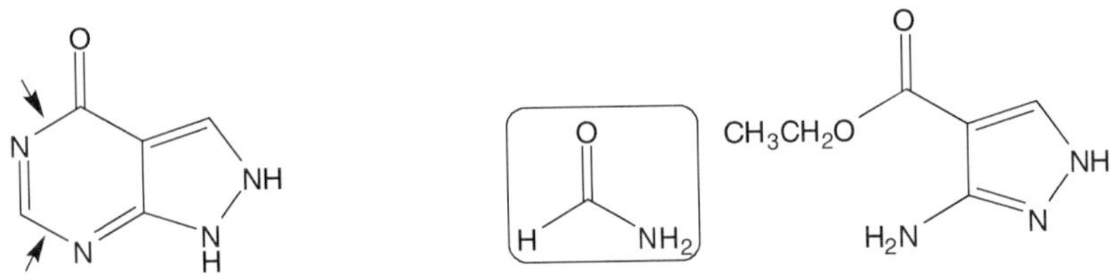

Amidotriazoate

Diagnostic Agents/Radiocontrast Media

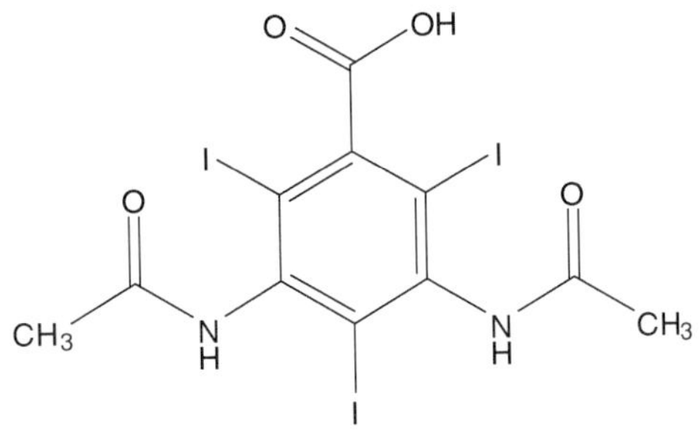

For a symmetrical molecule, symmetrical disconnections lead back to symmetrical intermediates and are likely associated with the shortest route.

Discussion. Since some amide hydrolysis is likely under iodination conditions, the diamide is formed in the final step by reaction of the diamine with acetic anhydride. The triiodide is formed by iodination of 3,5-diaminobenzoic acid.

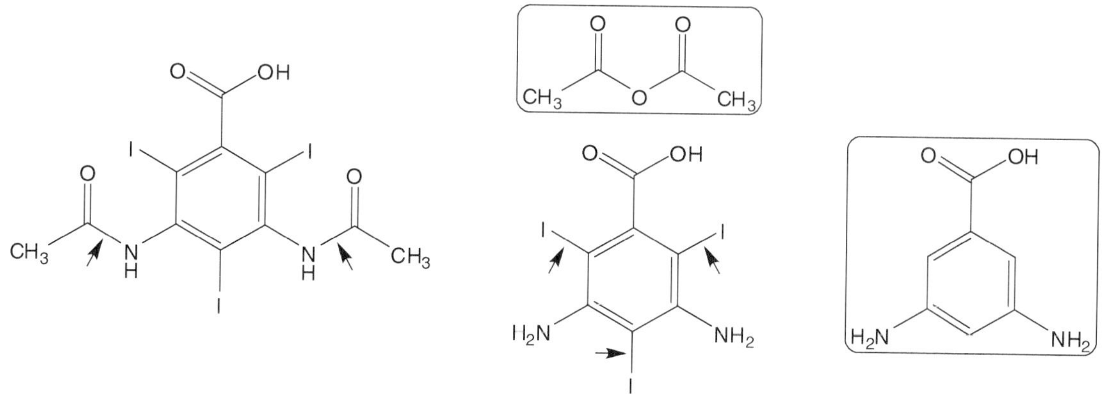

Extended Discussion

List reagents or reagent combinations used for direct iodination of an aromatic ring.

Amikacin

Anti-Infective Medicines/Antibacterials/Other Antibacterials
Anti-Infective Medicines/Antibacterials/Antituberculosis Medicines

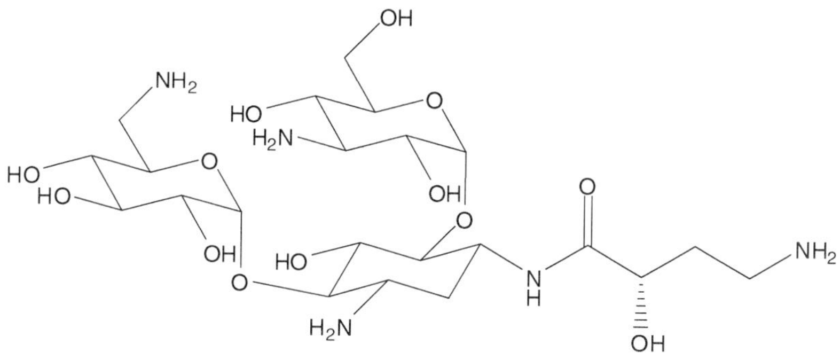

A single-enantiomer molecule with multiple chiral carbons is often made by modification of a natural product which has most or all of the chiral carbons already in place.

Discussion. Amikacin is semisynthetic. Amikacin is formed by acylation of the amino group at C1 of kanamycin A. This selective acylation requires a protection–deprotection strategy since kanamycin A has four amino groups and the amino group at C1 is not the most reactive.

Three of the amino groups of amikacin are released in the final step by benzyl carbamate hydrogenolysis. The amide at C1 is formed by reaction of the amino group with an *N*-hydroxysuccinimide ester. Amino groups at C3 and C6′ of kanamycin A are protected as benzyl carbamates (Cbz). Kanamycin A is produced by fermentation.

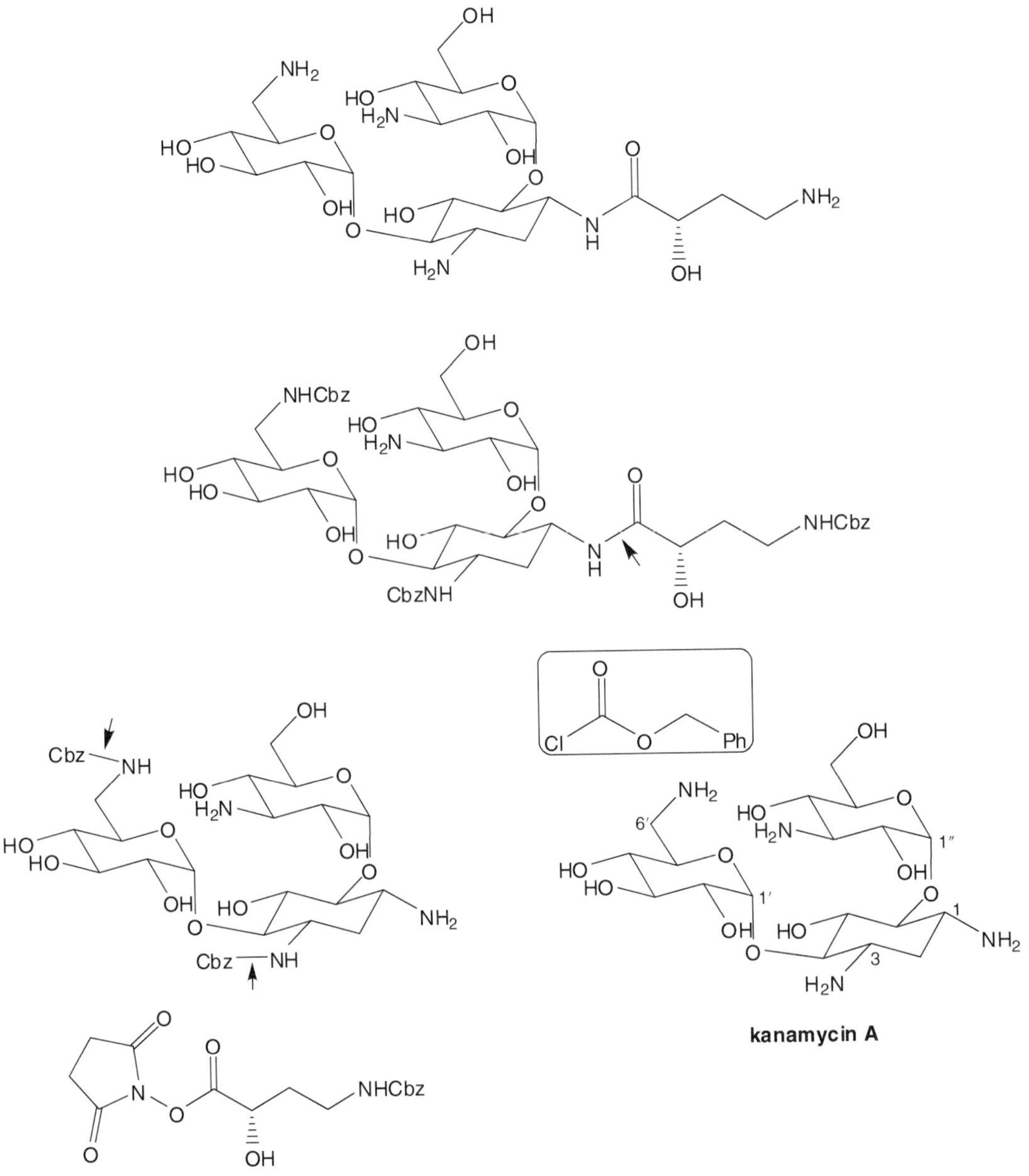

The *N*-hydroxysuccinimide ester is formed from the carboxylic acid. The amino group of the 4-amino-2-hydroxybutanoic acid is protected as the benzyl carbamate. (*S*)-4-Amino-2-hydroxybutanoic acid is formed from (S)-2-hydroxyglutaramic acid (**Hofmann Rearrangement**). The amide is formed from the lactone. (*S*)-5-Oxotetrahydrofuran-2-carboxylic acid lactone is formed by diazotization of L-glutamic acid. L-Glutamic acid is produced by fermentation.

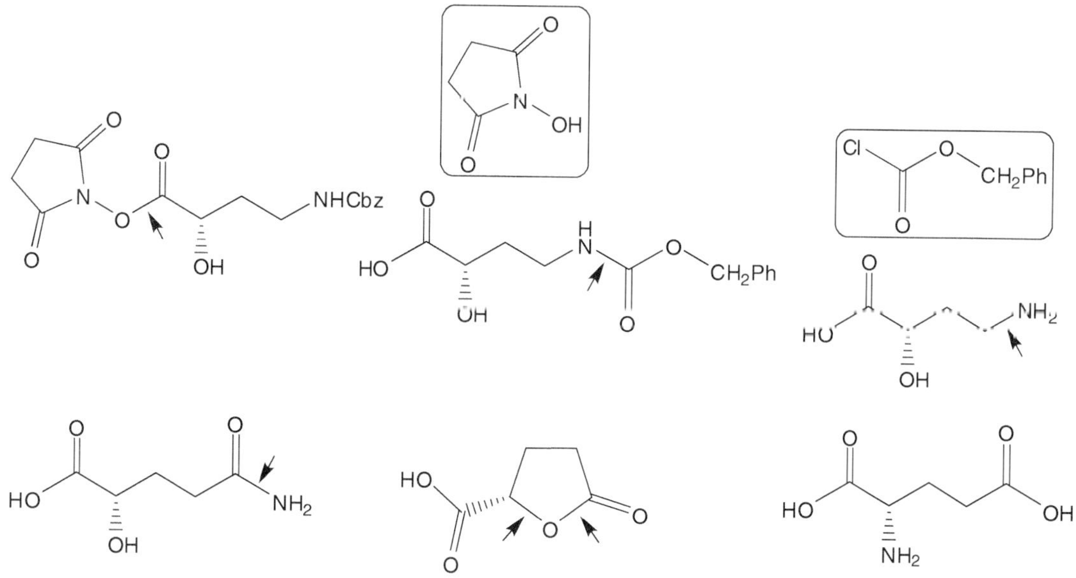

L-glutamic acid

Extended Discussion

Draw the structures of three side products which are likely to be formed in the reaction of kanamycin A with two equivalents of benzyl chloroformate. Draw the structure(s) of likely impurities in amikacin as each side product is carried through the amide formation and carbamate hydrogenolysis.

or

Draw the structures of the retrosynthetic analysis of the alternative route to (*S*)-4-amino-2-hydroxybutanoic acid from L-asparagine. List the pros and cons for both routes and select one route as the preferred route.

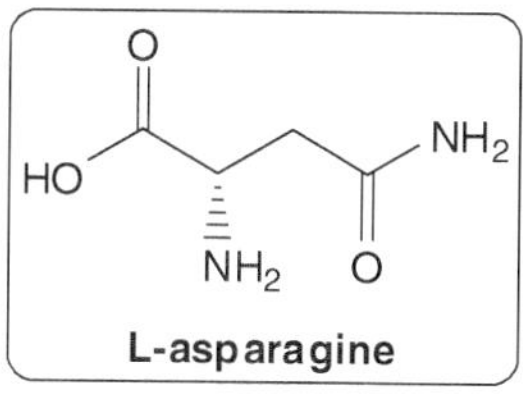

Amiloride

Diuretic

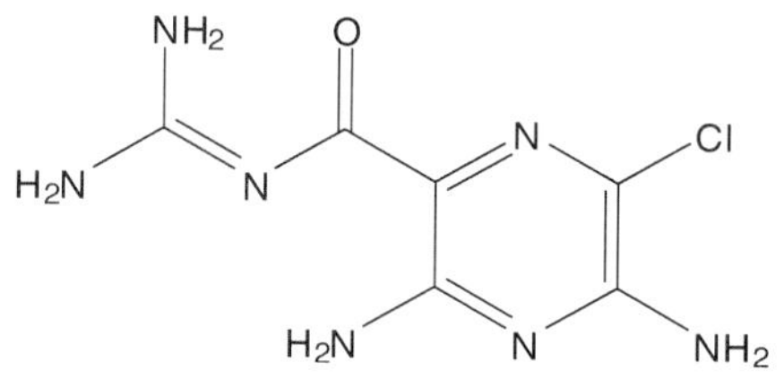

A nitrogen substituent on a pyrazine ring carbon is often introduced by displacement of chloride. The substitution is facilitated by the adjacent ring nitrogen and can be further facilitated by an electron-withdrawing group (NO_2, SO_2R, COOR, CN) on a *para* ring carbon.

Discussion. The guanidine group is introduced in the final step by reaction of guanidine with the methyl ester. Chloride at the 5-position of the 5,6-dichloropyrazine is displaced by ammonia. The 5,6-dichloropyrazine is formed by chlorination of methyl 3-aminopyrazine-2-carboxylate. The methyl ester is formed from the carboxylic acid (**Fischer Esterification**).

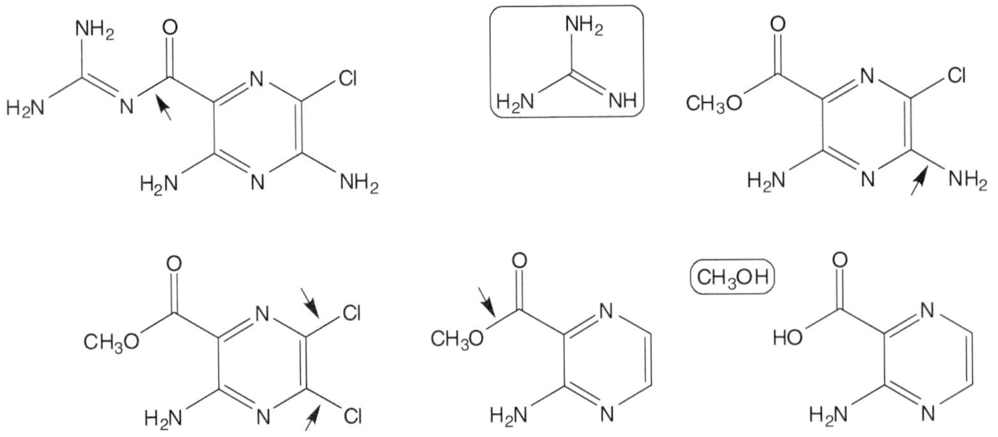

3-Aminopyrazine-2-carboxylic acid is formed by hydrolysis of the pyrimidine ring of lumazine (1H-pteridine-2,4-dione). The pyrazine ring of lumazine is formed by reaction of 5,6-diaminouracil with glyoxal. The amino group at the 5-position of 5,6-diaminouracil is formed by reduction of a nitroso group. The nitroso group is introduced by nitrosation of 6-aminouracil. 6-Aminouracil is formed from ethyl cyanoacetate and urea.

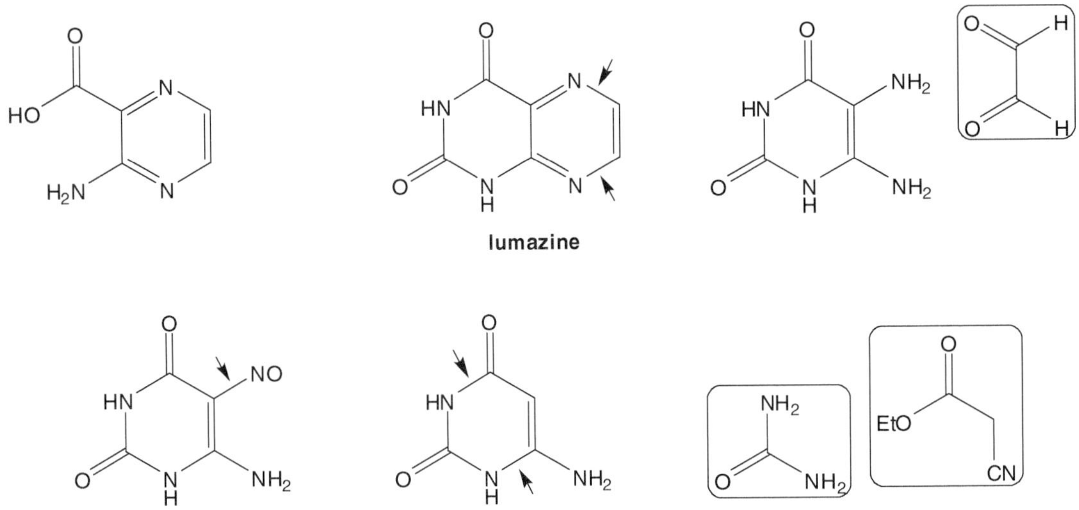

Extended Discussion

Draw the structures of the retrosynthetic analysis of one alternative route to 3-aminopyrazine-2-carboxylic acid. Include the structures of the retrosynthetic analysis of any organic starting material(s) from petrochemical or biochemical raw materials. List the pros and cons for both routes and select one route as the preferred route.

4-Aminosalicylic Acid

Anti-Infective Medicines/Antibacterials/Antituberculosis Medicines

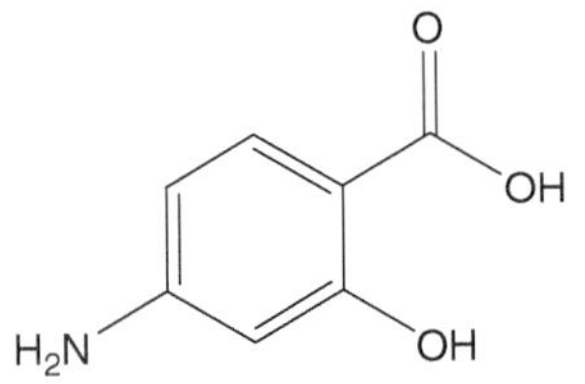

A 2-hydroxybenzoic acid is often formed by carboxylation of the phenol (Kolbe–Schmitt Reaction).

Discussion. 4-Aminosalicylic acid is formed from 3-aminophenol and carbon dioxide (**Kolbe–Schmitt Reaction**) (Draw the structure of one side product formed in this reaction. How is pure 4-aminosalicylic acid isolated from the product mixture?).

Extended Discussion

A preferred route to 3-aminophenol is from benzene via resorcinol. Draw the structures of a retrosynthetic analysis of one alternative route to 3-aminophenol. Include the structures of the retrosynthetic analysis of any organic starting material(s) from petrochemical or biochemical raw materials. List pros and cons for the two routes and select one route as the preferred route.

Amiodarone

Cardiovascular Medicines/Antihypertensive Medicines

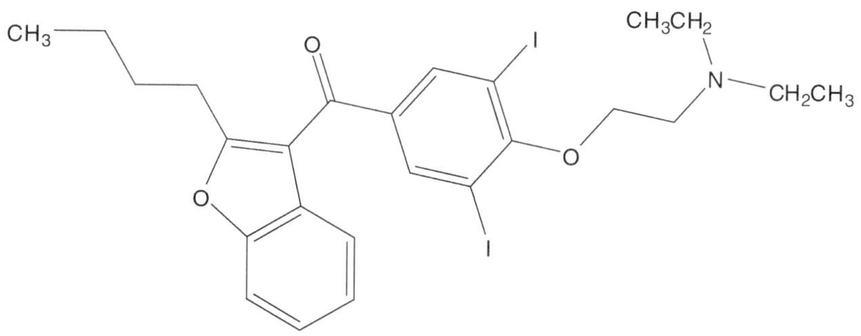

An aromatic ketone is often formed by Friedel–Crafts Acylation.

Discussion. The ether is formed in the final step by displacement of the chloride of 2-chloro-*N,N*-diethylethanamine by the phenol (**Williamson Ether Synthesis**).

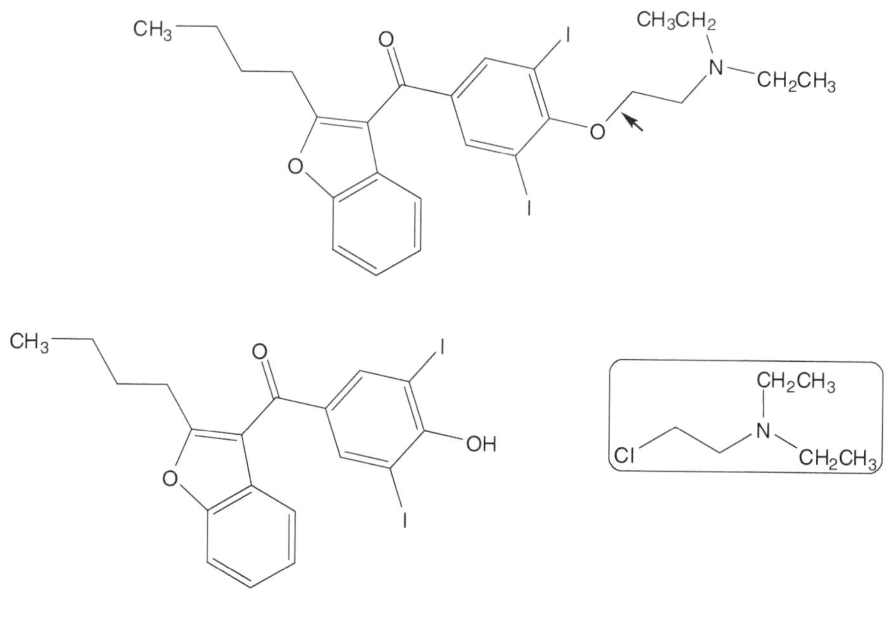

The phenol is iodinated in both *ortho* positions. The phenol is formed by demethylation of the ether. The ketone is formed by acylation of 2-butyl-2,3-benzofuran with 4-methoxybenzoyl chloride (**Friedel–Crafts Acylation**). 4-Methoxybenzoyl chloride is formed from *para*-anisic acid.

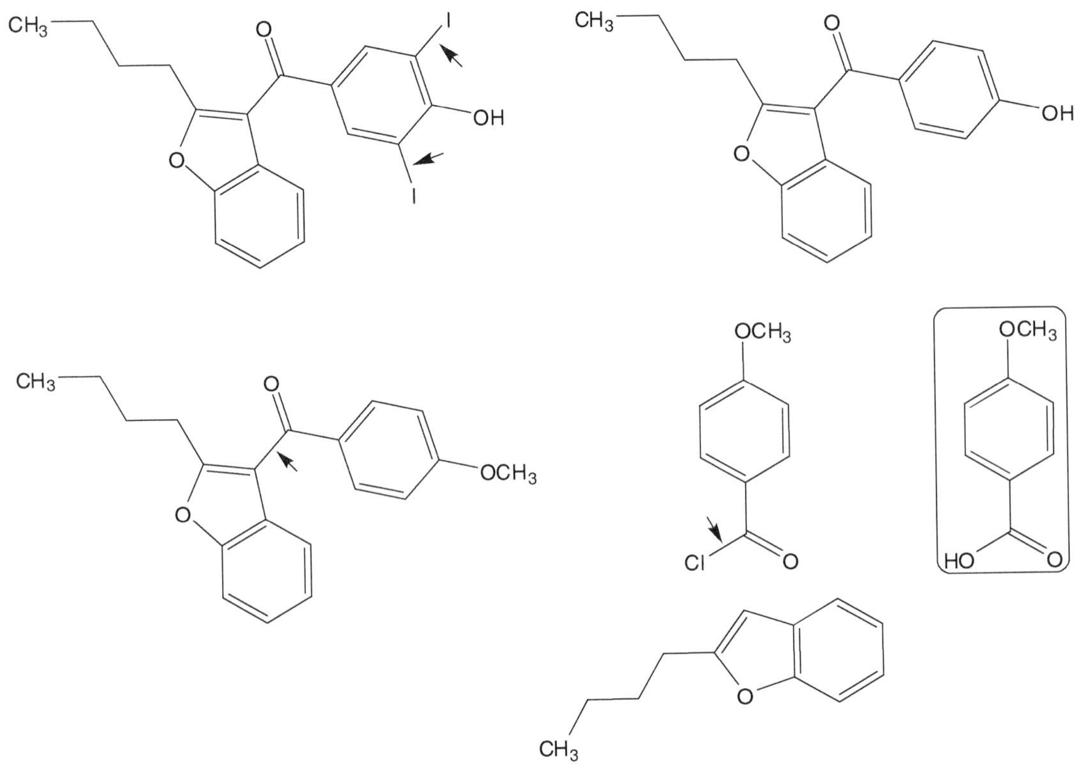

2-Butylbenzofuran is formed by rearrangement of the chlorohydrin. The tertiary alcohol of the chlorohydrin is formed by addition of butylmagnesium chloride to the ketone (**Grignard Reaction**). Butylmagnesium chloride is formed from 1-chlorobutane. 2-Chloro-2′-hydroxyacetophenone is formed from phenol and chloroacetonitrile (**Sugasawa Reaction**).

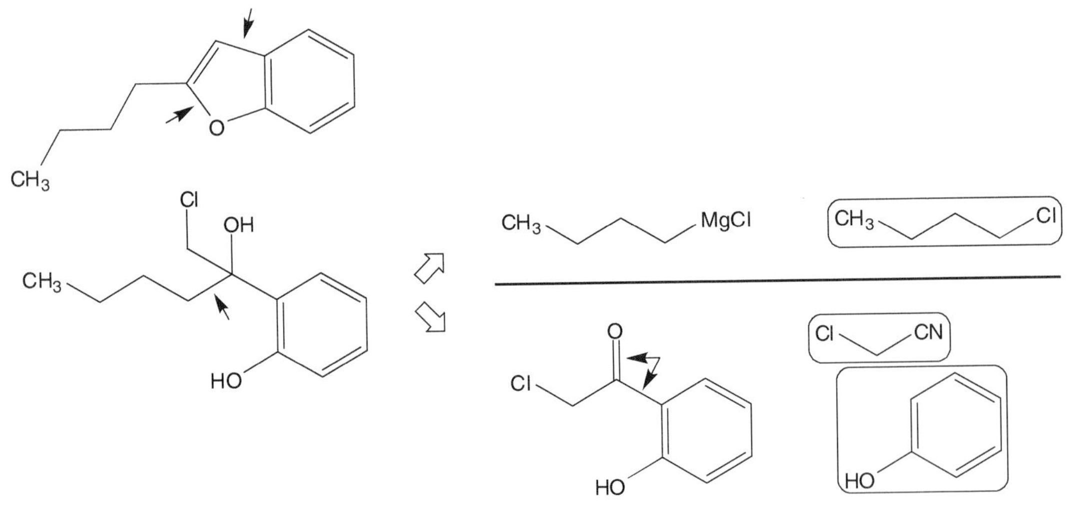

Extended Discussion

Draw the structures of the retrosynthetic analysis of one alternative route to 2-butylbenzofuran. Include the structures of the retrosynthetic analysis of any organic starting material(s) from petrochemical or biochemical raw materials. List the pros and cons for both routes and select one route as the preferred route.

Amitriptyline

Medicines for Pain and Palliative Care/Medicines for Other Common Symptoms in Palliative Care
Medicines for Mental and Behavioral Disorders/Medicines Used in Mood Disorders/Medicines Used in Depressive Disorders

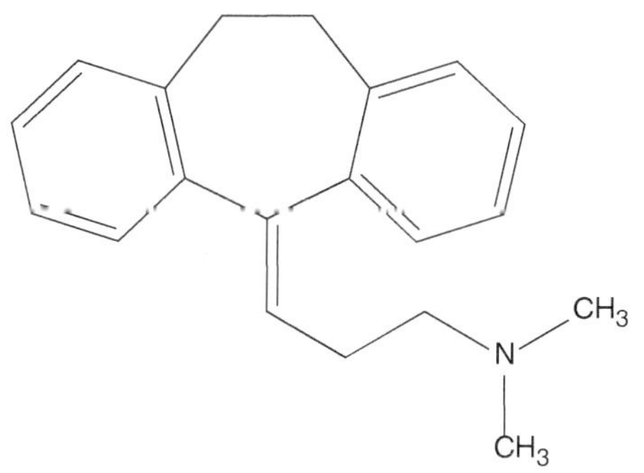

An alkene conjugated to two aromatic rings is often formed by dehydration of an alcohol.

Discussion. The alkene is formed in the final step by dehydration of the tertiary alcohol. The tertiary alcohol is formed by addition of the alkylmagnesium chloride to dibenzosuberone (**Grignard Reaction**). The alkylmagnesium chloride is formed from 3-chloro-*N,N*-dimethylpropan-1-amine. Dibenzosuberone is formed by cyclization of 2-phenethylbenzoic acid (**Friedel–Crafts Acylation**). 2-Phenethylbenzoic acid is formed by reduction of benzalphthalide. Benzalphthalide is formed from phthalic anhydride and phenylacetic acid.

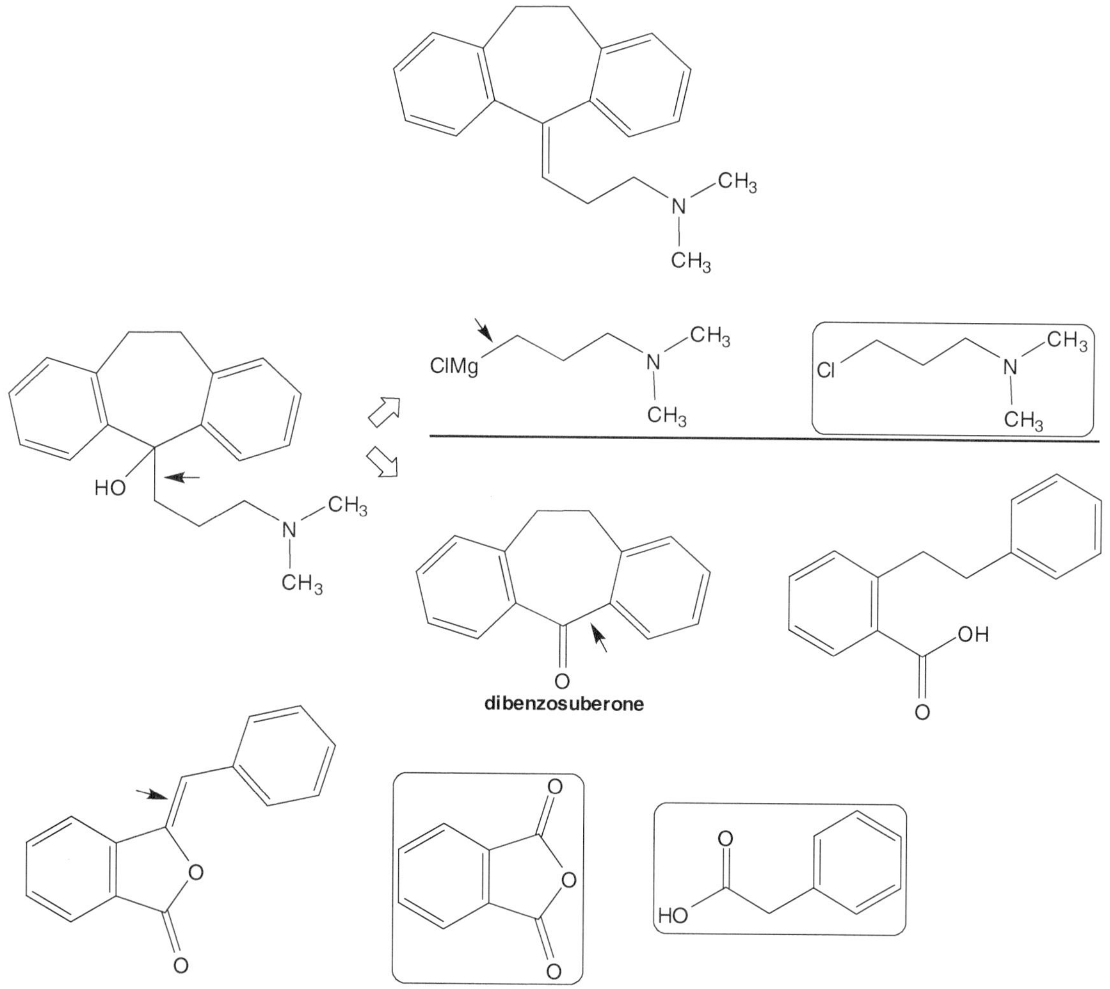

Extended Discussion

Dibenzosuberone is also formed from 2-bromobenzyl bromide and carbon dioxide (**Parham Cyclization**). List the pros and cons for the two dibenzosuberone routes and select one route as the preferred route.

Amlodipine

Cardiovascular Medicines/Antihypertensive Medicines

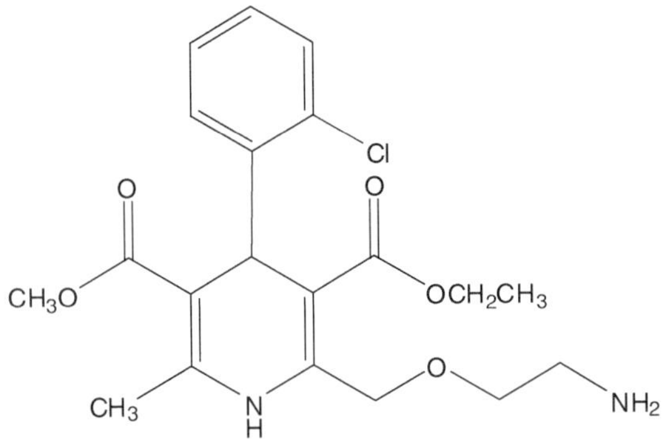

A dihydropyridine is often formed by Hantzsch Synthesis. The key step in the Hantzsch Synthesis is formation of a C3—C4 bond of the ring by Michael Addition of an enamine to an α,β-unsaturated ketone or ester.

Discussion. While the (*S*)-enantiomer is a thousand times more active than the (*R*)-enantiomer, amlodipine is sold as the racemate. Racemic amlodipine is constructed in just four steps! The primary amine is formed using a **Gabriel Synthesis**. The final step is release of the primary amine from the phthalimide. The 1,4-dihydropyridine is formed from methyl 3-aminocrotonate and an enone by C—C bond formation (**Michael Addition**) followed by C—N bond formation to close the ring (**Hantzsch Dihydropyridine Synthesis**). The enone is formed by condensation of a β-ketoester with 2-chlorobenzaldehyde (**Knoevenagel Condensation**). The ether on C4 of the β-ketoester is formed by chloride displacement from ethyl 4-chloro-3-oxobutanoate by the alcohol of *N*-(2-hydroxyethyl)phthalimide (**Williamson Ether Synthesis**).

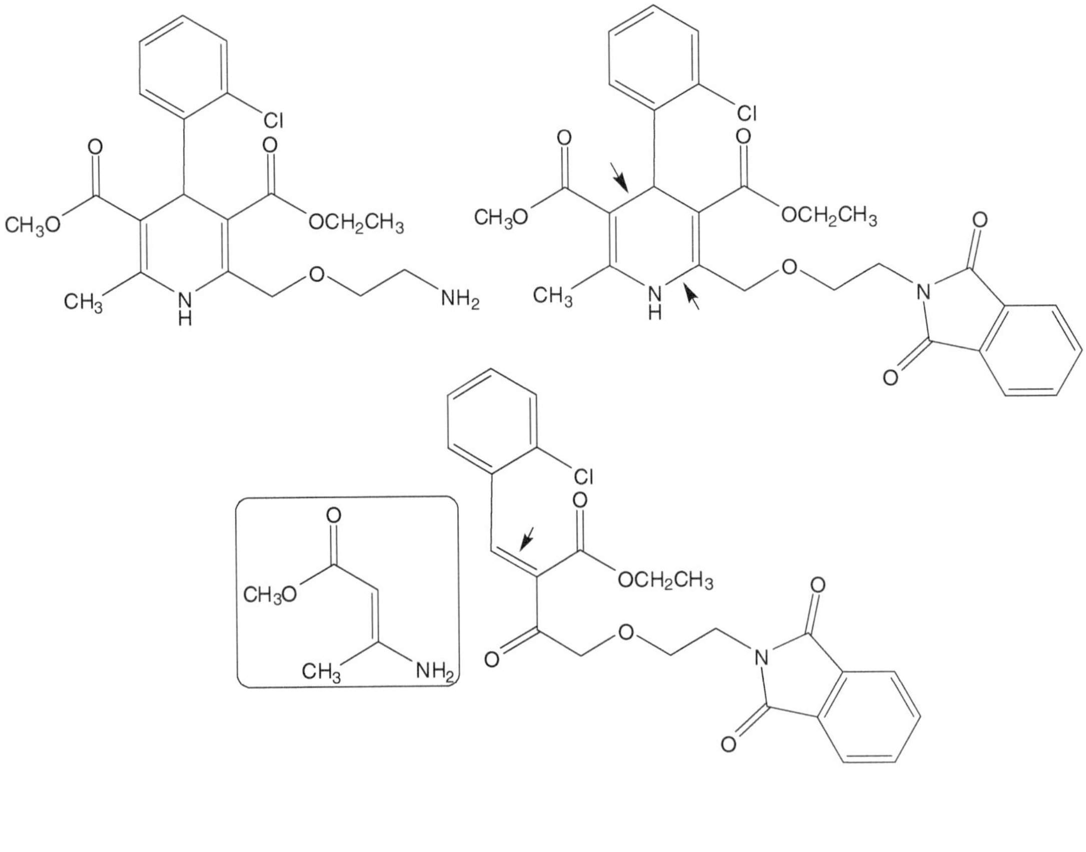

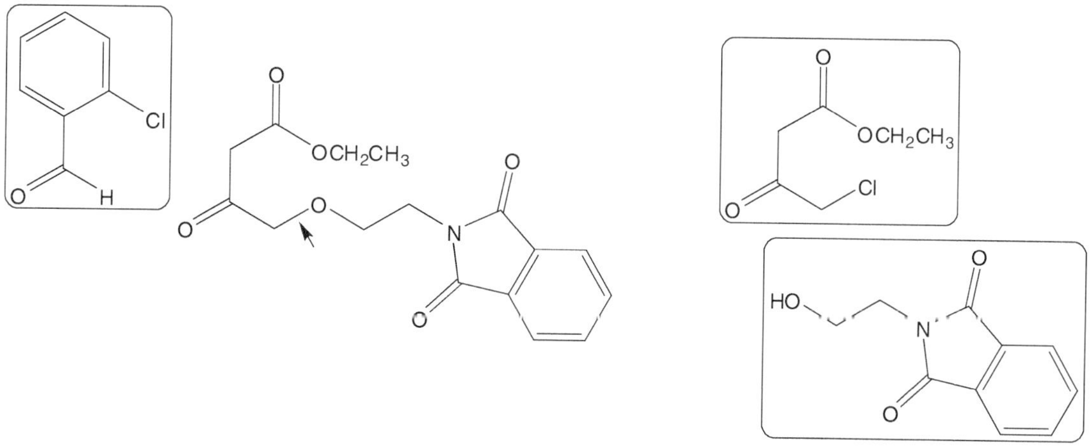

Extended Discussion

List the amine protecting groups that have been used in alternative syntheses of amlodipine. List the conditions used and amlodipine yields obtained in deprotection of each group in the final step. Why is the phthalimide group preferred?

Amodiaquine

Anti-Infective Medicines/Antiprotozoal Medicines/Antimalarial Medicines/For Curative Treatment

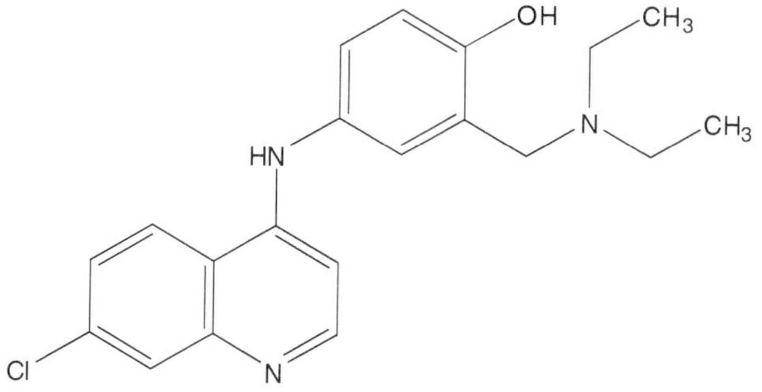

A nitrogen substituent at C2 or C4 on a quinoline ring is often introduced by displacement of chloride. The substitution is facilitated by the quinoline ring nitrogen and can be further facilitated by an electron-withdrawing group (NO_2, SO_2R, COOR, CN) on C3.

Discussion. A C—N bond is formed by displacement of a chloride at the quinoline 4-position by nitrogen of the 4-aminophenol.

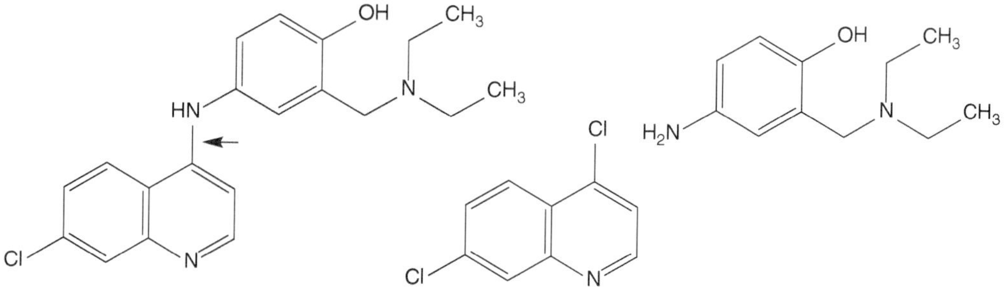

4,7-Dichloroquinoline is formed from 7-chloro- 4-hydroxyquinoline (7-chloroquinolin-4-one). 7-Chloro- 4-hydroxyquinoline is formed by thermolysis/decarboxylation of the 4-hydroxyquinoline-3-carboxylic acid. The carboxylic acid is formed by ester hydrolysis. The quinoline ring is formed by an intramolecular acylation at C6 of the 3-chloroaniline. An enamine is formed by reaction of the enol ether of ethyl ethoxymethylenemalonate with 3-chloroaniline (The four-step sequence from 3-chloroaniline to 7-chloro-4-hydroxyquinoline is an example of the **Gould–Jacobs Reaction**).

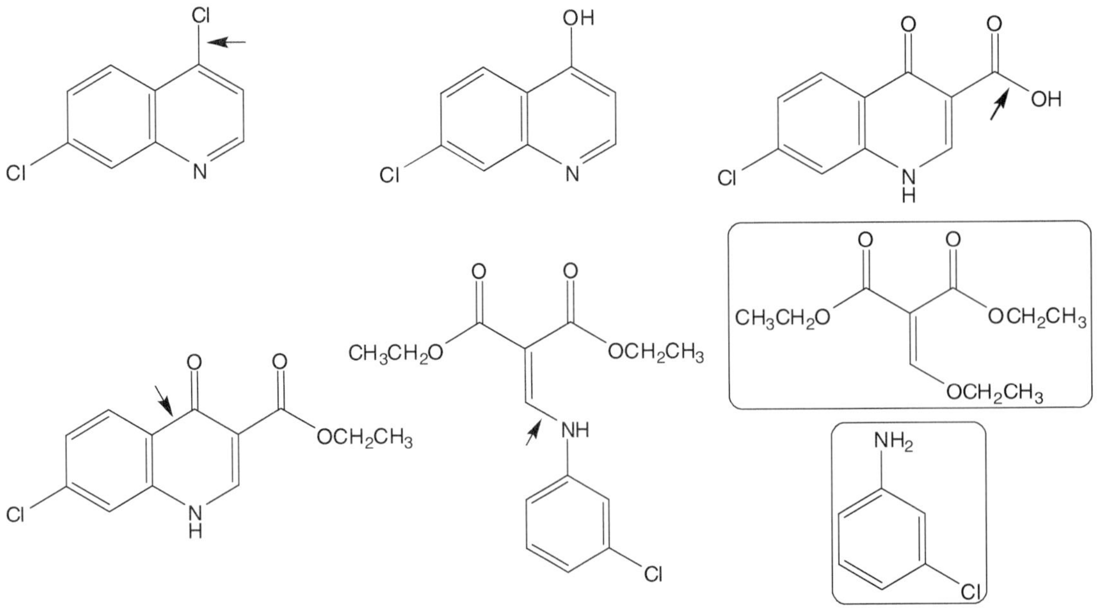

The 4-aminophenol is formed by hydrolysis of the acetanilide. Another essential medicine, 4-acetamidophenol (*para*-acetamol or acetaminophen) is alkylated by reaction with formaldehyde and *N,N*-diethylamine (**Betti Reaction**).

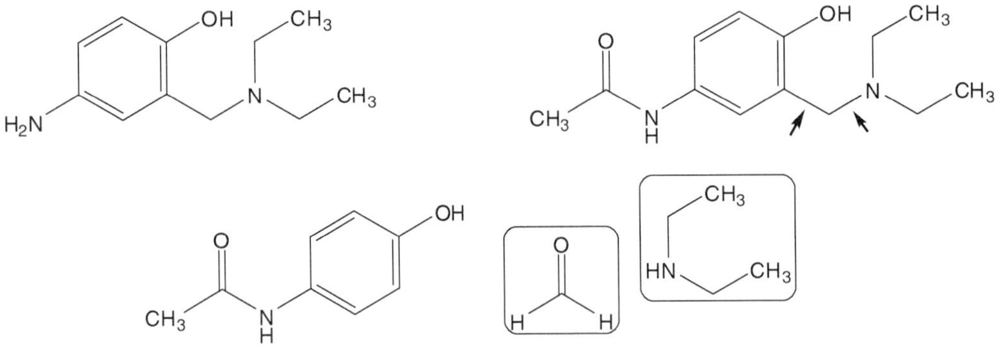

acetaminophen

Extended Discussion

Draw the structures of the retrosynthetic analysis of an alternative route from 3-chloroaniline to 7-chloro-4-hydroxyquinoline utilizing a **Conrad–Limpach Reaction**. Include the structures of the retrosynthetic analysis of any organic starting material(s) from petrochemical or biochemical raw materials. List pros and cons for the two routes and select one route as the preferred route.

Amoxicillin

Anti-Infective Medicines/Antibacterials/Beta-Lactam Medicines

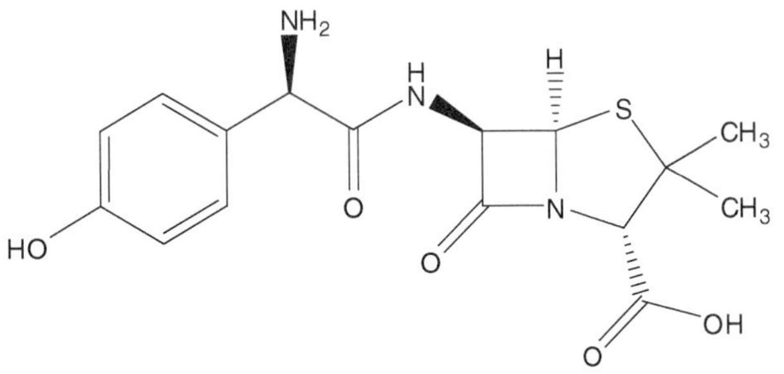

Penicillins are produced by fermentation or are semisynthetic. A semisynthetic penicillin is often formed by acylation of the amine of 6-aminopenicillanic acid (6-APA). 6-APA is produced from penicillin G (benzylpenicillin) by enzyme-mediated hydrolysis of the side-chain amide.

Discussion. Amoxicillin is a semisynthetic penicillin. The final step is enzyme-mediated formation of the side-chain amide by reaction of an amine with an ester. The amine, 6-aminopenicillanic acid (6-APA), is formed by enzyme-mediated hydrolysis of the side-chain amide of penicillin G. Penicillin G (benzylpenicillin) is produced by the fungus *Penicillium chrysogenum*.

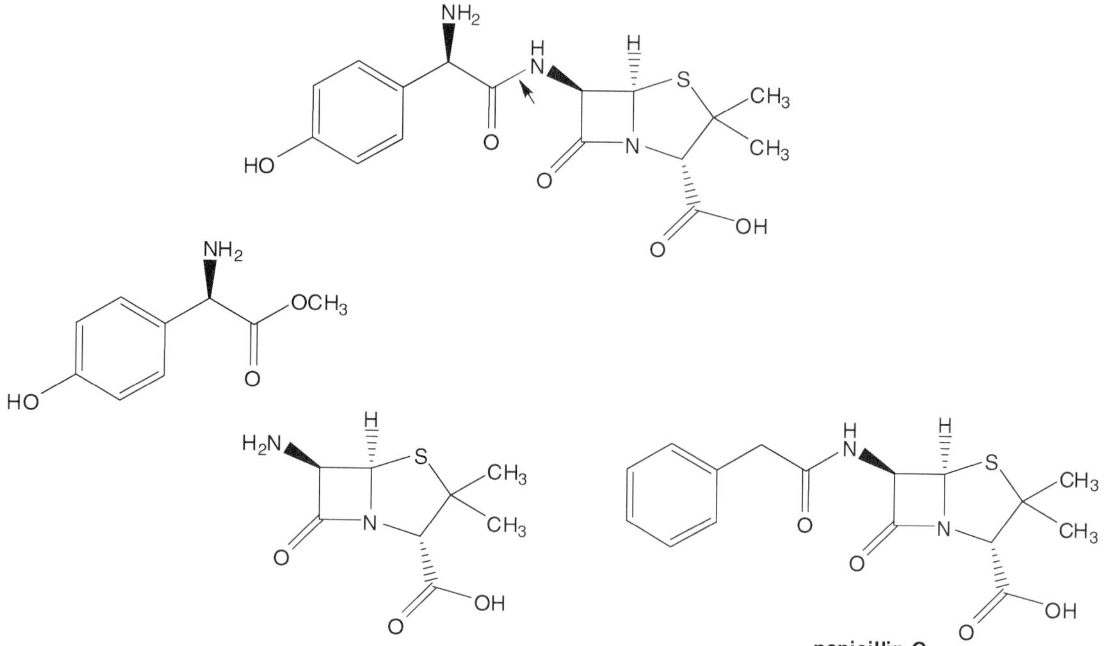

6-APA **penicillin G**

The ester is formed from the carboxylic acid (**Fischer Esterification**). The α-amino acid, (*R*)-α-(4-hydroxyphenyl)glycine, is formed by enzyme-mediated hydrolysis of the *N*-carbamoyl α-amino acid. The (*R*)-*N*-carbamoyl α-amino acid is formed by enzyme-mediated hydrolysis of the (*R*)-hydantoin in a mixture of the (*R*)- and (*S*)-hydantoins. Since the (*R*)- and (*S*)-hydantoins interconvert under the hydrolysis conditions, the (*S*)-hydantoin is also converted to the (*R*)-*N*-carbamoyl α-amino acid. The mixture of (*R*)- and (*S*)-hydantoins is formed from phenol, glyoxylic acid, and urea.

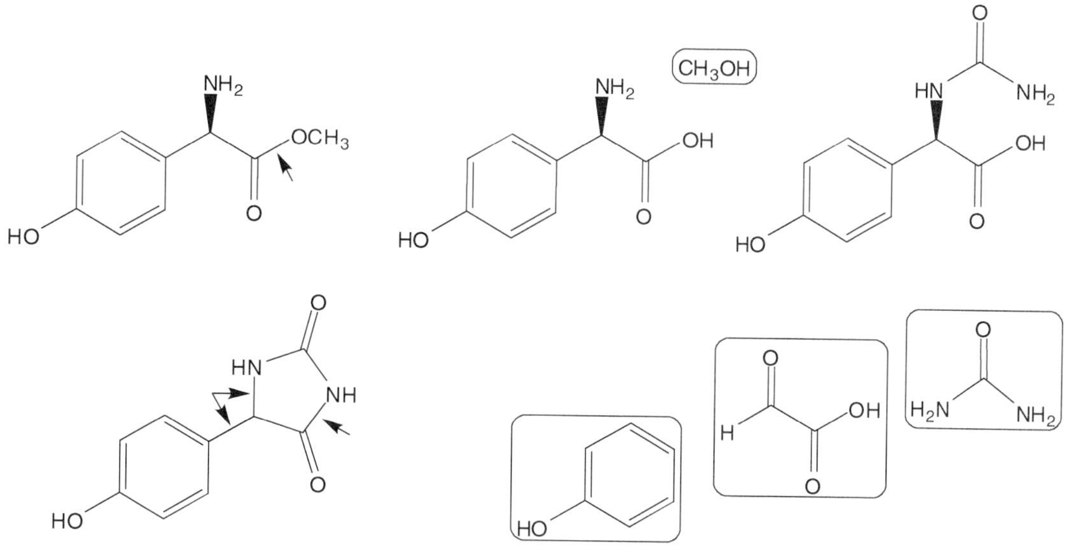

Extended Discussion

Draw the structures of the retrosynthetic analysis of an alternative route to amoxicillin from (*R*)-α-(4-hydroxyphenyl)glycine and 6-APA which does not utilize an enzyme to mediate the formation of the side-chain amide bond.

Ampicillin

Anti-Infective Medicines/Antibacterials/Beta-Lactam Medicines

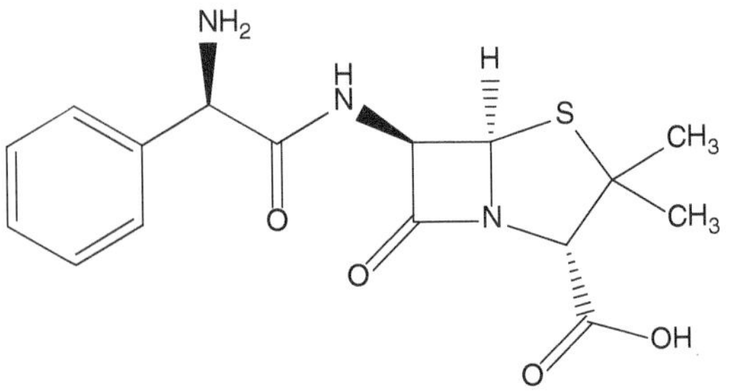

Penicillins are produced by fermentation or are semisynthetic. A semisynthetic penicillin is often formed by acylation of the amine of 6-aminopenicillanic acid (6-APA). 6-APA is produced from penicillin G (benzylpenicillin) by enzyme-mediated hydrolysis of the side-chain amide.

Discussion. Ampicillin is a semisynthetic penicillin. The final step is release of the amine by hydrolysis of an enamine. The side-chain amide is formed by enzyme-mediated formation of the reaction of 6-aminopenicillanic acid (6-APA) with a mixed anhydride. 6-Aminopenicillanic acid is formed by enzyme-mediated hydrolysis of the side-chain amide of penicillin G. Penicillin G (benzylpenicillin) is produced by the fungus *P. chrysogenum*.

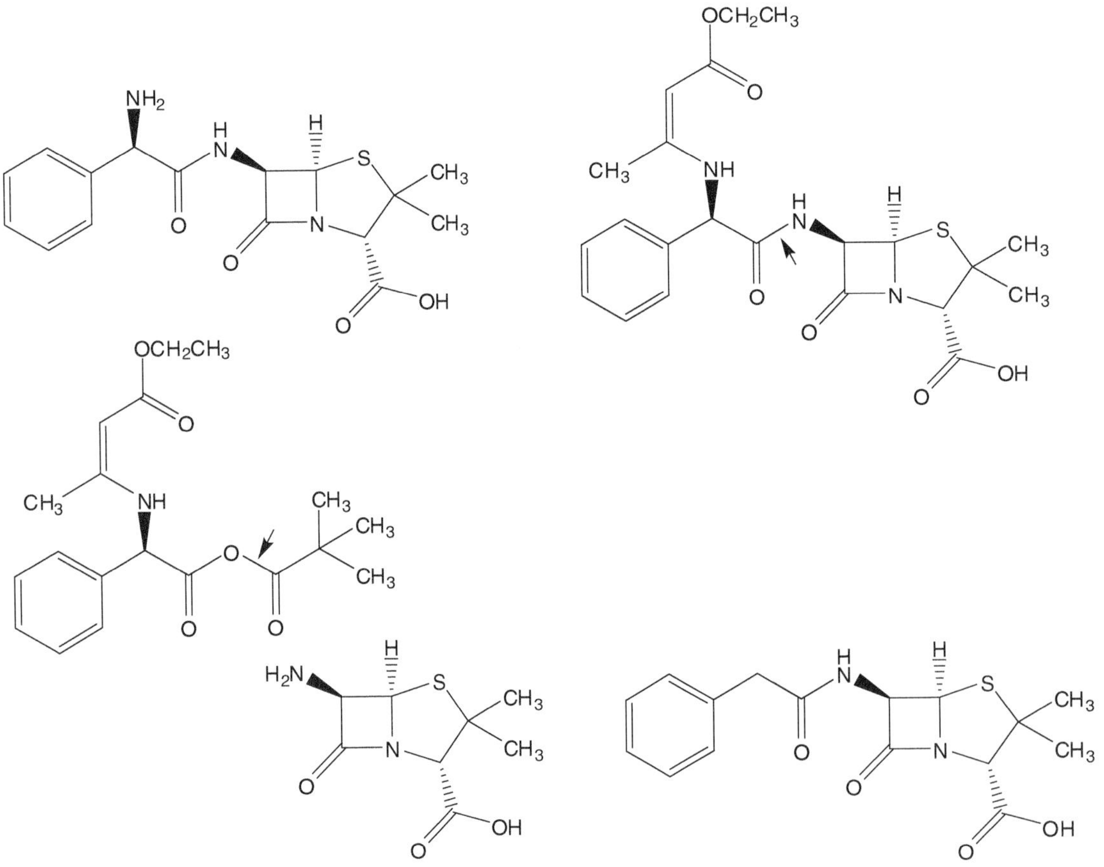

6-APA **penicillin G**

The mixed anhydride is formed from the potassium carboxylate salt and pivaloyl chloride. The N-protected potassium carboxylate salt of the amino acid (known as a **Dane Salt**) is formed from (*R*)-α-phenylglycine and ethyl acetoacetate.

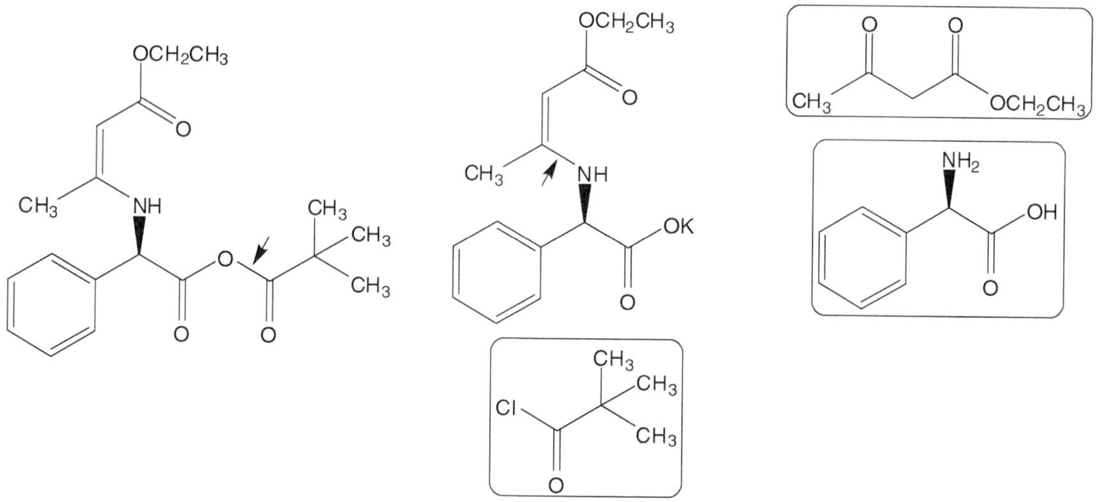

Extended Discussion

Ampicillin is also manufactured from 6-APA and the acid chloride, (*R*)-α-phenylglycine chloride hydrochloride. List the pros and cons for both routes and select one route as the preferred route.

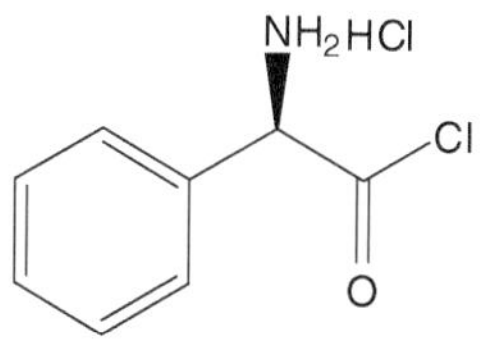

Anastrozole

Antineoplastics and Immunosuppressives/Hormones and Antihormones

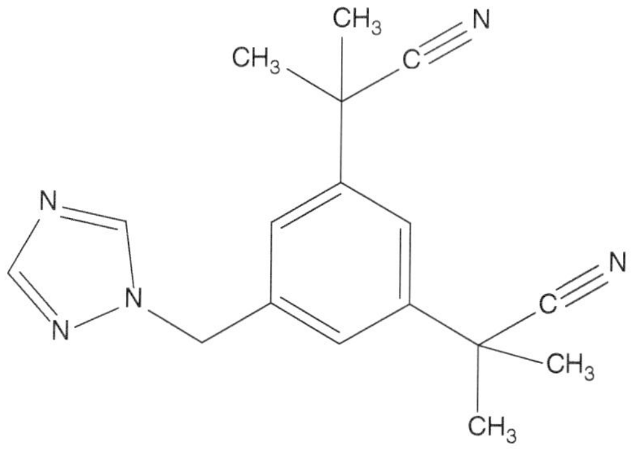

A phenylacetonitrile is often formed by displacement of a benzyl chloride or bromide by cyanide.

Discussion. Each substituent on the central ring of anastrozole has a functional group (cyanide or 1,2,4-triazole) which is likely introduced as a nucleophile. These features suggest disconnection strategies which have statistical product distribution problems. In a preferred strategy, the most significant problem is addressed in the preparation of the starting material.

Bromide is displaced by 1,2,4-triazole in the final step. The bromomethyl group is formed by bromination of the methyl group.

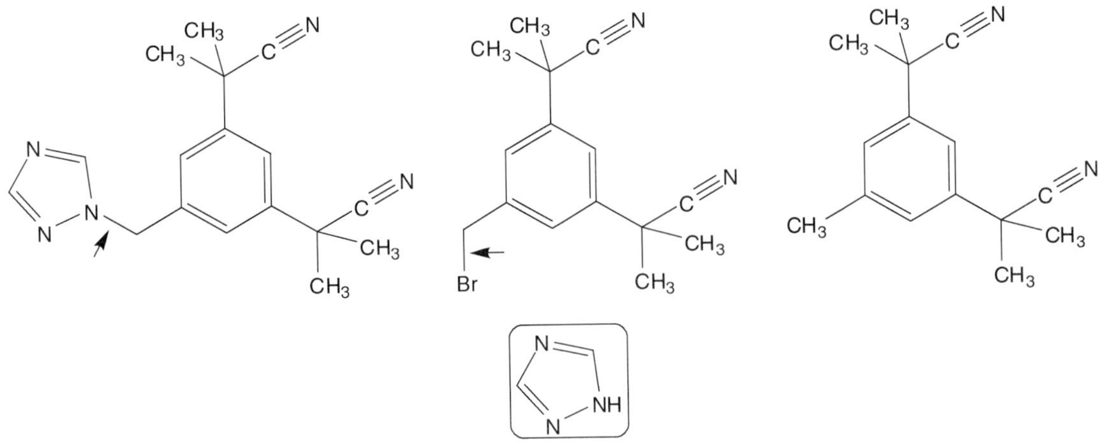

Four methyl groups are added by α-alkylation of the nitriles with iodomethane. The nitriles are formed by bromide displacement by sodium cyanide. The dibromide is formed by bromination of mesitylene.

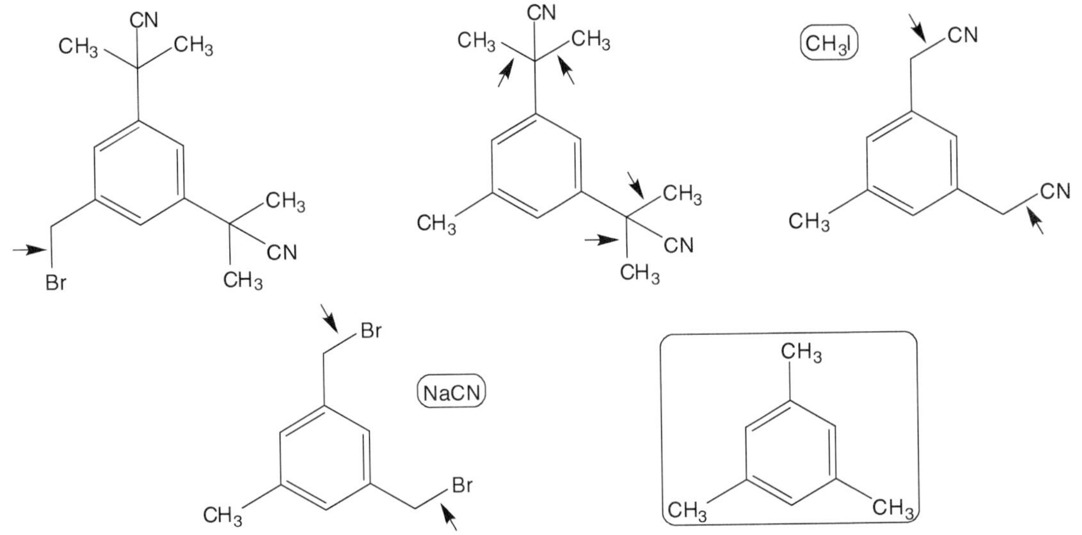

Extended Discussion

Final product purity is critical when manufacturing a drug substance. To ensure high purity of the product, no side products which are difficult to separate from the product should form in the final step. This is not the case for the anastrozole process. Explain. List the process details which ensure that anastrozole meets high purity specifications.

Artemether

Anti-Infective Medicines/Antiprotozoal Medicines/Antimalarial Medicines/For Curative Treatment

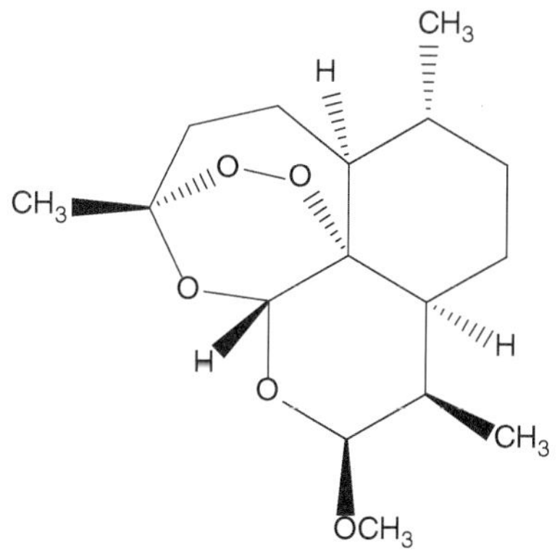

A single-enantiomer molecule with multiple chiral carbons is often formed by modification of a natural product which has most or all of the chiral carbons already in place.

Discussion. Artemether (β-artemether) is semisynthetic, it is manufactured in two steps from artemisinin. The methyl acetal of β-artemether is formed by the acid-catalyzed reaction of the hemiacetal (dihydroartemisinin) with methanol. The hemiacetal of dihydroartemisinin is formed by reduction of the ester of artemisinin. Artemisinin is a natural product isolated from the plant *Artemisia annua* or sweet wormwood.

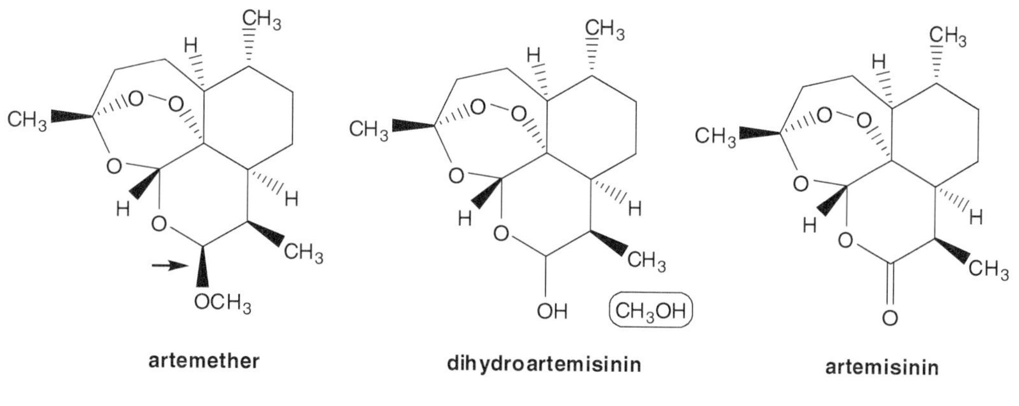

Artemisinin can also be manufactured in four steps from artemisinic acid. In the last step, a hydroperoxide is formed by α-oxidation of an aldehyde with triplet oxygen. The aldehyde, hydroperoxide, ketone, and carboxylic acid then assemble to form artemisinin. The aldehyde and ketone are formed by cleavage of an allylic hydroperoxide (**Hock Rearrangement**). The allylic hydroperoxide is formed from dihydroartemisinic acid (**Ene Reaction**). Dihydroartemisinic acid is formed by reduction of artemisinic acid. Artemisinic acid is a natural product also isolated from the plant *A. annua* or sweet wormwood. Artemisinic acid is also produced by fermentation.

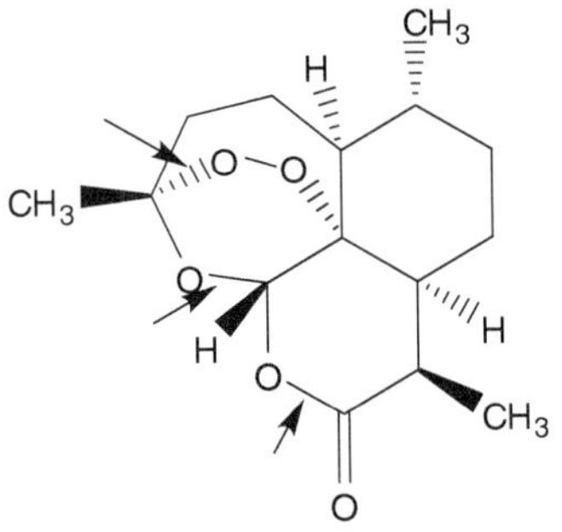

artemisinin

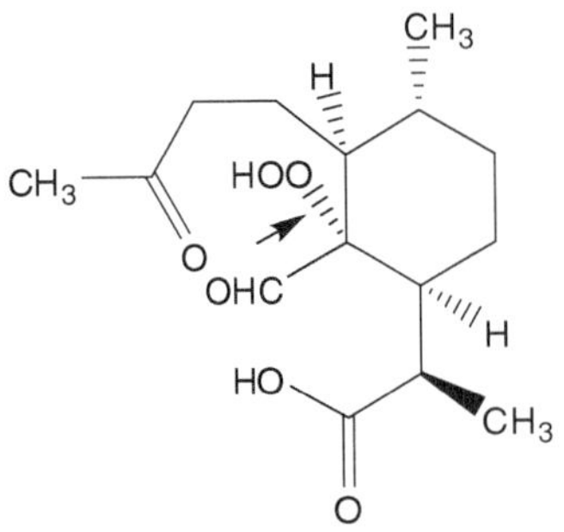

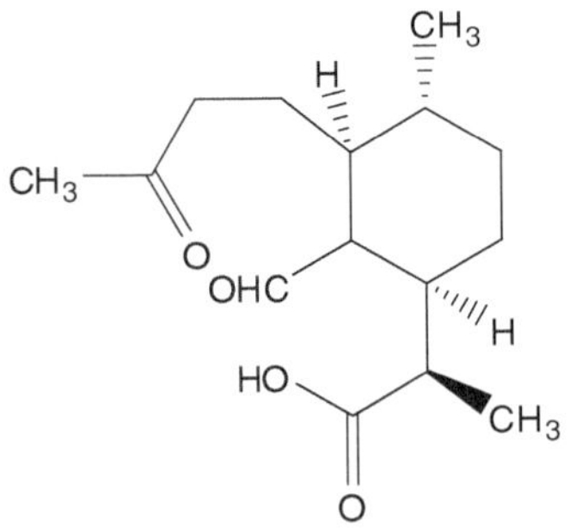

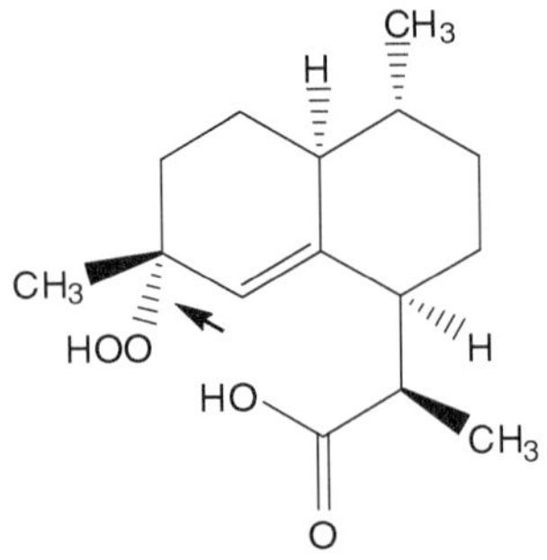

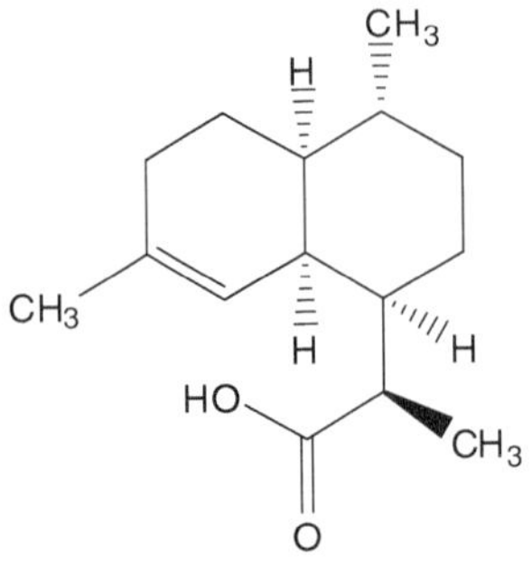

dihydroartemisinic acid

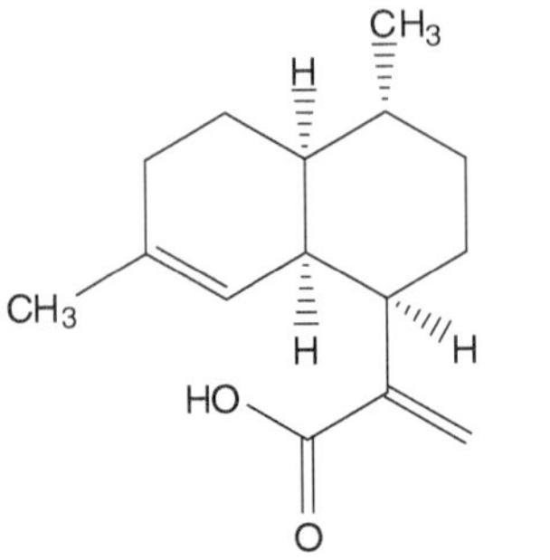

artemisinic acid

Extended Discussion

Draw the structures of four impurities which are likely to form in the conversion of dihydroartemisinin to artemether.

Artesunate

Anti-Infective Medicines/Antiprotozoal Medicines/Antimalarial Medicines/For Curative Treatment

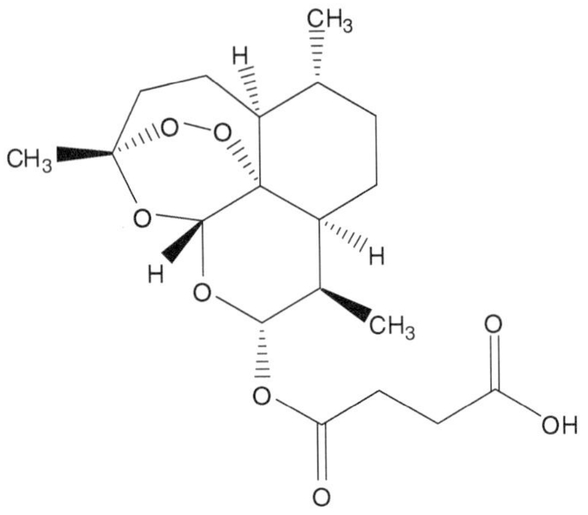

A single-enantiomer molecule with multiple chiral carbons is often formed by modification of a natural product which has most or all of the chiral carbons already in place.

Discussion. Artesunate is semisynthetic, and it is manufactured in two steps from artemisinin. The ester is formed by reaction of the hemiacetal (dihydroartemisinin) with succinic anhydride. The hemiacetal of dihydroartemisinin is formed by reduction of the ester of artemisinin. Artemisinin is a natural product isolated from the plant *A. annua* or sweet wormwood.

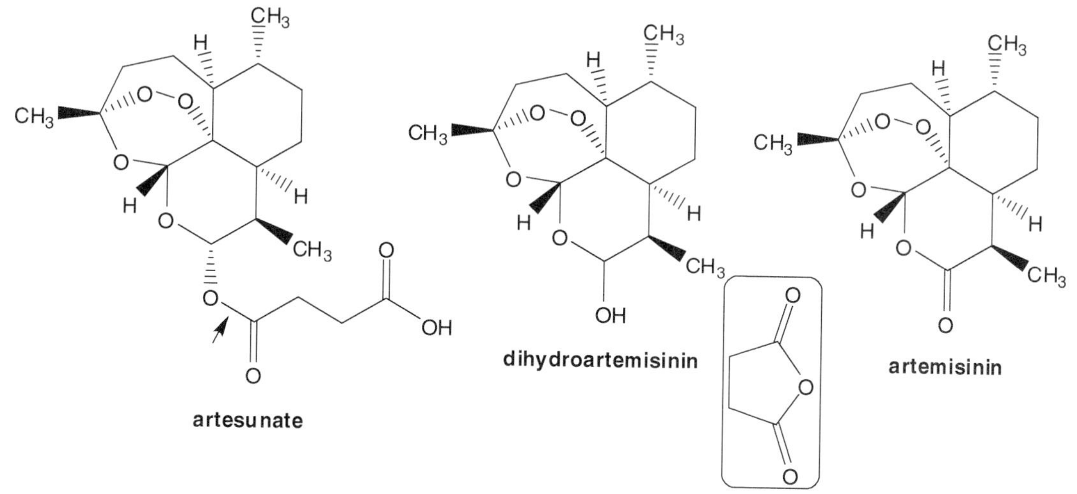

Artemisinin is also manufactured in four steps from artemisinic acid. In the last step, a hydroperoxide is formed by α-oxidation of an aldehyde with triplet oxygen. The aldehyde, hydroperoxide, ketone, and carboxylic acid then assemble to form artemisinin. The aldehyde and ketone are formed by cleavage of an allylic hydroperoxide (**Hock Rearrangement**). The allylic hydroperoxide is formed from the alkene (**Ene Reaction**). The alkene, dihydroartemisinic acid, is formed by reduction of artemisinic acid. Artemisinic acid is a natural product also isolated from the plant *A. annua* or sweet wormwood. Artemisinic acid is also produced by fermentation.

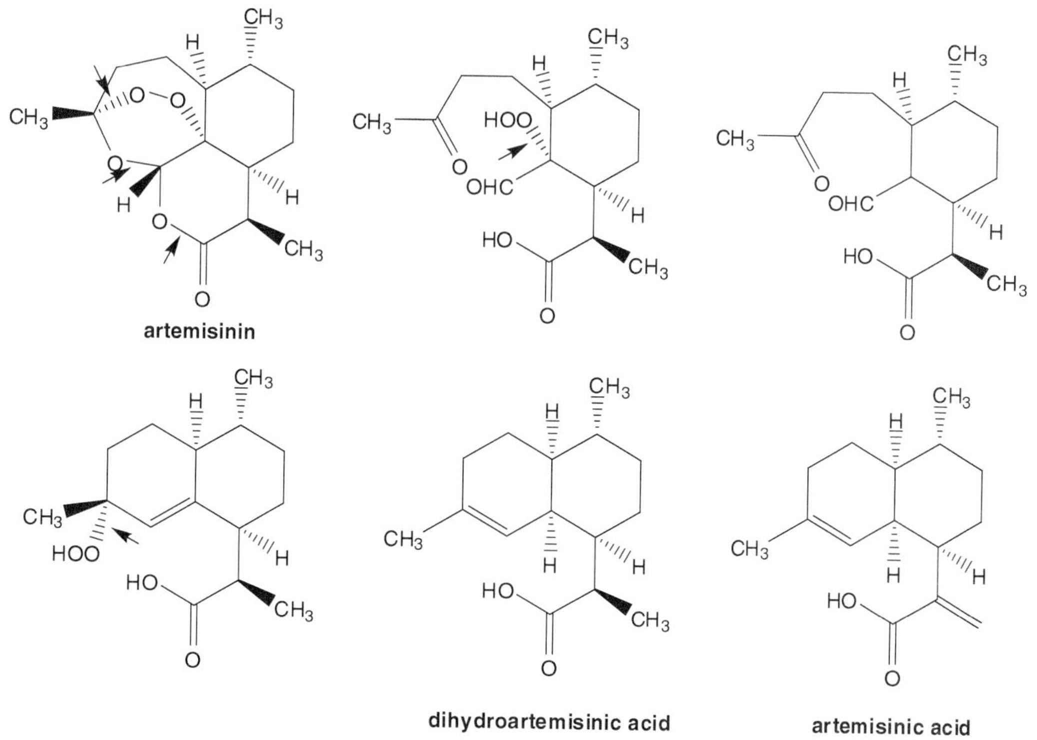

Extended Discussion

β-Artemether and α-artesunate are both formed from dihydroartemisinin. Draw the structures of a retrosynthetic analysis of β-artesunate.

Ascorbic Acid

Vitamins and Minerals

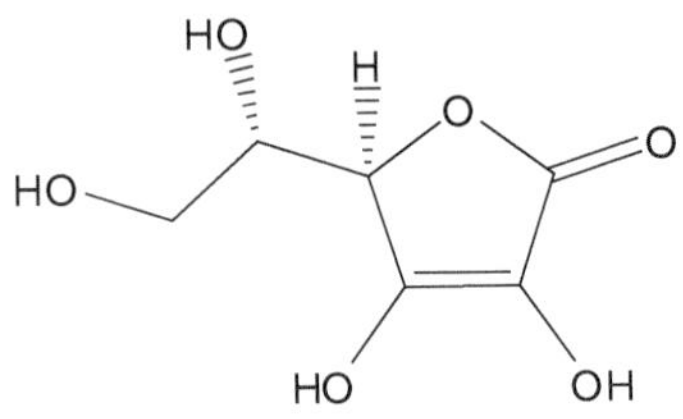

A single-enantiomer molecule with multiple chiral carbons is often formed by modification of a natural product which has most or all of the chiral carbons already in place.

Discussion. Ascorbic acid (vitamin C) is semisynthetic. Ascorbic acid is formed from 2-keto-L-gulonic acid. 2-Keto-L-gulonic acid is produced by fermentation from L-sorbose. L-Sorbose is produced by fermentation from D-sorbitol.

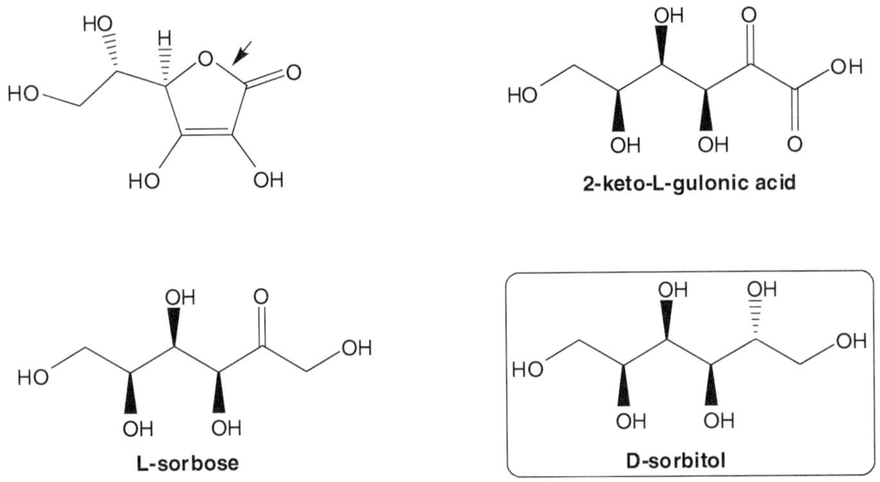

Extended Discussion

Draw the structures of the retrosynthetic analysis of an alternative non-fermentation route to 2-keto-L-gulonic acid from L-sorbose.

Atazanavir

Anti-Infective Medicines/Antiviral Medicines/Antiretrovirals/Protease Inhibitors

A β-amino alcohol with a primary β-C is often formed by ring-opening of an epoxide by an amine.

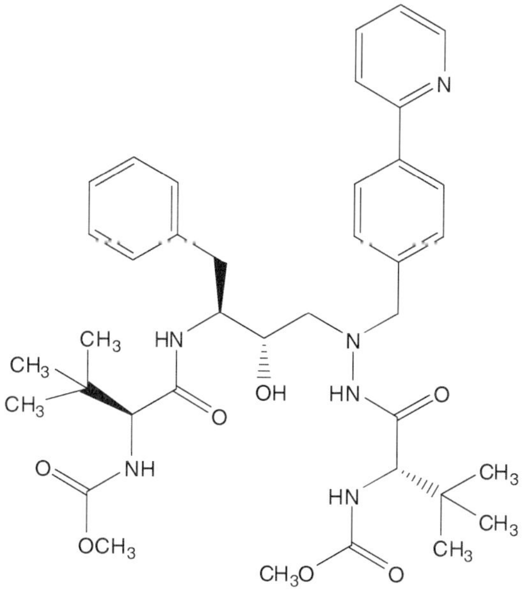

Discussion. Two amides are formed with expensive *N*-(methoxycarbonyl)-L-tertleucine (Moc-L-tertleucine) in the final step. The amine and hydrazine needed to form the amides are released by hydrolysis of *tert*-butoxycarbonyl (Boc) protecting groups. A key C—N bond near the center of the molecule is formed by ring-opening of an epoxide with a Boc-protected hydrazine.

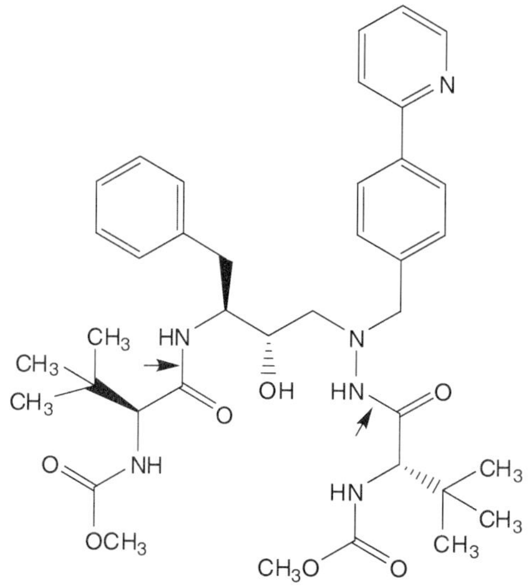

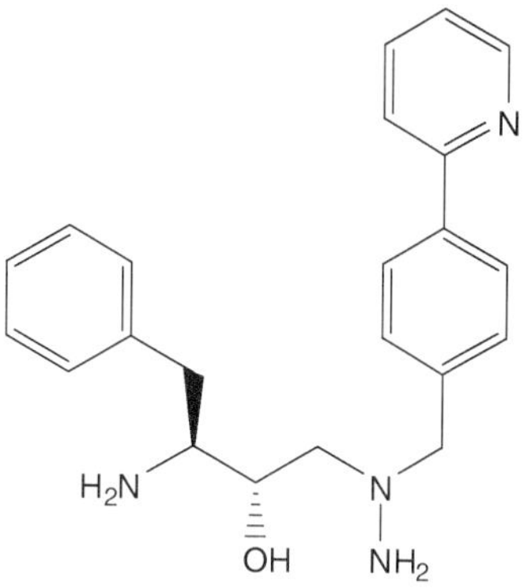

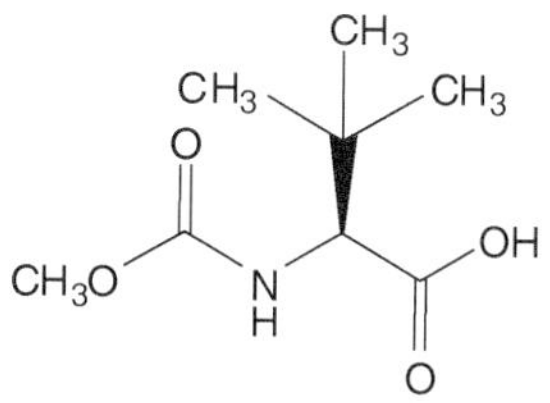

Moc-L-tertleucine

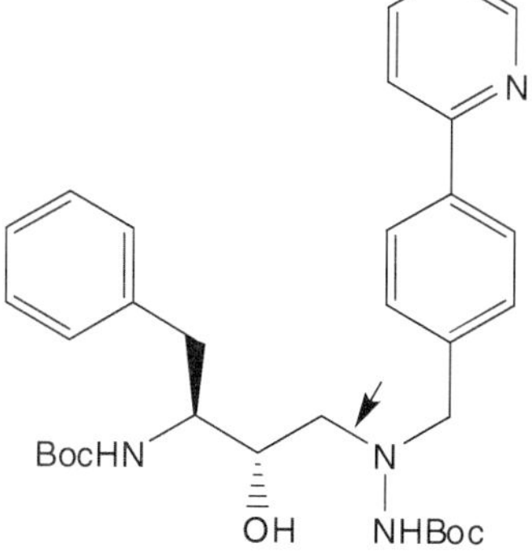

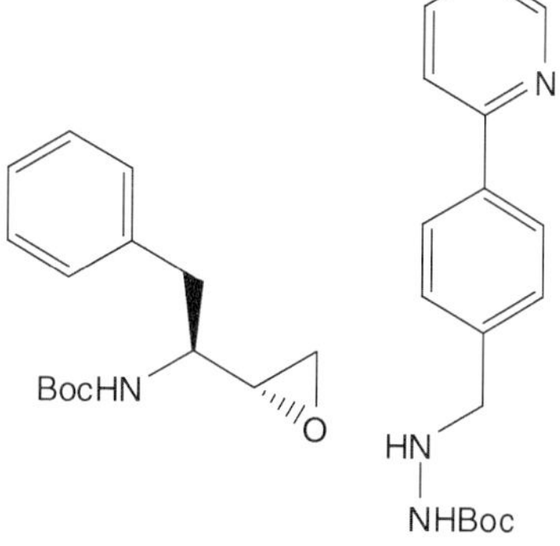

Moc-L-Tertleucine is formed from L-tertleucine and methyl chloroformate. L-Tertleucine is formed by an enzyme-mediated reductive amination of trimethylpyruvic acid. The pyruvic acid is formed by oxidation of the α-hydroxyacid. The α-hydroxyacid is formed from 1,1-dichloropinacolone by rearrangement and hydrolysis. 1,1-Dichloropinacolone is formed by α–chlorination of pinacolone.

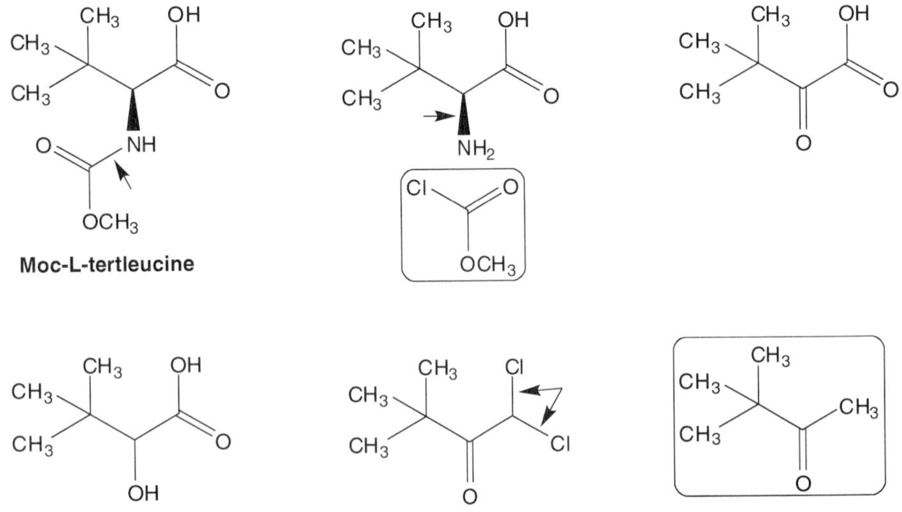

The epoxide is formed from the chlorohydrin by nucleophilic displacement of chloride by oxygen. The chlorohydrin is formed by reduction of the α-chloroketone, *N*-(*tert*-butoxycarbonyl)-3(*S*)-amino-1-chloro-4-phenyl-2-butanone. The α-chloroketone is formed by reduction of the α,α-dichloroketone. The α,α-dichloroketone is formed from *N*-(*tert*-butoxycarbonyl)-L-phenylalanine methyl ester and dichloromethane.

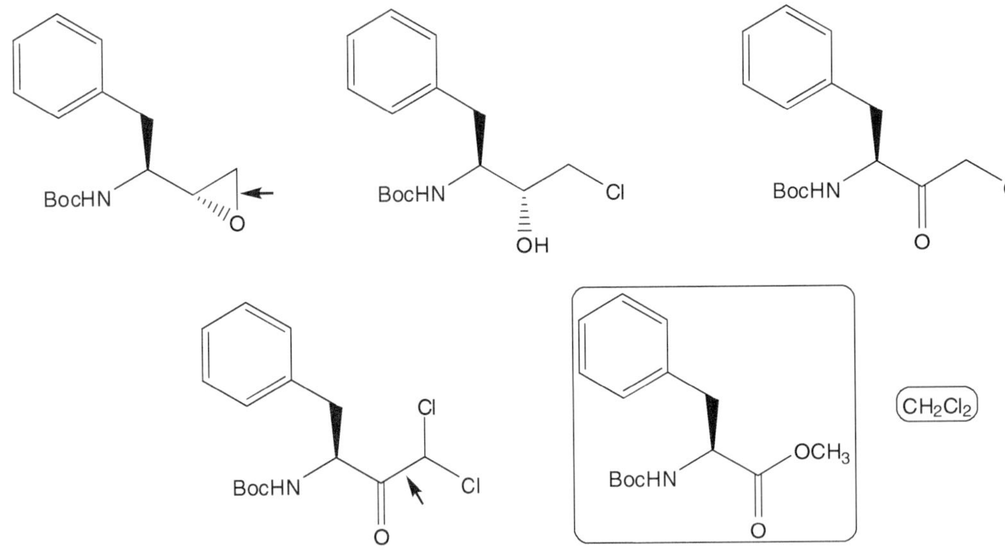

The Boc-protected hydrazine is formed by reduction of the hydrazone. The hydrazone is formed by reaction of the aldehyde with Boc-hydrazine. The aldehyde, 4-(2-pyridyl)benzaldehyde, is formed from 2-bromopyridine and 4-formylbenzeneboronic acid (**Suzuki–Miyaura Coupling**). The boronic acid is formed from 4-chlorobenzaldehyde and trimethylborate via an acetal.

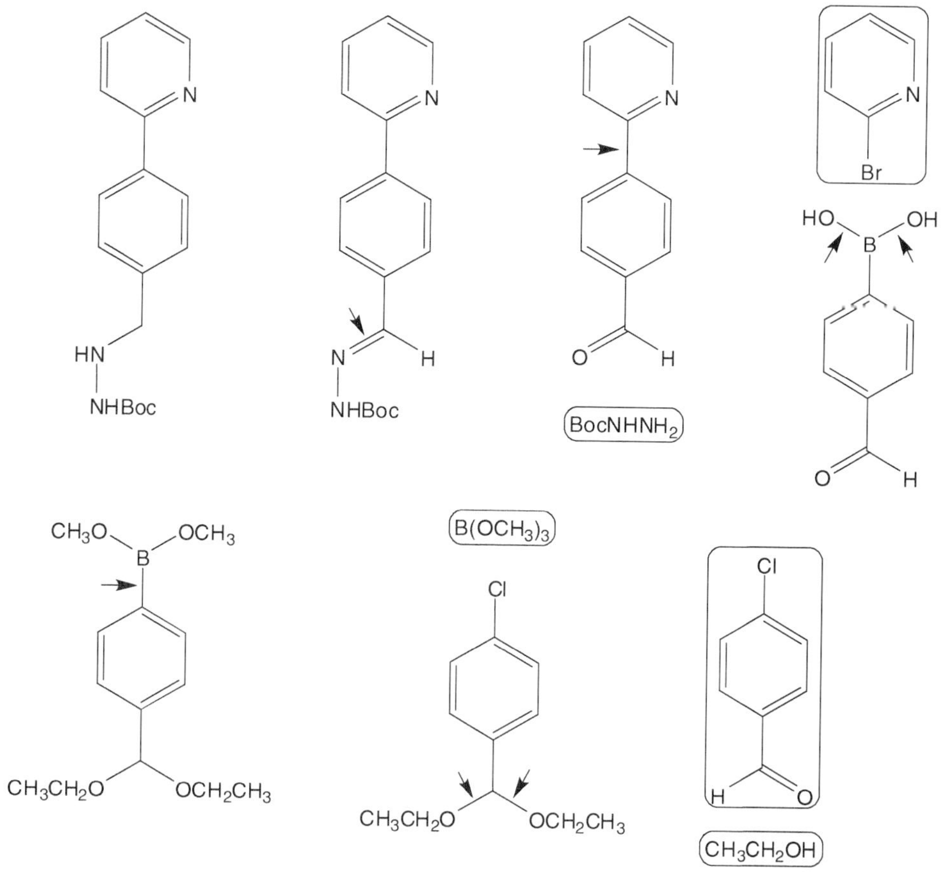

Extended Discussion

In the route presented, the key C—N bond is formed by ring-opening of an epoxide. Draw the structures of the retrosynthetic analysis of an alternative route to atazanavir which forms the same key C—N bond but does not involve an epoxide intermediate. List the pros and cons for both routes and select one route as the preferred route.

Atracurium Besylate

Muscle Relaxants (Peripherally Acting) and Cholinesterase Inhibitors

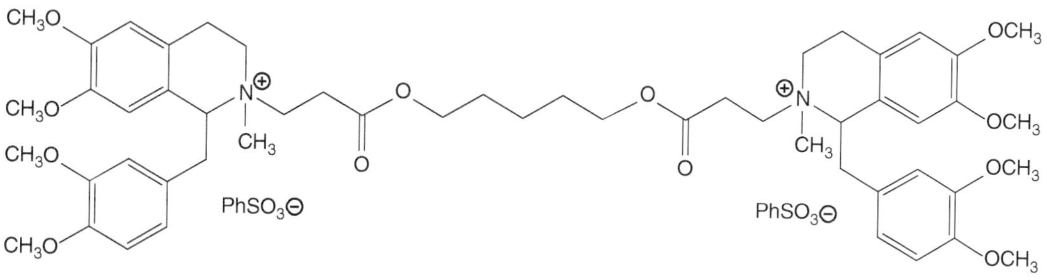

A tertiary amine is often formed by alkylation of a secondary amine.

Discussion. Atracurium besylate is produced as a mixture of 10 stereoisomers (label the four chiral centers in atracurium besylate). The discussion will be limited to routes to one of the 10 stereoisomers, the (1*R*-*cis*, 1′*R*-*cis*)-isomer known as cisatracurium besylate.

In one preferred route, the quaternary salts are formed in the final step by reaction of the tertiary amines with methyl benzenesulfonate (draw structures for two side products formed in this reaction. How is cisatracurium besylate separated from the side products?) The tertiary amines are formed by conjugate addition of (*R*)-tetrahydropapaverine to the acrylate.

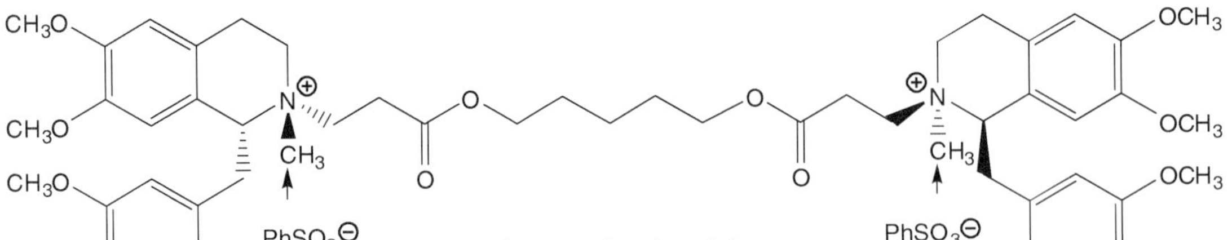

cisatracurium besylate

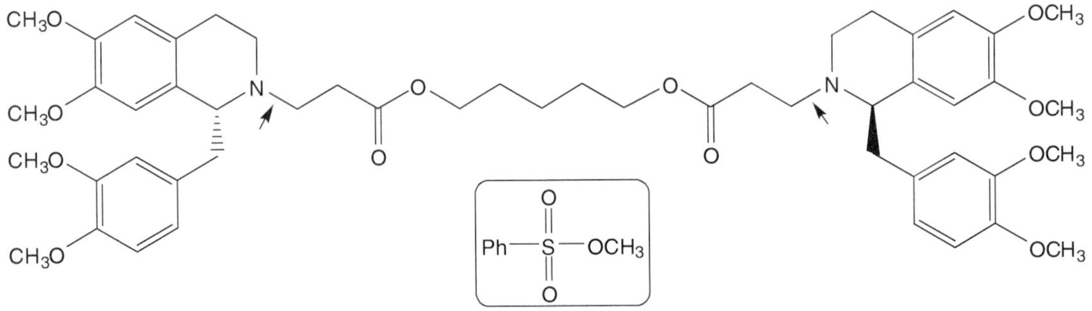

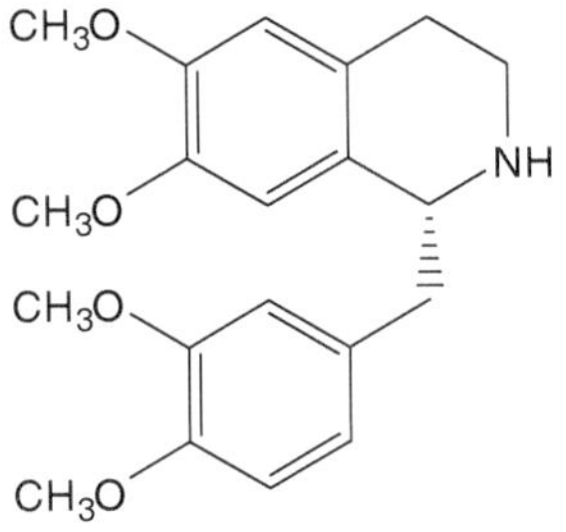

(R)-tetrahydropapaverine

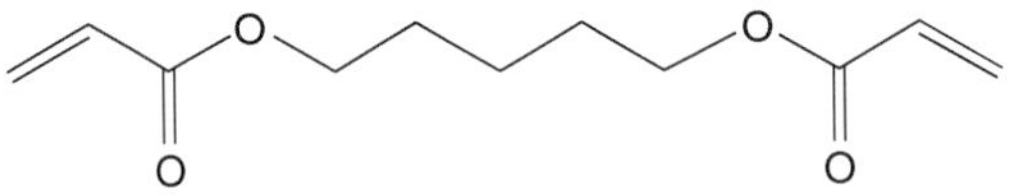

(*R*)-Tetrahydropapaverine is formed by resolution of tetrahydropapaverine. Tetrahydropapaverine is formed by reduction of dihydropapaverine. The dihydroisoquinoline ring of dihydropapaverine is formed from the amide (**Bischler–Napieralski Reaction**). The amide is formed from 3,4-dimethoxyphenylacetic acid and 3,4-dimethoxyphenethylamine.

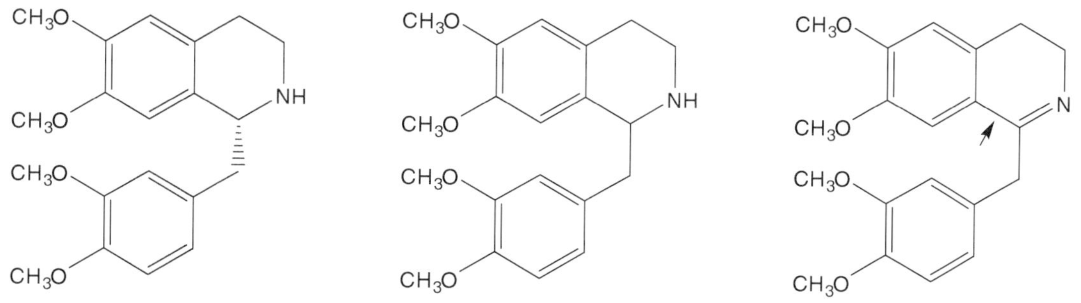

(R)-tetrahydropapaverine

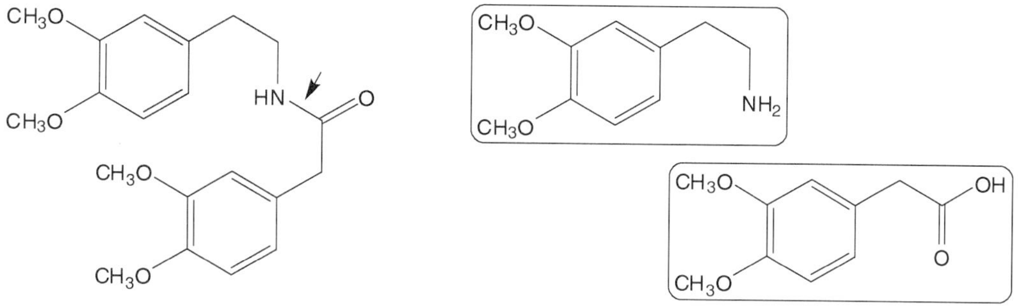

The acrylate is formed by elimination of hydrogen bromide. The diester is formed from 1,5-pentanediol and 3-bromopropanoic acid (**Fischer Esterification**).

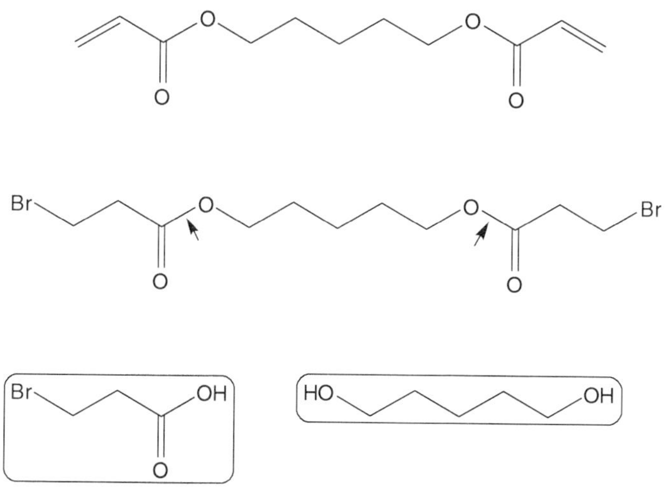

Extended Discussion

Draw the structures of the retrosynthetic analysis of one alternative route to cisatracurium besylate. List the pros and cons for both routes. Is one route preferred?

Atropine

Anesthetics, Preoperative Medicines and Medical Gases/Preoperative Medication and Sedation for Short-Term Procedures
Antidotes and Other Substances Used in Poisonings/Specific
Ophthalmological Preparations/Mydriatics

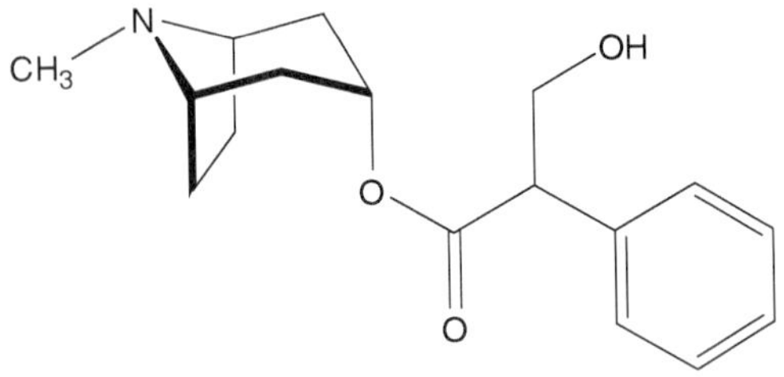

A rigid bicyclic structure is often used to direct the formation of a new chiral carbon.

Discussion. Atropine, a 1 : 1 mixture of the tropane alkaloids (*R*)-hyoscyamine and (*S*)-hyoscyamine, is usually produced by extraction from the plants *Atropa belladonna*, *Datura stramonium*, or *Duboisis myoporoides*.

Atropine can also be synthesized from tropic acid and tropinone. In the final step of the synthesis, the primary alcohol is released by acetate ester hydrolysis. The tropic acid ester is formed from the acid chloride and the alcohol, 3-tropanol (tropine).

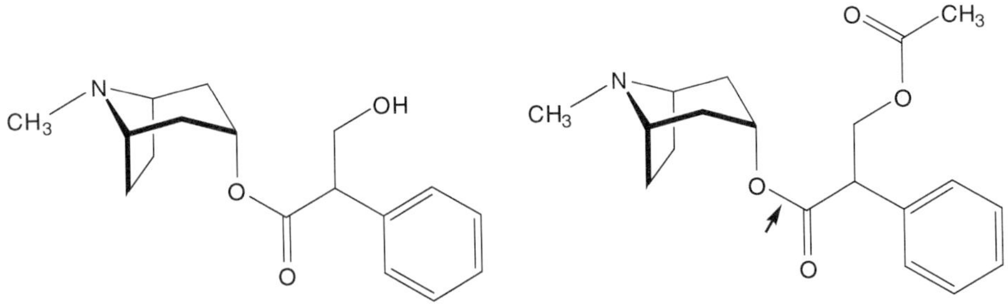

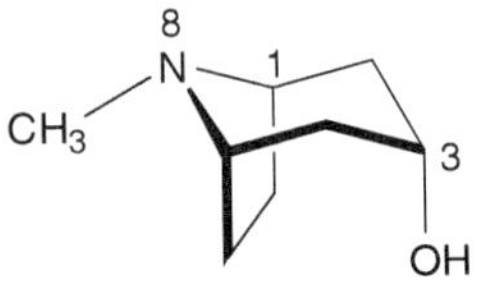

tropine

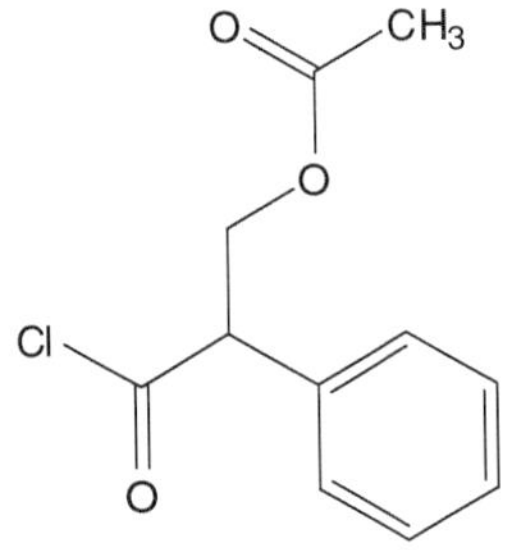

In a one-pot procedure, the acetate ester is formed from acetyl chloride and the primary alcohol of tropic acid then the acid chloride is then formed from the carboxylic acid. Tropic acid is formed by hydrolysis of the ethyl ester. The α-hydroxymethyl ester is formed by reduction of the α-formyl ester. The α-formyl ester is formed from ethyl phenylacetate and ethyl formate (mixed **Claisen Condensation**).

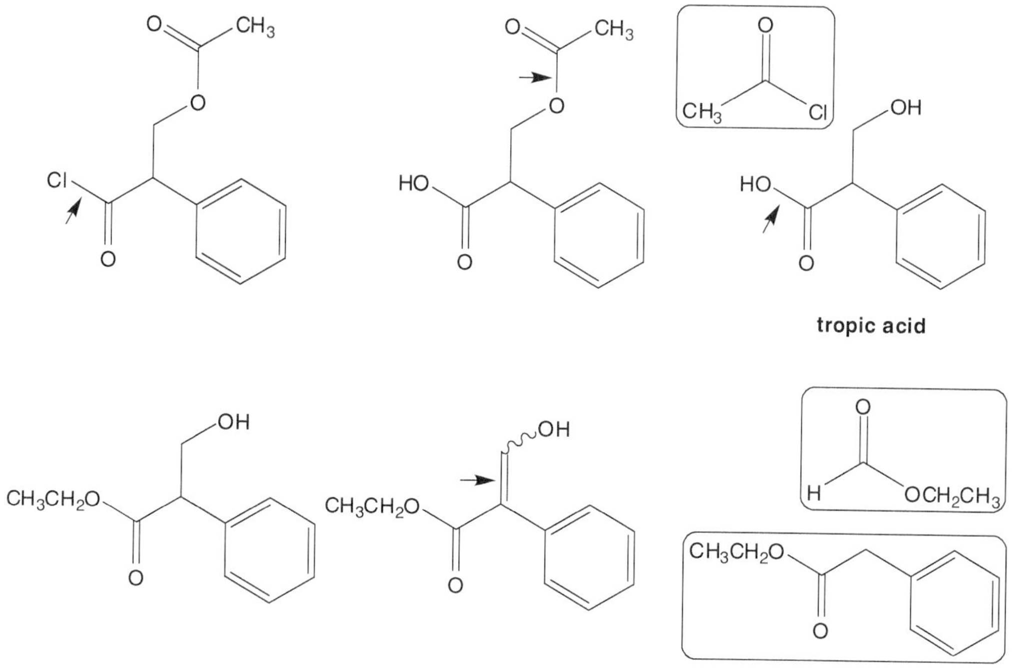

The axial alcohol of tropine is formed by reduction of the ketone. Tropinone is efficiently assembled in a single step from methylamine, 2,5-dimethoxytetrahydrofuran, and 1,3-acetonedicarboxylic acid (**Robinson–Schopf Reaction**). 1,3-Acetonedicarboxlic acid is formed by oxidative decarboxylation of citric acid. Citric acid is produced by fermentation.

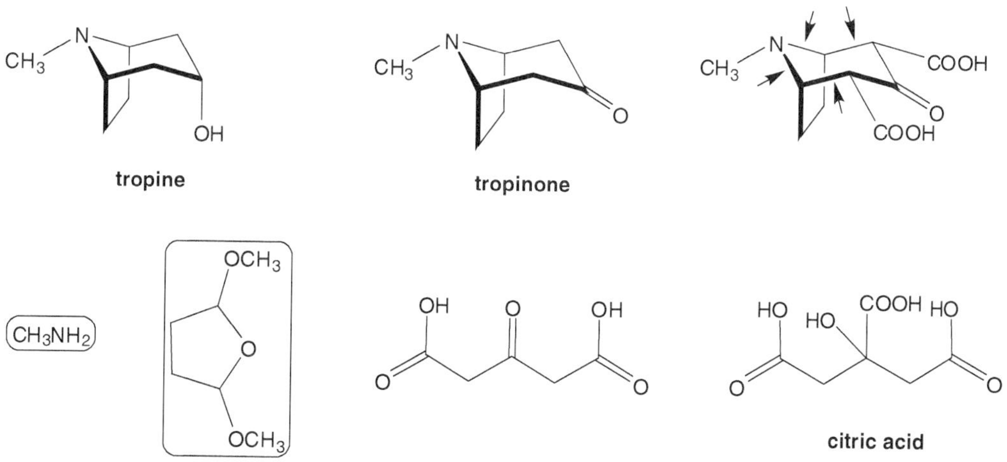

Extended Discussion

The yield in the first one-pot synthesis of tropinone, described by Robinson in 1917, was just 17%. After a century of process development, the yield for the one-pot synthesis is now 90%! List the references for the available procedures, the process modification(s) made, and the tropinone yield. Which process modification had the greatest impact on the yield?

Azathioprine

Antineoplastics and Immunosuppressives/Immunosuppressive Medicines
Medicines for Diseases of Joints/Disease-Modifying Agents Used in Rheumatoid Disorders

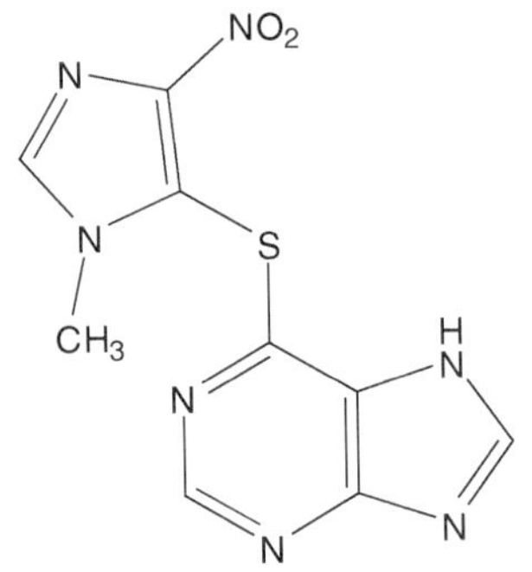

An aromatic thioether is often formed by displacement of chloride or bromide by a thiol.

Discussion. The thioether is formed in the final step by displacement of chloride from 5-chloro-1-methyl-4-nitroimidazole by 6-mercaptopurine. 5-Chloro-1-methyl-4-nitroimidazole is formed by nitration of 5-chloro-1-methylimidazole. 5-Chloro-1-methylimidazole is formed from *N,N'*-dimethyloxamide (**Wallach Imidazole Synthesis**). The oxamide is formed from diethyl oxalate and methylamine. 6-Mercaptopurine is formed from hypoxanthine.

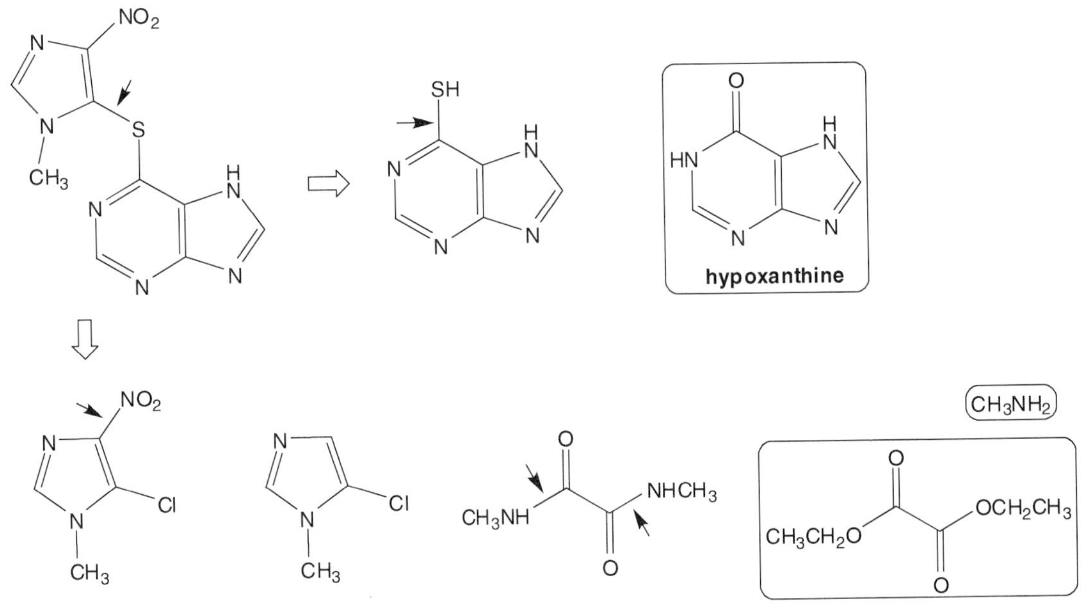

Extended Discussion

Draw the structures of a retrosynthetic analysis of one alternative route to the thioether by nucleophilic aromatic substitution of chloride from 6-chloropurine. Compare the two routes and select one route as the preferred route.

Azithromycin

Anti-Infective Medicines/Antibacterials/Other Antibacterials
Ophthalmological Medicines/Anti-Infective Agents

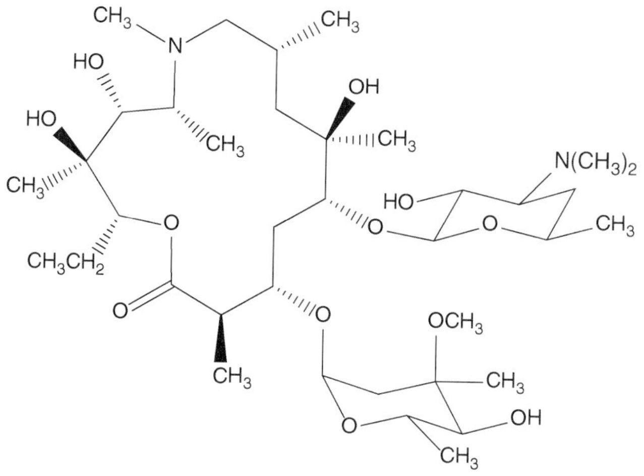

Macrolide antibiotics are produced by fermentation or are semisynthetic. The process for manufacture of a semisynthetic macrolide antibiotic often begins with conversion of the C9 ketone of erythromycin A to the oxime.

Discussion. Azithromycin is semisynthetic. Azithromycin is formed from the essential medicine erythromycin A which is produced by fermentation.

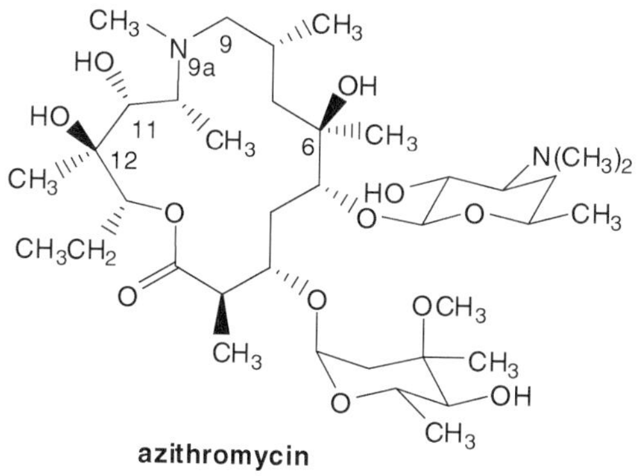

azithromycin

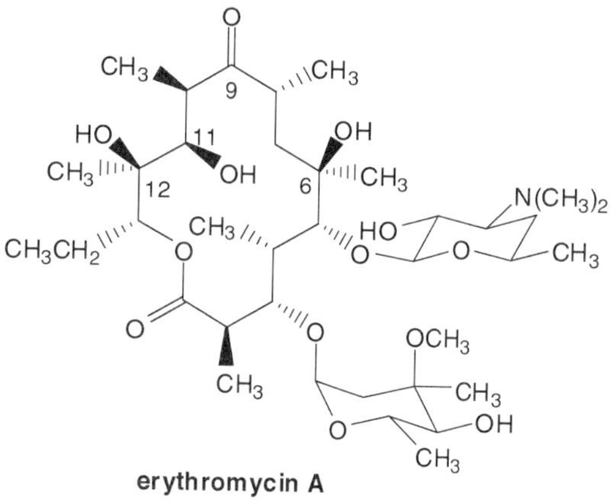

erythromycin A

The tertiary amine of azithromycin is formed in the final step by methylation of the secondary amine with formaldehyde and formic acid (**Eschweiler–Clarke Reaction**). The secondary amine (9-deoxo-9a-aza-homoerythromycin) is formed by reduction of an iminoether. The 6,9-iminoether (the iminoether involving the hydroxyl group at C6 and carbon at C9) is formed by rearrangement of an (*E*)-*O*-arylsulfonyl oxime (**Beckmann Rearrangement**). The (*E*)-*O*-arylsulfonyloxime is formed from the (*E*)-oxime and an arylsulfonyl chloride (List the arylsulfonyl chlorides and (*O*)-arylsulfonyloxime yields). (*E*)-Erythromycin A oxime is formed from erythromycin A. (How is (*E*)-erythromycin A oxime separated from (*Z*)-erythromycin A oxime?)

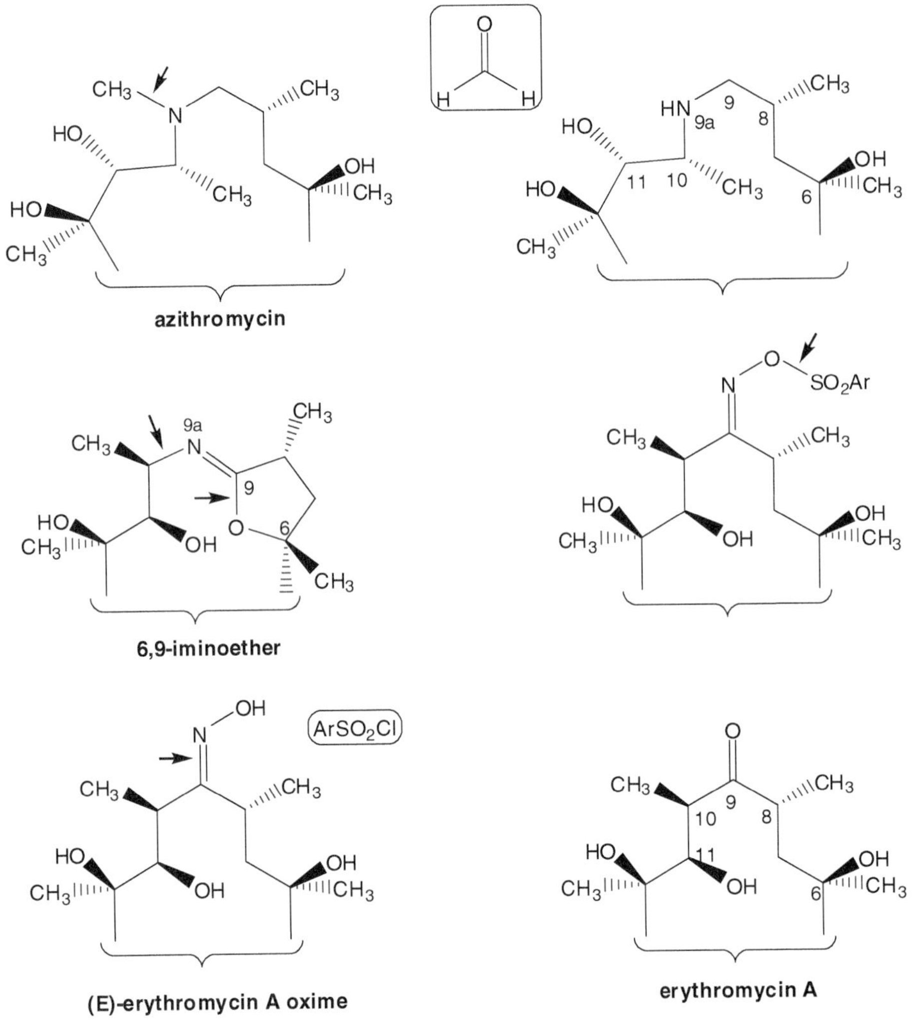

Extended Discussion

(Z)-Erythromycin A oxime can be prepared from (E)-erythromycin A oxime. Draw structures for the products formed as (Z)-erythromycin A oxime is carried through the same sequence used to convert (E)-erythromycin A oxime to azithromycin.

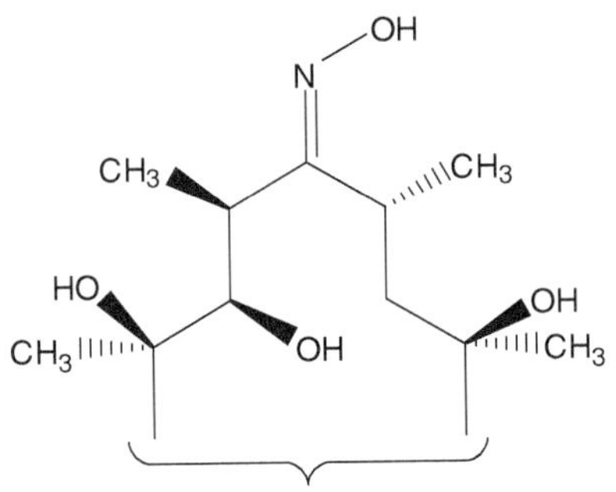

(E)-erythromycin A oxime

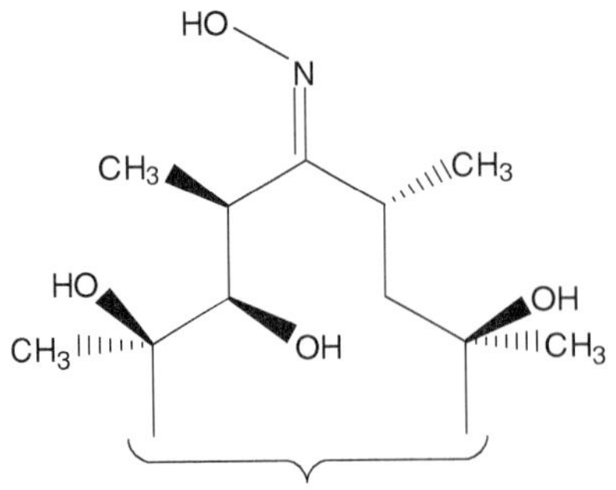

(Z)-erythromycin A oxime

Aztreonam

Anti-Infective Medicines/Antibacterials/β-Lactam Medicines

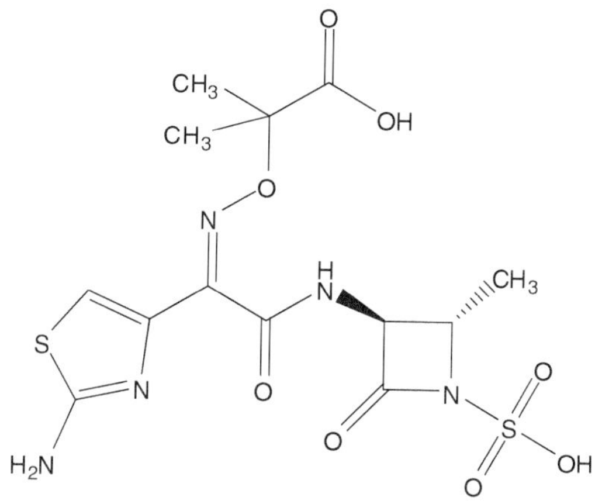

A single-enantiomer molecule with multiple chiral carbons is often formed by modification of a natural product which has most or all of the chiral carbons already in place.

Discussion. In the final step, the carboxylic acid of aztreonam is released by hydrolysis of the *tert*-butyl ester. The amide bond near the center of the molecule is formed by reaction of the amine of the ammonium sulfamate zwitterion with a thioester.

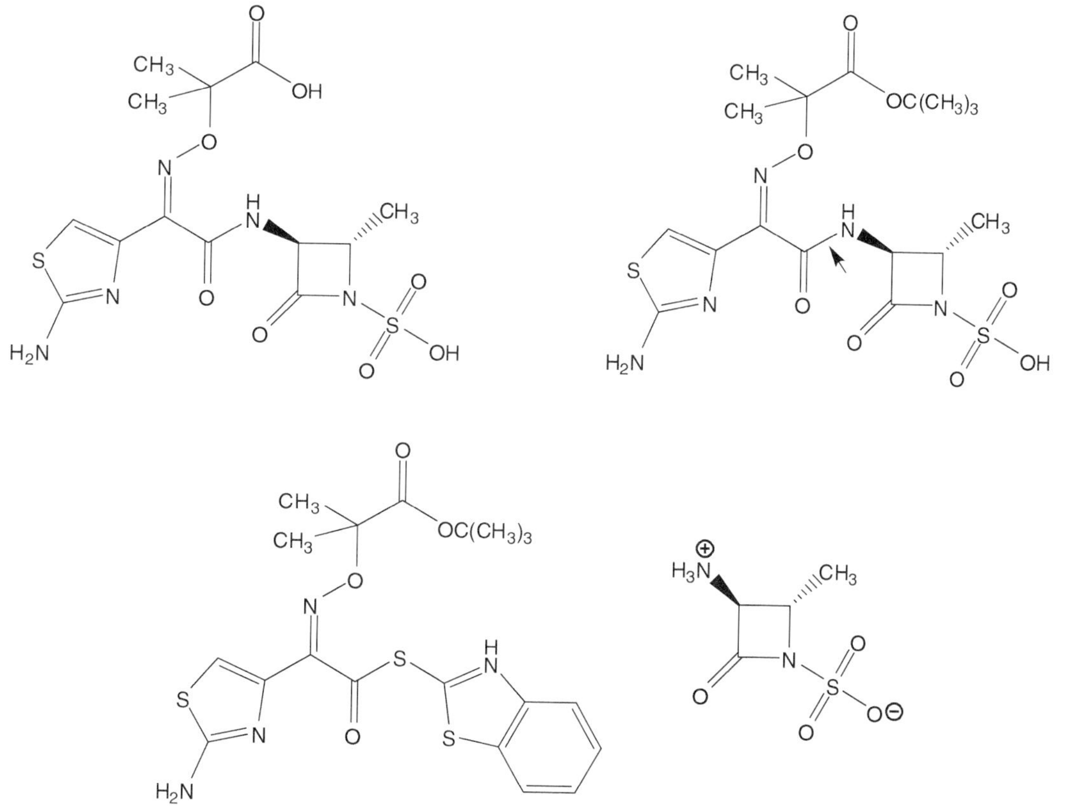

The amine of the ammonium sulfamate zwitterion is released by cleavage of the protecting group. The β-lactam ring is formed by displacement of the secondary methanesulfonate by the *N*-acylsulfamate nitrogen. The *N*-acyl sulfamate is formed from the amide. The methanesulfonate is formed from the alcohol. The amine of L-threonine amide is protected (List the protecting groups. Select one protecting group to use in the analysis). L-Threonine amide is formed from L-threonine methyl ester. L-Threonine methyl ester hydrochloride is formed from L-threonine (**Fischer Esterification**). L-Threonine is produced by fermentation.

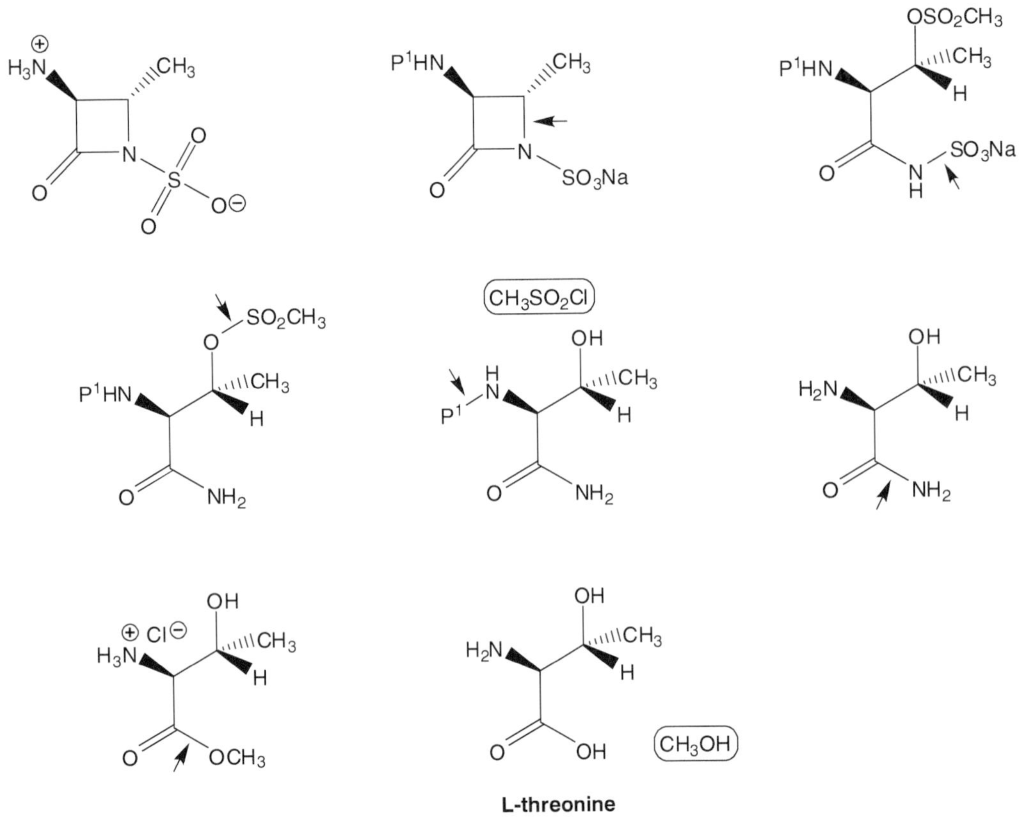

The thioester is formed from the carboxylic acid and 2,2′-dithiobis(benzothiazole). The carboxylic acid is formed by hydrolysis of the ethyl ester. The *O*-alkyloxime is formed by *O*-alkylation of the oxime with *tert*-butyl α-bromoisobutyrate. The thiazole ring is formed by reaction of an α-bromoketone with thiourea (**Hantzsch Thiazole Synthesis**). The α-bromoketone is formed by bromination of the ketone. The oxime is formed by nitrosation of ethyl acetoacetate.

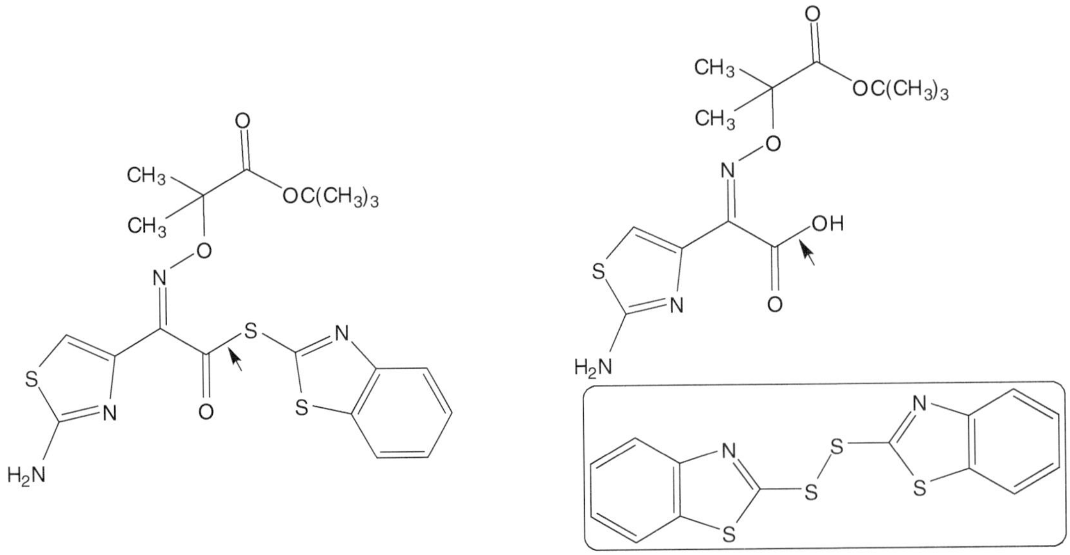

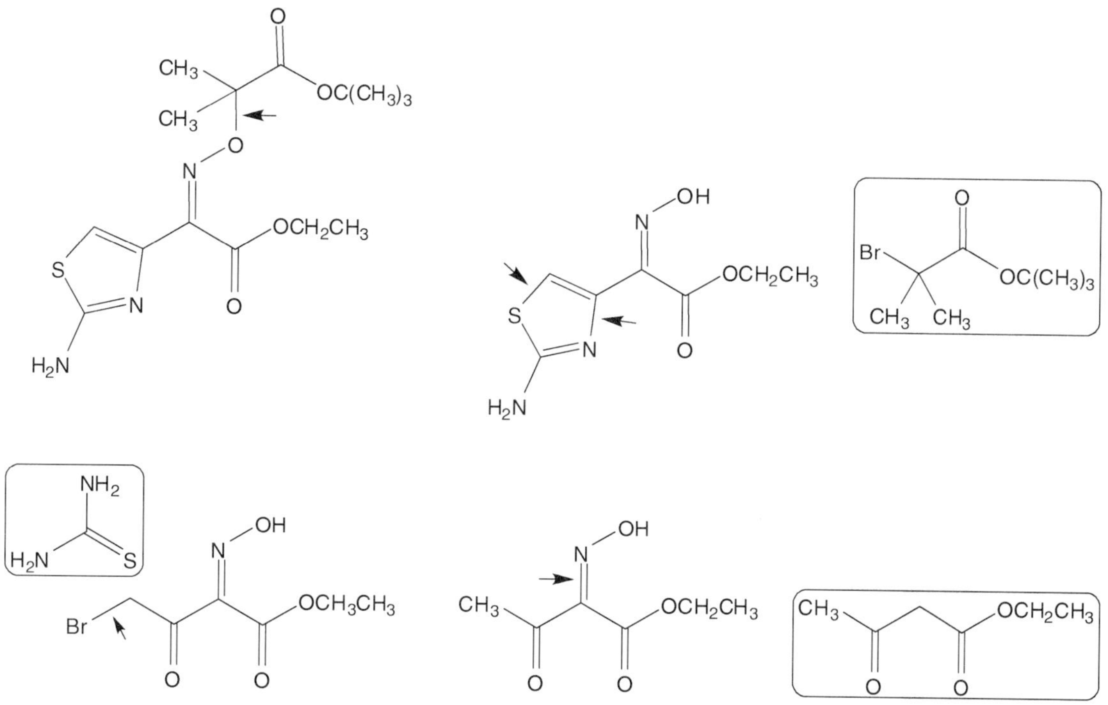

Extended Discussion

Draw the structures of the retrosynthetic analysis of an alternative route to the ammonium sulfamate zwitterion. Include the structures of the retrosynthetic analysis of any organic starting material(s) from petrochemical or biochemical raw materials. List the pros and cons for both routes and select one route to the ammonium sulfamate zwitterion as the preferred route.

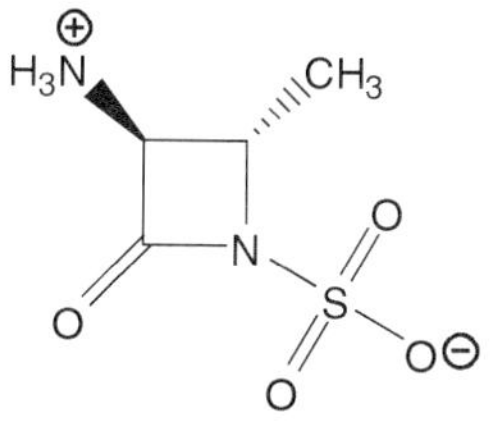

B

Beclomethasone Dipropionate

Medicines Acting on the Respiratory Tract/Antiasthmatic and Medicines for Chronic Obstructive Pulmonary Disease

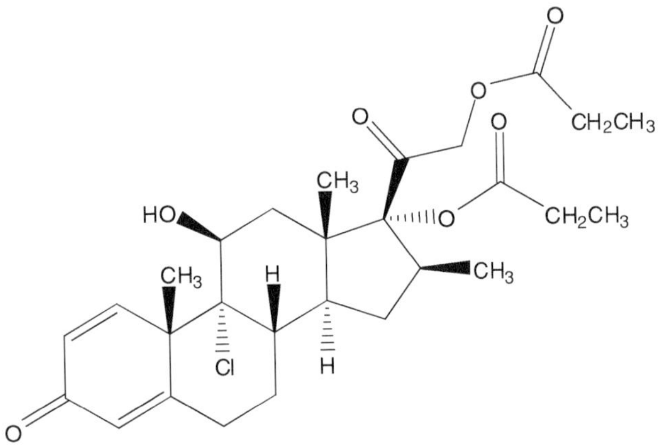

A single-enantiomer molecule with multiple chiral carbons is often formed by modification of a natural product which has most or all of the chiral carbons already in place. A steroid with 16β-methyl and 17α-hydroxy substituents is often formed by ring-opening of a 16α,17-epoxide with a methylmagnesium halide.

Discussion. Beclomethasone dipropionate is manufactured in 3 steps from beclomethasone, 12 steps from 16β-methylpregnane-3β,17α,21-triol 21-acetate, 19 steps from 16-dehydropregnenolone acetate, and 22 steps from diosgenin. Diosgenin is a phytosteroid sapogenin isolated from the tubers of *Dioscorea* wild yam.

Beclomethasone 17α,21-dipropionate is formed from the 17α-propionate and propanoic anhydride. The 17α-propionate is formed via the 17α,21-orthoester. The 17α,21-orthoester is formed by reaction of beclomethasone and trimethyl orthopropionate (1,1,1-trimethoxypropane).

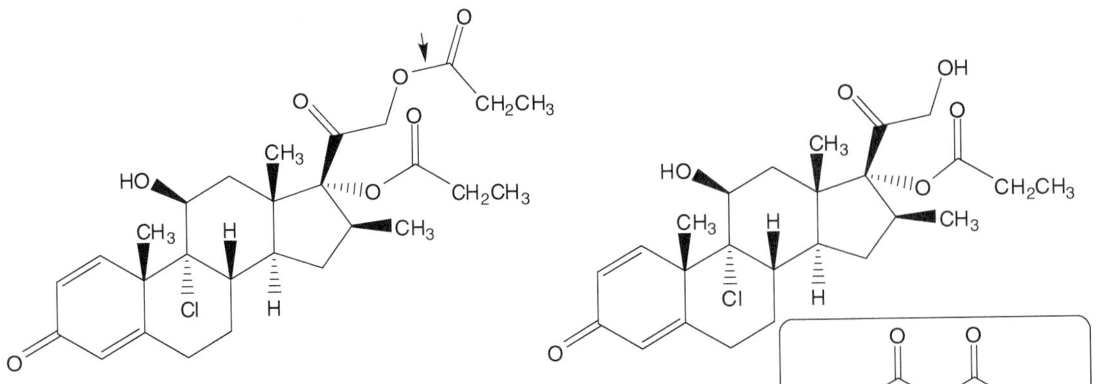

Routes to Essential Medicines: A Workbook for Organic Synthesis, First Edition. Peter J. Harrington.

Companion website: www.wiley.com/go/Harrington/routes_essential_medicine

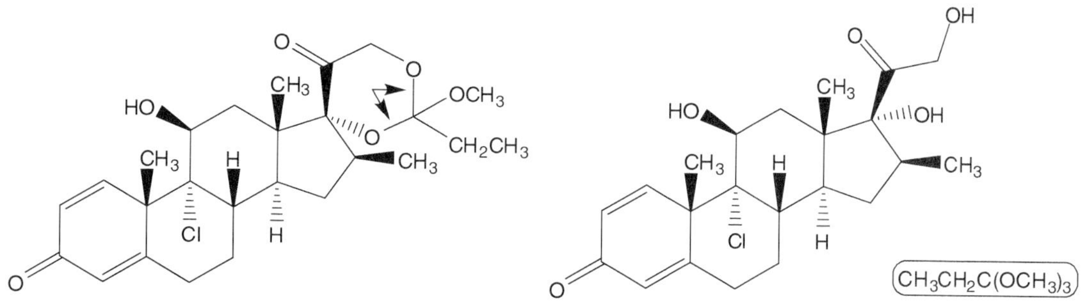

Chlorine is introduced at C9 on the α–face by ring-opening of the 9β,11-epoxide with hydrogen chloride. The 9β,11-epoxide is formed from the bromohydrin. The 11β-alcohol of the bromohydrin is formed in situ by hydrolysis of the formate ester. The 21-alcohol is also released by hydrolysis of the carbonate ester. The bromohydrin formate is formed from the 9(11)-alkene. The 9(11)-alkene is formed by dehydration of the 11α-alcohol. The 21-alcohol is protected as a carbonate.

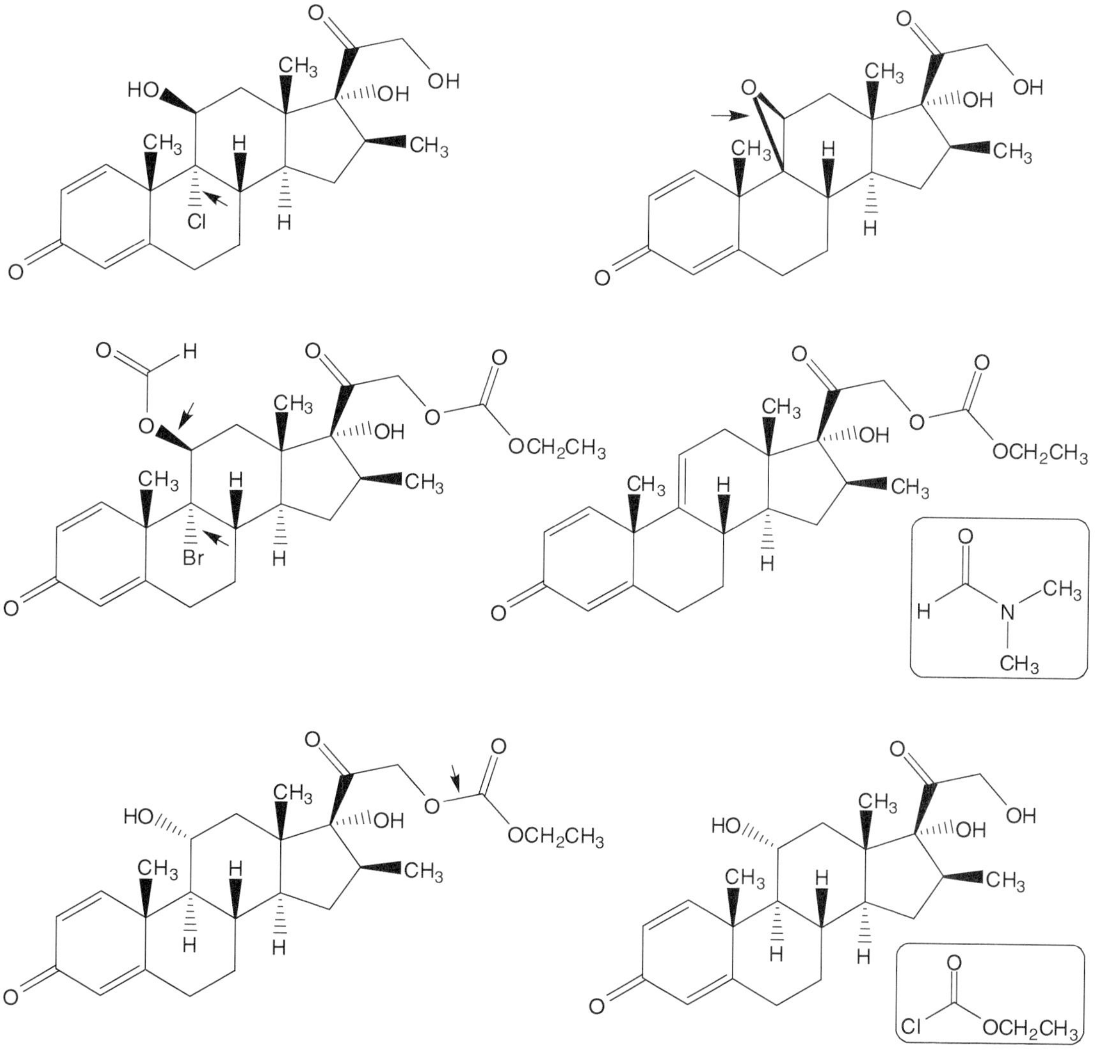

The 11α-alcohol is formed by microbial oxidation. The C21 acetate ester is hydrolyzed during the microbial oxidation. The 1,4-diene-3-one is formed from the 2α,4α-dibromo-3-one by double elimination. The 2α,4α-dibromo-3-one is formed by α-bromination of the C3 ketone. The C3 ketone is formed by oxidation of the 3β-alcohol of 16β-methylpregnane-3β,17α,21-triol 21-acetate.

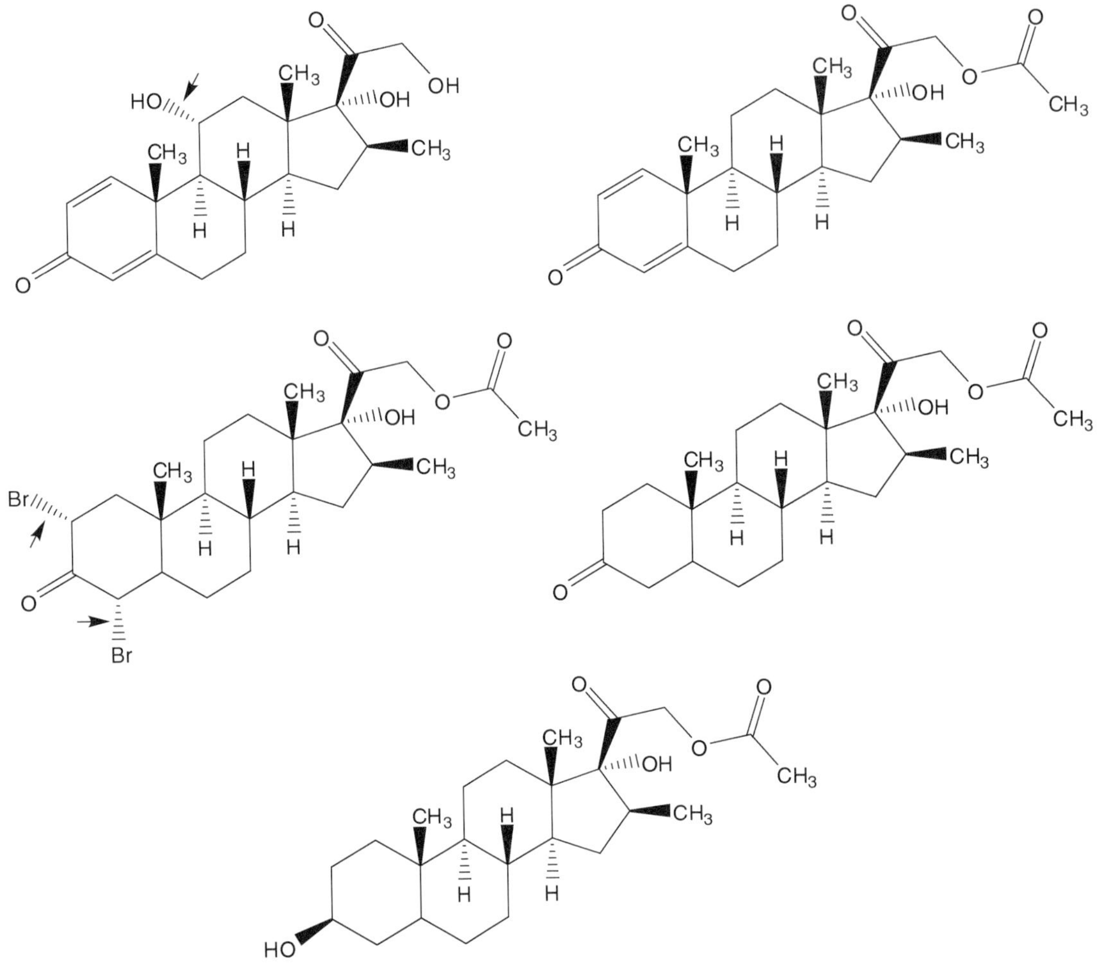

The acetate ester is formed by displacement of bromide by acetate. The α-bromoketone is formed by α-bromination of the C20 ketone. The C20 ketone is released by acetal hydrolysis. The 16β-methyl substituent is introduced by ring-opening of the 16α,17-epoxide with methylmagnesium bromide (**Grignard Reaction**). Methylmagnesium bromide is formed from bromomethane. The 5,6-alkene is reduced by catalytic hydrogenation. (What is the stereochemistry of the new chiral carbon at C5? Add this stereochemical feature to the structures in the retrosynthetic analysis.) The C20 ketone is protected as an acetal by reaction with ethylene glycol. The 16α,17-epoxide is formed by epoxidation of the 16-alkene of 16-dehydropregnenolone acetate. The acetate ester is hydrolyzed under the epoxidation conditions.

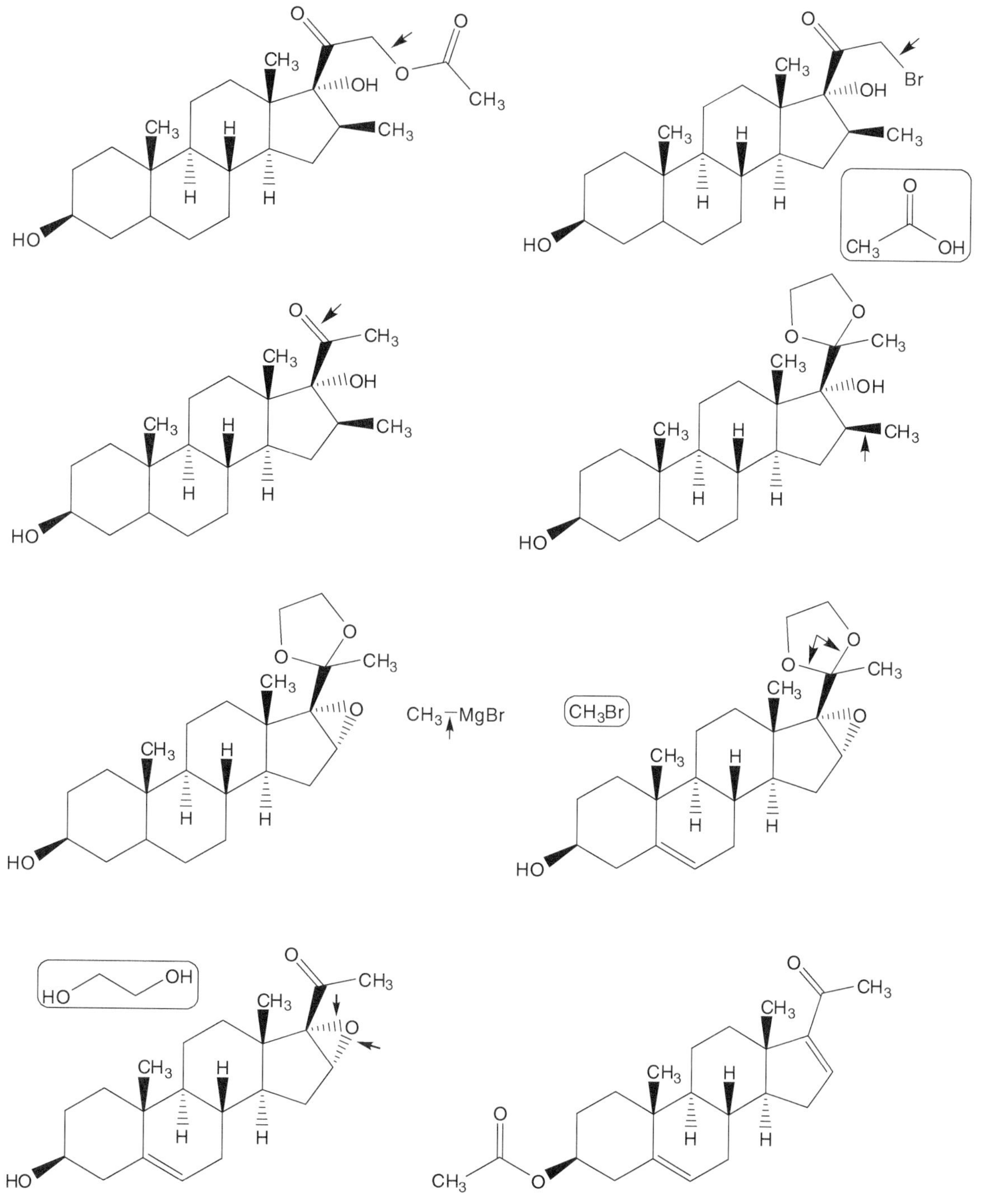

16-dehydropregnenolone acetate

The 16-alkene of 16-dehydropregnenolone acetate is formed from diosone by β-elimination. Diosone is formed by oxidation of the 20(22)-alkene of pseudodiosgenin-3,26-diacetate. Pseudodiosgenin 3,26-diacetate is formed by reaction of diosgenin with acetic anhydride. The three-step synthesis of 16-dehydropregnenolone acetate from diosgenin by acetylation, oxidation, and elimination is known as the **Marker Degradation**.

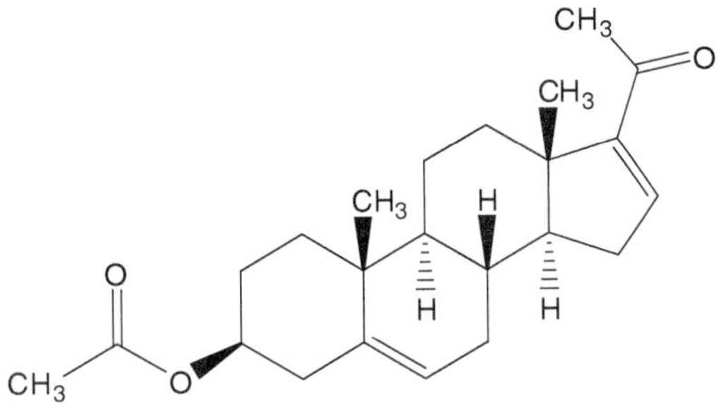

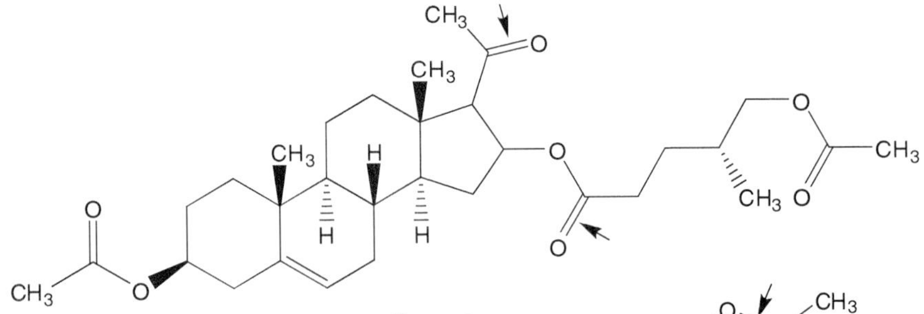

diosone

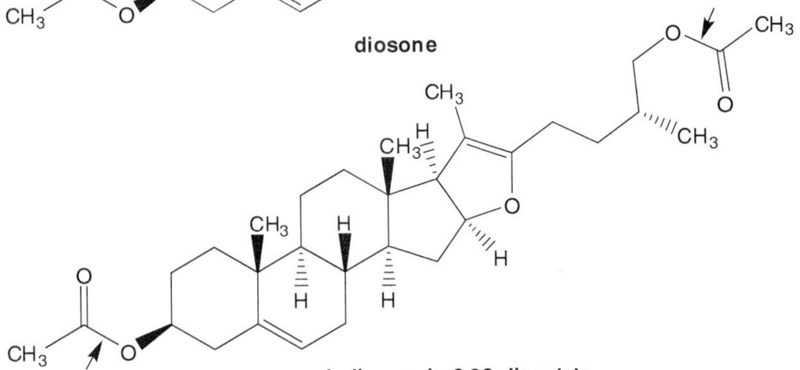

pseudodiosgenin-3,26-diacetate

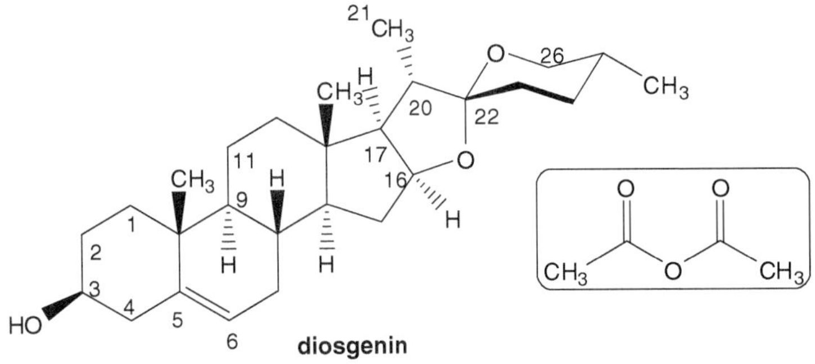

diosgenin

Extended Discussion

Draw the structures of the retrosynthetic analysis of one alternative route to beclomethasone dipropionate via the 11β-trifluoroacetate intermediate.

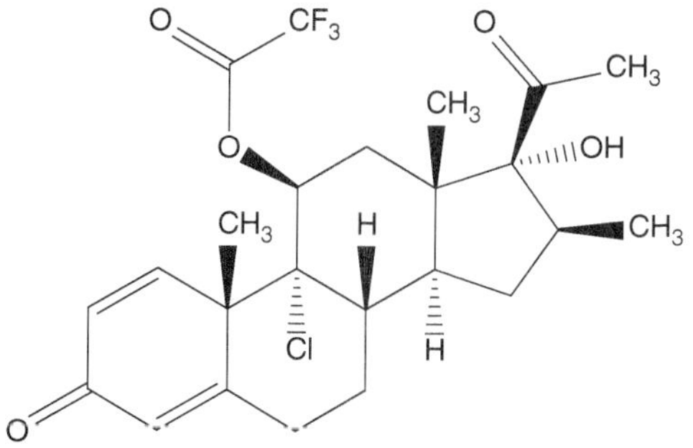

9α-chloro-11β,17α-dihydroxy-16β-methylpregna-1,4-diene-3,20-dione11-trifluoroacetate

Bedaquiline

Anti-infective Medicines/Antibacterials/Antituberculosis Medicines

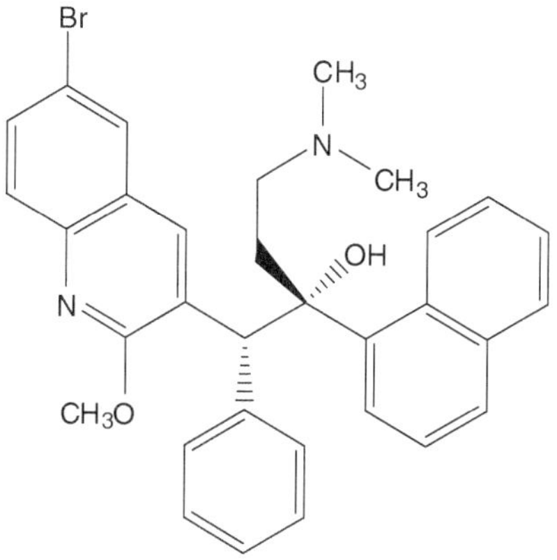

A tertiary alcohol is often formed by addition of an organometallic reagent RLi or RMX (M = Mg, Zn, X = Cl, Br, I) to a ketone.

Discussion. Bedaquiline, the (1*R*,2*S*)-enantiomer, is separated from a 1 : 1 mixture of (1*R*,2*S*)- and (1*S*,2*R*)-enantiomers by resolution in the final step. (Draw the structures of the chiral acids used for the resolution. What is the yield of bedaquiline for the two-step resolution process associated with each chiral acid?) The addition of an alkyllithium to a ketone results in a mixture of four stereoisomers. The undesired (1*S*,2*S*)- and (1*R*,2*R*)-enantiomers are separated from the mixture by crystallization. The desired (1*R*,2*S*)- and (1*S*,2*R*)-enantiomers are then isolated by crystallization. (What is the highest diastereoselectivity for the addition reaction? What reagents and reaction conditions are associated with the highest diastereoselectivity?) The alkyllithium is formed from 3-benzyl-6-bromo-2-methoxyquinoline. The β–dimethylamino ketone is formed from 1-acetonaphthone, formaldehyde, and dimethylamine (**Mannich Reaction**).

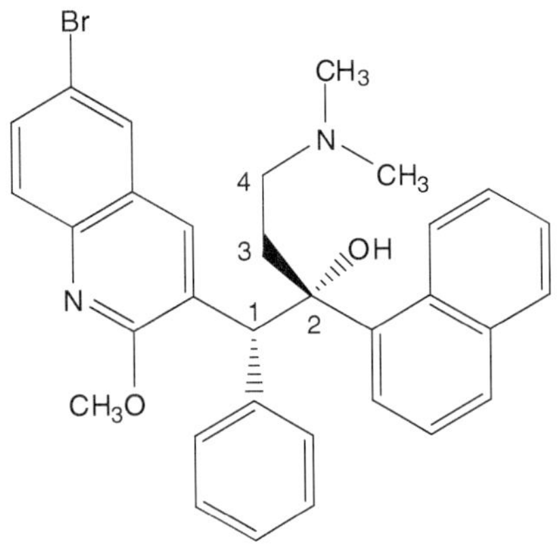

(1R, 2S)

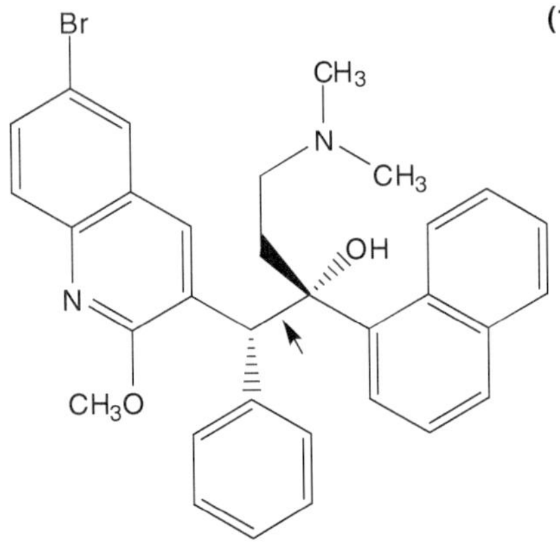

(1R, 2S)

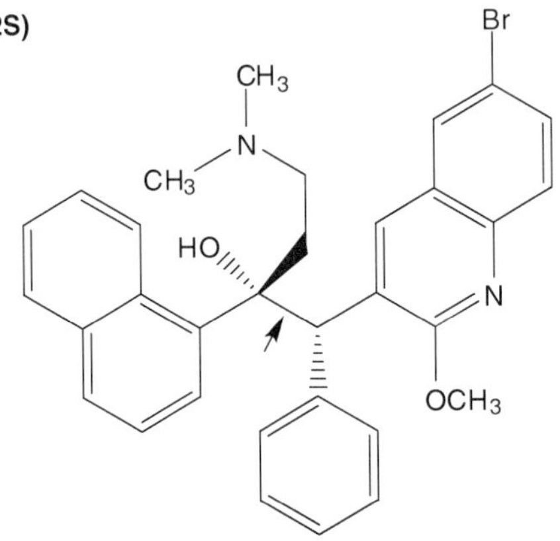

(1S, 2R)

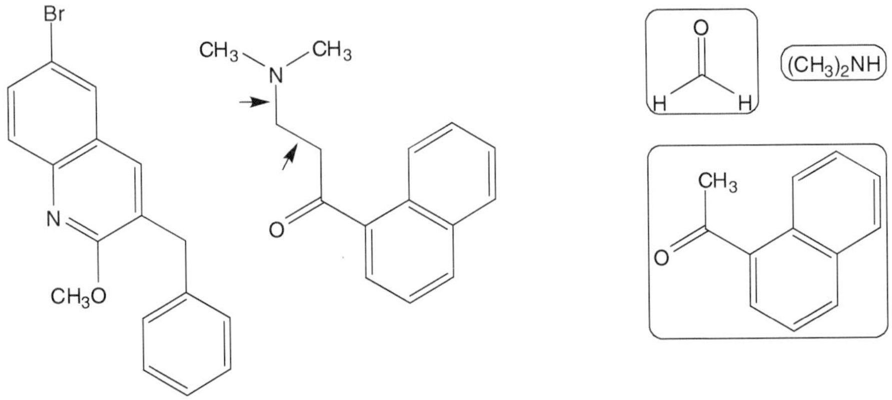

The 2-methoxyquinoline is formed from the 2-chloroquinoline by chloride displacement. 3-Benzyl-6-bromo-2-chloroquinoline is formed from *N*-(4-bromophenyl)-3-phenylpropanamide, and *N,N*-dimethylformamide (**Vilsmeier–Haack Reaction**). *N*-(4-Bromophenyl)-3-phenylpropanamide is formed from 4-bromoaniline and the acid chloride. The acid chloride is formed from 3-phenylpropanoic acid (dihydrocinnamic acid).

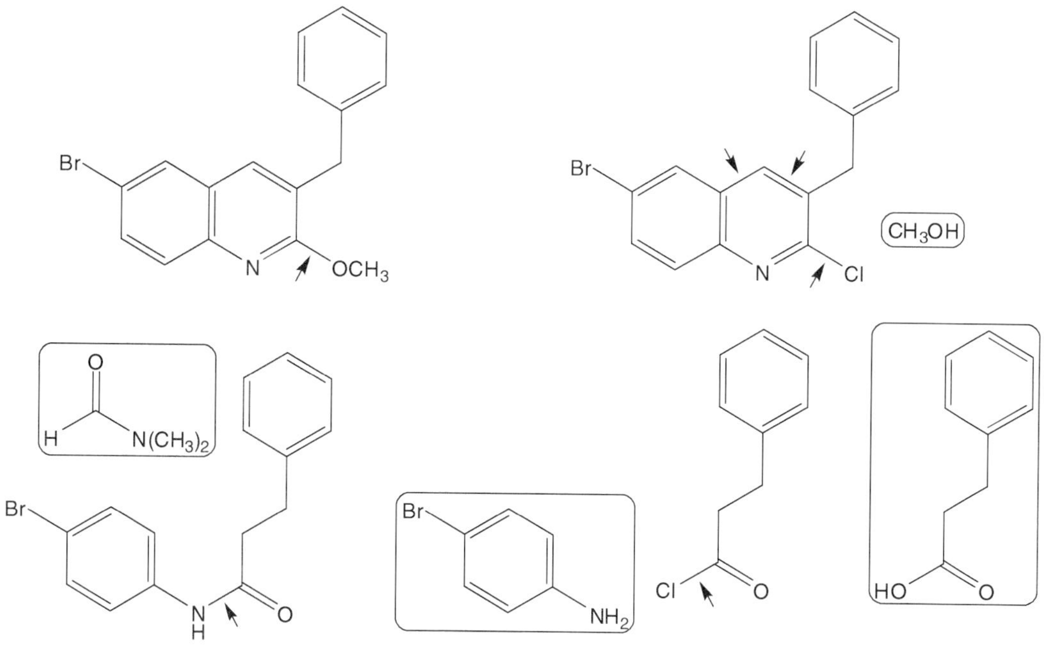

Extended Discussion

Bedaquiline is converted to a 1 : 1 salt with fumaric acid in the final step of the manufacturing process. Calculate the overall yield of bedaquiline fumarate (1 : 1) from 4-bromoaniline. (What percent of the 4-bromoaniline starting material is lost to waste?)

Bendamustine

Antineoplastics and Immunosuppressives/Cytotoxic and Adjuvant Medicines

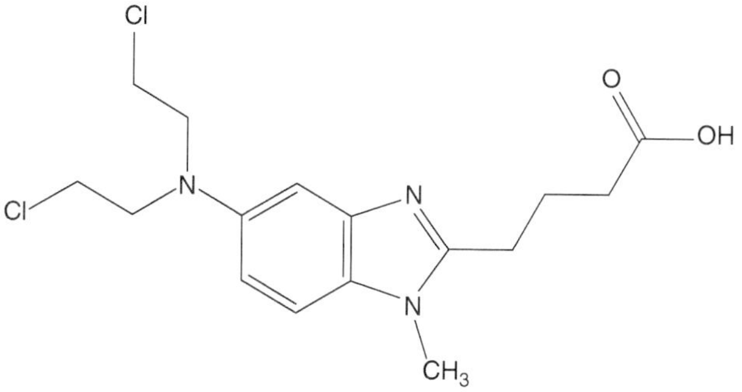

A 2-alkylbenzimidazole is often formed by the reaction of a 1,2-phenylenediamine with an acid chloride, anhydride, ester or carboxylic acid.

Discussion. The carboxylic acid is formed by ester hydrolysis in the final step. The bis-(2-chloroethyl)amine is formed from the bis-(2-hydroxyethyl)amine. The bis-(2-hydroxyethyl)amine is formed by ring-opening of ethylene oxide by the 5-aminobenzimidazole.

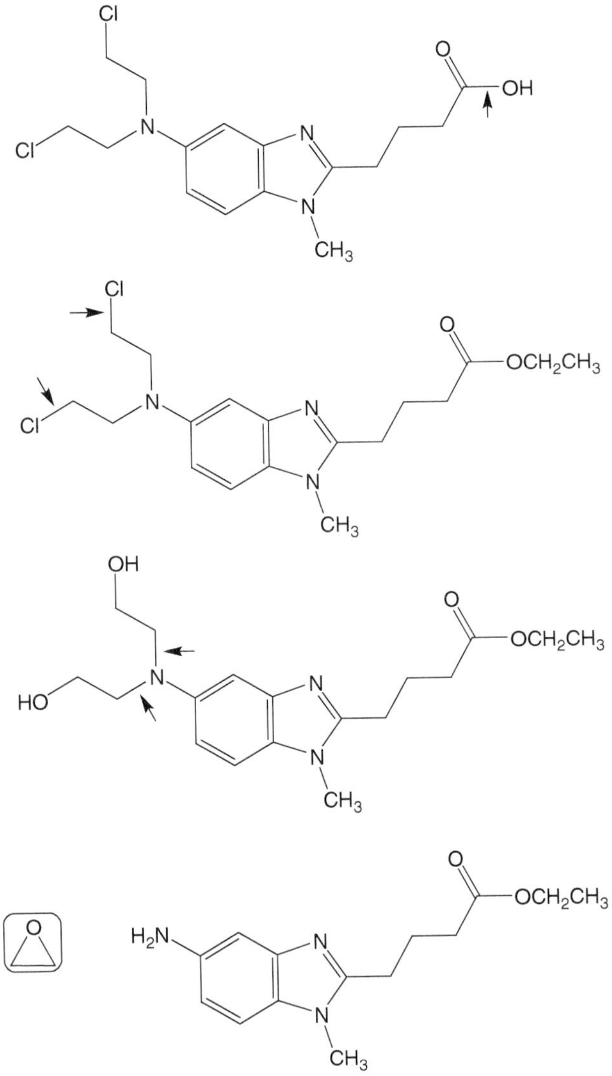

The 5-aminobenzimidazole is formed by reduction of the 5-nitrobenzimidazole. The ester is formed from the carboxylic acid and ethanol (**Fischer Esterification**). The benzimidazole ring and the carboxylic acid both form in the reaction of N1-methyl-4-nitro-1,2-phenylenediamine with glutaric anhydride. The phenylenediamine is formed by a selective reduction of *N*-methyl-2,4-dinitroaniline. The dinitroaniline is formed by the displacement of chloride from 1-chloro-2,4-dinitrobenzene by methylamine.

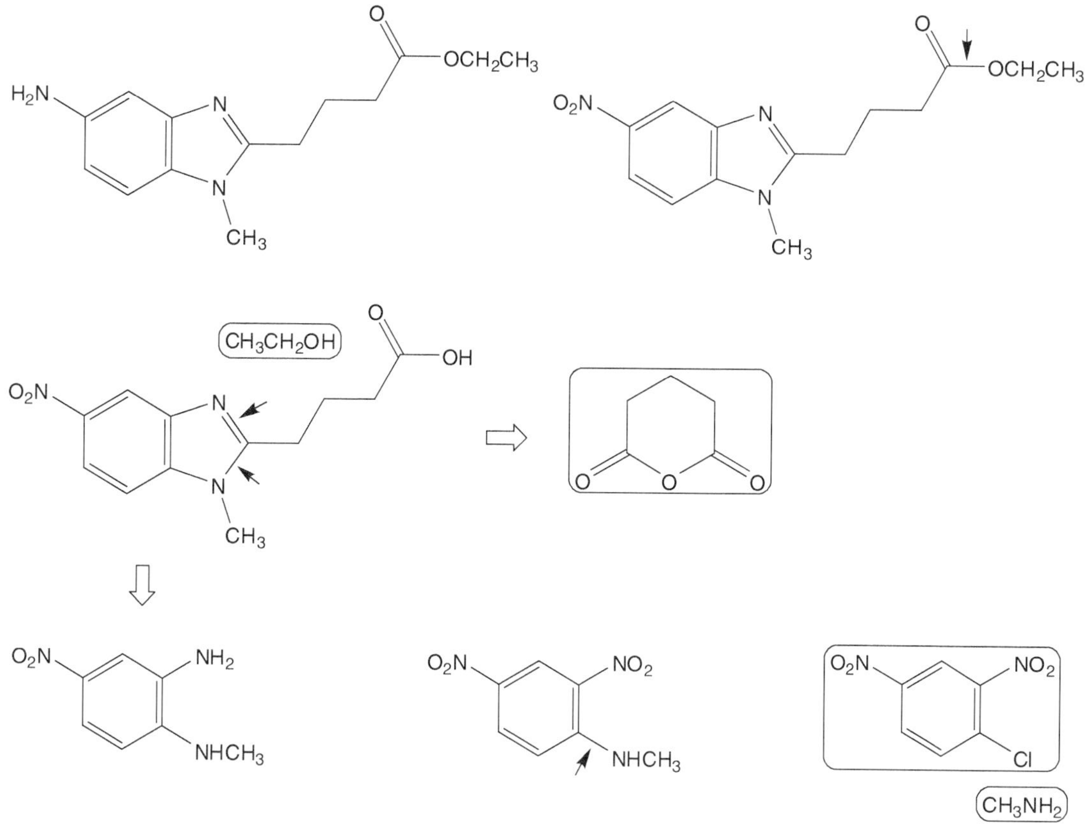

Extended Discussion

The selective reduction of one nitro group in a 2,4-dinitroaniline is a key step in the synthesis of bendamustine. (How selective is this reduction?) Draw the structures of the retrosynthetic analysis of an alternative route to bendamustine which is based on reduction of both nitro groups of a dinitroaromatic in the same step. List the pros and cons for both routes and select one route as the preferred route.

Benznidazole

Anti-Infective Medicines/Antiprotozoal Medicines/Antitrypanosomal Medicines/American Trypanosomiasis

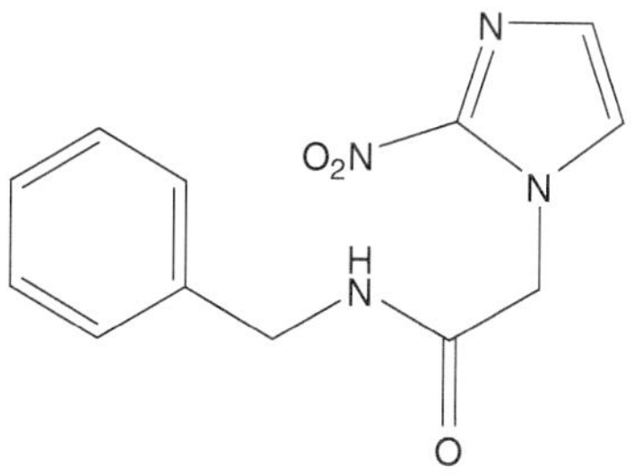

A 2-nitroimidazole is often formed from a 2-aminoimidazole via the diazonium salt.

Discussion. The amide is formed in the final step by reaction of the ethyl ester with benzylamine. Ethyl 2-nitroimidazole-1-ylacetate is formed from ethyl bromoacetate by bromide displacement by 2-nitroimidazole. 2-Nitroimidazole is formed from 2-aminoimidazole via the diazonium salt. 2-Aminoimidazole is formed from *O*-methylisourea and 2,2-dimethoxyethanamine. *O*-Methylisourea is formed from cyanamide and methanol.

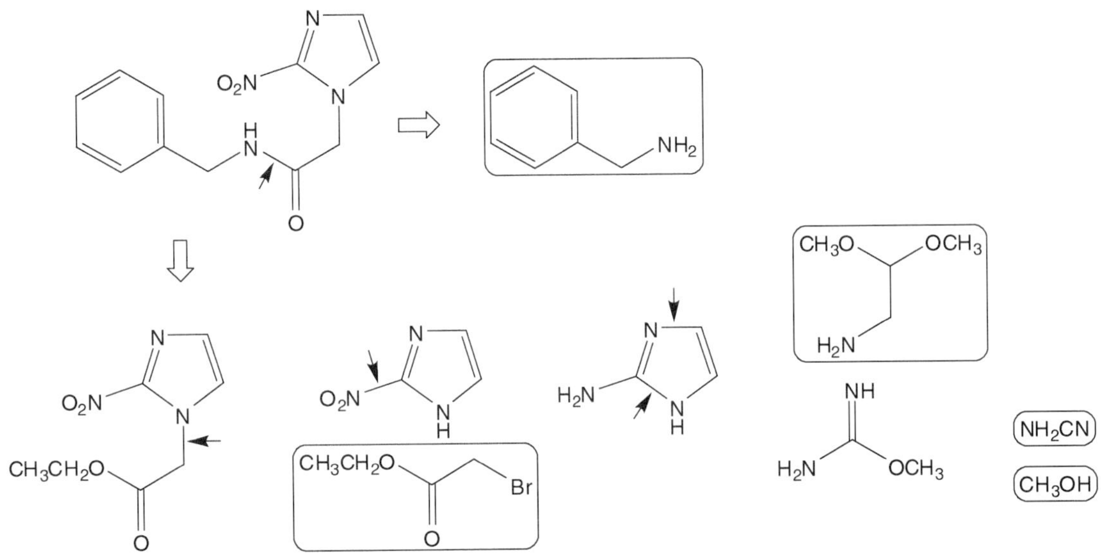

Extended Discussion

Why are anhydrous conditions preferred for the final two steps of the synthesis?

Benzoyl Peroxide

Dermatological Medicines/Medicines Affecting Skin Differentiation and Proliferation

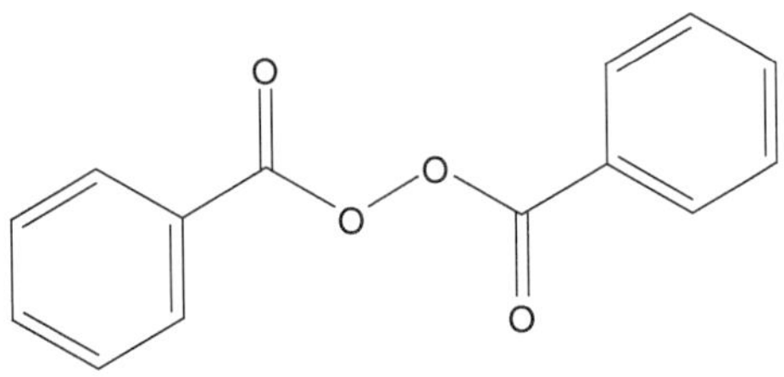

Discussion. Benzoyl peroxide is formed by reaction of benzoyl chloride with hydrogen peroxide.

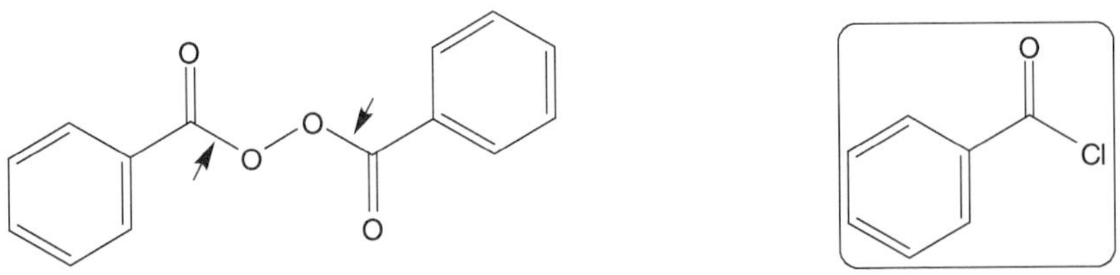

Dry (98%) benzoyl peroxide is a very dangerous material!
Benzoyl peroxide is not purified by crystallization.
Benzoyl peroxide melts with decomposition and may explode at 106° C.
Benzoyl peroxide may explosively decompose on shock, friction, or concussion.
Benzoyl peroxide is a strong oxidant and reacts violently with combustible or reducing materials.
Benzoyl peroxide reacts violently with inorganic acids, organic acids, alcohols, and amines.
Benzoyl peroxide (75%, with 25% added water) has a better safety profile.

Benzyl Benzoate

Dermatological Medicines/Scabicides and Pediculicides

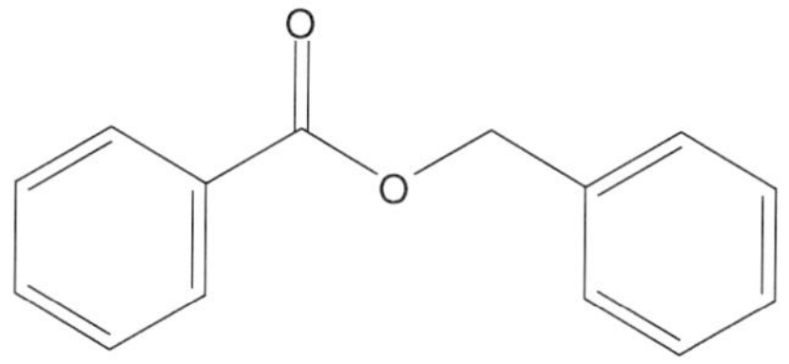

When a target molecule is accessible by more than one efficient route, the preferred route is likely to have the lowest raw material costs.

Discussion. Disconnection of the C—O bond suggests benzyl benzoate is manufactured from benzyl alcohol and the acid chloride, anhydride, ester, or carboxylic acid. Methyl benzoate, a byproduct from the manufacture of polyethylene terephthalate (PET or PETE) is used in one manufacturing route. Benzaldehyde is converted to benzyl benzoate in an alternative manufacturing route (**Claisen–Tishchenko Reaction**).

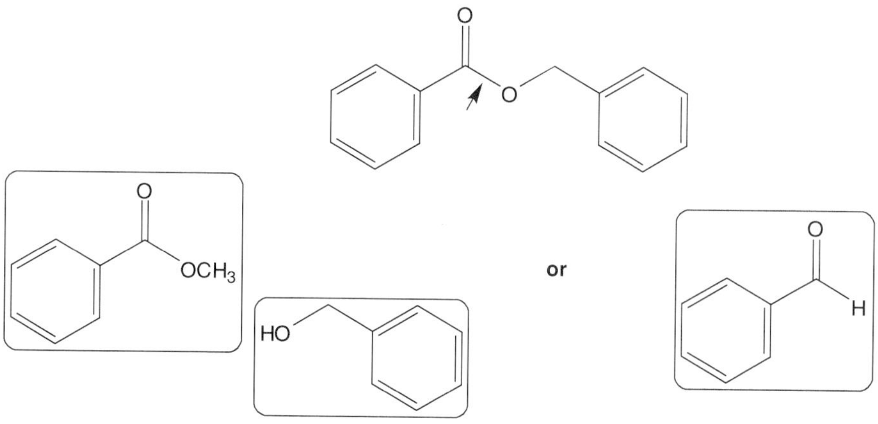

Extended Discussion

Estimate a lowest cost per mole for methyl benzoate, benzyl alcohol, and benzaldehyde. Explain the estimation process used. Use the estimates to calculate a raw material cost to produce 1 mol of benzyl benzoate by both manufacturing routes.

Betamethasone

Dermatological Medicines/Anti-Inflammatory and Antipruritic Medicines

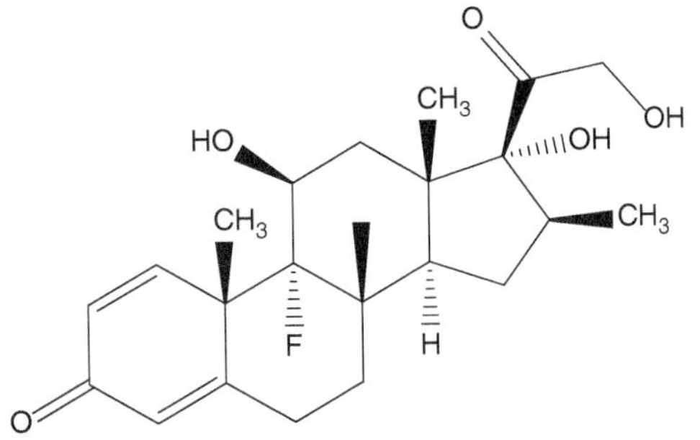

A single-enantiomer molecule with multiple chiral carbons is often formed by modification of a natural product which has most or all of the chiral carbons already in place. A steroid with 16β-methyl and 17α-hydroxy substituents is often formed by ring-opening of a 16α,17-epoxide with a methylmagnesium halide.

Discussion. Betamethasone is manufactured in 9 steps from 16β-methylpregnane-3β,17α,21-triol 21-acetate, 16 steps from 16-dehydropregnenolone acetate, and 19 steps from diosgenin. Diosgenin is a phytosteroid sapogenin isolated from the tubers of *Dioscorea* wild yam.

Fluorine is introduced at C9 on the α–face in the final step by ring-opening of the 9β,11-epoxide with hydrogen fluoride. The 9β,11-epoxide is formed from the bromohydrin. The 11β-alcohol of the bromohydrin is formed in situ by hydrolysis of the formate ester. The 21-alcohol is also released by hydrolysis of the carbonate ester. The bromohydrin formate is formed from the 9(11)-alkene. The 9(11)-alkene is formed by dehydration of the 11α-alcohol. The C21 alcohol is protected as a carbonate.

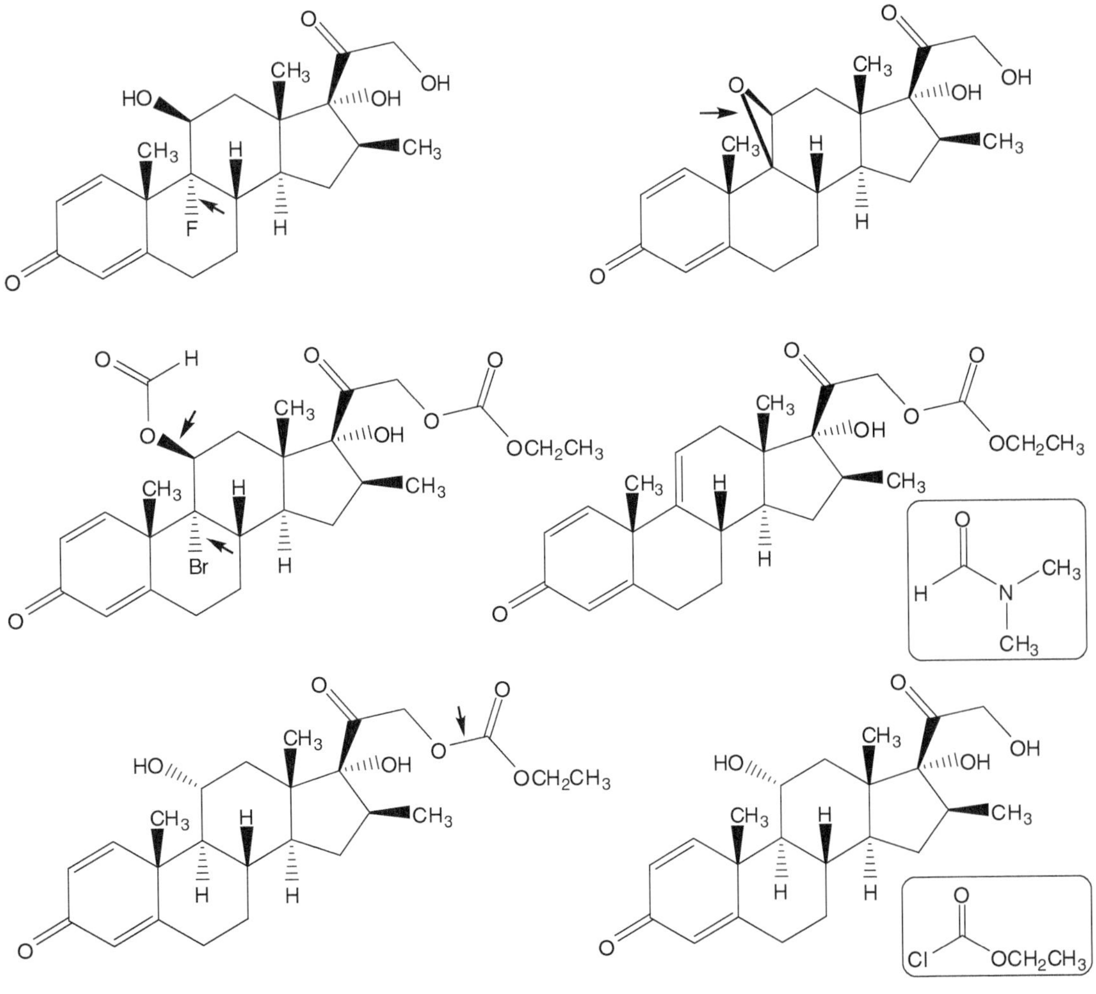

The 11α-alcohol is formed by microbial oxidation. The acetate ester is hydrolyzed during the microbial oxidation. The 1,4-diene-3-one is formed from the 2α,4α-dibromo-3-one by double elimination. The 2α,4α-dibromo-3-one is formed by α-bromination of the C3 ketone. The C3 ketone is formed by oxidation of the C3 alcohol of 16β-methylpregnane-3β,17α,21-triol 21-acetate.

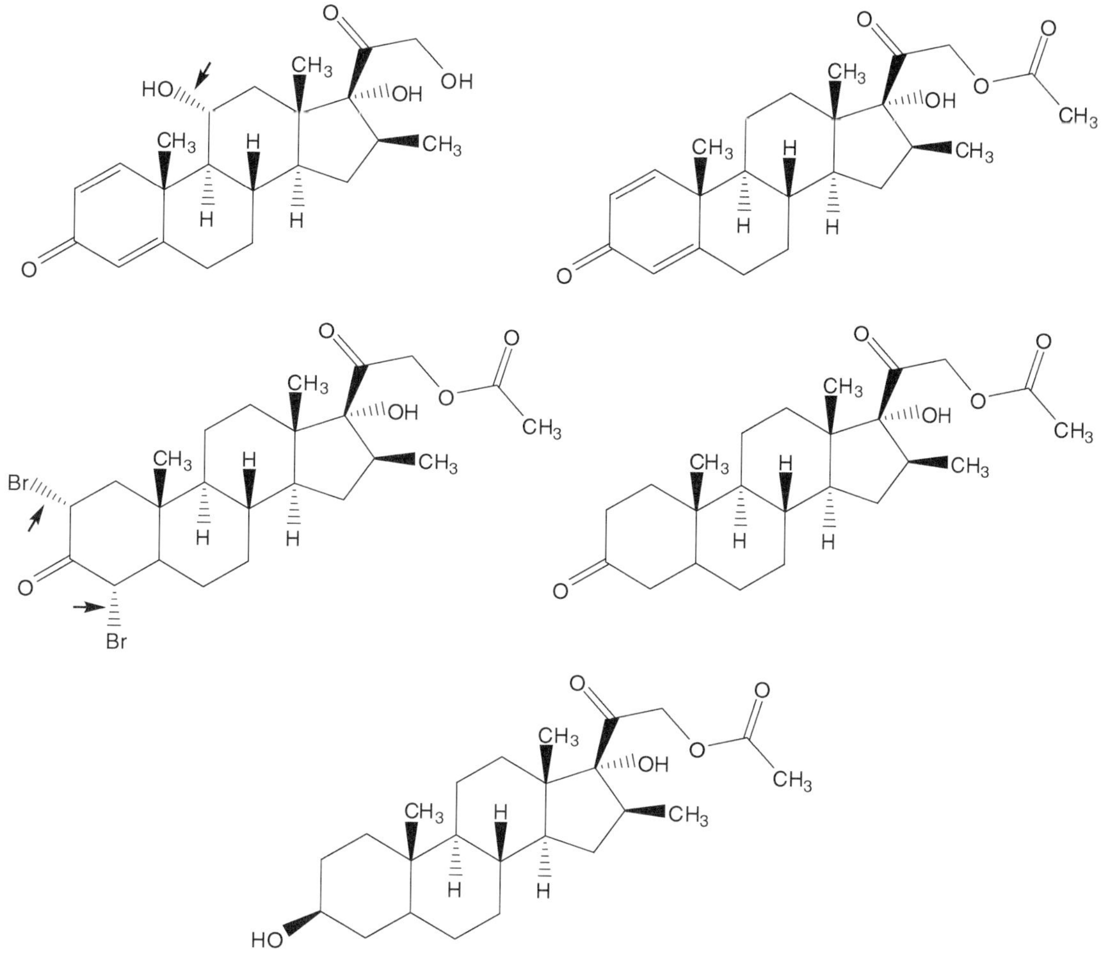

The acetate ester is formed by displacement of bromide by acetate. The α-bromoketone is formed by bromination of the C20 ketone. The C20 ketone is released by acetal hydrolysis. The 16β-methyl substituent is introduced by ring-opening of the 16α,17-epoxide with methylmagnesium bromide (**Grignard Reaction**). Methylmagnesium bromide is formed from bromomethane. The 5,6-alkene is reduced by catalytic hydrogenation. (What is the stereochemistry of the new chiral carbon at C5? Add this stereochemical feature to the structures in the retrosynthetic analysis.) The C20 ketone is protected as an acetal by reaction with ethylene glycol. The 16α,17-epoxide is formed by epoxidation of the 16-alkene of 16-dehydropregnenolone acetate. The acetate ester is hydrolyzed under the epoxidation conditions.

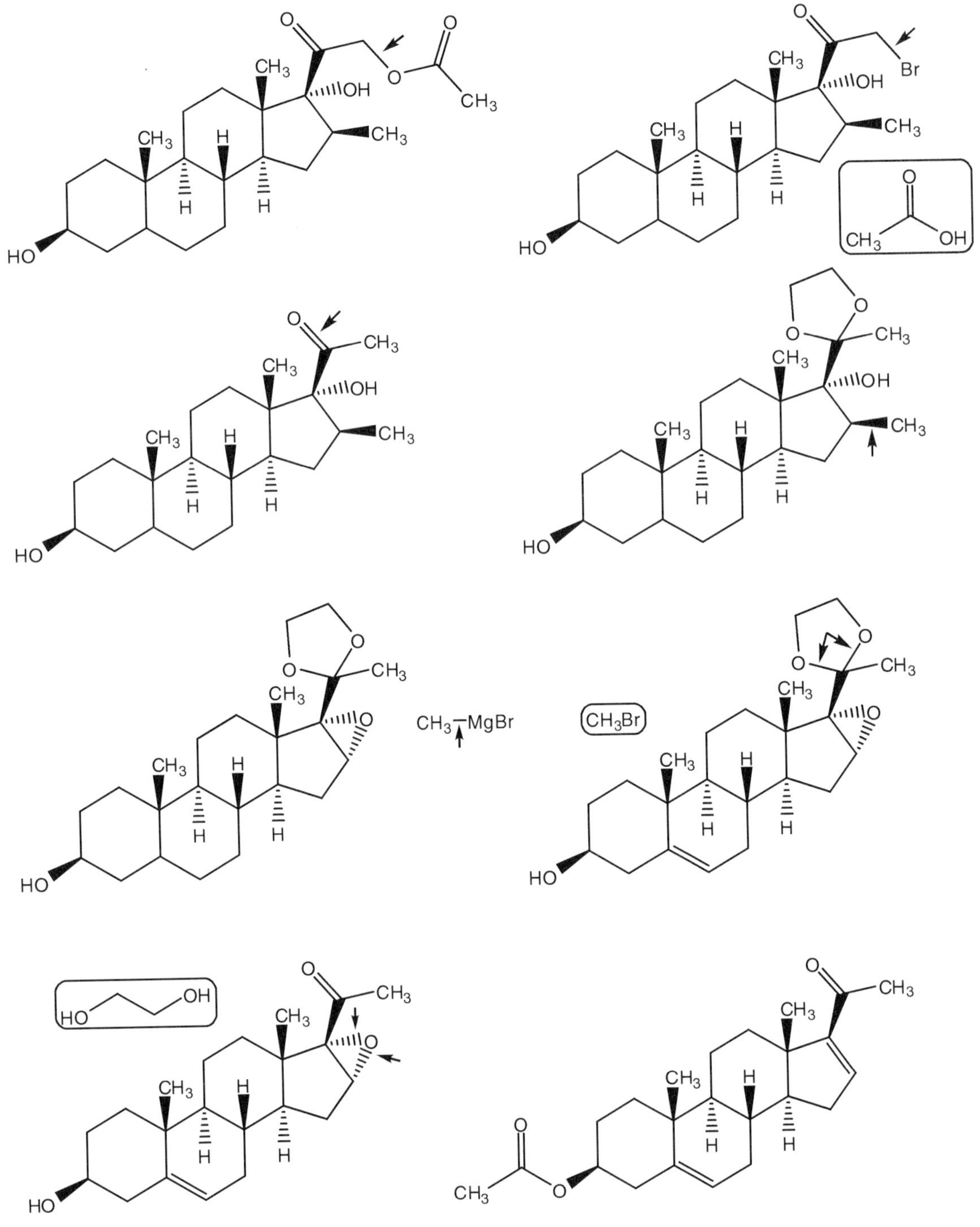

16-dehydropregnenolone acetate

The 16-alkene of 16-dehydropregnenolone acetate is formed from diosone by β-elimination. Diosone is formed by oxidation of the 20(22)-alkene of pseudodiosgenin-3,26-diacetate. Pseudodiosgenin 3,26-diacetate is formed by reaction of diosgenin with acetic anhydride. The three-step synthesis of 16-dehydropregnenolone acetate from diosgenin by acetylation, oxidation, and elimination is known as the **Marker Degradation**.

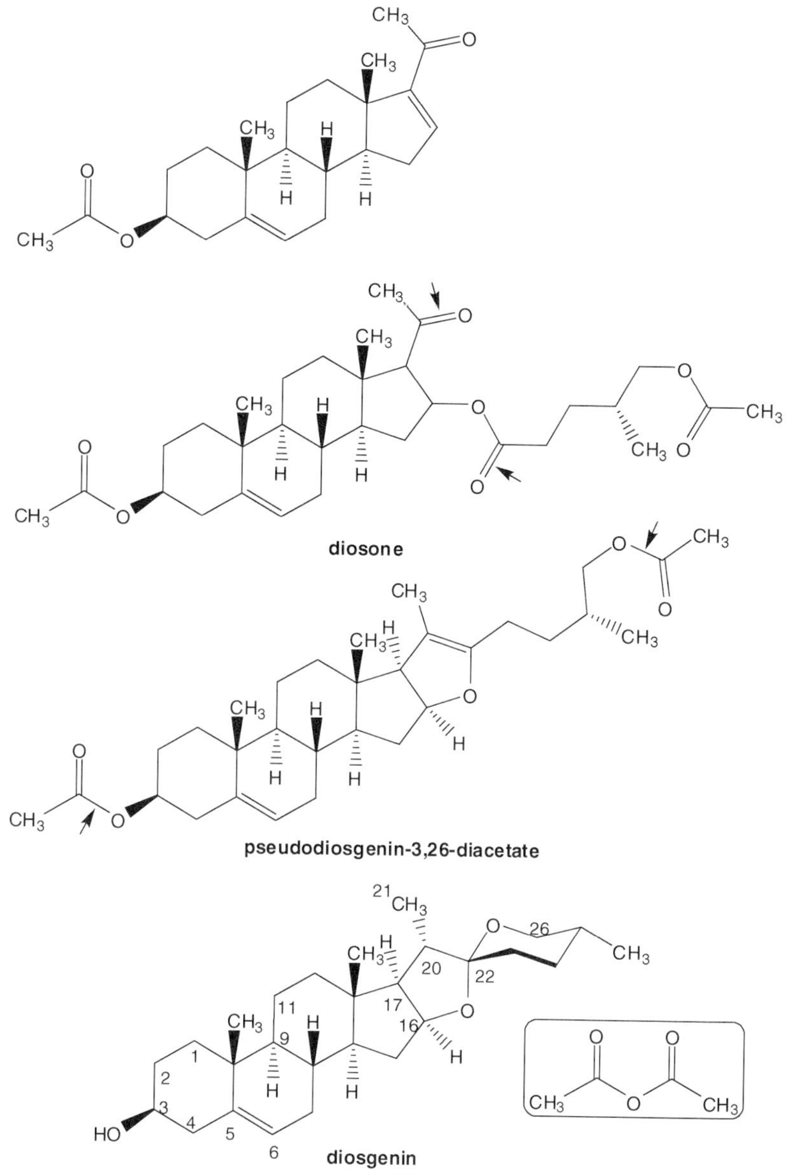

Extended Discussion

Hecogenin is a phytosteroid sapogenin isolated from the sisal plant, *Agave sisalana*. Draw the structures of the retrosynthetic analysis of an alternative route to betamethasone from hecogenin via 17α-hydroxy-16β-methylpregn-9(11)-ene-3,20-dione.

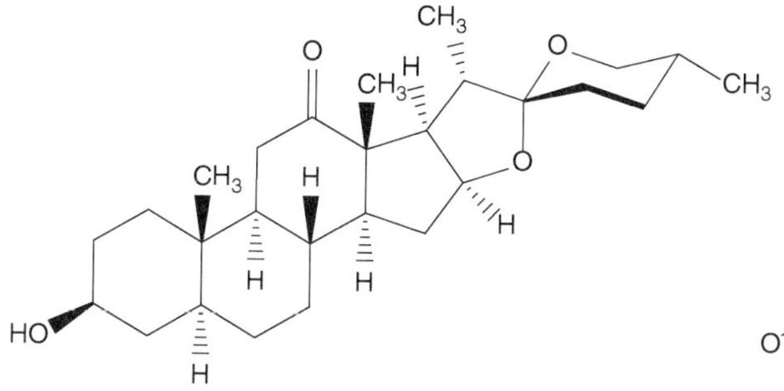

hecogenin

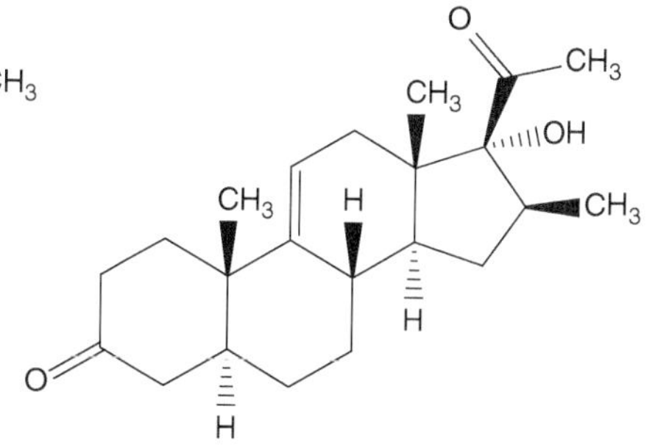

17α-hydroxy-16β-methylpregn-9(11)-ene-3,20-dione

Bicalutamide

Antineoplastics and Immunosuppressives/Hormones and Antihormones

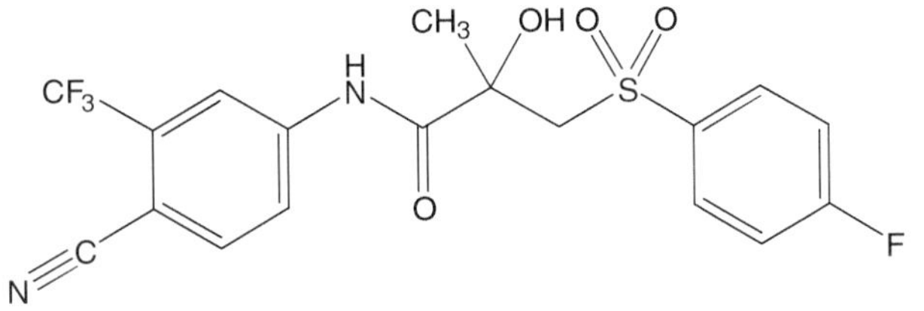

A β-hydroxythioether is often formed by ring opening of an epoxide by a thiol.

Discussion. Bicalutamide is a 1 : 1 mixture of (*R*)- and (*S*)-enantiomers. The sulfone is formed in the final step by oxidation of the thioether. The thioether is formed by ring opening of the epoxide by 4-fluorothiophenol.

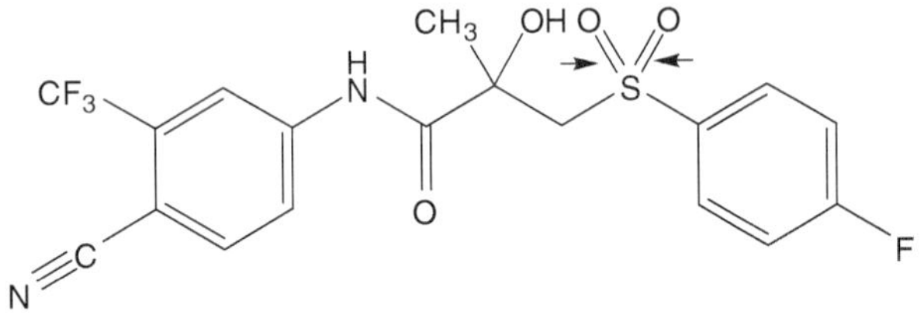

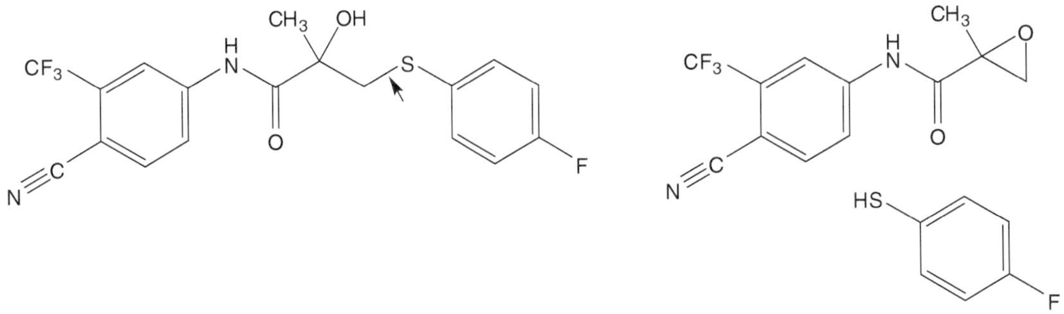

The epoxide is formed by epoxidation of the acrylamide. The acrylamide is formed by reaction of methacryloyl chloride with the amine. Methacryloyl chloride is formed from methacrylic acid. The amine, 4-amino-2-(trifluoromethyl)benzonitrile, is formed by reaction of 4-chloro-3-(trifluoromethyl)aniline with cuprous cyanide. 4-Chloro-3-(trifluoromethyl)aniline is formed by chlorination of 3-(trifluoromethyl)aniline. (How selective is the chlorination? What conditions are associated with the highest selectivity for formation of 4-chloro-3-(trifluoromethyl)aniline?)

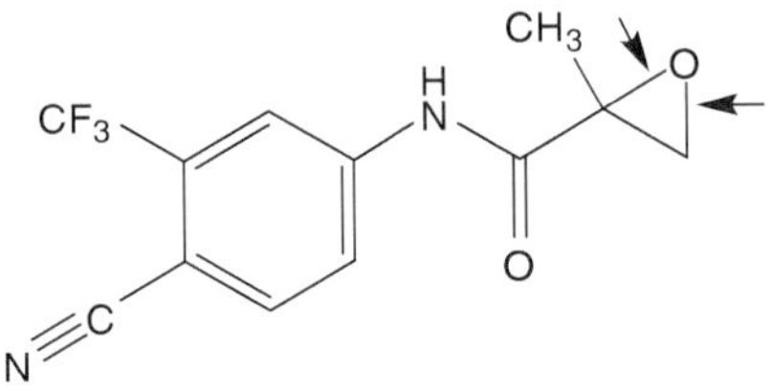

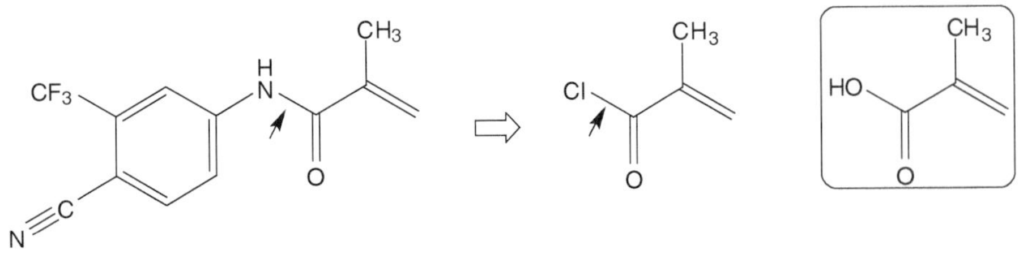

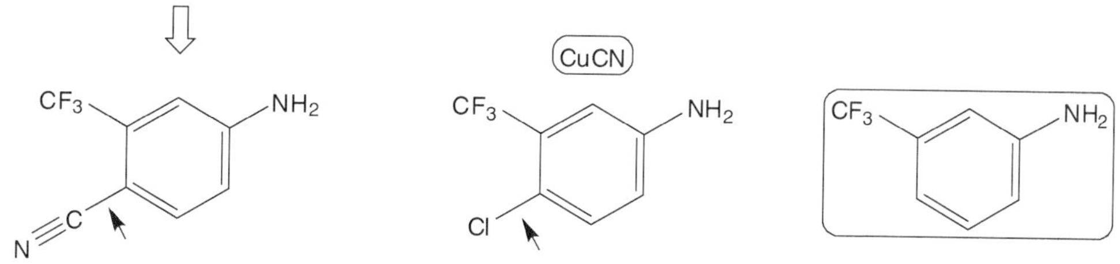

4-Fluorothiophenol is formed by reduction of the disulfide. The disulfide is formed by reduction of the sulfinate salt. The sulfinate salt is formed by reduction of 4-fluorobenzenesulfonyl chloride.

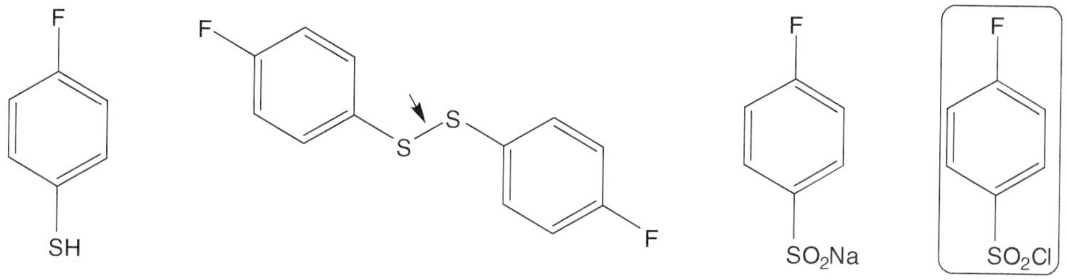

Extended Discussion

Draw the structures of the retrosynthetic analysis of one alternative route to bicalutamide which forms the C—S bond before forming the amide C—N bond. List pros and cons for both routes and select one route as the preferred route.

Biperiden

Antiparkinsonism Medicines

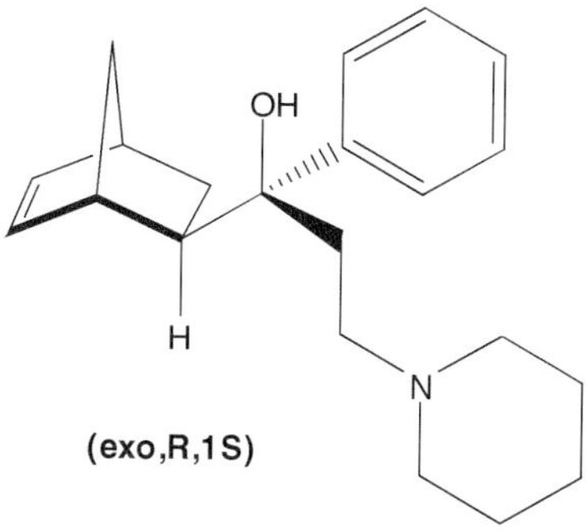

(exo,R,1S)

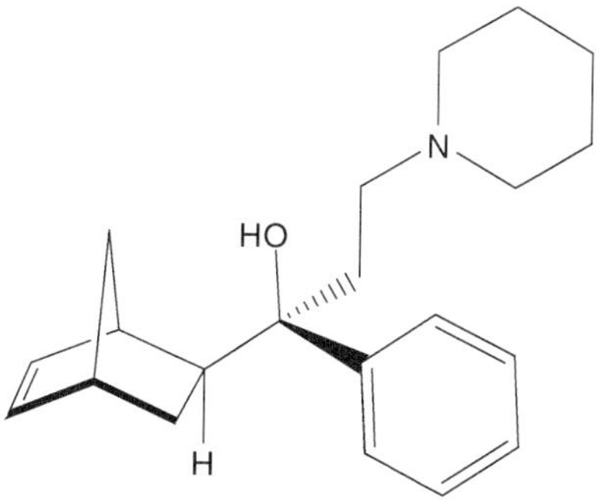

(exo,S,1R)

A tertiary alcohol is often formed by addition of an alkylmagnesium halide or arylmagnesium halide to a ketone (Grignard Reaction).

Discussion. Biperiden is a mixture of (*exo*,*R*,1*S*)- and (*exo*,*S*,1*R*)-enantiomers. The retrosynthetic analysis of the (*exo*,*R*,1*S*)-enantiomer is shown. The tertiary alcohol is formed from phenylmagnesium chloride and the ketone (**Grignard Reaction**). (How is biperiden separated from the mixture of stereoisomers that is formed in the **Grignard Reaction**?) Phenylmagnesium chloride is formed from chlorobenzene. The β-piperidinoethyl ketone is formed by the reaction of the methyl ketone with formaldehyde and piperidine (**Mannich Reaction**). (Some *exo*-to-*endo* isomerization is observed in the **Mannich Reaction**. What reaction conditions are associated with the minimum amount of *exo*-to-*endo* isomerization?) A mixture of four stereoisomers of 2-acetyl-5-norbornene is formed in the [4+2]-cycloaddition of cyclopentadiene with methyl vinyl ketone (**Diels–Alder Reaction**). (Draw the structures of the stereoisomers.) The endo-stereoisomers are isomerized to the exo-stereoisomers.

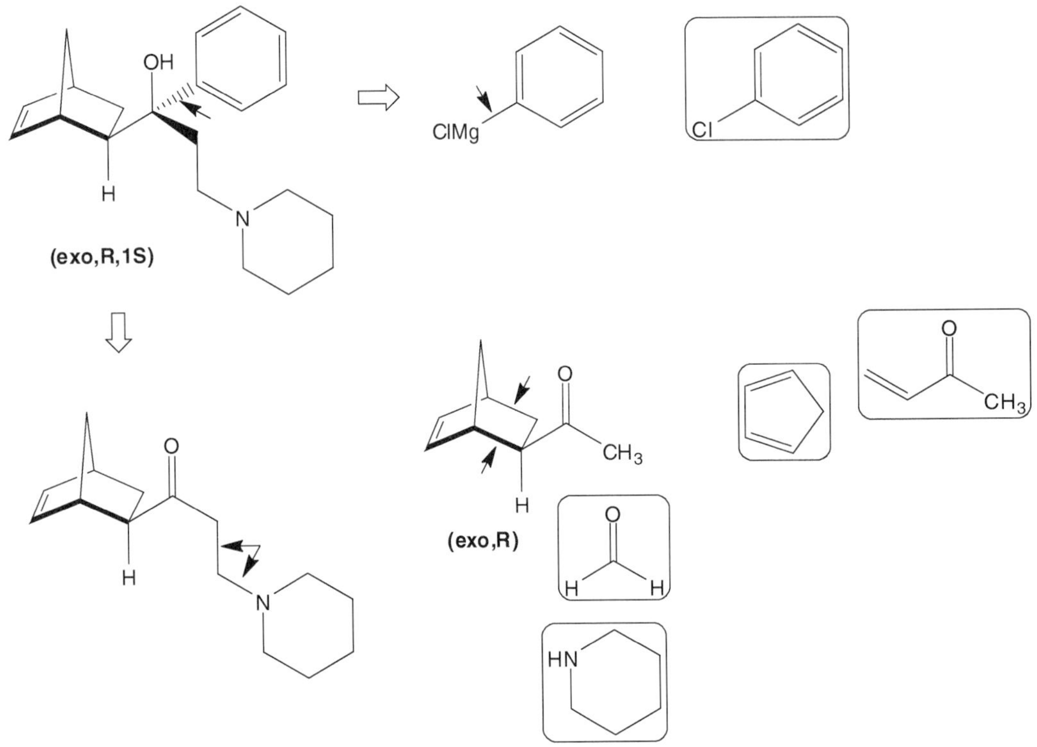

Extended Discussion

Draw the structures of the retrosynthetic analysis of one alternative route to biperiden. List the pros and cons for both routes and select one route as the preferred route.

Bisoprolol

Cardiovascular Medicines/Antiarrhythmic Medicines
Cardiovascular Medicines/Medicines Used in Heart Failure

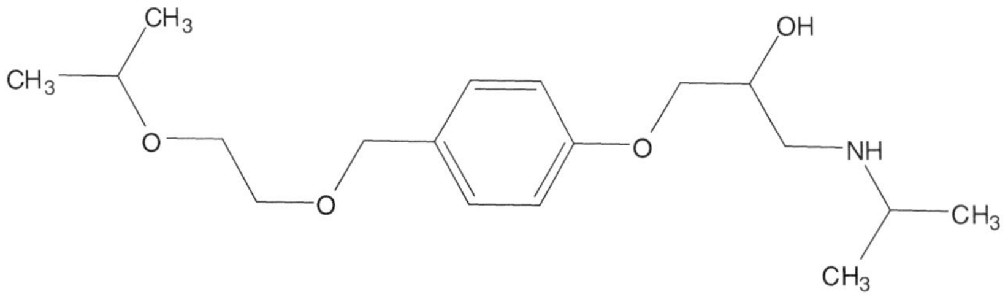

A β-amino alcohol is often formed by ring-opening an epoxide with an amine.

Discussion. Bisoprolol is a 1 : 1 mixture of (*R*)- and (*S*)-enantiomers. The β–amino alcohol is formed in the final step by ring-opening of the epoxide with isopropylamine. The epoxide is formed from the chlorohydrin. The chlorohydrin is formed by the ring-opening of epichlorohydrin by the phenol. The benzyl ether is formed by the reaction of 4-hydroxybenzyl alcohol with 2-isopropoxyethanol.

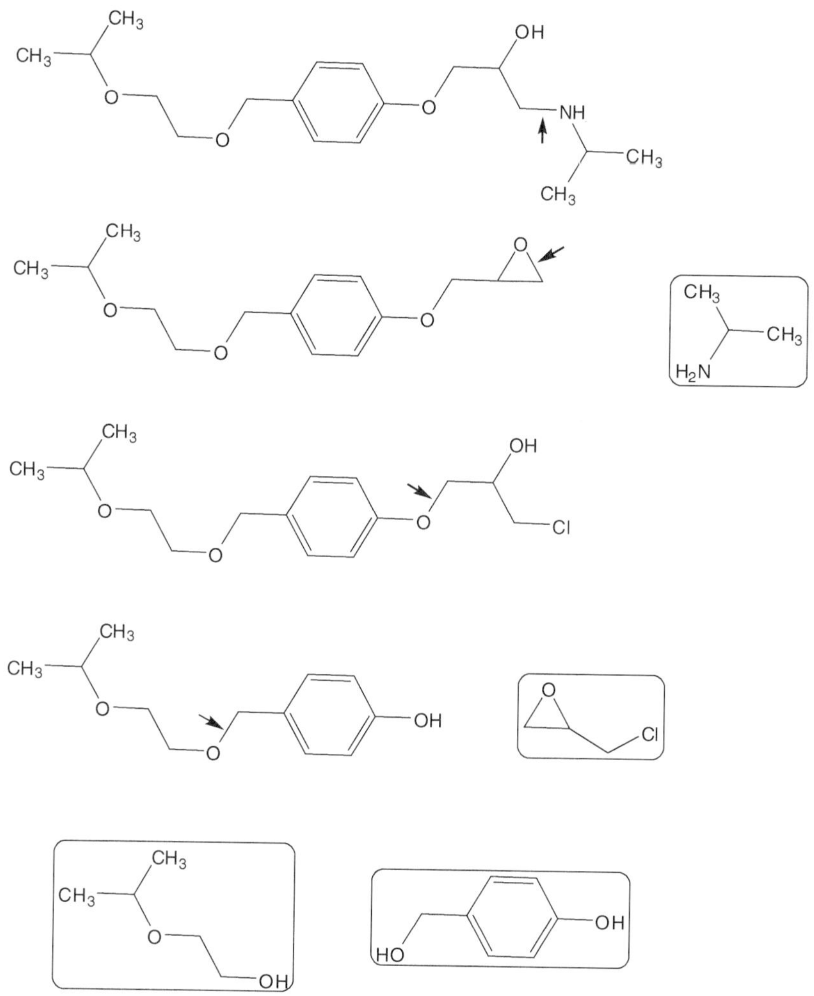

Extended Discussion

Draw the structures of the retrosynthetic analysis of one alternative route to bisoprolol from phenol that does not have 4-hydroxybenzaldehyde as an intermediate. List the pros and cons for both routes. Is one route preferred?

Budesonide

Medicines Acting on the Respiratory Tract/Antiasthmatic and Medicines for Chronic Obstructive Pulmonary Disease
Ear, Nose, and Throat Medicines

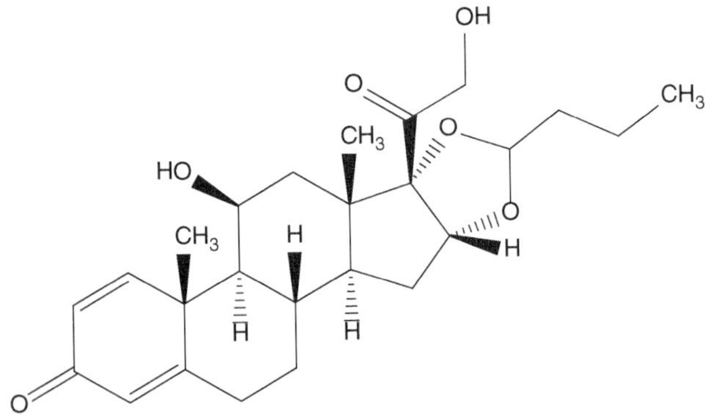

A single-enantiomer molecule with multiple chiral carbons is often formed by modification of a natural product which has most or all of the chiral carbons already in place. A 16α,17α-dihydroxy-20-one steroid is often formed by dihydroxylation of a 16-ene-20-one.

Discussion. Budesonide is manufactured in five steps from 21-acetoxypregna-1,4,9(11),16-tetraene-3,20-dione and in 10 steps from another essential medicine, prednisolone.

Budesonide is a mixture of the 22*R*- and 22*S*-diastereomers. The acetal is formed in the final step by reaction of the 16α,17α-diol with butanal. The C21 alcohol is released by hydrolysis of the acetate ester. The 9α-hydrogen is introduced by reduction of the 9α-bromide. The bromohydrin is formed from the 9(11)-ene. The 16α,17α-diol is formed by dihydroxylation of the 16-alkene of 21-acetoxypregna-1,4,9(11),16-tetraene-3,20-dione.

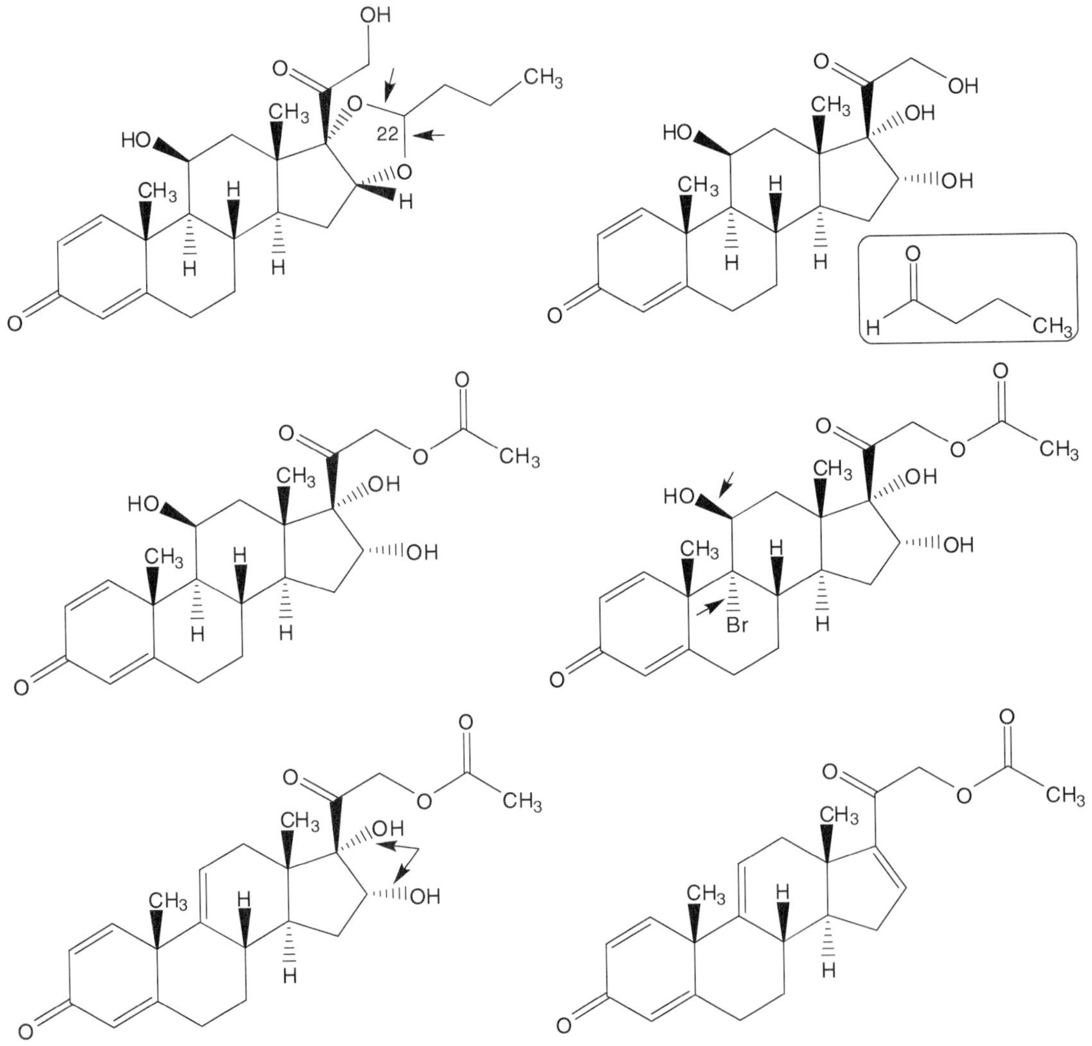

The 9(11)-ene of the tetraene is formed by elimination of the C11β mesylate which is formed in situ. The 16-alkene is formed by elimination of the 17α-acetate of prednisolone 17α,21-diacetate. Prednisolone 17α,21-diacetate is formed by reaction of the 17α-acetate with acetic anhydride. The 17α-acetate is formed from prednisolone 17α,21-ethyl orthoacetate. The orthoester is formed from prednisolone and triethyl orthoacetate.

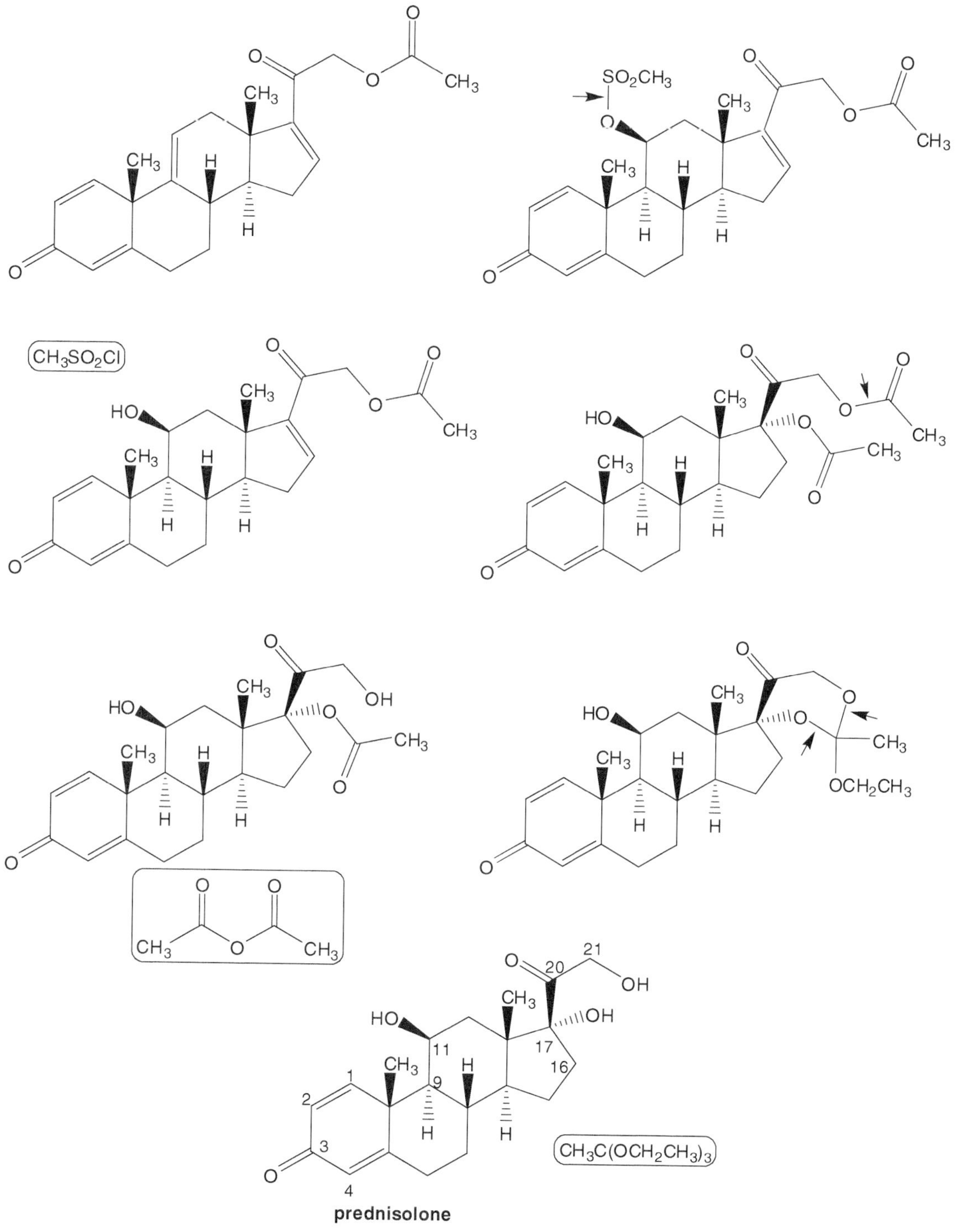

Extended Discussion

Draw the structures of the retrosynthetic analysis of one alternative route to the key tetraene intermediate.

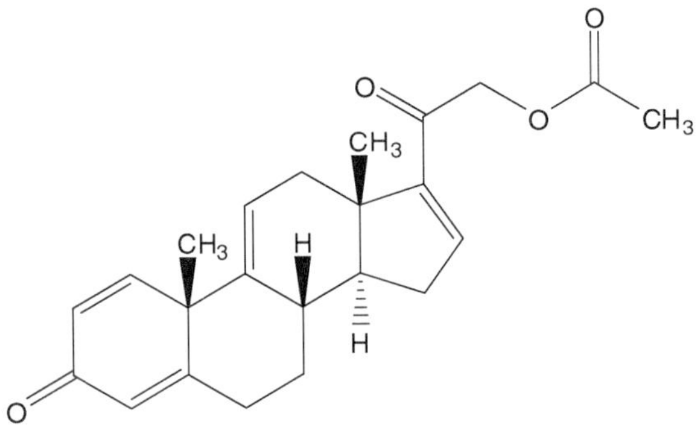

21-acetoxypregna-1,4,9(11),16-tetraene-3,20-dione

Bupivacaine

Anesthetics, Preoperative Medicines, and Medical Gases/Local Anesthetics

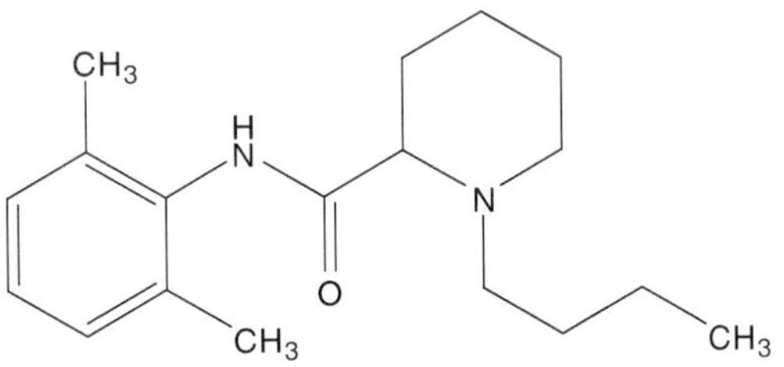

A piperidine is often formed by catalytic hydrogenation of a pyridine. Palladium, platinum, rhodium, ruthenium, and nickel catalysts have all been used. The hydrogenation is often run under acidic conditions to prevent catalyst deactivation by the piperidine product.

Discussion. Bupivacaine is a 1 : 1 mixture of enantiomers. Levobupivacaine, the (*S*)-enantiomer, is also manufactured for clinical use. The tertiary amine of bupivacaine is formed in the final step by *N*-alkylation of the secondary amine with 1-bromobutane. The amide is formed from 2,6-dimethylaniline (2,6-xylidene) and the acid chloride. Pipecolic acid chloride hydrochloride is formed from pipecolic acid. Pipecolic acid is formed by catalytic hydrogenation of 2-picolinic acid.

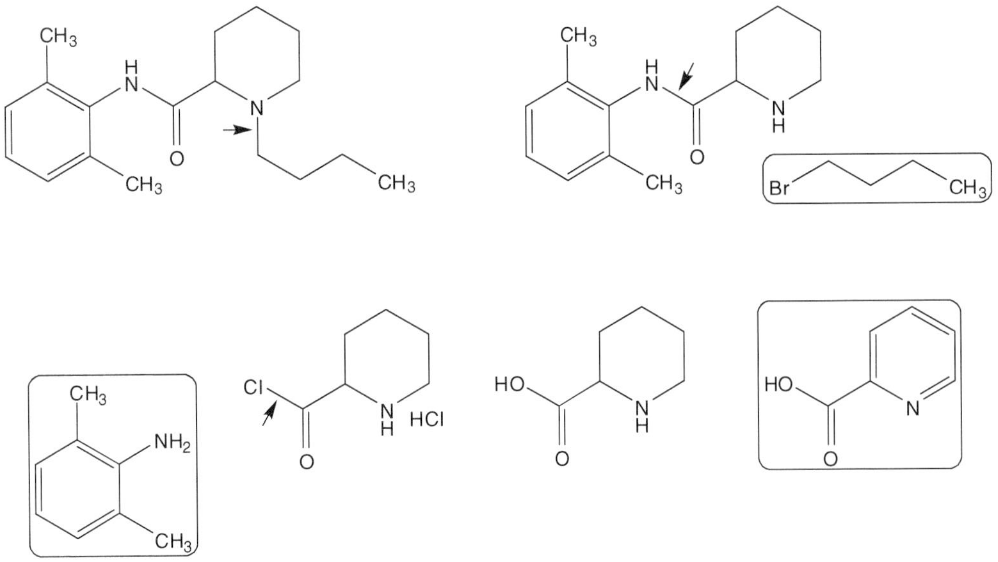

Extended Discussion

Draw the structures of the retrosynthetic analyses for three alternative routes to bupivacaine from the same starting materials (2,6-dimethylaniline, picolinic acid, 1-bromobutane) using the same transformations (catalytic hydrogenation, amide C—N bond formation, *N*-alkylation) in a different order.

C

Caffeine

Specific Medicines for Neonatal Care/Medicines Administered to the Neonate

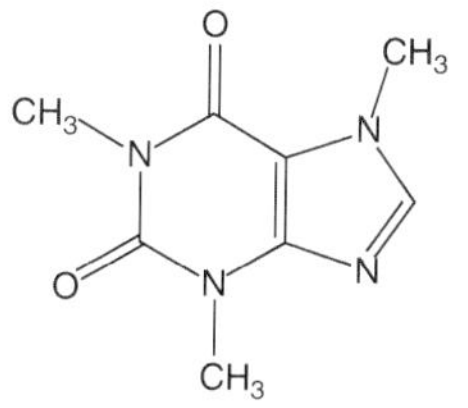

A urea, thiourea, or amidine often provides the carbon at position 2 of a pyrimidine.

Discussion. Caffeine is formed by methylation of theophylline. (List the methylating reagents and caffeine yield associated with each reagent.) Theophylline is formed by dehydration of 5,6-diamino-1,3-dimethyluracil 5-formamide. The 5-formamide is formed by reaction of the 5,6-diamine with formic acid. The amine at position 5 results from reduction of the oxime. The oxime is formed by nitrosation of 6-amino-1,3-dimethyluracil. 6-Amino-1,3-dimethyluracil is formed by condensation of cyanoacetic acid with *N,N′*-dimethylurea.

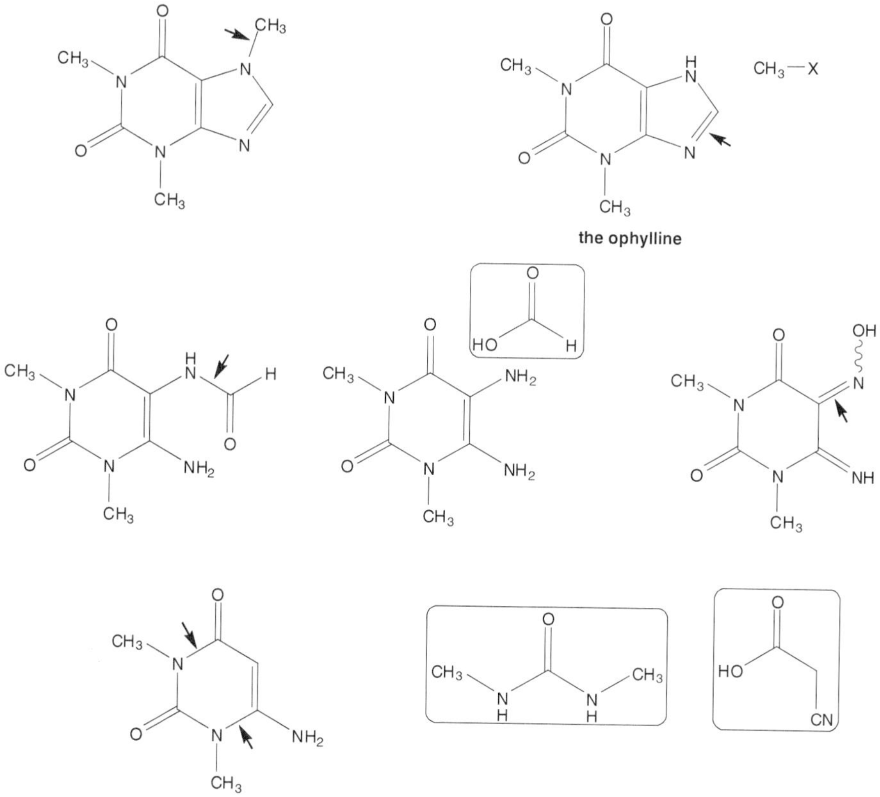

Routes to Essential Medicines: A Workbook for Organic Synthesis, First Edition. Peter J. Harrington.

Companion website: www.wiley.com/go/Harrington/routes_essential_medicine

Extended Discussion

In one alternative route, theophylline is formed by the reaction of 5,6-diamino-1,3-dimethyluracil with 1,3,5-triazine. Is this alternative route preferred?

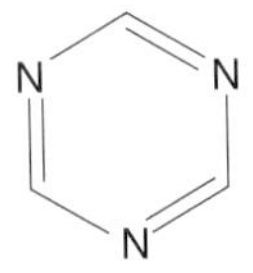

1,3,5-triazine

Calcium Folinate/Folinic Acid

Antineoplastics and Immunosuppressives/Cytotoxic and Adjuvant Medicines

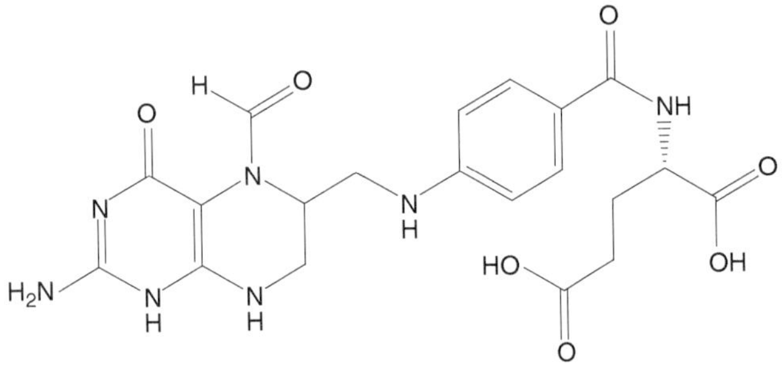

Oxidative instability of a intermediate presents a significant challenge when the intermediate is isolated. A folinic acid synthesis avoiding the isolation of 5,6,7,8-tetrahydrofolic acid and/or 10-formyl-5,6,7,8-tetrahydrofolic acid is preferred.

Discussion. Folinic acid (5-formyl-5,6,7,8-tetrahydrofolic acid, leucovorin) is manufactured from another essential medicine, folic acid, as a 1:1 mixture of two diastereomers [(6S,S) and (6R,S)]. Folinic acid is formed by hydrolysis of 5,10-methenyl-5,6,7,8-tetrahydrofolic acid in the final step. (10-Formyl-5,6,7,8,-tetrahydrofolic acid also forms in the hydrolysis. What are preferred conditions for formation of folinic acid?) 5,10-Methenyl-5,6,7,8-tetrahydrofolic acid is formed by reduction of 10-formylfolic acid and in situ dehydration of 10-formyl-5,6,7,8,-tetrahydrofolic acid. 10-Formylfolic acid is formed from folic acid and formic acid.

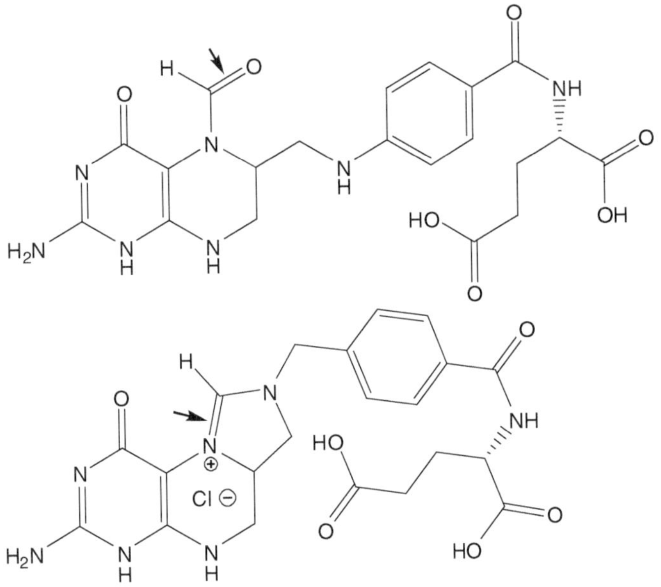

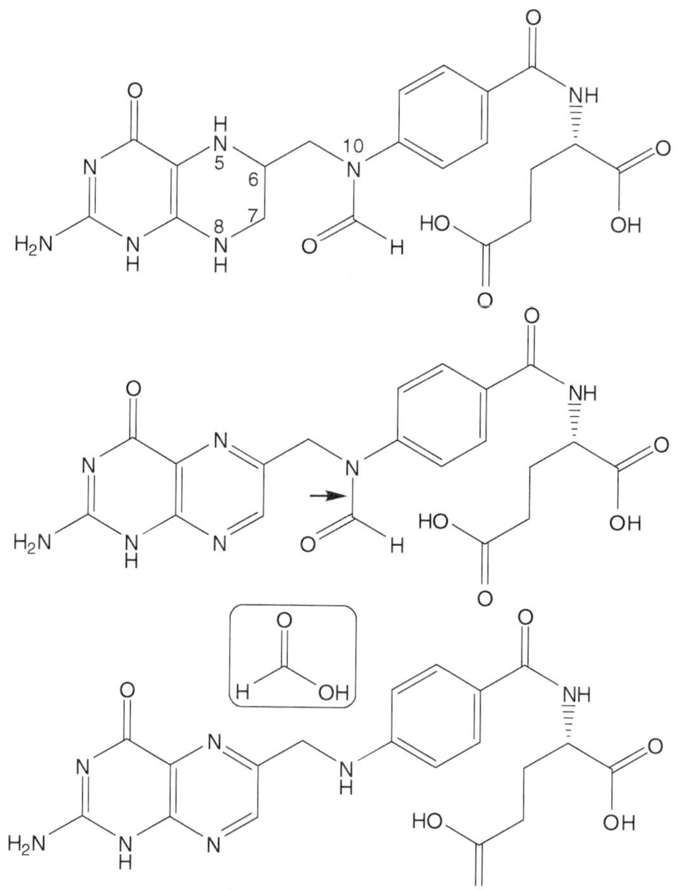

folicacid

Extended Discussion

The (6*S*,*S*)-diastereomer, levoleucovorin, is the biochemical cofactor associated with the desired pharmacological activity. Draw the structures of a retrosynthetic scheme or draw a flow diagram for one route to levoleucovorin.

Capecitabine

Antineoplastics and Immunosuppressives/Cytotoxic and Adjuvant Medicines

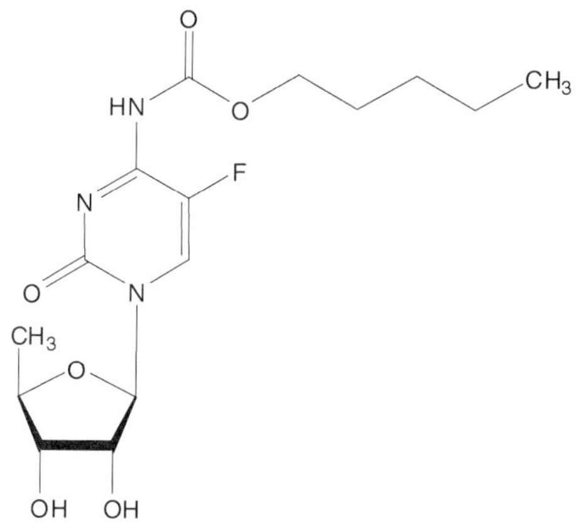

A nucleoside is often formed by displacement of a leaving group on the sugar by nitrogen of the heterocycle. The displacement usually results in a mixture of two products, with the heterocycle on the top face (β) or the bottom face (α) of the sugar. Factors which influence the β-product to α-product ratio include the heterocycle, the sugar, the leaving group, and the reaction conditions.

Discussion. The hydroxyl groups are released by ester hydrolysis in the final step. The carbamate is formed from pentyl chloroformate and the amine. The nucleoside C—N bond is formed by displacement of the C1 acetate of 1,2,3-triacetyl-5-deoxyribose by N1 of the essential medicine flucytosine (5-fluorocytosine) (**silyl–Hilbert–Johnson Reaction**). (What is the highest reported yield of the β–anomer? What reaction conditions are associated with the highest yield? How are the α– and β-anomers separated?)

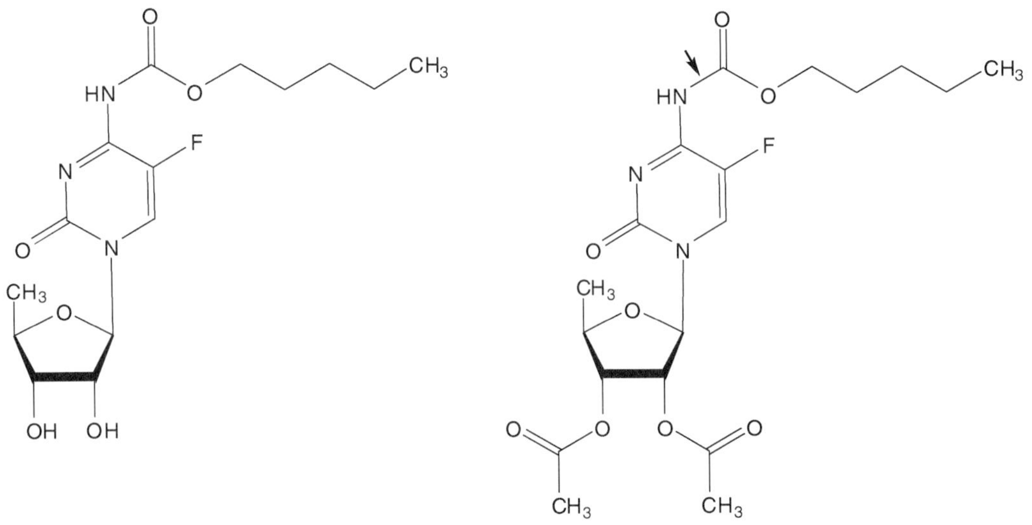

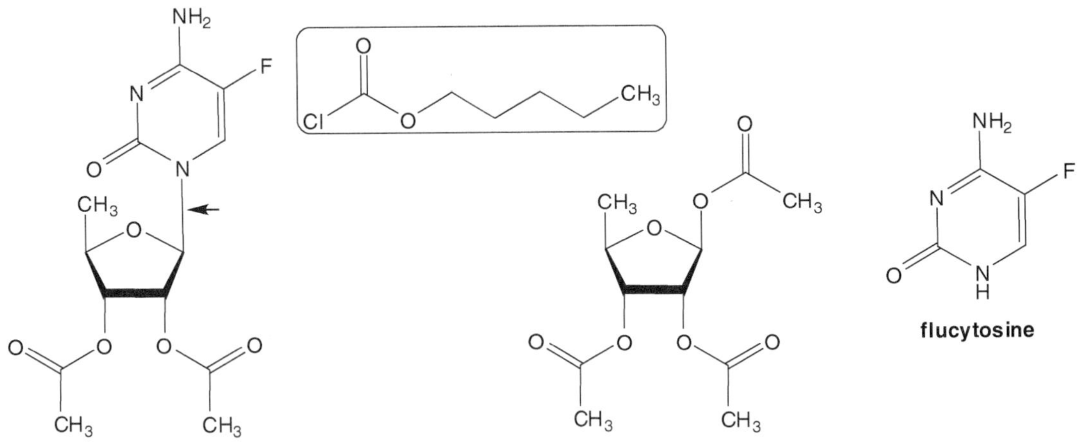

1,2,3-Triacetyl-5-deoxy-D-ribose is formed by reaction of 5-deoxy-D-ribose with acetic anhydride. The three hydroxyl groups of 5-deoxy-D-ribose are released by acetal hydrolysis. The methyl group at C5 is formed by reduction of a *p*-toluenesulfonate ester. The *p*-toluenesulfonate ester is formed by reaction of *p*-toluenesulfonyl chloride with the C5 alcohol of methyl 2,3-*O*-isopropylidene-D-ribofuranoside. Methyl 2,3-*O*-isopropylidene-D-ribofuranoside is formed from D-ribofuranose, acetone, and methanol. D-Ribofuranose is produced by fermentation.

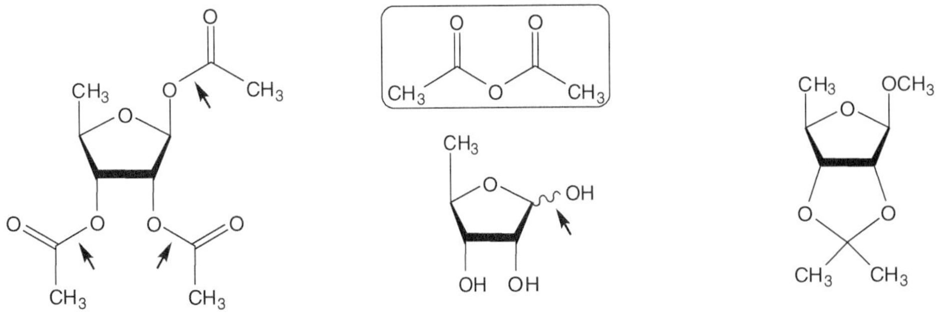

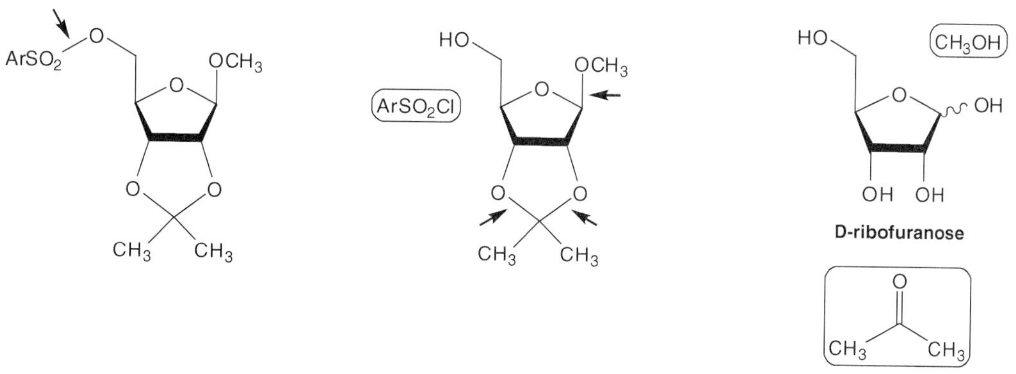

Carbamazepine

Anticonvulsants/Antiepileptics
Medicines for Mental and Behavioral Disorders/Medicines Used in Mood Disorders/Medicines Used in Bipolar Disorders

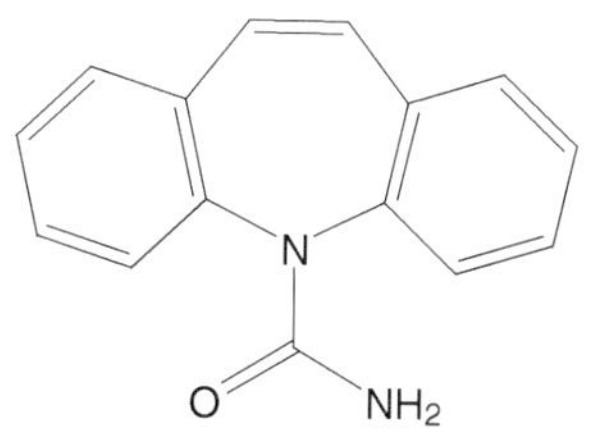

Diphenylamine and ammonia are formed by the reaction of two molecules of aniline under strongly acidic conditions at elevated temperature.

Discussion. Carbamazepine is formed by the reaction of 5*H*-dibenzo[b,f]azepine with potassium cyanate. 5*H*-Dibenzo [b,f]azepine is formed from 10,11-dihydro-5*H*-dibenzo[b,f]azepine. (These compounds are often called iminostilbene and iminodibenzyl.) The azepine ring is formed by cyclization of 1,2-bis(2-aminophenyl)ethane. The diamine is formed by reduction of 1,2-bis(2-nitrophenyl)ethane. 1,2-bis(2-Nitrophenyl)ethane is formed by dimerization of 2-nitrotoluene.

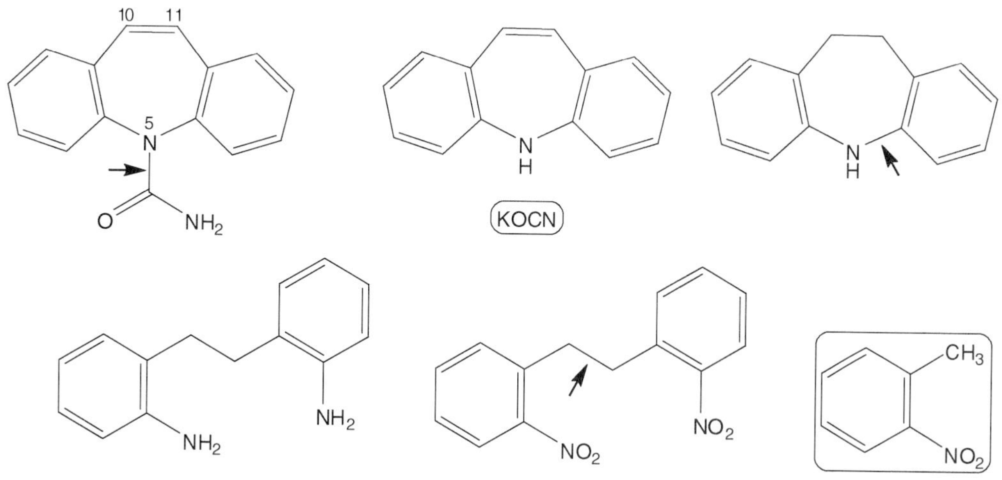

Extended Discussion

List the reagent(s) used in one alternative route from 5*H*-dibenzo[b,f]azepine to carbamazepine. List the pros and cons for both routes. Is one route preferred?

Cefalexin

Anti-Infective Medicines/Antibacterials/Beta-Lactam Medicines

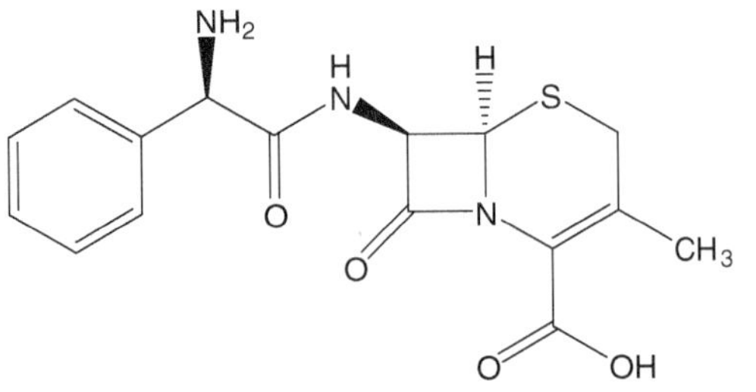

Many cephalosporin antibiotics are semisynthetic. When the cephalosporin target has a methyl substituent at position 3, the core structure is often formed from penicillin G (benzylpenicillin).

Discussion. The side chain amide is formed by the enzyme-mediated acylation of 7-aminodeacetoxycephalosporanic acid (7-ADCA) with D-α-phenylglycine methyl ester. The amine of 7-ADCA is released by enzyme-mediated deacylation of the side chain phenacetyl amide. Ring expansion from the five-membered ring of penicillin G to the six-membered ring of phenacetyl 7-ADCA is also mediated by an enzyme. Penicillin G is produced by the fungus *Penicillium chrysogenum*.

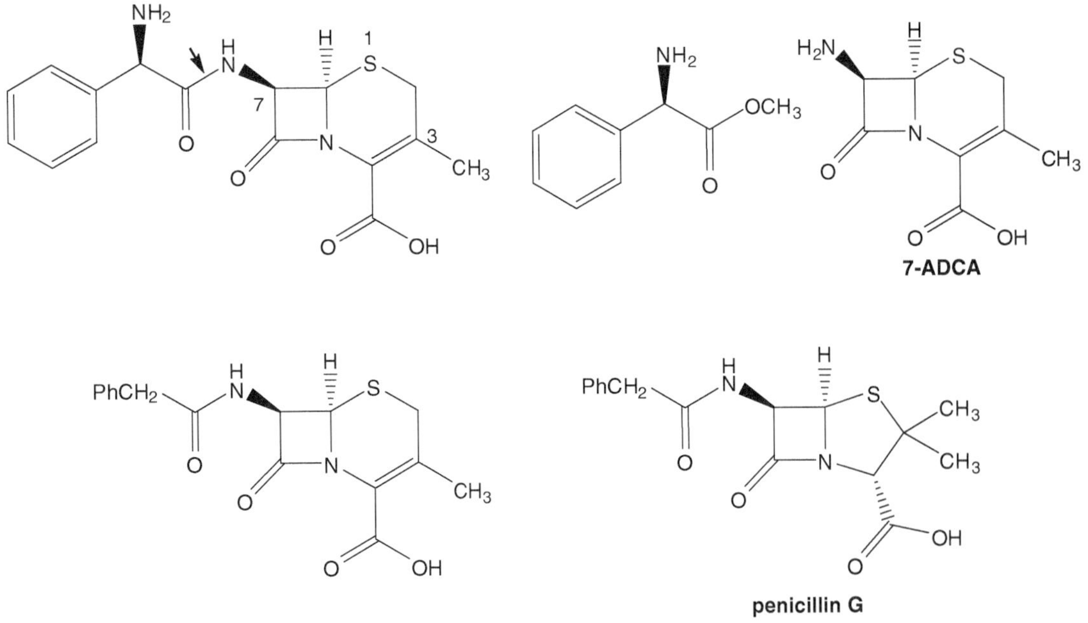

D-α-Phenylglycine methyl ester is produced by resolution. DL-α-Phenylglycine methyl ester is formed from the carboxylic acid and methanol (**Fischer Esterification**).

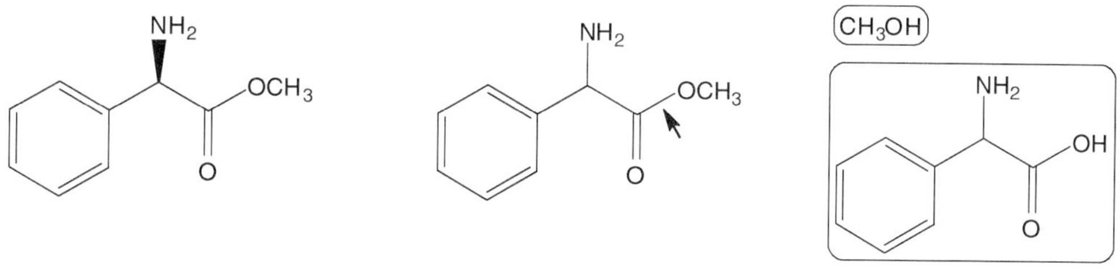

Extended Discussion

Draw the structures of a retrosynthetic analysis for an alternative nonenzyme route to phenacetyl 7-ADCA from penicillin G. List the byproducts associated with each reagent used in the alternative route.

Cefazolin

Anti-Infective Medicines/Antibacterials/Beta-Lactam Medicines

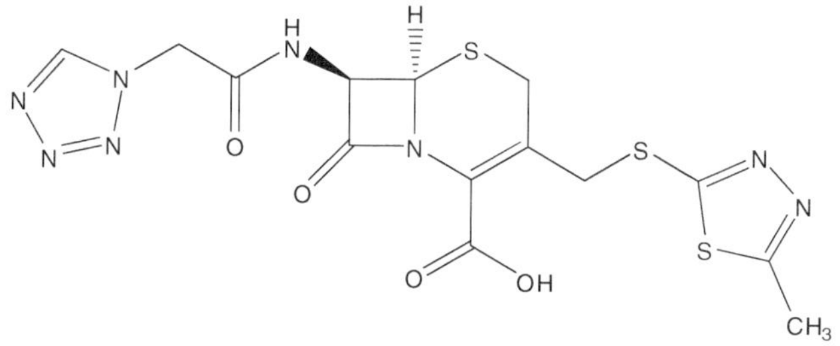

Many cephalosporin antibiotics are semisynthetic. When the cephalosporin target has a substituted methyl or an alkene substituent at position 3, it is often formed from cephalosporin C.

Discussion. The side chain amide of cefazolin is formed in the final step from an amine and a mixed anhydride formed from 1*H*-tetrazole-1-acetic acid. The side chain thioether is formed by displacement of the allylic acetate of 7-aminocephalosporanic acid (7-ACA) by a thiol. 7-ACA is formed by enzyme-mediated hydrolysis of the side chain amide of cephalosporin C. Cephalosporin C is produced by the fungus *Acremonium chrysogenum*.

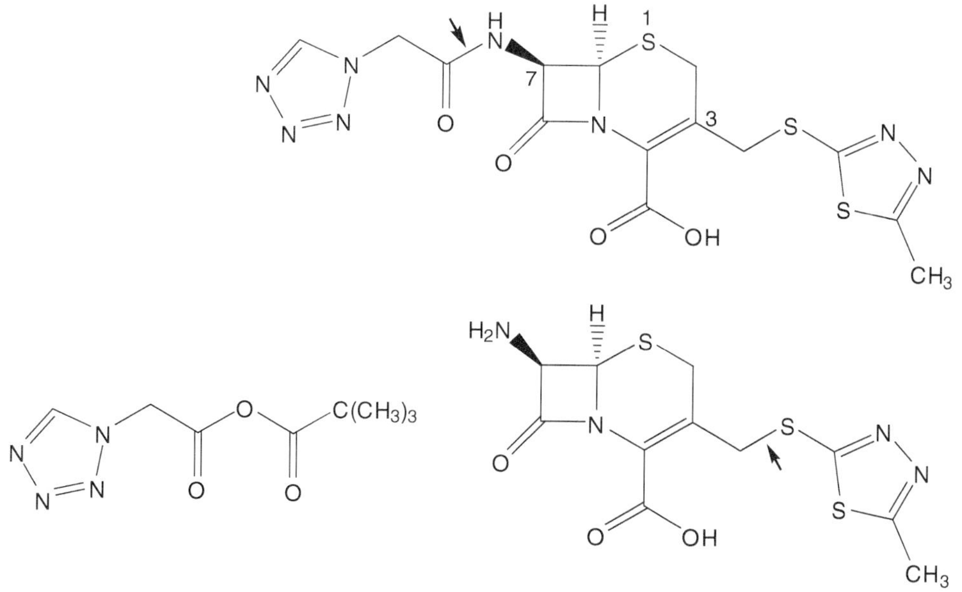

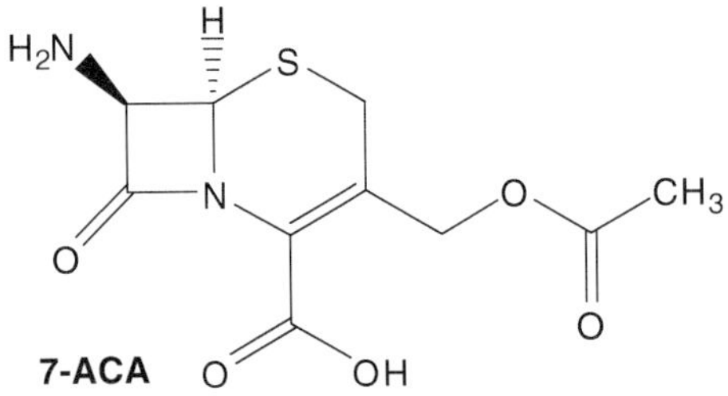

7-ACA

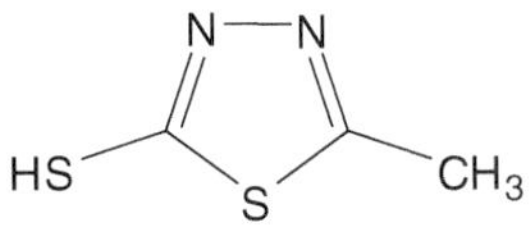

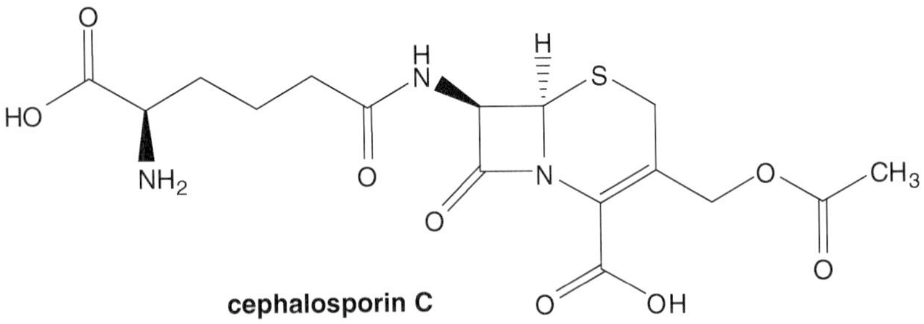

cephalosporin C

The mixed anhydride is formed from the carboxylic acid and pivaloyl chloride. 1*H*-Tetrazole-1-acetic acid is formed in a single step from glycine, triethyl orthoformate, and azidotrimethylsilane. Azidotrimethylsilane is formed from chlorotrimethylsilane and sodium azide.

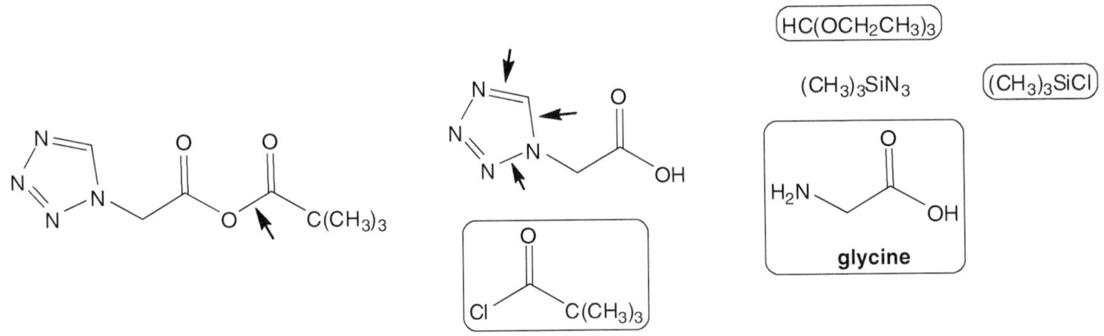

The thiol, 2-mercapto-5-methyl-1,3,4-thiadiazole is assembled in a single step from thioacetamide, hydrazine, and carbon disulfide.

Extended Discussion

Draw the structures of a retrosynthetic analysis of one alternative route to 1*H*-tetrazole-1-acetic acid. List and discuss the reactivity hazards(s) associated with both routes to 1*H*-tetrazole-1-acetic acid.

Cefepime

Anti-Infective Medicines/Antibacterials/Beta-Lactam Medicines

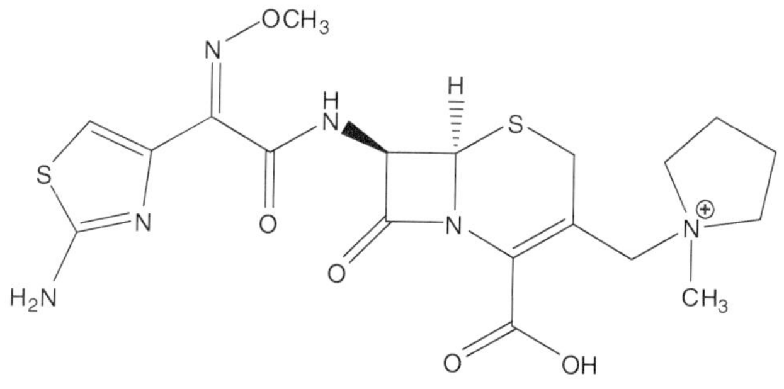

Many cephalosporin antibiotics are semisynthetic. When the cephalosporin target has a substituted methyl substituent or an alkene at position 3, it is often formed from cephalosporin C.

Discussion. The side chain amide is formed in the final step from the amine and a thioester. The amine and carboxylic acid are released by cleavage of N—Si and O—Si bonds. The quaternary ammonium salt is formed by displacement of iodide by *N*-methylpyrrolidine. Iodide displaces the allylic acetate in a silylated derivative of 7-aminocephalosporanic acid (7-ACA). 7-ACA is formed by enzyme-mediated hydrolysis of the side chain amide of cephalosporin C. Cephalosporin C is produced by the fungus *Acremonium chrysogenum*.

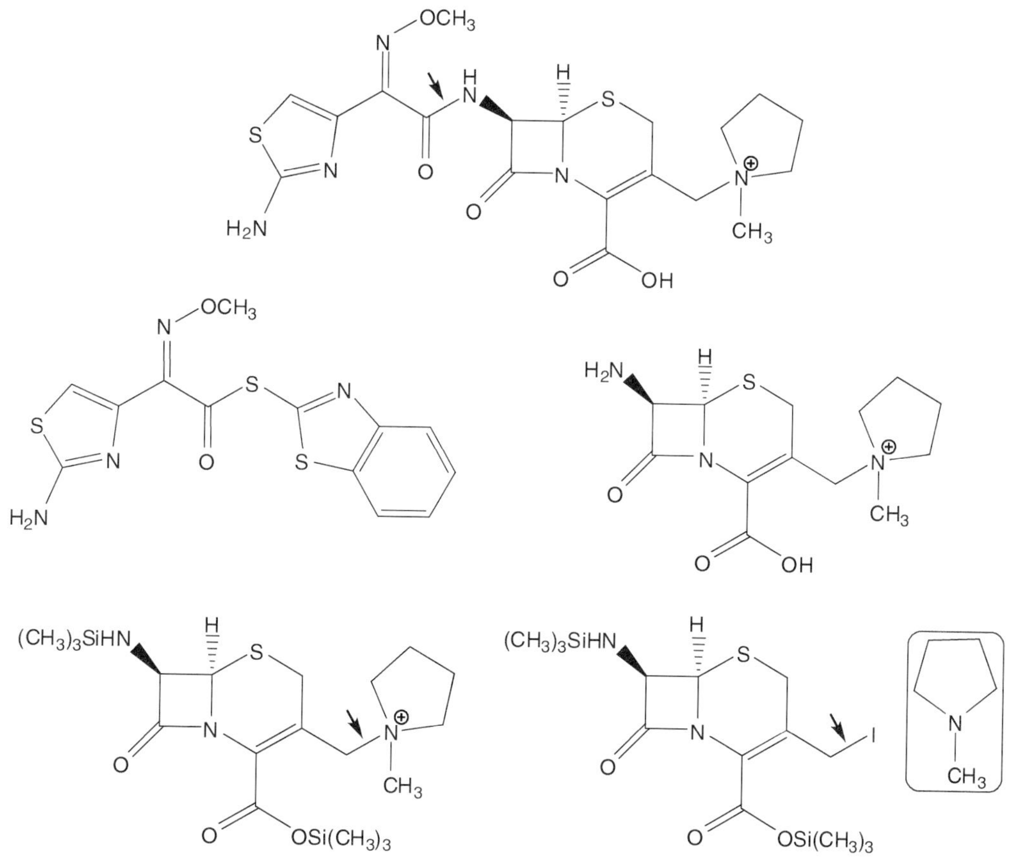

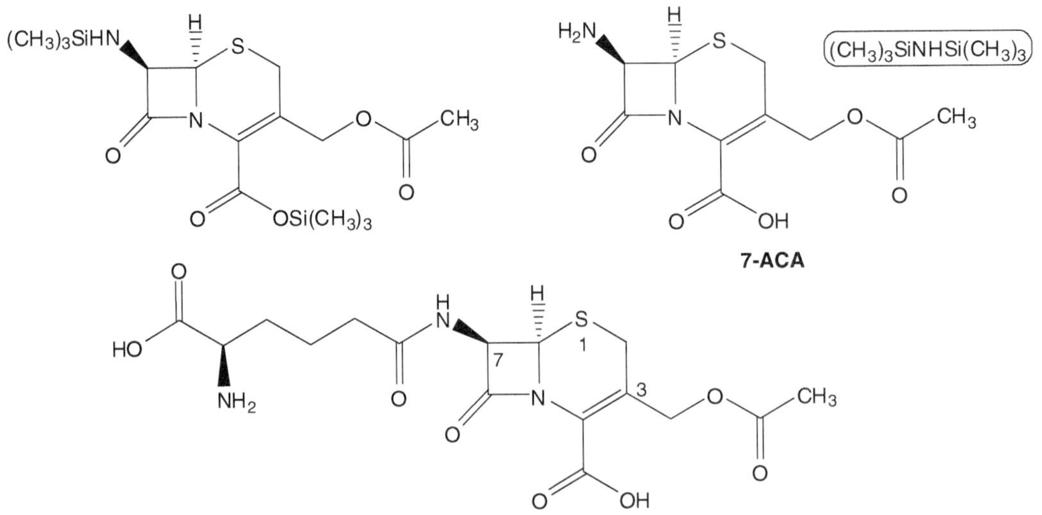

cephalosporin C

The thioester is formed from the carboxylic acid and 2,2′-dithiobis(benzothiazole). The carboxylic acid is formed by hydrolysis of the ethyl ester. The thiazole ring is formed by reaction of an α-bromoketone with thiourea (**Hantzsch Thiazole Synthesis**). The α-bromoketone is formed by bromination of the ketone. The *O*-methyloxime is formed by *O*-alkylation of the oxime. The oxime is formed by nitrosation of ethyl acetoacetate.

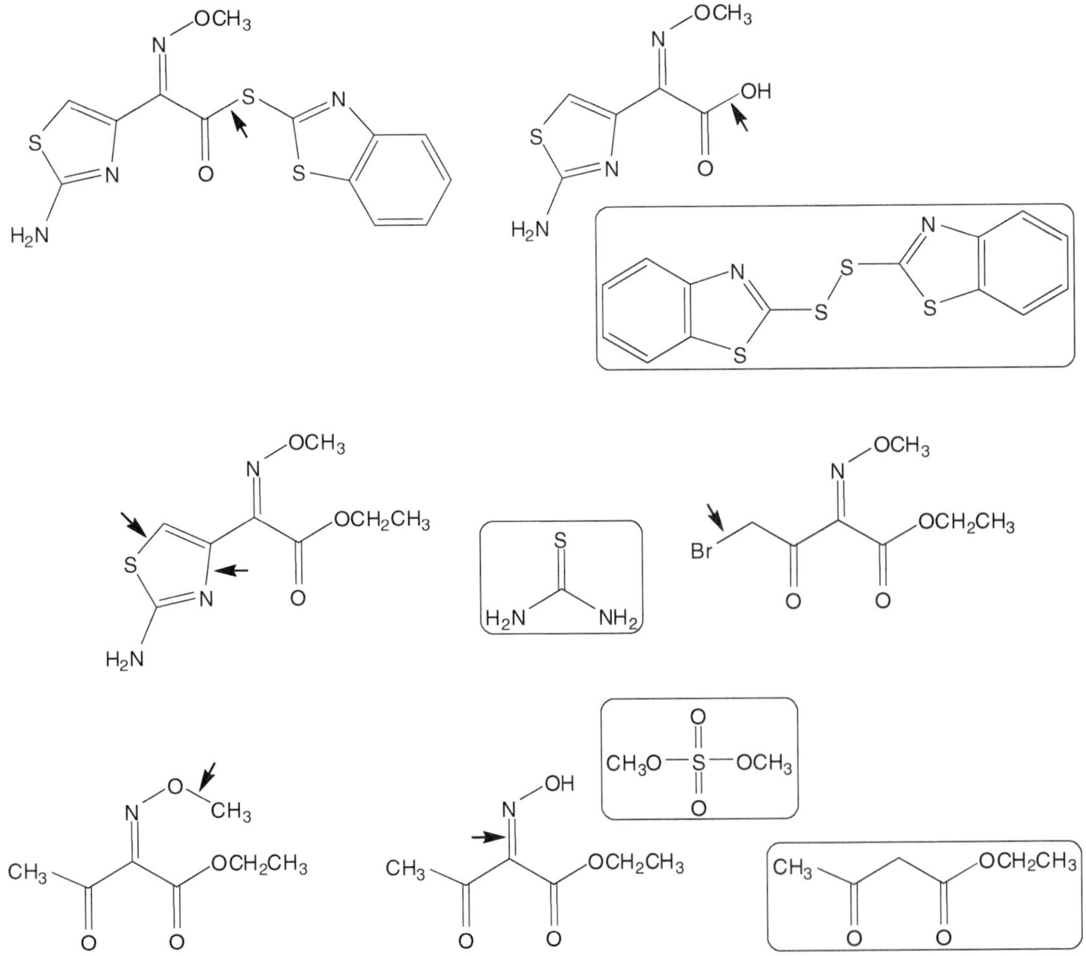

Extended Discussion

Draw the structures of a retrosynthetic analysis of one alternative route which utilizes a different protection–deprotection strategy in the conversion of 7-ACA to the amine used in the final acylation step. List pros and cons for both routes and select one route as the preferred route.

Cefixime

Anti-Infective Medicines/Antibacterials/Beta-Lactam Medicines

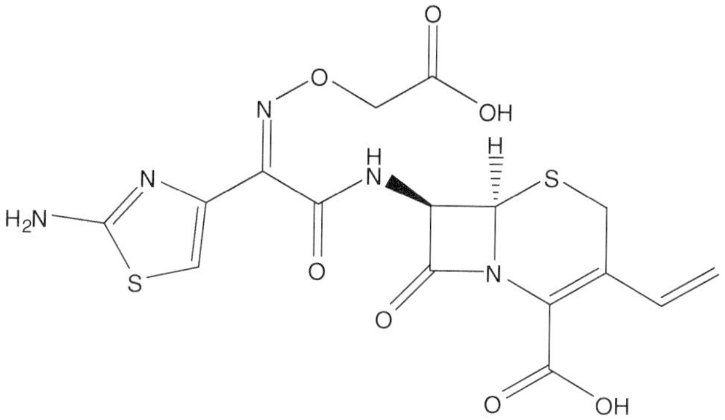

Many cephalosporin antibiotics are semisynthetic. When the cephalosporin target has a substituted methyl or an alkene substituent at position 3, it is often formed from cephalosporin C.

Discussion. The side chain carboxylic acid is released in the final step by ester hydrolysis. The side chain amide is formed from the amine, 7-amino-3-vinylcephem-4-carboxylic acid (7-AVCA), and a thioester.

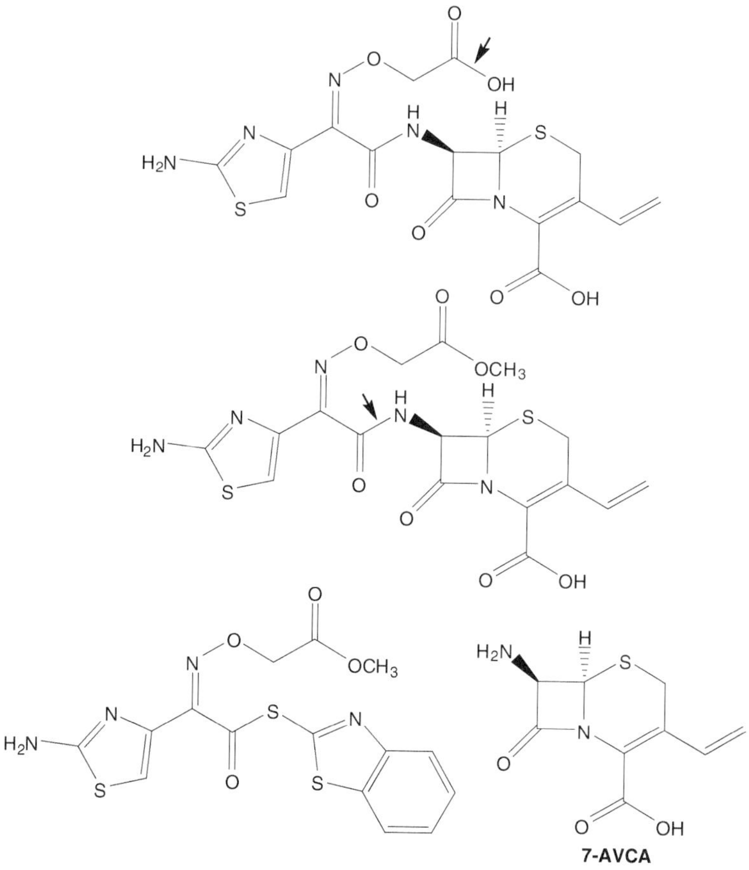

7-AVCA is formed from 7-aminocephalosporanic acid (7-ACA) in an eight-step linear sequence. Four steps in the sequence are protections or deprotections. The protecting groups are not specified in the analysis since there are two manufacturing routes which use the same sequence. The carboxylic acid is released by cleavage of protecting group P_2. The amine is released by cleavage of a protecting group P_1. The vinyl group is formed from the phosphonium salt and formaldehyde (**Wittig Reaction**). The phosphonium salt is formed by leaving group (I, Br, Cl) displacement by triphenylphosphine. The halide is formed from the alcohol. The carboxylic acid is protected. The amine is protected. The alcohol is formed from 7-aminocephalosporanic acid (7-ACA) by hydrolysis of the ester. 7-ACA is formed by enzyme-mediated hydrolysis of the side chain amide of cephalosporin C. Cephalosporin C is produced by the fungus *A. chrysogenum*.

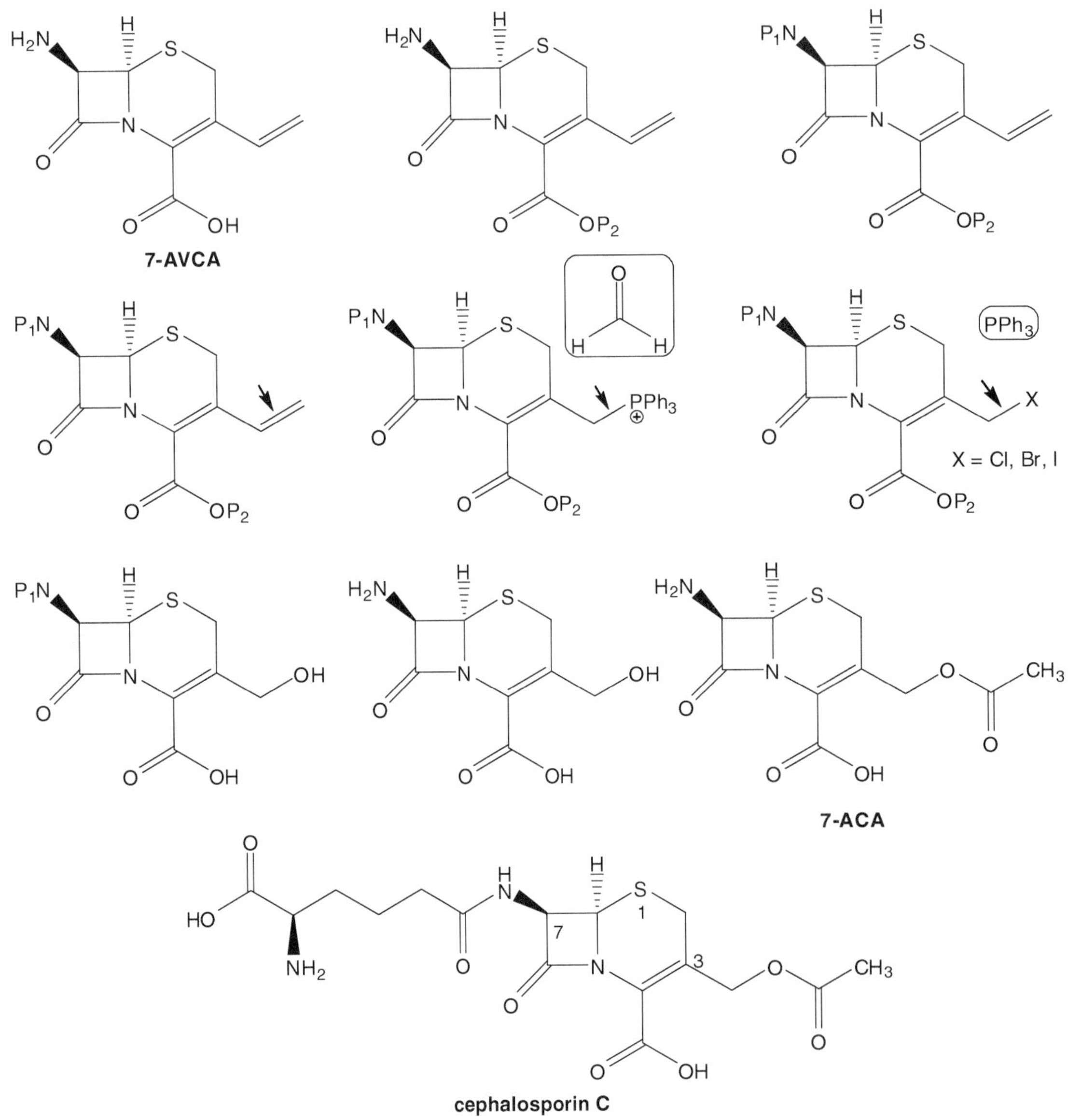

The thioester is formed from the carboxylic acid and 2,2′-dithiobis(benzothiazole). The thiazole ring is formed by reaction of an α–bromoketone with thiourea (**Hantzsch Thiazole Synthesis**). The α-bromoketone is formed from the ketone and the carboxylic acid is formed by hydrolysis of the *tert*-butyl ester. The *O*-alkyloxime is formed by reaction of the oxime with methyl chloroacetate. The oxime is formed by nitrosation of *tert*-butyl acetoacetate.

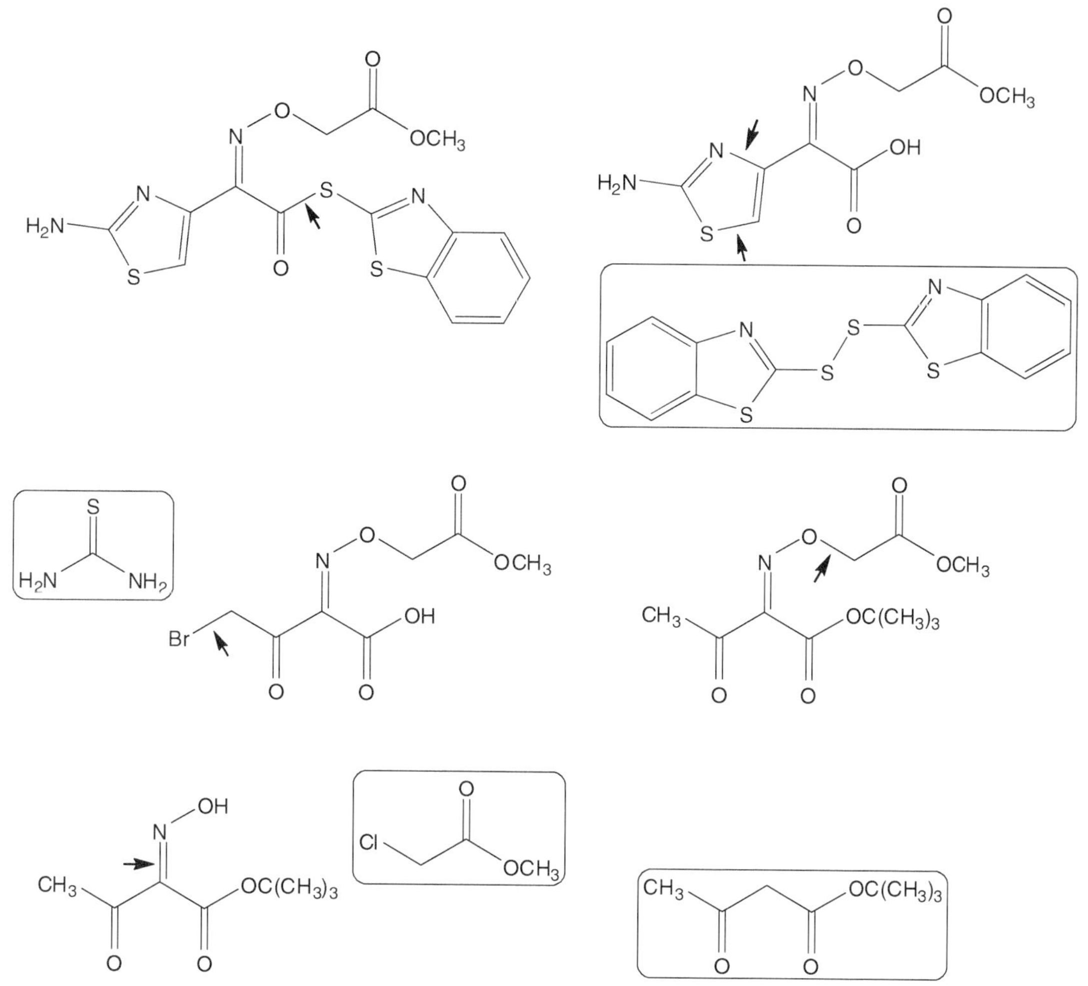

Extended Discussion

Identify the protecting groups P_1 and P_2 used in both routes to 7-amino-3-vinylcephem-4-carboxylic acid (7-AVCA). List the pros and cons for both routes. Is one route preferred?

Cefotaxime

Anti-Infective Medicines/Antibacterials/Beta-Lactam Medicines

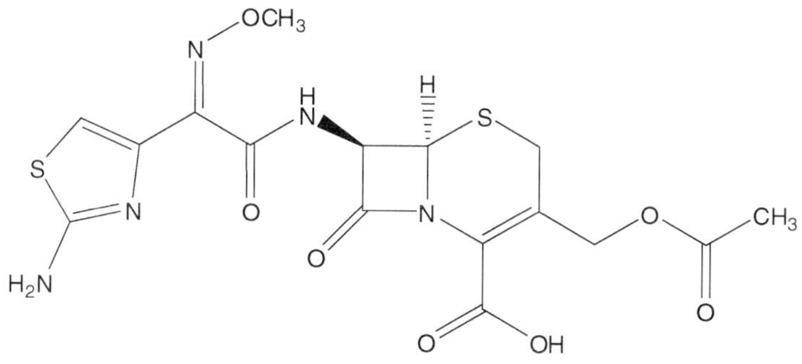

Many cephalosporin antibiotics are semisynthetic. When the cephalosporin target has a substituted methyl or an alkene substituent at position 3, it is often formed from cephalosporin C.

Discussion. The side chain amide is formed in the final step from 7-aminocephalosporanic acid (7-ACA) and a thioester. 7-ACA is formed by enzyme-mediated hydrolysis of the side chain amide of cephalosporin C. Cephalosporin C is produced by the fungus *A. chrysogenum*.

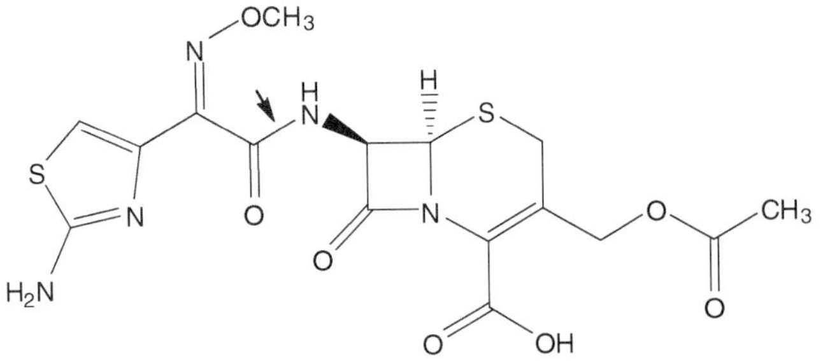

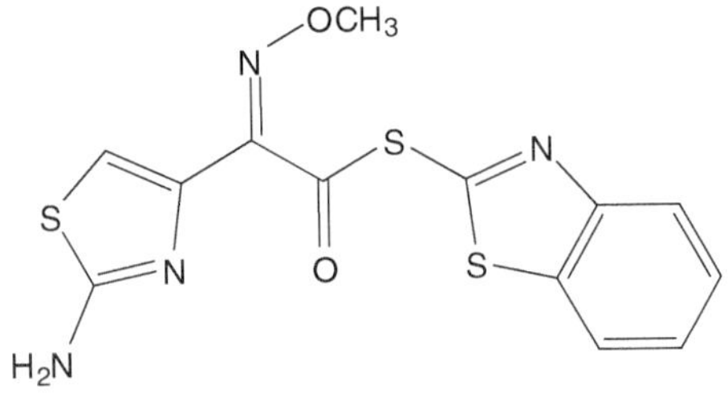

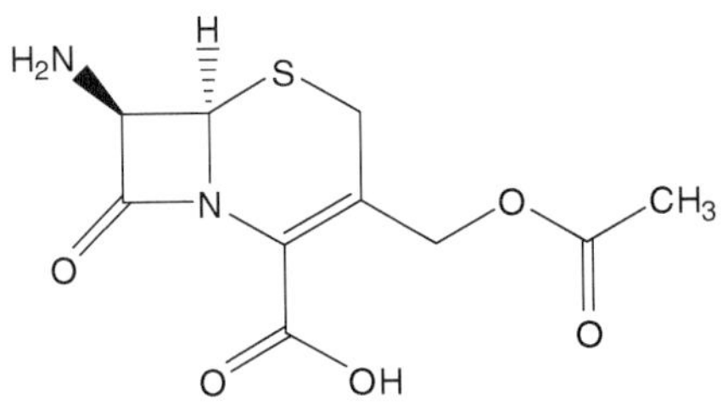

7-ACA

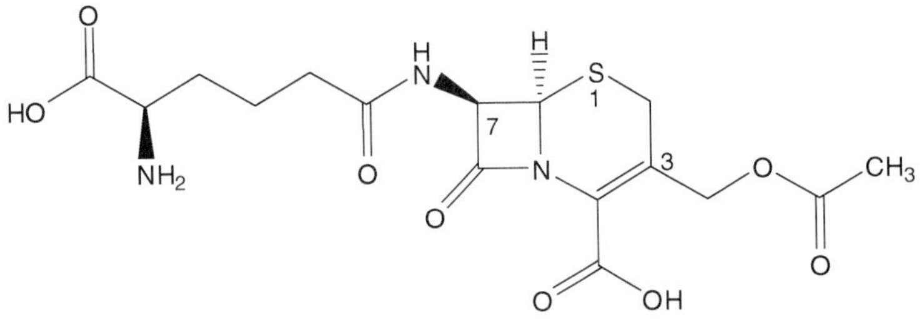

cephalosporin C

The thioester is formed from the carboxylic acid and 2,2′-dithiobis(benzothiazole). The carboxylic acid is formed by hydrolysis of the ethyl ester. The thiazole ring is formed by reaction of an α–bromoketone with thiourea (**Hantzsch Thiazole Synthesis**). The α-bromoketone is formed by bromination of the ketone. The *O*-methyloxime is formed by *O*-alkylation of the oxime. The oxime is formed by nitrosation of ethyl acetoacetate.

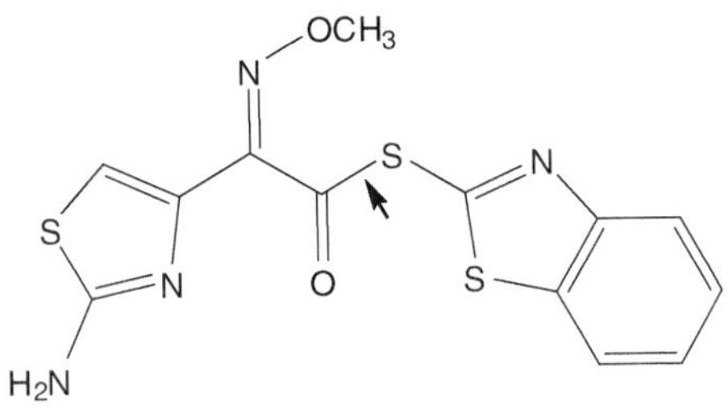

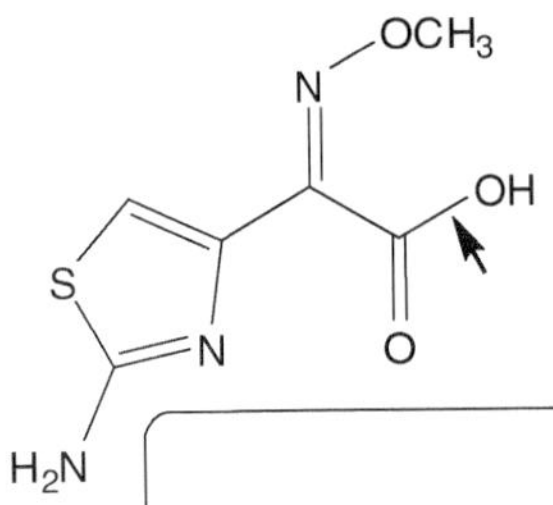

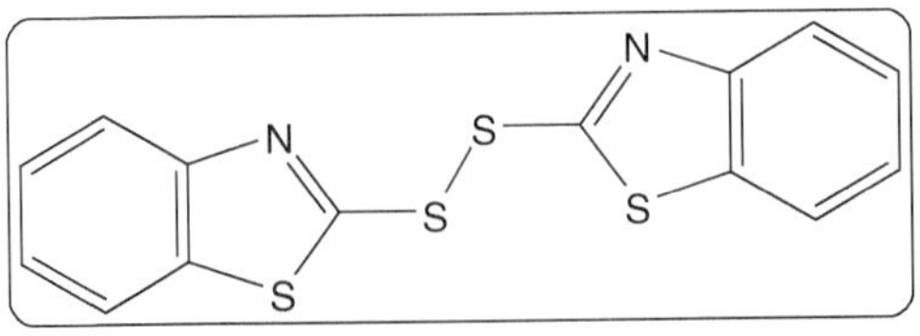

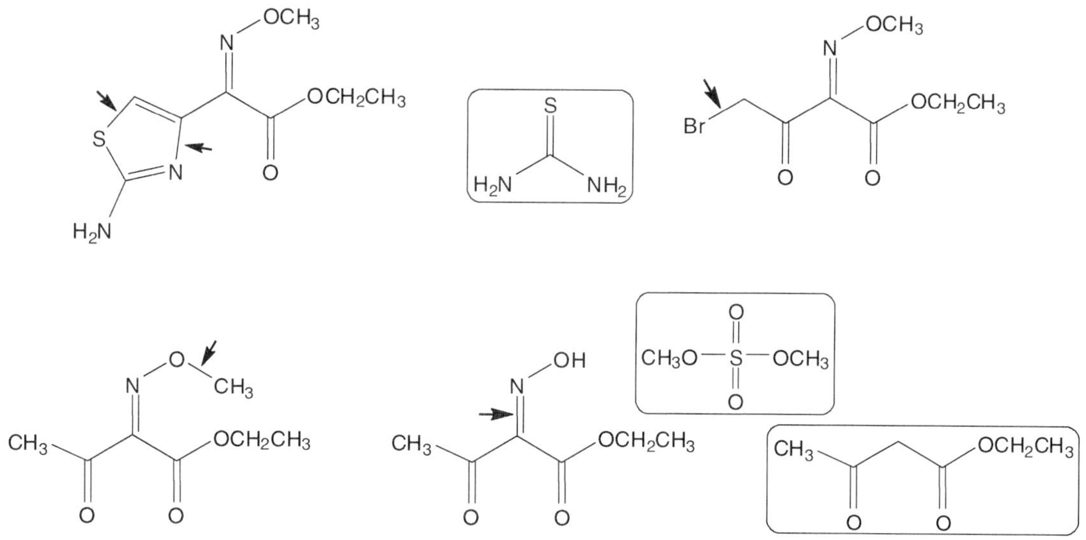

Extended Discussion

There are many alternative active esters which react with 7-ACA to form cefotaxime. Select one of these alternative active esters.

1) Draw the structures of the reagent(s) used to prepare the active ester. Draw the structures of the retrosynthetic analysis of each reagent from petrochemical or biochemical raw materials.
2) What is the yield of active ester?
3) What is the yield of cefotaxime using this active ester?
4) What byproduct(s) form in the amide formation?

Ceftaroline

Anti-Infective Medicines/Antibacterials/Beta-Lactam Medicines

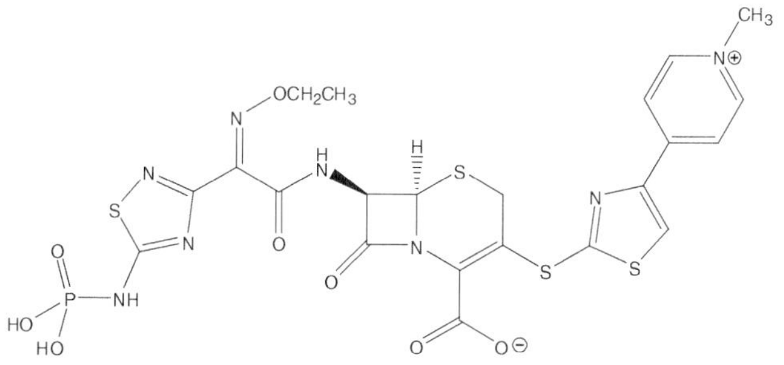

Many cephalosporin antibiotics are semisynthetic. An oxygen or sulfur substituent at the 3-position of a cephalosporin is often formed from a 3-exomethylene substituent. The cephalosporin with a 3-exomethylene substituent is often formed from penicillin G (benzylpenicillin).

Discussion. The side chain amide is formed in the final step from an amine and an acid chloride. The dichlorophosporamide is hydrolyzed in the workup of the final step. The carboxylic acid at cepham C4 is released from the diphenylmethyl (benzhydryl) ester. The amine is released by cleavage of the side chain phenacetyl amide. The pyridine is alkylated by reaction with iodomethane. The thioether at cepham C3 is formed by displacement of methanesulfonate by the thiol.

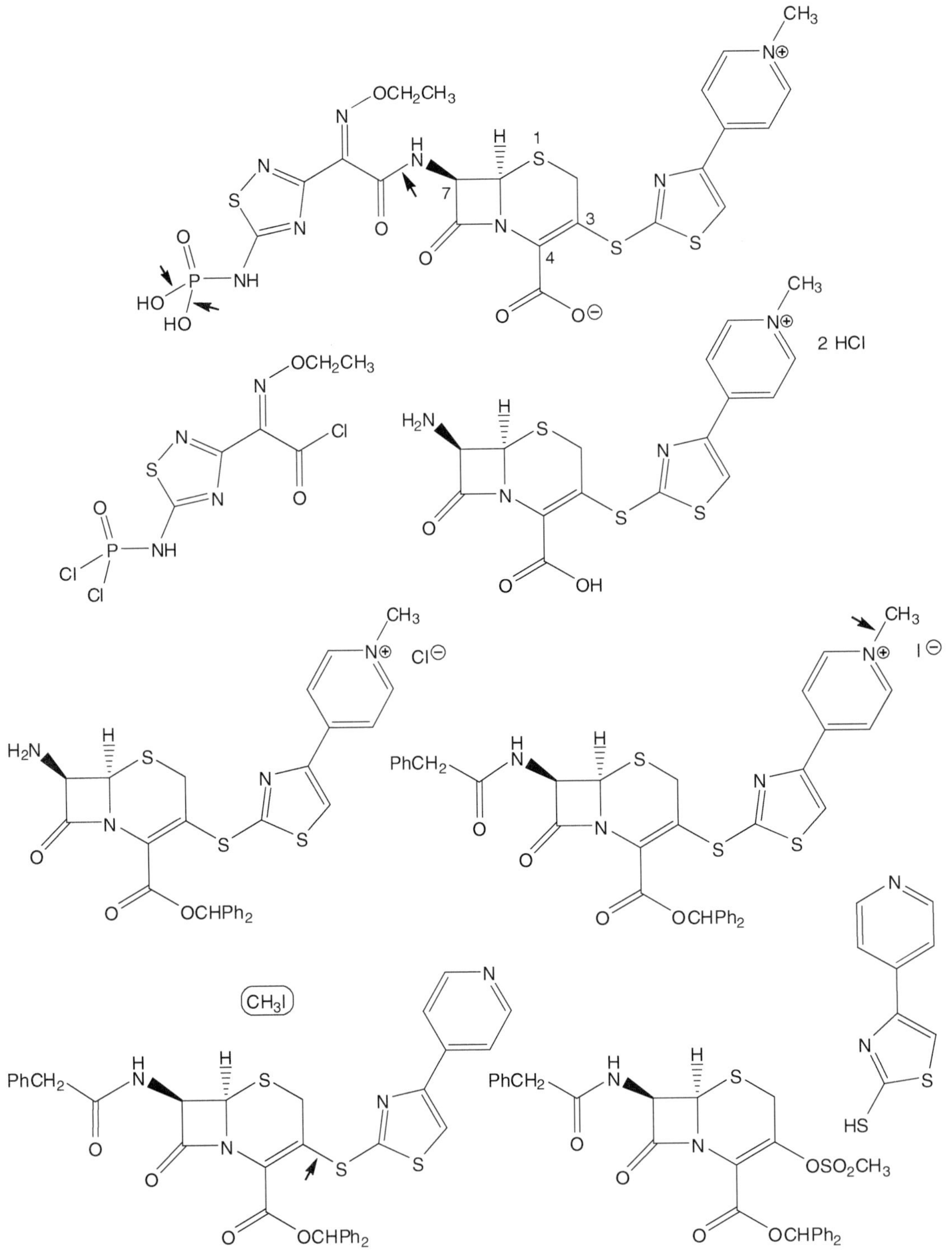

The methanesulfonate is formed by reaction of the enol with methanesulfonyl chloride. The sulfide is formed by reduction of the sulfoxide. The enol is formed by ozonolysis of the cephem C3 exomethylene substituent. The cephem with the 3-exomethylene substituent is formed by reaction of a sulfinyl chloride with an alkene. The sulfinyl chloride and alkene are both formed by ring-opening of the penam sulfoxide. The penam sulfoxide is formed by oxidation of the penam sulfide. The benzhydryl ester is formed by *O*-alkylation of the carboxylic acid of penicillin G. Penicillin G is produced by the fungus *P. chrysogenum*.

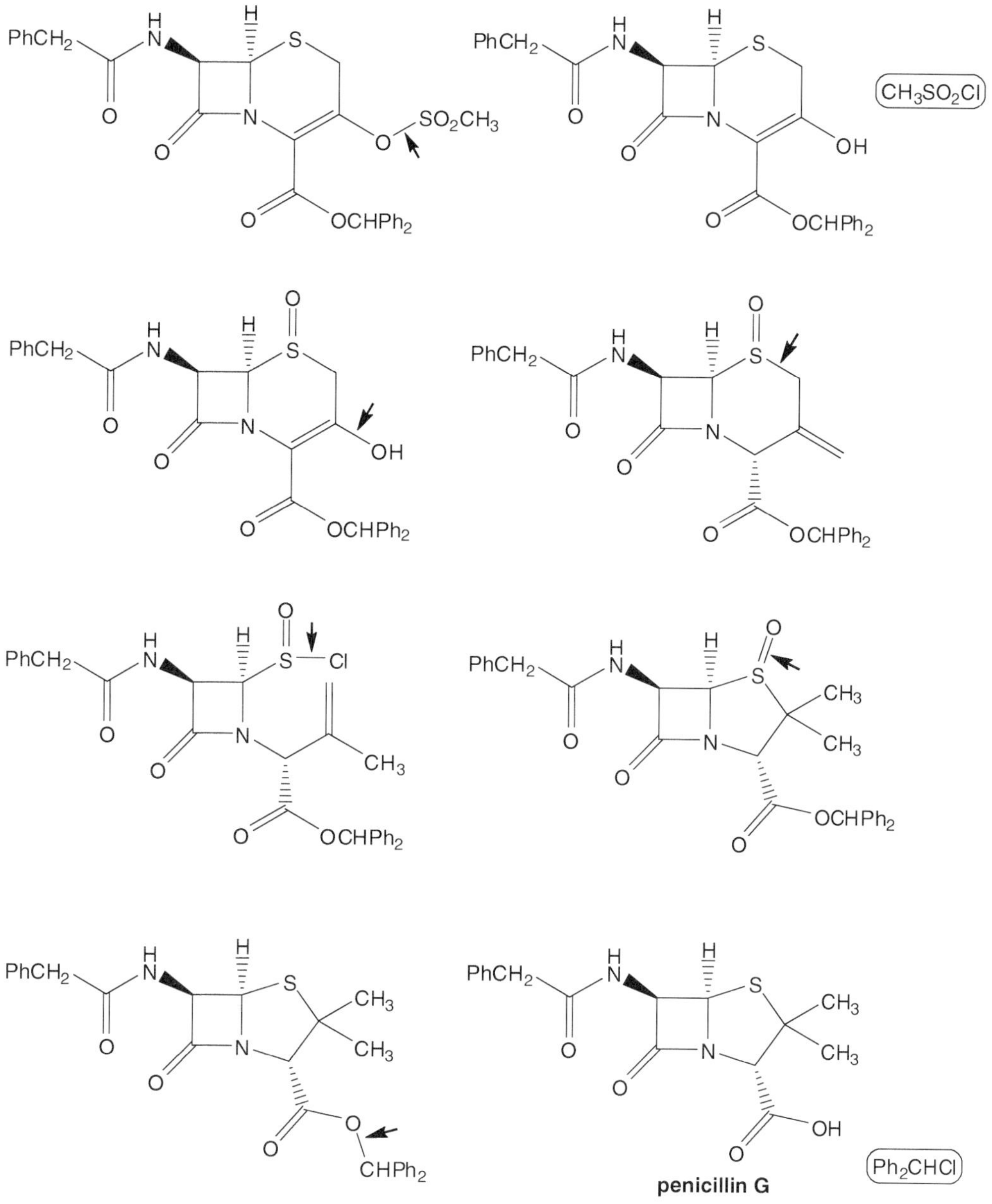

The acid chloride is formed by reaction of the carboxylic acid with phosphorus trichloride. The dichlorophosphoramide is also formed. The carboxylic acid is formed by hydrolysis of the nitrile. The 1,2,4-thiadiazole ring is formed by reaction of an amidine with potassium thiocyanate. The amidine is formed from a nitrile. (The desired diastereomer is the major product.) The C—O bond is formed by *O*-alkylation of the oxime. The oxime is formed by nitrosation of malononitrile.

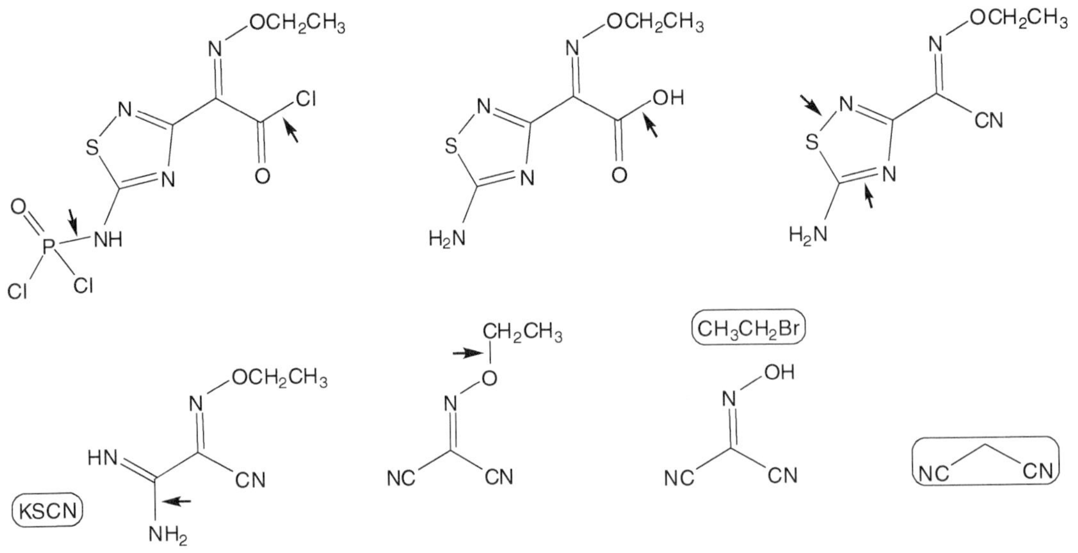

4-(4-Pyridinyl)thiazole-2-thiol is formed by the reaction of ammonium dithiocarbamate with the α–bromoketone. Ammonium dithiocarbamate is formed from carbon disulfide. The α–bromoketone is formed by bromination of the ketone, 4-acetylpyridine.

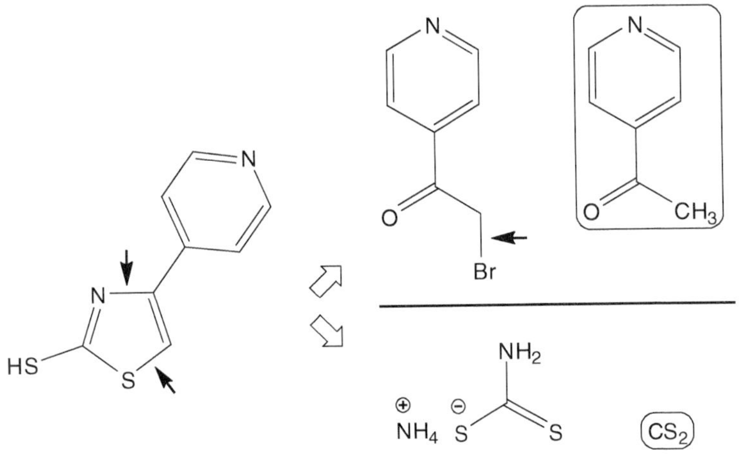

Extended Discussion

Draw the structures of a retrosynthetic analysis of a more convergent alternative route where methylation of the pyridine nitrogen is accomplished before formation of the thioether. List the pros and cons for both routes. Is one route preferred?

Ceftazidime

Anti-Infective Medicines/Antibacterials/Beta-Lactam Medicines

Many cephalosporin antibiotics are semisynthetic. When the cephalosporin target has a substituted methyl or an alkene substituent at position 3, it is often formed from cephalosporin C.

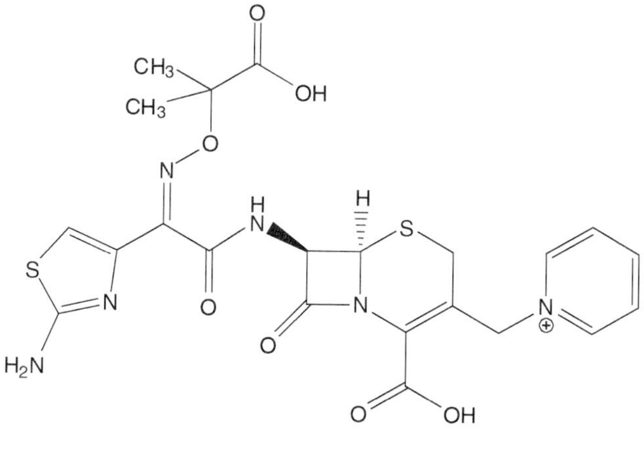

Discussion. Hydrolysis of a *tert*-butyl ester releases the side chain carboxylic acid in the final step. The side chain amide is formed from the amine (7-PyCA) and a thioester.

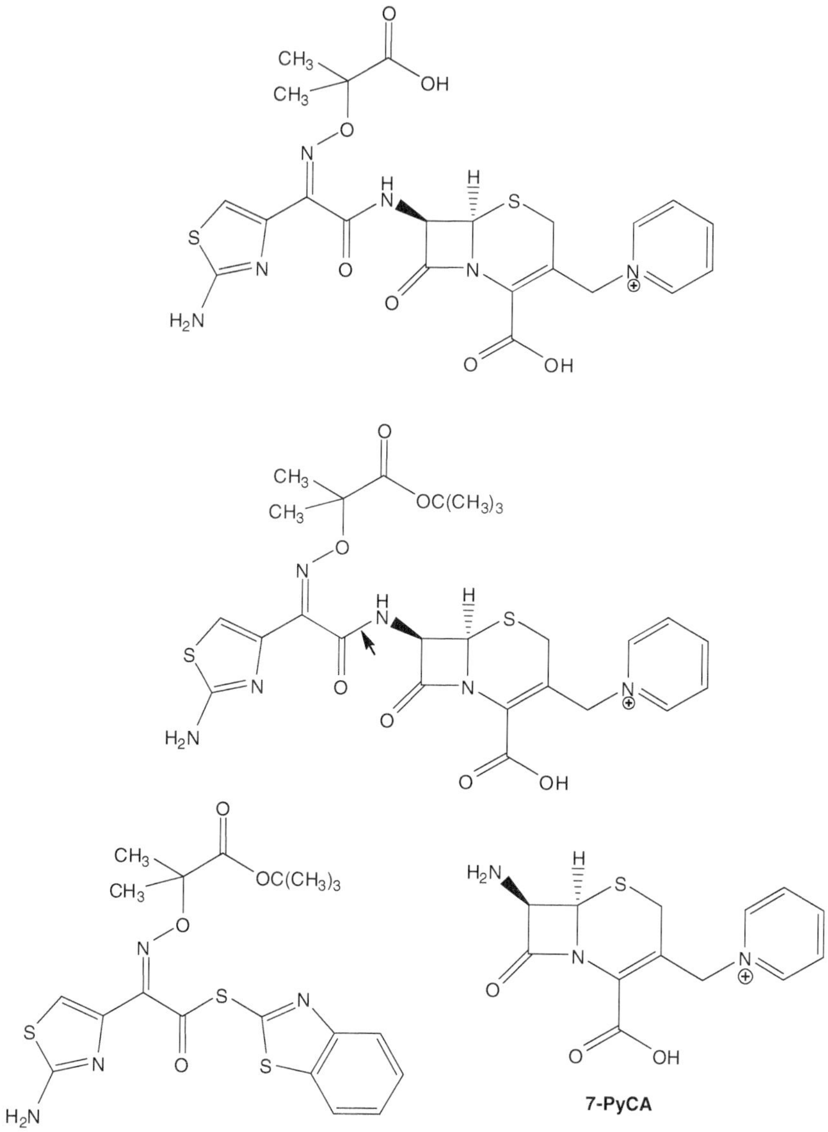

The amine and carboxylic acid of 7-PyCA are released by cleavage of N-Si and O-Si bonds. The pyridinium salt is formed by displacement of iodide by pyridine. The iodide is introduced by displacement of the allylic acetate in a derivative produced by *N* and *O*-silylation of 7-aminocephalosporanic acid (7-ACA). 7-ACA is formed by enzyme-mediated hydrolysis of the side chain amide of cephalosporin C. Cephalosporin C is produced by the fungus *A. chrysogenum*.

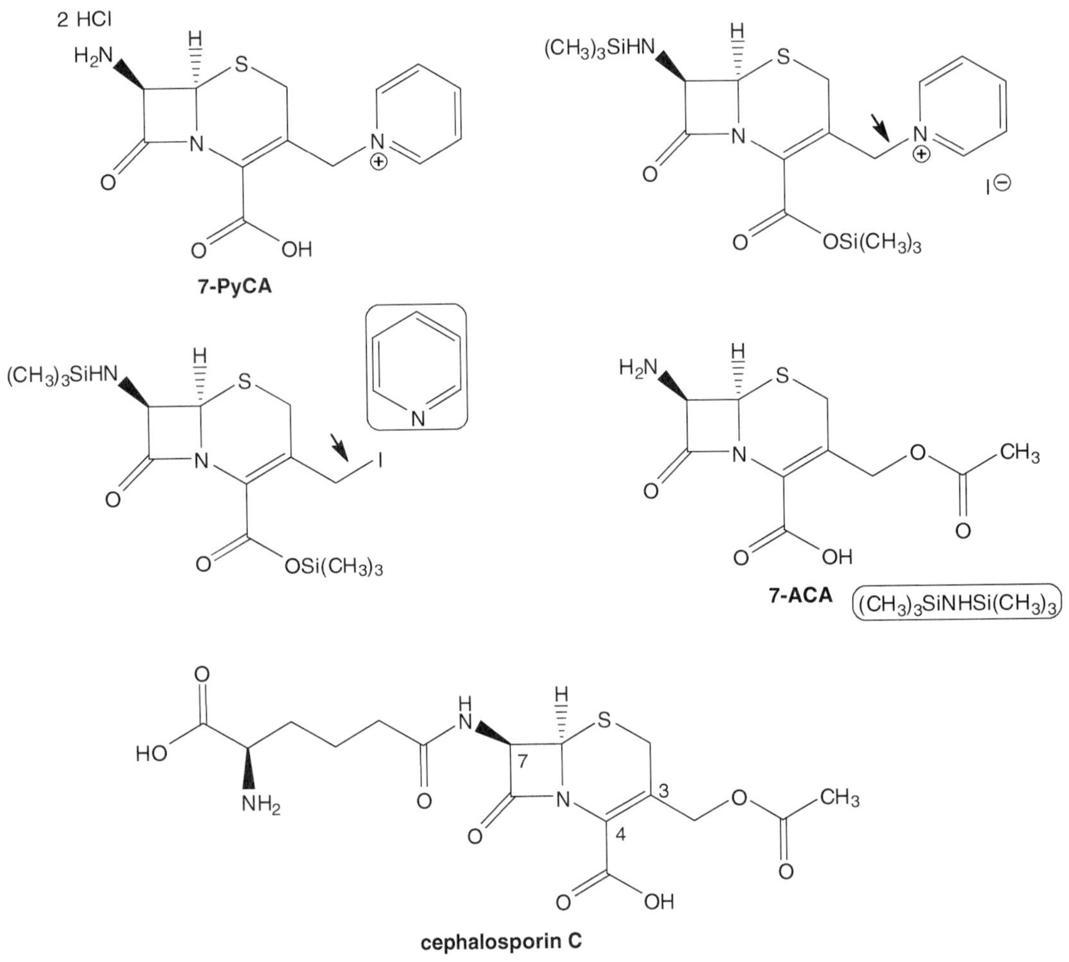

The thioester is formed from the carboxylic acid and 2,2′-dithiobis(benzothiazole). The carboxylic acid is formed by hydrolysis of the ethyl ester. The *O*-alkyloxime is formed by *O*-alkylation of the oxime with *tert*-butyl α-bromoisobutyrate. The thiazole ring of is formed by reaction of the α–bromoketone with thiourea (**Hantzsch Thiazole Synthesis**). The α-bromoketone is formed by bromination of the ketone. The oxime is formed by nitrosation of ethyl acetoacetate.

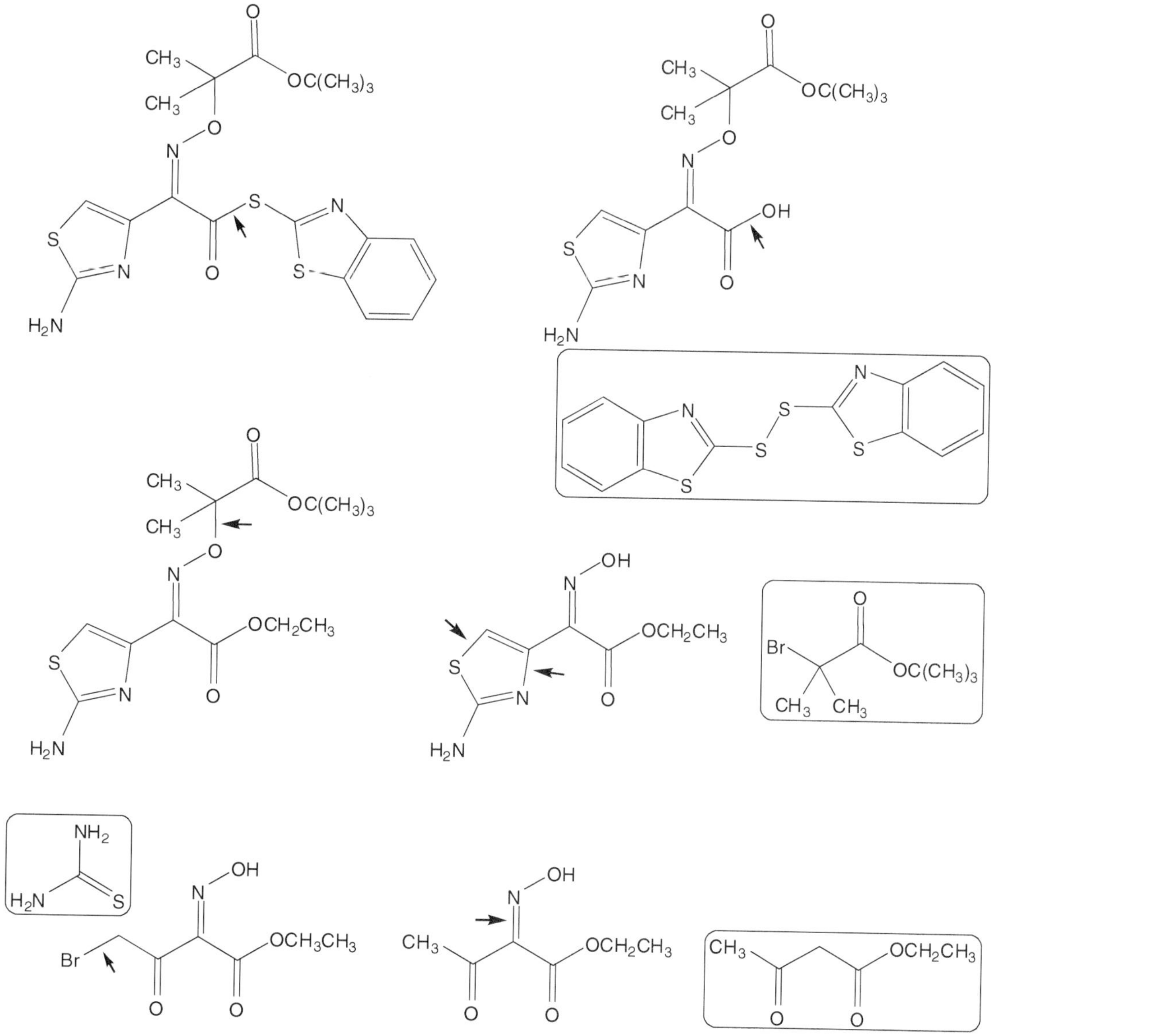

Extended Discussion

Ceftazidime and cefixime both have a carboxylic acid on the amide side chain. In the ceftazidime retrosynthetic analysis, the carboxylic acid precursor is a *tert*-butyl ester. In the cefixime retrosynthetic analysis, the carboxylic acid precursor is a methyl ester. Why are different esters used?

Ceftriaxone

Anti-Infective Medicines/Antibacterials/Beta-Lactam Medicines

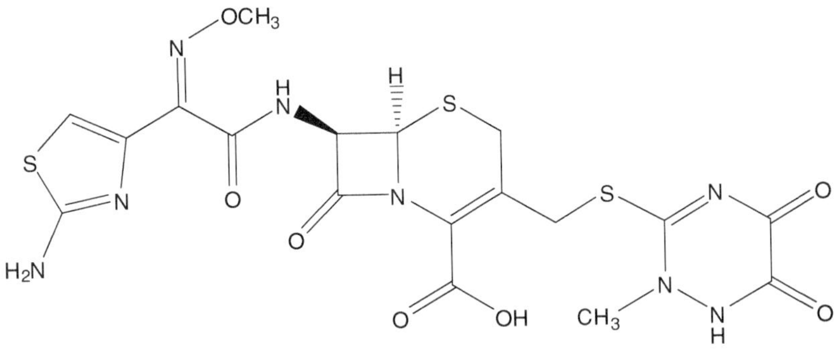

Many cephalosporin antibiotics are semisynthetic. When the cephalosporin target has a substituted methyl or an alkene substituent at position 3, it is often formed from cephalosporin C.

Discussion. The side chain amide is formed in the final step from an amine (7-ACT) and a thioester. The side chain thioether of 7-ACT is formed by displacement of the allylic acetate of 7-aminocephalosporanic acid (7-ACA) by the thiol. 7-ACA is formed by enzyme-mediated hydrolysis of the side chain amide of cephalosporin C. Cephalosporin C is produced by the fungus *A. chrysogenum*.

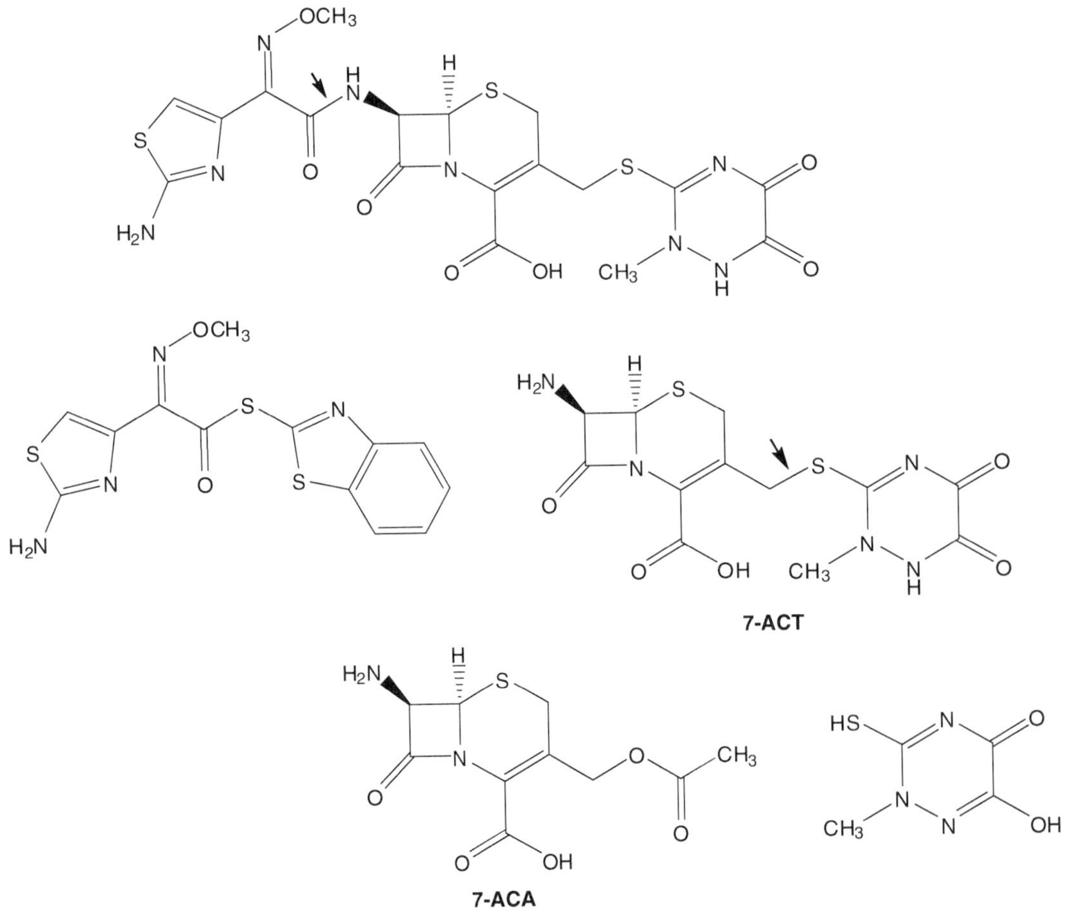

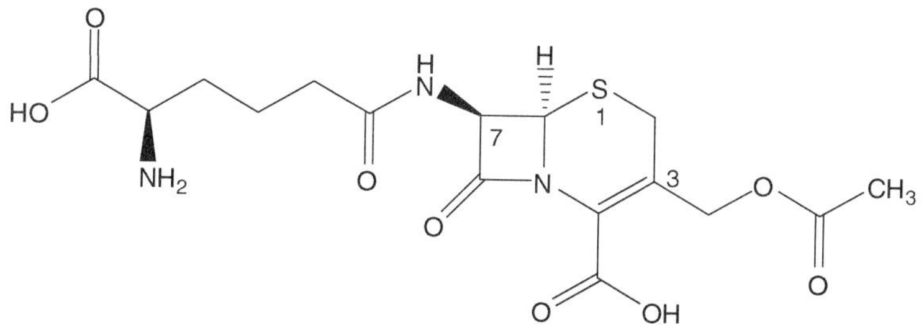

cephalosporin C

The thioester is formed from the carboxylic acid and 2,2′-dithiobis(benzothiazole). The carboxylic acid is formed by hydrolysis of the ethyl ester. The thiazole ring is formed by reaction of an α–bromoketone with thiourea (**Hantzsch Thiazole Synthesis**). The α-bromoketone is formed by bromination of the ketone. The *O*-methyloxime is formed by *O*-alkylation of the oxime. The oxime is formed by nitrosation of ethyl acetoacetate.

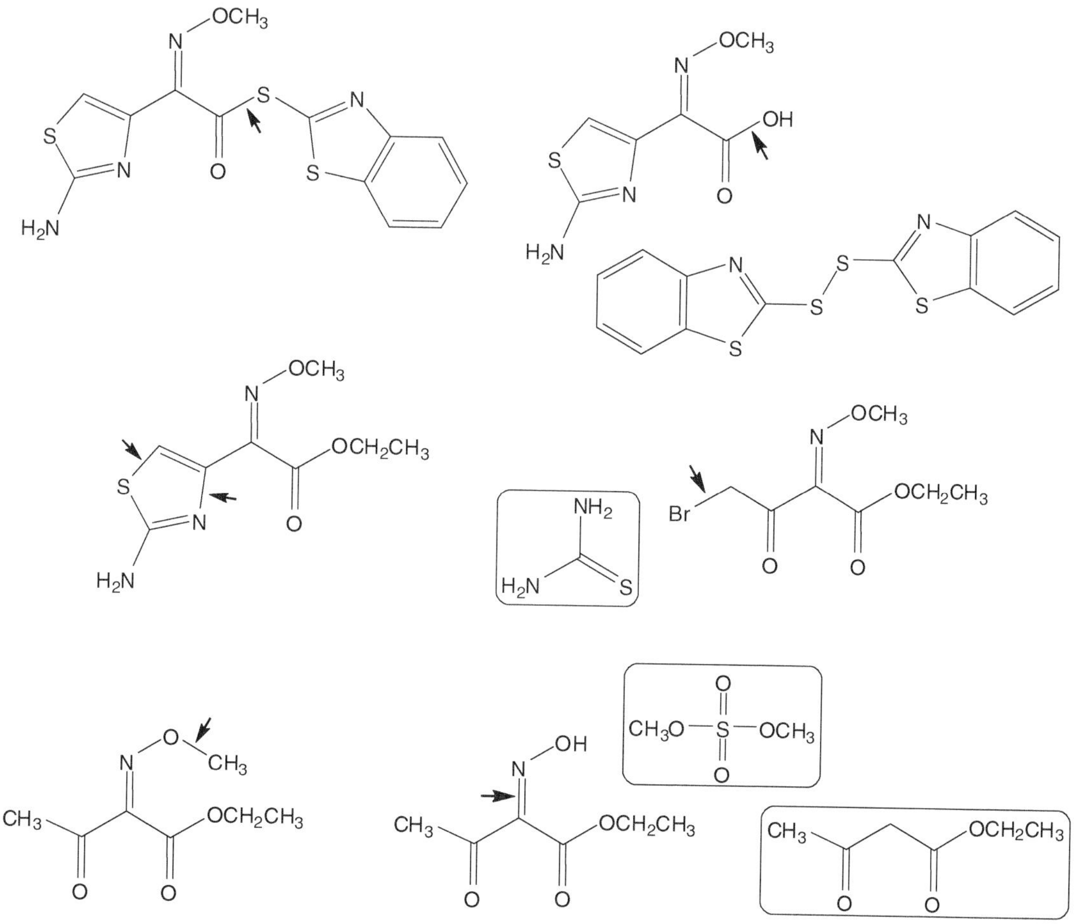

The thiol 1,2,4-triazineone ring is formed by the reaction of 1-methylhydrazine-1-carbothioamide with dimethyl oxalate. 1-Methylhydrazine-1-carbothioamide is formed by the reaction of methylhydrazine with potassium thiocyanate.

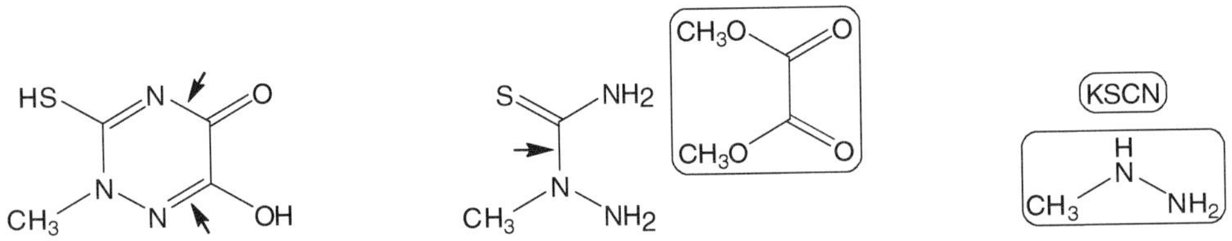

Extended Discussion

The thiazole ring is formed in the final step in several alternative routes to ceftriaxone. Draw the structures for a retrosynthetic analysis for one of these alternative routes. List the pros and cons for both routes and select one route as the preferred route.

Chlorambucil

Antineoplastics and Immunosuppressives/Cytotoxic and Adjuvant Medicines

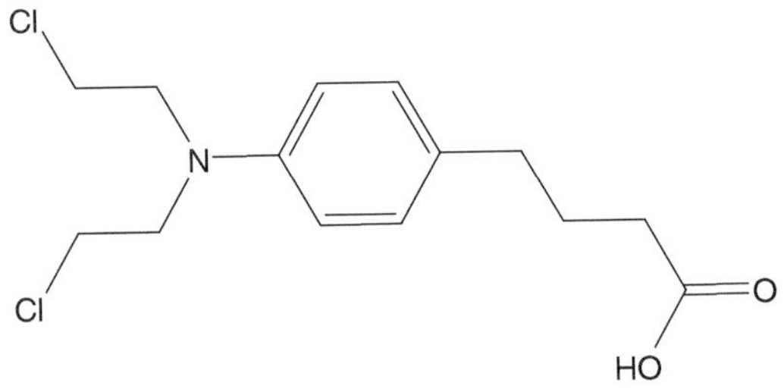

A β-chloroethylamine is a very reactive alkylating agent. A β-chloroethylamine is often formed from a β-amino alcohol in the final step of the synthesis.

Discussion. The bis-(2-chloroethyl)amine is formed from the bis-(2-hydroxyethyl)amine in the final step. The carboxylic acid is formed by hydrolysis of the methyl ester during the aqueous workup of the final step. The bis-(2-hydroxyethyl) amine is formed by ring-opening of ethylene oxide by the amine. The methylene directly attached to the ring is formed by reduction of the ketone. Under the ketone reduction conditions, the aniline is released from the acetanilide and the ester is formed from the carboxylic acid. The aryl ketone is formed by the reaction of acetanilide with succinic anhydride (**Friedel–Crafts Acylation**).

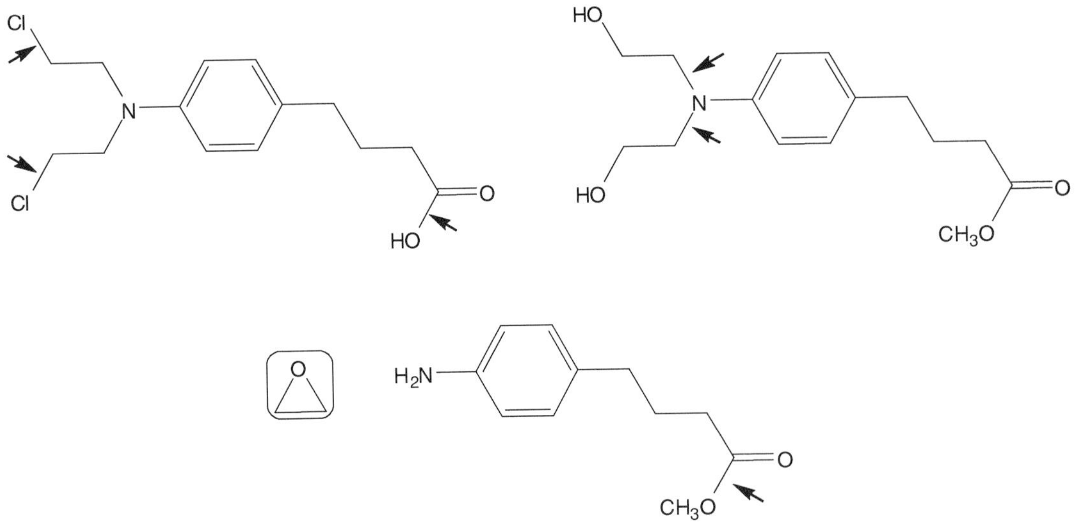

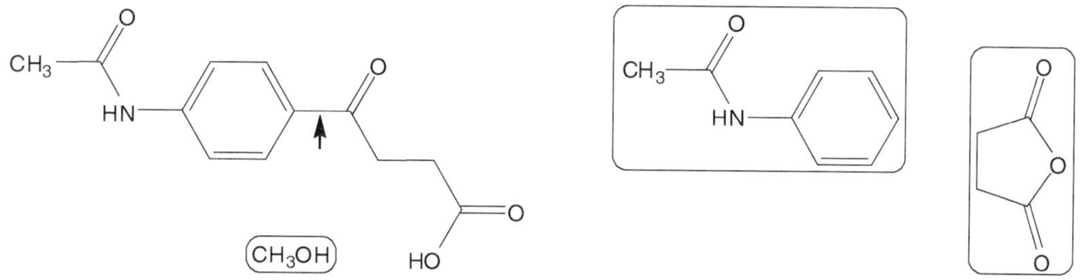

Extended Discussion

Draw the structures of the retrosynthetic analysis of one alternative route to chlorambucil from benzene. List the pros and cons for both routes and select one route as the preferred route.

Chloramphenicol

Anti-Infective Medicines/Antibacterials/Other Antibacterials

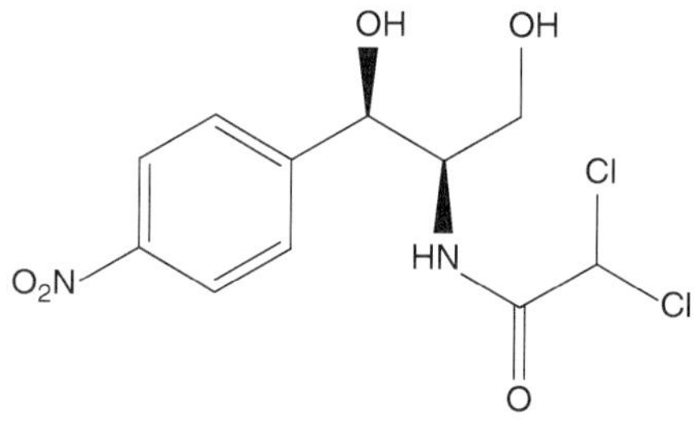

When a target molecule has two adjacent chiral carbons, one chiral carbon is often used to direct the formation of the other.

Discussion. Chloramphenicol has been manufactured by fermentation and by at least 3 unique synthetic routes published more than 50 years ago. These synthetic routes rely on a resolution of an intermediate to separate one enantiomer from a racemic mixture.

In one preferred route utilizing a resolution, the amide is formed in the final step from the amine and methyl dichloroacetate. The amine, (*R,R*)-2-amino-1-(4-nitrophenyl)-1,3-propanediol, is separated from the (*S,S*)-enantiomer by resolution. The amine to be resolved is released by hydrolysis of the acetamide. The 2-acetamido-1,3-propanediol is formed by reduction of the 2-acetamido-3-hydroxyketone (**Meerwein–Pondorf–Verley Reduction**). Four stereoisomeric products are formed in this reduction. The (*R,R*)- and (*S,S*)-enantiomer pair crystallizes from the mixture. 2-Acetamido-3-hydroxy-1-(4-nitrophenyl)-1-propanone is formed by the reaction of 2-acetamido-4′-nitroacetophenone with formaldehyde (**Aldol Reaction**). The acetamide is formed from the amine and acetic anhydride. The amine is formed by bromide displacement from 2-bromo-4′-nitroacetophenone by hexamethylenetetramine (**Delepine Reaction**). The α-bromoketone is formed by bromination of 4′-nitroacetophenone.

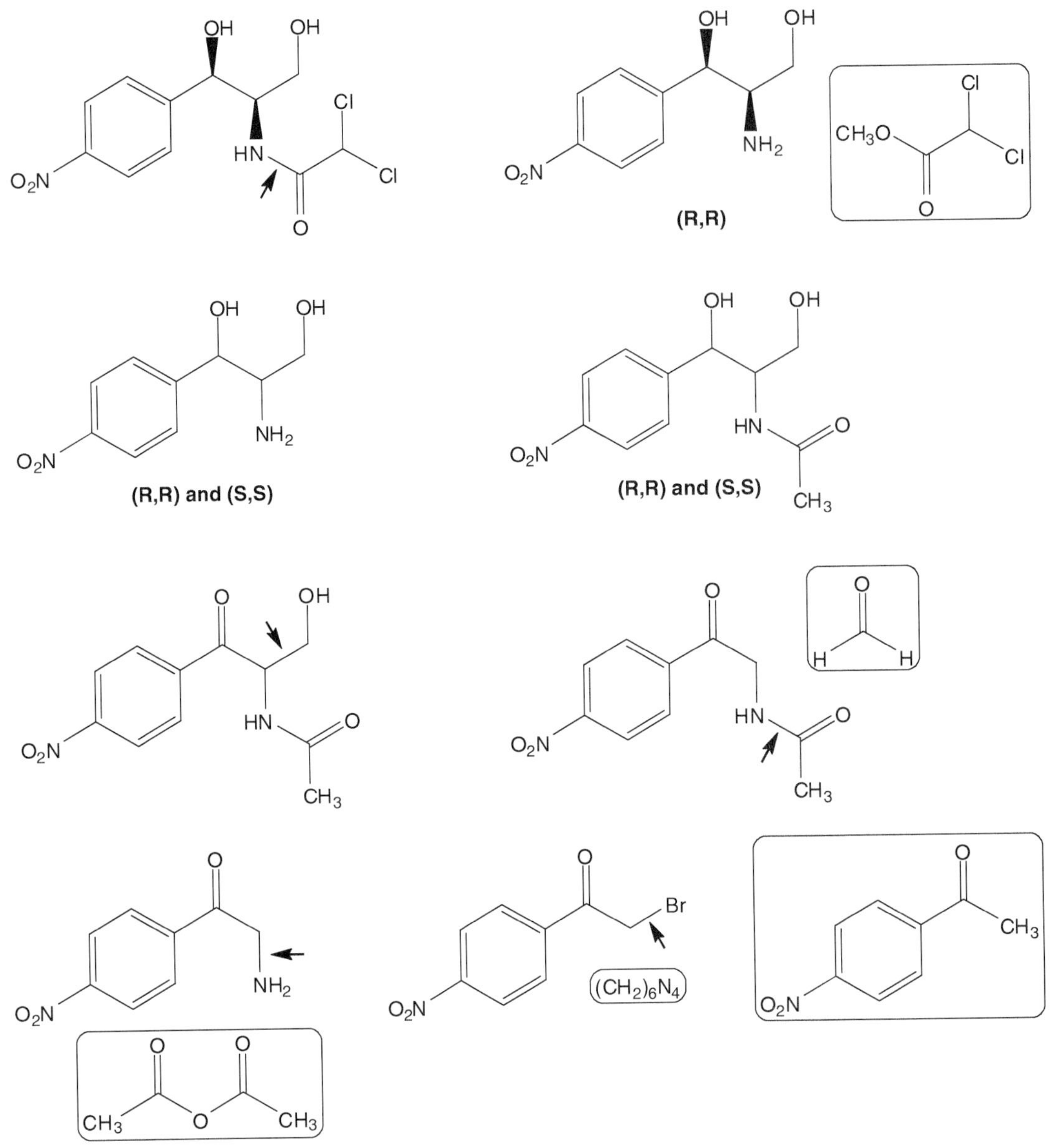

Extended Discussion

Draw the structures of the retrosynthetic analysis of one alternative route to chloramphenicol which does not utilize a resolution of an intermediate. Include the structures of the retrosynthetic analysis of any organic starting material(s) from petrochemical or biochemical raw materials. List the pros and cons for both routes.

Chlorhexidine

Disinfectants and Antiseptics/Antiseptics

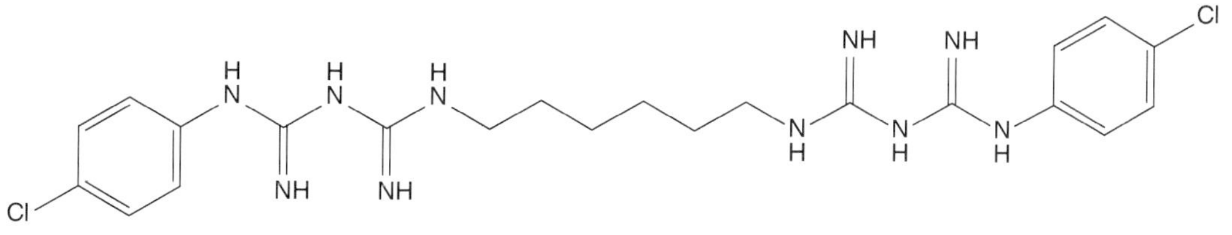

A biguanide is formed by addition of an amine to a cyanoguanidine.

Discussion. The functional group on the left and right of symmetrical chlorhexidine is called a biguanide. The biguanide is formed by addition of 4-chloroaniline to the nitrile of a cyanoguanidine. The cyanoguanidine is formed by addition an amine (1,6-hexanediamine) to dicyanamide. Dicyanamide is formed in situ from sodium dicyanamide.

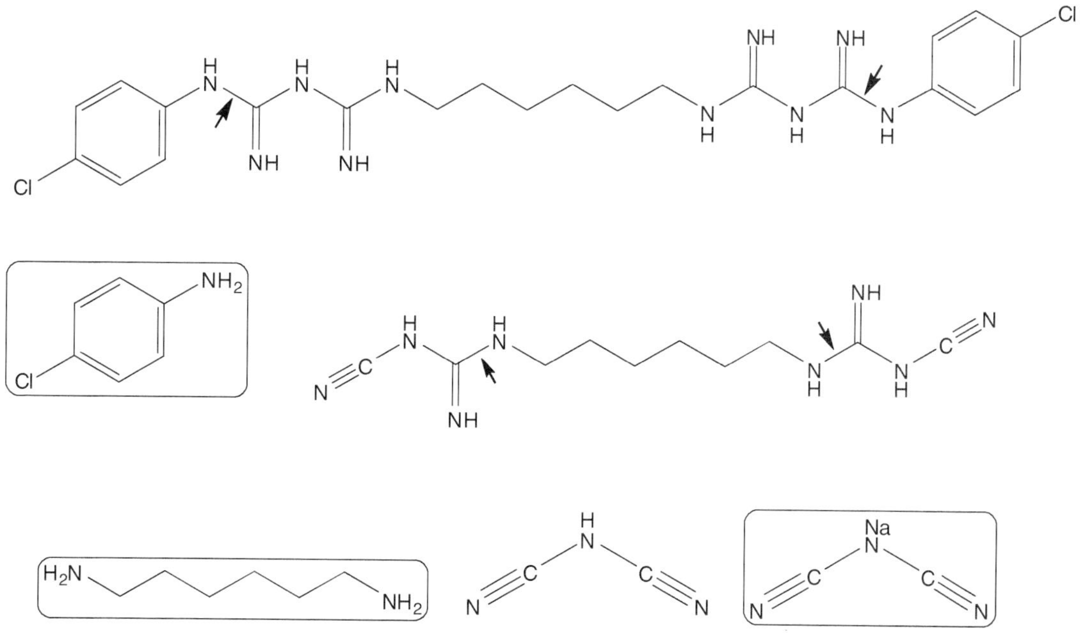

Extended Discussion

Draw the structures of the retrosynthetic analysis of an alternative route to chlorhexidine from the same starting materials. List the pros and cons for both routes. Is one route preferred?

Chloroquine

Anti-Infective Medicines/Antiprotozoal Medicines/Antimalarial Medicines/For Curative Treatment
Anti-Infective Medicines/Antiprotozoal Medicines/Antimalarial Medicines/For Prophylaxis
Medicines for Diseases of Joints/Disease-Modifying Agents Used in Rheumatoid Disorders

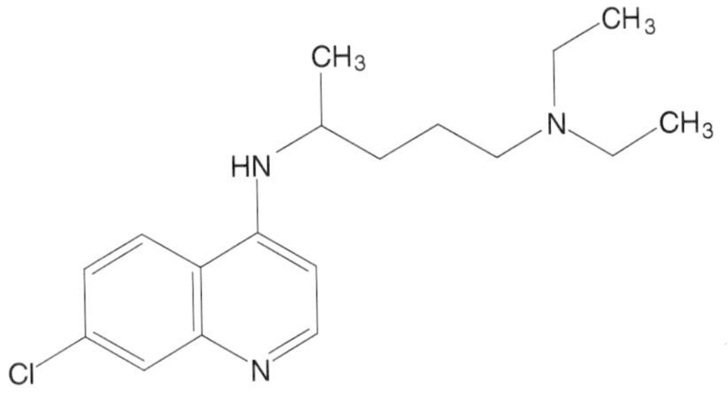

A nitrogen substituent at C2 or C4 on a quinoline ring is often introduced by displacement of chloride.

Discussion. Chloroquine is a 1 : 1 mixture of the (*R*)- and (*S*)-enantiomers. Chloride at C4 of 4,7-dichloroquinoline is displaced by N^1,N^1-diethyl-1,4-pentanediamine (novoldiamine) in the final step.

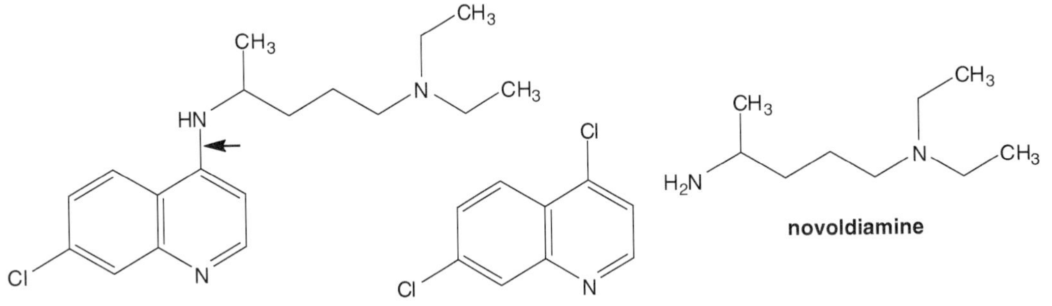

4,7-Dichloroquinoline is formed from 7-chloro-4-hydroxyquinoline (7-chloroquinolin-4-one). 7-Chloroquinolin-4-one is formed by thermolysis/decarboxylation of the quinolone-3-carboxylic acid. The carboxylic acid is formed by ester hydrolysis. The quinoline ring is formed by *ortho*-acylation of the aniline. The enamine is formed by reaction of ethyl ethoxymethylenemalonate with 3-chloroaniline. The four-step sequence from 3-chloroaniline to 7-chloro-4-hydroxyquinoline is an example of the **Gould-Jacobs Reaction**.

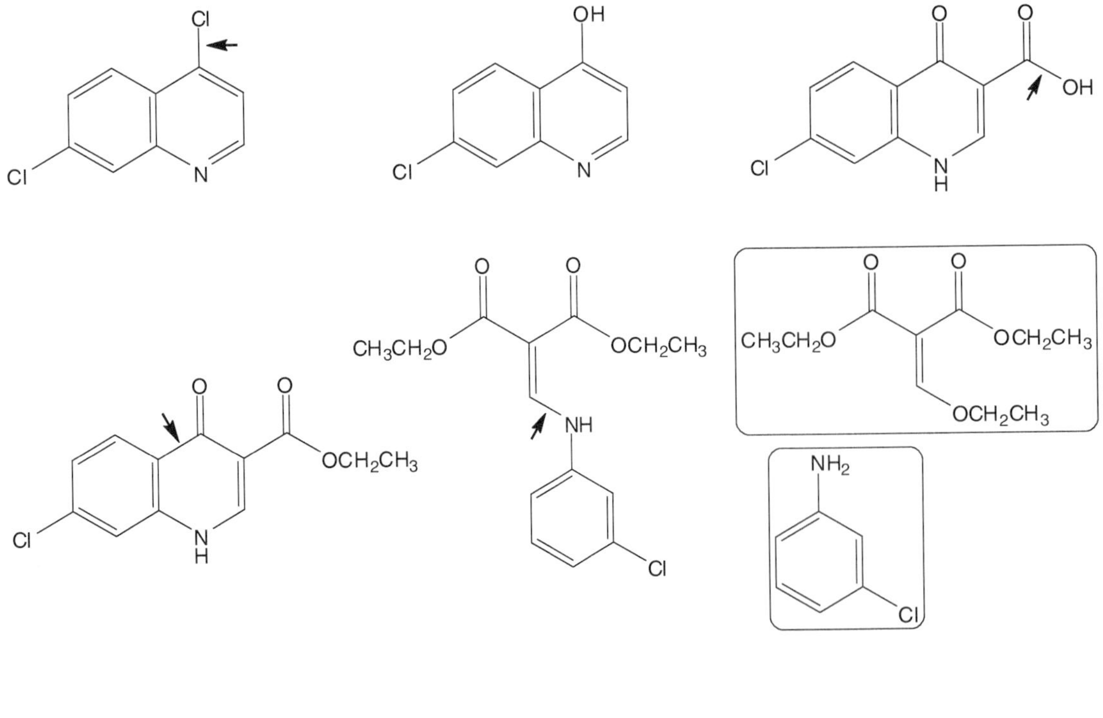

The primary amine of novoldiamine is formed by reductive amination of the ketone with ammonia. The tertiary amine is formed by chloride displacement from 5-chloro-2-pentanone with *N,N*-diethylamine. 5-Chloro-2-pentanone is formed from α–acetyl-γ-butyrolactone.

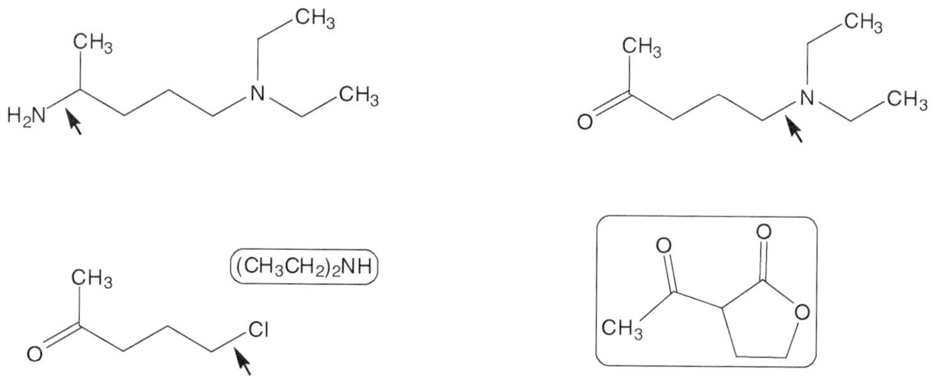

Extended Discussion

Draw the structures of the retrosynthetic analysis of one alternative route to novoldiamine.

Chloroxylenol

Disinfectants and Antiseptics/Disinfectants

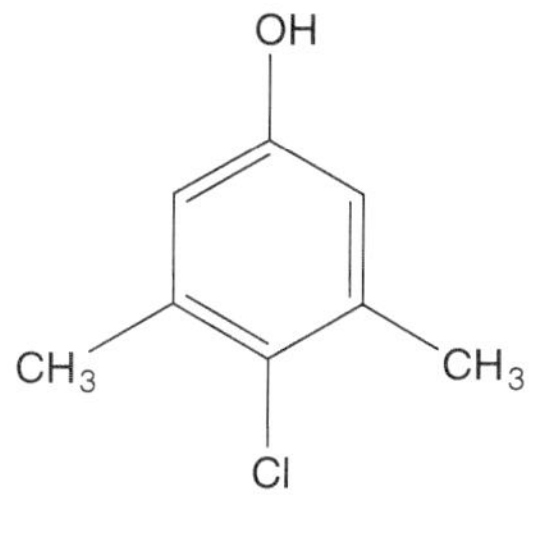

3,5-Dimethylphenol (*sym-meta*-xylenol) is produced by catalytic cracking of isophorone.

Discussion. Chloroxylenol is manufactured by chlorination of 3,5-dimethylphenol. (List all the options for the chlorinating agent and reaction conditions. Which of these options is associated with the highest selectivity for chlorination at the 4-position?) 3,5-Dimethylphenol (*sym-meta*-xylenol) is produced by catalytic cracking of isophorone.

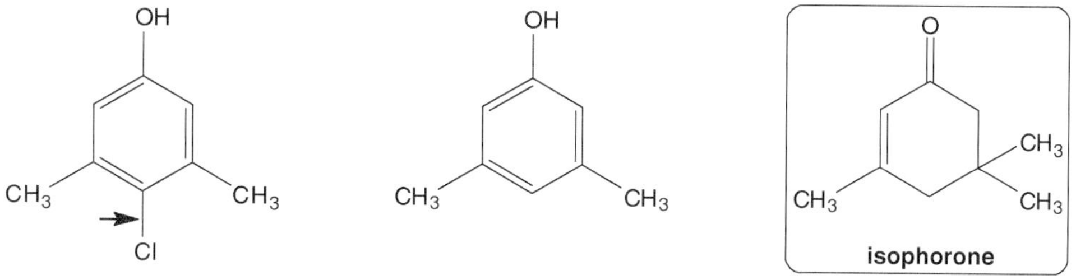

Chlorpromazine

Medicines for Mental and Behavioral Disorders/Medicines Used in Psychotic Disorders

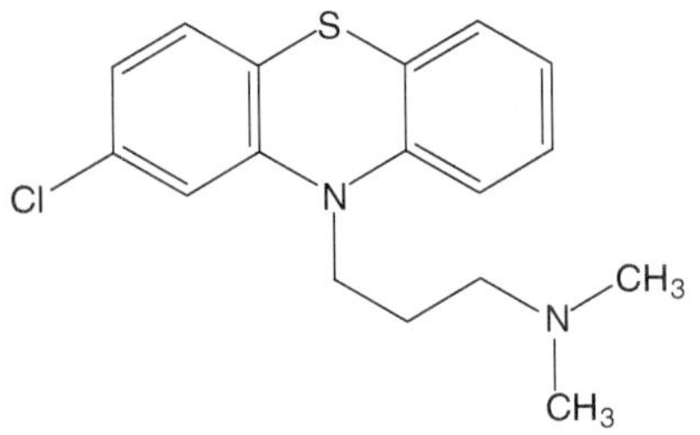

Nucleophilic aromatic substitution is often facilitated by an electron-withdrawing group (NO_2, SO_2R, COOR, CN) on an *ortho* or *para* ring carbon. No electron-withdrawing group is required when the displacement results in formation of a five- or six-membered ring. Leaving groups for nucleophilic aromatic substitution include fluorine, chlorine, and nitro.

Discussion. 2-Chloro-10*H*-phenothiazine displaces chloride from 3-chloro-*N*,*N*-dimethylpropan-1-amine in the final step. The 10*H*-phenothiazine ring is formed by intramolecular displacement of a nitro group by the formamide nitrogen. After the nitro group displacement, the formamide is hydrolyzed in situ. The amide is formed by reaction of the amine with formic acid. The thioether is formed by chloride displacement from 1-chloro-2-nitrobenzene by 2-amino-4-chlorobenzenethiol. 2-Amino-4-chlorobenzenethiol is formed from 1,4-dichloro-2-nitrobenzene by chloride displacement by hydrogen sulfide followed by reduction of the nitro group.

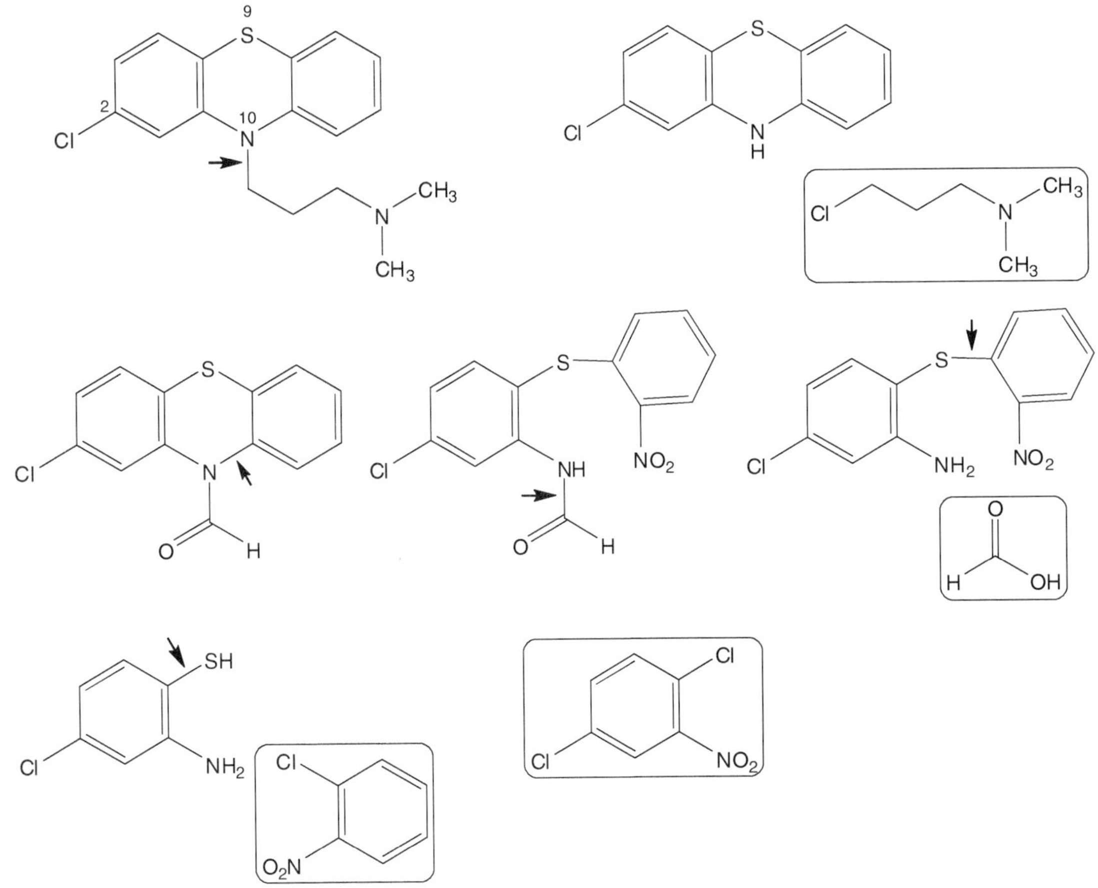

Extended Discussion

List the alternative reagents which react with 2-amino-4-chlorobenzenethiol to form 2-chloro-10*H*-phenothiazine. Indicate the yield for the phenothiazine ring formation for each reagent.

Ciprofloxacin

Anti-Infective Medicines/Antibacterials/Other Antibacterials
Ear, Nose, and Throat Medicines

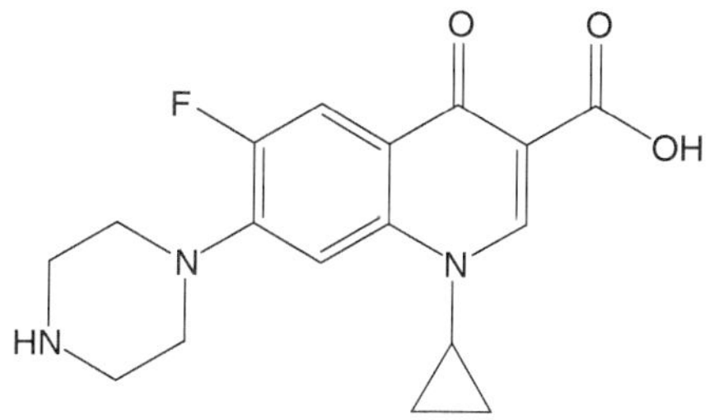

Nucleophilic aromatic substitution is often facilitated by an electron-withdrawing group (NO_2, SO_2R, COOR, CN) on an *ortho* or *para* ring carbon. No electron-withdrawing group is required when the substitution results in formation of a five- or six-membered ring. Leaving groups for nucleophilic aromatic substitution include fluorine, chlorine, and nitro.

Discussion. Chlorine at C7 of the quinoline ring is displaced by piperazine in the final step. The carboxylic acid is released by ester hydrolysis.

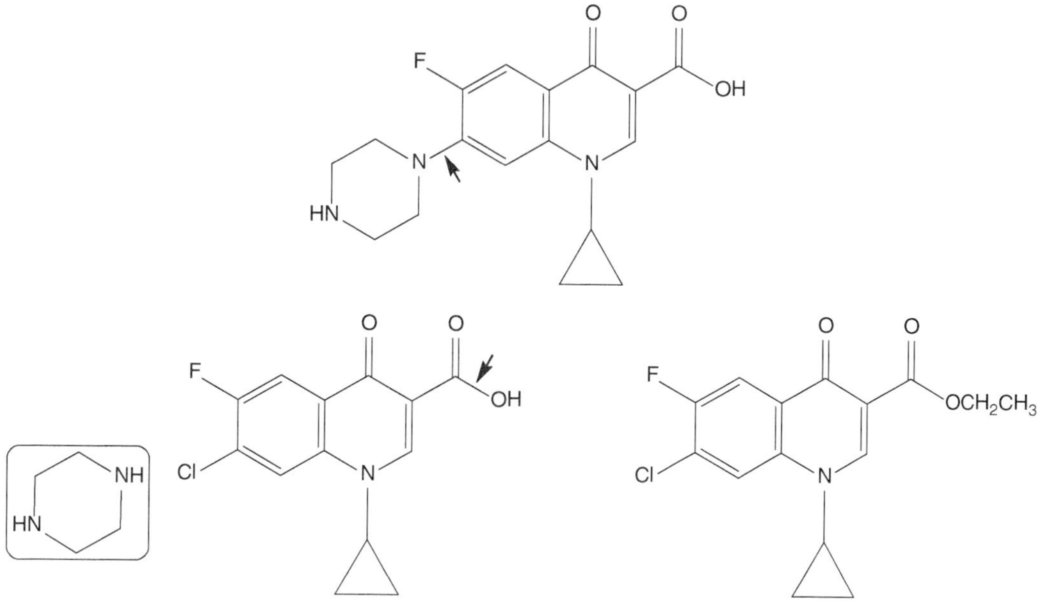

Intramolecular chloride displacement by nitrogen of the cyclopropanamine forms the quinoline ring. The enamine is formed by displacement of ethanol from the enol ether by cyclopropanamine. The enol ether is formed by reaction of the β–ketoester with triethyl orthoformate. The β–ketoester is formed from 2,4-dichloro-5-fluorobenzoyl chloride and diethyl malonate. The four-step conversion of the 2-chlorobenzoyl chloride to the quinolone is known as the **Grohe–Heitzer Sequence**. The acid chloride is formed from the carboxylic acid. The carboxylic acid is formed by oxidation of the acetophenone.

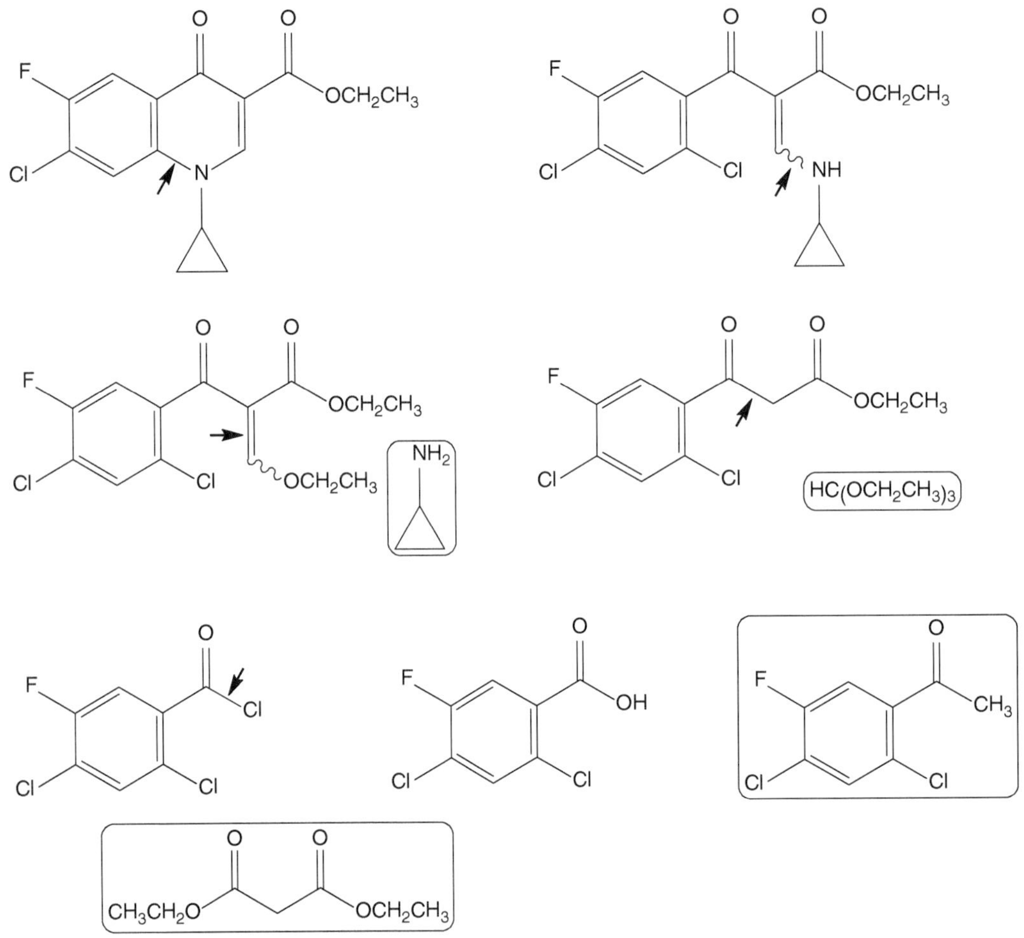

Extended Discussion

Ciprofloxacin can also be formed from 2,4,5-trifluorobenzoic acid. List the pros and cons for the two routes to ciprofloxacin. Is one route preferred?

Cisplatin, Carboplatin, Oxaliplatin

Antineoplastics and Immunosuppressives/Cytotoxic and Adjuvant Medicines

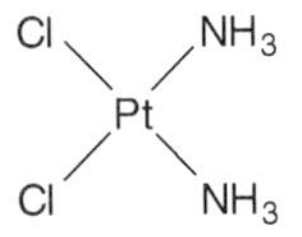

cisplatin

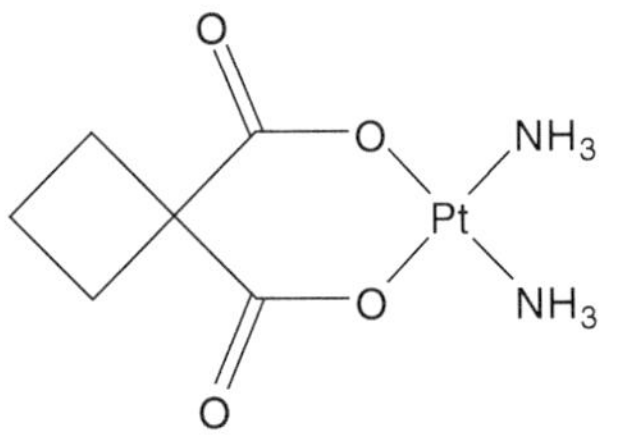

carboplatin

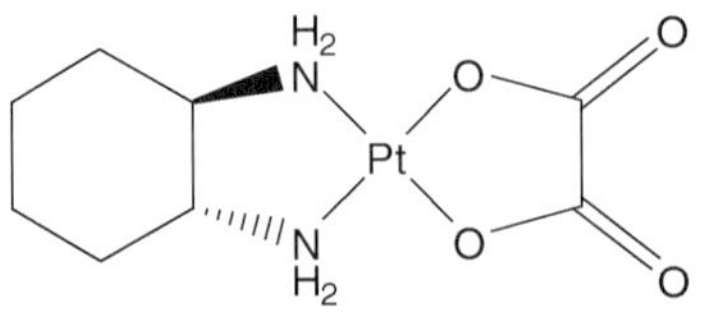

oxaliplatin

A d^8 square-planar 16-electron platinum(II) complex is often formed from potassium tetrachloroplatinate by ligand substitution reactions. A chloride or iodide ligand is often substituted with retention of configuration at platinum. The *trans*-positioned ligand influences the rate of substitution of a chloride or iodide ligand.

Discussion. Cisplatin precipitates when potassium chloride is added to a solution of the *cis*-diaminediaquaplatinum(II) cation. The aqueous solution of *cis*-diaminediaquaplatinum(II) cation is formed by reaction of *cis*-diamino(diiodo)platinum with silver nitrate. *cis*-Diamino(diiodo)platinum precipitates when ammonium hydroxide is added to a solution of potassium tetraiodoplatinate. An aqueous solution of potassium tetraiodoplatinate is formed by reaction of potassium tetrachloroplatinate with potassium iodide.

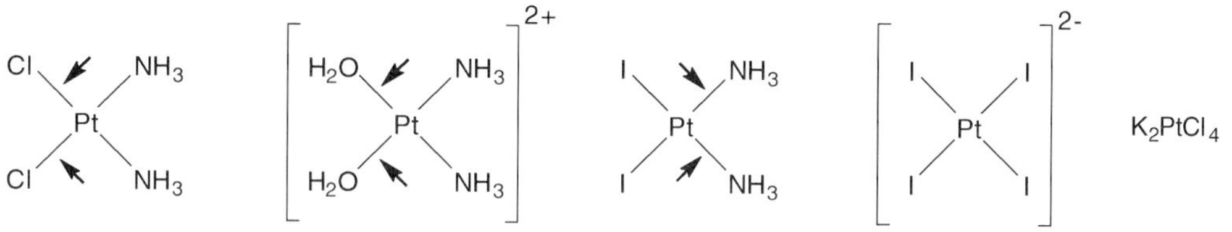

Carboplatin precipitates when 1,1-cyclobutanedicarboxylic acid is added to the solution of the same *cis*-diaminediaquaplatinum(II) cation. 1,1-Cyclobutanedicarboxylic acid is formed by hydrolysis of the diester. Diethyl 1,1-cyclobutanedicarboxylate is formed from diethyl malonate and 1,3-dibromopropane. (Draw the structure of a side product formed in the reaction of diethyl malonate and 1,3-dibromopropane. How is diethyl 1,1-cyclobutanedicarboxylate separated from the side product?)

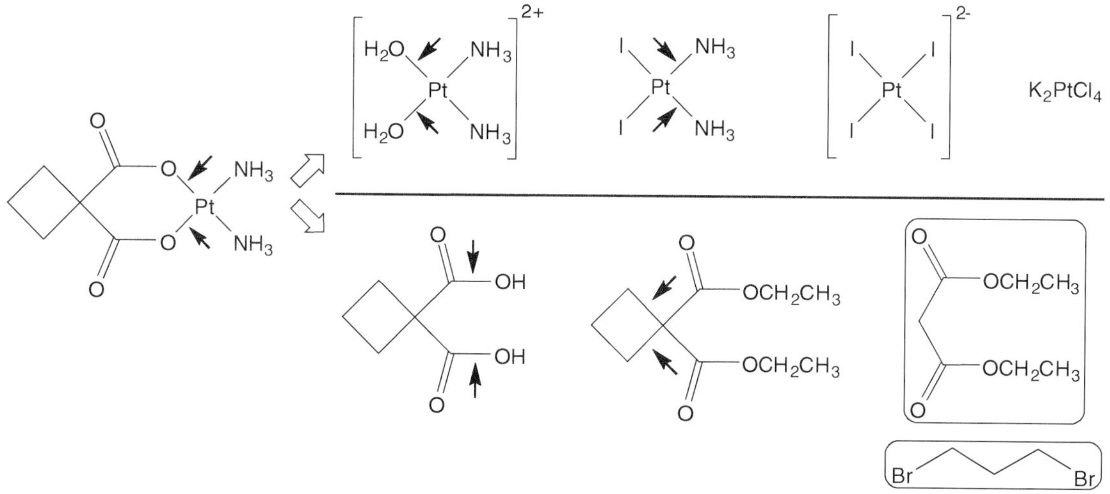

Oxaliplatin precipitates when oxalic acid is added to a solution of the diaminediaquaplatinum(II) cation formed from (1*R*,2*R*)-1,2-diaminocyclohexane. The aqueous solution of the diaminediaquaplatinum(II) cation is formed by reaction of the diiodoplatinum complex with silver nitrate. Diiodo[(1*R*,2*R*)-*trans*-1,2-diaminocyclohexane]platinum precipitates when (1*R*,2*R*)-1,2-diaminocyclohexane is added to a solution of potassium tetraiodoplatinate. An aqueous solution of potassium tetraiodoplatinate is formed by reaction of potassium tetrachloroplatinate with potassium iodide.

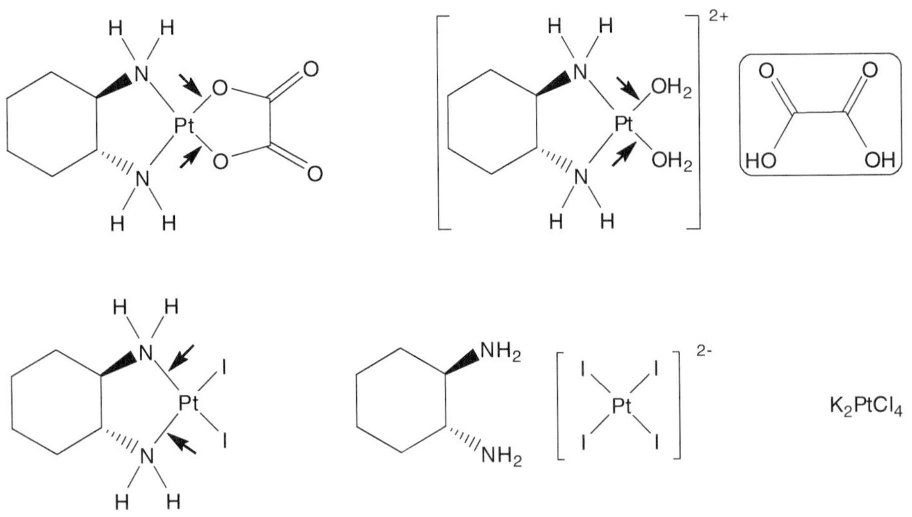

(1*R*,2*R*)-1,2-Diaminocyclohexane is separated from (±)-*trans*-1,2-diaminocyclohexane by resolution. (±)-*trans*-1,2-Diaminocyclohexane is separated from *cis*-1,2-diaminocyclohexane. The mixture of *cis* and *trans*-1,2-diaminocyclohexanes is formed by hydrogenation of *ortho*-phenylenediamine.

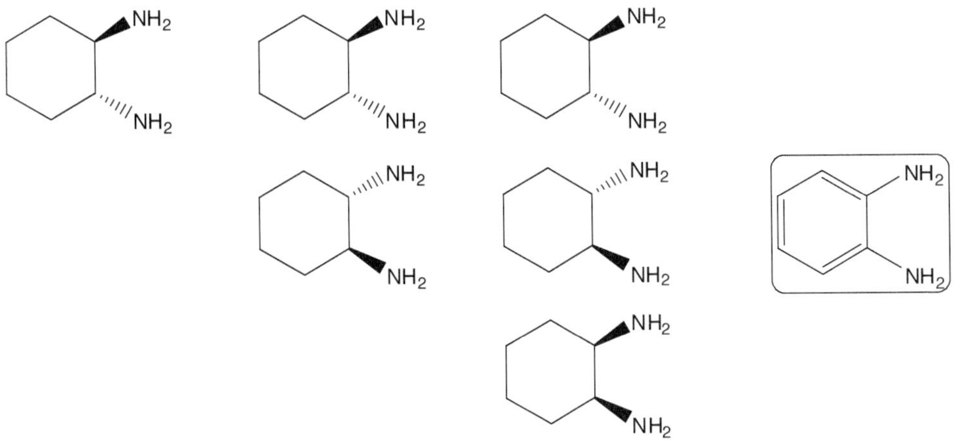

Extended Discussion

Draw the structures of the retrosynthetic analysis of one alternative route to (1*R*,2*R*)-1,2-diaminocyclohexane. Include the structures of the retrosynthetic analysis of any organic starting material(s) from petrochemical or biochemical raw materials. List the pros and cons for both routes.

Clarithromycin

Anti-Infective Medicines/Antibacterials/Other Antibacterials

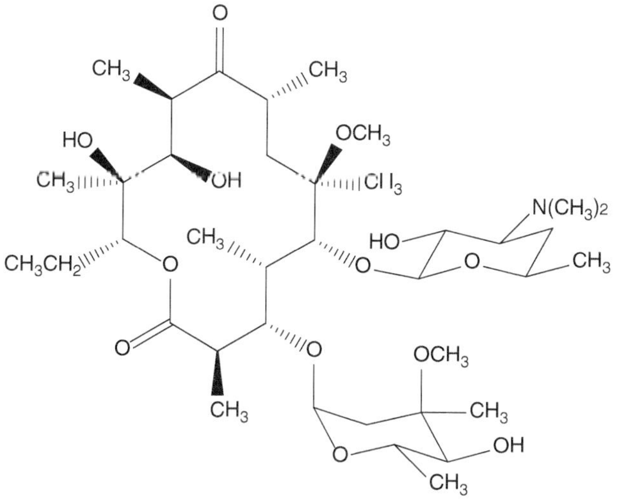

Macrolide antibiotics are produced by fermentation or are semisynthetic. The process for manufacture of a semisynthetic macrolide antibiotic often begins with conversion of the C9 ketone of erythromycin A to the oxime.

Discussion. Clarithromycin is semisynthetic. A preferred synthesis of clarithromycin (OCH_3 at C6) from erythromycin A (OH at C6) was developed by trial-and-error over decades. Shorthand forms for clarithromycin and erythromycin A which highlight the ketone at C9 and the functional groups at positions 6, 11, 12, 2′, and 4″ will be used in the retrosynthetic analysis.

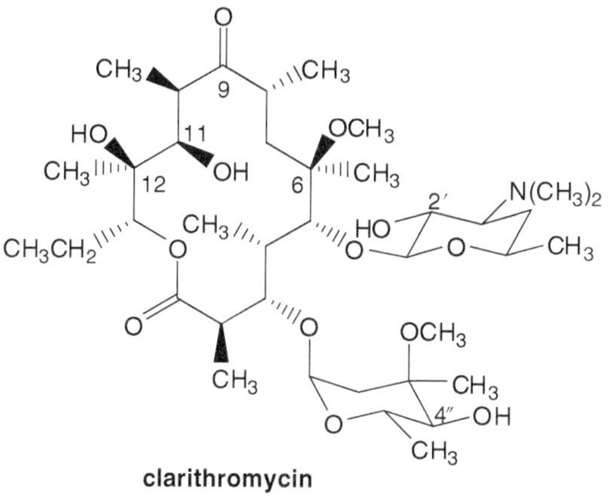

clarithromycin

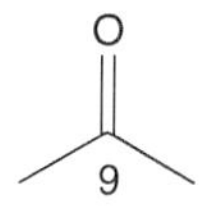

6 **OCH_3**
11 OH
12 OH
2′ OH
4″ OH

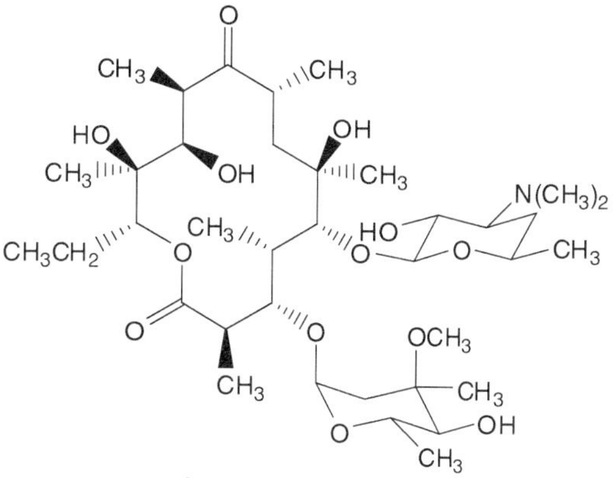

erythromycin A

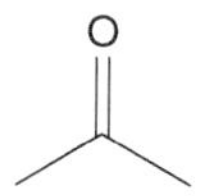

6 **OH**
11 OH
12 OH
2′ OH
4″ OH

Clarithromycin is formed from erythromycin A in seven steps. The sequence begins with three protection steps and ends with three deprotection steps. In the final step, the ketone of clarithromycin is released from the ketone bisulfite addition product. The ketone bisulfite addition product is formed in situ from the oxime. The oxime hydroxyl group is released by hydrolysis of an acetal. The secondary alcohols at C2′ and C4″ are released by trimethylsilyl ether hydrolysis. With the oxime hydroxyl group and secondary alcohols at C2′ and C4″ protected, methylation with iodomethane is highly selective for the tertiary alcohol at C6 (**Williamson Ether Synthesis**). Trimethylsilyl ethers are formed on the secondary alcohols at the C2′ (of D-desosamine) and C4″ (of L-cladinose). The oxime hydroxyl group of (E)-erythromycin A oxime is protected as an acetal by reaction with 2-methoxypropene. (E)-Erythromycin A oxime is formed from erythromycin A. Erythromycin A is produced by fermentation.

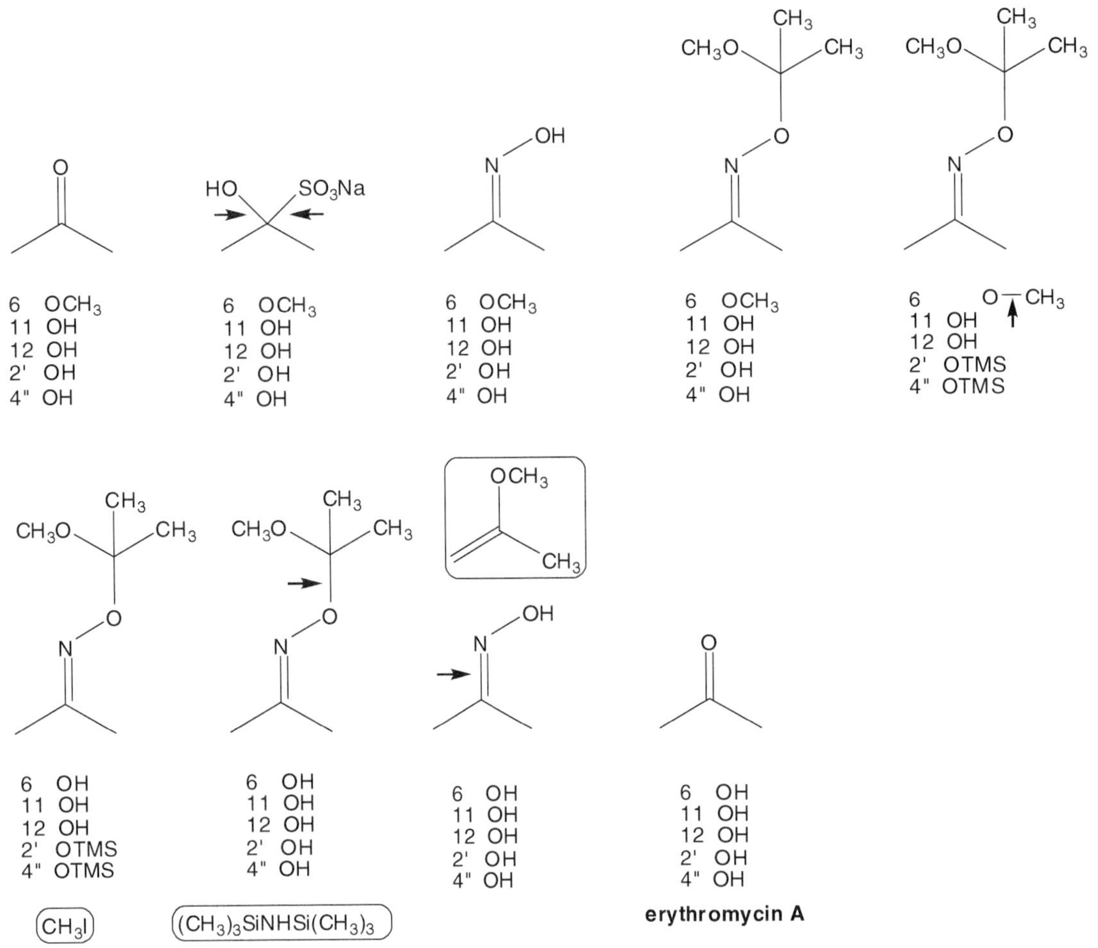

Extended Discussion

List the alternative groups used to protect the oxime hydroxyl group. Select one alternative oxime hydroxyl protecting group and draw shorthand forms for the retrosynthetic analysis of the alternative route to clarithromycin from erythromycin A using this group.

Clindamycin

Anti-Infective Medicines/Antibacterials/Other Antibacterials

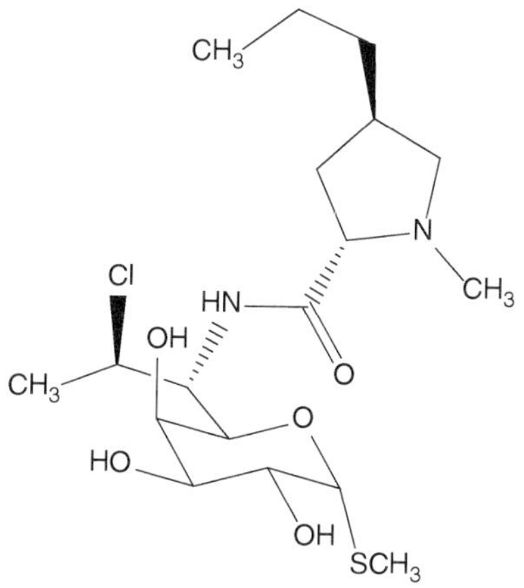

A single-enantiomer molecule with multiple chiral carbons is often formed by modification of a natural product which has most or all of the chiral carbons already in place.

Discussion. Clindamycin (Cl at C7) is formed from lincomycin (OH at C7). Lincomycin is produced by fermentation. (List all the reagents and reagent combinations used for the conversion of lincomycin to clindamycin. Which reagent or reagent combination gives the highest yield of clindamycin?)

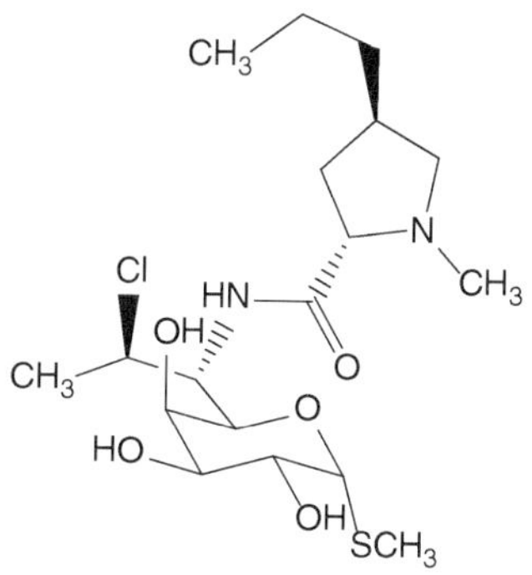

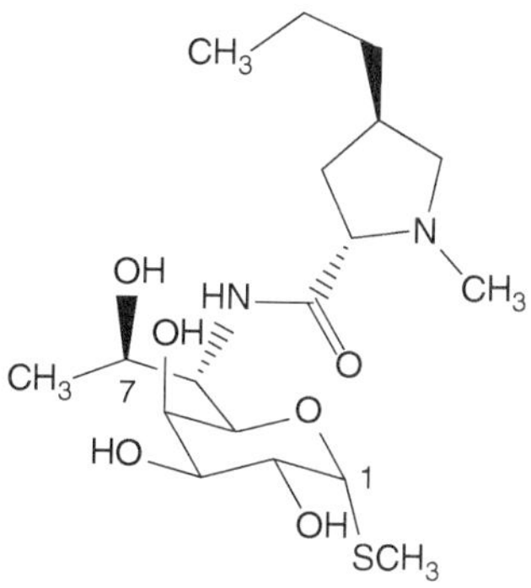

lincomycin

Extended Discussion

Triphenylphosphine is often one of the reagents used in reagent combinations for the conversion of lincomycin to clindamycin. List the options for the recycle/reuse of triphenylphosphine oxide.

Clofazimine

Anti-Infective Medicines/Antibacterials/Antileprosy Medicines
Anti-Infective Medicines/Antibacterials/Antituberculosis Medicines

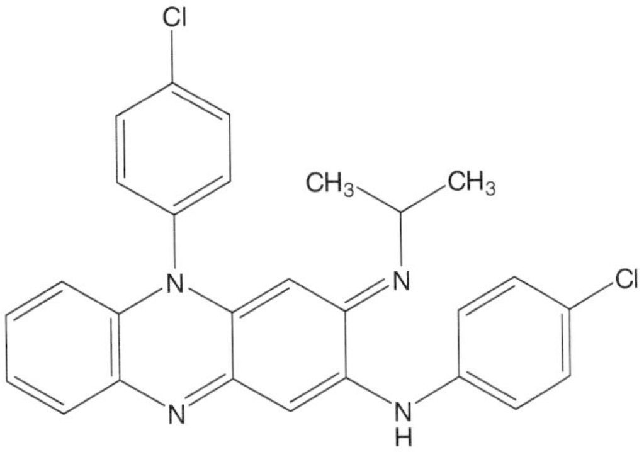

Nucleophilic aromatic substitution is often facilitated by an electron-withdrawing group (NO_2, SO_2R, COOR, CN) on an *ortho* or *para* ring carbon. Leaving groups for nucleophilic aromatic substitution include fluorine, chlorine, and nitro.

Discussion. Clofazimine, an R-iminophenazine (riminophenazine), is formed by an imine exchange with isopropylamine in the last step. Disconnection of two C—N bonds suggests the iminophenazine is a dimer formed by oxidation of *N*-(4-chlorophenyl)-1,2-benzenediamine. The 1,2-benzenediamine is formed by reduction of the 2-nitroaniline. The 2-nitroaniline is formed by fluoride displacement from 1-fluoro-2-nitrobenzene by 4-chloroaniline.

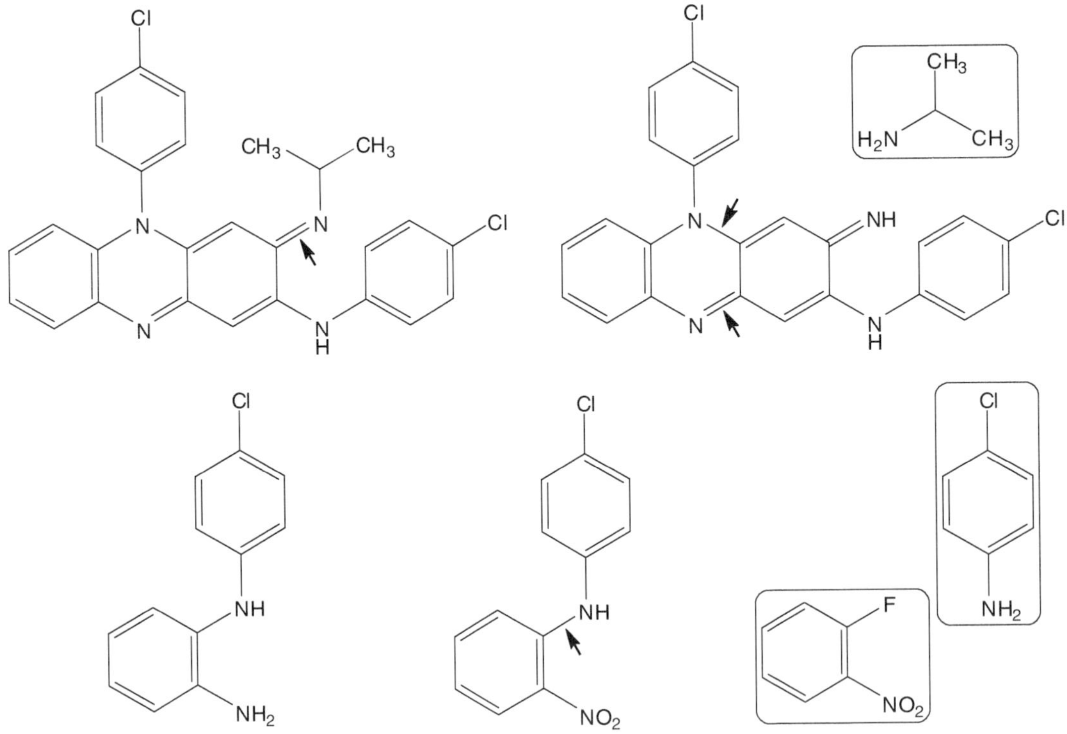

Extended Discussion

The oxidative dimerization results in a iminophenazine with two identical aryl groups (Ar = 4-chlorophenyl). Draw the structures of a retrosynthetic analysis of one alternative route to clofazimine which could be used to prepare analogs **which have two different aryl groups**. Include the structures of the retrosynthetic analysis of any organic starting material(s) from petrochemical or biochemical raw materials.

Clomifene

Hormones, Other Endocrine Medicines and Contraceptives/Ovulation Inducers

An alkene conjugated to three aromatic rings is often formed by dehydration of an alcohol.

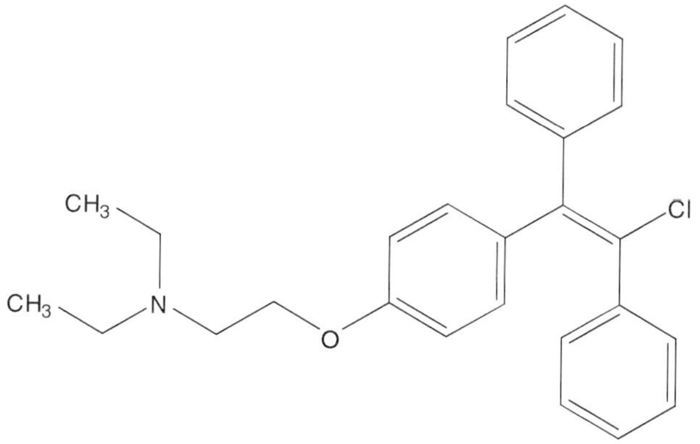

Discussion. Clomifene is manufactured as a mixture of the (E)- and (Z)-alkenes. The (E)-alkene, also called the *trans*-alkene (refers to the relationship of the unsubstituted phenyl substituents), has the desired estrogenic activity. The chloro-alkene is formed by chlorination and dehydrochlorination of the alkene in the final step. The alkene is formed by dehydration of the tertiary alcohol. The tertiary alcohol is formed by addition of benzylmagnesium chloride to the ketone (**Grignard Reaction**). Benzylmagnesium chloride is formed from benzyl chloride. The ether is formed by displacement of chloride from 2-chloro-*N,N*-diethylethanamine by 4-hydroxybenzophenone (**Williamson Ether Synthesis**).

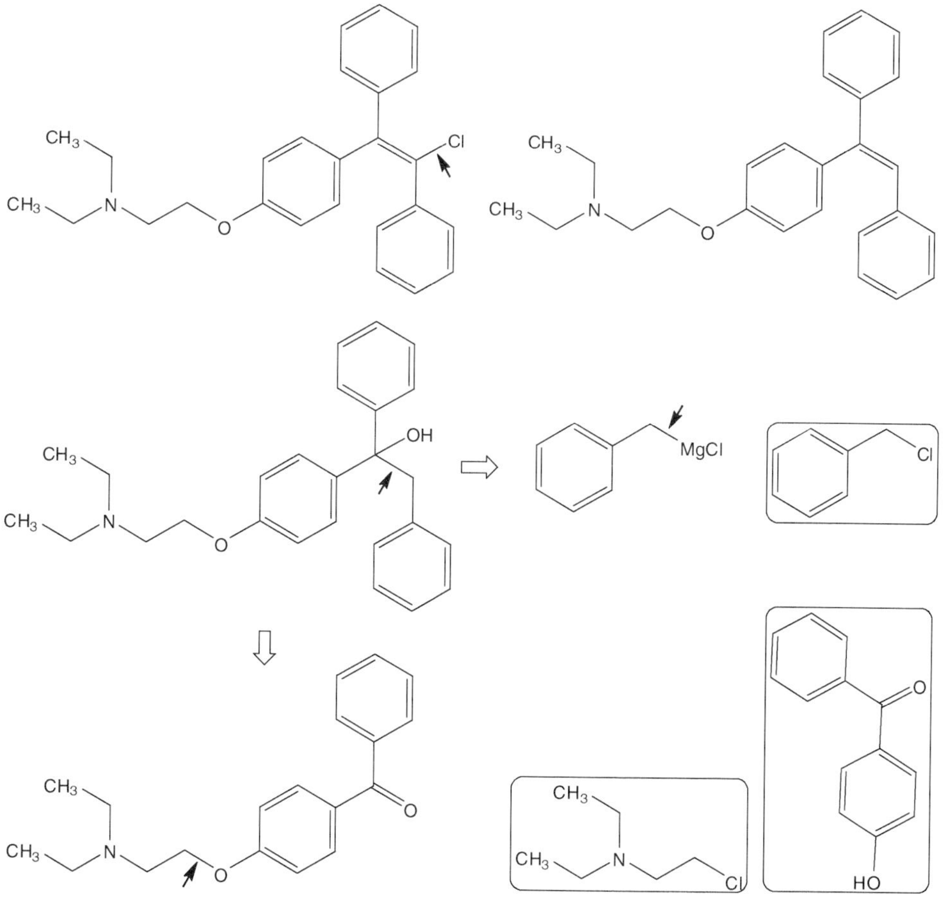

Extended Discussion

Draw the structures of the retrosynthetic analysis of one alternative route to clomifene that does not involve a **Grignard Reaction**. Include the structures of the retrosynthetic analysis of any organic starting material(s) from petrochemical or biochemical raw materials.

Clomipramine

Medicines for Mental and Behavioral Disorders/Medicines Used for Obsessive Compulsive Disorders

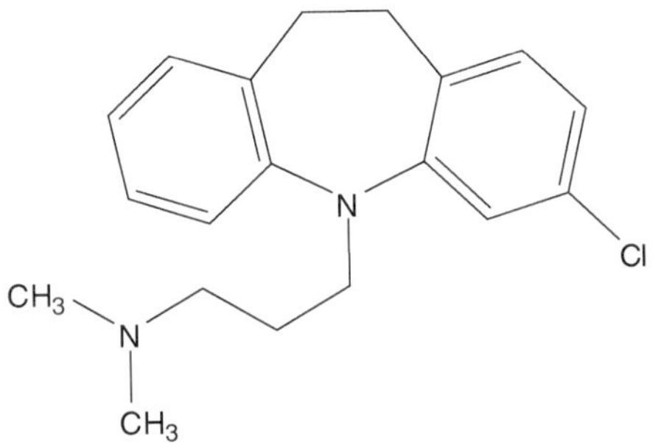

Seven-membered rings are often formed by ring expansion of six-membered rings.

Discussion. 3-Chloro-10,11-dihydrodibenzo[b,f]azepine displaces chloride from 3-chloro-*N*,*N*-dimethylpropan-1-amine in the final step. The 10,11-dihydrodibenzo[b,f]azepine is formed by reduction of 3-chlorodibenzo[b,f]azepine.

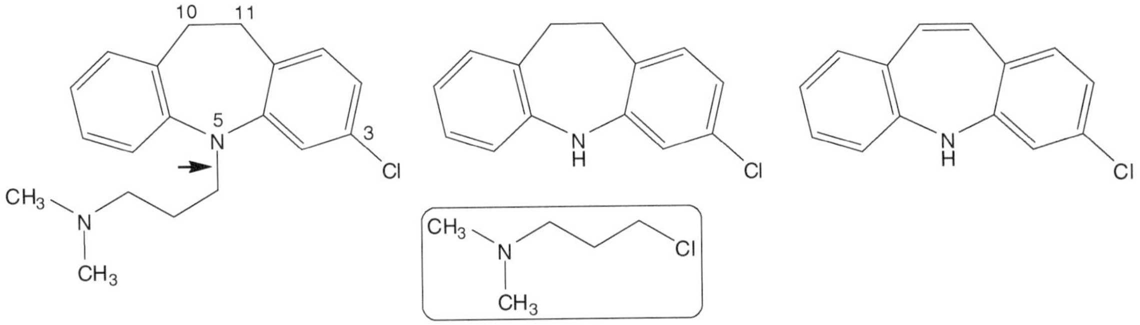

The azepine ring is formed from (3-chloro-9,10-dihydroacridin-9-yl)methanol by a ring expansion. (3-Chloro-9,10-dihyd roacridin-9-yl)methanol is formed by reduction of the ethyl 3-chloroacridine-9-carboxylate. The ester is formed from the acid chloride, the acid chloride from the acid, and the acid from the nitrile. The nitrile is formed by displacement of the chloride at C9 of 3,9-dichloroacridine by cyanide. The reaction of 2-anilino-4-chlorobenzoic with phosphorous oxychloride results in formation of the acridine ring and replacement of the oxygen at C9 by chlorine. 2-Anilino-4-chlorobenzoic acid is formed from by displacement of the chloride at C2 of 2,4-dichlorobenzoic acid by aniline (**Ullmann–Goldberg Reaction**).

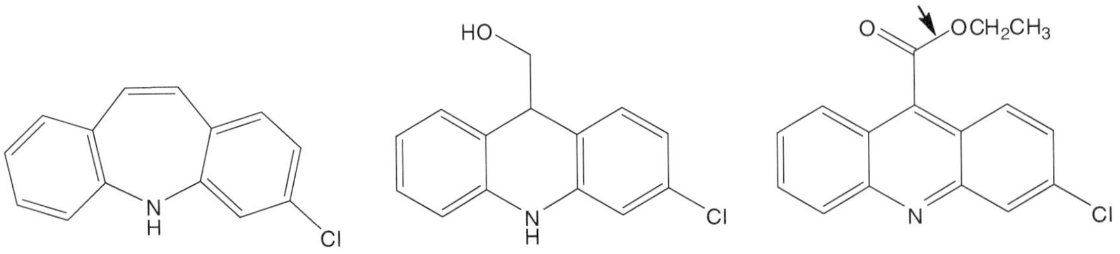

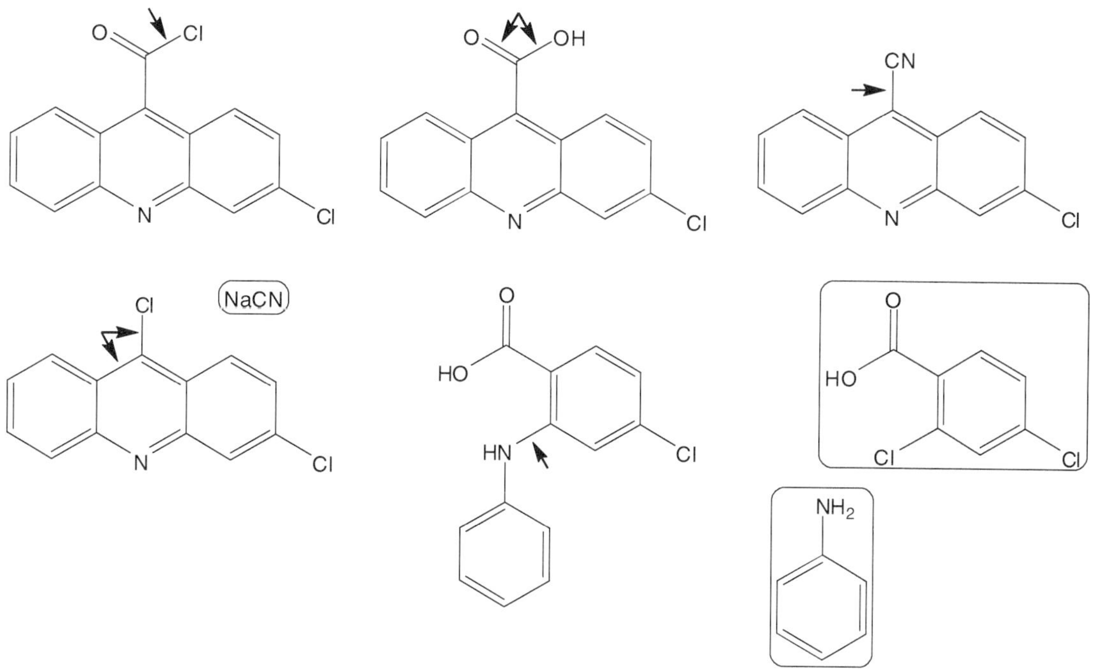

Extended Discussion

Design a **new synthesis** of clomipramine using a transition metal-catalyzed reaction to form the azepine ring. Draw the structures of the retrosynthetic analysis. Include the structures of the retrosynthetic analysis of any organic starting material(s) from petrochemical or biochemical raw materials.

Clopidogrel

Cardiovascular Medicines/Antithrombotic Medicines/Anti-Platelet Medicines

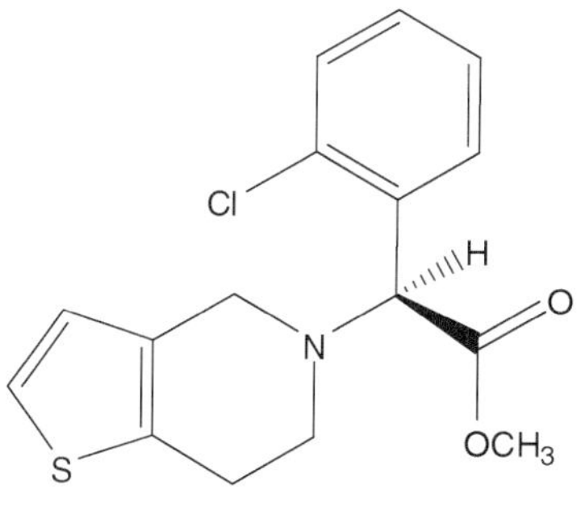

Thiophene reacts with *n*-butyllithium to form 2-lithiothiophene. A 2-substituted thiophene is often formed by the reaction of 2-lithiothiophene with an electrophile.

Discussion. Clopidogrel, the (*S*)-enantiomer, is separated from the racemic mixture by resolution in the final step. The byproduct mixture rich in the (*R*)-enantiomer is recycled: the (*R*)-enantiomer is converted back to the racemic mixture. The α-amino ester is formed from the α-bromo ester by bromide displacement with 4,5,6,7-tetrahydrothieno[3,2-c]pyridine.

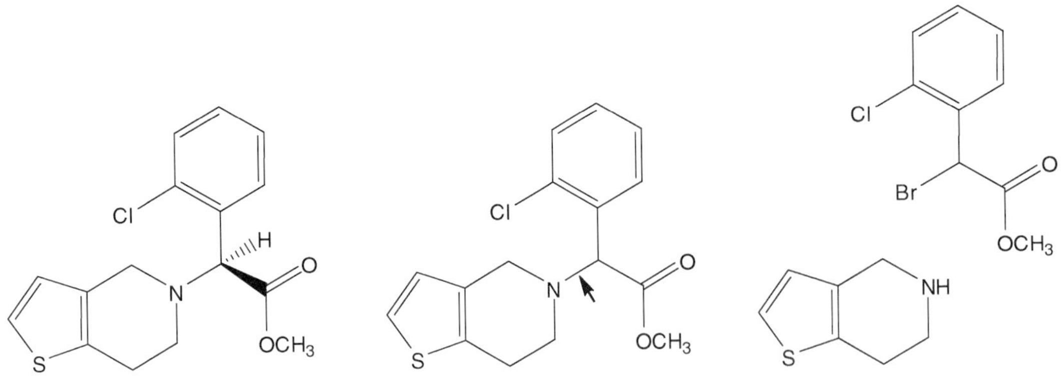

The α-bromo ester is formed by bromination of the ester. Methyl 2-chlorophenylacetate is formed by methoxycarbonylation of 2-chlorobenzyl chloride.

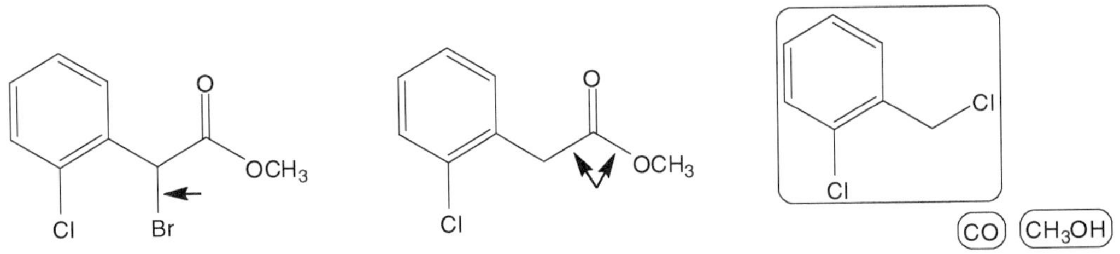

4,5,6,7-Tetrahydrothieno[3,2-c]pyridine is formed from 2-(thiophen-2-yl)ethanamine and formaldehyde (**Pictet–Spengler Reaction**). The amine is formed from the alcohol. (List all the options for reagents and conditions for this conversion. List the yield associated with each option.) 2-(Thiophen-2-yl)ethanol is formed by reaction of thiophene with ethylene oxide.

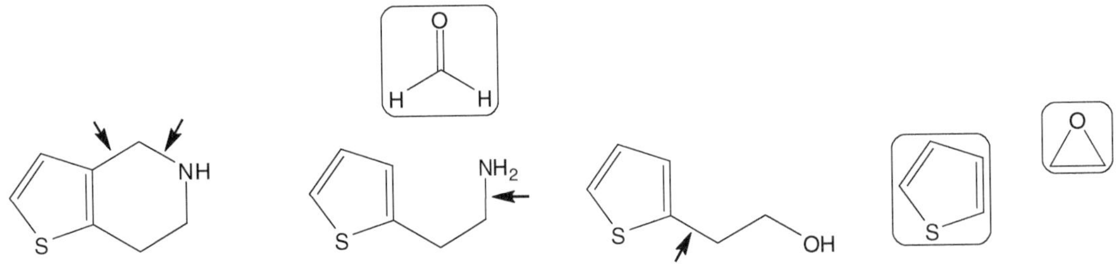

Extended Discussion

Draw the structures of the retrosynthetic analysis of one alternative route to clopidogrel. List the pros and cons for both approaches and select one route as the preferred route.

Cloxacillin

Anti-Infective Medicines/Antibacterials/Beta-Lactam Medicines

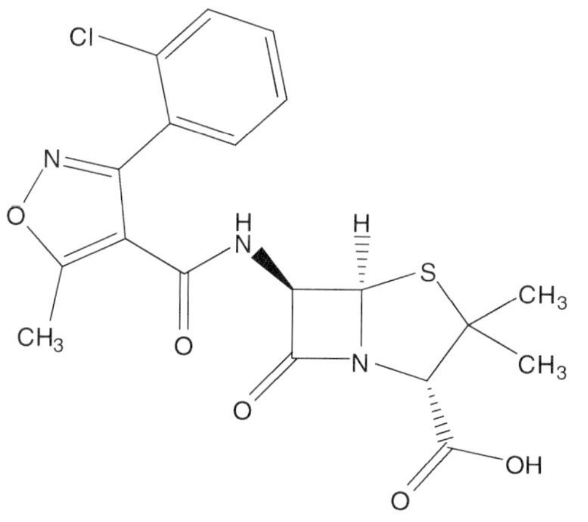

Penicillins are produced by fermentation or are semisynthetic. A semisynthetic penicillin is often formed by acylation of the amine of 6-aminopenicillanic acid (6-APA). 6-APA is produced from penicillin G (benzylpenicillin) by enzyme-mediated hydrolysis of the side-chain amide.

Discussion. Cloxacillin is semisynthetic. The side-chain amide bond is formed in the final step by reaction of an amine with an acid chloride. The amine, 6-aminopenicillanic acid (6-APA), is formed by enzyme-mediated hydrolysis of the side chain amide of penicillin G. Penicillin G (benzylpenicillin) is produced by the fungus *P. chrysogenum*.

The acid chloride is formed from the carboxylic acid. The carboxylic acid is formed by ester hydrolysis. Ethyl 3-(2-chlorophenyl)-5-methylisoxazole-4-carboxylate is formed by condensation of 2-chloro-*N*-hydroxybenzimidoyl chloride with ethyl acetoacetate. The benzimidoyl chloride is formed by chlorination of the oxime. The oxime is formed from 2-chlorobenzaldehyde.

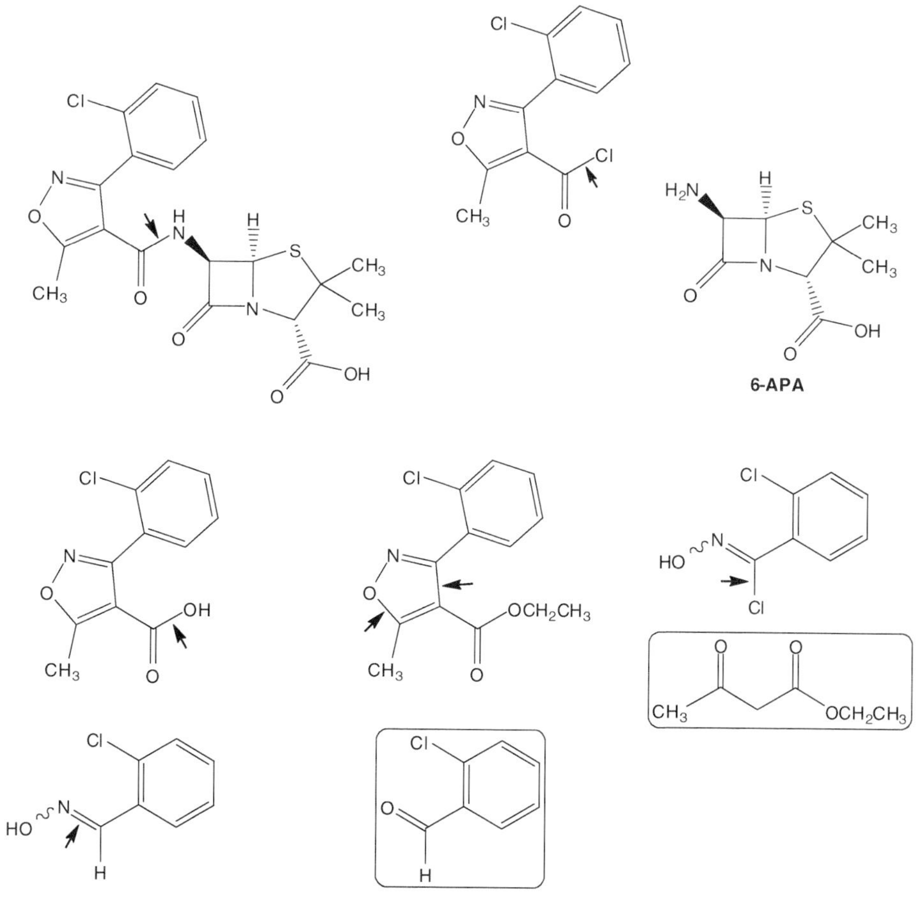

Extended Discussion

Draw the structures of the retrosynthetic analysis of one alternative route to 3-(2-chlorophenyl)-5-methylisoxazole-4-carb oxylic acid. Include the structures of the retrosynthetic analysis of any organic starting material(s) from petrochemical or biochemical raw materials. List the pros and cons for both routes and select one route as the preferred route.

Clotrimazole

Anti-Infective Medicines/Antifungal Medicines

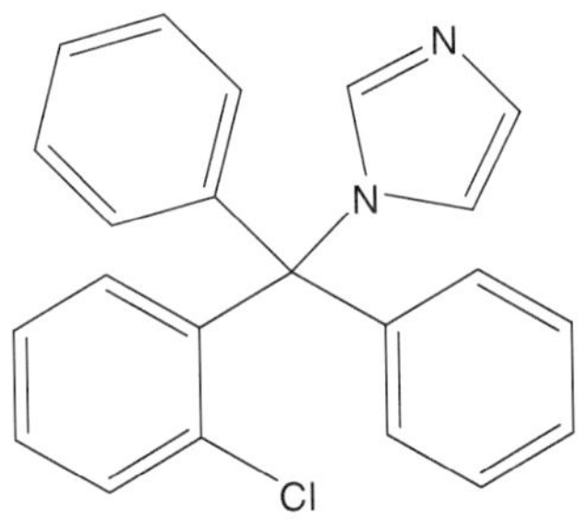

A benzylic C—N bond is often formed by displacement of a leaving group by a nitrogen nucleophile.

Discussion. Clotrimazole is manufactured in just two steps. A benzylic hydroxyl group is replaced by imidazole in the final step. The benzyl alcohol, (2-chlorophenyl)diphenylmethanol, is formed from the benzyl chloride in the water workup of the reaction of 2-chlorobenzotrichloride with benzene (**Friedel–Crafts Alkylation**).

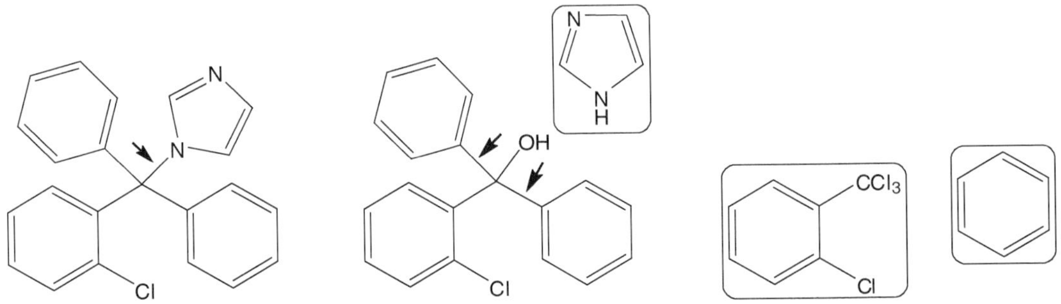

Extended Discussion

Draw the structures of the retrosynthetic analysis of one alternative route to (2-chlorophenyl)diphenylmethanol. List the pros and cons for both routes and select one route as the preferred route.

Clozapine

Medicines for Mental and Behavioral Disorders/Medicines Used in Psychotic Disorders

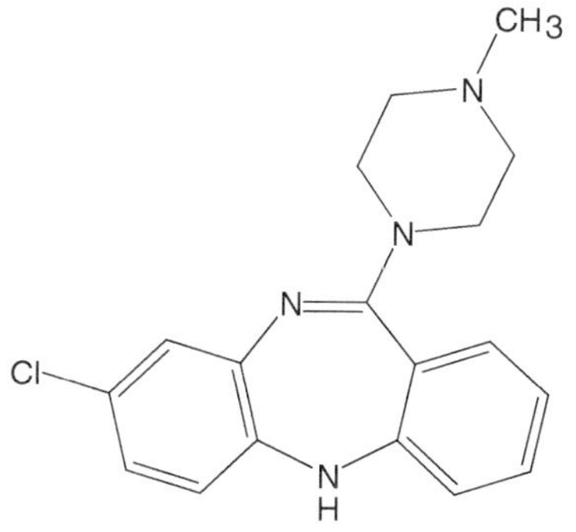

A 5H-dibenzo[b.e][1,4]diazepine is often formed from a diphenylamine.

Discussion. The C11 chloride of 8,11-dichloro-5H-dibenzo[b.e][1,4]diazepine is displaced by 1-methylpiperazine in the final step. The imino chloride is formed from the amide. The diazepine ring is formed when the amide is formed from the amine and carboxylic acid. The amine is formed by nitro group reduction. The diphenylamine is formed by displacement of halide (X = Cl, Br) from 1,4-dichloro-2-nitrobenzene or 1-bromo-4-chloro-2-nitrobenzene by anthranilic acid (**Ullmann–Goldberg Reaction**). (Draw structures for the retrosynthetic analysis of 1-bromo-4-chloro-2-nitrobenzene from petrochemical or biochemical raw materials.)

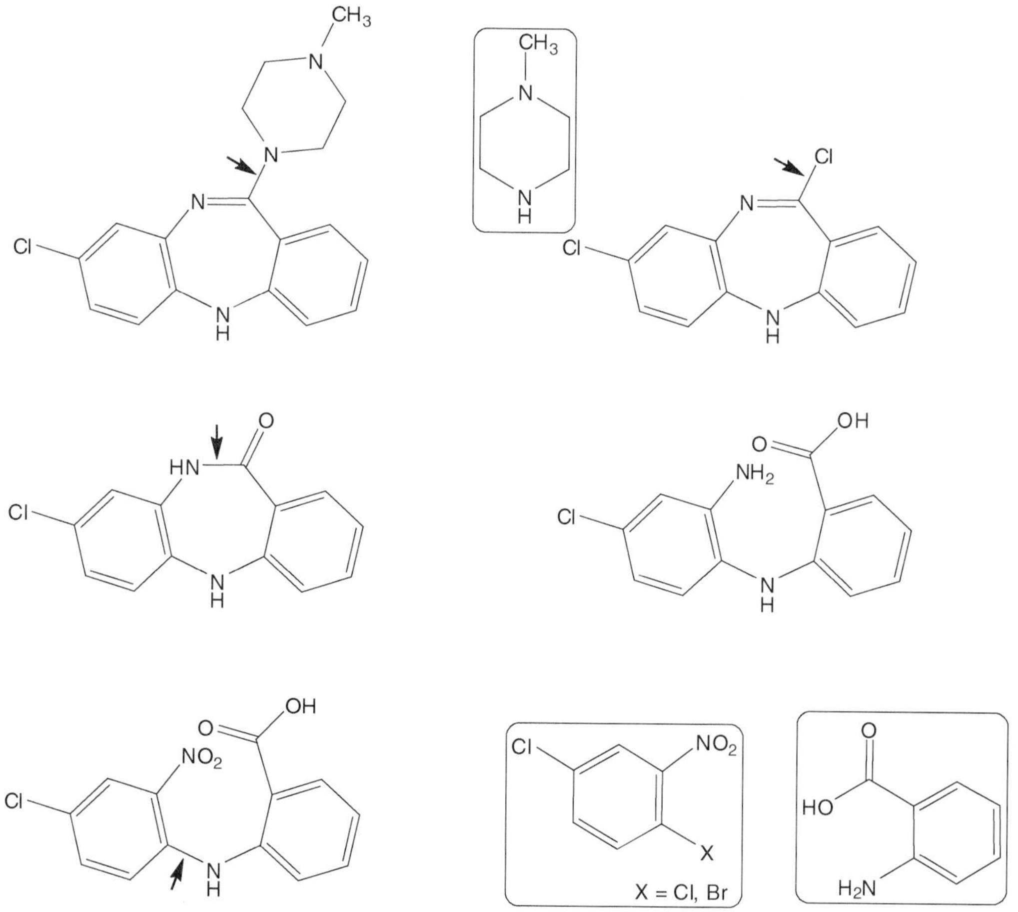

Extended Discussion

Draw the structures of the retrosynthetic analysis of one alternative route to clozapine which does not use anthranilic acid as a starting material. Include the structures of the retrosynthetic analysis of any organic starting material(s) from petrochemical or biochemical raw materials. List the pros and cons for both routes and select one route as the preferred route.

Codeine

Medicines for Pain and Palliative Care/Opioid Analgesics

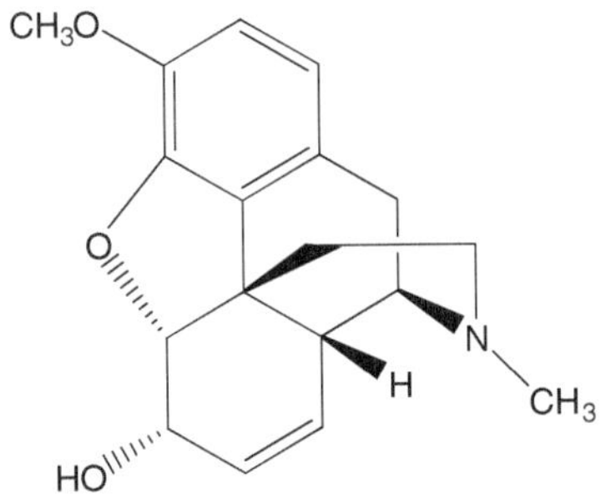

A single-enantiomer molecule with multiple chiral carbons is often formed by modification of a natural product which has most or all of the chiral carbons already in place.

Discussion. Codeine is an opium poppy alkaloid but most codeine is manufactured by *O*-methylation of another opium poppy alkaloid, morphine. Morphine has three sites likely to react with iodomethane or dimethyl sulfate. Selective *O*-methylation is accomplished with trimethylphenylammonium chloride. (*N,N*-Dimethylaniline is a byproduct of the *O*-methylation. How is *N,N*-dimethylaniline separated from codeine?) Trimethylphenylammonium chloride is formed from *N,N*-dimethylaniline and chloromethane.

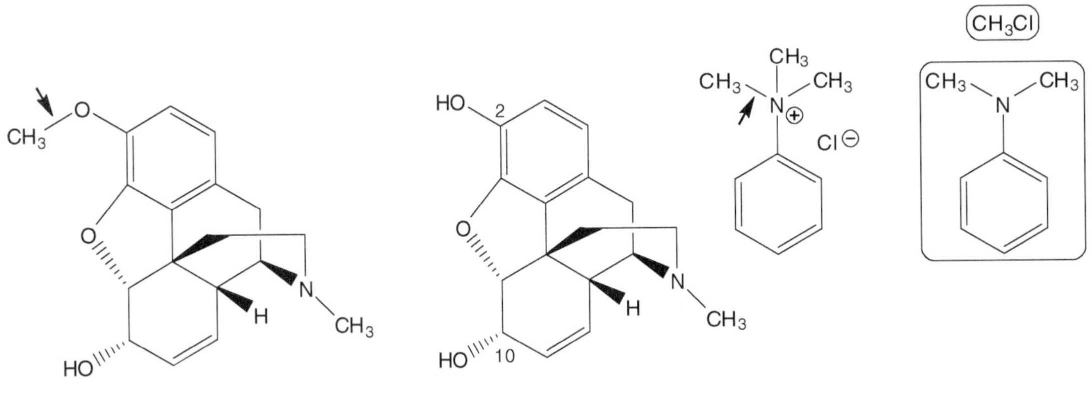

morphine

Extended Discussion

Morphine is prone to oxidation. Draw the structures of three impurities in codeine which result from morphine oxidation.

Colecalciferol

Vitamins and Minerals

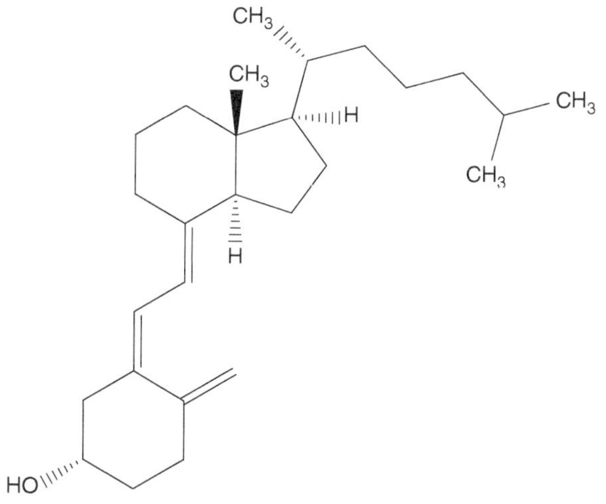

A single-enantiomer molecule with multiple chiral carbons is often formed by modification of a natural product which has most or all of the chiral carbons already in place.

Discussion. Colecalciferol (cholecalciferol, vitamin D_3) is a semisynthetic secosteroid (seco from Latin *secare* "to cut"). Colecalciferol is produced by cleavage of the B-ring of the steroid natural product cholesterol.

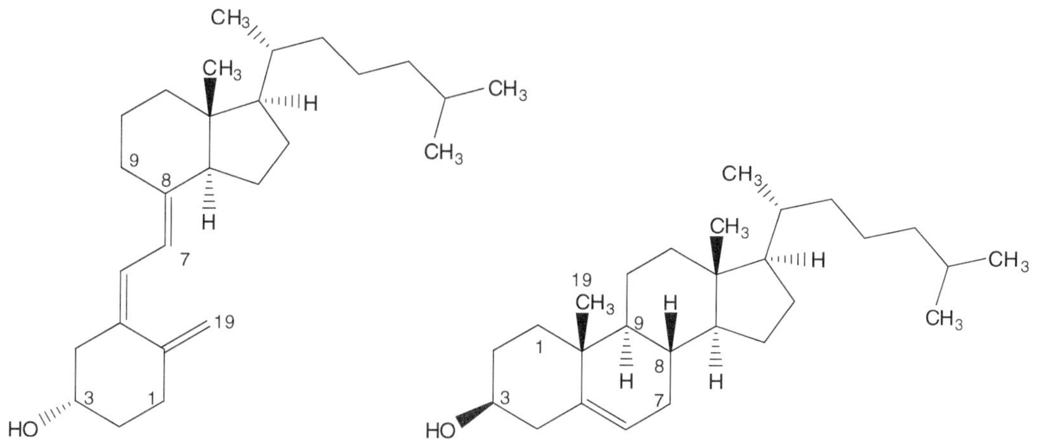

cholesterol

Colecalciferol (vitamin D_3) is formed from pre-vitamin D_3. Pre-vitamin D_3 is formed by UV-photoirradiation of pro-vitamin D_3 (7-dehydrocholesterol). Low conversion in the UV-photoirradiation step minimizes the formation of two side products. (Draw structures for and name these side products. How is 7-dehydrocholesterol separated from the UV-photoirradiation product mixture?) 7-Dehydrocholesterol is formed by hydrolysis of the ester of 7-dehydrocholesteryl benzoate.

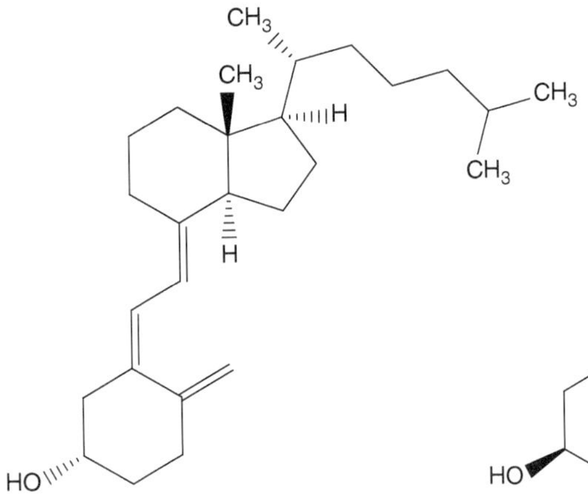

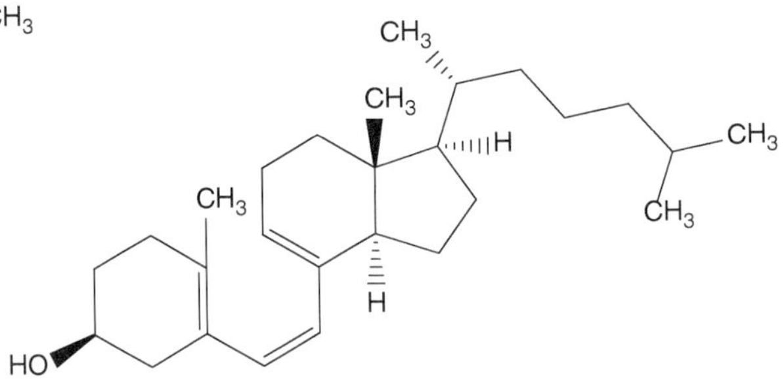

pro-vitamin D_3

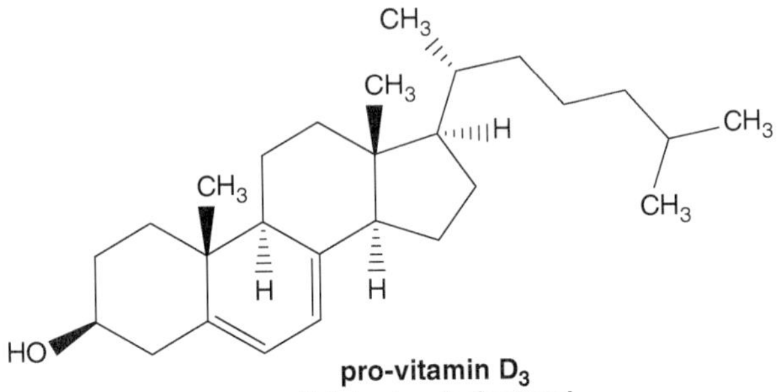

pro-vitamin D_3
7-dehydrocholesterol

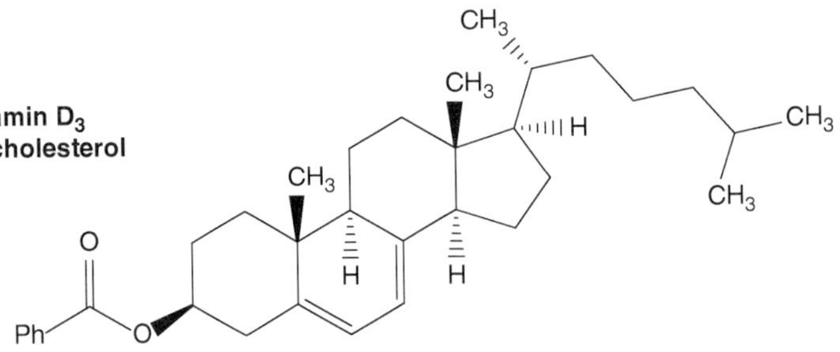

There are two routes to 7-dehydrocholesteryl benzoate from cholesterol. In Route A, the 7,8-alkene is formed from the 7-*p*-toluenesulfonyl hydrazone (**Bamford–Stevens Reaction**). The hydrazone is formed from 7-oxocholesteryl benzoate (7-ketocholesteryl benzoate) and *p*-toluenesulfonyl hydrazide. 7-Oxocholesteryl benzoate is formed by oxidation of cholesteryl benzoate. Cholesteryl benzoate is formed from cholesterol.

RouteA

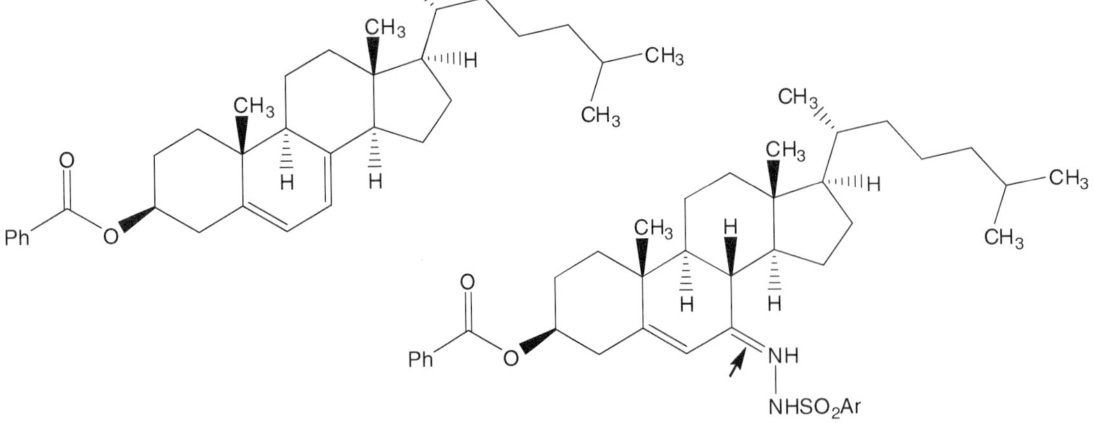

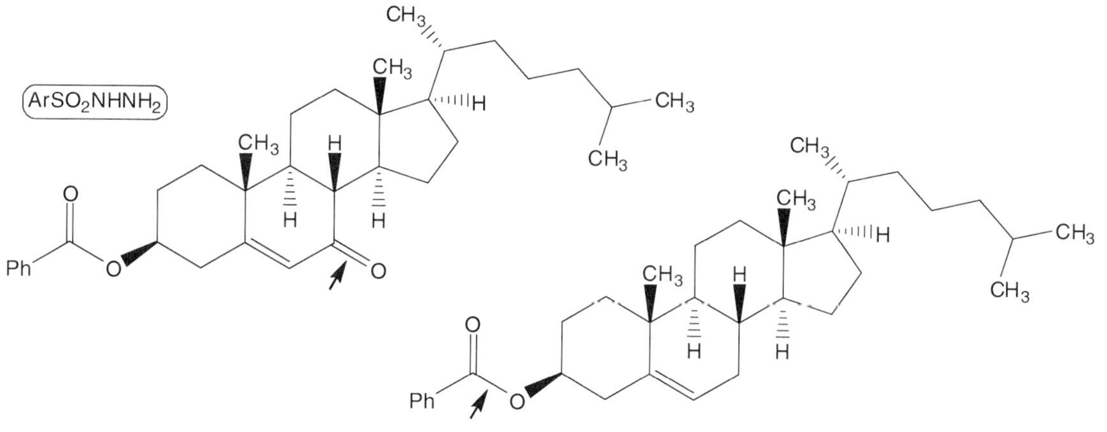

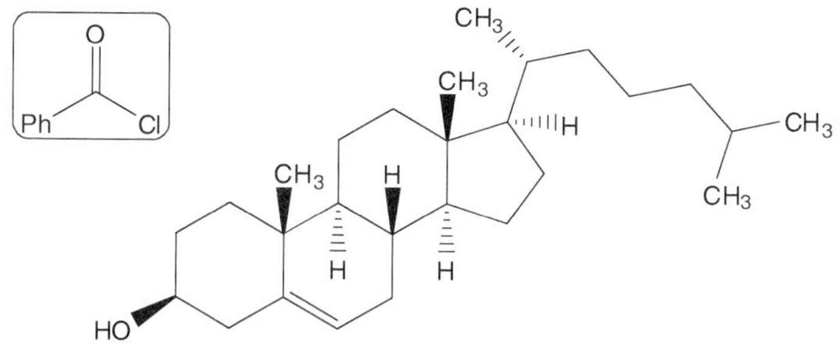

cholesterol

In Route B, the 7,8-alkene is formed by elimination of hydrogen bromide from 7-bromocholesteryl benzoate. (Draw the structure of a side product which is also formed in this elimination.) 7-Bromocholesteryl benzoate is formed by bromination of cholesteryl benzoate. (List the options for brominating reagent and reaction conditions and the ratio of 7α-bromocholesteryl benzoate to 7β-bromocholesteryl benzoate associated with each option.) Cholesteryl benzoate is formed from cholesterol. Cholesterol is isolated from sheep wool grease.

Route B

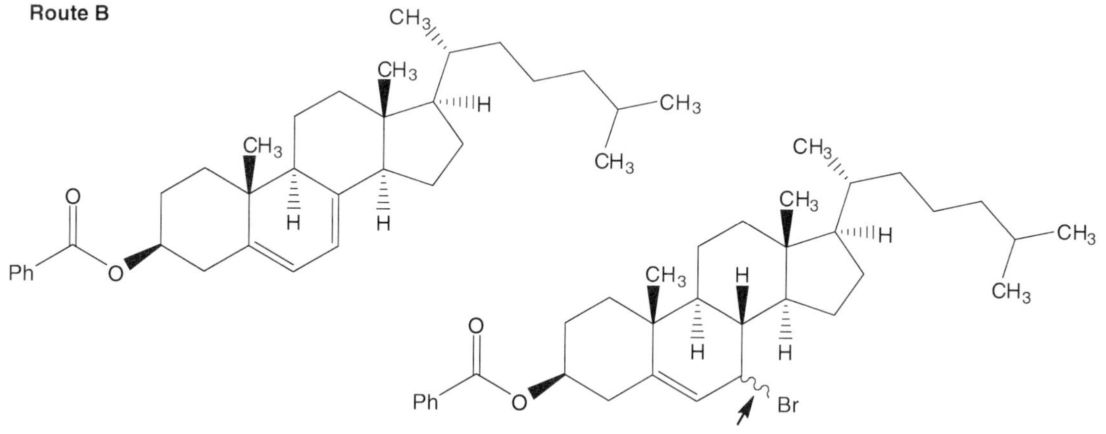

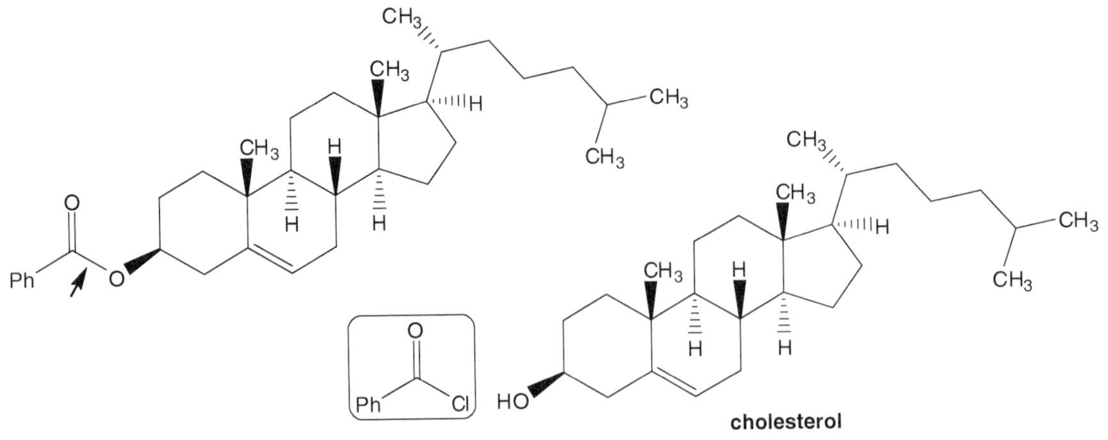

Extended Discussion

7-Dehydrocholesterol can also be formed by fermentation (Route C). List the pros and cons for the three routes to 7-dehydrocholesterol.

Cyclizine

Medicines for Pain and Palliative Care/Medicines for Other Common Symptoms in Palliative Care

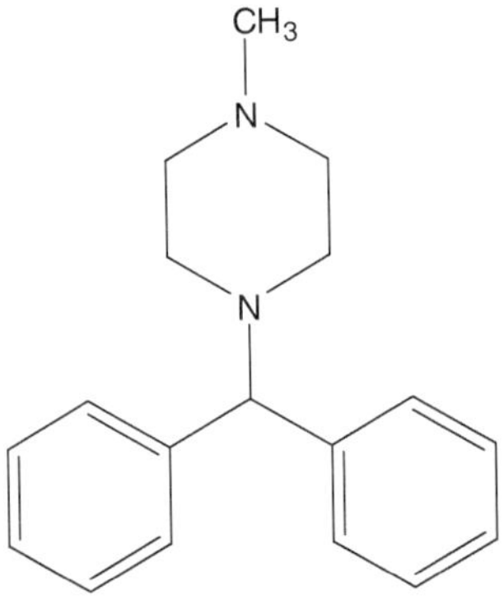

Tertiary amines are ubiquitous in drug structures. A tertiary amine is often formed by alkylation of a secondary amine. The alkylation is most efficient when the amine is used in excess and the carbon with the leaving group (Cl, Br, I, OTs, OMs) is primary or benzylic.

Discussion. Cyclizine is manufactured in one step. The tertiary amine near the center of the molecule is formed by chloride displacement from chlorodiphenylmethane (benzhydryl chloride) by 1-methylpiperazine.

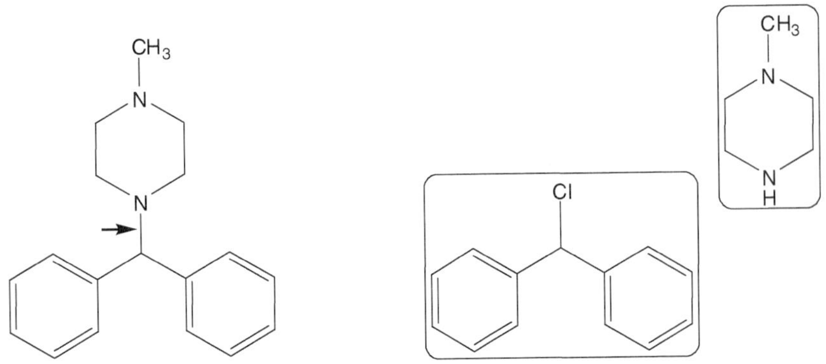

Cyclophosphamide

Antineoplastics and Immunosuppressives/Cytotoxic and Adjuvant Medicines

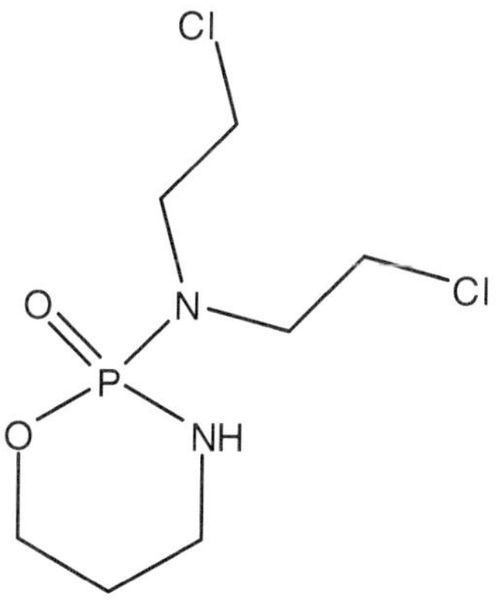

A phosphoric acid amide is often formed from a phosphoryl chloride and a primary or secondary amine. A phosphoric acid ester is often formed from a phosphoryl chloride and an alcohol.

Discussion. Cyclophosphamide is a 1 : 1 mixture of the (*R*)- and (*S*)-enantiomers. The (*S*)-enantiomer has the higher therapeutic index. The tetrahydro-2*H*-1,3,2-oxazaphosphorine-2-oxide is formed in the final step by the reaction of the phosphoramidic dichloride with 3-amino-1-propanol. The phosphoramidic dichloride is formed by the reaction of phosphorus oxychloride with bis(2-chloroethyl)amine hydrochloride.

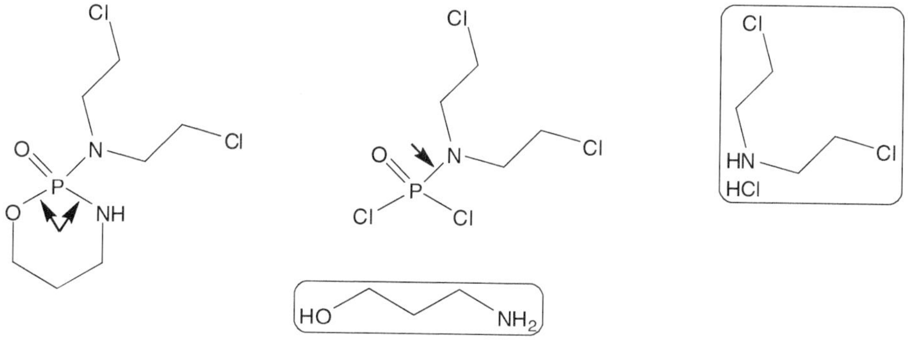

Extended Discussion

Cyclophosphamide is also formed from the same starting materials using the same disconnections in the reverse order. List the pros and cons for both routes. Is one route preferred?

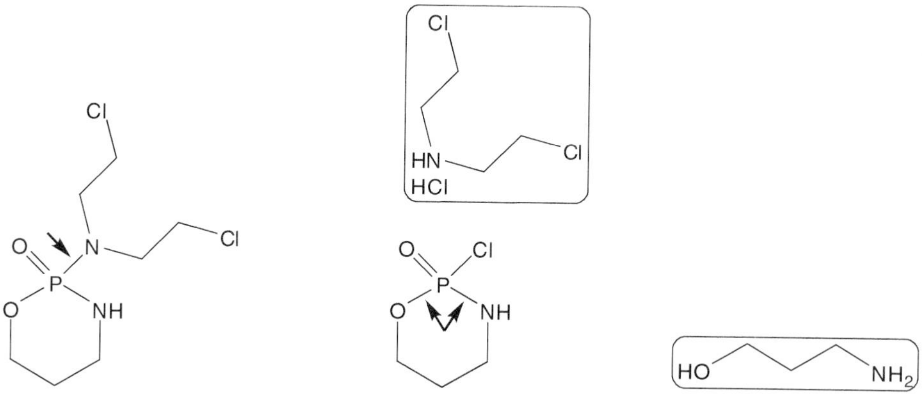

Cycloserine

Anti-Infective Medicines/Antibacterials/Antituberculosis Medicines

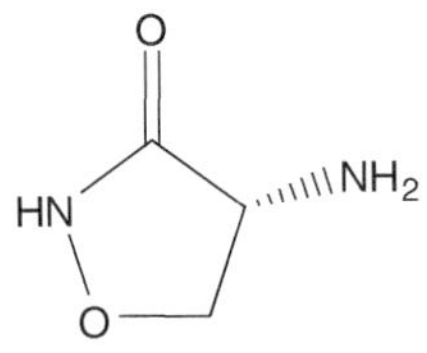

A chiral carbon in a single-enantiomer molecule is often delivered in a starting material.

Discussion. Cycloserine has been manufactured by fermentation (*Streptomyces garyphalussive orchidaceus*) and by chemical synthesis from D-serine. In the chemical synthesis, the isoxazolidinone ring is formed in the final step by the reaction of 3-chloro-D-alanine methyl ester hydrochloride with hydroxylamine hydrochloride. 3–Chloro-D-alanine methyl ester hydrochloride is formed from D-serine methyl ester hydrochloride. D-Serine methyl ester hydrochloride is formed from D-serine (**Fischer Esterification**). D-Serine is produced by fermentation.

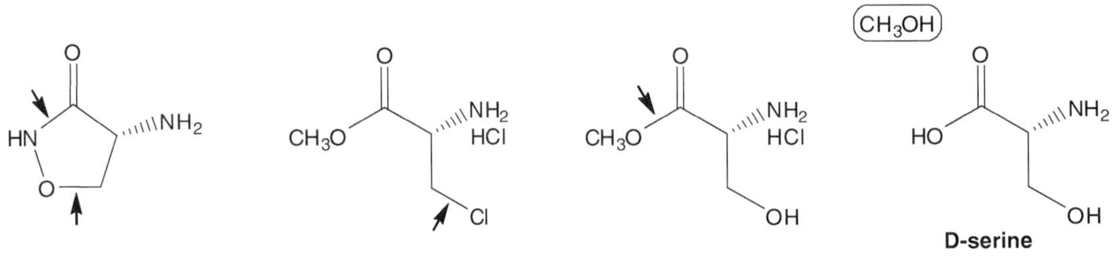

Extended Discussion

Alternative routes to cycloserine utilize protection and deprotection of the amino group in the conversion of D-serine to cycloserine. Draw the structures of the retrosynthetic analysis of one of these alternative routes. List the pros and cons for both routes.

Cytarabine

Antineoplastics and Immunosuppressives/Cytotoxic and Adjuvant Medicines

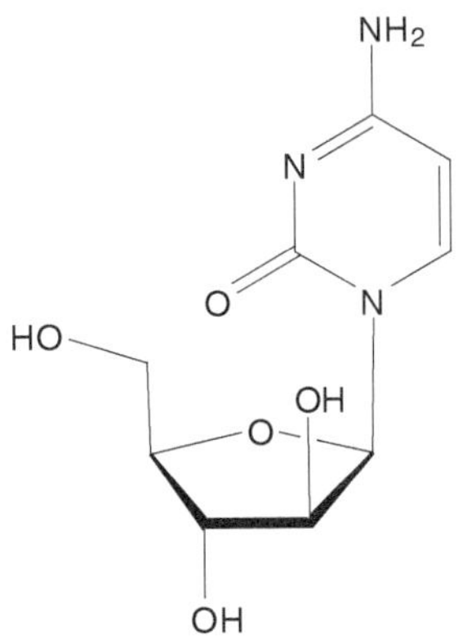

A single-enantiomer molecule with multiple chiral carbons is often formed by modification of a natural product which has most or all of the chiral carbons already in place.

Discussion. Cytarabine is formed by hydrolysis of 2,2′-*O*-anhydro(1-β-D-arabinofuranosyl)cytosine. The anhydronucleoside is formed from cytidine.

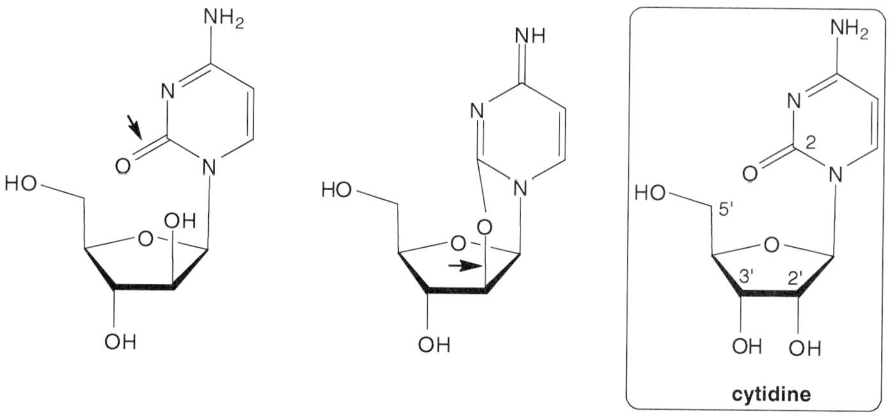

Extended Discussion

Cytidine can be converted to a 2,2′-*O*-anhydro(1-β-D-arabinofuranosyl)cytosine by other routes which require an additional step to release the 3′ and 5′-hydroxyl groups of cytarabine. Draw the structures of the retrosynthetic analysis of one alternative route from cytidine to cytarabine. List the pros and cons for both routes and select one route as the preferred route.

D

Dacarbazine

Antineoplastics and Immunosuppressives/Cytotoxic and Adjuvant Medicines

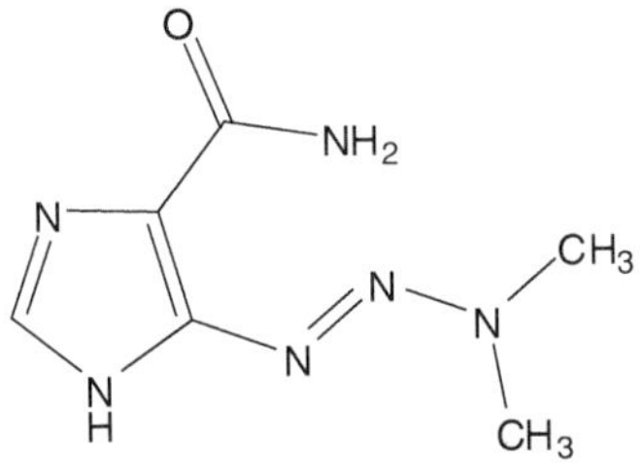

A highly reactive/unstable functional group in the target molecule is best formed last in the synthetic sequence.

Discussion. The unusual triazene functional group is unstable/highly reactive. Triazenes must be protected from light and fragment to form diazonium salts and amines in the presence of Bronsted acids or alkylating agents.

The *N*-aryltriazene is formed by reaction of the aryldiazonium salt with dimethylamine. The diazonium salt is formed by diazotization of 5-aminoimidazole-4-carboxamide.

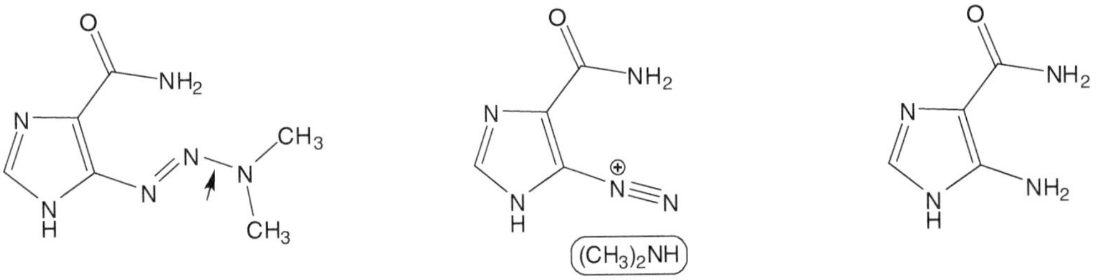

The carboxamide is formed by hydrolysis of the nitrile. The amine is formed from the carboxamide (**Hofmann Rearrangement**). 4-Cyano-1*H*-imidazole-5-carboxamide is formed by hydrolysis of 4,5-dicyanoimidazole. 4,5-Dicyanoimidazole is formed from triethyl orthoformate and diaminomaleonitrile.

Routes to Essential Medicines: A Workbook for Organic Synthesis, First Edition. Peter J. Harrington.

Companion website: www.wiley.com/go/Harrington/routes_essential_medicine

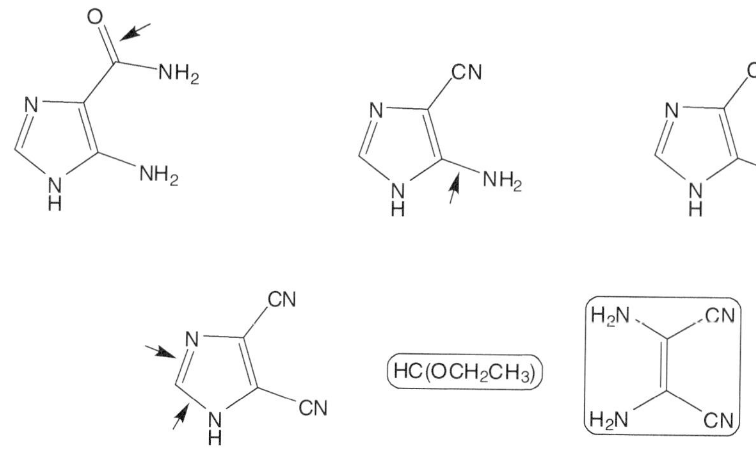

Extended Discussion

Temozolomide is used to treat some brain cancers and is the first-line treatment for glioblastoma multiforme. Draw the structures of a retrosynthetic analysis of this important drug.

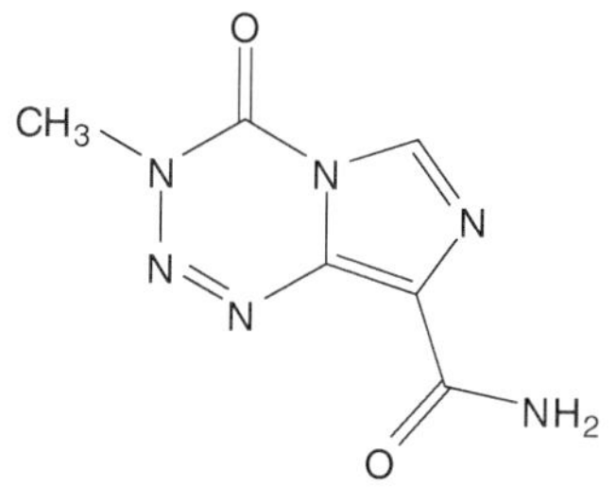

temozolomide

Daclatasvir

Anti-infective Medicines/Antiviral Medicines/Antihepatitis Medicines/Medicines for Hepatitis C/NS5A Inhibitors

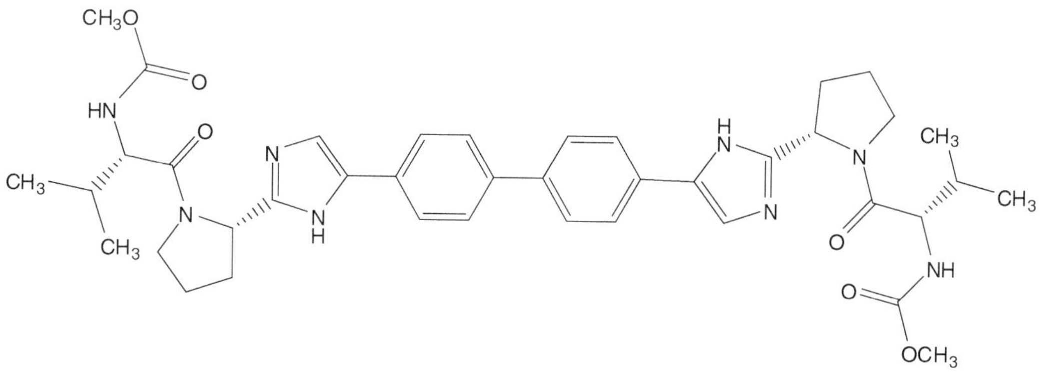

A 2,5-disubstituted imidazole is formed by reaction of an α-acyloxyketone with ammonium acetate. The four C-N bonds of the imidazole ring are formed in the reaction.

Discussion. In a preferred route, the identical components on the left and right of the central C-C bond are introduced in the same step. The amide is formed by reaction of the pyrrolidine with *N*-(methoxycarbonyl)-L-valine (Moc-L-valine) in the final step. The pyrrolidine is released by deprotection of the *tert*-butyl (Boc) carbamate. The imidazole ring is formed by reaction of the α-acyloxyketone with ammonium acetate. The α-acyloxyketone is formed by bromide displacement from the α-bromoketone by *N*-(*tert*-butoxycarbonyl)-L-proline (Boc-L-proline).

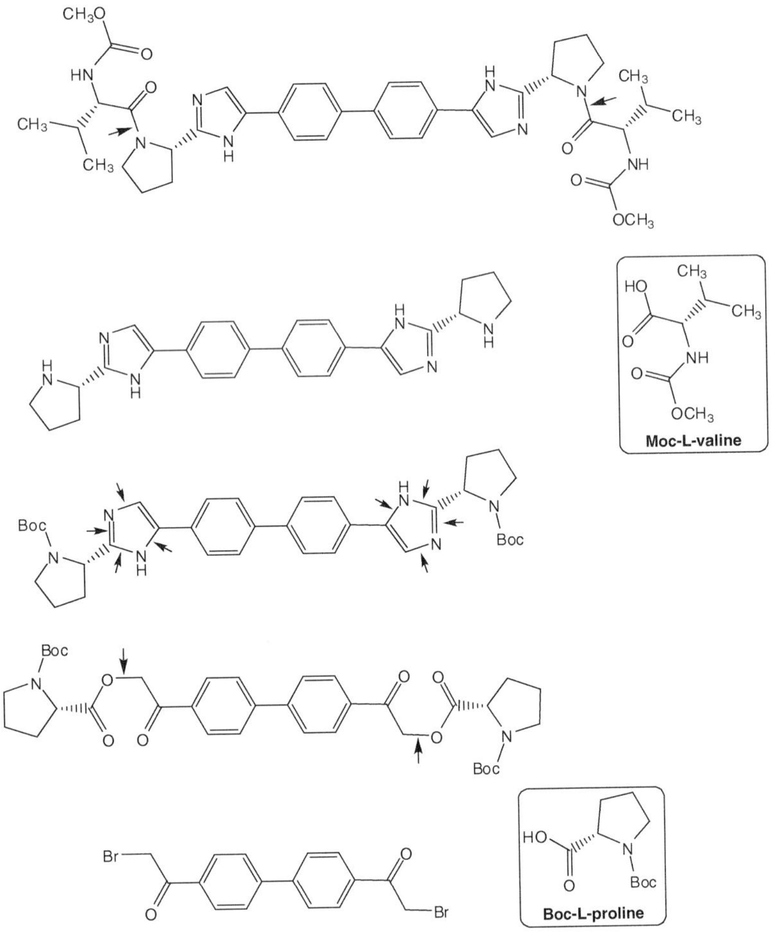

The α-bromoketone is formed by bromination of the ketone. 4,4′-Diacetylbiphenyl is formed by the palladium-catalyzed coupling of 4′-bromoacetophenone and the arylboronic acid or arylboronate ester derived from 4′-bromoacetophenone (**Suzuki–Miyaura Coupling**). (Evaluate the acylation of biphenyl as an alternative route to 4,4′-diacetylbiphenyl.)

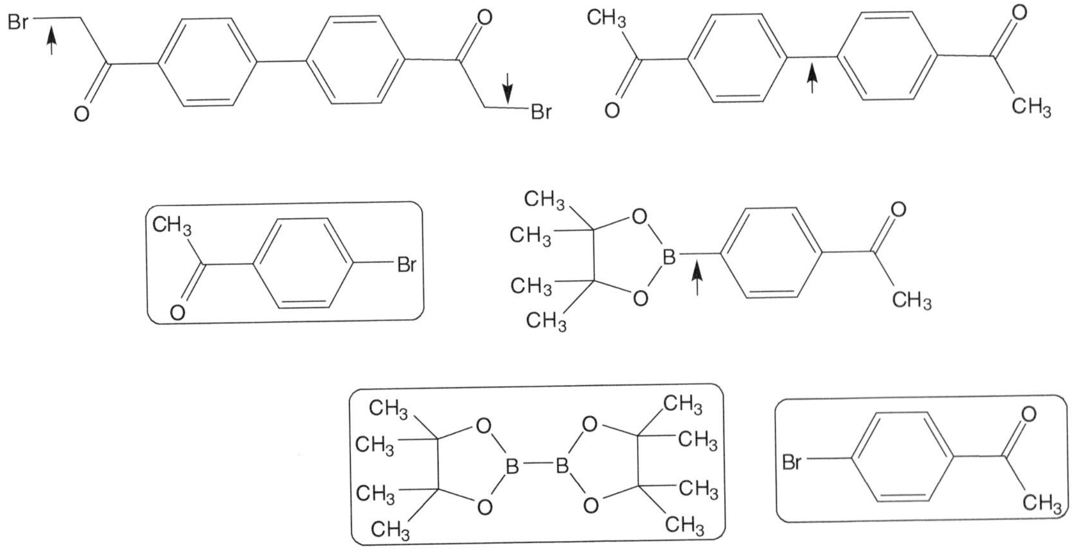

Extended Discussion

Draw the structures of a retrosynthetic analysis of an alternative **Suzuki–Miyaura Coupling** route to daclatasvir.

Dapsone

Anti-infective Medicines/Antibacterials/Antileprosy Medicines

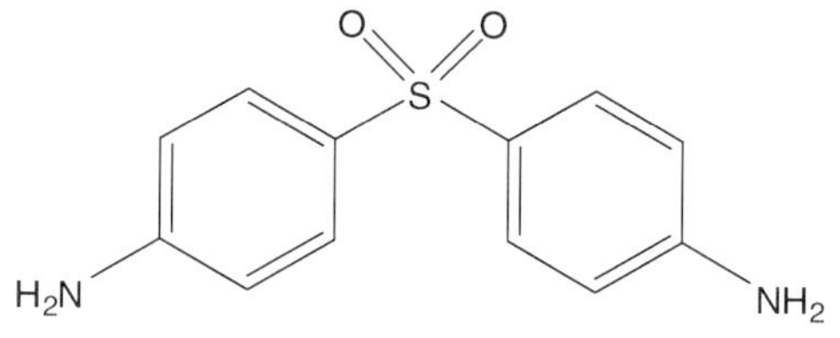

Sulfur is encountered in organic synthesis in oxidation states ranging from S^{-2} to S^{+6}. Diarylsulfones (S^{+2}) are often formed by oxidation of diarylsulfides (S^{-2}).

Discussion. The second amino group is formed by reduction of the nitro group in the final step. The amino group is released by hydrolysis of the acetamide. The diarylsulfone is formed by oxidation of the diarylsulfide. The amino group of 4-amino-4′-nitrodiphenylsulfide is protected as the acetamide. 4-Amino-4′-nitrodiphenylsulfide is formed by displacement of chloride from 1-chloro-4-nitrobenzene by 4-aminothiophenol. 4-Aminothiophenol is formed by chloride displacement and nitro group reduction from 1-chloro-4-nitrobenzene.

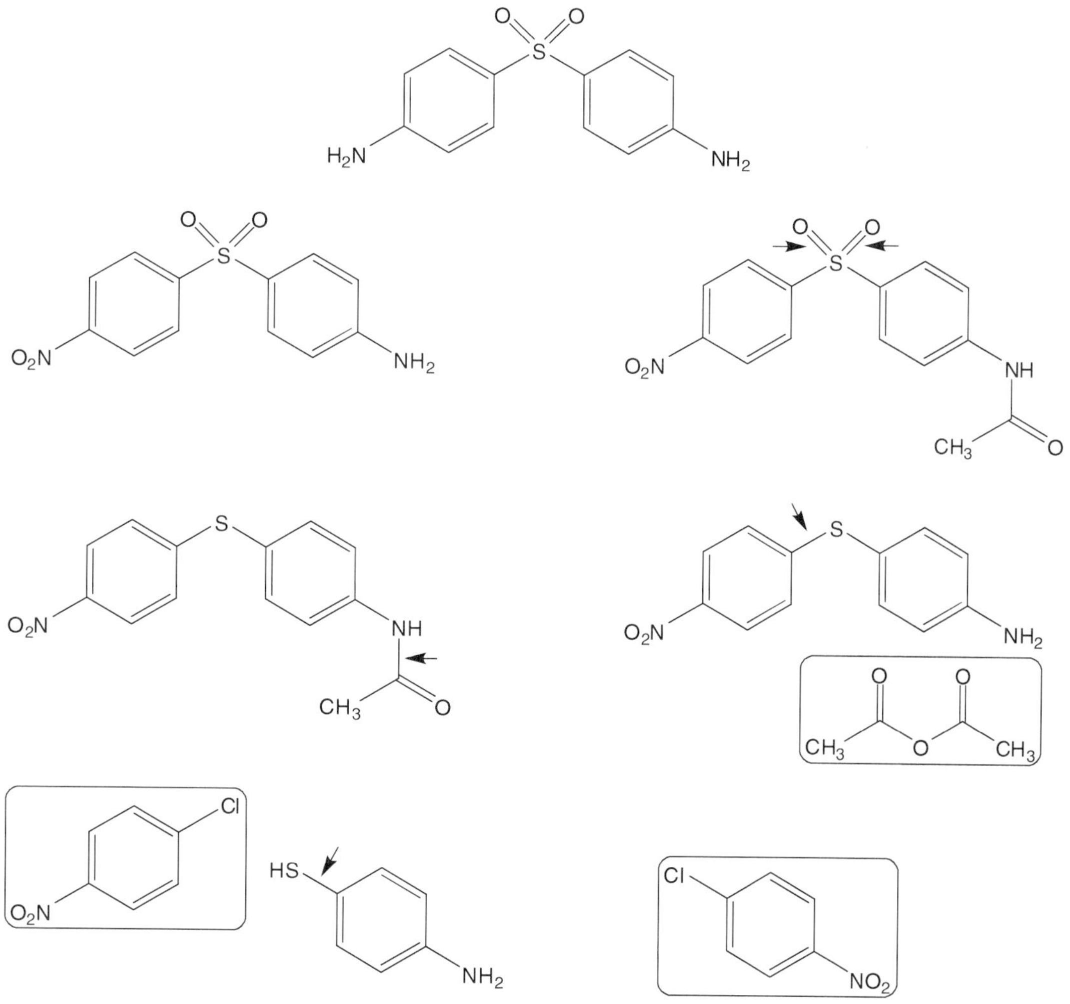

Extended Discussion

Draw the structures of the retrosynthetic analysis of one alternative route to dapsone. List the pros and cons for both routes and select one route as the preferred route.

Darunavir

Anti-infective Medicines/Antiviral Medicines/Antiretrovirals/Protease Inhibitors

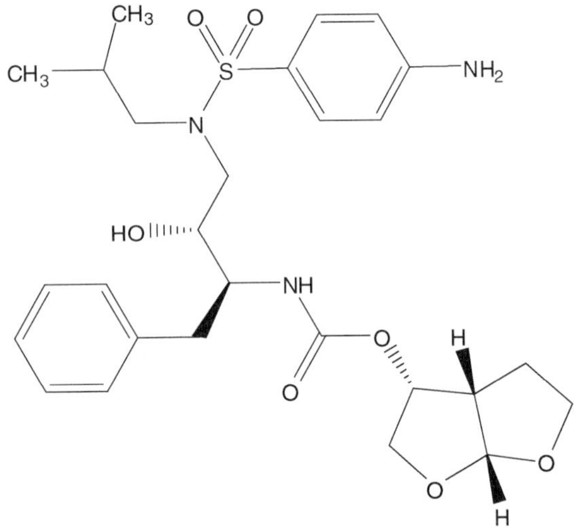

A β-amino alcohol with a chiral β-C is often formed from an α-amino acid if the configuration and β-substituent of the β-amino alcohol match the configuration and α-substituent of a common α-amino acid.

Discussion. The carbamate is formed from the amine and the *N*-succinimidyl carbonate in the final step. The amine is released by cleavage of the *tert*-butoxycarbonyl (Boc) protecting group. The 4-aminosulfonamide is formed by reduction of the 4-nitrosulfonamide.

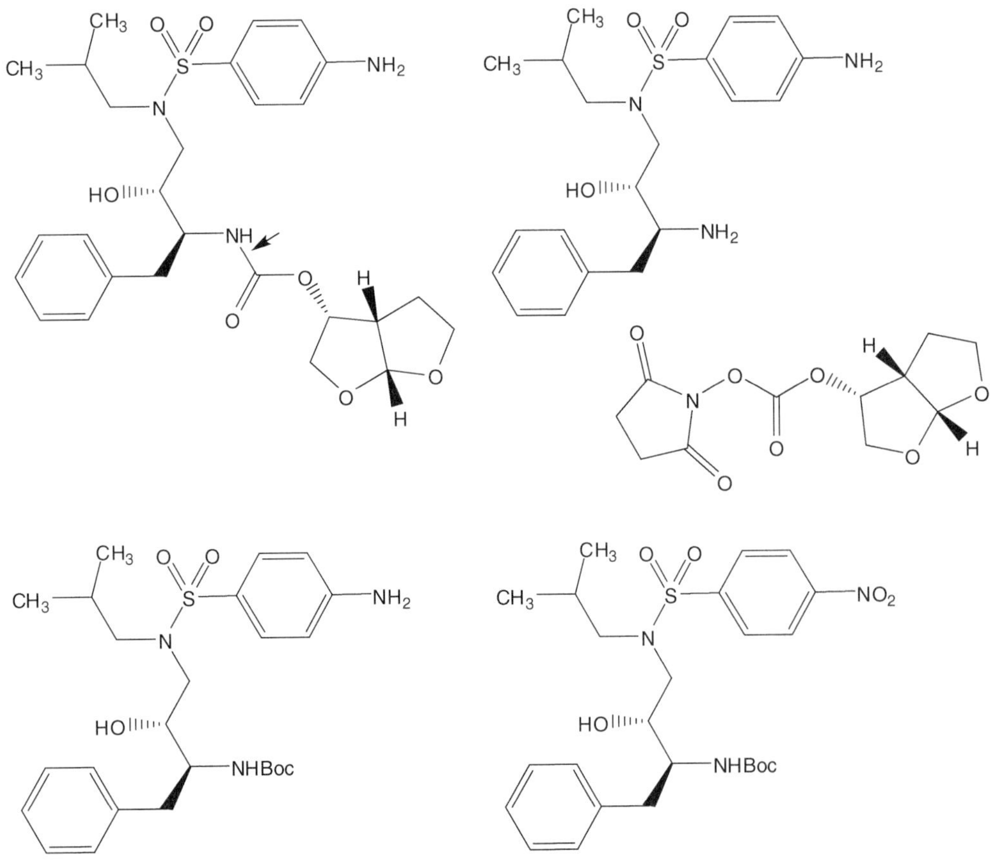

The 4-nitrosulfonamide is formed by reaction of the amine with 4-nitrobenzenesulfonyl chloride. The amine is formed by ring-opening of the epoxide by isobutylamine.

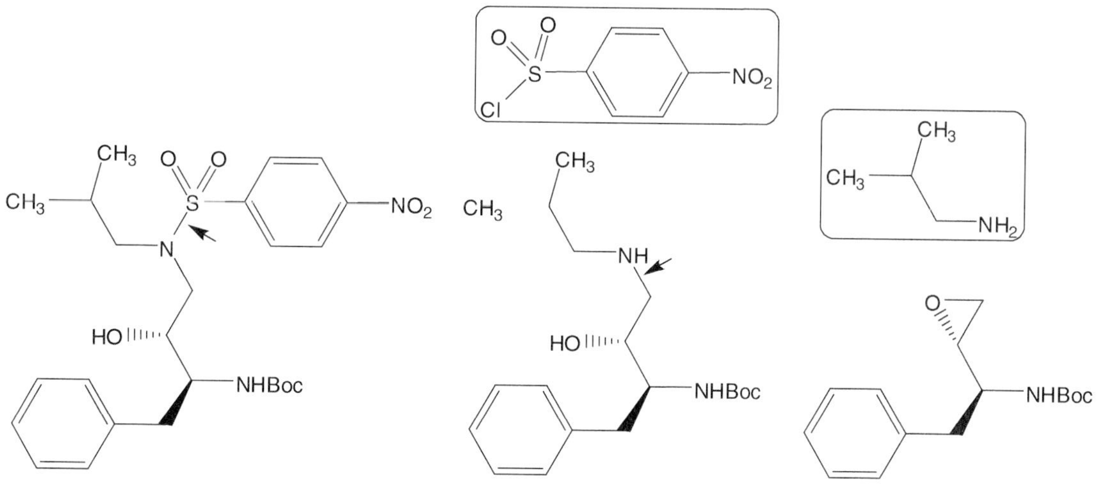

The epoxide is formed from the chlorohydrin. The chlorohydrin is formed by reduction of the α-chloroketone. The α-chloroketone is formed from (*N-tert*-butoxycarbonyl)-L-phenylalanine methyl ester and chloroacetic acid (**Claisen Condensation**).

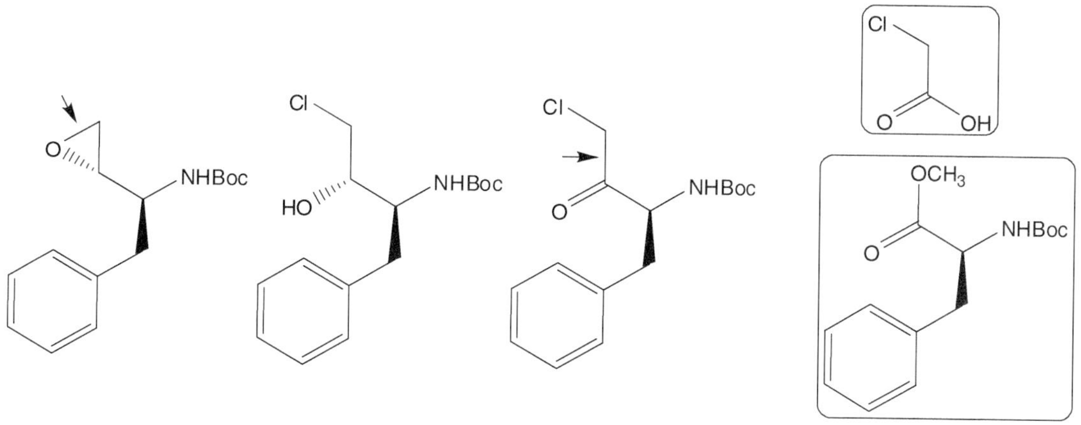

The *N*-succinimidyl carbonate for the final step is formed by reaction of (3*R*,3a*S*,6a*R*)-3-hydroxyhexahydrofuro[2,3-*b*] furan with *N*,*N*′-disuccinimidyl carbonate. The (3*R*,3a*S*,6a*R*)-alcohol is separated from the (3*S*,3a*R*,6a*S*)-enantiomer. (List methods used for the separation.) The alcohol is formed by reduction of the ketone. The ketone is formed by ozonolysis of the alkene. Hexahydro-3-methylenefuro[2,3-*b*]furan is formed by radical cyclization. A 1 : 1 mixture of (2*R*,3*S*)- and (2*S*,3*R*)-enantiomers of (*trans*)-3-iodo-2-(propynyloxy)tetrahydrofuran is formed from 2,3-dihydrofuran and propargyl alcohol.

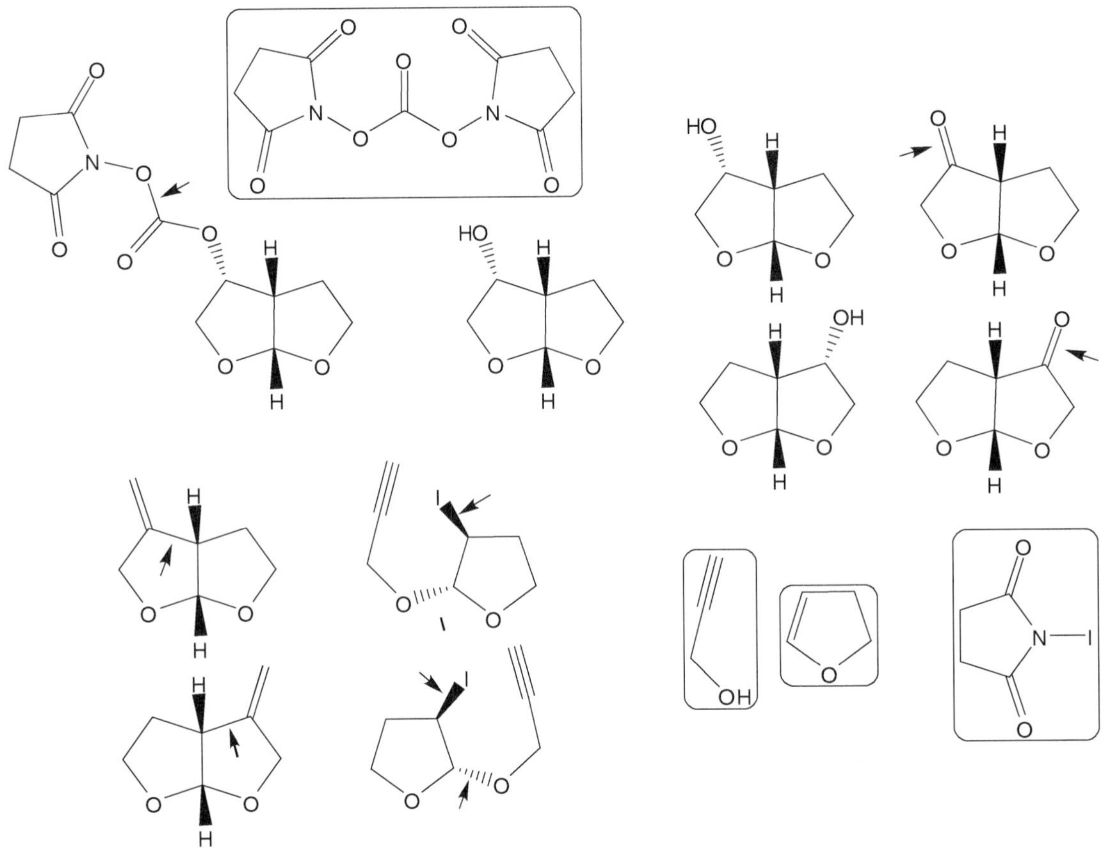

Extended Discussion

Draw the structures of a retrosynthetic analysis of one alternative route to (3*R*,3a*S*,6a*R*)-3-hydroxyhexahydrofuro[2,3-*b*]furan. Include the structures of the retrosynthetic analysis of any organic starting material(s) from petrochemical or biochemical raw materials. List the pros and cons for both routes and select one route as the preferred route to this key intermediate.

Dasabuvir

Anti-infective Medicines/Antiviral Medicines/Antihepatitis Medicines/Medicines for Hepatitis C/Non-nucleoside Polymerase Inhibitors

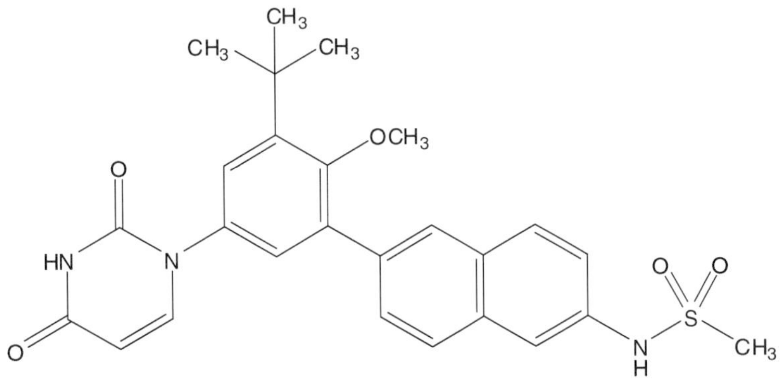

A biaryl is often formed by palladium-catalyzed coupling of an aryl iodide and arylboronate ester (Suzuki–Miyaura Coupling).

Discussion. The C-C bond joining the rings is formed in the final step by palladium-catalyzed coupling of an aryl iodide and an arylboronate ester (**Suzuki–Miyaura Coupling**).

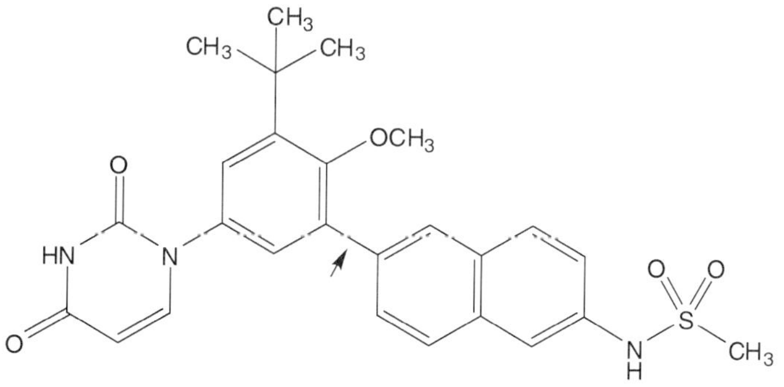

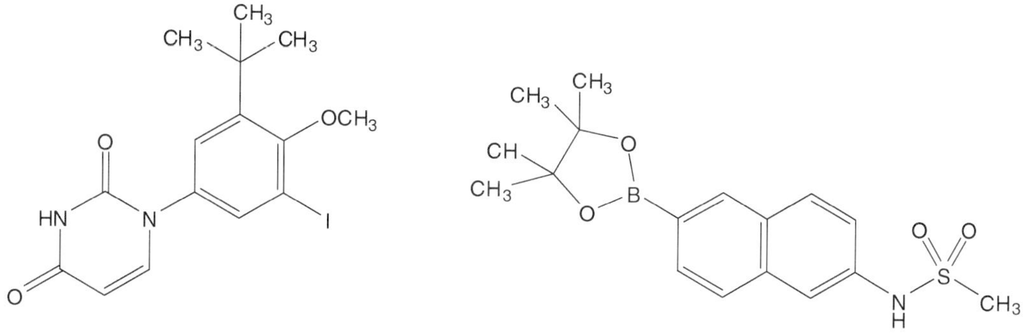

The C-N bond joining the rings of the aryl iodide is formed by displacement of the C4 iodide in 2-*tert*-butyl-4,6-diiodoanisole by N1 of uracil. The anisole methoxy group is formed by *O*-methylation of the phenol (**Williamson Ether Synthesis**). (List the methylating agents used and the anisole yield associated with each methylating agent.) The diiodophenol is formed by iodination of 2-*tert*-butylphenol.

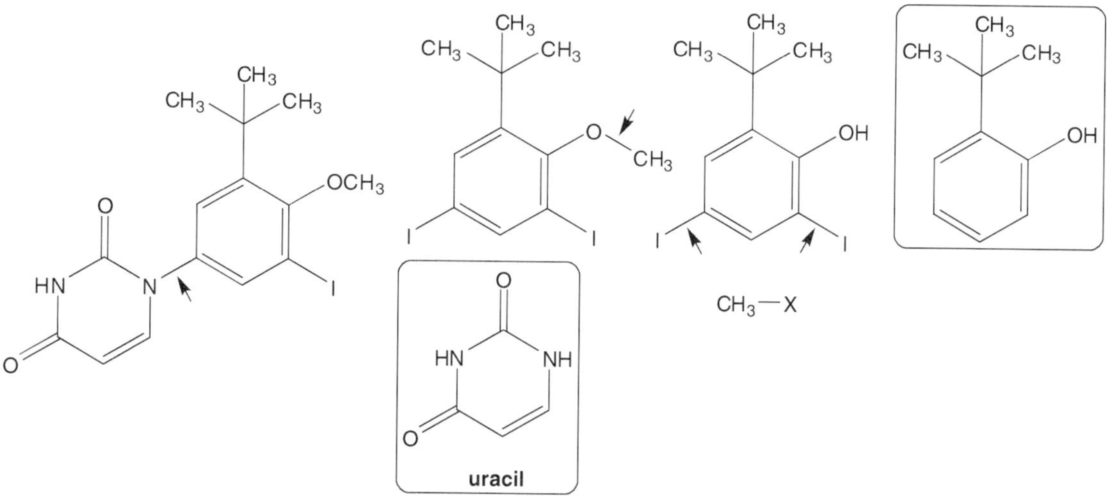

The arylboronate ester is formed by the palladium-catalyzed reaction of the aryl bromide with bis(pinacolato)diboron. The sulfonamide is formed by reaction of 6-bromo-2-naphthylamine with methanesulfonyl chloride. The naphthylamine is formed from the naphthol (**Bucherer Reaction**). 6-Bromo-2-naphthol is formed by debromination of 1,6-dibromo-2-naphthol. The dibromonaphthol is formed from 2-naphthol.

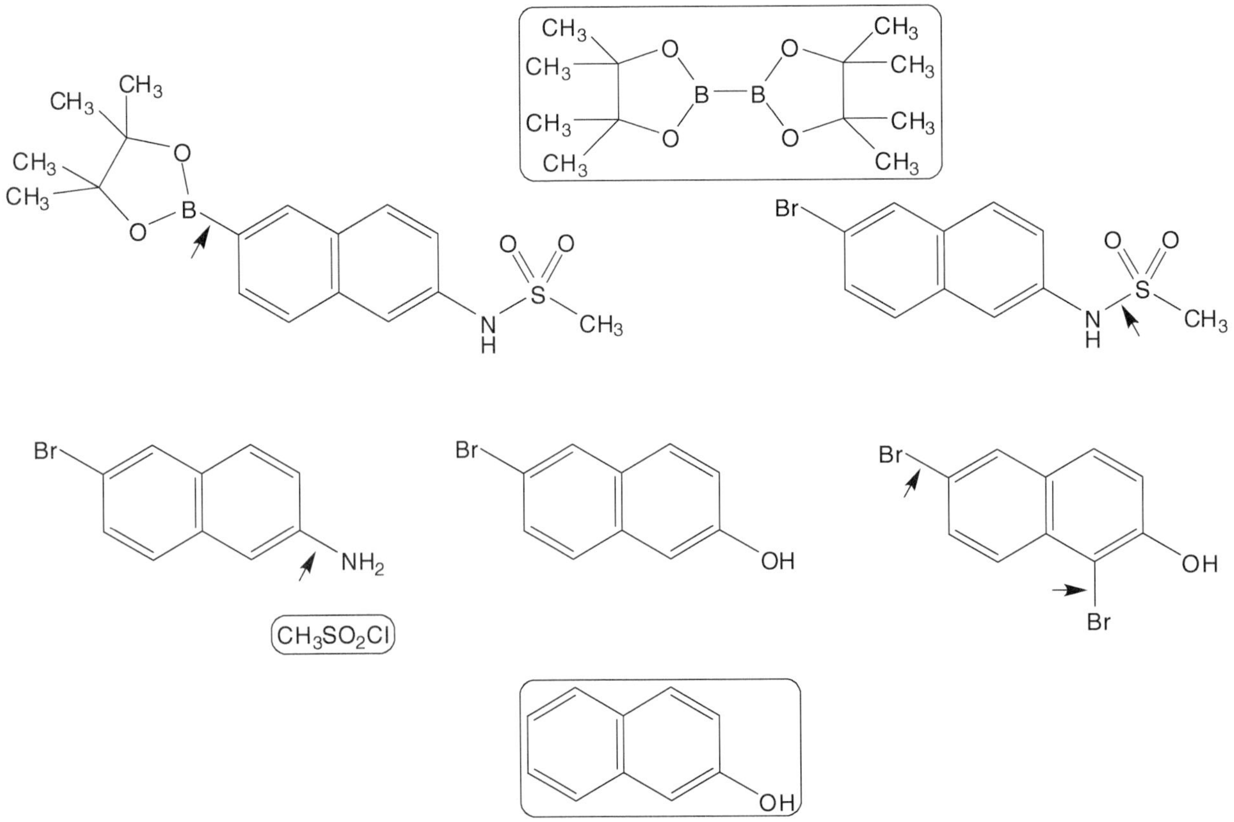

Extended Discussion

Draw the structures of the retrosynthetic analysis of one alternative route to the aryl iodide used in the final **Suzuki–Miyaura Coupling**. Include structures of the retrosynthetic analysis of any organic starting material(s) from petrochemical or biochemical raw materials.

Dasatinib

Antineoplastics and Immunosuppressives/Cytotoxic and Adjuvant Medicines

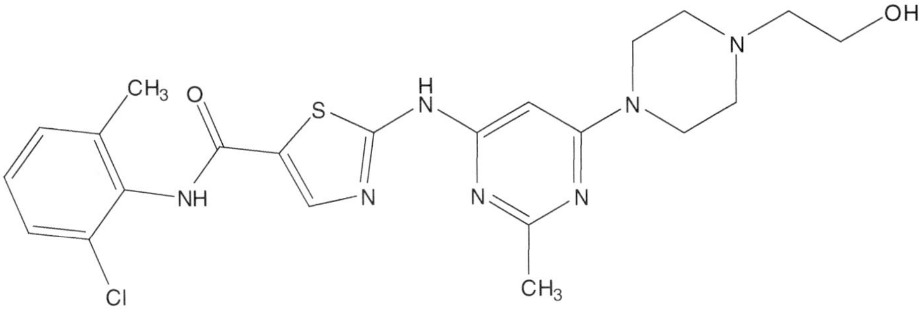

A nitrogen substituent at C4 or C6 of a pyrimidine ring is often introduced by displacement of chloride by an amine.

Discussion. The C-N bond to C6 of the pyrimidine is formed by displacement of chloride by the 2-thiazolamine in the final step. The C-N bond to C4 of the pyrimidine is formed by displacement of chloride by 2-(1-piperazinyl)ethanol. The 4,6-dichloro-2-methylpyrimidine is formed from 2-methylpyrimidine-4,6-dione.

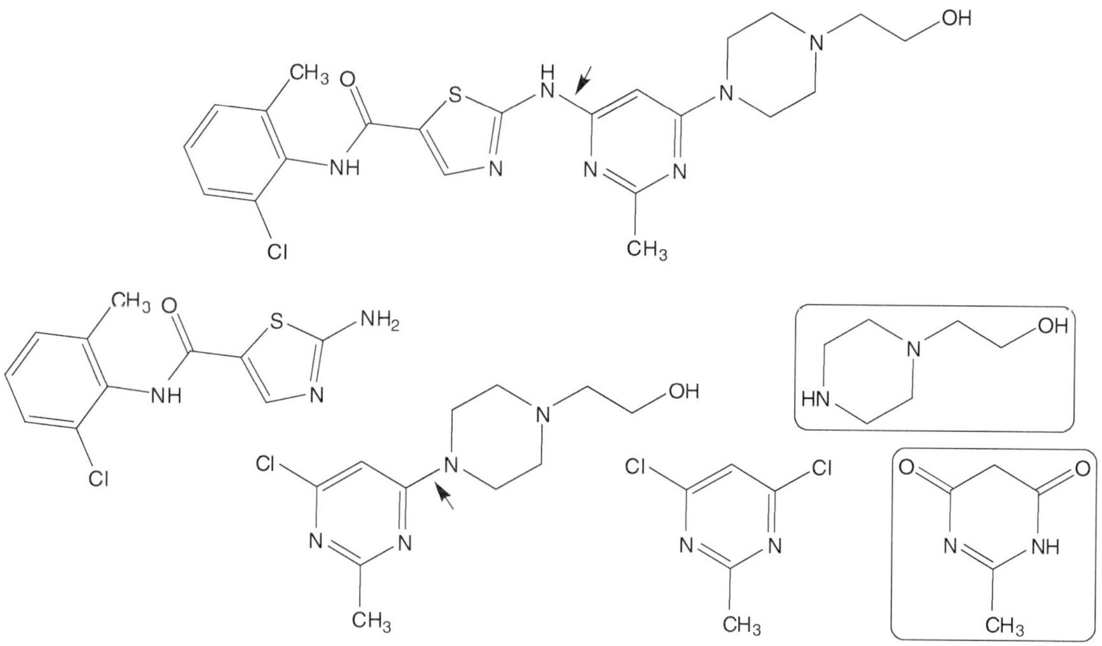

The thiazole ring of the 2-thiazolamine is formed by reaction of thiourea with the 2-chloro-3,3-dimethoxypropanamide (**Hantzsch Thiazole Synthesis**). The 2-chloro-3,3-dimethoxypropanamide is formed by reaction of the 2,3-dichloroacrylamide with methanol. The acrylamide is formed from the amine and the acid chloride. The acid chloride is formed from the carboxylic acid. 2,3-Dichloroacrylic acid is formed from mucochloric acid.

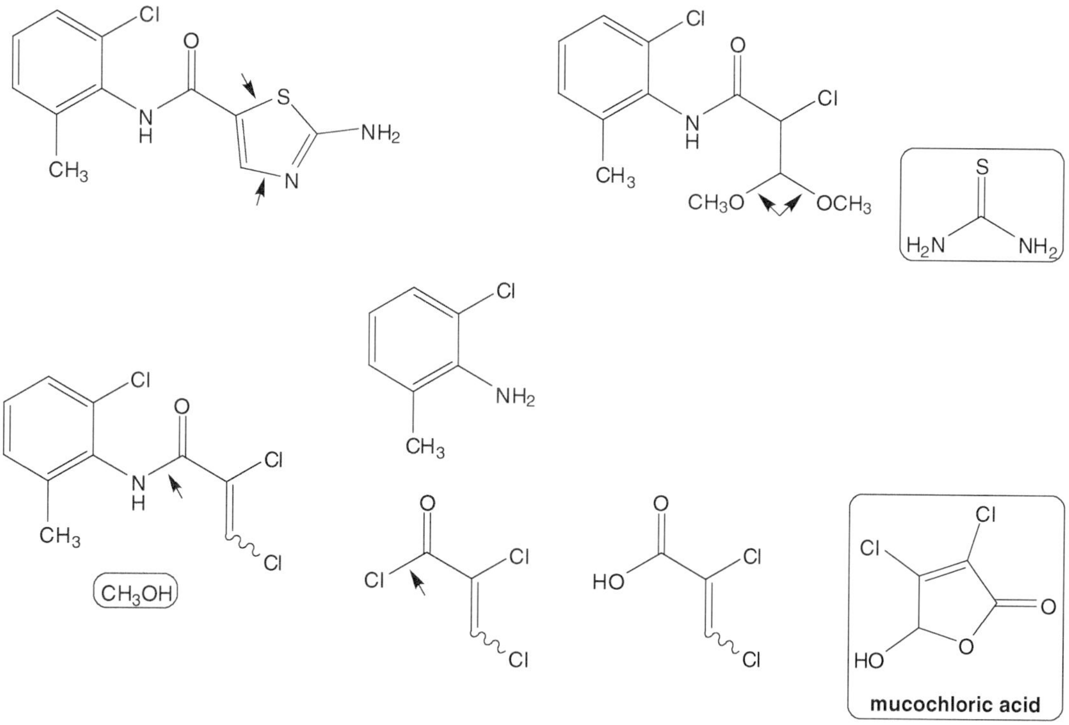

Chlorination at the position *ortho* to the amino group is best accomplished when the position *para* to the amino group is blocked. 2-Chloro-6-methylaniline is produced by removing the blocking sulfonic acid group. *ortho*-Toluidine is converted to the symmetrical urea which is then used in the sulfonation and chlorination.

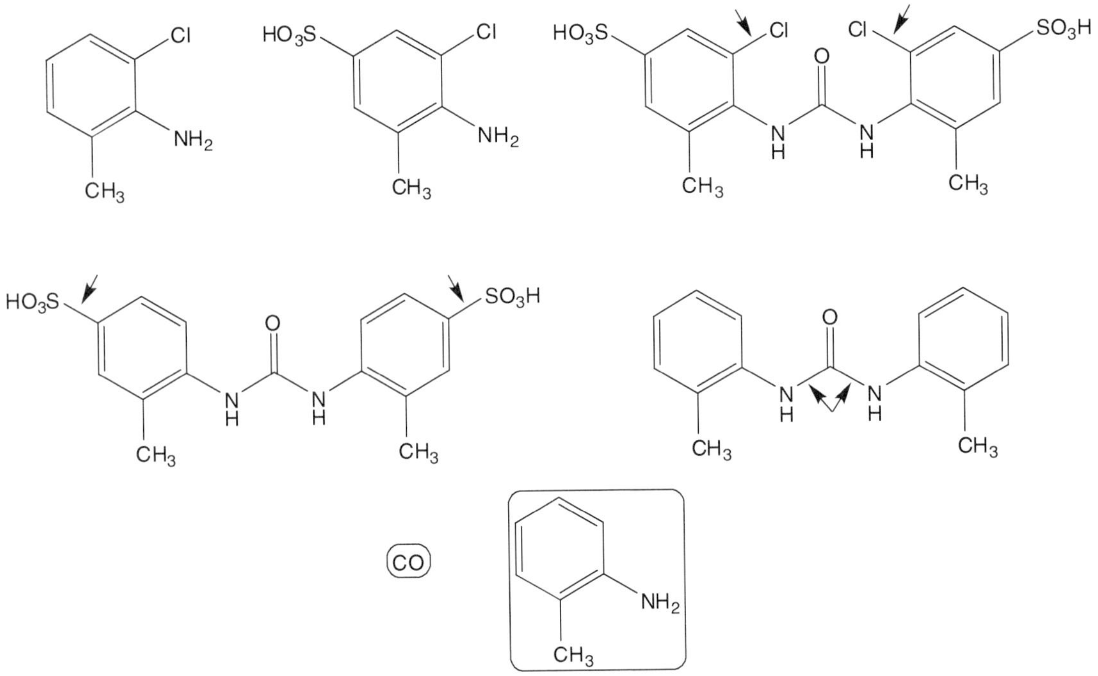

Extended Discussion

Draw the structures of the retrosynthetic analysis of one alternative route to 2-chloro-6-methylaniline. List the pros and cons of both routes and select one route as the preferred route to this intermediate.

Delamanid

Anti-infective Medicines/Antibacterials/Antituberculosis Medicines

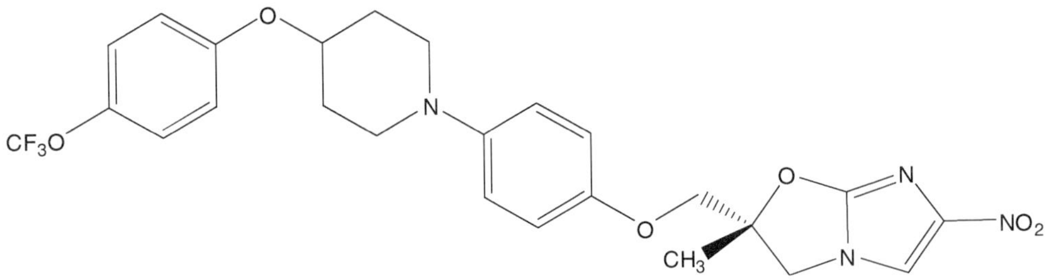

A nitrogen or oxygen substituent on a primary β-C of an alcohol is often introduced by ring-opening of an epoxide.

Discussion. The oxazoline ring is formed by ring-opening of the (*R*)-epoxide with 2-chloro-4-nitroimidazole followed by chloride displacement by the alcohol formed.

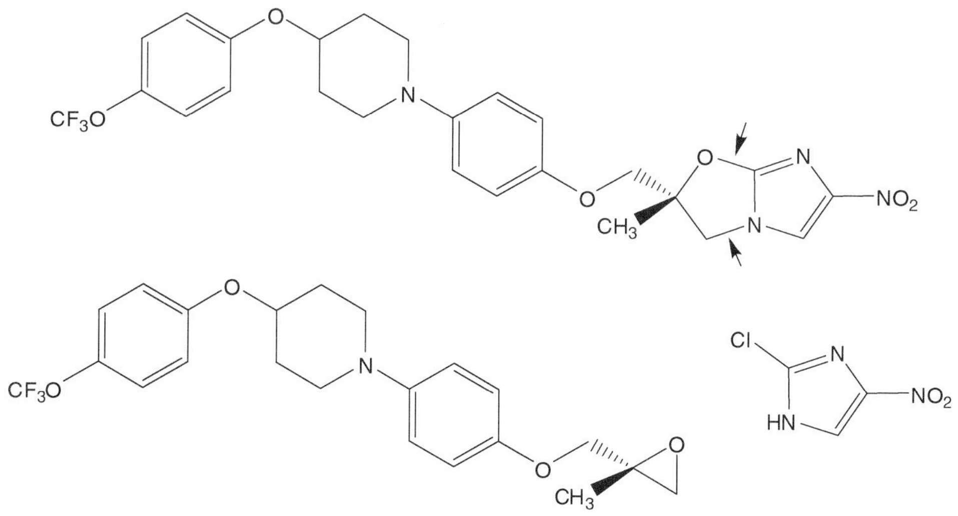

The epoxide is formed by displacement of the methanesulfonate by the alcohol. The methanesulfonate is formed by reaction of the primary alcohol of the diol with methanesulfonyl chloride.

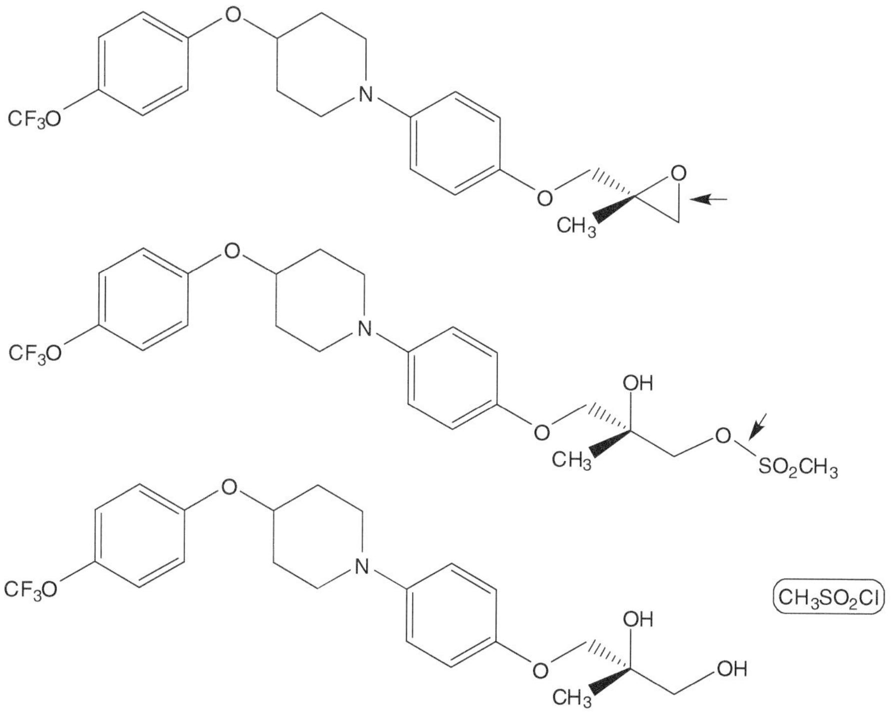

The bond joining nitrogen to the aromatic ring is formed by displacement of bromide by 4-[4-(trifluoromethoxy)phenoxy]piperidine (**Buchwald–Hartwig Amination**). The diol side chain on the aryl bromide is formed by ring-opening of the (*R*)-epoxide with 4-bromophenol. The epoxide is formed by epoxidation of 2-methyl-2-propen-1-ol (**Sharpless Epoxidation**).

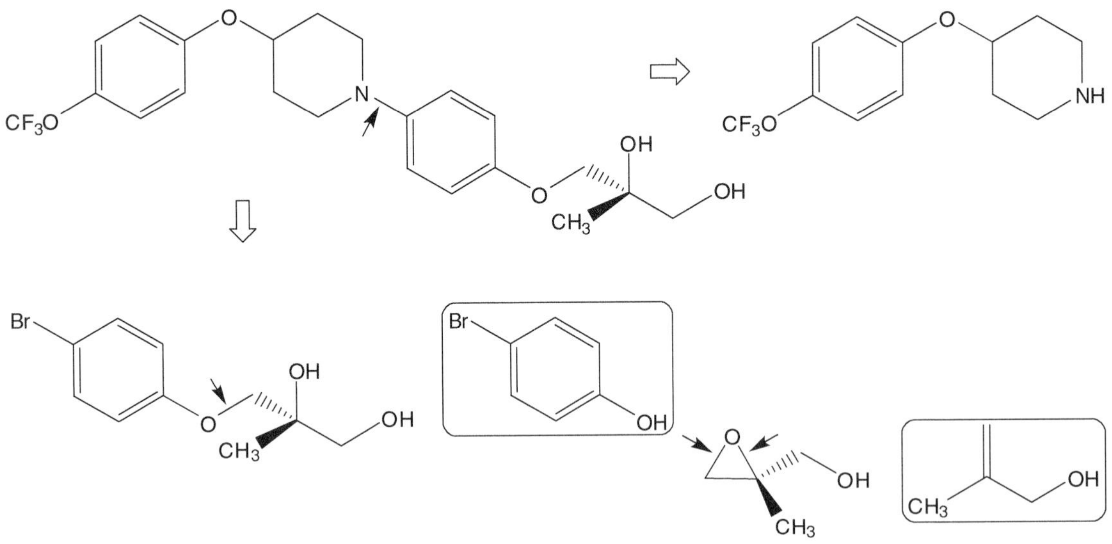

The amine of 4-[4-(trifluoromethoxy)phenoxy]piperidine is released by cleavage of the carbamate. The ether is formed from 4-(trifluoromethoxy)phenol and a methanesulfonate or alcohol (**Mitsunobu Reaction**). The piperidin-4-ol is formed by reduction of the 4-piperidinone. The carbamate is formed by reaction of 4-piperidinone with di-*tert*-butyl dicarbonate.

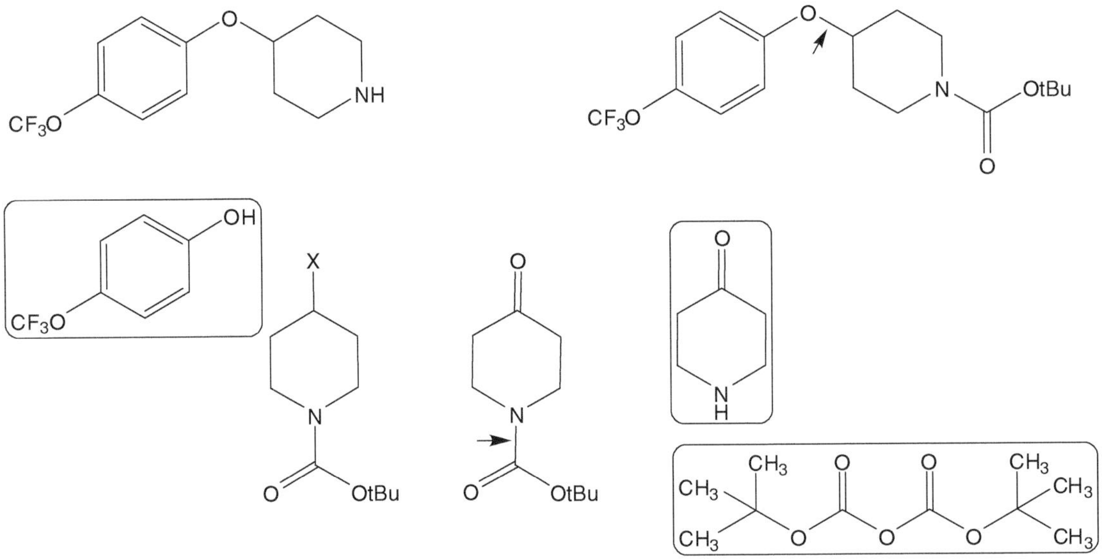

There are several routes to 2-chloro-4-nitroimidazole from imidazole. In a four-step sequence developed by experts in the handing of potentially explosive intermediates, 2-chloro-4-nitroimidazole is formed from 2,4-dinitroimidazole. 2,4-Dinitroimidazole is formed by thermal rearrangement of 1,4-dinitroimidazole. 1,4-Dinitroimidazole is formed by nitration of 4-nitroimidazole. 4-Nitroimidazole is formed by nitration of imidazole.

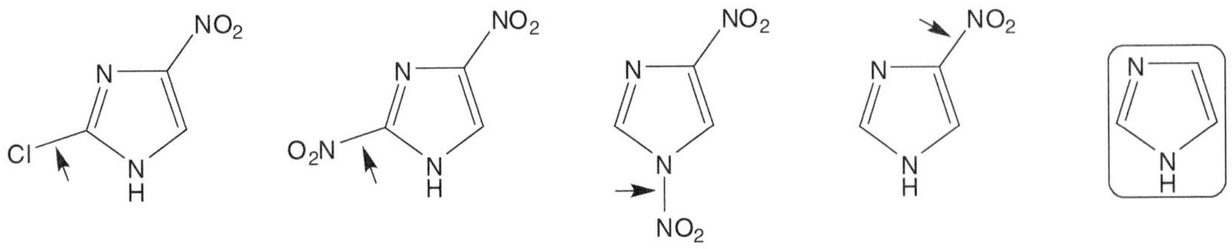

In the alternative five-step sequence, 2-chloro-4-nitro-imidazole is formed from 2-bromo-1-methoxymethyl-4-nitroimid azole. 2-Bromo-1-methoxymethyl-4-nitroimidazole is formed by reduction of the dibromoimidazole. The imidazole N1 is protected by reaction with dimethoxymethane. 2,5-Dibromo-4-nitroimidazole is formed by bromination of 4-nitroimidazole. 4-Nitroimidazole is formed by nitration of imidazole.

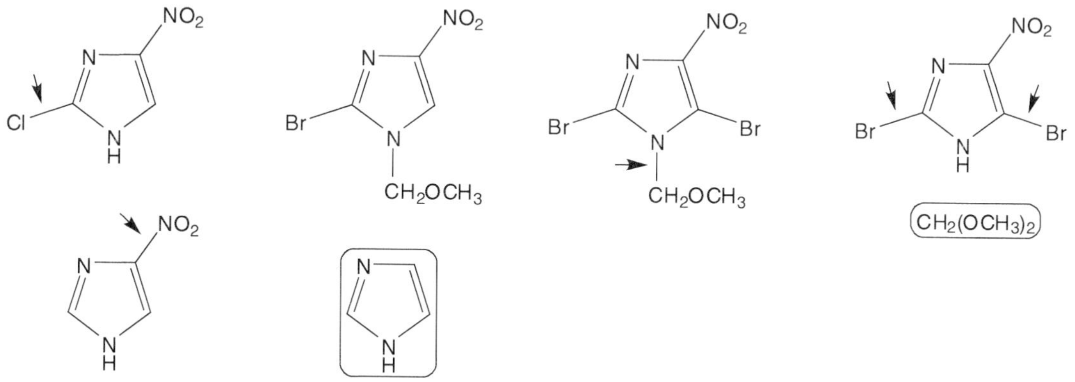

Extended Discussion

List the pros and cons for the routes to the ether via displacement of the methanesulfonate or directly from the alcohol (**Mitsunobu Reaction**). Select one route as the preferred route to this intermediate.

Desmopressin

Medicines Affecting the Blood/Medicines Affecting Coagulation

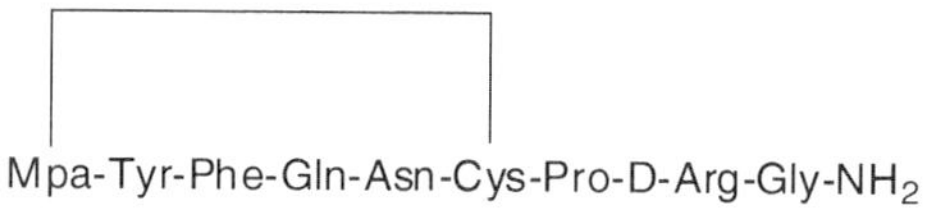

A polypeptide is often constructed from the constituent amino acids by forming the amide bonds. The amide bonds are formed via a solid phase peptide synthesis, a solution phase peptide synthesis, or a hybrid approach using both solid phase and solution phase methods.

Discussion. To simplify polypeptide retrosynthetic schemes, the amino acids are represented by acronyms and are assumed to be in the L-configuration. Protecting groups used in construction of the polypeptide are also represented by acronyms. A disulfide is represented by a bridge connecting the amino acids involved in the bridge. Polypeptide amino acid sequences are drawn with the C-terminal amino acid (carboxylic acid) on the right and the *N*-terminal amino acid (amino group) on the left. (Create a Table of acronyms used in the analysis. Draw the structure associated with each acronym.)

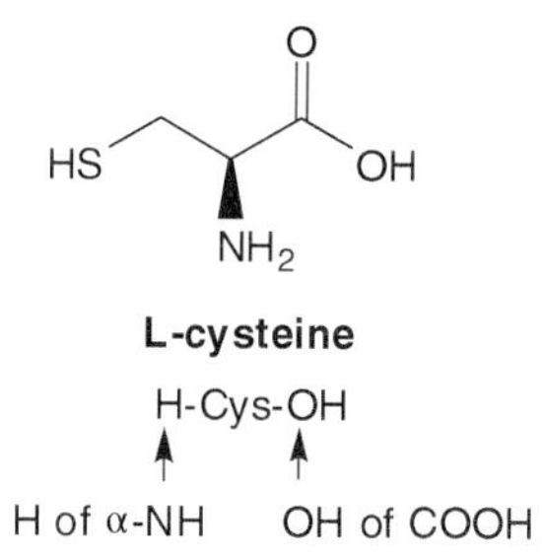

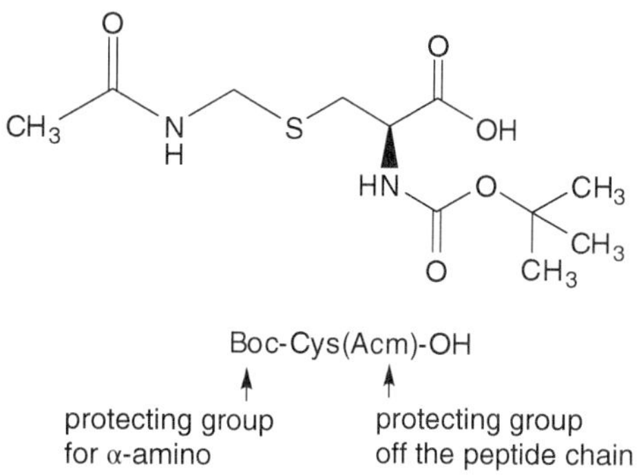

In one solution-phase synthesis, desmopressin is formed from *N-tert*-butoxycarbonyl-D-arginine in 10 steps. The disulfide bridge is formed in the final step. The last amide bond of the nonapeptide chain is formed by reaction of the carboxylic acid of the 1–3 tripeptide (amino acids 1–3 of the chain numbered left-to-right) and amine of the 4–9 hexapeptide.

The 1–3 tripeptide is formed by reaction of the ester Mpa(Acm)-OPFP with the amine of the 2–3 dipeptide H-Tyr-Phe-OH. The pentafluorophenyl (PFP) ester is formed from the carboxylic acid and pentafluorophenol. The 2–3 dipeptide is formed from Z-Tyr-Phe-OBn by release of the amino and carboxylic acid groups by hydrogenolysis. The protected 2–3 dipeptide is formed from Z-Tyr-OH and H-Phe-OBn.

The 4–9 hexapeptide is formed by cleavage of the Boc carbamate to release of the terminal amino group. The Boc-hexapeptide is formed from Boc-Gly-OH and the 5–9 pentapeptide. The 5–9 pentapeptide is formed by cleavage of the Boc carbamate to release of the terminal amino group. The Boc-pentapeptide is formed from the ester Boc-Asn-ONp and the 6–9 tetrapeptide. The ester is formed from Boc-Asn-OH and 4-nitrophenol (Np). The 6–9 tetrapeptide is formed by cleavage of the Boc carbamate to release of the terminal amino group. The Boc-tetrapeptide is formed from the Boc-67 dipeptide and the 89 dipeptide.

The Boc-67 dipeptide is formed by reaction of the ester Boc-Cys(Acm)-OPFP with L-proline. The pentafluorophenyl ester is formed from the carboxylic acid and pentafluorophenol. The 89 dipeptide is formed by cleavage of the Boc carbamate to release of the terminal amino group. The Boc-89 dipeptide is formed from Boc-D-Arg-OH and glycinamide (H-Gly-NH_2).

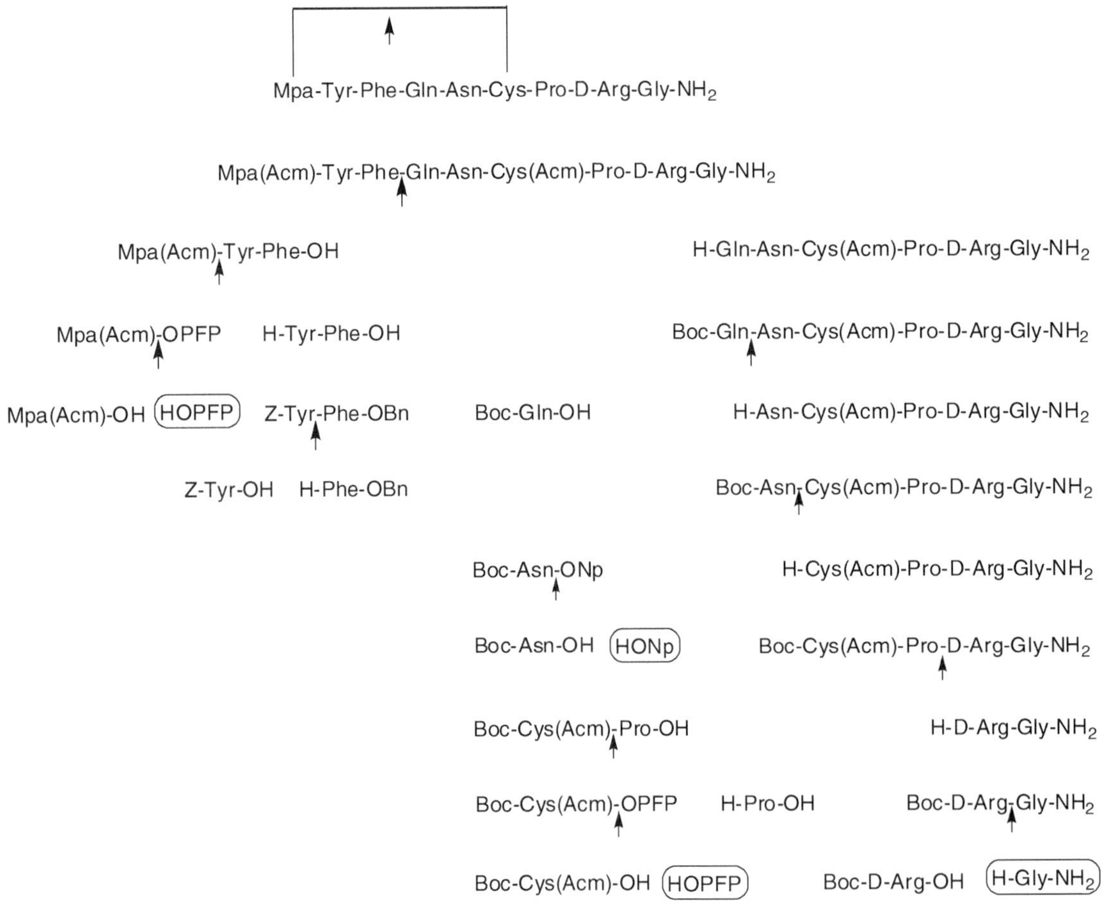

Routes to the protected amino acid starting materials are presented. The amino acids tyrosine, phenylalanine, glutamine, cysteine, proline, and L-arginine are produced by fermentation.

Mpa(Acm)-OH	Mpa	Acm-OH		
Z-Tyr-OH	H-Tyr-OH	Z-Cl		
H-Phe-OBn	H-Phe-OH	Bn-OH		
Boc-Gln-OH	H-Gln-OH	$(Boc)_2O$		
Boc-Asn-OH	H-Asn-OH	$(Boc)_2O$		
Boc-Cys(Acm)-OH	H-Cys(Acm)-OH	$(Boc)_2O$	H-Cys-OH	Acm-OH
Boc-D-Arg-OH	H-D-Arg-OH	$(Boc)_2O$	H-DL-Arg-OH	H-Arg-OH

Extended Discussion

Draw a scheme for the retrosynthetic analysis of one alternative solution-phase synthesis of desmopressin. Add any new acronyms used in the alternative synthesis to the table. List the pros and cons for both of the solution-phase syntheses.

Dexamethasone

Medicines for Pain and Palliative Care/Medicines for Other Common Symptoms in Palliative Care
Antiallergics and Medicines Used in Anaphylaxis
Antineoplastics and Immunosuppressives/Hormones and Antihormones
Gastrointestinal Medicines/Antiemetic Medicines
Specific Medicines for Neonatal Care/Medicines Administered to the Mother

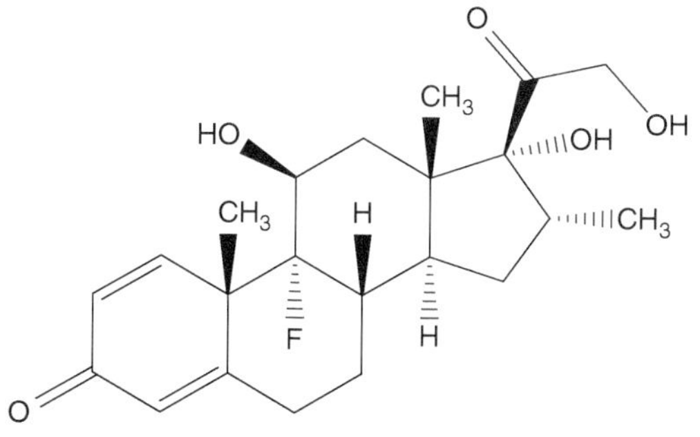

A single-enantiomer molecule with multiple chiral carbons is often formed by modification of a natural product which has most or all of the chiral carbons already in place. A steroid with a 16α-methyl substituent is often formed by copper-catalyzed conjugate addition of a methylmagnesium halide to a 16-ene-20-one.

Discussion. Dexamethasone is manufactured in seven steps from pregna-1,4,9(11),16-tetraene-3,20-dione, 13 steps from 16-dehydropregnenolone acetate, and 16 steps from diosgenin. Diosgenin is a phytosteroid sapogenin isolated from the tubers of *Dioscorea* wild yam.

The C21 alcohol is released by hydrolysis of the acetate ester in the final step. The acetate ester is formed by bromide displacement by acetate. The α-bromoketone is formed by α-bromination of the ketone.

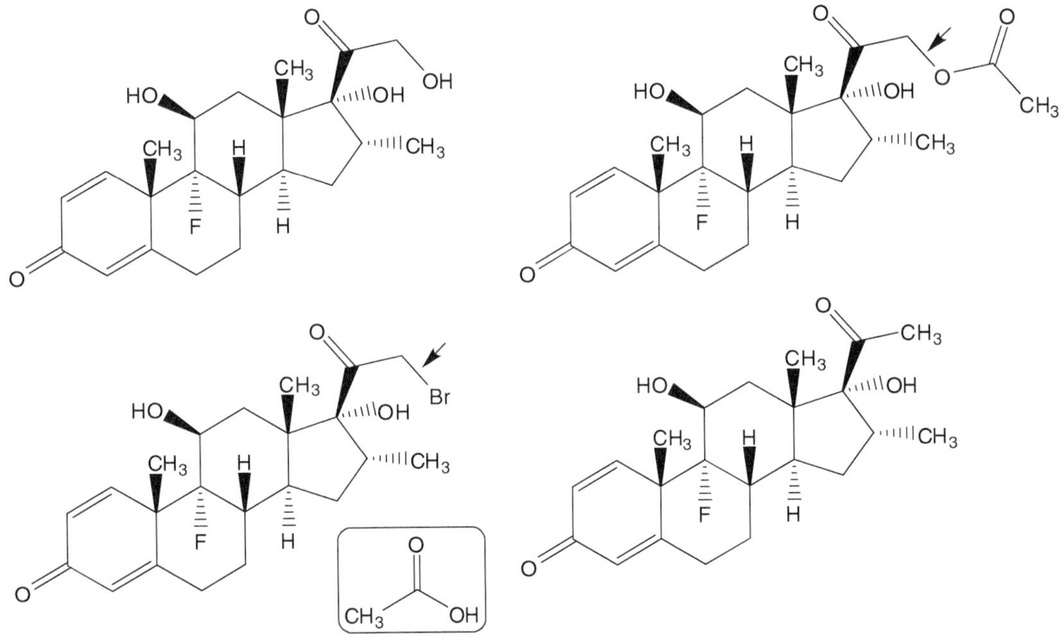

Fluorine is introduced at C9 on the α-face by ring-opening of the 9β,11-epoxide with hydrogen fluoride. The methyl group is introduced at C16 on the α-face by copper-catalyzed conjugate addition of a methylmagnesium bromide to the 16-alkene (**Grignard Reaction**). The α-alcohol at C17 is then formed by α-oxidation of the C20 ketone. Methylmagnesium bromide is formed from bromomethane. The 9β,11-epoxide is formed from the bromohydrin. The 11β-alcohol of the bromohydrin is formed in situ by hydrolysis of the formate ester. The bromohydrin formate is formed from the 9(11)-alkene of pregna-1,4,9(11),16-tetraene-3,20-dione.

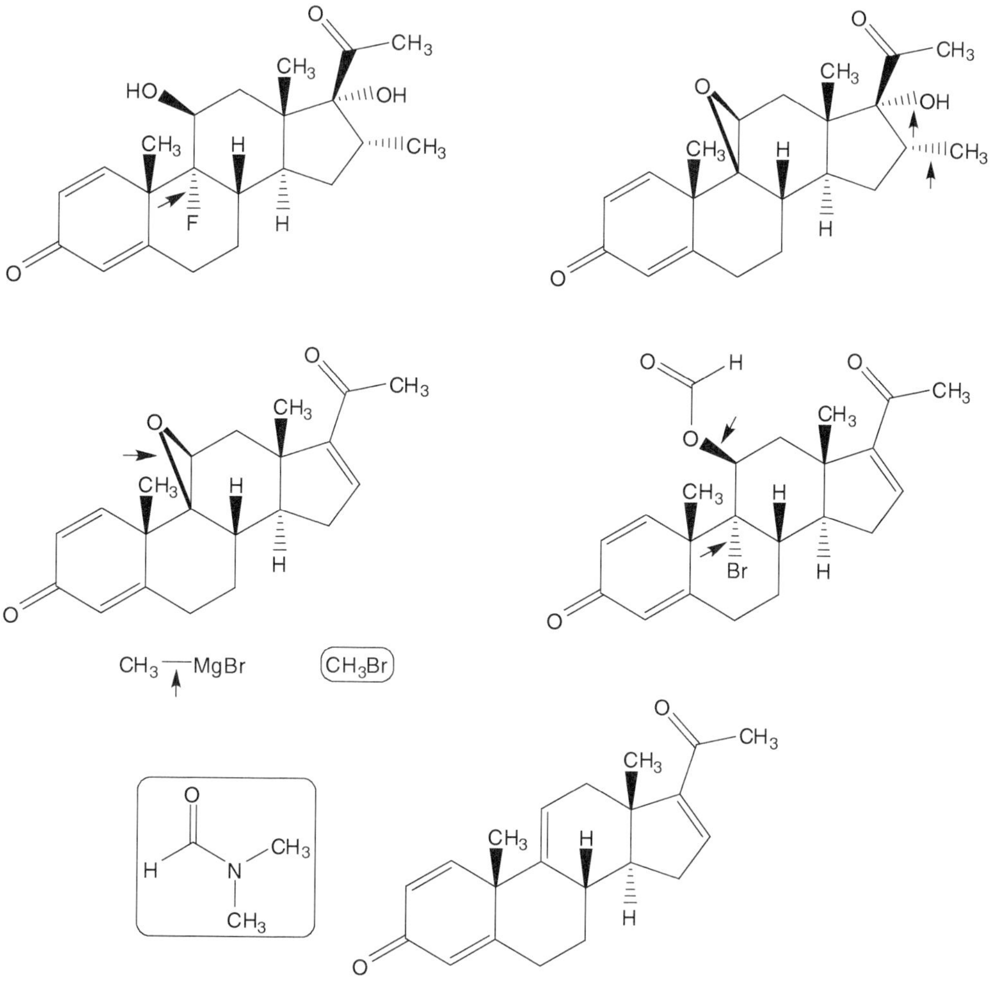

The 9(11)-alkene is formed by elimination of the methanesulfonate formed in situ from the 11α-alcohol. The 16-alkene is released by deoxygenation of the 16α,17-epoxide. The 1,4-diene-3-one is formed by microbial dehydrogenation of the 4-ene-3-one. The 11α-alcohol is formed by microbial oxidation. The 4-ene-3-one is formed by oxidation of the C3 alcohol followed by double bond migration. The 16α,17-epoxide is formed by epoxidation of the 16-alkene of 16-dehydropregnenolone acetate. The acetate ester is hydrolyzed under the epoxidation conditions.

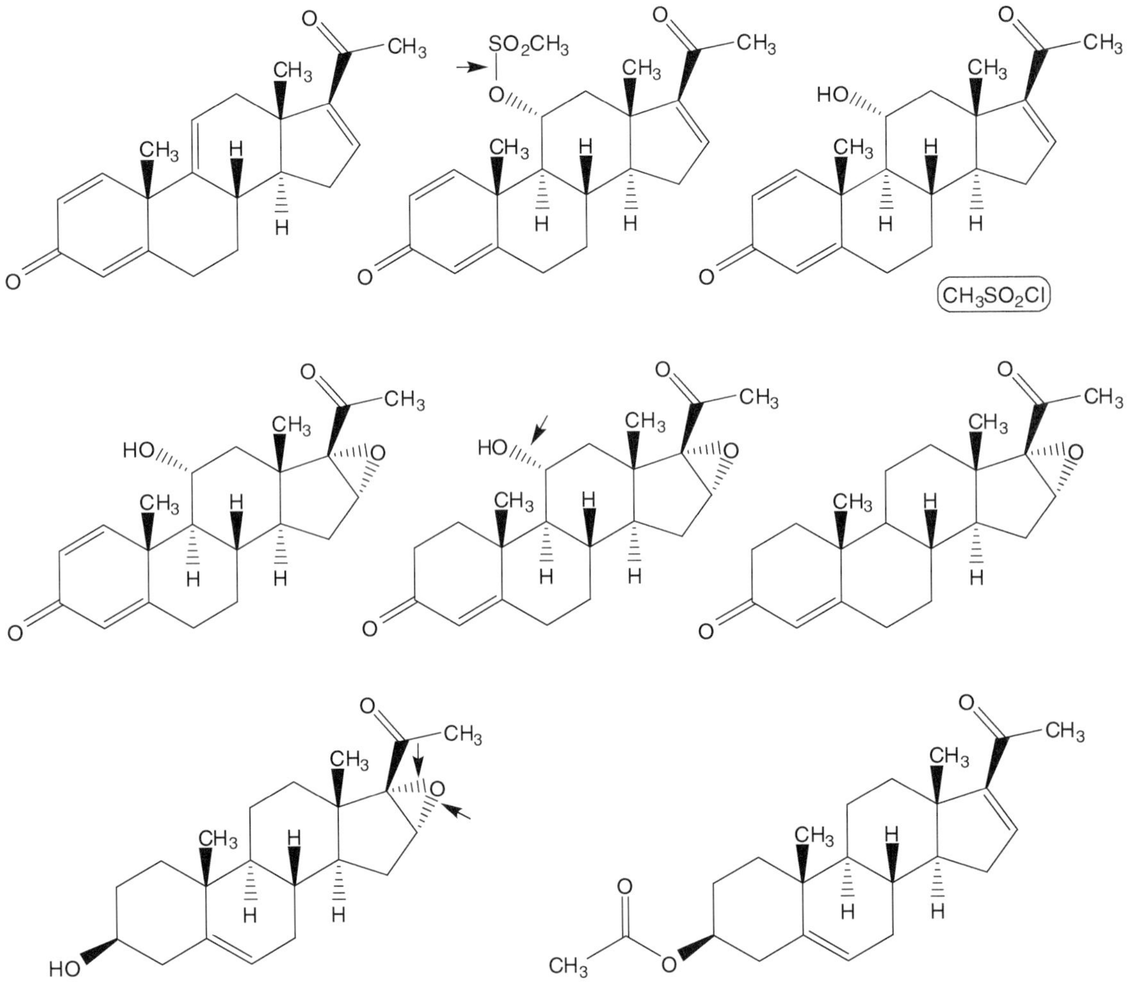

16-dehydropregnenolone acetate

The 16-alkene of 16-dehydropregnenolone acetate is formed from diosone by β-elimination. Diosone is formed by oxidation of the 20(22)-alkene of pseudodiosgenin-3,26-diacetate. Pseudodiosgenin 3,26-diacetate is formed by reaction of diosgenin with acetic anhydride. The three-step synthesis of 16-dehydropregnenolone acetate from diosgenin by acetylation, oxidation, and elimination is known as the **Marker Degradation**.

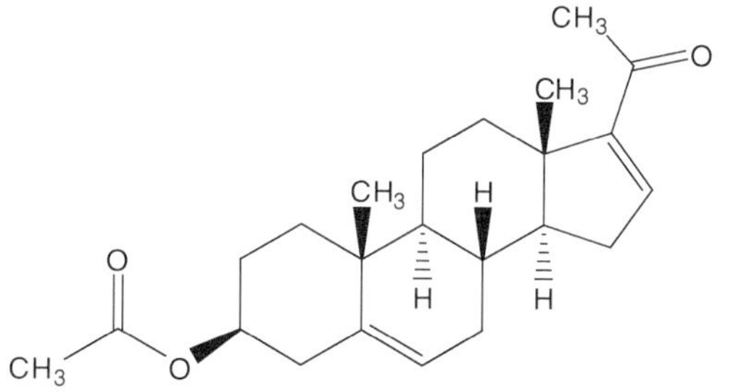

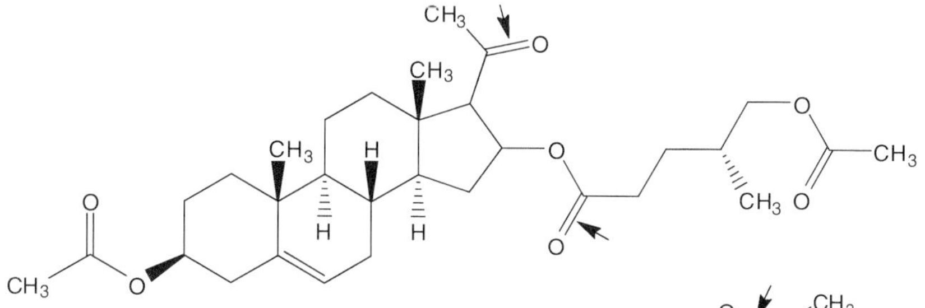

diosone

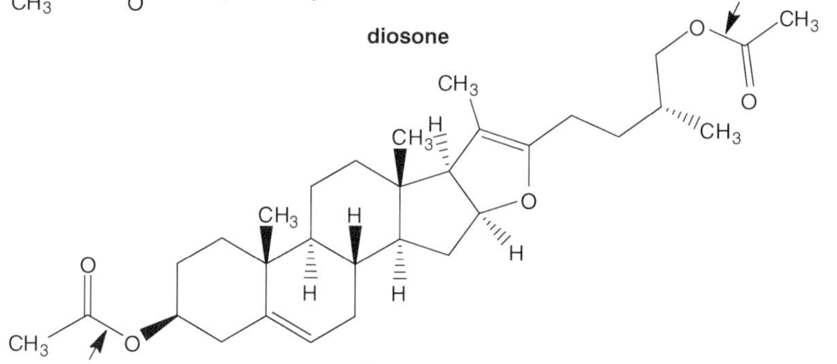

pseudodiosgenin-3,26-diacetate

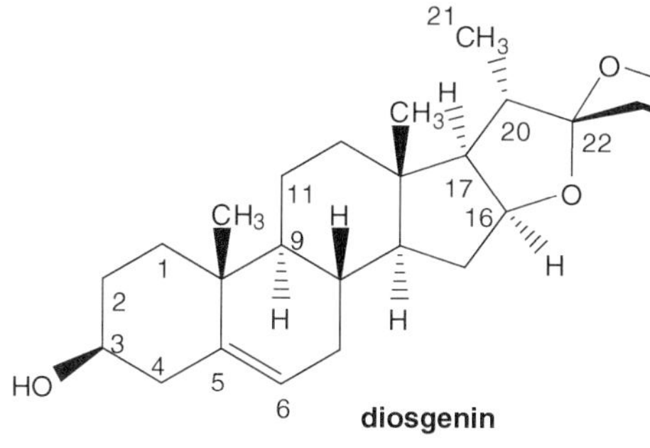

diosgenin

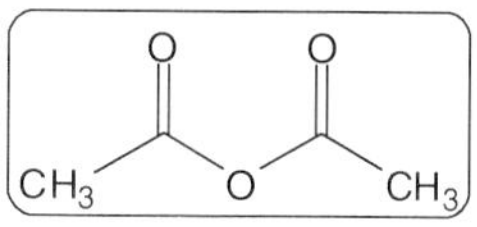

Extended Discussion

Draw the structures of a retrosynthetic analysis of an alternative route to dexamethasone via 3β,17α-dihydroxy-16α-methyl-5α-pregnan-20-one.

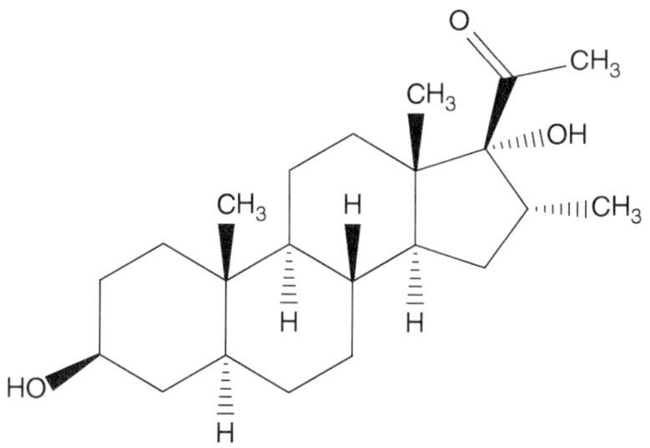

3β,17α-dihydroxy-16α-methyl-5α-pregnan-20-one

Diazepam

Medicines for Pain and Palliative Care/Medicines for Other Common Symptoms in Palliative Care
Anticonvulsants/Antiepileptics
Medicines for Mental and Behavioral Disorders/Medicines for Anxiety Disorders

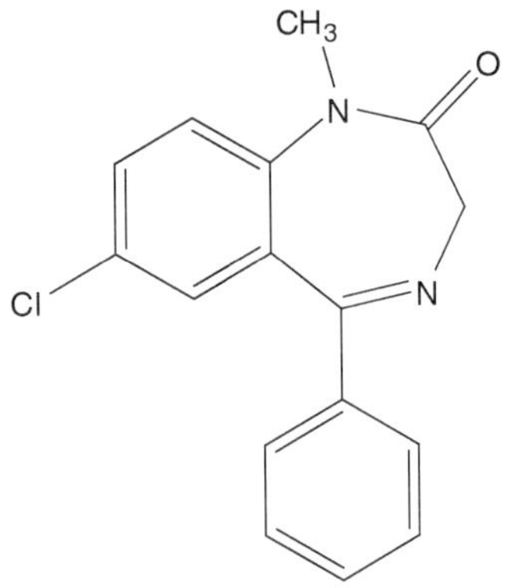

A 5-aryl-1,3-dihydro-2*H*-1,4-benzodiazepin-2-one is often formed in two steps from a 2-aminobenzophenone, chloroacetyl chloride, and ammonia/hexamethylenetetramine.

Discussion. The diazepine ring is closed in the final step by imine formation. The amine is formed by displacement of chloride from the α-chloroamide by ammonia. The α-chloroamide is formed from chloroacetyl chloride and 5-chloro-2-methylaminobenzophen one. 5-Chloro-2-methylaminobenzophenone is formed by *N*-methylation of 2-amino-5-chlorobenzophenone. (List the methylating agents, reaction conditions, and yields.)

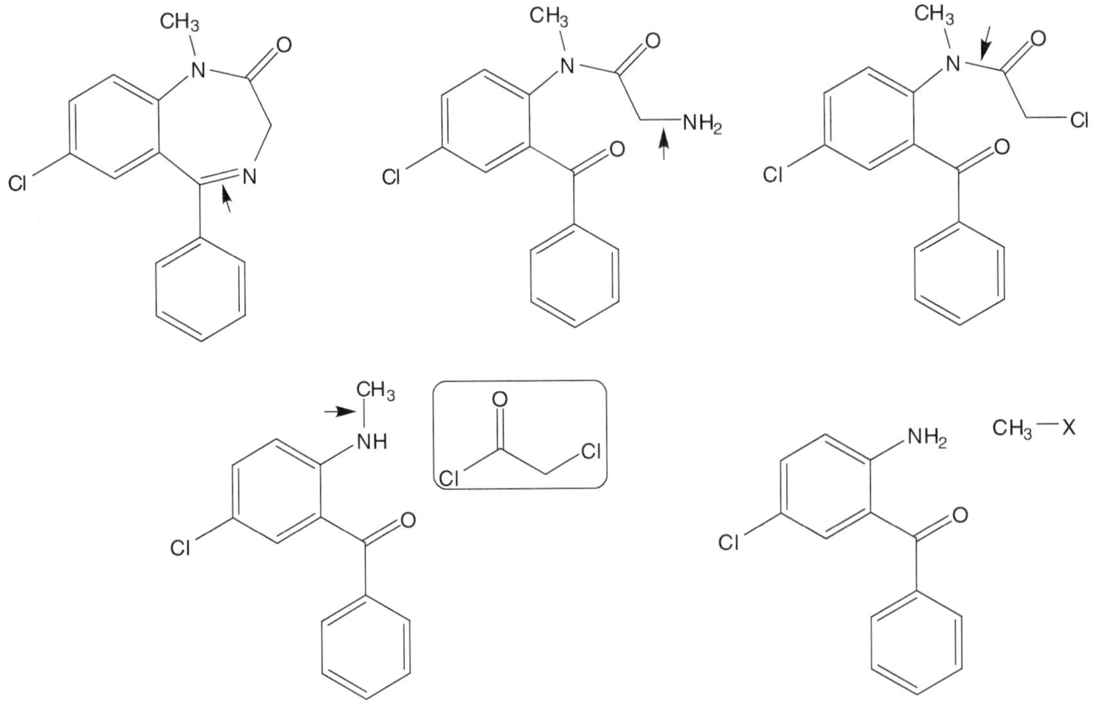

Three routes to 2-amino-5-chlorobenzophenone are well-known. In Route A, the benzophenone is formed from 4-chloroaniline and benzoyl chloride (**Friedel–Crafts Acylation**). In Route B, the benzophenone is formed from 4-chloroaniline and benzonitrile (**Sugasawa Reaction**). In Route C, the benzophenone is formed by reduction of 5-chloro-3-phenyl-2,1-benzisoxazole (5-chloro-3-phenylanthranil). 5-Chloro-3-phenylanthranil is formed from 4-chloronitrobenzene and phenylacetonitrile.

Route A

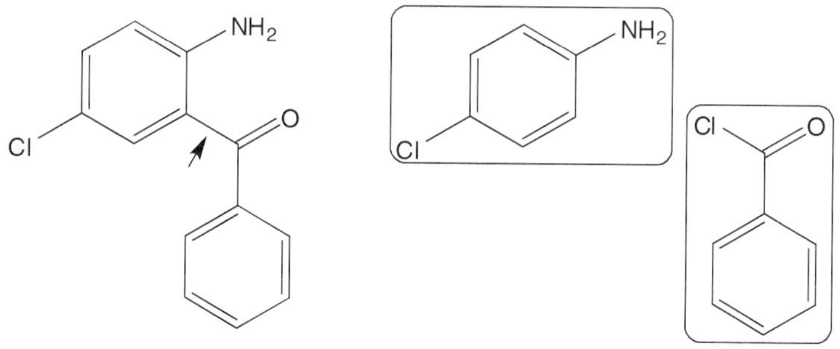

Route B

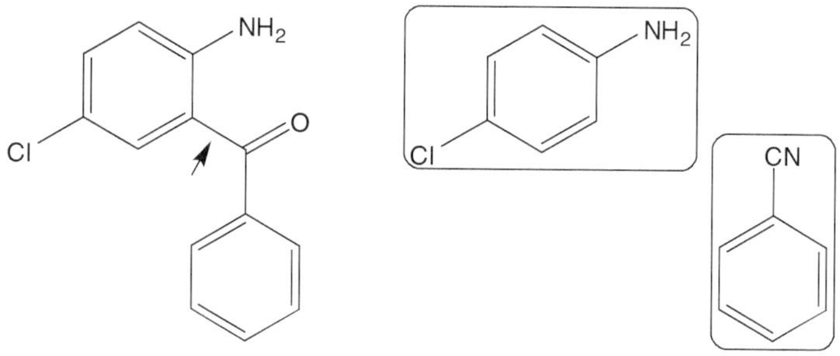

Route C

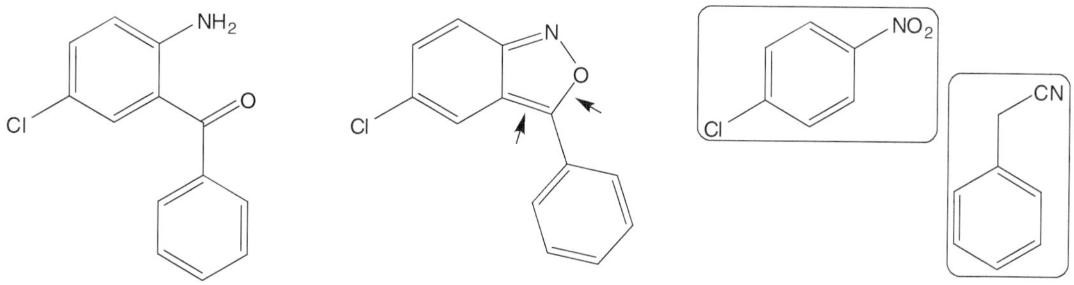

Extended Discussion

List the pros and cons for the three routes to 2-amino-5-chlorobenzophenone. Is one route preferred?

Diethylcarbamazine

Anti-infective Medicines/Anthelminthics/Antifilarials

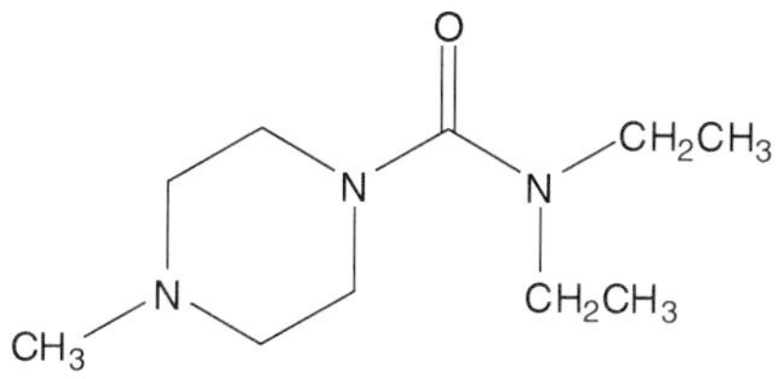

An unsymmetrical urea is often formed by the reaction of an amine with a carbamoyl chloride. The carbamoyl chloride is formed by the reaction of an amine with phosgene.

Discussion. The unsymmetrical urea is formed from 1-methylpiperazine and *N,N*-diethylcarbamoyl chloride (Route A) or from *N,N*-diethylamine and 4-methylpiperazinecarbonyl chloride (Route B). The carbamoyl chlorides are formed from the amines and phosgene.

Route A

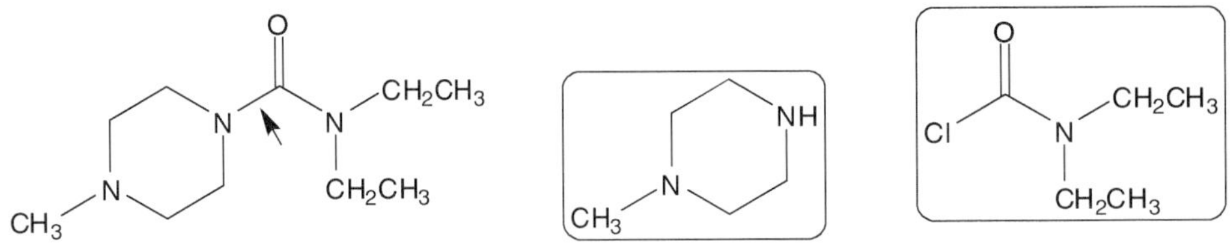

Route B

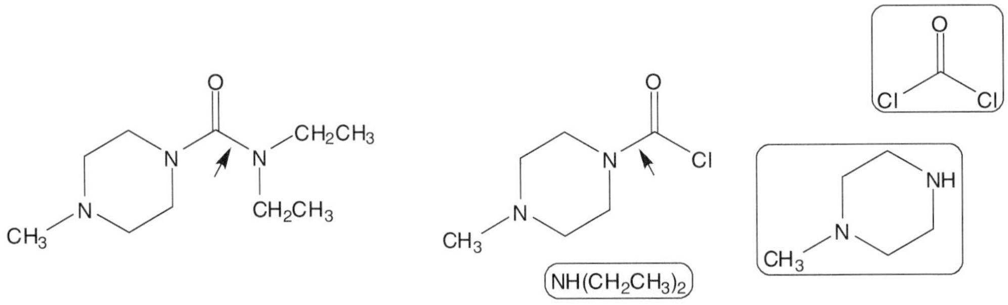

Diethylcarbamazine is formed by reductive methylation of *N,N*-diethylpiperazine-1-carboxamide in the final step of Route C. The carbamate is formed from piperazine and *N,N*-diethylcarbamoyl chloride.

Route C

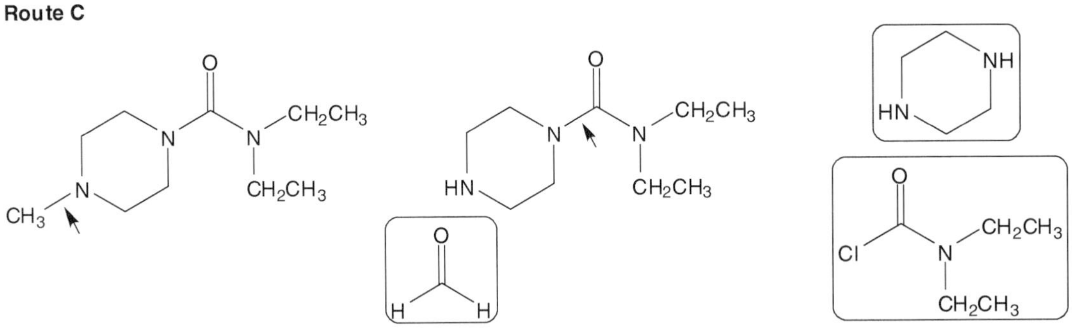

Extended Discussion

List the pros and cons for the three routes and select one route as the preferred route.

Dihydroartemisinin

Anti-infective Medicines/Antiprotozoal Medicines/Antimalarial Medicines/For Curative Treatment

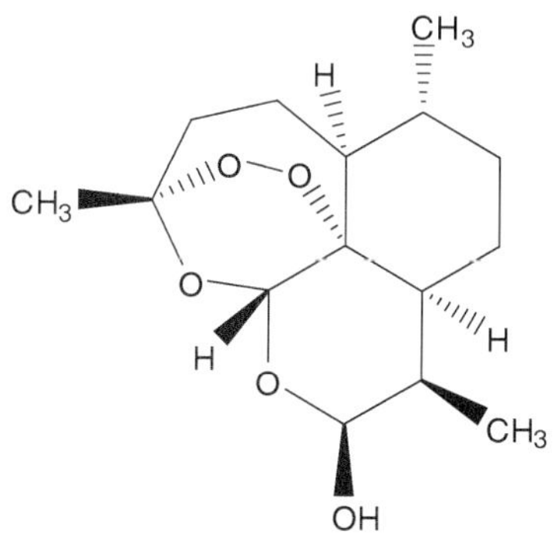

A cyclic hemiacetal can be formed by the reduction of the carbonyl carbon of a lactone. If the hemiacetal has a propensity to ring-open, further reduction of the aldehyde to a primary alcohol is likely.

Discussion. Dihydroartemisinin (artenimol or DHA) is semisynthetic. It is manufactured by reduction of artemisinin. Artemisinin is a natural product isolated from the plant *Artemisia annua* or sweet wormwood.

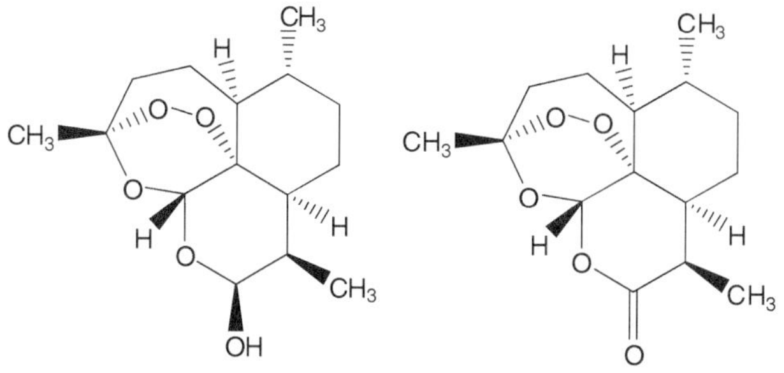

dihydroartemisinin **artemisinin**

An alternative supply of artemisinin is manufactured in four steps from artemisinic acid. In the last step, the aldehyde, hydroperoxide, ketone, and carboxylic acid assemble to form artemisinin. The hydroperoxide is formed in situ by reaction of the aldehyde with triplet oxygen. The aldehyde and ketone are formed by cleavage of an allylic hydroperoxide (**Hock Rearrangement**). The allylic hydroperoxide is formed from the alkene (**Ene Reaction**). The alkene, dihydroartemisinic acid, is formed by reduction of artemisinic acid. Artemisinic acid is a natural product also isolated from the plant *A. annua* or sweet wormwood. An alternative supply of artemisinic acid is produced from sugar by yeast fermentation.

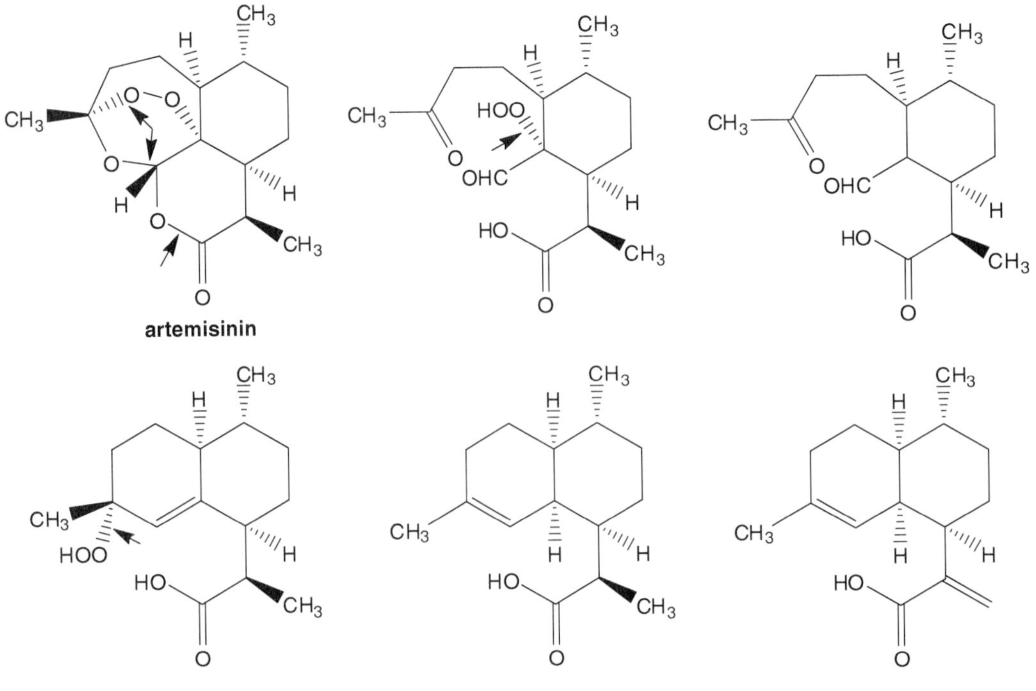

Extended Discussion

Dihydroartemisinin is usually isolated from methanol–water. Drying the wet solid under vacuum at 15–25 °C affords dihydroartemisinin of very high purity (>99.9%) but some degradation is observed when the drying temperature is increased to 50 °C. Draw the structures of two degradation products.

Diloxanide Furoate

Anti-infective Medicines/Antiprotozoal Medicines/Antiamoebic and Antigiardiasis Medicines

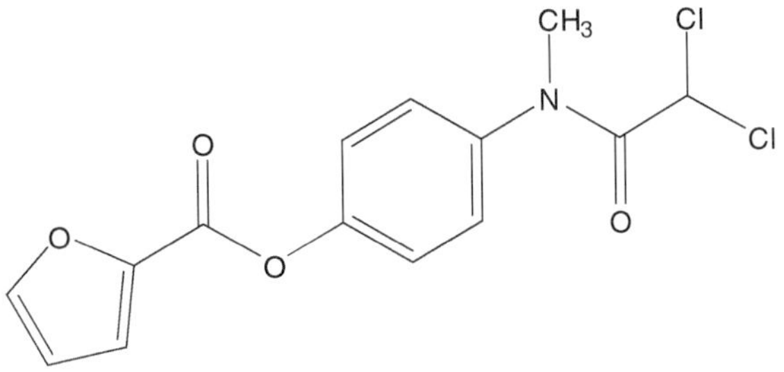

Amides are ubiquitous in drug structures. Amide formation from an amine and acid chloride, anhydride, ester, or carboxylic acid is often very efficient.

Discussion. There are just two steps in the synthesis of diloxanide furoate. The ester is formed in the final step from 2-furoyl chloride and the phenol. The acid chloride is formed from 2-furoic acid. The amide is formed from 4-methylaminophenol (the hydrogen sulfate salt is called metol) and dichloroacetyl chloride.

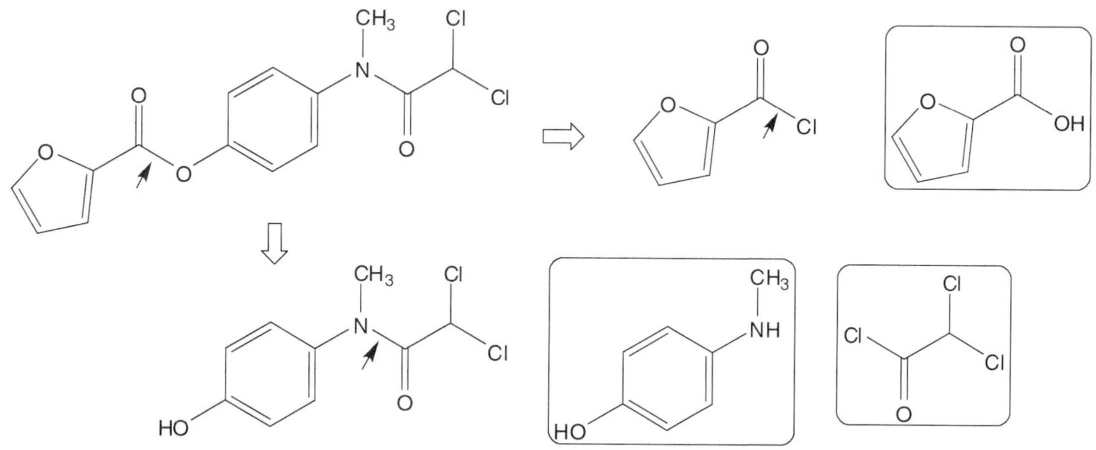

Dimercaprol

Antidotes and Other Substances Used in Poisonings/Specific

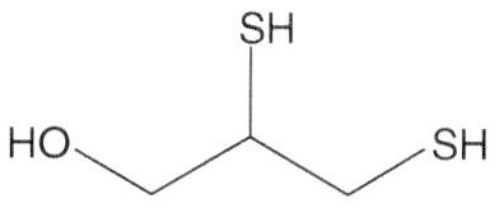

Thiols are often formed by displacement of a leaving group (Cl, Br, I, OTs, OMs) by hydrogen sulfide anion.

Discussion. Dimercaprol (also known as British anti-Lewisite) is a 1 : 1 mixture of (*R*)- and (*S*)-enantiomers. The 1,2-dithiol is formed from 2,3-dibromo-1-propanol by bromide displacement with hydrogen sulfide.

Docetaxel

Antineoplastics and Immunosuppressives/Cytotoxic and Adjuvant Medicines

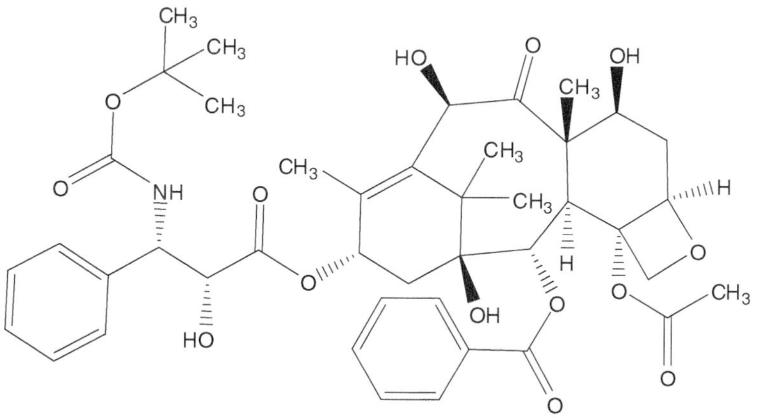

A single-enantiomer molecule with multiple chiral carbons is often formed by modification of a natural product which has most or all of the chiral carbons already in place. Taxanes are often formed by modification of 10-deacetylbaccatin III. 10-Deacetylbaccatin III is a diterpenoid isolated from the leaves of the European yew *Taxus baccata*.

Discussion. Docetaxel is semisynthetic. The final step is release of the secondary alcohols at C7 and C10 by cleavage of the trichloroethyl carbonates. (List other alcohol protecting groups used in alternative routes to docetaxel. How are the C7 and C10 alcohols released using the alternative protecting groups?) The alcohol at C2′ and carbamate NH at C3′ are released from the oxazolidine.

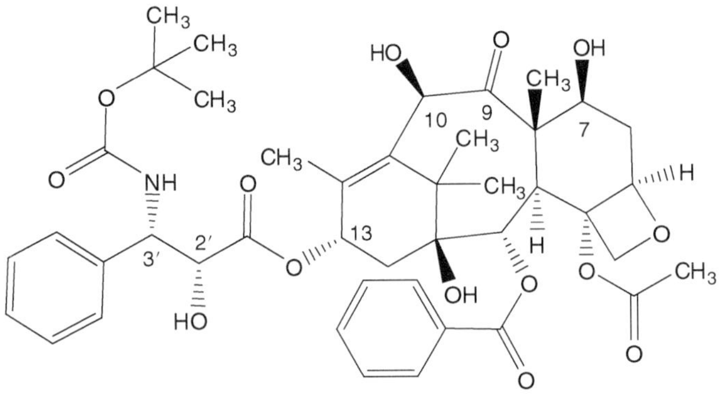

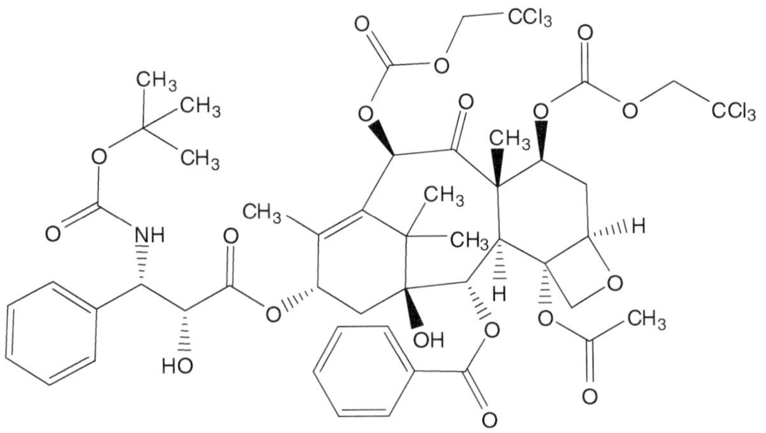

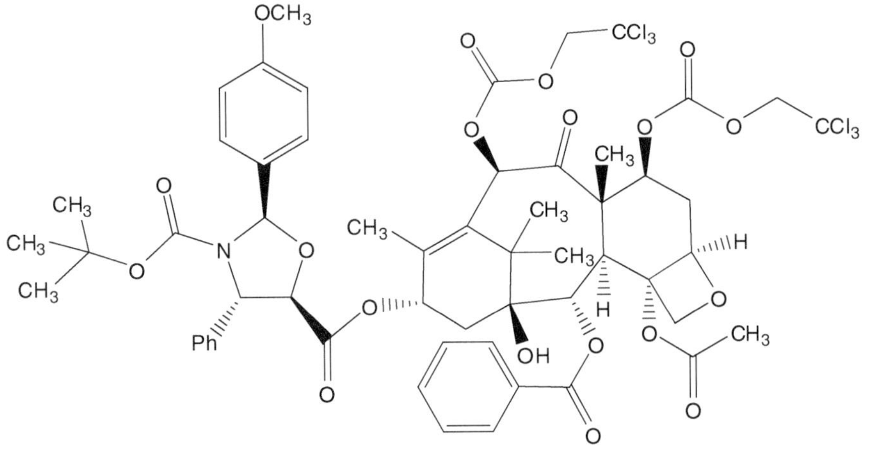

The ester at C13 is formed from the carboxylic acid and the secondary alcohol. The C7 and C10 secondary alcohols of 10-deacetylbaccatin III (10-DAB) are protected as trichloroethyl carbonates (troc). 10-Deacetylbaccatin III is a diterpenoid isolated from the leaves of the European yew *T. baccata*.

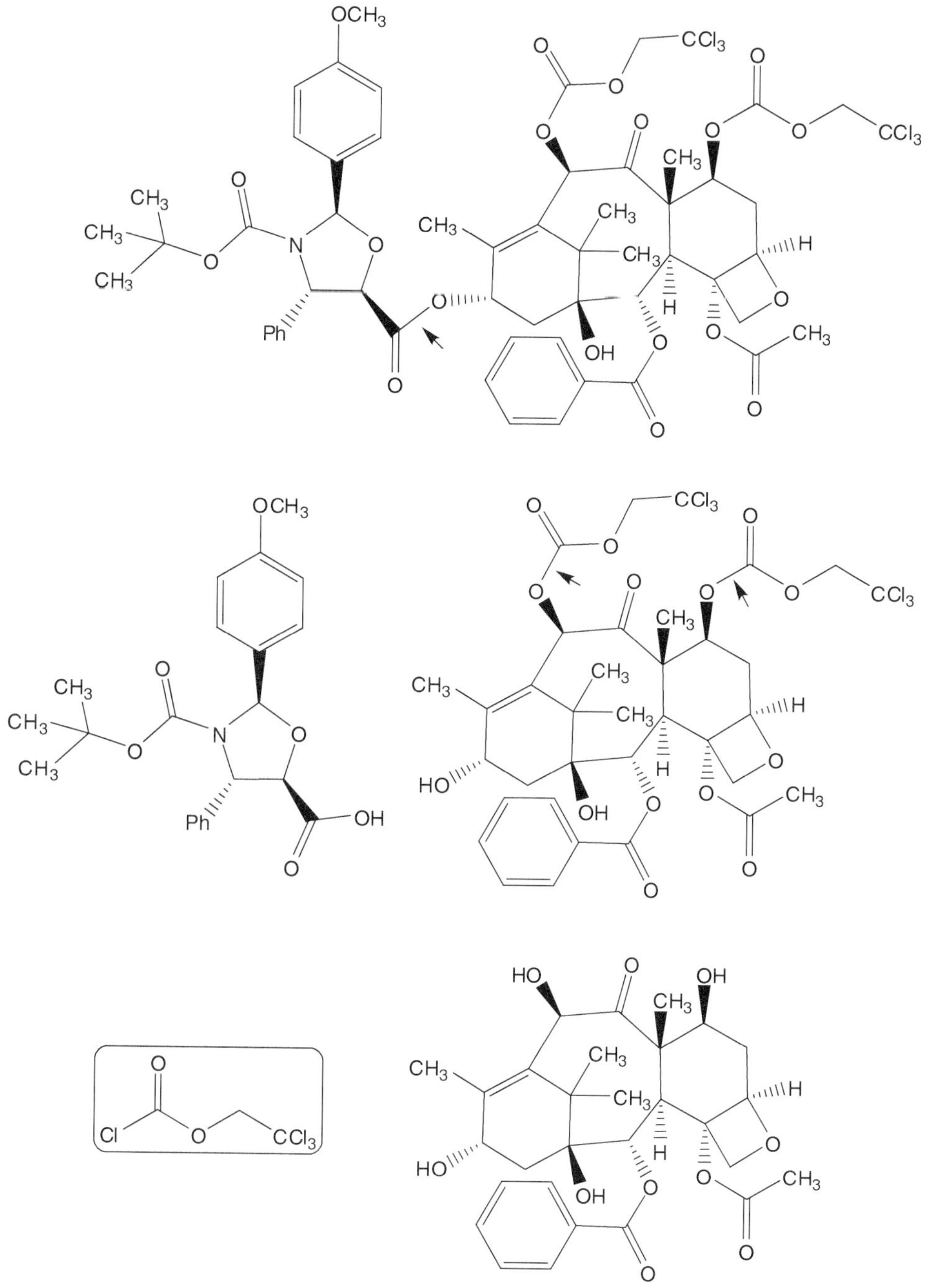

10-deacetylbaccatin III

The carboxylic acid used in formation of the C13 ester is formed by hydrolysis of the methyl ester. The oxazolidine is formed from (2*R*,3*S*)-*N*-(*tert*-butoxycarbonyl)-3-phenylisoserine methyl ester and *para*-anisaldehyde. (List other aldehydes and ketones used to form oxazolidines in alternative routes to docetaxel. How are the alcohol at C2′ and carbamate NH at C3′ released from these alternative oxazolidines?) The carbamate is formed by reaction of (2*R*,3*S*)-3-phenylisoserine methyl ester with di-*tert*-butyl dicarbonate.

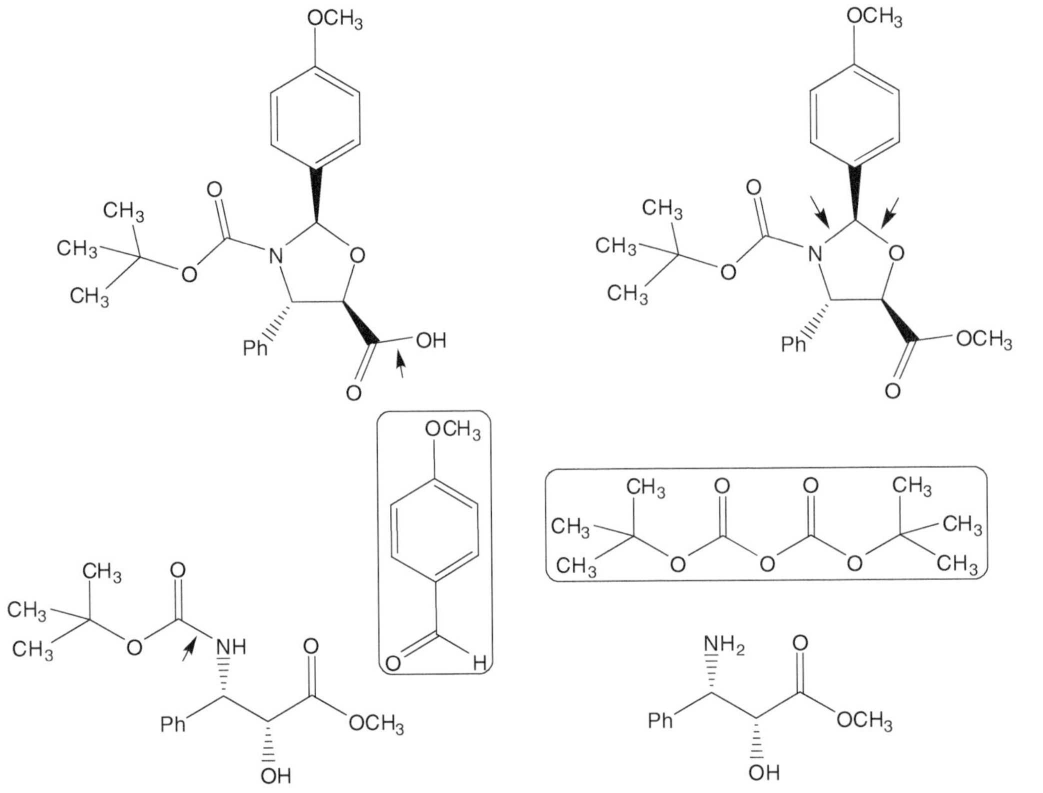

In one preferred route to (2*R*,3*S*)-3-phenylisoserine methyl ester, the ester is formed from the amide. The (2*R*,3*S*)- and (2*S*,3*R*)-3-phenylisoserine amides are separated by resolution. The β-amino alcohol is formed by ring-opening of the epoxide of methyl *cis*-3-phenylglycidate with ammonia. The amide is formed by the reaction of the methyl ester with ammonia in the same pot. The epoxide is formed from the *threo*-bromohydrin. The *threo*-bromohydrin is formed by the ring-opening of the epoxide of methyl *trans*-3-phenylglycidate by hydrogen bromide. Methyl *trans*-3-phenylglycidate is formed by condensation of benzaldehyde and methyl chloroacetate (**Darzens Reaction**). (Methyl *cis*-3-phenylglycidate is a side product of this condensation. How much of the *cis*-side product is present in the crude methyl 3-phenylglycidate? What is the fate of this side product in the next step?)

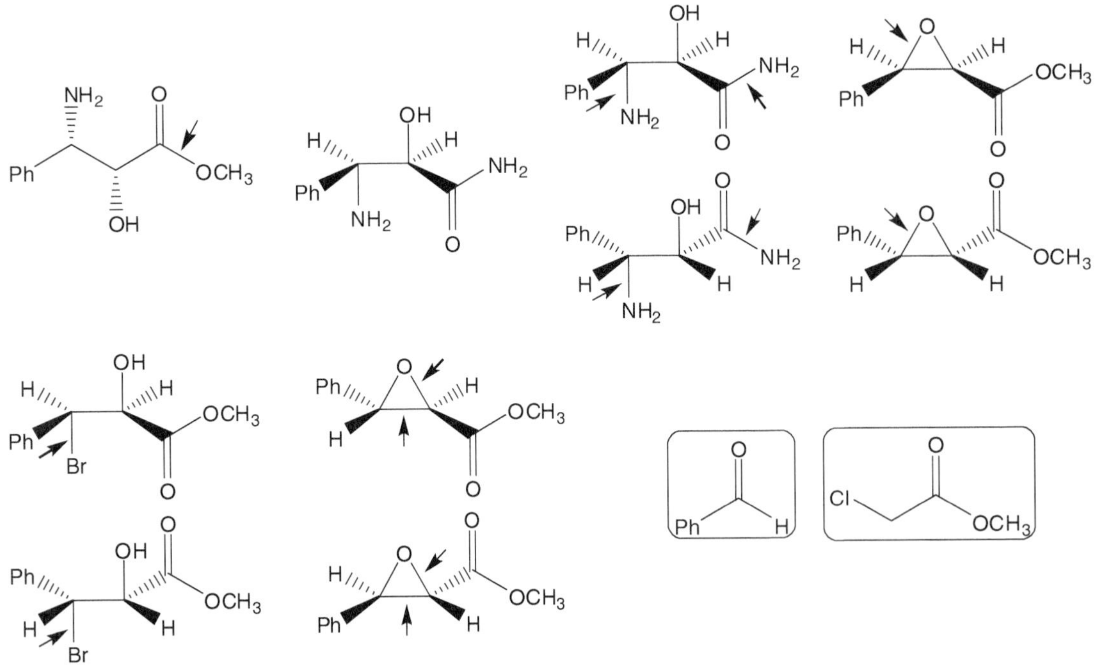

Extended Discussion

Draw the structures of the retrosynthetic analysis of one alternative route to (2*R*,3*S*)-3-phenylisoserine methyl ester utilizing a **Sharpless Dihydroxylation**. Include the structures of the retrosynthetic analysis of any organic starting material(s) from petrochemical or biochemical raw materials. List the pros and cons for both routes.

Docusate Sodium

Medicines for Pain and Palliative Care/Medicines for Other Common Symptoms in Palliative Care

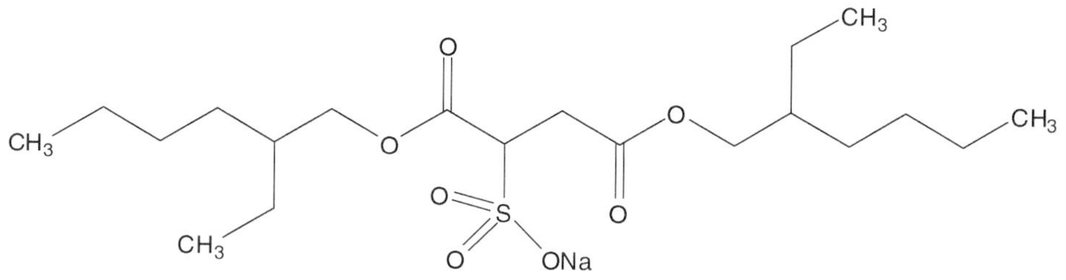

A sulfonic acid is often formed by conjugate addition of sodium bisulfite to an acrylate ester.

Discussion. The sulfonate of docusate (dioctylsulfosuccinate) is formed in the final step by sodium bisulfite addition to the maleate diester. The maleate diester is formed by reaction of maleic anhydride with 2-ethyl-1-hexanol.

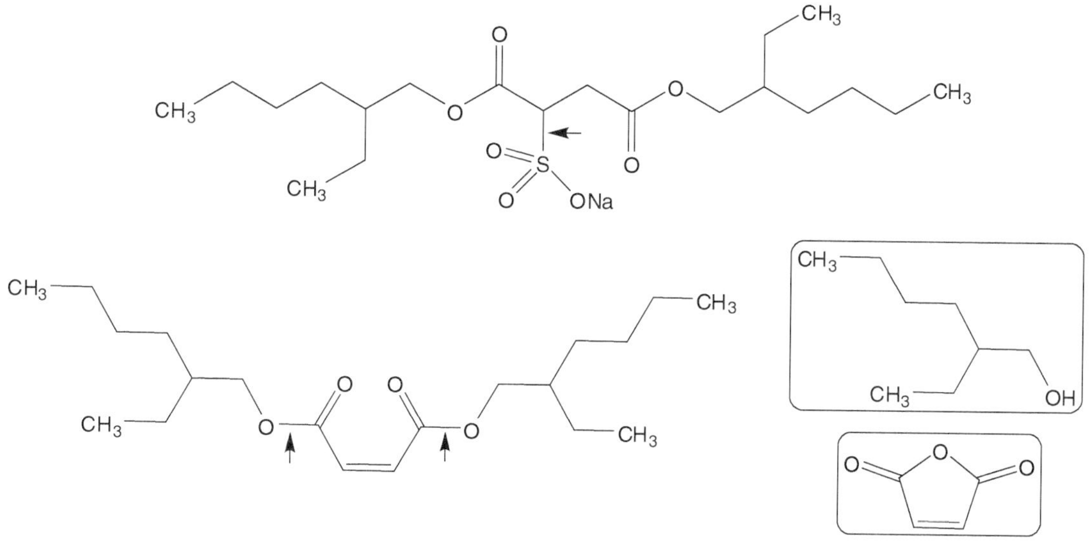

Dolutegravir

Anti-infective Medicines/Antiviral Medicines/Antiretrovirals/Integrase Inhibitors

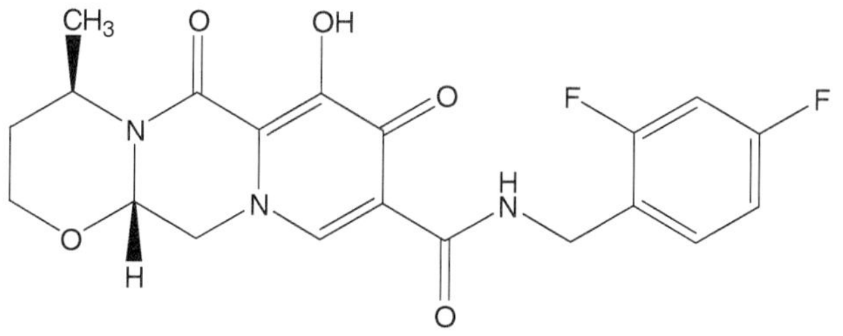

In a molecule with two chiral carbons, one chiral carbon is often used to direct the formation of the second.

Discussion. The hydroxyl group is protected through the entire sequence as the benzyl ether. The final step is hydrogenolysis of the benzyl ether. The 1,3-oxazine ring is assembled from the aldehyde, ester, and (*R*)-3-amino-1-butanol. The formation of the new chiral carbon at C2 of the oxazine ring is directed by the chiral carbon from the amino alcohol. The aldehyde is formed by hydrolysis of an acetal.

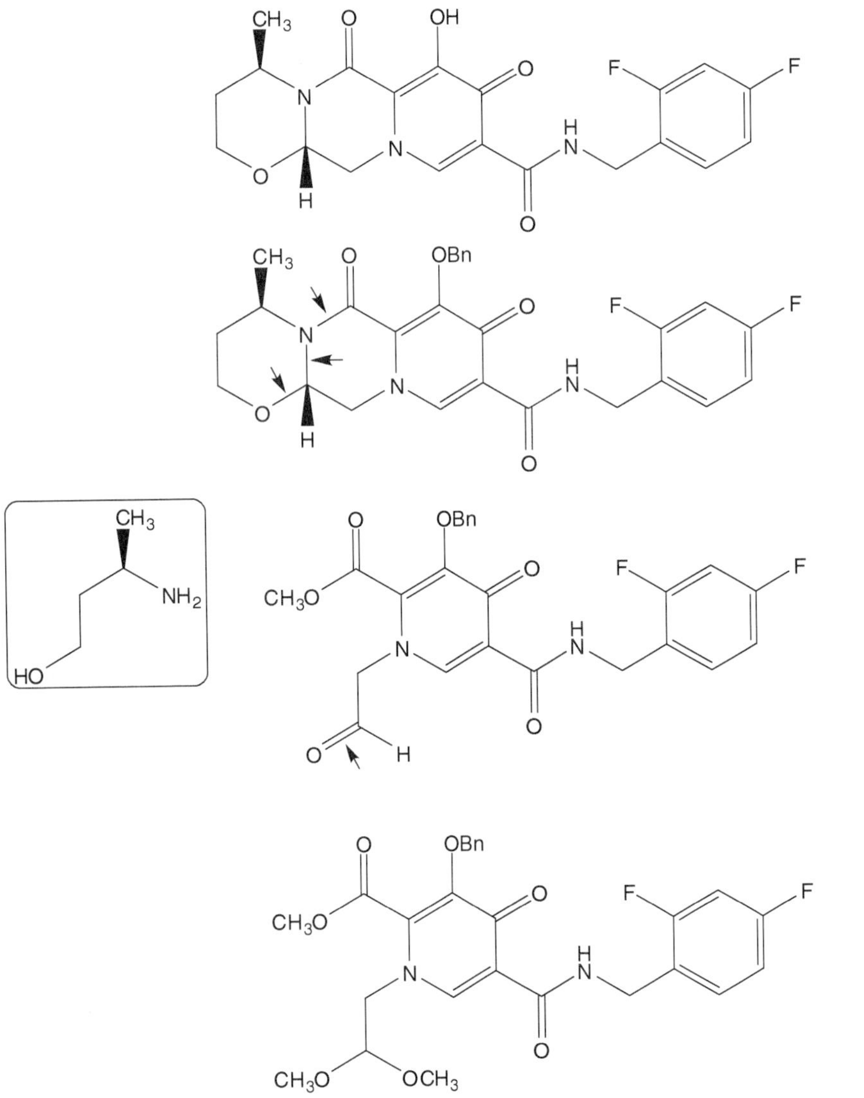

The amide is formed from the selective reaction of one of two esters with 2,4-difluorobenzylamine. (Why is this ester more reactive? Was this difference in reactivity known prior to the synthesis of dolutegravir?) The pyridine-2,5-dicarboxylate is formed by reaction of the 4-oxo-4*H*-pyran-2,5-dicarboxylate with 2-aminoacetaldehyde dimethyl acetal. The 4-oxo-4*H*-pyran-2,5-dicarboxylate is formed by condensation of the aminomethylene acetoacetate with dimethyl oxalate (mixed **Claisen Condensation**). The aminomethylene acetoacetate is formed from methyl 4-benzyloxy-3-oxobutanoate and dimethylformamide dimethyl acetal. The benzyl ether is formed by displacement of chloride from methyl 4-chloro-3-oxobutanoate by benzyl alcohol (**Williamson Ether Synthesis**).

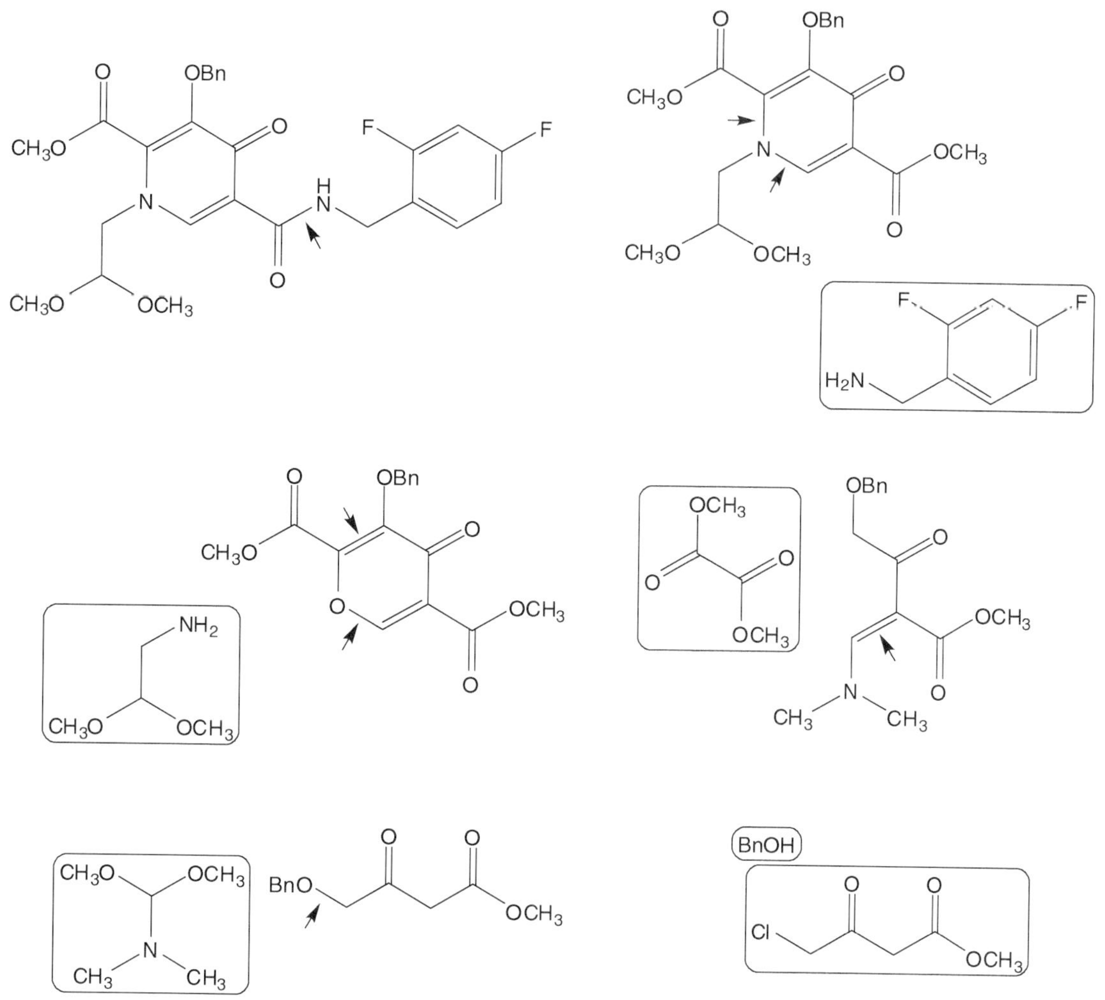

Extended Discussion

Sequences with the same disconnections but in a different order are often evaluated in the search for the optimal route to a target molecule. Draw the structures of the retrosynthetic analysis of an alternative route to dolutegravir which forms the amide with 2,4-difluorobenzylamine after formation of the 1,3-oxazine. List pros and cons for the two routes and select one route as the preferred route.

Dopamine

Cardiovascular Medications/Medicines Used in Heart Failure

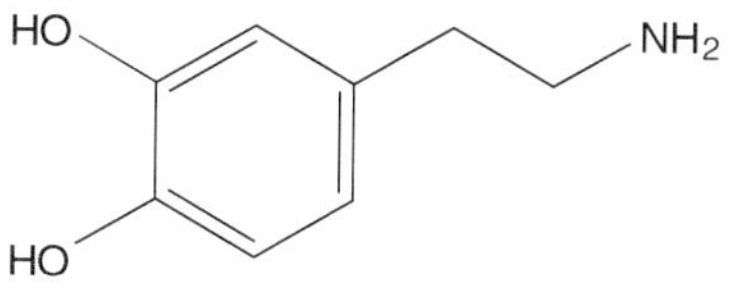

A primary alkanamine is often formed by catalytic hydrogenation of a nitrile.

Discussion. The hydroxyl groups are released by demethylation of the ethers in the final step. The primary amine is formed by reduction of the nitrile. The nitrile is formed by displacement of chloride by cyanide. 3,4-Dimethoxybenzyl chloride is formed by chloromethylation of veratrole.

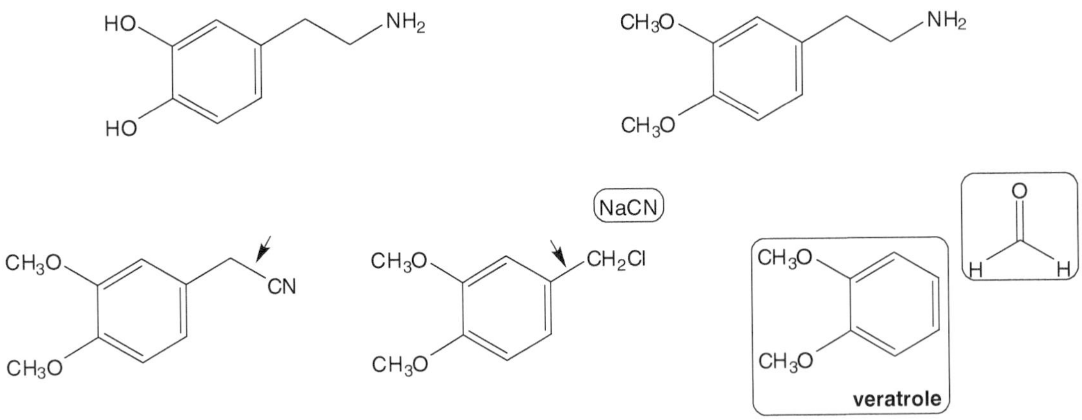

Extended Discussion

List the many challenges associated with the chloromethylation step. Discuss procedure details provided to address the challenges.

Doxycycline

Anti-infective Medicines/Antibacterials/Other Antibacterials

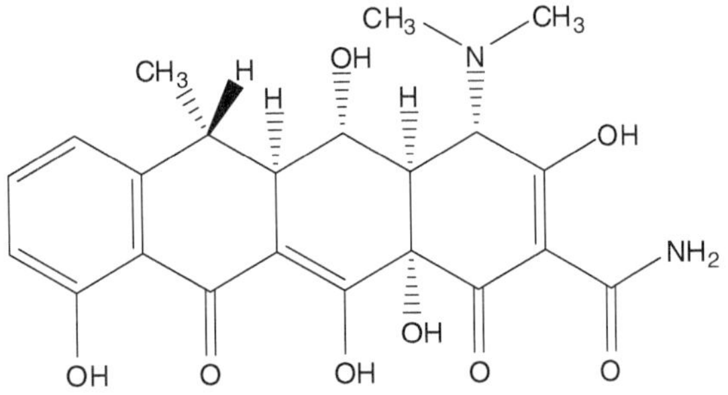

Many tetracyclines are semisynthetic, they are formed by modification of the C and/or D rings of a tetracycline natural product (tetracycline, chlorotetracycline, oxytetracycline) or a tetracycline produced by a mutant strain of *Streptomyces aureofaciens* (6-demethyltetracycline, 7-chloro-6-demethyltetracycline, 7-bromotetracycline).

Discussion. Doxycycline (H at the 6β-position) is produced in three steps from oxytetracycline (OH at the 6β-position). In the final step, hydrogen at the 6-position is introduced by hydrogenation of the 6-methylene group of 6-methylenetetracycline (methacycline). Methacycline is formed by reduction of the α-chloroketone (α-chlorine at C11a). The 6-methylene group and the α-chloroketone are formed in the dehydration of an α-chlorohemiacetal. The α-chlorohemiacetal is formed by the reaction of oxytetracycline with a chlorinating agent.

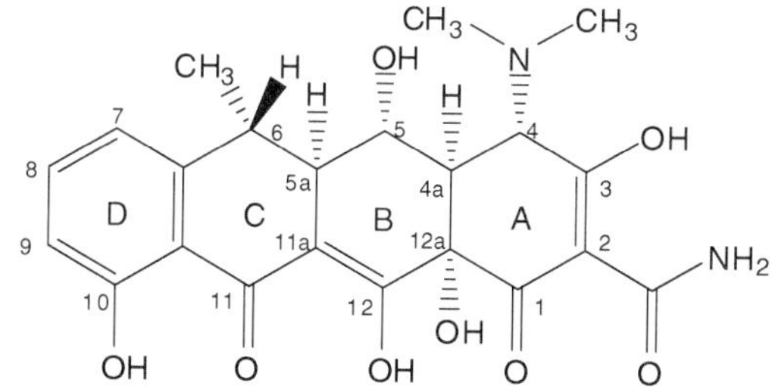

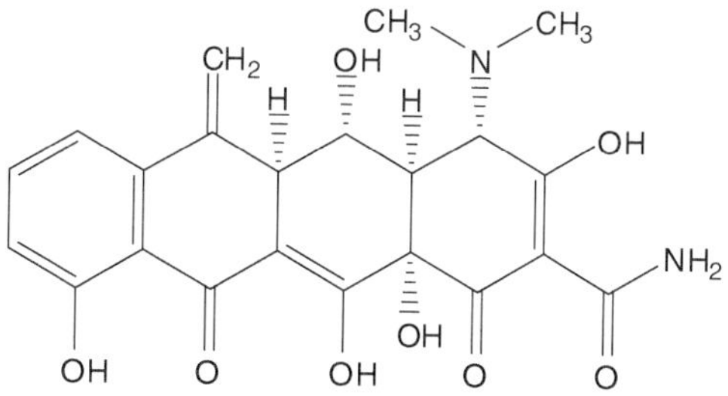

methacycline

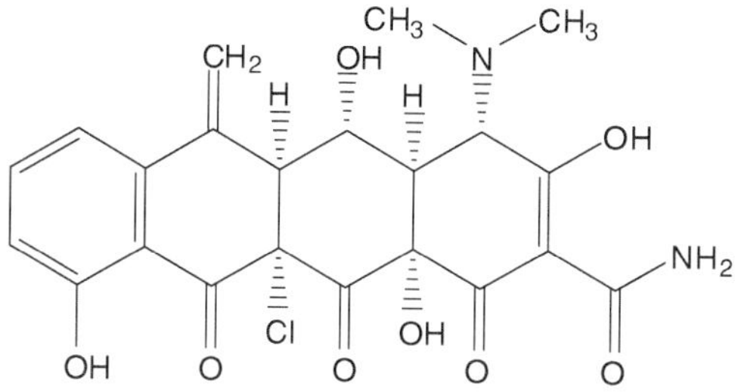

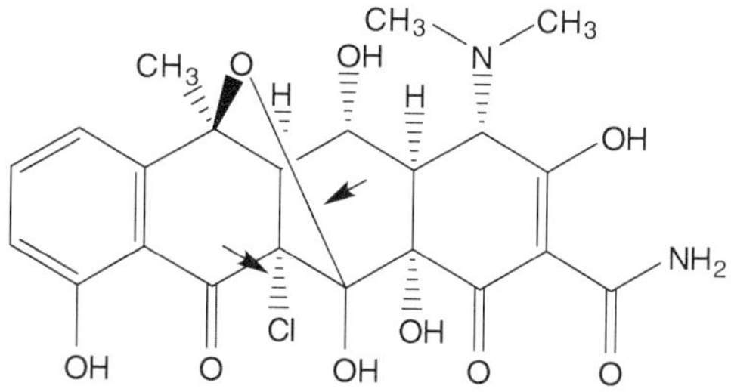

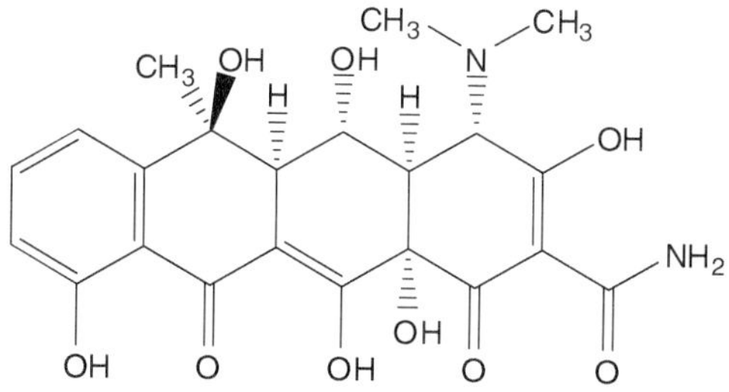

oxytetracycline

Extended Discussion

Draw the structure of another 6-deoxytetracycline formed in a one-step hydrogenolysis of the 6β-hydroxyl group of oxytetracycline.

E

Efavirenz

Anti-infective Medicines/Antiviral Medicines/Antiretrovirals/Non-nucleoside Reverse Transcriptase Inhibitors
Anti-infective Medicines/Antiviral Medicines/Antiretrovirals/Integrase Inhibitors

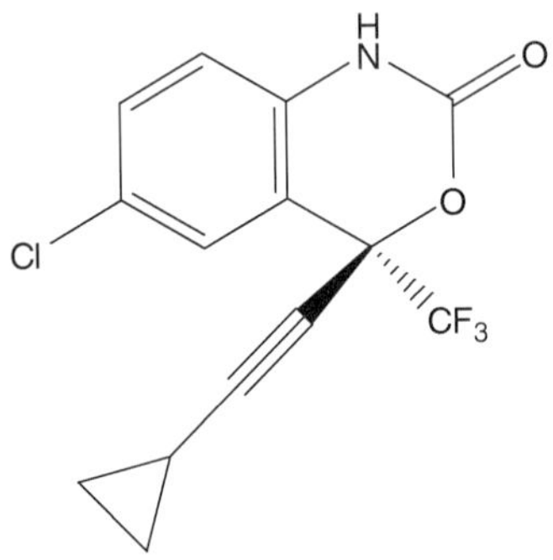

A tertiary alcohol is often formed by addition of an organometallic reagent RLi or RMX (M = Mg, Zn, X = Cl, Br, I) to a ketone.

Discussion. The carbamate is formed from the amino alcohol in the final step. (List all the known reagents for the carbamate formation. Which reagent is preferred?) The tertiary alcohol is formed by enantioselective addition of the acetylide from cyclopropylacetylene to the ketone. (List the reagents used to form the metal acetylide and the associated yield of the tertiary alcohol.) The amine is released by hydrolysis of the pivalamide. The ketone is formed by ortholithiation of the *N*-(4-chlorophenyl) pivalamide and capture of the aryllithium by ethyl trifluoroacetate. The pivalamide is formed from 4-chloroaniline and pivaloyl chloride.

Routes to Essential Medicines: A Workbook for Organic Synthesis, First Edition. Peter J. Harrington.

Companion website: www.wiley.com/go/Harrington/routes_essential_medicine

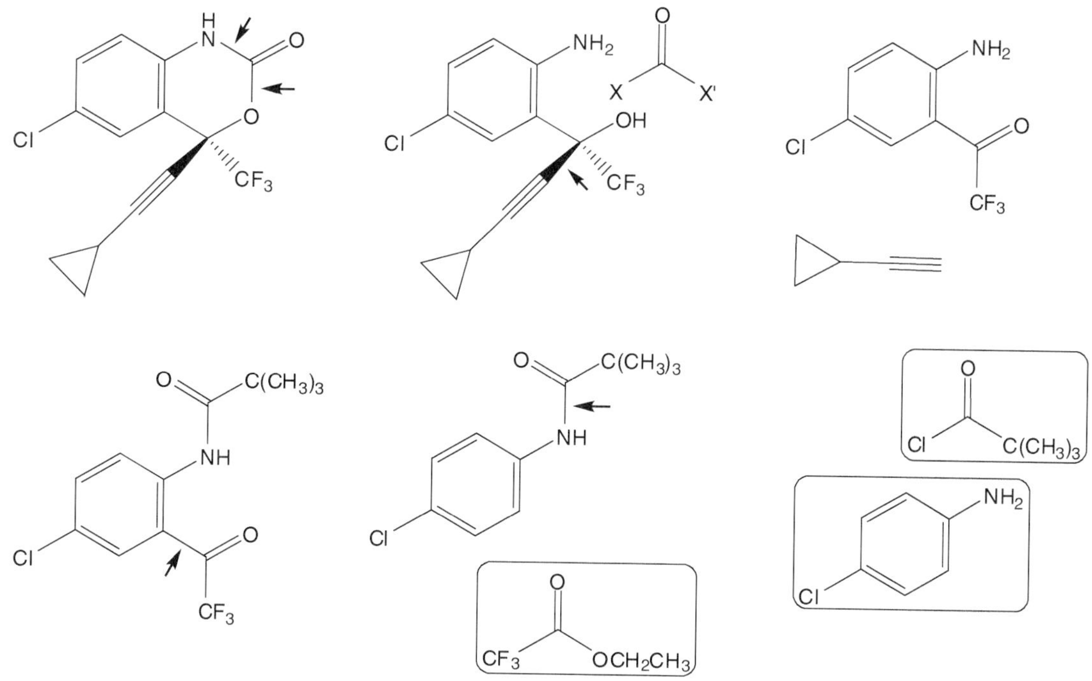

Cyclopropylacetylene is formed from cyclopropyl methyl ketone.

Extended Discussion

(1*R*,2*S*)-*N*-Pyrrolidinylnorephedrine is the preferred ligand for the enantioselective addition of the Grignard reagent to the ketone. Draw the structures of a retrosynthetic analysis of this ligand. Include the structures of the retrosynthetic analysis of any organic starting material(s) from petrochemical or biochemical raw materials.

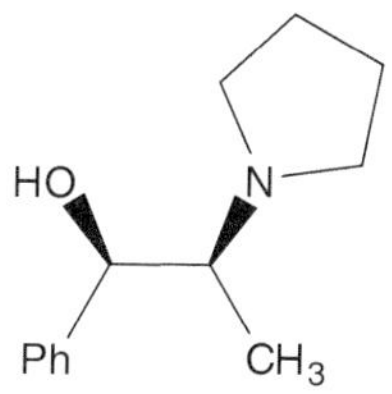

(1R,2S)-N-pyrrolidinylnorephedrine

Eflornithine

Anti-infective Medicines/Antiprotozoal Medicines/Antitrypanosomal Medicines/African Trypanosomiasis

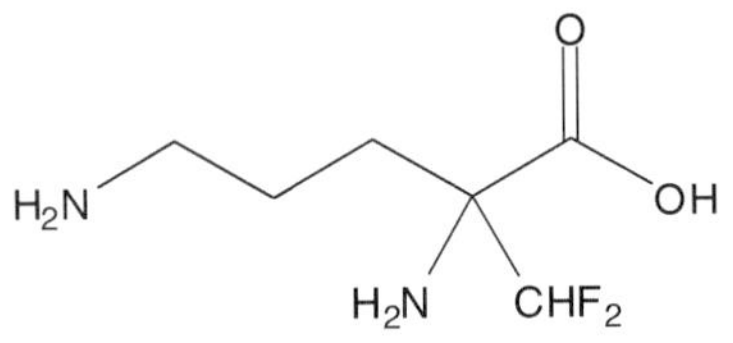

A difluoromethyl group is often introduced by reaction of chlorodifluoromethane (HCFC-22) with a nucleophile.

Discussion. Eflornithine is a 1 : 1 mixture of (*R*)- and (*S*)-enantiomers. The carboxylic acid is released in the final step by hydrolysis of the ester. The amine at C5 is formed by hydrogenation of the nitrile. The amine at C2 is released by hydrolysis of the benzaldehyde imine. The *N*-protected α-amino ester is alkylated by reaction with chlorodifluoromethane (HCFC-22). The *N*-protected α-amino ester is alkylated by reaction with acrylonitrile (**Michael Addition**). The *N*-protected α-amino ester is formed from glycine ethyl ester and benzaldehyde. Benzophenone is used in place of benzaldehyde in the parallel sequence in alternative Route B.

Route A

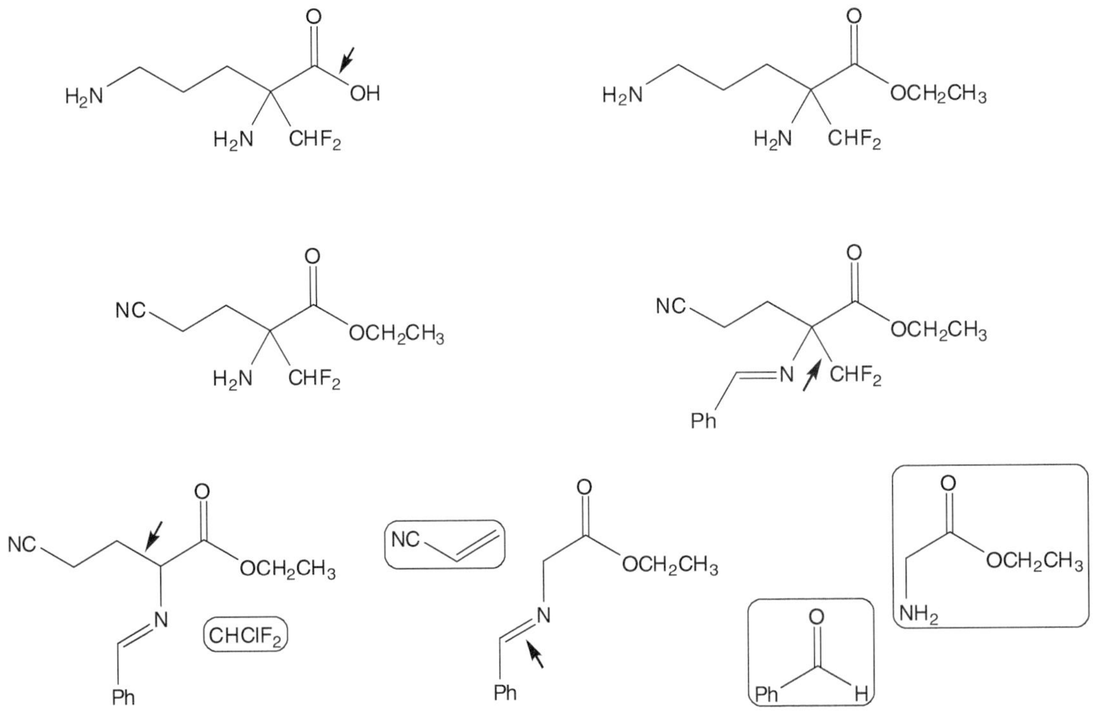

In Route C, the carboxylic acid is released in the final step by hydrolysis of the ester. Both amines are released by hydrolysis of the benzaldehyde imines. The *N*-protected α-amino ester is alkylated by reaction with fluoroform (HFC-23). The *N*-protected α-amino ester is formed from L-ornithine methyl ester and benzaldehyde. L-Ornithine methyl ester is formed from L-ornithine (**Fischer Esterification**). L-Ornithine is produced by microbial fermentation. This may be the only example of a racemic drug manufactured from a single enantiomer starting material!

Route C

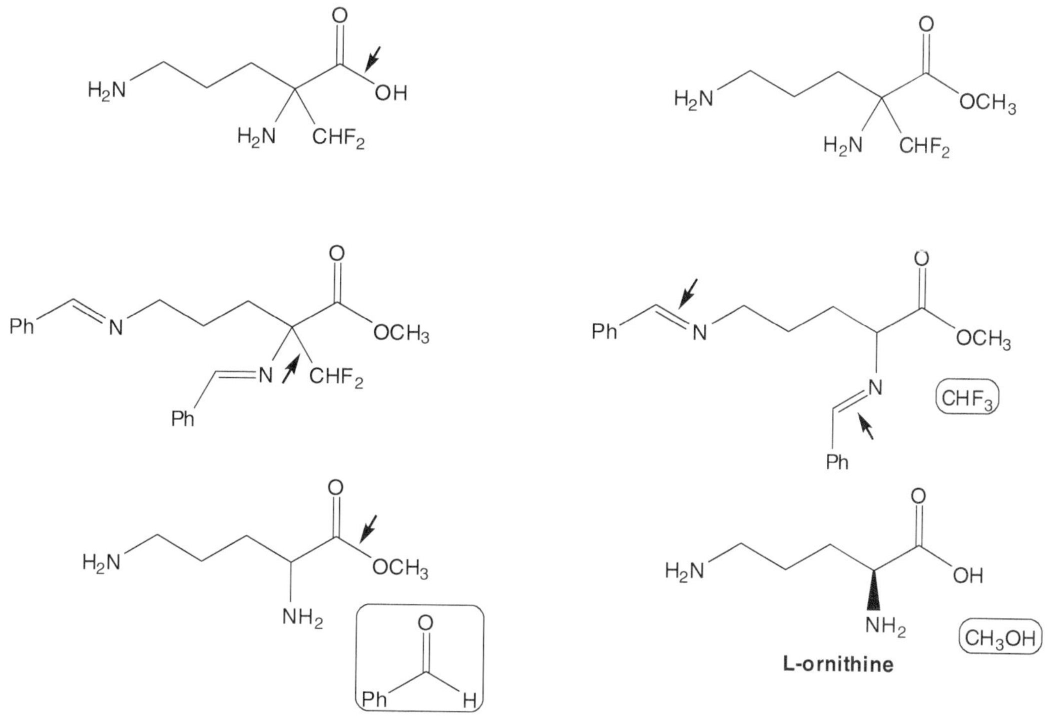

Extended Discussion

List the pros and cons for Routes A, B, and C. Is one route preferred?

Emtricitabine

Anti-infective Medicines/Antiviral Medicines/Antiretrovirals/Integrase Inhibitors

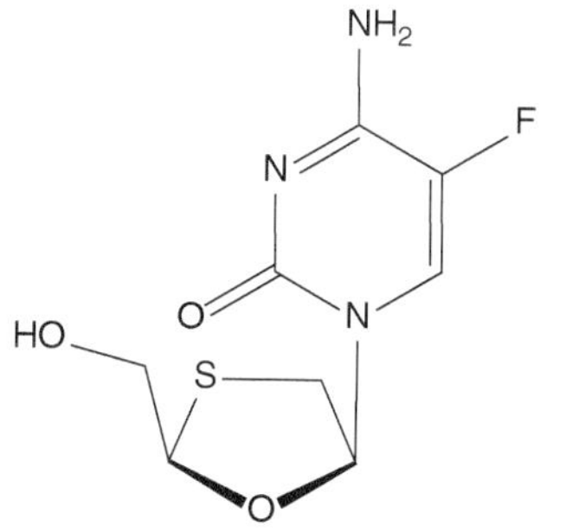

A chiral auxiliary can be used to control the stereochemical outcome of a reaction which forms a nearby chiral carbon. A chiral auxiliary strategy for creating a chiral carbon is cost-effective when the auxiliary is inexpensive and when the auxiliary can be recovered after the chiral carbon is formed.

Discussion. The primary alcohol of emtricitabine is formed by reduction of the (−)-menthyl ester in the final step. (How is (−)-menthol separated from emtricitabine?) The rings are joined by nucleophilic displacement of the acetate from the (5*R*)-acetoxy-1,3-oxathiolane-(2*R*)-carboxylate by N1 of the essential medicine flucytosine (5-fluorocytosine) (**Silyl–Hilbert–Johnson Reaction**). (The desired product, with both oxathiolane ring substituents *cis*, and the *trans*-side product are both formed. What is the highest ratio of *cis*-product to *trans*-side product? What reaction conditions are associated with the highest ratio of *cis*-product to *trans*-side product?) The acetate is formed from the alcohol and acetyl chloride.

(How is (1*R*-(−)-menthyl (5*R*)-acetoxy-1,3-oxathiolane-(2*R*)-carboxylate separated from the mixture of stereoisomers produced?) The (5*R*)-hydroxy-1,3-oxathiolane-(2*R*)-carboxylate is formed by the condensation of (1*R*)-(−)-menthyl glyoxalate hydrate with 1,4-dithiane-2,5-diol.

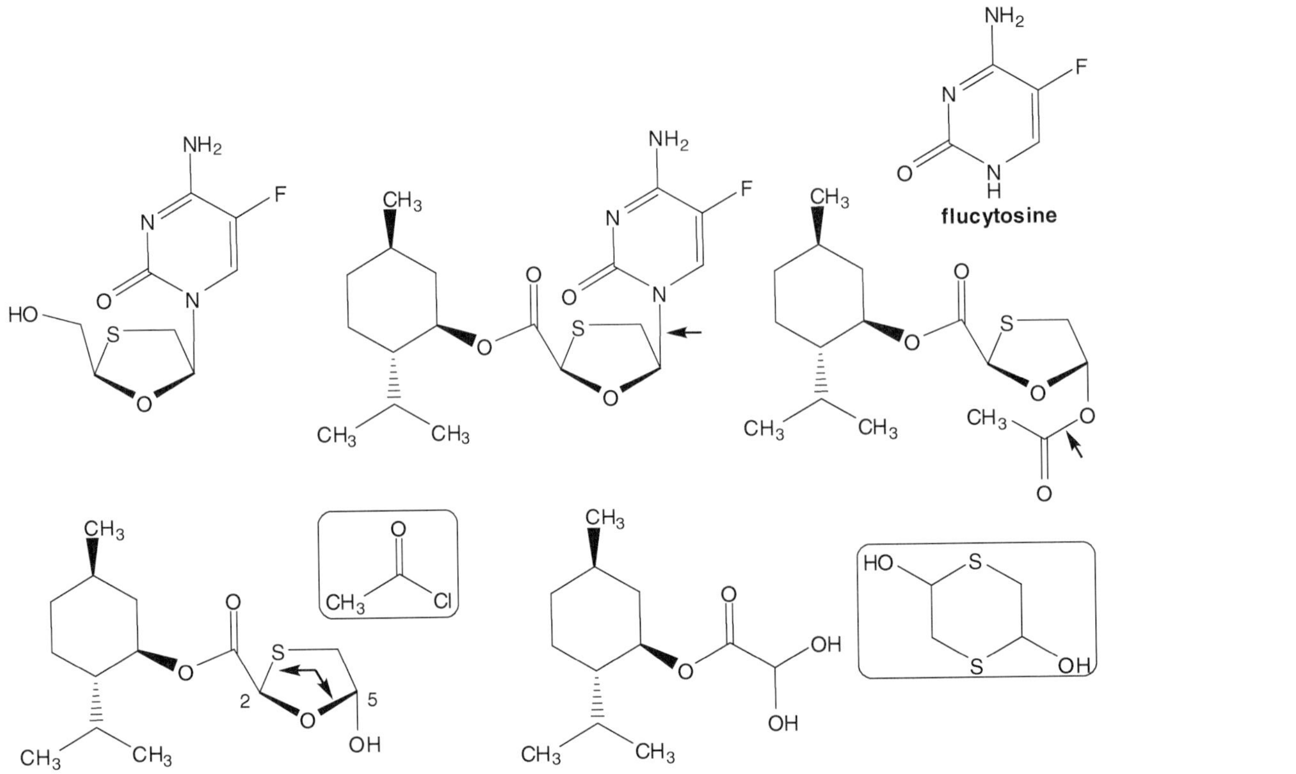

(1*R*)-(−)-Menthyl glyoxalate hydrate is formed by ozololysis of (−)-dimenthyl fumarate. The fumarate diester is formed from maleic anhydride and (−)-menthol.

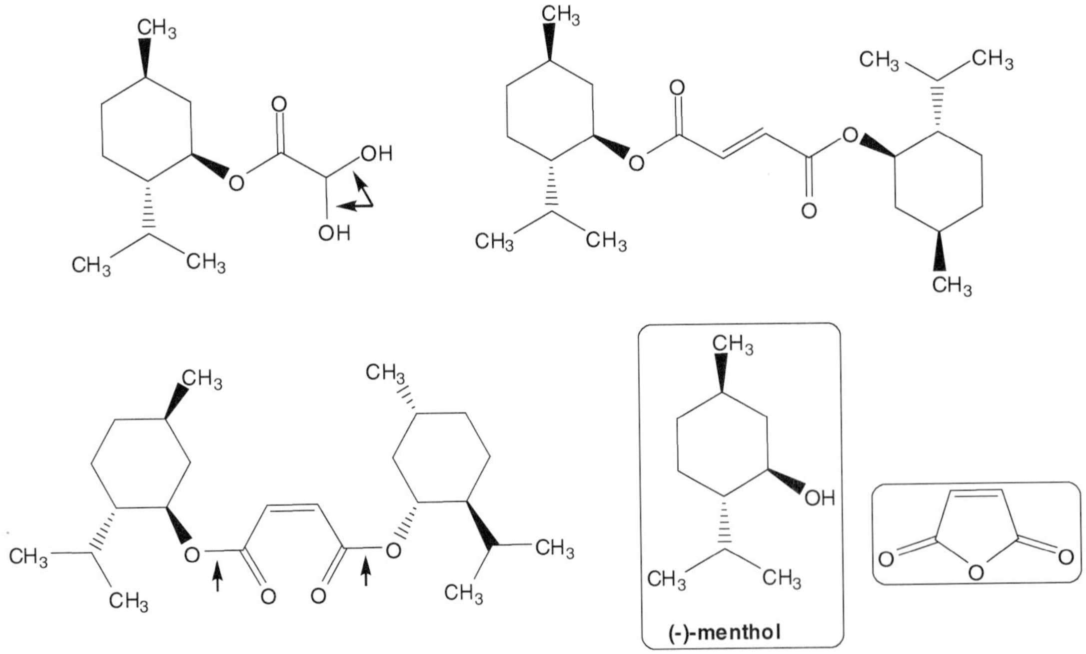

Extended Discussion

The key C-N bond formation in the synthesis of emtricitabine can be accomplished using **other leaving groups**. Select one alternative leaving group. What is the highest ratio of *cis*-product to *trans*-product and what reaction conditions are associated with the highest ratio?

Enalapril

Cardiovascular Medicines/Medicines Used in Heart Failure

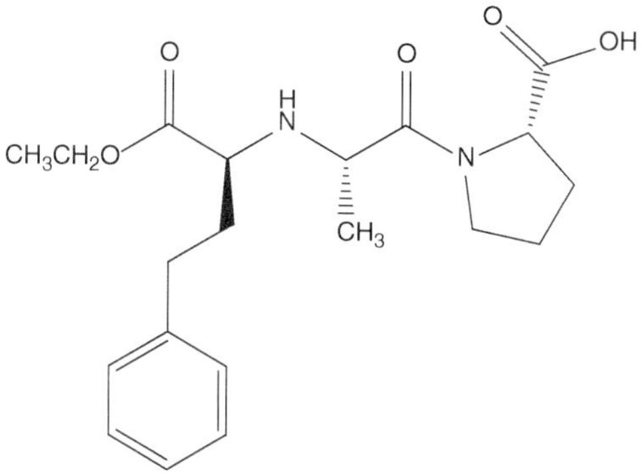

An amide linkage between two α-amino acids in a peptide is often formed by converting one α-amino acid to an *N*-carboxyanhydride (NCA) then using the NCA to *N*-acylate nitrogen of the other α-amino acid.

Discussion. Enalapril is manufactured from the amino acids L-alanine and L-proline by two routes. In Route A, the amide is formed in the final step by reaction of the *N*-carboxyanhydride with L-proline. L-Proline is produced by fermentation. The *N*-carboxyanhydride is formed from the α-amino acid (*S*,*S*)-*N*-(1-ethoxycarbonyl-3-phenylpropyl)alanine. (List the reagents and conditions used to make this NCA.) (S,S)-N-(1-ethoxycarbonyl-3-phenylpropyl)alanine is separated from the (*R*,*S*)-diastereomer by crystallization. The mixture of (*S*,*S*)- and (*R*,*S*)-diastereomers is formed by hydrogenation of the (*S*,*S*)- and (*R*,*S*)-diastereomers of *N*-(1-ethoxycarbonyl-3-oxo-3-phenylpropyl) alanine. The (*S*,*S*)- and (*R*,*S*)-diastereomers of *N*-(1-ethoxycarbonyl-3-oxo-3-phenylpropyl)alanine are formed by conjugate addition of L-alanine to ethyl 3-benzoylacrylate. (What is the highest *SS* to *RS* ratio reported for the conjugate addition? What conditions are associated with this high ratio?) Ethyl 3-benzoylacrylate is formed from glyoxylic acid, acetophenone, and ethanol (**Aldol Condensation**).

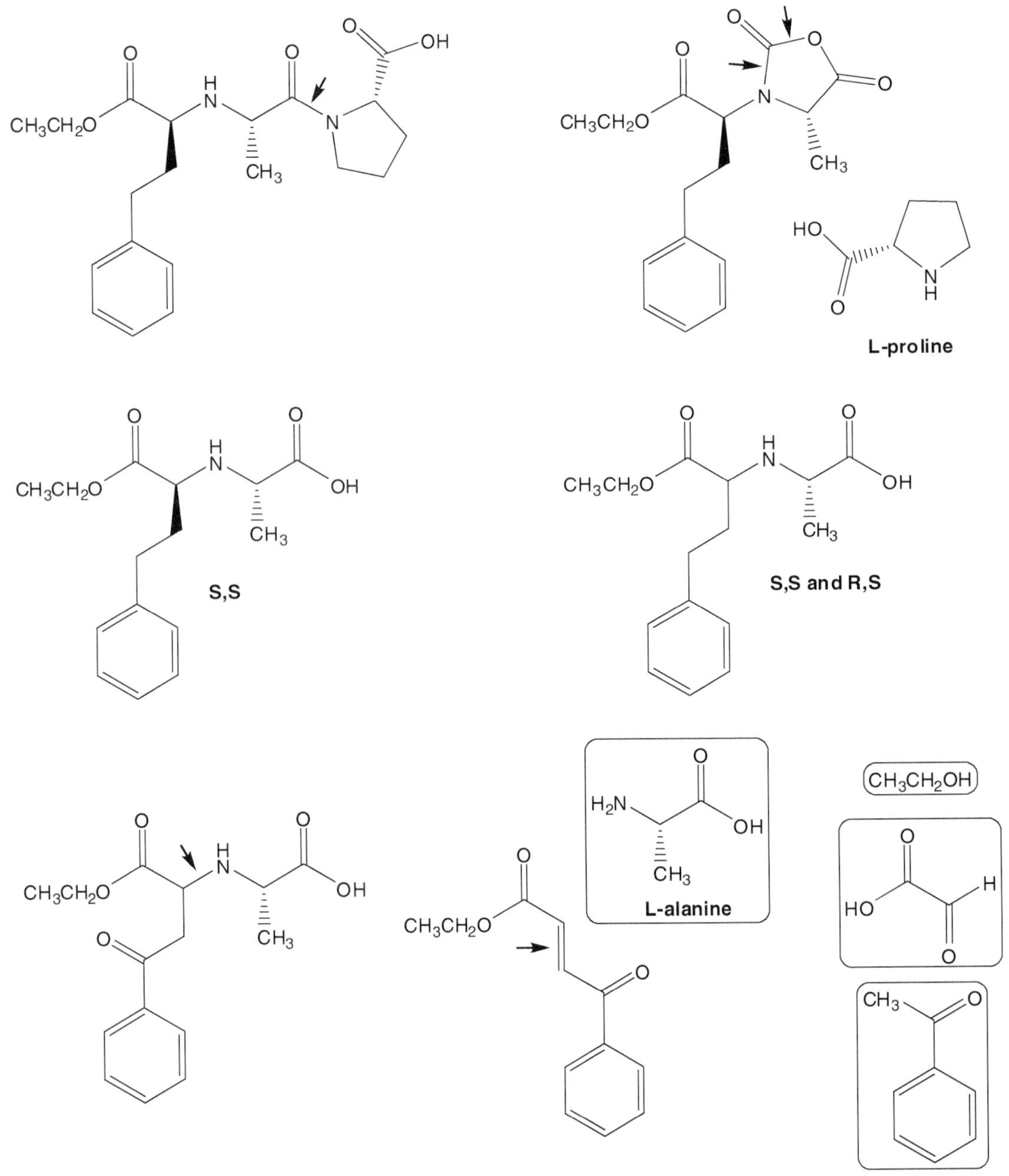

In Route B, the (*S*,*S*,*S*)-diastereomer is separated from the mixture of (*S*,*S*,*S*)- and (*R*,*S*,*S*)-diastereomers in the final step by crystallization of the maleate salt. Both diastereomers are formed in the reductive amination of ethyl 2-oxo-4-phenylbutanoate with the dipeptide L-ala-L-pro. (What is the highest *SSS* to *RSS* ratio reported for the reductive amination? What conditions are associated with this high ratio?) The amide in dipeptide L-ala-L-pro is formed by reaction of L-alanine *N*-carboxyanhydride with L-proline. L-Alanine *N*-carboxyanhydride is formed from L-alanine. (List the reagents and conditions used to make this NCA.)

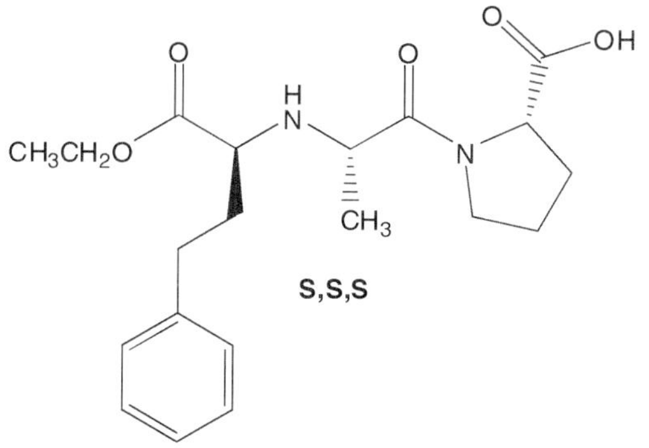

S,S,S

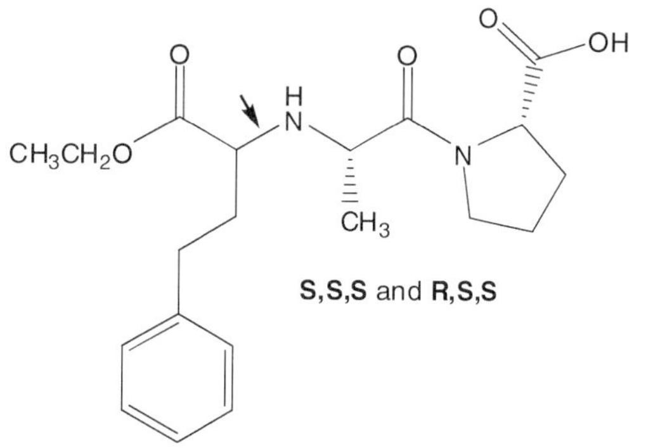

S,S,S and R,S,S

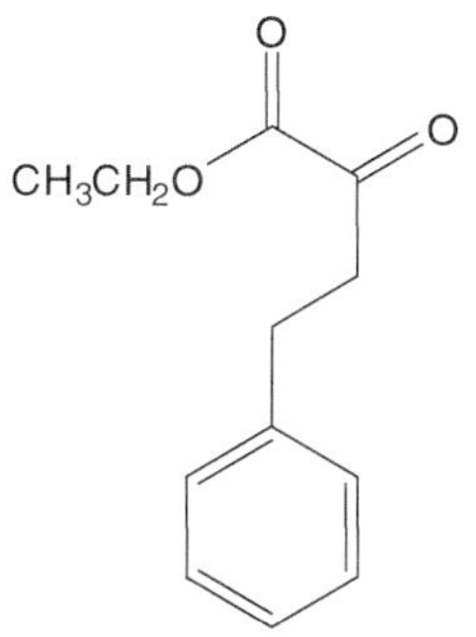

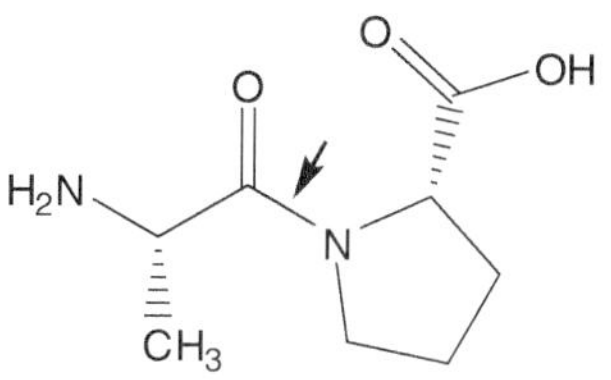

L-ala-L-pro

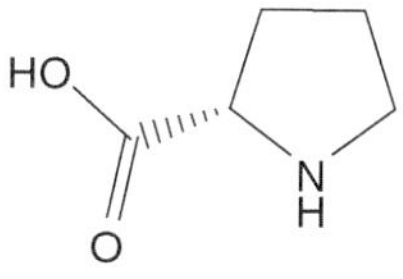

L-proline

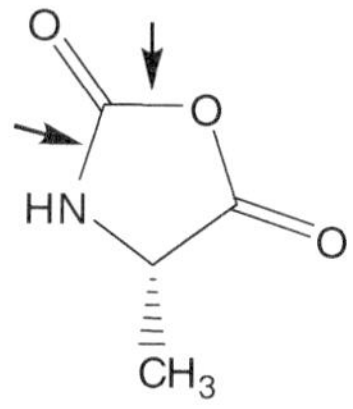

L-ala NCA

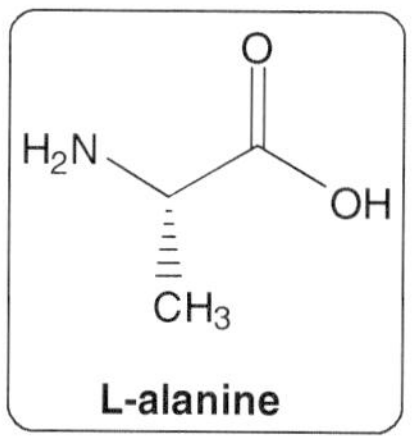

L-alanine

Ethyl 2-oxo-4-phenylbutanoate is formed by addition of 2-phenylethylmagnesium bromide to diethyl oxalate (**Grignard Reaction**). (Draw the structures of two side products formed in this reaction. How is ethyl 2-oxo-4-phenylbutanoate separated from these side products?) The Grignard reagent is formed from (2-bromoethyl)benzene.

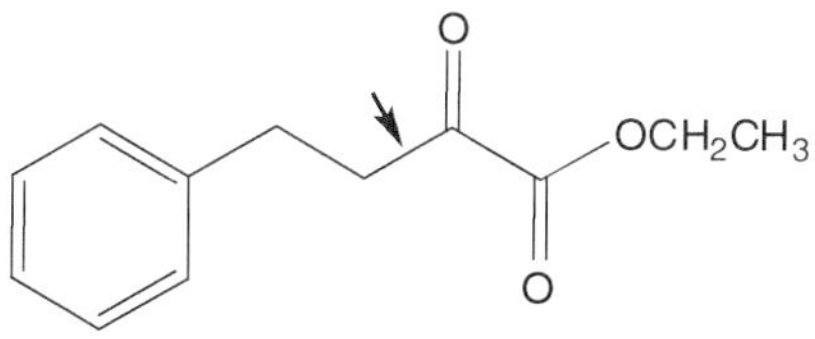

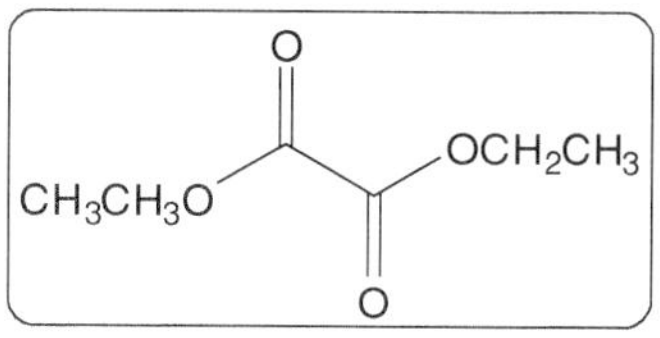

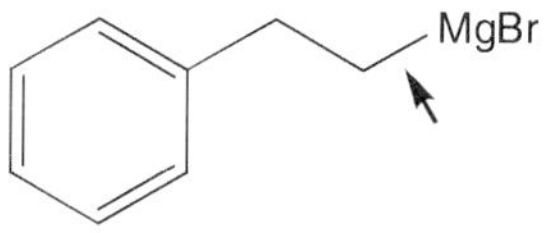

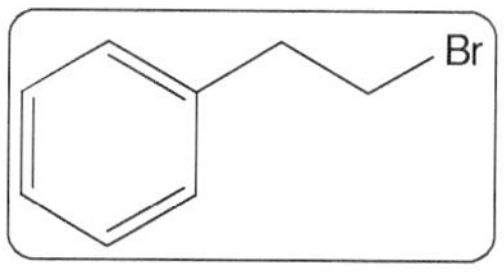

Extended Discussion

List the pros and cons for both routes. Is one route preferred?

Entecavir

Anti-infective Medicines/Antiviral Medicines/Antihepatitis Medicines/Medicines for Hepatitis B/Nucleoside/Nucleotide Reverse Transcriptase Inhibitors

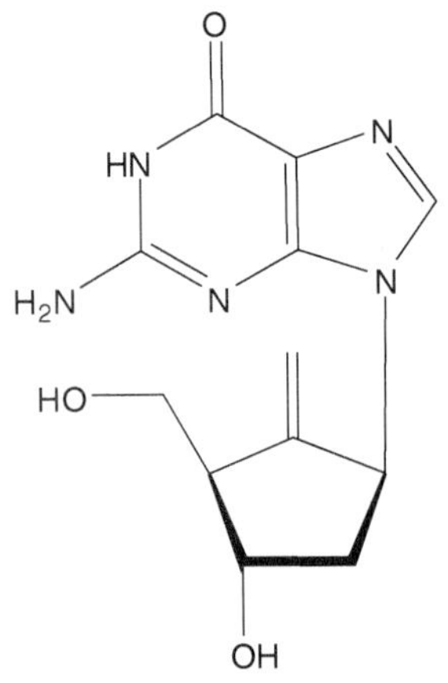

A nucleoside analog is often formed by displacement of a leaving group on the sugar by nitrogen of the heterocycle. Oxygen-based leaving groups include water, alcohols, and carboxylic acids. The leaving group is usually geminal to another oxygen atom, allylic, or is the oxygen of an epoxide.

Discussion. The alcohols are released in the final step by cleavage of the benzyl ethers. Cleavage of benzyl and 4-monomethoxytrityl (MMTr) protecting groups reveals the guanine. The alkene is formed by reaction of the cyclopentanone with a reagent produced from dibromomethane, zinc, and tetrahydrofuran (**Nysted Reagent**). The cyclopentanone is formed by oxidation of the secondary alcohol. The purine amino group is protected by reaction with 4-monomethoxytrityl chloride prior to the alcohol oxidation. The secondary alcohol and C-N bond joining the two rings form in the key step by ring-opening of the α-face epoxide by N9 of 2-amino-6-(benzyloxy)purine.

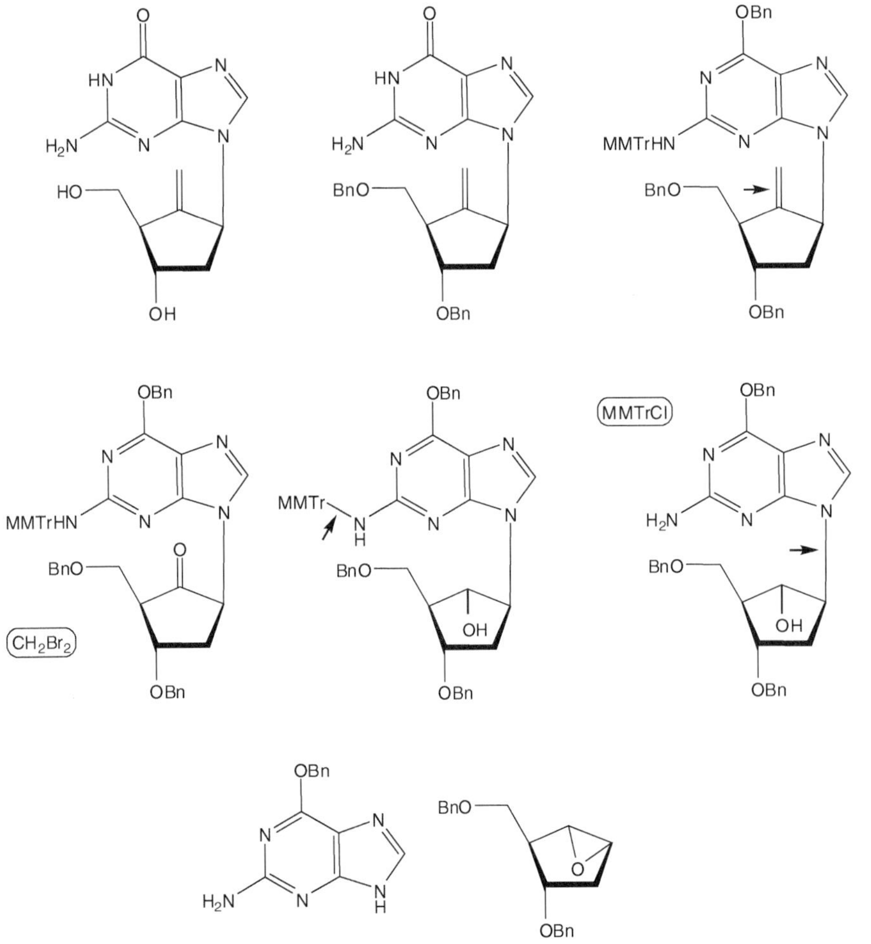

The benzyl ether of 2-amino-6-(benzyloxy)purine is formed by displacement of chloride by benzyl alcohol. 2-Amino-6-chloropurine is formed from guanine.

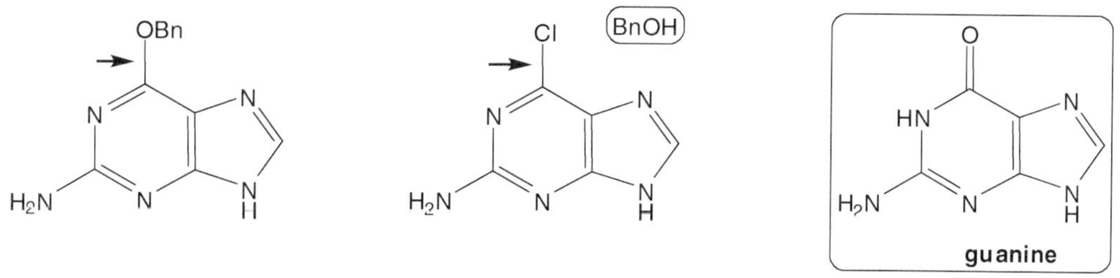

On the epoxide reaction partner, the α-face benzyl ether is formed by *O*-alkylation of the alcohol (**Williamson Ether Synthesis**). The α-face epoxide is formed epoxidation of the alkene. The α-face alcohol is formed by hydroboration-oxidation of the 5-benzyloxymethyl-1,3-cyclopentadiene. The 5-substituted-1,3-cyclopentadiene is formed from cyclopentadiene and benzyl chloromethyl ether.

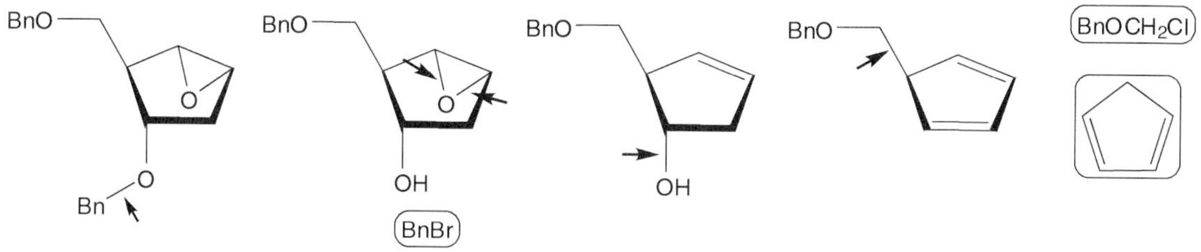

Extended Discussion

Draw the structures of a retrosynthetic analysis of one alternative route to entecavir via displacement of an allylic oxygen leaving group. Include the structures of the retrosynthetic analysis of any organic starting material(s) from petrochemical or biochemical raw materials.

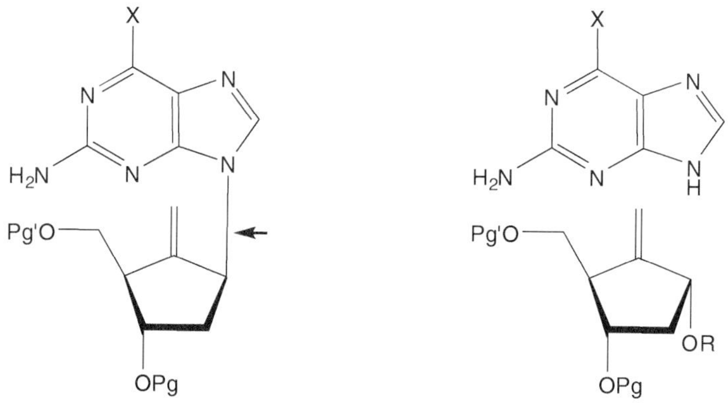

Pg, Pg' = hydroxyl protecting groups

Ephedrine

Anesthetics, Preoperative Medicines and Medical Gases/Local Anesthetics

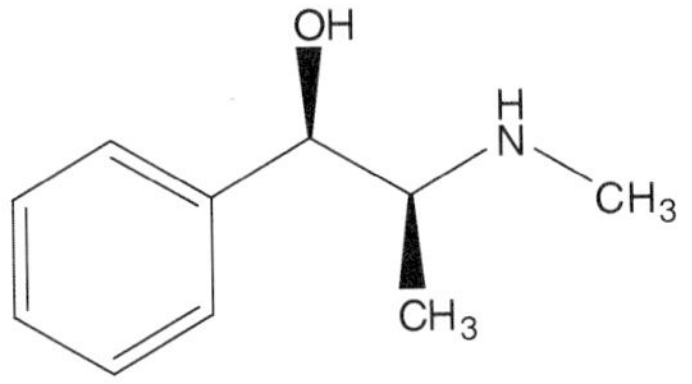

In a molecule with two adjacent chiral carbons, one chiral carbon is often used to direct the formation of the second.

Discussion. Ephedrine is an alkaloid isolated from plants of the genus *Ephedra* (*Ephedra equisetina*, *Ephedra vulgaris*). Ephedrine is also manufactured by both semisynthetic and synthetic routes. In the preferred semisynthetic route, ephedrine is formed in a single step from (*R*)-1-hydroxy-1-phenylacetone (also known as (*R*)-phenylacetylcarbinol) and methylamine by reductive amination. (*R*)-1-Hydroxy-1-phenylacetone is produced from benzaldehyde by fermentation.

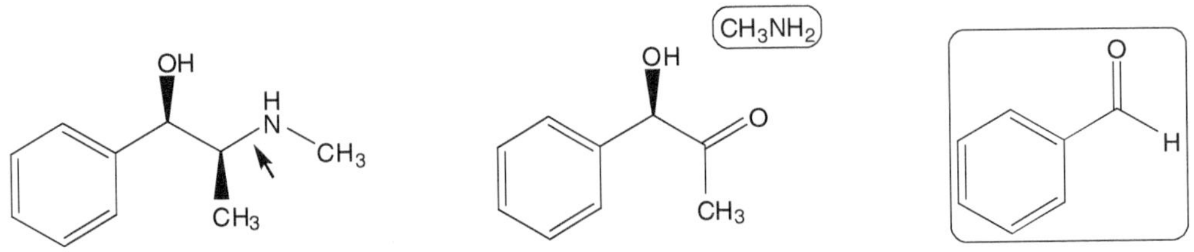

Extended Discussion

Draw the structures of the retrosynthetic analysis of one synthetic route to ephedrine. Include the structures of the retrosynthetic analysis of any organic starting material(s) from petrochemical or biochemical raw materials.

Epinephrine

Antiallergics and Medicines Used in Anaphylaxis
Cardiovascular Medicines/Antiarrhythmic Medicines
Opthalmological Preparations/Mydriatics
Medicines Acting on the Respiratory Tract/Antiasthmatic and Medicines for Chronic Obstructive Pulmonary Disease

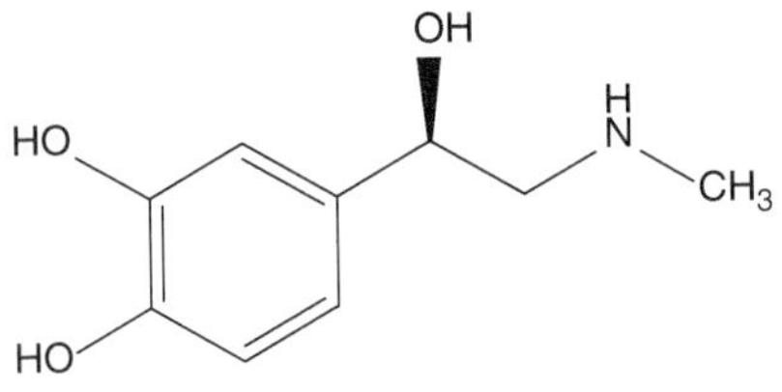

A secondary benzylic alcohol is often formed by hydrogenation of a ketone.

Discussion. Epinephrine is produced from 2-chloro-3′,4′-dihydroxyacetophenone by three routes. In Route A, the (*R*)-alcohol is formed by asymmetric hydrogenation of the ketone. The α-aminoketone is formed by chloride displacement by methylamine.

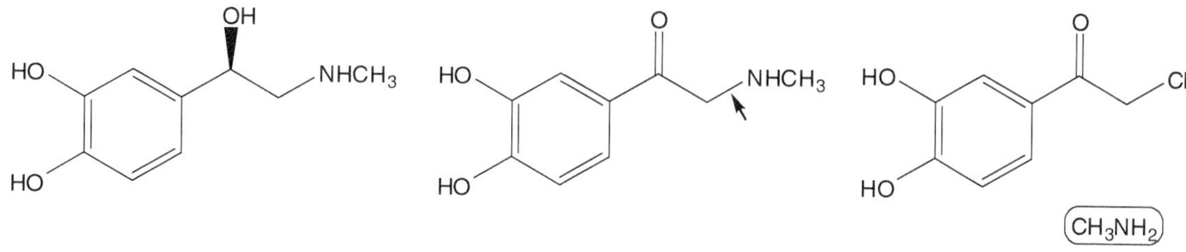

In Route B, the secondary amine is released by hydrogenolysis of the benzyl group of the tertiary amine in the final step. (What reaction conditions are required to avoid hydrogenolysis of the benzyl alcohol?) The (*R*)-alcohol is formed by asymmetric hydrogenation of the ketone. The α-aminoketone is formed by chloride displacement by *N*-benzylmethylamine.

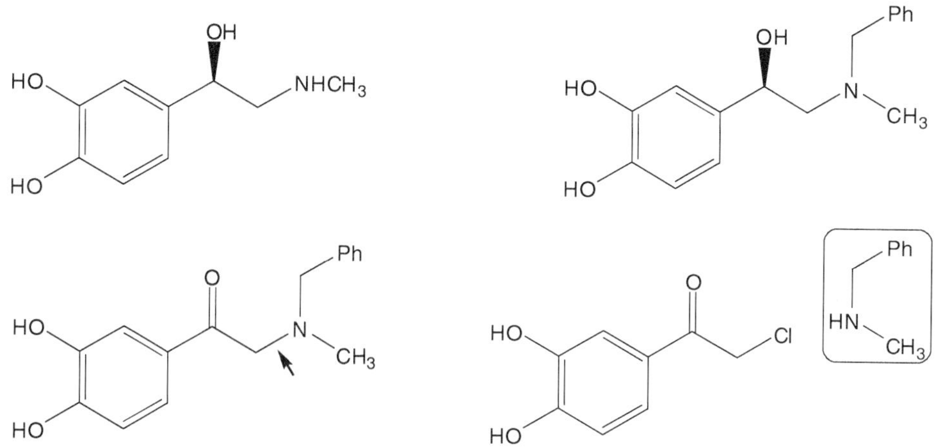

In Route C, the (*R*)-alcohol is formed by resolution of the racemic alcohol. (How is the undesired (*S*)-alcohol converted to the racemic alcohol?) The racemic alcohol is formed by hydrogenation of the ketone. The α-aminoketone is formed by chloride displacement by methylamine.

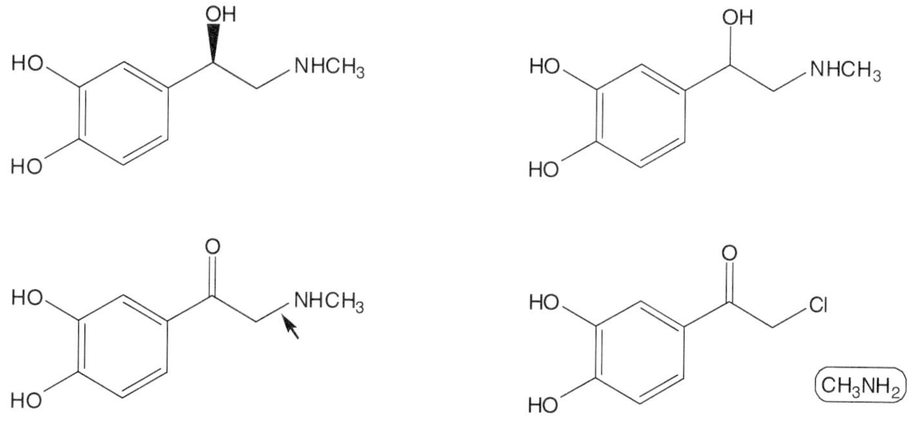

2-Chloro-3′,4′-dihydroxyacetophenone is formed from catechol and chloroacetyl chloride (**Fries Rearrangement**).

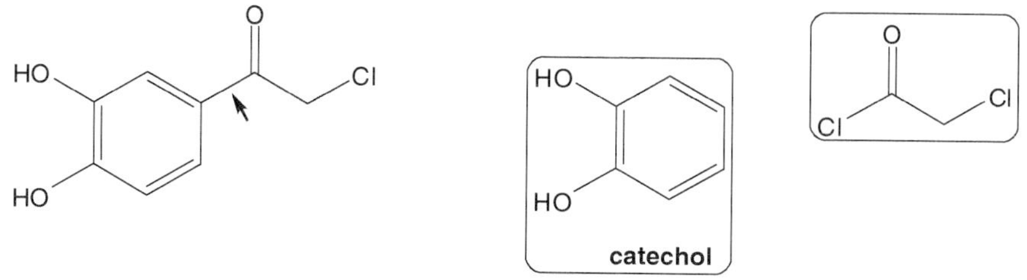

Extended Discussion

List the pros and cons for the three routes to epinephrine.

Ergocalciferol

Vitamins and Minerals

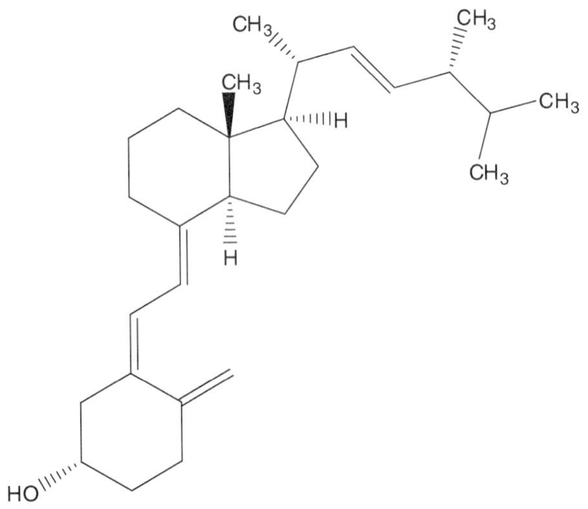

A single-enantiomer molecule with multiple chiral carbons is often formed by modification of a natural product which has most or all of the chiral carbons already in place.

Discussion. Ergocalciferol (calciferol, vitamin D_2) is semisynthetic. The *secosteroid* ergocalciferol is produced by cleavage of the B-ring of the steroid natural product ergosterol.

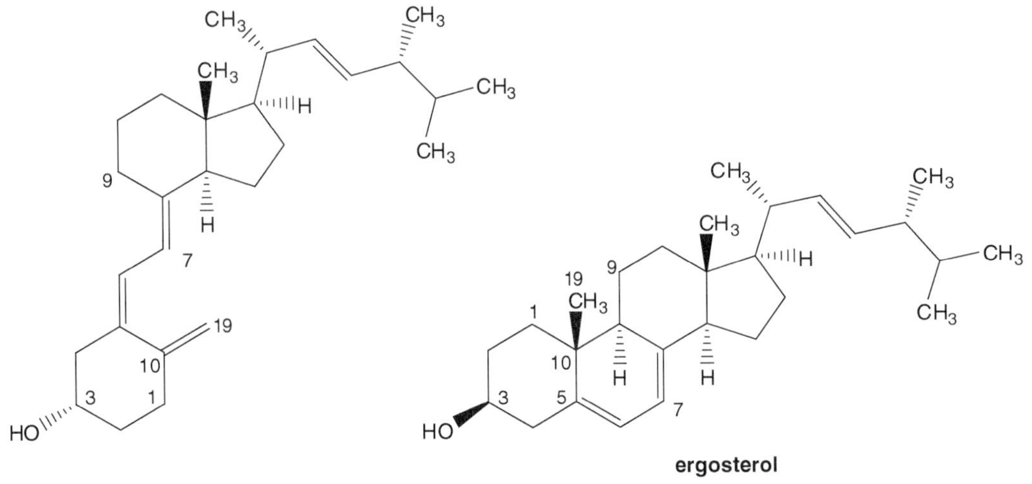

Ergocalciferol (vitamin D_2) is formed from pre-vitamin D_2. Pre-vitamin D_2 is formed by UV-photoirradiation of pro-vitamin D_2 (ergosterol). Low conversion in the UV-photoirradiation step minimizes the formation of two side products. (Draw structures for and name these side products. How is ergosterol separated from the UV-photoirradiation product mixture?) Ergosterol is produced by fermentation.

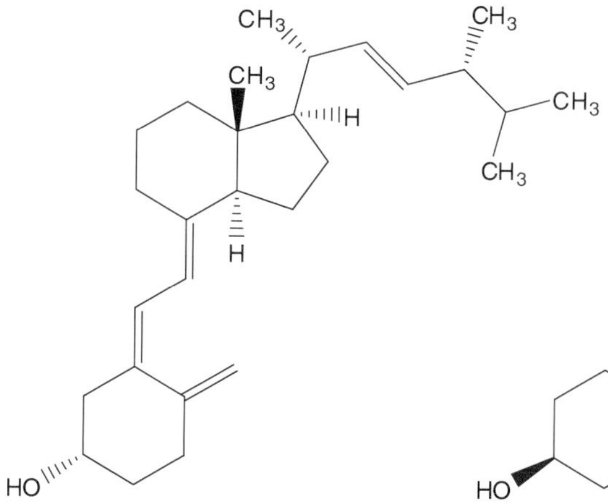

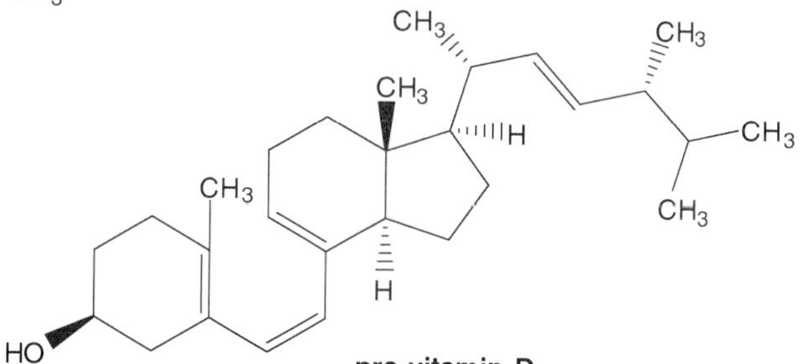

pre-vitamin D_2

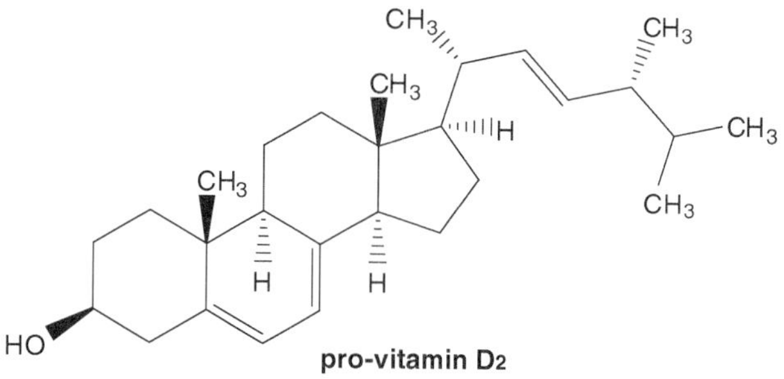

pro-vitamin D2
ergosterol

Ergometrine

Oxytocics and Antioxytocics/Oxytocics

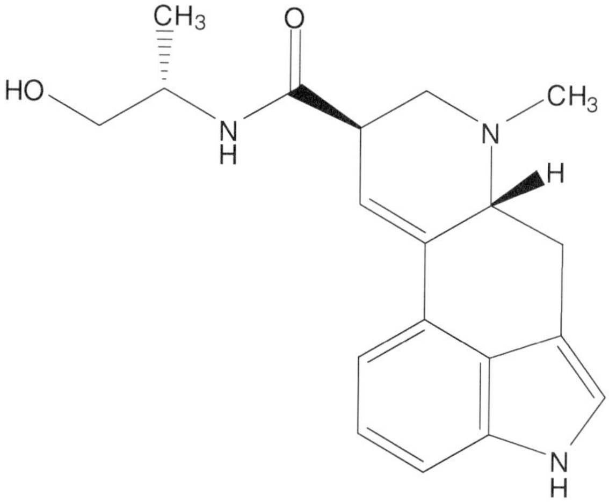

A single-enantiomer molecule with multiple chiral carbons is often formed by modification of a natural product which has most or all of the chiral carbons already in place.

Discussion. Ergometrine (ergonovine) is accessible by isolation of the alkaloid from field ergot, by fermentation, and by semisynthesis. In the semisynthesis, the amide is formed in the final step from the carboxylic acid (lysergic acid) and (*S*)-2-amino-1-propanol (L-alaninol). (List the coupling agents used and the yield of ergometrine associated with each coupling agent.) Lysergic acid is formed from an ergot alkaloid by amide hydrolysis or from the fermentation product paspalic acid by isomerization. (Draw the structure of side product isolysergic acid also formed by paspalic acid isomerization. How is isolysergic acid separated from lysergic acid?)

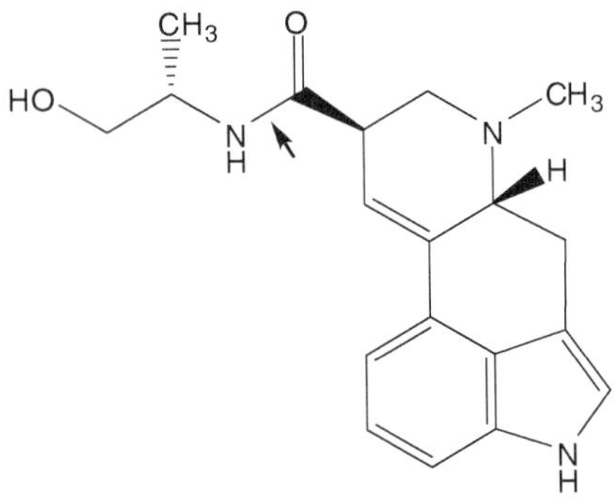

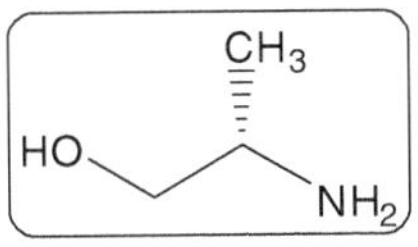

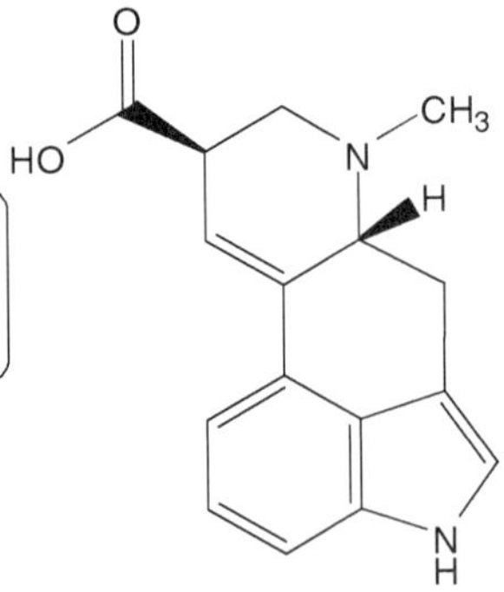

lysergic acid

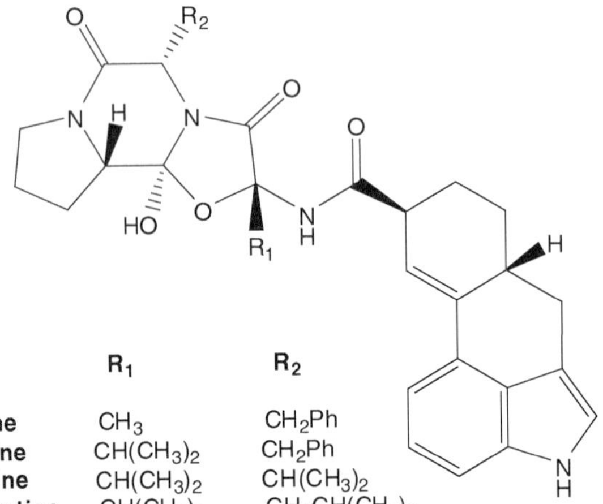

or

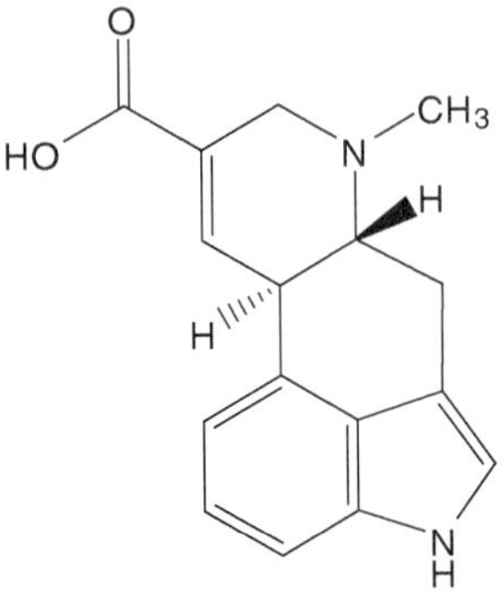

	R1	R2
ergotamine	CH_3	CH_2Ph
ergocristine	$CH(CH_3)_2$	CH_2Ph
ergocornine	$CH(CH_3)_2$	$CH(CH_3)_2$
α-ergocryptine	$CH(CH_3)_2$	$CH_2CH(CH_3)_2$
β-ergocryptine	$CH(CH_3)_2$	$CH(CH_3)CH_2CH_3$

paspalic acid

Estradiol Cypionate

Hormones, Other Endocrine Medicines and Contraceptives/Contraceptives/Injectable Hormonal Contraceptives

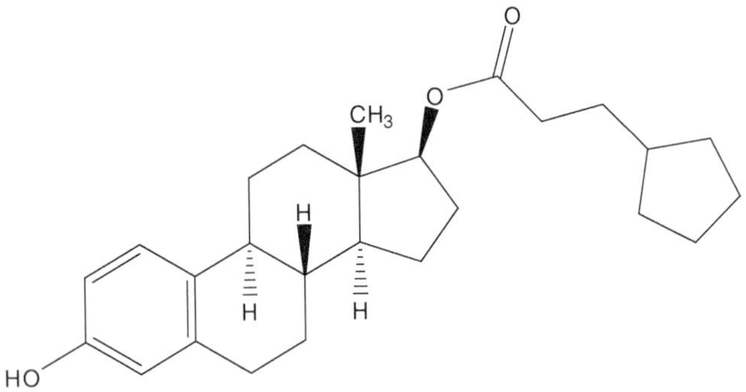

A single-enantiomer molecule with multiple chiral carbons is often formed by modification of a natural product which has most or all of the chiral carbons already in place. Steroids with an aromatic A ring are often formed from androsta-1,4-diene-3,17-dione which is produced by microbial oxidation/side chain degradation of a phytosterol.

Discussion. Estradiol cypionate is manufactured in three steps from estrone and in six steps from the plant sterol β-sitosterol. The final step is hydrolysis of the C3 ester of the 3,17-diester. The 3,17-diester is formed by reaction of estradiol with 3-cyclopentylpropionyl chloride. Estradiol is formed by reduction of estrone. Estrone is formed by A-ring aromatization from androsta-1,4-diene-3,17-dione ethylene glycol ketal (acetal). The acetal is formed from androsta-1,4-diene-3,17-dione (boldione). Androsta-1,4-diene-3,17-dione is formed by microbial oxidation/side chain degradation of phytosterols (plant sterols) including β-sitosterol.

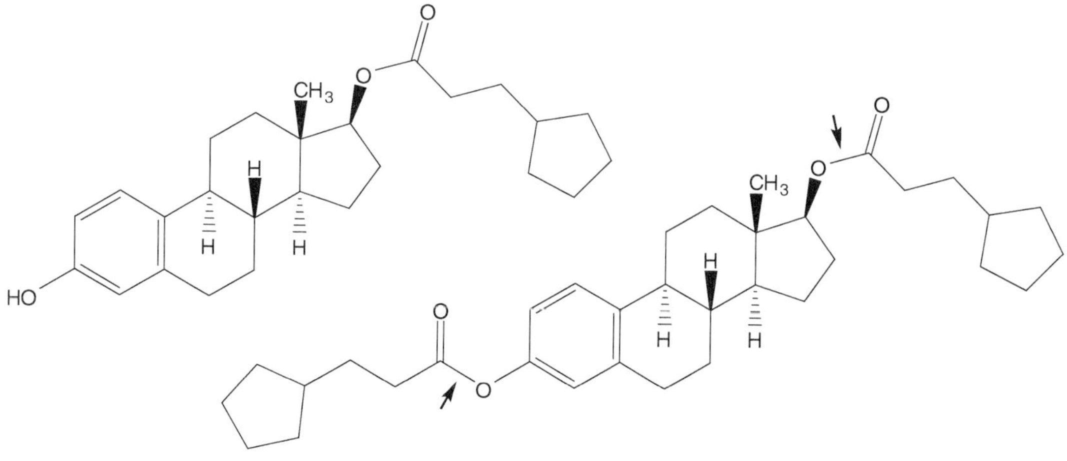

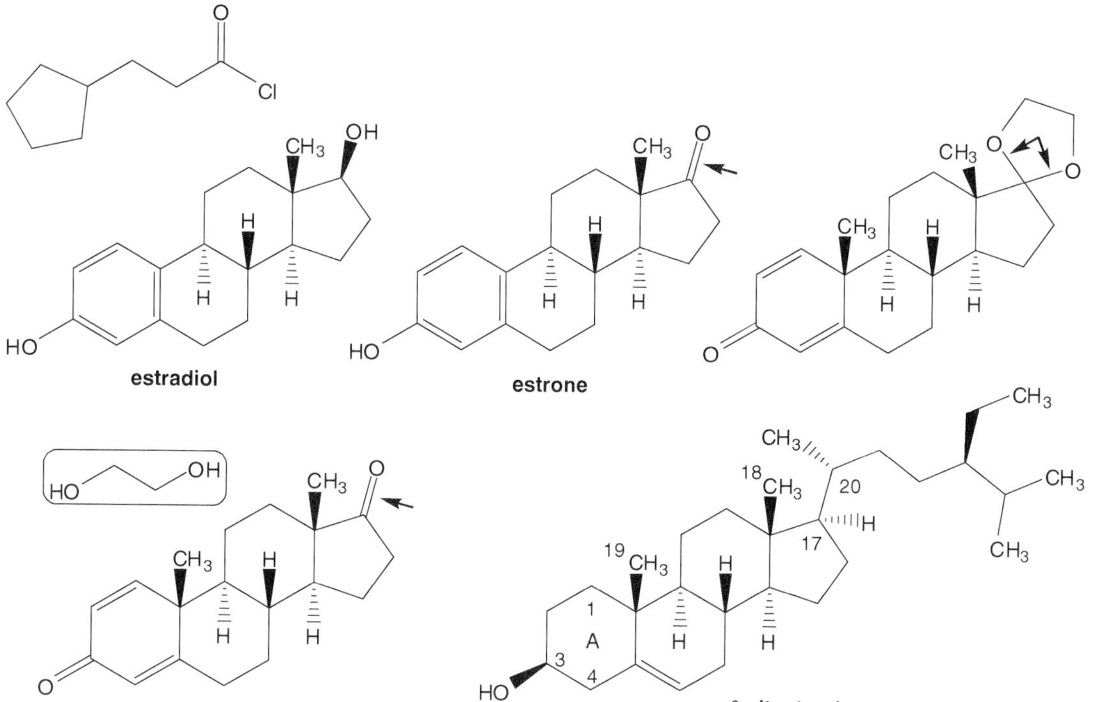

estradiol

estrone

β-sitosterol

3-Cyclopentylpropionyl chloride is formed from the carboxylic acid. 3-Cyclopentylpropionic acid (cypionic acid) is formed from 3-(2-oxocyclopentyl)propionitrile by reduction of the ketone and hydrolysis of the nitrile. 3-(2-Oxocyclopentyl)propionitrile is formed by addition of an imine or enamine of cyclopentanone to acrylonitrile (**Michael Addition**).

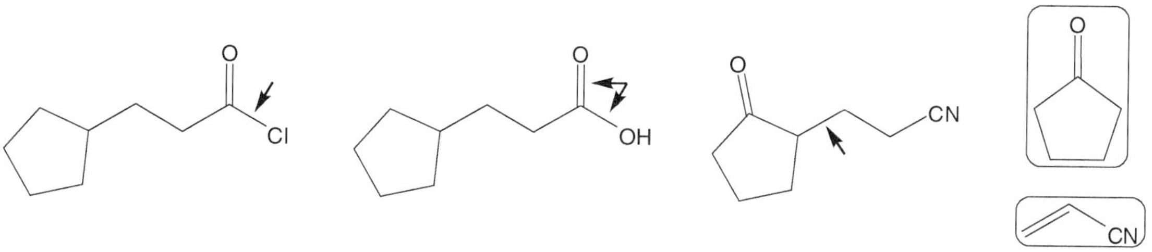

Extended Discussion

Draw the structures of the retrosynthetic analysis of one alternative route to 3-cyclopentylpropionic acid.

Ethambutol

Anti-infective Medicines/Antibacterials/Antituberculosis Medicines

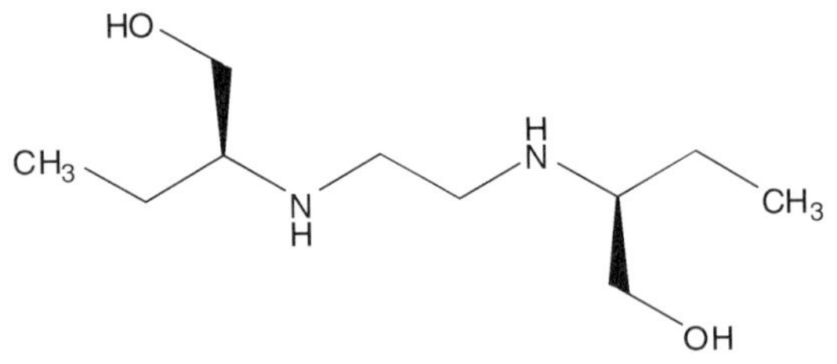

A beta-amino alcohol with a primary α-C is often formed by reduction of a beta-nitro alcohol.

Discussion. The 1,2-diamine is formed by chloride displacement from 1,2-dichloroethane by the primary amine (*S*)-2-amino-1-butanol. (Draw the structures of likely side products of this reaction. How is the procedure designed to minimize the formation of these side products? How are these side products separated from ethambutol?) (*S*)-2-Amino-1-butanol is produced by resolution. The unwanted (*R*)-2-amino-1-butanol can be racemized and the racemate returned to the resolution.

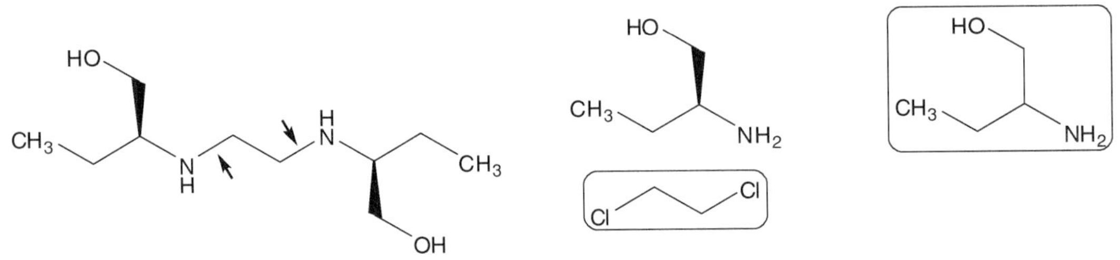

Extended Discussion

(*S*)-2-Amino-1-butanol is also formed from an inexpensive proteinogenic amino acid. Identify one amino acid and outline the route from the amino acid to (*S*)-2-amino-1-butanol.

Ethinylestradiol

Hormones, Other Endocrine Medicines and Contraceptives/Contraceptives/Oral Hormone Contraceptives

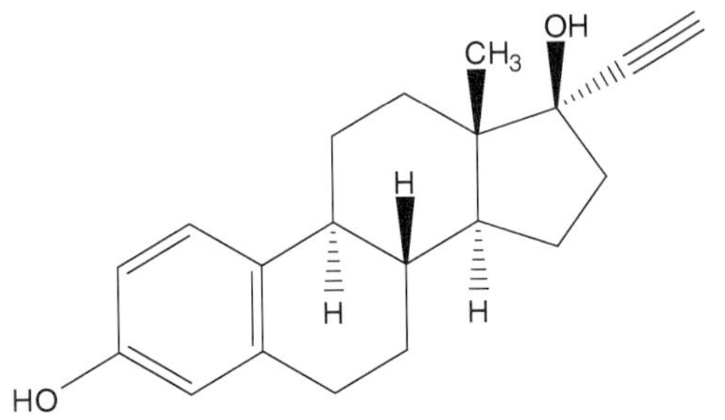

A single-enantiomer molecule with multiple chiral carbons is often formed by modification of a natural product which has most or all of the chiral carbons already in place. Steroids with an aromatic A ring are often formed from androsta-1,4-diene-3,17-dione which is produced by microbial oxidation/side chain degradation of a phytosterol.

Discussion. Ethinylestradiol is manufactured in four steps from the plant sterol β-sitosterol. The tertiary alcohol at C17 is formed by addition of a metal acetylide to the C17 ketone of estrone in the final step. (List the reagents and conditions used to form the metal acetylide and the associated yield of ethinylestradiol.) Estrone is formed by A-ring aromatization from androsta-1,4-diene-3,17-dione ethylene glycol ketal (acetal). The acetal is formed from androsta-1,4-diene-3,17-dione (boldione). Androsta-1,4-diene-3,17-dione is formed by microbial oxidation/side chain degradation of phytosterols (plant sterols) including β-sitosterol.

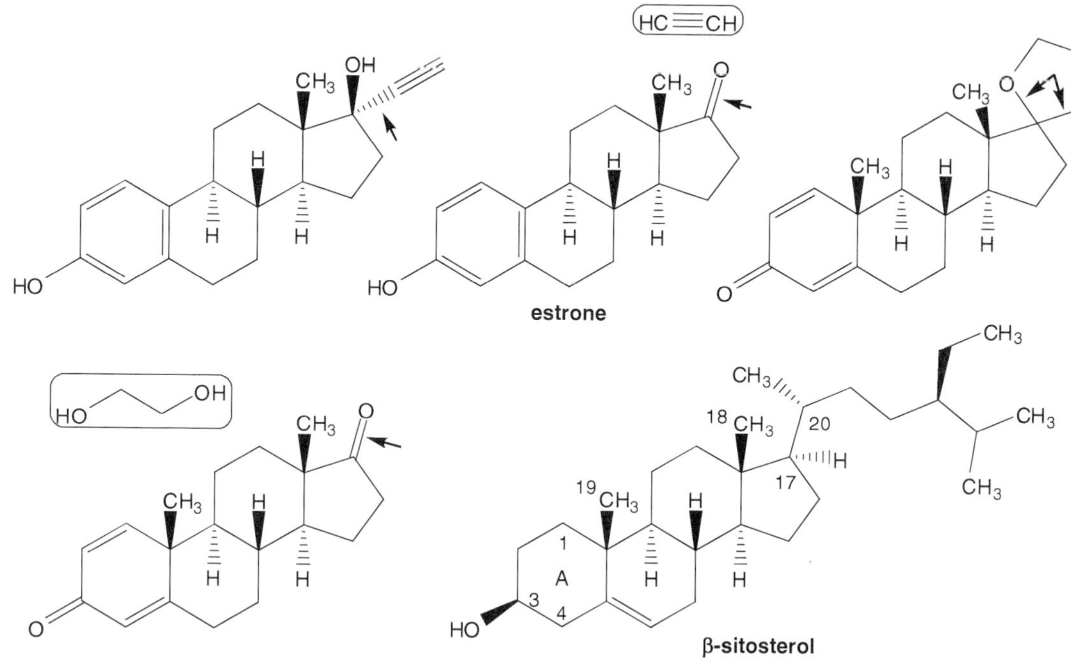

Extended Discussion

Draw the structures of the retrosynthetic analysis of one alternative route to estrone. List the pros and cons for both routes.

Ethionamide

Anti-infective Medicines/Antibacterials/Antituberculosis Medicines

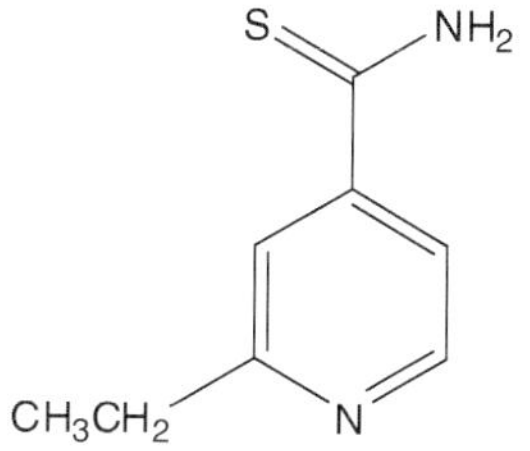

Hydrogen at C2 and/or C6 of a pyridine ring is often introduced by hydrogenolysis of a chlorine substituent. The chloropyridine is formed from the pyridone.

Discussion. Three functional group interconversions at the end of the sequence convert the ester to the thioamide. The thioamide is formed from the nitrile, the nitrile is formed by dehydration of the amide, and the amide is formed from the ester.

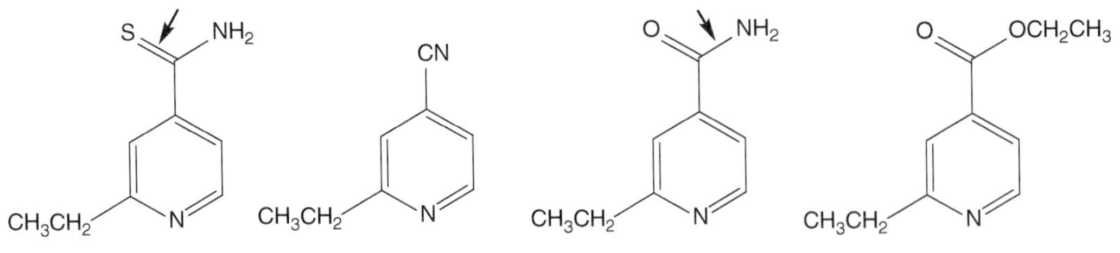

The simplicity of ethionamide and the ester precursor suggests an unusual retrosynthetic strategy: *functional groups are added to construct the pyridine ring*. After the pyridine ring is formed, the functional groups are replaced by hydrogen. Chlorine at C6 is removed by hydrogenolysis. The ester is formed from the acid chloride. The chloropyridine and acid chloride are formed from the pyridone and carboxylic acid. The 2-pyridone-4-carboxylic acid is formed by decarboxylation of the 2-pyridone-3,4-dicarboxylic acid. The dicarboxylic acid is formed by hydrolysis of an ester at C4 and a nitrile at C3. The pyridine ring is formed by reaction of cyanoacetamide and a β-diketone. The β-diketone is formed from 2-butanone and diethyl oxalate (mixed **Claisen Condensation**).

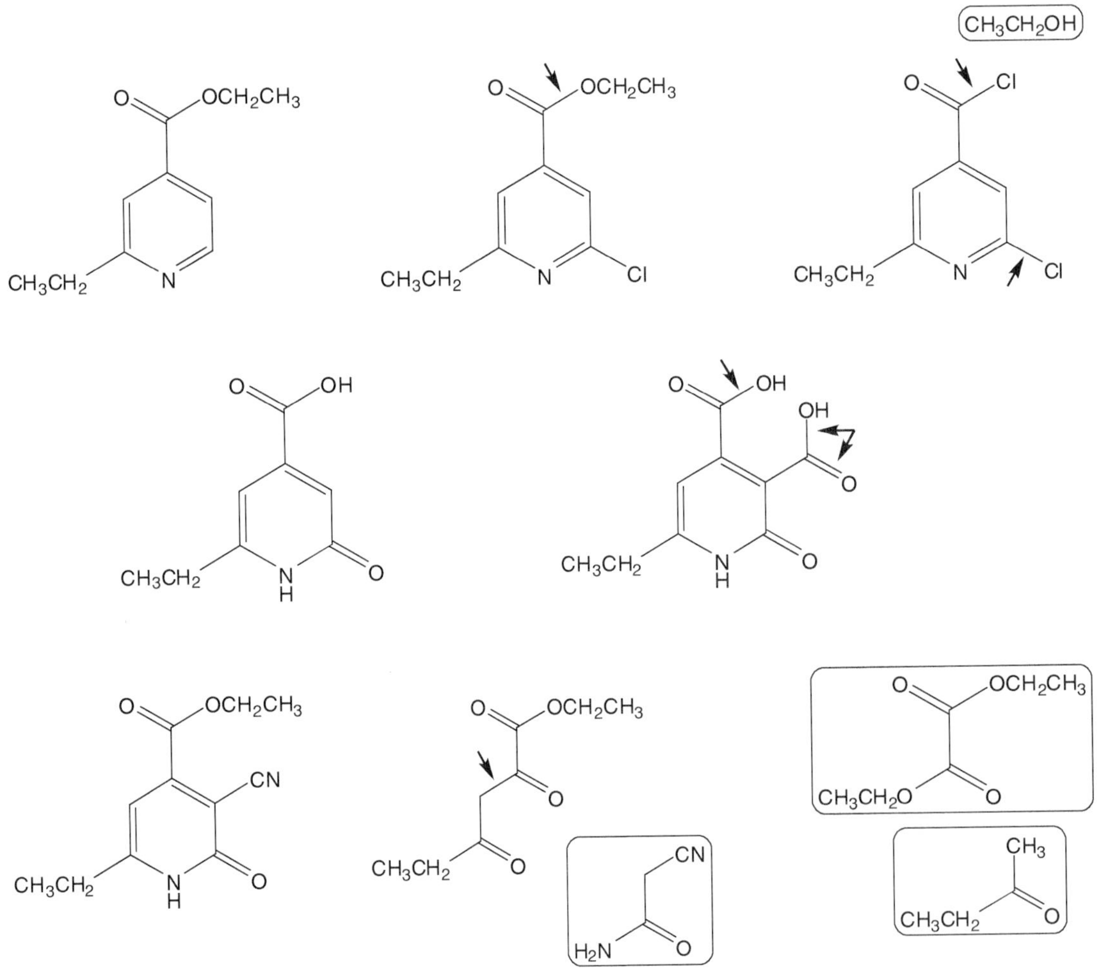

Extended Discussion

Adding functional groups which are not in the target molecule results in a nine-step synthesis of this simple molecule. Draw the structures of the retrosynthetic analysis of an alternative route to ethionamide from 2-ethylpyridine. Include the structures of the retrosynthetic analysis of 2-ethylpyridine from petrochemical or biochemical raw materials. List the pros and cons for the two routes and select one route as the preferred route.

Ethosuximide

Anticonvulsants/Antiepileptics

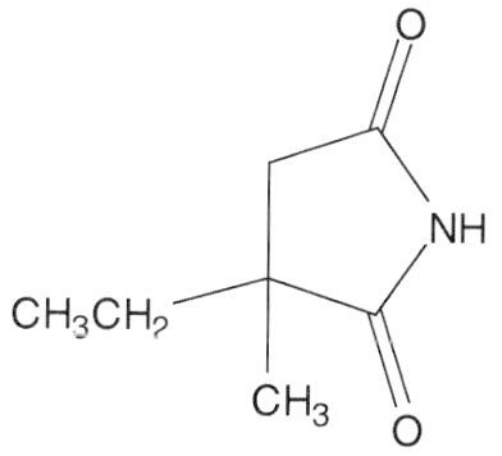

A succinimide is often formed from the succinic acid and ammonia.

Discussion. Ethosuximide is a 1 : 1 mixture of (*R*)- and (*S*)-enantiomers. The succinimide is formed by heating the ammonium succinate. (Ethosuximide is often distilled at reduced pressure. List the reported boiling points and pressures. What is the melting point of ethosuximide?) The ammonium succinate is formed in situ from the succinic acid. The succinic acid is formed by hydrolysis and decarboxylation of ethyl 2,3-dicyano-3-methylpentanoate. Ethyl 2,3-dicyano-3-methylpentanoate is formed by addition of cyanide to ethyl 2-cyano-3-methyl-2-pentenoate (**Michael Addition**). Ethyl 2-cyano-3-methyl-2-pentenoate (*E*/*Z* mixture) is formed by condensation of ethyl cyanoacetate and 2-butanone (**Knoevenagel Condensation**).

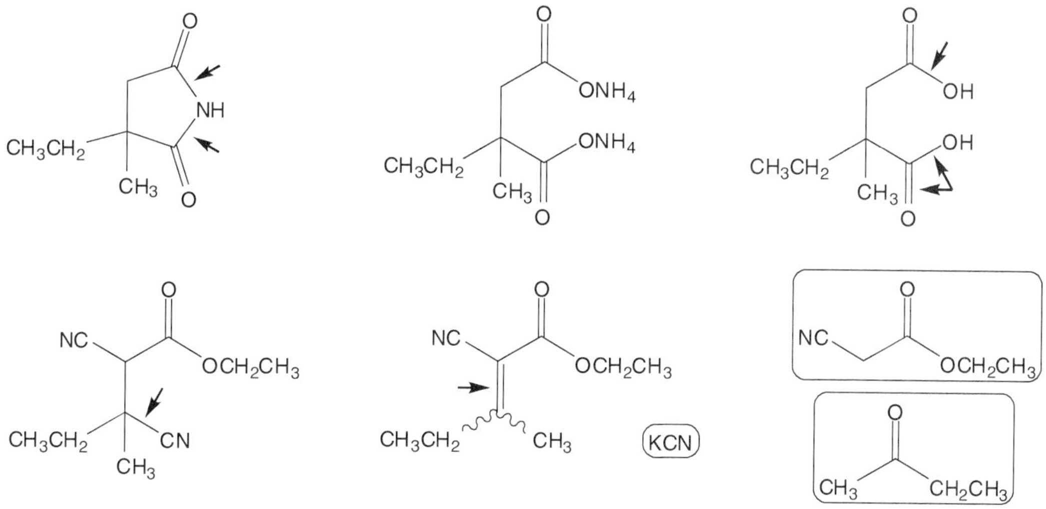

Etonogestrel

Hormones, Other Endocrine Medicines and Contraceptives/Contraceptives/Implantable Contraceptives

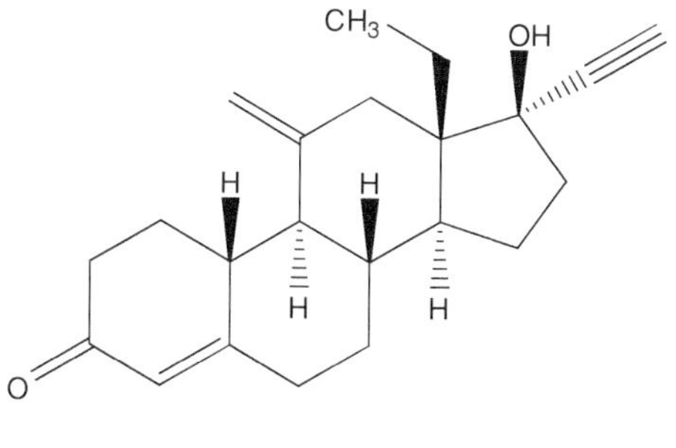

Steroids with an ethyl group at C13 on the gonane structure are often manufactured by total synthesis. The gonane is constructed from 6-methoxy-1-tetralone, vinyl chloride, and 2-ethyl-1,3-cyclopentanedione (Torgov–Smith Synthesis). Chirality at C13 is established by microbial reduction of a 2,2-disubstituted-1,3-cyclopentanedione intermediate.

Discussion. Etonogestrel is manufactured in eight steps from 13β-ethylgon-4-ene-3,17-dione, in 13 steps from 13-ethyl-3-methoxygona-1,3,5(10),8,14-pentaen-17β-ol acetate, and in 18 steps from 6-methoxy-1-tetralone.

The 4-ene-3-one of etonogestrel is released in the final step by hydrolysis of the C3 acetal and migration of the double bond. The tertiary alcohol at C17 is formed by addition of a metal acetylide to the C17 ketone. The C17 ketone is formed by oxidation of the secondary alcohol. The C11 ketone is converted to the C11 methylene (**Wittig Reaction**). (List other reagents and conditions used to convert the C11 ketone to the C11 methylene.) The C17 ketone of the 11,17-dione is protected as/reduced to the secondary alcohol. The C3 ketone of the 3,11,17-trione is protected as an acetal. The C11 ketone is formed by oxidation of the C11 α-alcohol. The C11 α-alcohol is formed by microbial oxidation of 13β-ethylgon-4-ene-3,17-dione.

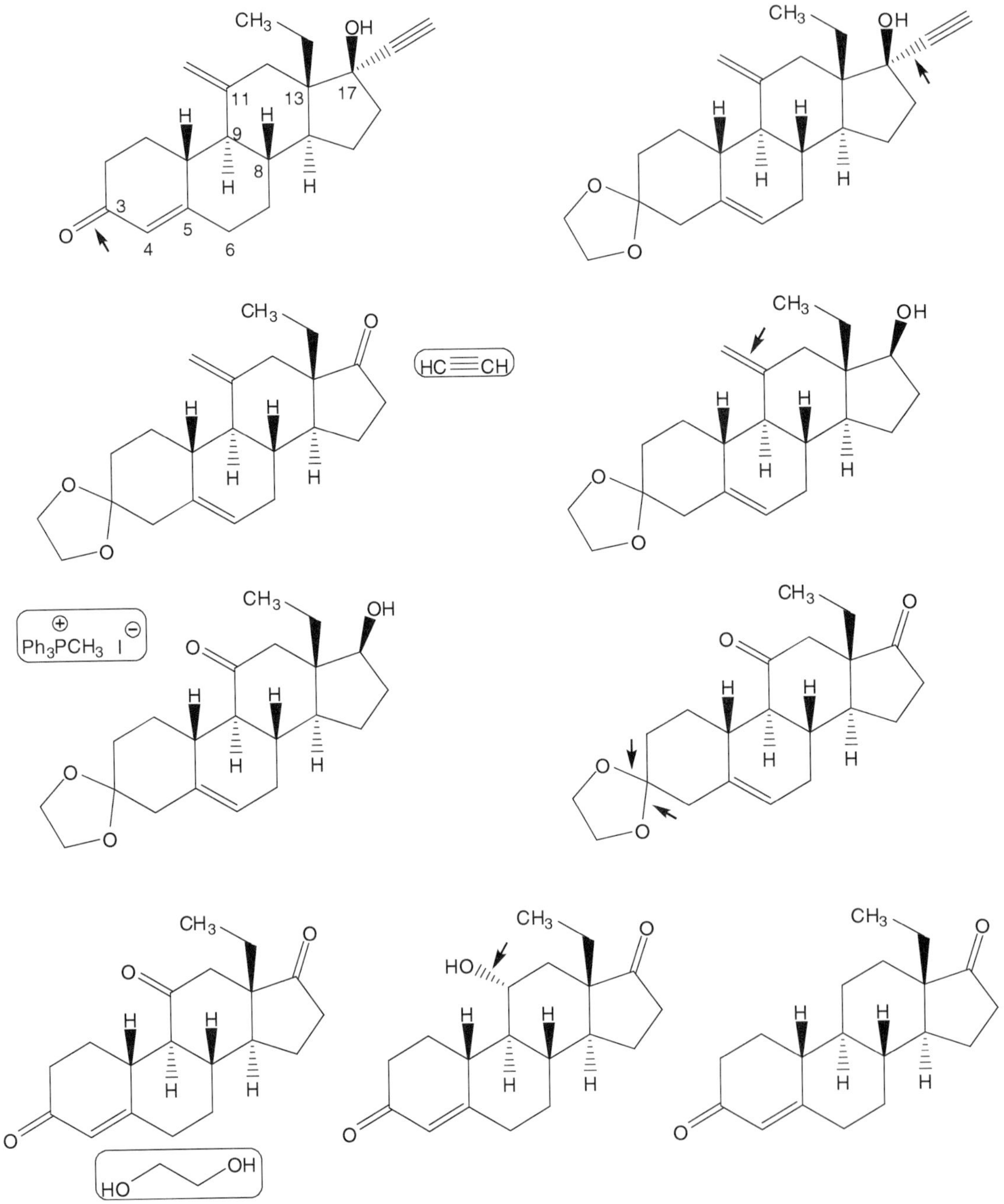

The 4-ene-3-one is formed by hydrolysis of the enol ether and migration of the 5(10)-alkene. The C17 ketone is formed by oxidation of the C17 secondary alcohol (**Oppenauer Oxidation**). The 3-methoxy-2,5(10)-diene is formed by reduction of the aromatic ring (**Birch Reduction**). The 8-alkene is also reduced under the conditions used for reduction of the aromatic ring. The C17 secondary alcohol is released by ester hydrolysis. The 14-alkene of the gona-1,3,5(10),8,14-pentaene is reduced by catalytic hydrogenation.

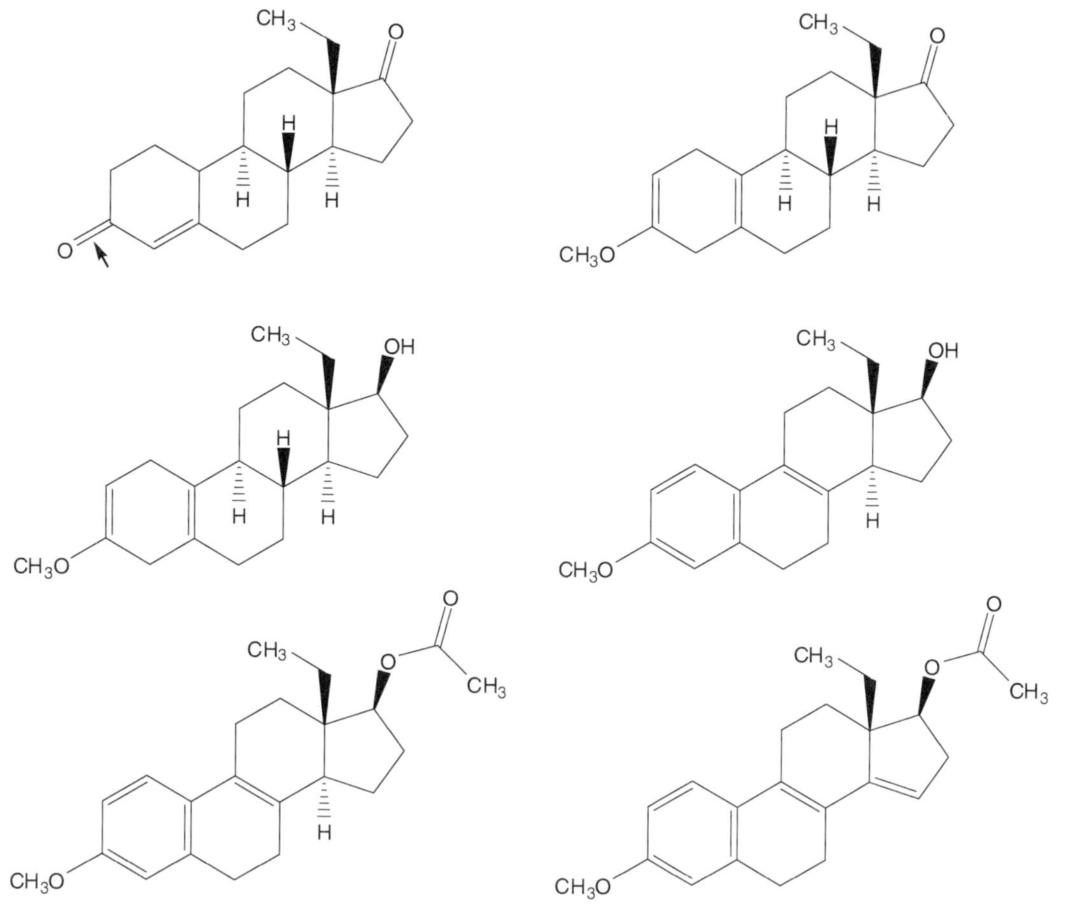

The gona-1,3,5(10),8,14-pentaene is formed from the secosteroid. The acetate ester of the secosteroid is formed by reaction of the C17 alcohol with acetic anhydride. The chiral carbons at C13 and C17 are formed by microbiological reduction of the secodione. The C12—C13 bond of the secodione is formed by the reaction of 6-methoxy-1-vinyl-1,2,3,4-tetrahydronaphthalen-1-ol with 2-ethyl-1,3-cyclopentanedione. 6-Methoxy-1-vinyl-1,2,3,4-tetrahydronaphthalen-1-ol is formed by addition of vinylmagnesium chloride to 6-methoxy-1-tetralone (**Grignard Reaction**). Vinylmagnesium chloride is formed from vinyl chloride.

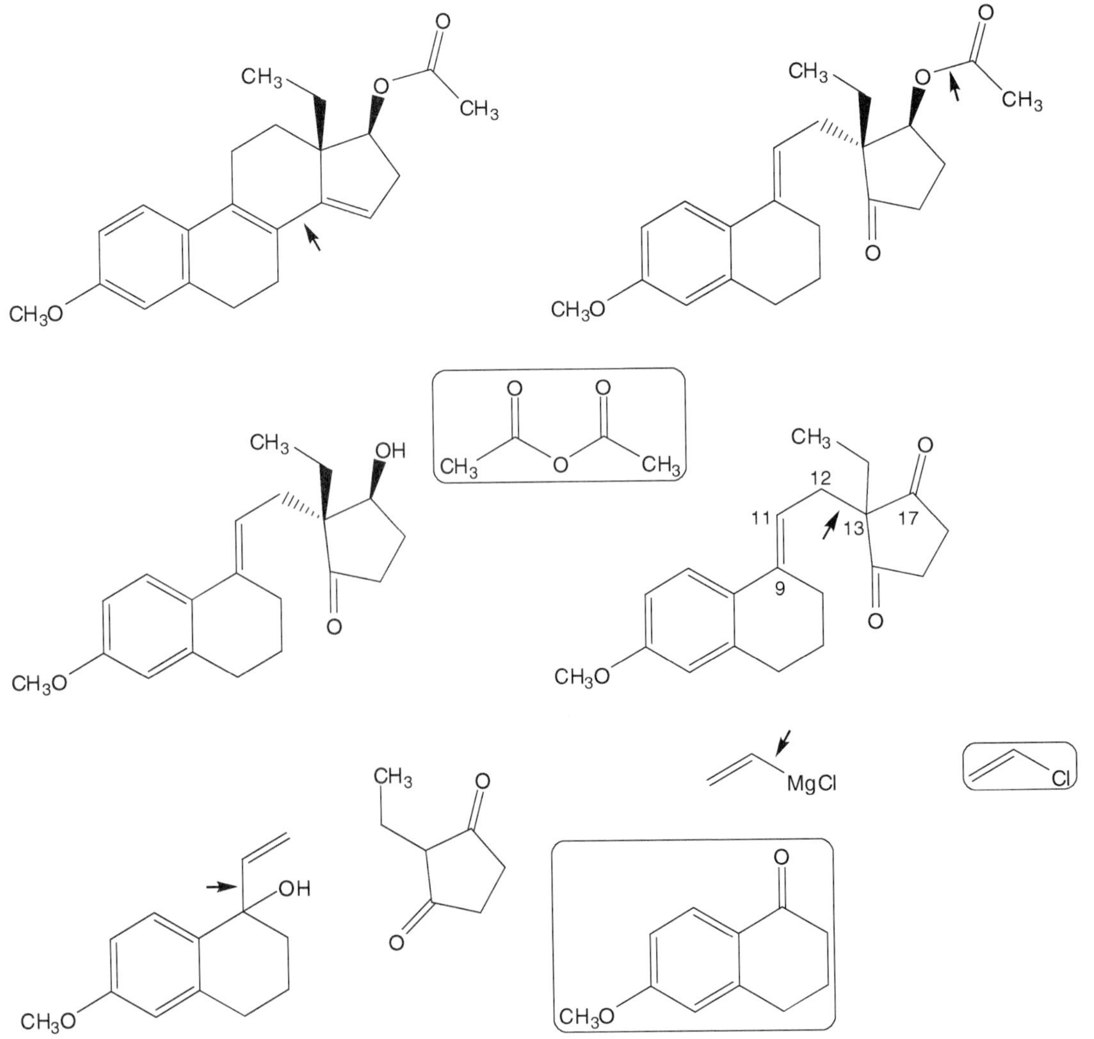

2-Ethyl-1,3-cyclopentanedione is formed from succinic anhydride and butanoic anhydride.

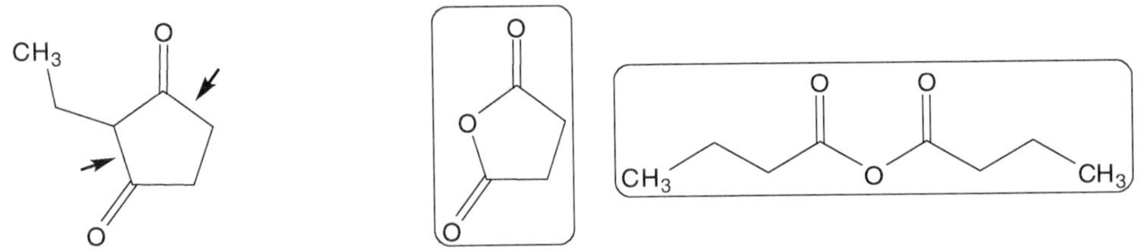

Extended Discussion

Draw the structures of the retrosynthetic analysis of one alternative route to etonogestrel from 6-methoxy-1-tetralone. Include the structures of the retrosynthetic analysis of any organic starting material(s) from petrochemical or biochemical raw materials. How many steps are there in the linear sequence from 6-methoxy-1-tetralone to etonogestrel in the alternative route?

Etoposide

Antineoplastics and Immunosuppressives/Cytotoxic and Adjuvant Medicines

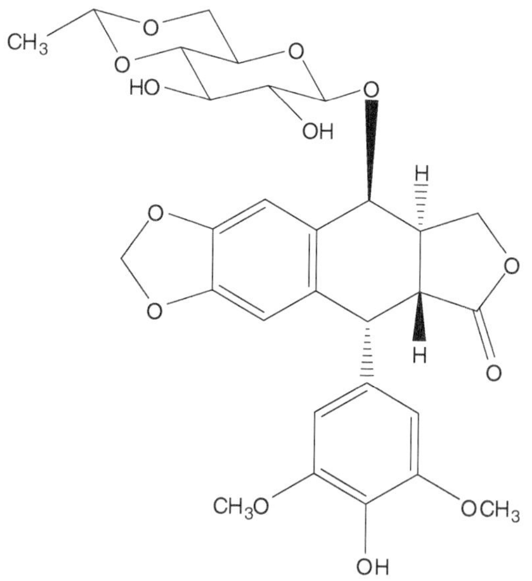

A single-enantiomer molecule with multiple chiral carbons is often formed by modification of a natural product which has most or all of the chiral carbons already in place.

Discussion. Etoposide is semisynthetic. In the final step of one preferred route, the two secondary alcohols are released by cleavage of dichloroacetate esters. The ether link between the sugar and the aglycone is formed by reaction of the β-D-glucopyranose with the alcohol at C4 of 4′-demethylepipodophyllotoxin.

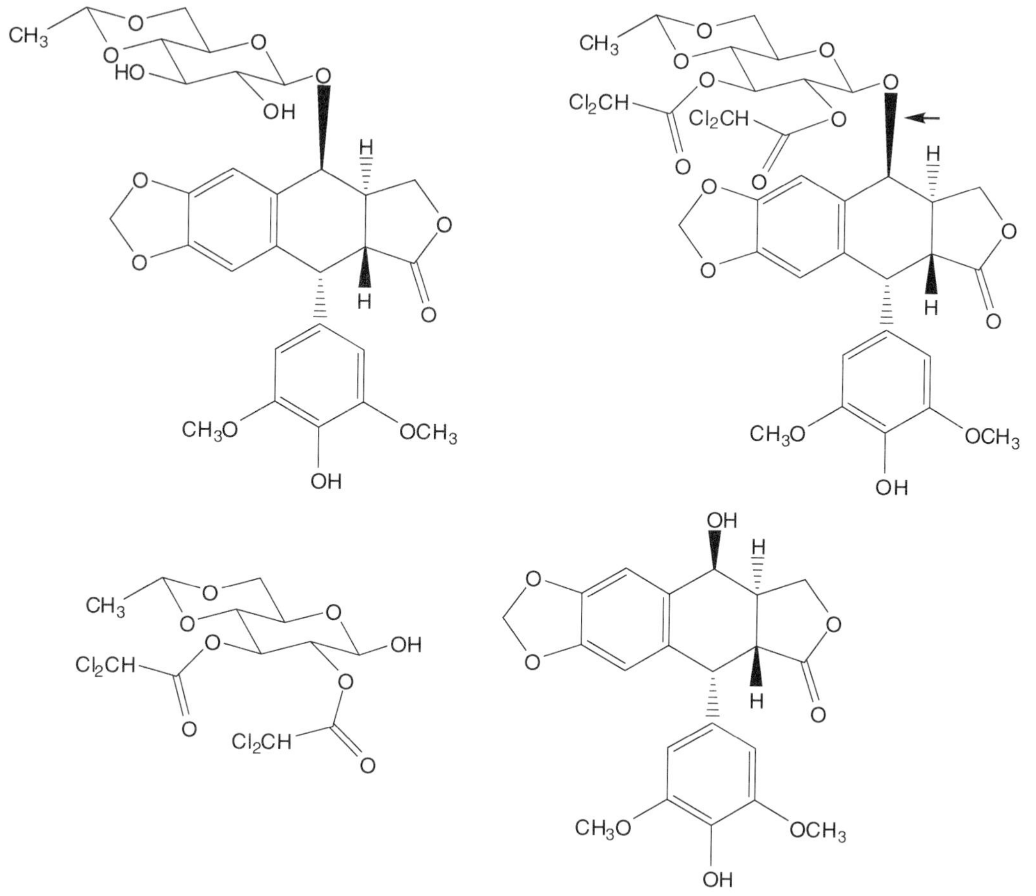

The alcohol at C4 of 4′-demethylepipodophyllotoxin is formed by hydrolysis of the benzylic bromide. Reaction of podophyllotoxin with hydrogen bromide results in formation of the benzylic bromide and demethylation of the ether at C4′. Podophyllotoxin is a non-alkaloid lignan isolated from the rhizomes of *Podophyllum peltatum*, the North American Mayapple.

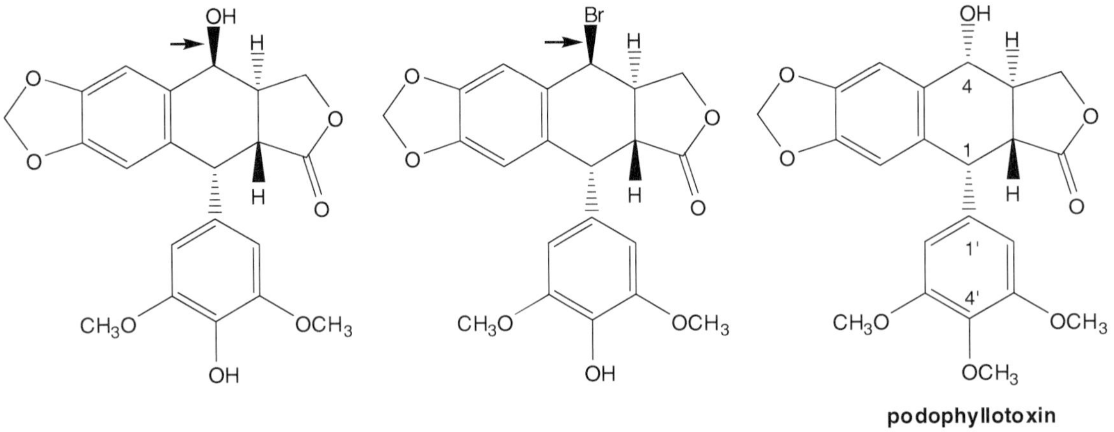

The hemiacetal at C1 of the β-D-glucopyranose is released by hydrogenolysis of the benzyl carbonate. The alcohols at C2 and C3 are protected as dichloroacetate esters. The benzyl carbonate is formed by reaction of 4,6-*O*-ethylidene-D-glucopyr anose with benzyl chloroformate. The acetal of 4,6-*O*-ethylidene-D-glucopyranose is formed by reaction of D-glucose with acetaldehyde.

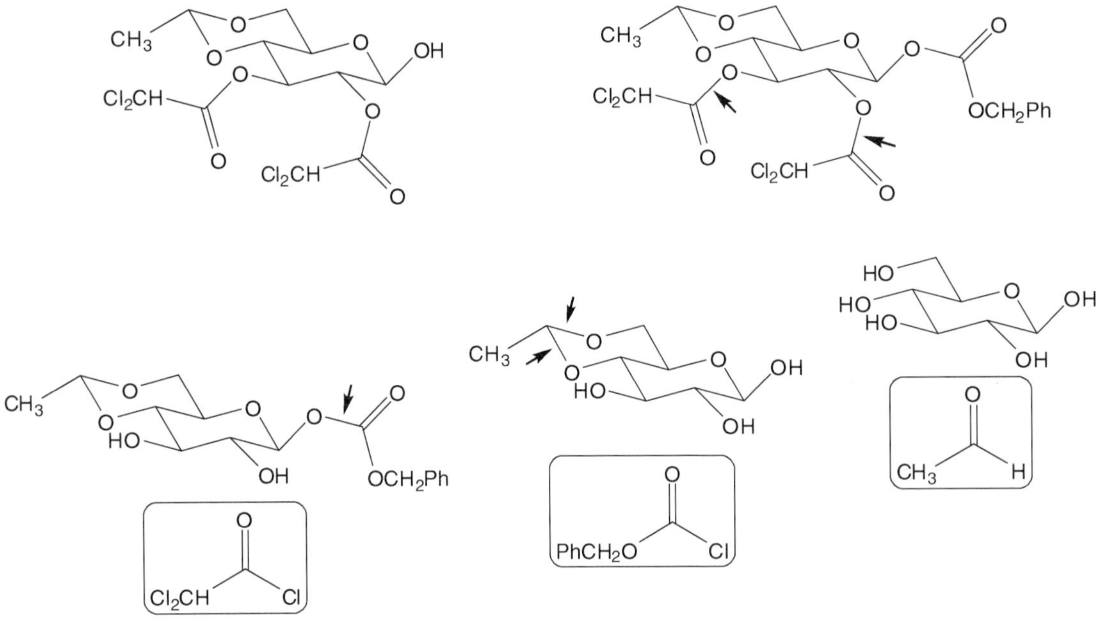

Extended Discussion

Draw the structures of the retrosynthetic analysis of one alternative route to etoposide from podophyllotoxin. List the pros and cons for both routes and select one route as the preferred route.

F

Fentanyl

Medicines for Pain and Palliative Care/Opioid Analgesics

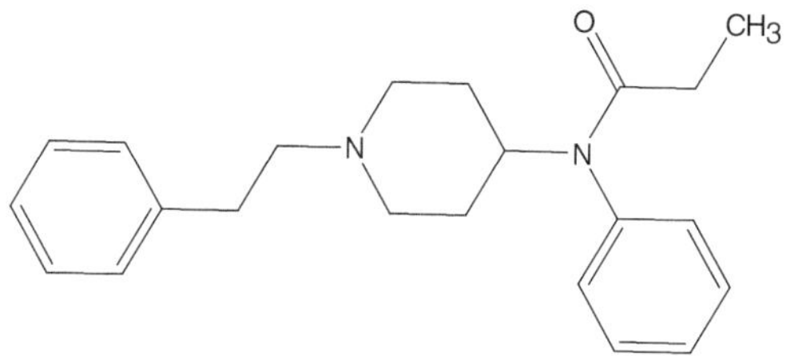

A tertiary amine is often formed by alkylation of a secondary amine. The alkylation is most efficient when the amine is used in excess and when the carbon with the leaving group (Cl, Br, I, OTs, OMs) is primary or benzylic.

Discussion. The fentanyl amide is formed in the final step from propanoyl chloride and the secondary amine, *N*-[1-(2-phenylethyl)-4-piperidinyl]aniline. The secondary amine is formed by reductive amination from the piperidinone and aniline. *N*-Phenylethylpiperidin-4-one is formed by *N*-alkylation of piperidin-4-one with (2-bromoethyl)benzene.

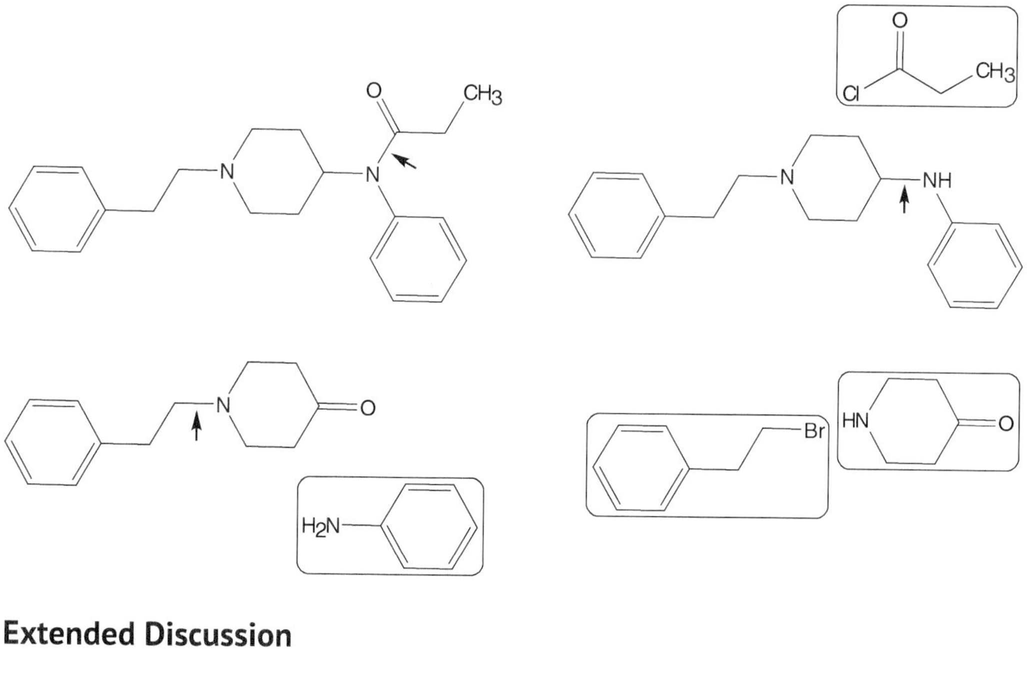

Extended Discussion

Draw the structures of the retrosynthetic analysis of one alternative route to fentanyl. List the pros and cons for both routes and select one route as the preferred route.

Routes to Essential Medicines: A Workbook for Organic Synthesis, First Edition. Peter J. Harrington.

Companion website: www.wiley.com/go/Harrington/routes_essential_medicine

Fluconazole

Antifungal Medicines

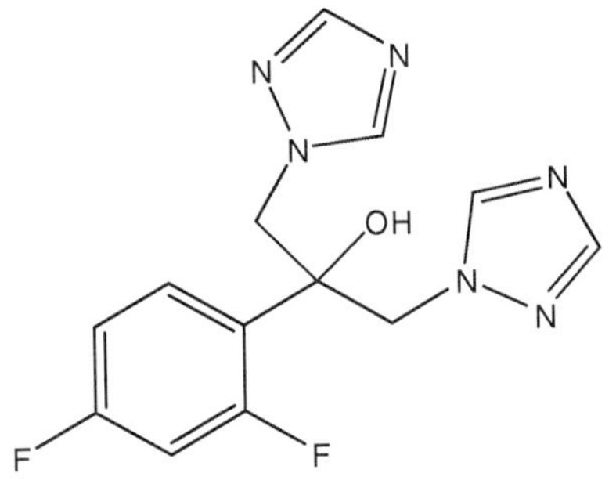

Fluorine on an aromatic ring is often delivered in a fluorinated starting material.

Discussion. The tertiary alcohol is formed in the final step by ring-opening of the epoxide with 1,2,4-triazole. The epoxide is formed by reaction of the ketone with trimethylsulfoxonium iodide (**Corey-Chaykovsky Reaction**). The ketone, 2′,4′-difluoro-2-(1,2,4-triazolyl)acetophenone, is formed by displacement of chloride by 1,2,4-triazole. 2-Chloro-2′,4′-difluoroa cetophenone is formed from 1,3-difluorobenzene and chloroacetyl chloride (**Friedel–Crafts Acylation**).

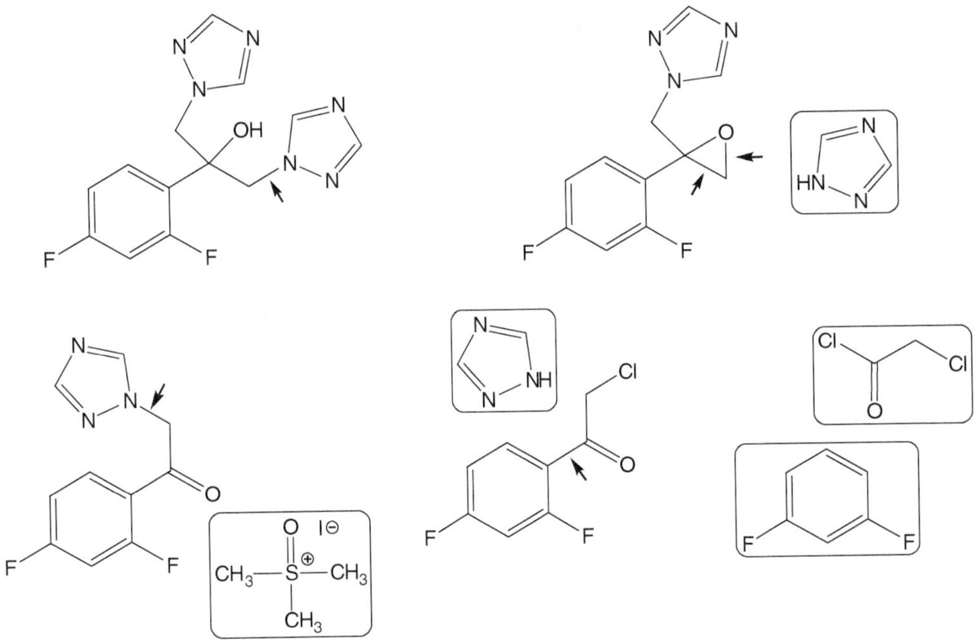

Extended Discussion

Draw the structures of a retrosynthetic analysis of a route to fluconazole starting with 1,3-dichloroacetone. List the pros and cons for the two routes and select one route as the preferred route.

Flucytosine

Anti-infective Medicines/Antifungal Medicines

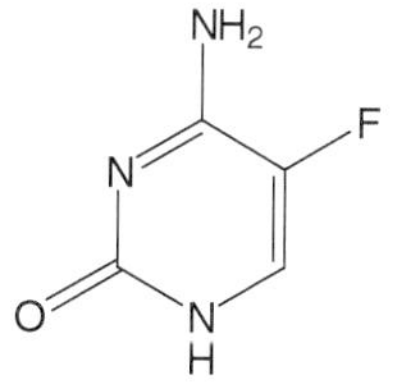

An oxygen or nitrogen substituent on C2 or C4 of a pyrimidine is often introduced by displacement of chloride.

Discussion. Flucytosine (5-fluorocytosine) is manufactured in three steps from the essential medicine 5-fluorouracil. The final step is hydrolysis of 4-amino-2-chloro-5-fluoropyrimidine. The 4-amino-2-chloropyrimidineis formed from the 2,4-dichloropyrimidine. The 2,4-dichloropyrimidine is formed from the pyrimidine-2,4-dione of 5-fluorouracil.

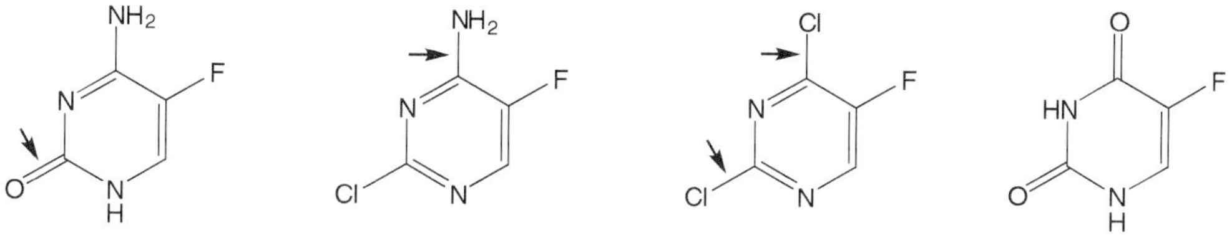

Extended Discussion

Flucytosine is also formed by the direct fluorination of cytosine. This alternative route is attractive when cytosine is comparable in cost to uracil. Draw the structures of a retrosynthetic analysis of one route to cytosine. Include the structures of the retrosynthetic analysis of any organic starting material(s) from petrochemical or biochemical raw materials.

Fludarabine Phosphate

Antineoplastics and Immunosuppressives/Cytotoxic and Adjuvant Medicines

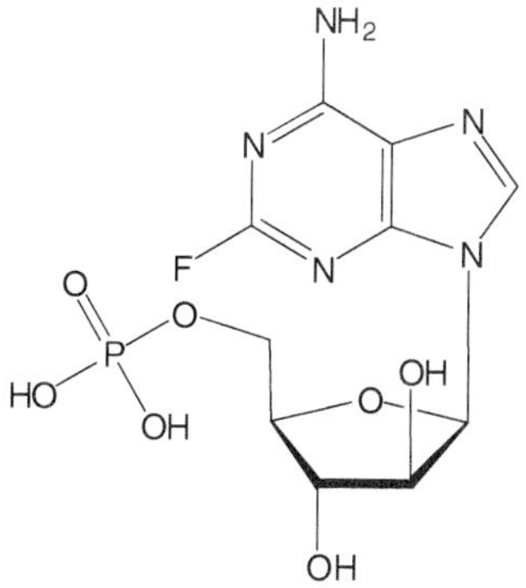

A single-enantiomer molecule with multiple chiral carbons is often formed by modification of a natural product which has most or all of the chiral carbons already in place.

Discussion. Fludarabine phosphate is formed from fludarabine in the final step. The alcohols at C2′, C3′, and C5′ are released by hydrolysis of 2′,3′,5′-tri-*O*-acetylfludarabine. The β-face acetate at the 2′-position is formed by displacement of a trifluoromethanesulfonate on the α-face. The α-trifluoromethanesulfonate is formed from the α-alcohol of 3′,5′-di-*O*-acetyl-2-fluoroadenosine. 3′,5′-Di-*O*-acetyl-2-fluoroadenosine is formed from 2′,3′,5′-tri-*O*-acetyl-2-fluoroadenosine. (Draw the structures for all the products formed in this reaction. What reaction conditions are associated with the highest yield of 3′,5′-di-*O*-acetyl-2-fluoroadenosine? How is 3′,5′-di-*O*-acetyl-2-fluoroadenosine separated from the other products?)

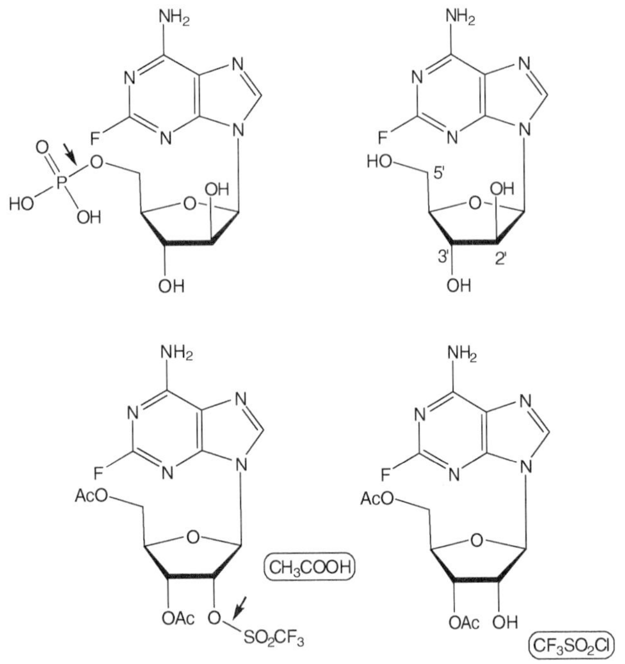

2′,3′,5′-Tri-*O*-acetyl-2-fluoroadenosine is formed from 2-fluoroadenosine. 2-Fluoroadenosine is formed from 2-aminoadenosine via the diazonium salt (**Balz–Schiemann Reaction**). 2-Aminoadenosine is formed from guanosine. Guanosine is produced by fermentation.

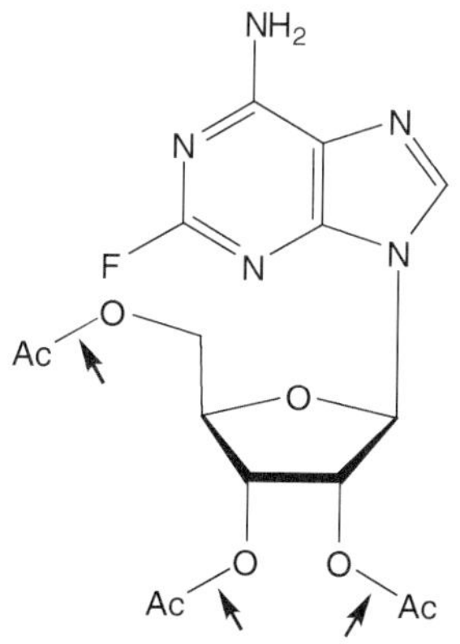

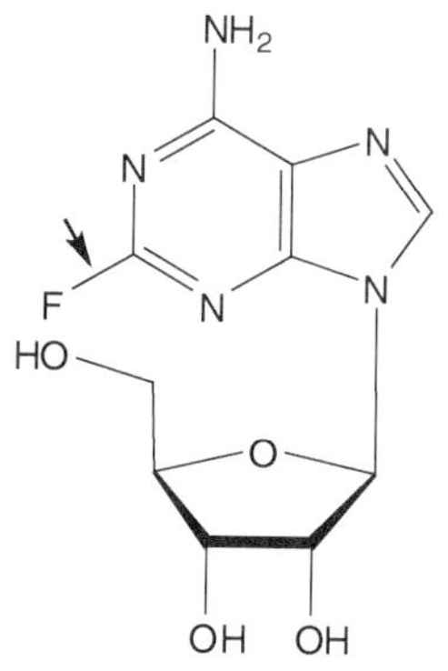

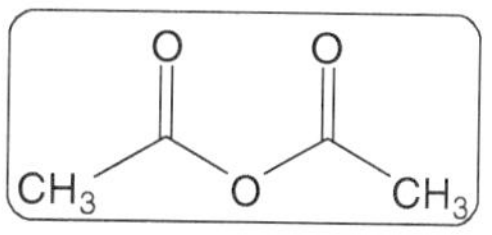

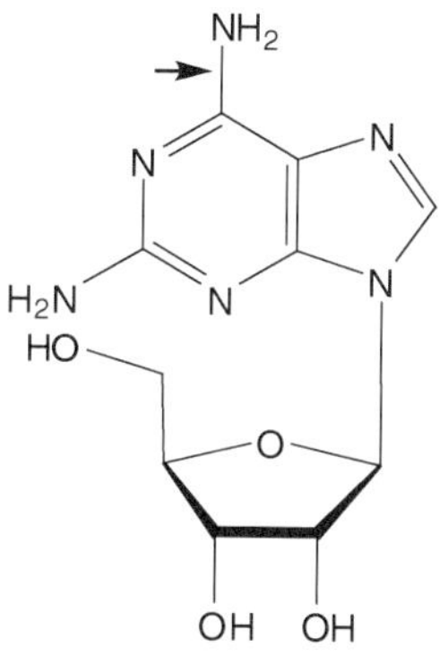

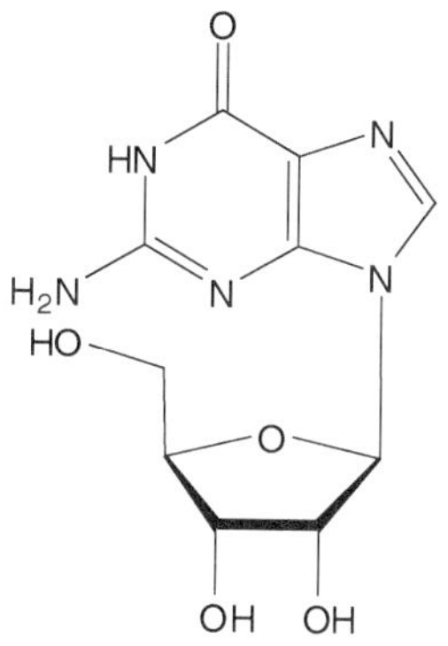

guanosine

Extended Discussion

Draw the structures of a retrosynthetic analysis of an alternative route to fludarabine starting with D-arabinose. Include the structures of the retrosynthetic analysis of any organic starting material(s) from petrochemical or biochemical raw materials.

Fludrocortisone Acetate

Hormones, Other Endocrine Medicines and Contraceptives/Adrenal Hormones and Synthetic Substitutes

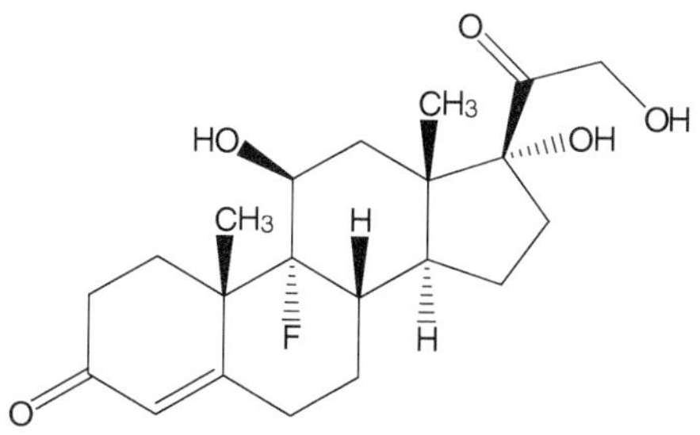

A single-enantiomer molecule with multiple chiral carbons is often formed by modification of a natural product which has most or all of the chiral carbons already in place. A halogen atom at C9 on the α-face of a corticosteroid is often introduced by ring-opening of a 9(11)-epoxide on the β-face.

Discussion. Fludrocortisone acetate is manufactured in five steps from another essential medicine, hydrocortisone. Fluorine is introduced at C9 on the α-face in the final step by ring-opening of the 9(11)-epoxide with hydrogen fluoride. The 9(11)-epoxide is formed from the bromohydrin. The bromohydrin is formed from the 9(11)-alkene. The 9(11)-alkene is formed by dehydration of hydrocortisone acetate. Hydrocortisone acetate is formed by reaction of the C21 alcohol of hydrocortisone with acetic anhydride.

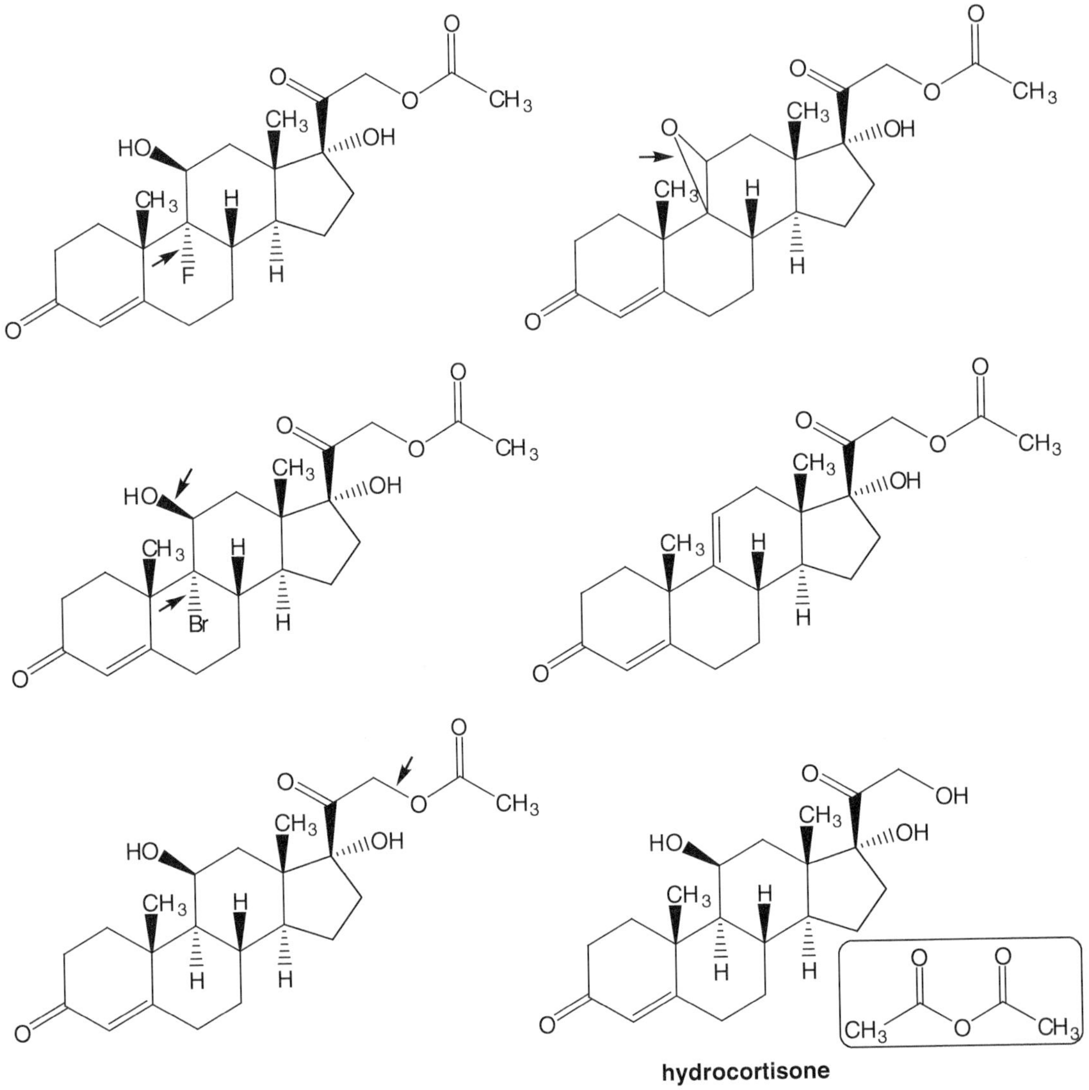

Fluorouracil

Anti-infective Medicines/Antifungal Medications

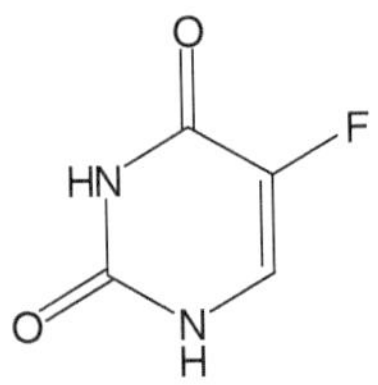

A fluorine substituent on a heterocycle is often delivered in a starting material used to construct the heterocycle.

Discussion. Manufacturers with the process equipment and expertise to work with electrophilic fluorinating agents produce 5-fluorouracil by fluorination of uracil or fluorination of cytosine followed by hydrolysis of the amino group. List the options for starting material (uracil or cytosine), fluorinating agent, solvent, reaction temperature, reaction time, and the yield of 5-fluorouracil associated with each option.

Route A

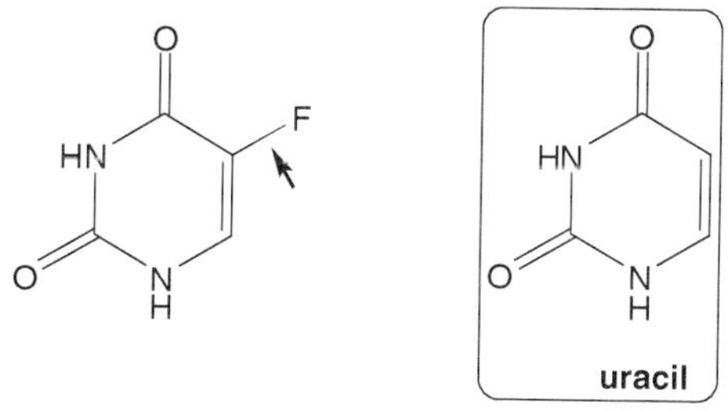

Route B

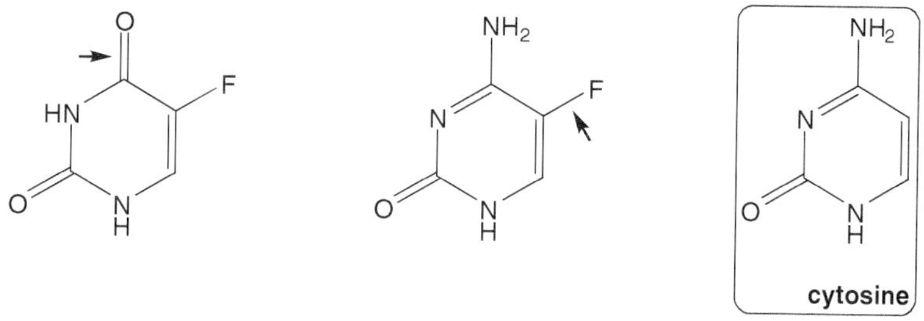

Extended Discussion

Draw the structures of a retrosynthetic analysis of one alternative route to 5-fluorouracil from sodium fluoroacetate. Include the structures of the retrosynthetic analysis of sodium fluoroacetate and any other organic starting material(s) from petrochemical or biochemical raw materials.

Fluoxetine

Medicines for Pain and Palliative Care/Medicines for Other Common Symptoms in Palliative Care
Medicines for Mental and Behavioral Disorders/Medicines Used in Mood Disorders/Medicines Used in Depressive Disorders

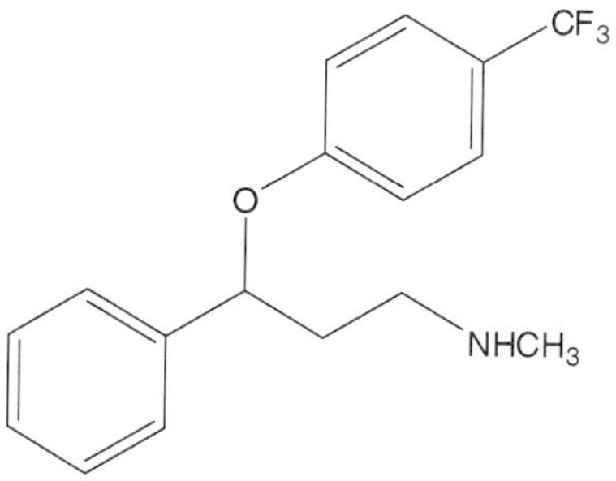

The trifluoromethyl group is often formed from a trichloromethyl group by halogen exchange (Halex Reaction).

Discussion. Fluoxetine is a 1 : 1 mixture of (*R*)- and (*S*)-enantiomers. The ether is formed in the final step by displacement of chloride from 4-chlorobenzotrifluoride by 3-methylamino-1-phenyl-1-propanol. (List all the options for the base, solvent, catalyst, reaction temperature, and time. Which reaction conditions are preferred?) The alcohol and the amine are released by carbonate and carbamate hydrolysis. The carbonate and carbamate are formed by reaction of 3-dimethylamino-1-phenyl-1-propanol with ethyl chloroformate. The secondary alcohol is formed by reduction of the ketone. 3-Dimethylamino-1-phenyl-1-propanone is formed from acetophenone, formaldehyde, and dimethylamine (**Mannich Reaction**).

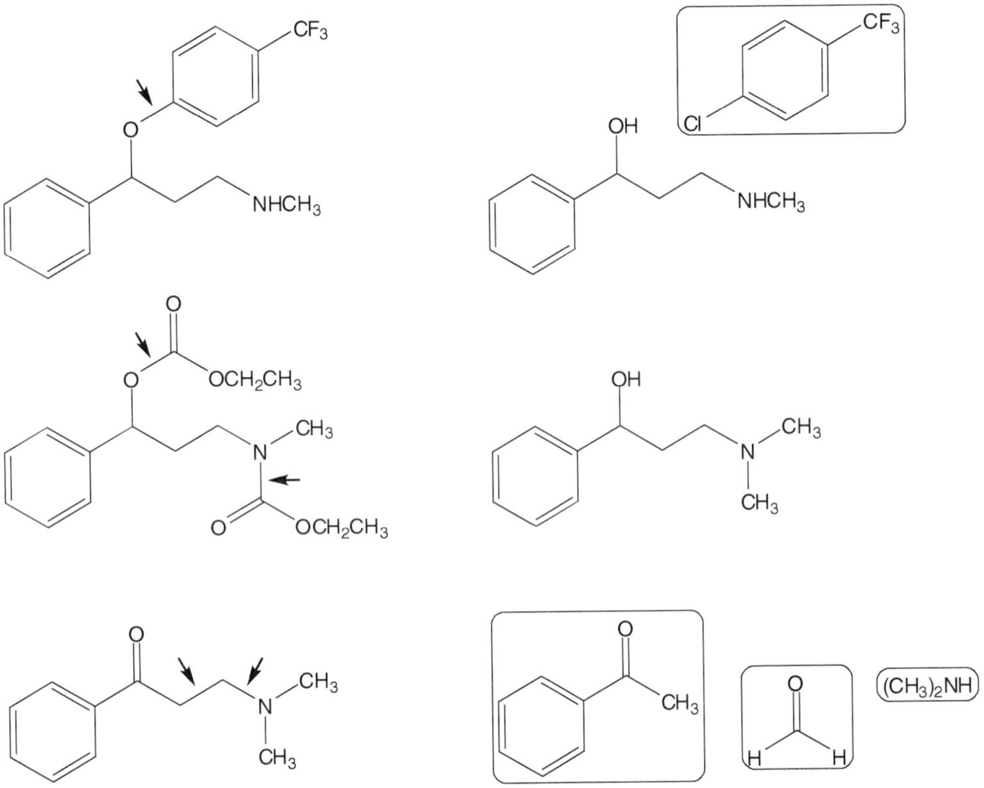

Extended Discussion

Draw the structures for the retrosynthetic analysis of one alternative route to 3-methylamino-1-phenyl-1-propanol. Include the structures of the retrosynthetic analysis of any organic starting material(s) from petrochemical or biochemical raw materials. List the pros and cons for both routes and select one route as the preferred route to this intermediate.

Fluphenazine

Medicines for Mental and Behavioral Disorders/Medicines Used in Psychotic Disorders

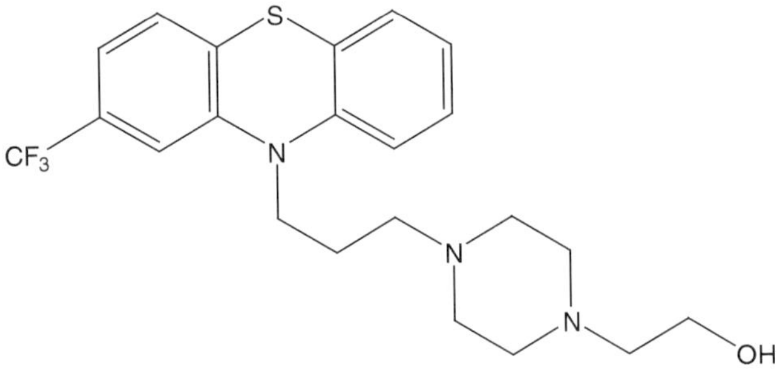

Nucleophilic aromatic substitution is often facilitated by an electron-withdrawing group (NO_2, SO_2R, COOR, CN) on an *ortho* or *para* ring carbon. No electron-withdrawing group is required when the displacement results in the formation of a five- or six-membered ring. Leaving groups for a nucleophilic aromatic substitution include fluorine, chlorine, and nitro.

Discussion. A tertiary amine is formed in the final step by displacement of chloride by 2-(1-piperazinyl)ethanol. Another tertiary amine is formed by the displacement of bromide in 1-bromo-3-chloropropane by 2-trifluoromethyl-10*H*-phenoth iazine. 2-(1-Piperazinyl)ethanol is formed by ring-opening of ethylene oxide with piperazine.

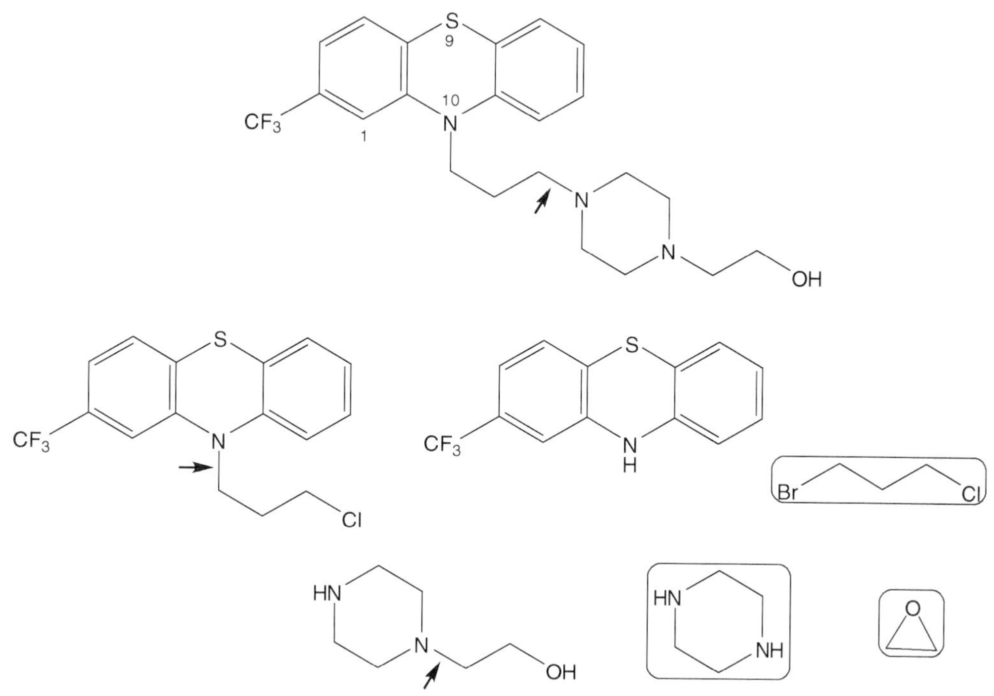

2-Trifluoromethyl-10*H*-phenothiazine is formed by intramolecular displacement of a nitro group by a formamide followed by amide hydrolysis under the ring-formation conditions. The formamide is formed from the amine and formic acid. The thioether is formed by chloride displacement by the thiol. The thiol, 2-amino-4-(trifluoromethyl)benzenethiol, is formed from 4-chloro-3-nitrobenzotrifluoride by displacement of chloride by sodium sulfide followed by reduction of the nitro group . 4-Chloro-3-nitrobenzotrifluoride is formed by nitration of 4-chlorobenzotrifluoride.

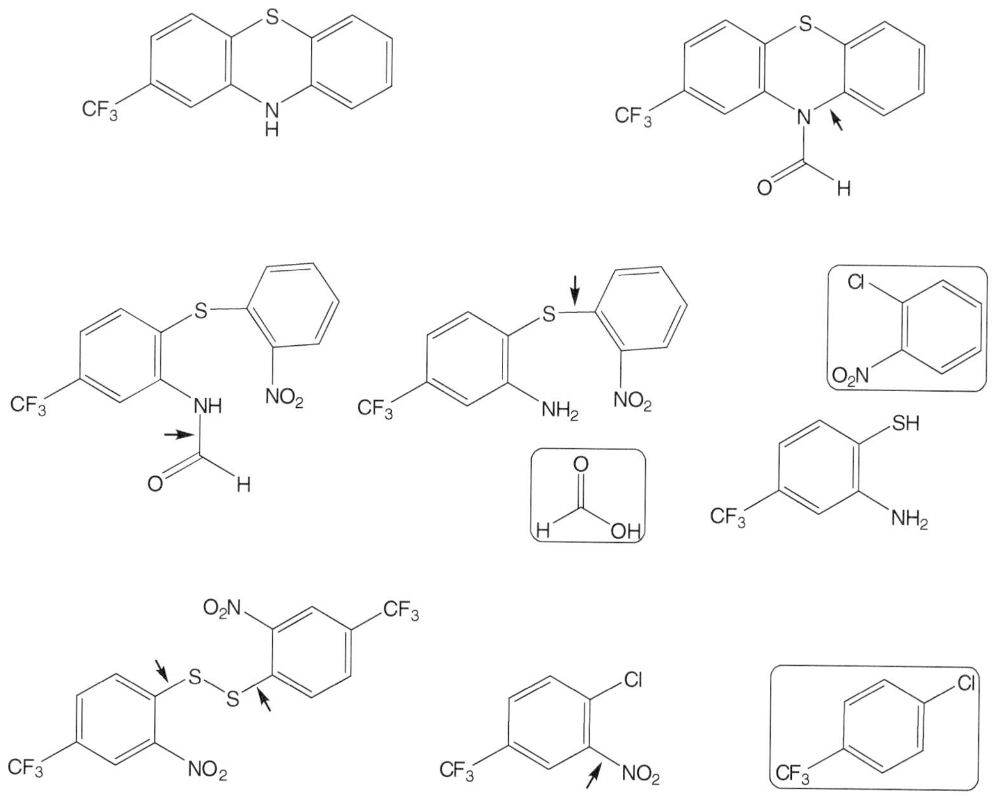

Extended Discussion

Draw the structures of the retrosynthetic analysis of an alternative route to 2-trifluoromethyl-10*H*-phenothiazine from 3-chloro-4-nitrobenzotrifluoride. Draw the structures for the retrosynthetic analysis for 3-chloro-4-nitrobenzotrifluori de from petrochemical or biochemical raw materials. List the pros and cons for both routes to 2-trifluoromethyl-10*H*-phenothiazine and select one route as the preferred route.

Folic Acid

Medicines Affecting the Blood/Antianaemia Medicines

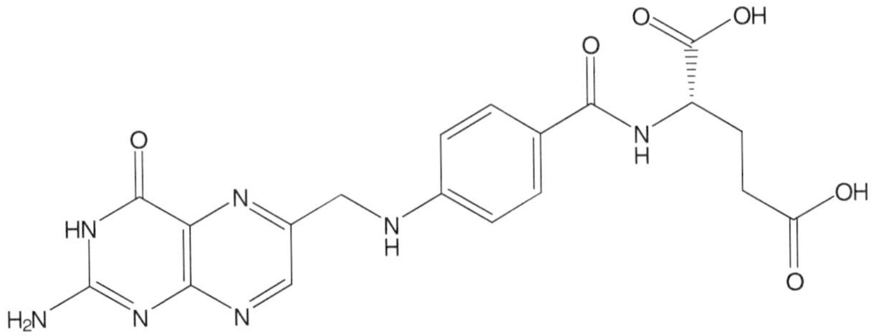

A 2-amino-4-hydroxypteridine is often formed by condensation of 2,4,5-tetraamino - 6-hydroxypyrimidine with an α-halo or α-hydroxy aldehyde or ketone.

Discussion. Folic acid has one chiral carbon with the (*S*)-configuration. The chiral carbon is delivered in the starting material L-glutamic acid. Folic acid is assembled in a single step from three components: 2,4,5-triamino-6-hydroxypyrimidine, *N*-(4-aminobenzoyl)-L-glutamic acid, and 1,1,3-trichloroacetone. 1,1,3-Trichloroacetone is formed by chlorination of acetone.

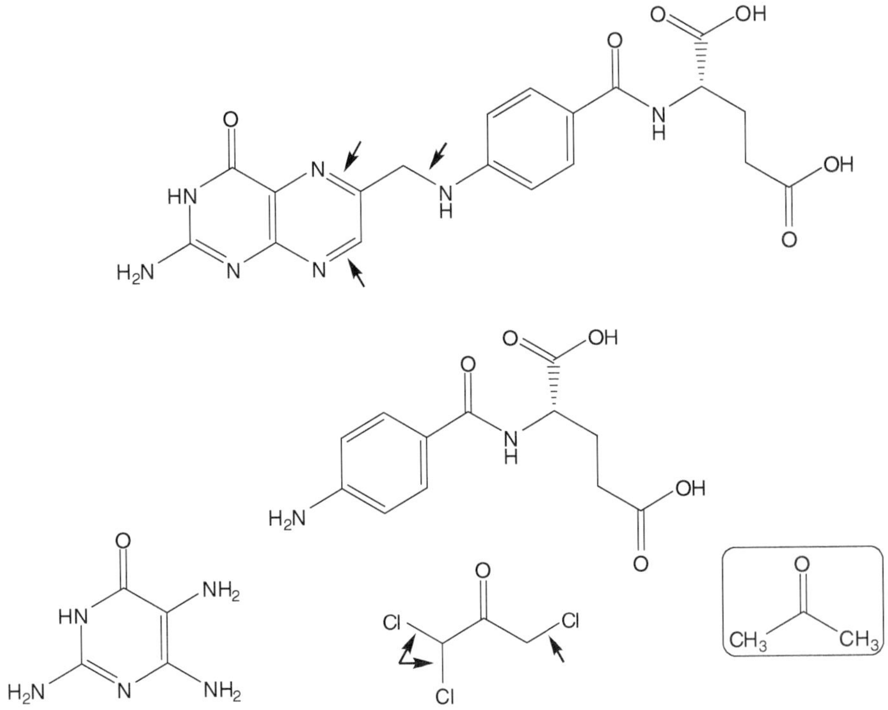

2,4,5-Triamino-6-hydroxypyrimidine is formed by reduction of the 5-nitrosopyrimidine. The Nitrosopyrimidine is formed by nitrosation of 2,4-diamino-6-hydroxypyrim-idine.

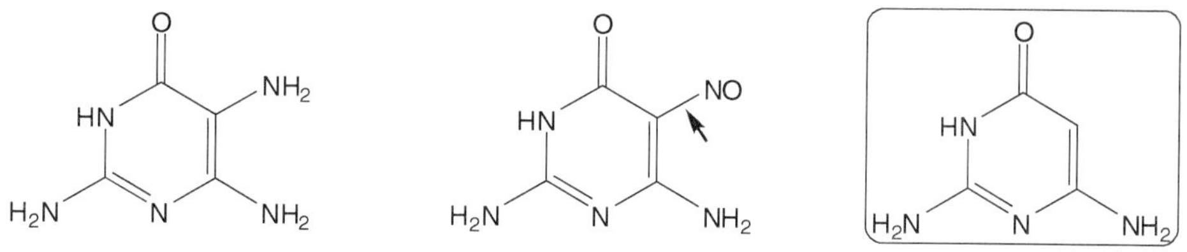

N-(4-Aminobenzoyl)-L-glutamic acid is formed by reduction of *N*-(4-nitrobenzoyl)-L-glutamic acid. The amide is formed from 4-nitrobenzoyl chloride and L-glutamic acid. L-Glutamic acid is produced by fermentation.

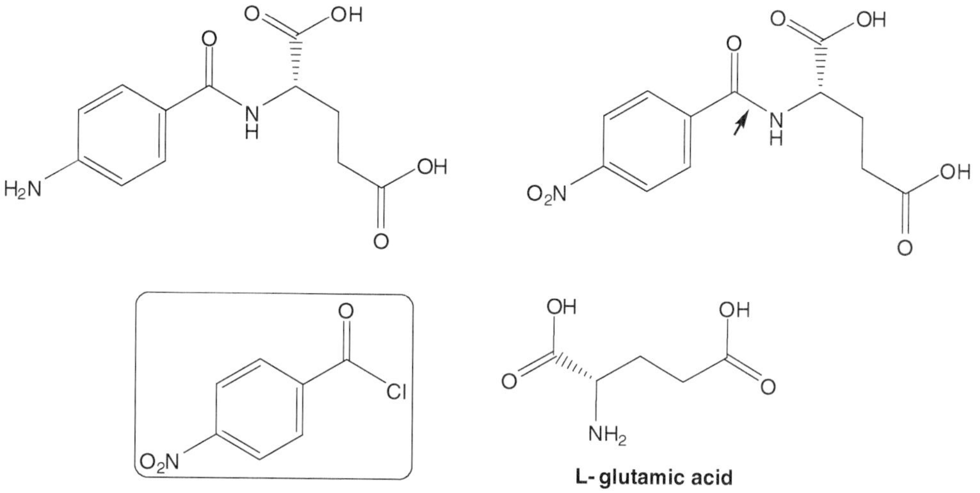

Extended Discussion

List the many challenges associated with manufacturing of high-purity 1,1,3-trichloroacetone. Other reagents have been used in a three-component condensation with 2,4,5-triamino-6-hydroxypyrimidine and *N*-(4-aminobenzoyl)-L-glutamic acid to produce folic acid. Draw structures for three of these reagents.

Fomepizole

Antidotes and Other Substances Used in Poisonings/Specific

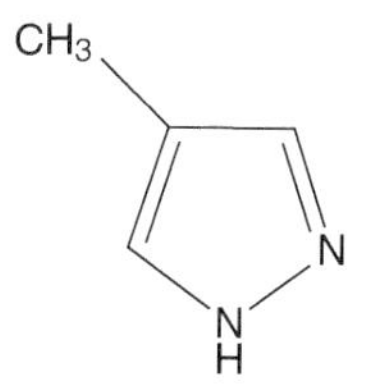

A pyrazole is often formed from hydrazine and a 1,3-dicarbonyl compound or its functional equivalent.

Discussion. Disconnections suggest fomepizole is formed from hydrazine and methylmalonaldehyde. The stable diacetal, 1,1,3,3-tetraethoxy-2-methylpropane, serves as a functional equivalent of the unstable dialdehyde. 1,1,3,3-Tetraethoxy-2-methylpropane is formed from 1-ethoxy-1-propene and triethyl orthoformate. 1-Ethoxy-1-propene is formed from 1,1-diethoxypropane by the elimination of ethanol.

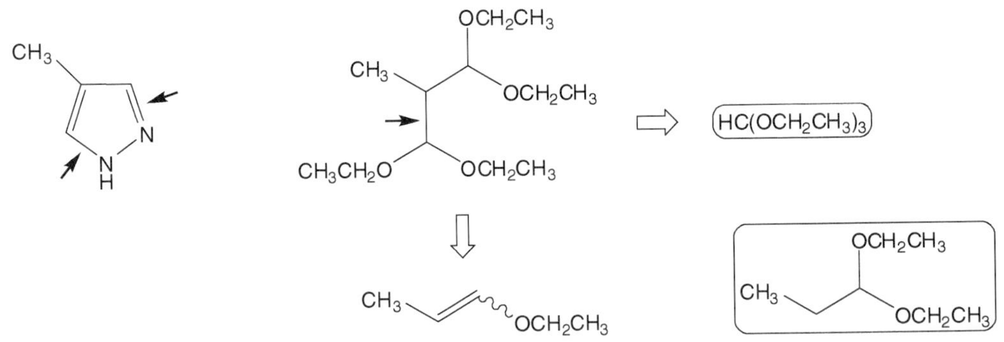

Extended Discussion

Alternative routes to fomepizole proceed via 4-methyl-4,5-dihydropyrazole, 4-methylpyrazole-3,5-dicarboxylic acid, 4-methylpyrazole-3-carboxylic acid, and 3,5-diamino-4-methylpyrazole. Draw a retrosynthetic scheme for one of these alternative routes. Include the structures of the retrosynthetic analysis of any organic starting material(s) from petrochemical or biochemical raw materials. Compare the two routes and select one route as the preferred route.

Formoterol

Medicines Acting on the Respiratory Tract/Antiasthmatic and Medicines for Chronic Obstructive Pulmonary Disease

A β-amino alcohol with a primary β-C is often formed by ring-opening of an epoxide by an amine.

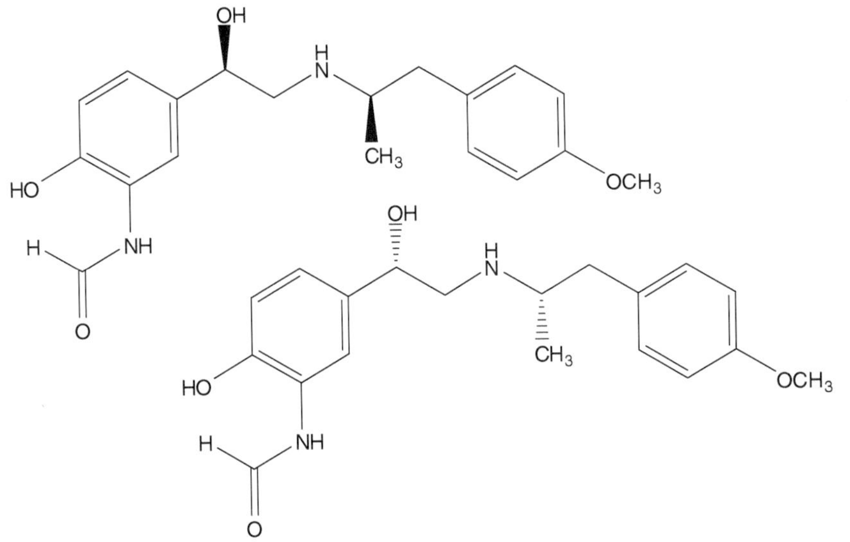

Discussion. Formoterol is a 1 : 1 mixture of the (*R*,*R*)- and (*S*,*S*)-enantiomers.

All four stereoisomers have been evaluated for bronchodilation activity. The potency order is: *R*,*R* > *R*,*S* = *S*,*R* > *S*,*S*. The (*R*,*R*)-enantiomer is 1000 times more potent than the (*S*,*S*)-enantiomer.

The phenol and the secondary amine are released in the final step by hydrogenolysis of the benzyl ether and benzyl amine. The (*R*,*R*)- and (*S*,*S*)-enantiomers of dibenzylformoterol are separated from the (*R*,*S*)- and (*S*,*R*)-enantiomers of dibenzylformoterol by crystallization of the fumarate salts.

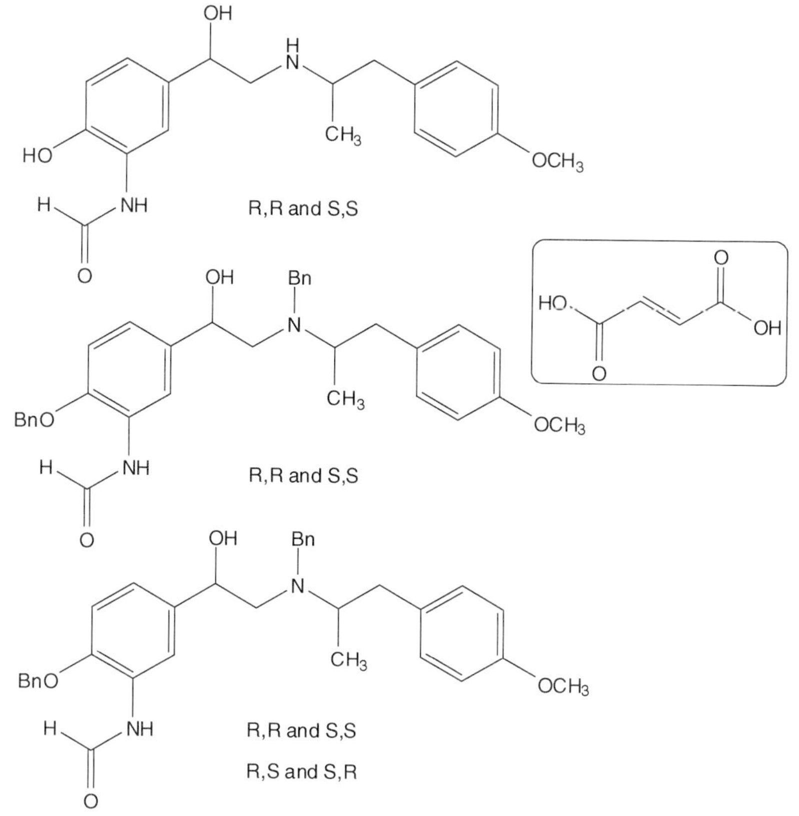

The amide is formed by acylation of the aniline with the mixed anhydride from formic acid and acetic anhydride. The aniline is formed by reduction of the nitroaromatic.

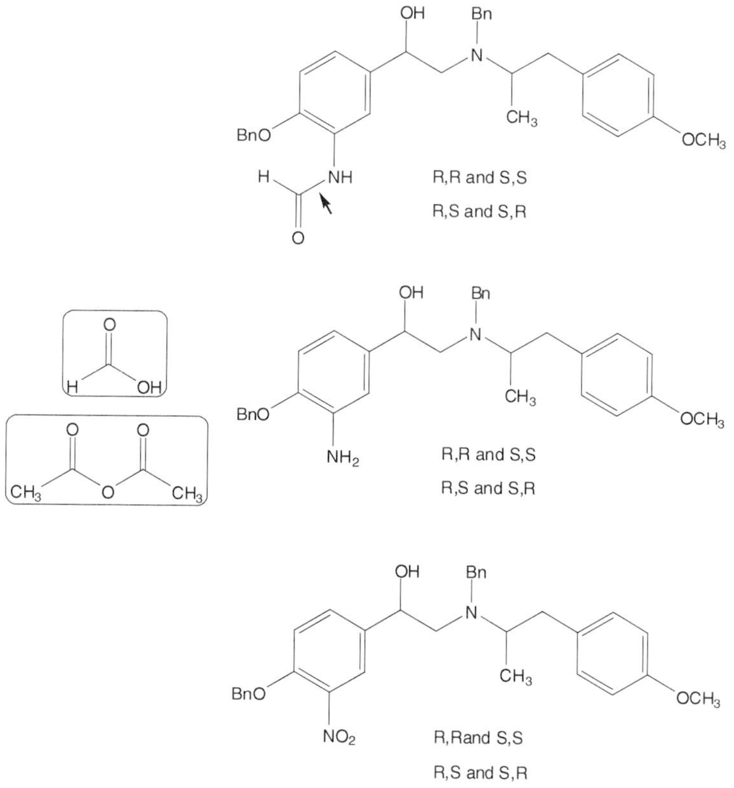

The alcohol is formed by reduction of the ketone. The tertiary amine is formed bromide displacement from the α-bromoketone by the secondary amine.

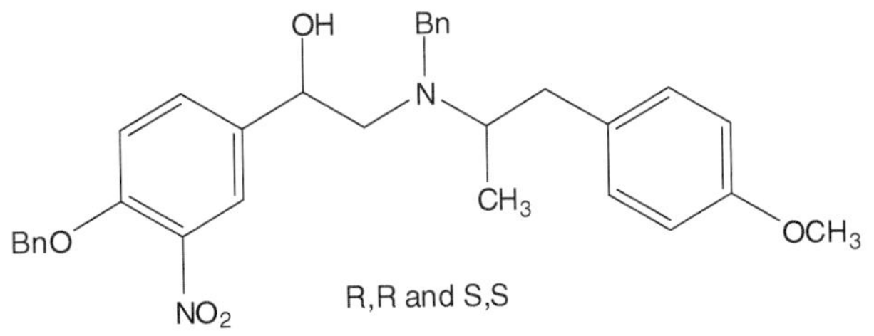

R,R and S,S

R,S and S,R

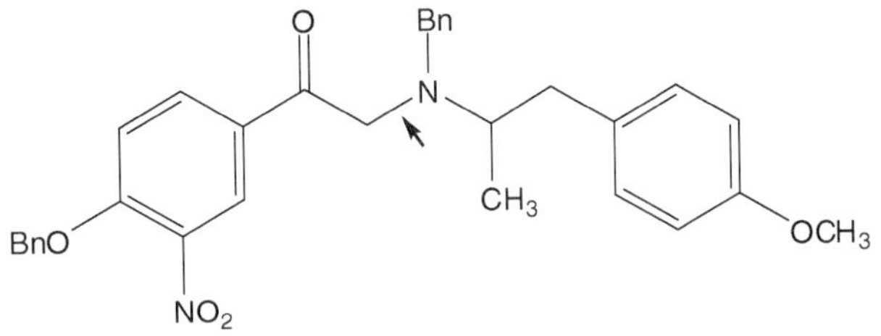

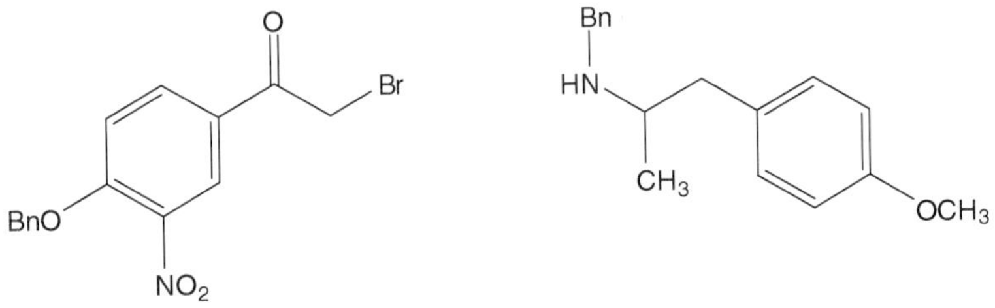

The α-bromoketone is formed by bromination of the ketone. The benzyl ether is formed by chloride displacement from benzyl chloride by the phenol (**Williamson Ether Synthesis**). 4′-Hydroxy-3′-nitroacetophenone is formed by nitration of 4′-hydroxyacetophenone.

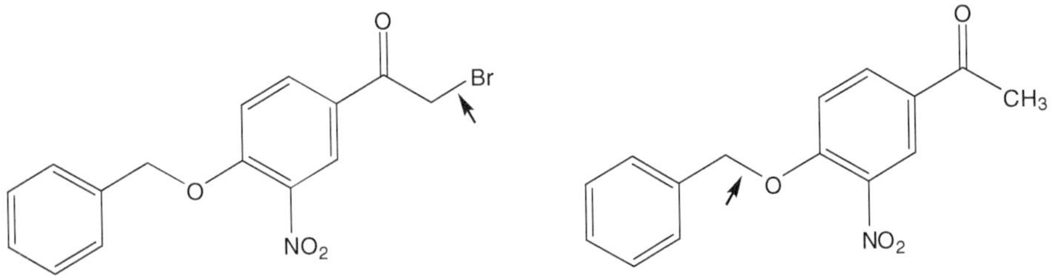

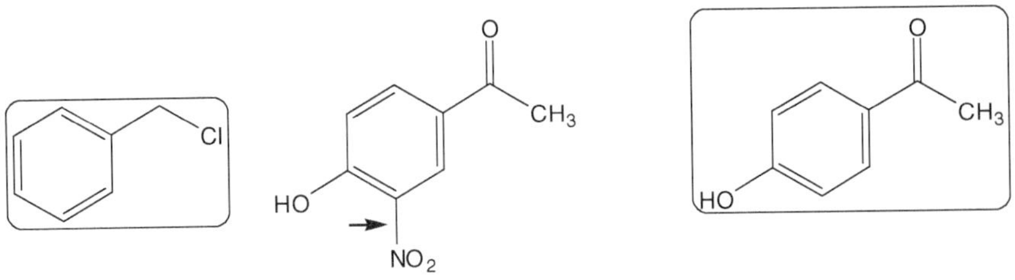

The secondary amine is formed by reductive amination from 4-methoxyphenylacetone and benzylamine. 4-Methoxyphenylacetone is formed from 4-methoxyphenylacetic acid and acetic anhydride.

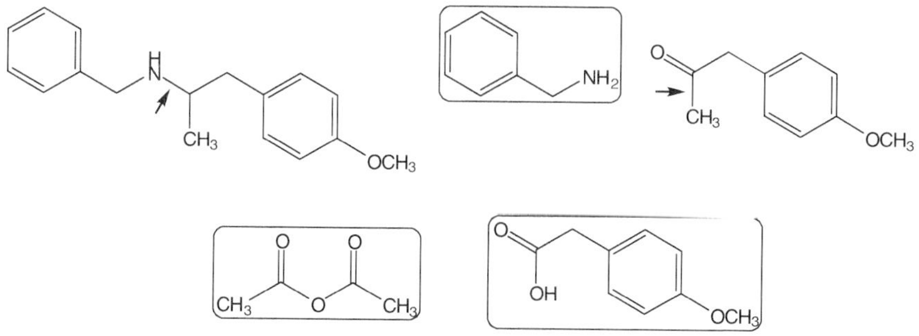

Extended Discussion

Arformoterol is the pure (*R*,*R*)-enantiomer. Draw the structures of the retrosynthetic analysis of one arformoterol process which has the β-amino alcohol formed by ring-opening of an epoxide by an amine.

Fosfomycin

Anti-infective Medicines/Antibacterials/Other Antibacterials

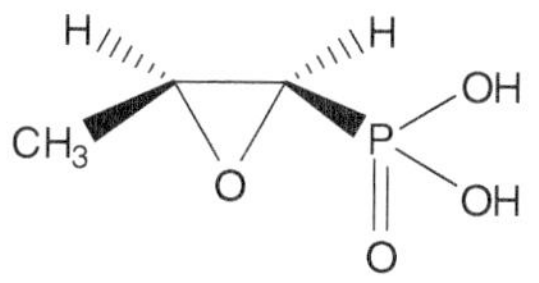

In a molecule with two adjacent chiral carbons, one chiral carbon is often used to direct the formation of the second.

Discussion. Two routes to fosfomycin, (1*R*,2*S*)-1,2-epoxypropylphosphonic acid, will be presented. Route A utilizes a resolution. Route B is an asymmetric synthesis.

In Route A, fosfomycin is separated from the 1 : 1 mixture of (1*R*,2*S*)- and (1*S*,2*R*)-enantiomers by resolution in the final step. The 1 : 1 mixture of enantiomeric epoxides is formed by epoxidation of *cis*-propenylphosphonic acid. The phosphonic acid is formed from di-*tert*-butyl *cis*-propenylphosphonate by acid-catalyzed cleavage of the esters. Di-*tert*-butyl *cis*-propenylphosphonate is formed by hydrogenation of di-*tert*-butyl propadienylphosphonate. The propadienylphosphonate is formed by the rearrangement of di-*tert*-butyl 2-propynyl phosphite. The phosphite is formed in situ from propargyl alcohol and di-*tert*-butyl phosphorochloridite. Di-*tert*-butyl phosphorochloridite is formed in situ from *tert*-butyl alcohol and phosphorus trichloride.

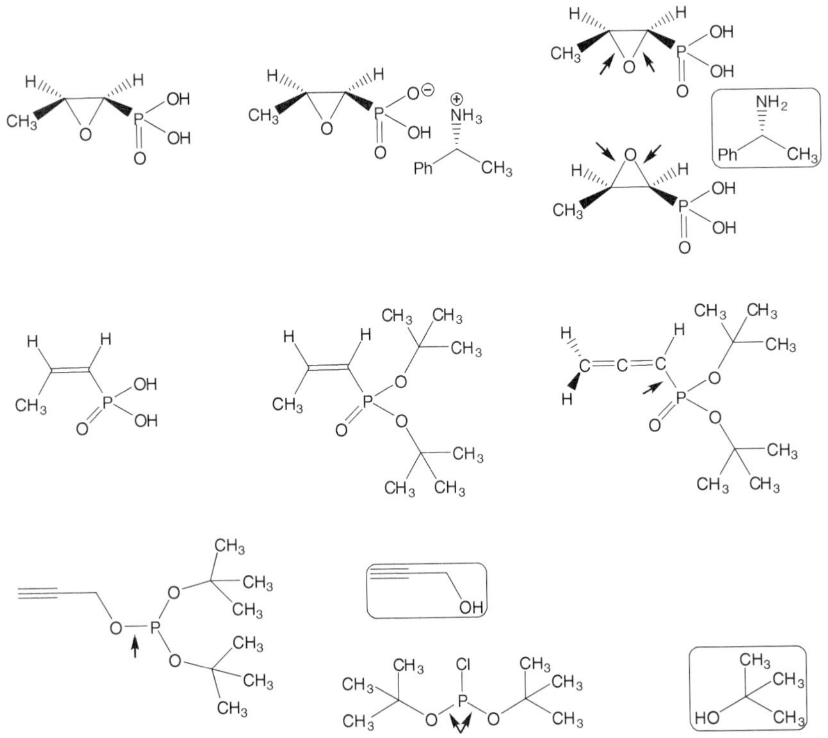

In Route B, the epoxide is formed in the final step by displacement of *para*-toluenesulfonate on C1 by the hydroxyl group on C2. The phosphonic acid is formed by hydrolysis of the dimethyl phosphonate. The *para*-toluenesulfonate is formed from the 1,2-diol. The C2 hydroxyl group is released by transesterification of the acetate ester. The C1 hydroxyl group is formed by reduction of the carbonyl group of the acyl phosphonate. [What is the highest ratio of the major (1*S*,2*S*)- and minor (1*R*,2*S*)-diastereomers formed in this reduction? What reaction conditions are associated with the highest ratio?] The acyl phosphonate is formed from the acid chloride and dimethylphosphite (**Michaelis–Arbuzov Reaction**). The acid chloride is formed from the carboxylic acid. (*S*)-2-Acetoxypropionic acid is formed from (*S*)-lactic acid and acetic acid. (*S*)-Lactic acid is manufactured by chemical synthesis and by fermentation.

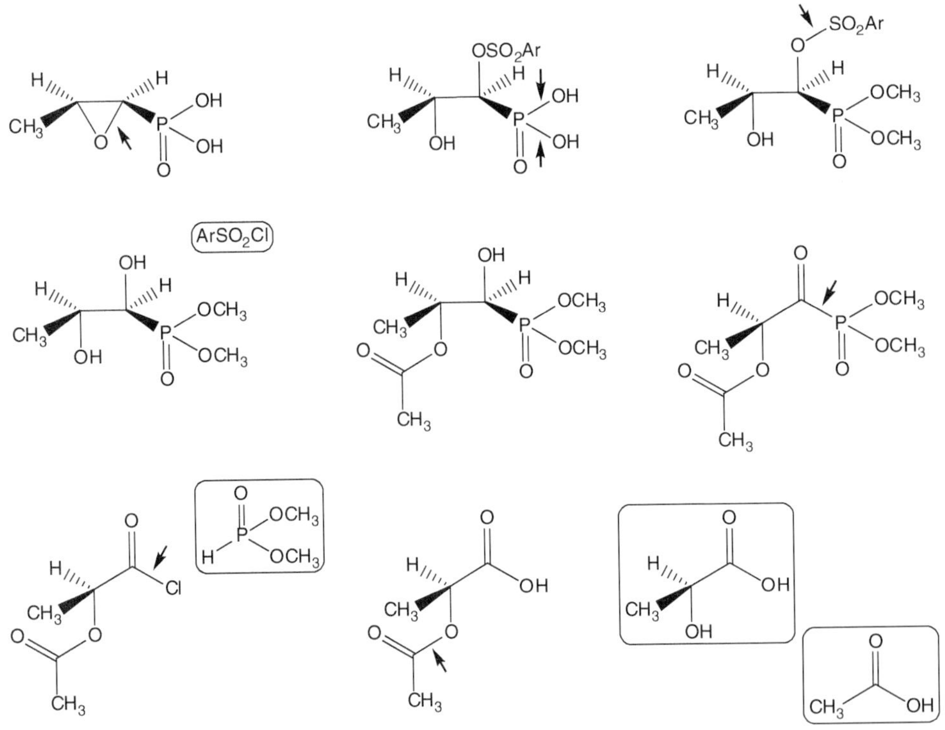

Extended Discussion

List the pros and cons for both routes to fosfomycin. Is one route preferred?

Furosemide

Cardiovascular Medicines/Medicines Used in Heart Failure
Diuretics

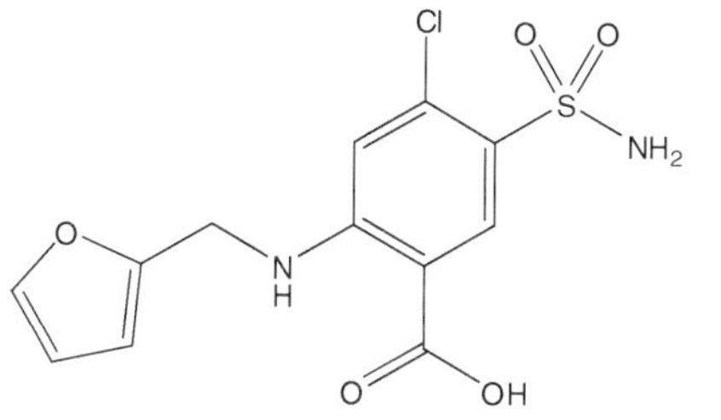

When the target molecule has a benzene ring with multiple substituents, it is likely that one or more of the substituents will be introduced by electrophilic aromatic substitution. Activating/deactivating and directing effects of substituents often determine the order of introduction of ring substituents in the preferred synthesis.

Discussion. Chloride is displaced by furfurylamine in the final step. (Why is this displacement selective for one of the two chlorides?) The sulfonamide is formed by reaction of the sulfonyl chloride with ammonia. The sulfonyl chloride is formed by chlorosulfonylation of 2,4-dichlorobenzoic acid.

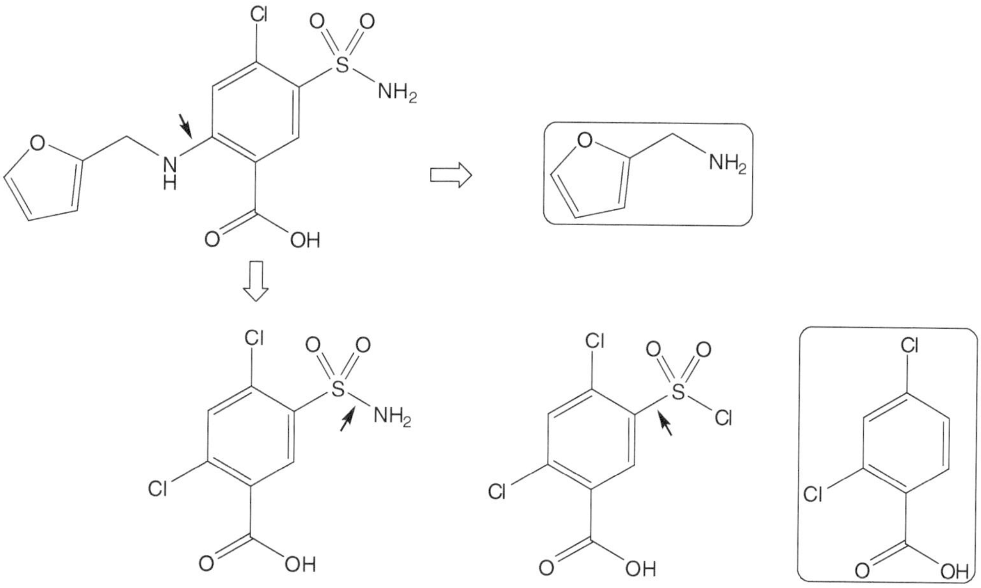

Extended Discussion

What is the highest yield reported for the **chloride** displacement by furfurylamine? by any primary amine? What is the best yield reported for the analogous **fluoride** displacement by furfurylamine? Draw the structures of the retrosynthetic analysis of 4-chloro-2-fluoro-5-sulfamoylbenzoic acid. Include the structures of the retrosynthetic analysis of any organic starting material(s) from petrochemical or biochemical raw materials.

G

Gemcitabine

Antineoplastics and Immunosuppressives/Cytotoxic and Adjuvant Medicines

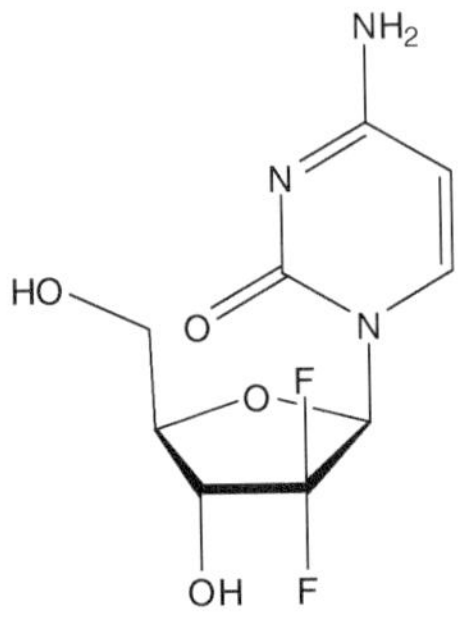

A nucleoside is often formed by displacement of a leaving group on the sugar by nitrogen of the heterocycle. The displacement usually results in a mixture of two products, with the heterocycle on the top face (β) or the bottom face (α) of the sugar. Factors which influence the β-product to α-product ratio include the heterocycle, the sugar, the leaving group, and the reaction conditions.

Discussion. Gemcitabine is crystallized from the mixture of β-anomer and α-anomer. The mixture of anomers is formed in the final step by release of the alcohols by ammonolysis of the benzoate esters. A mixture of the β-anomer (major) and α-anomer (minor) is formed in the displacement of the α-methanesulfonate at C1 by N1 of cytosine. (What is the highest reported β to α ratio? What reaction conditions are associated with the highest ratio?)

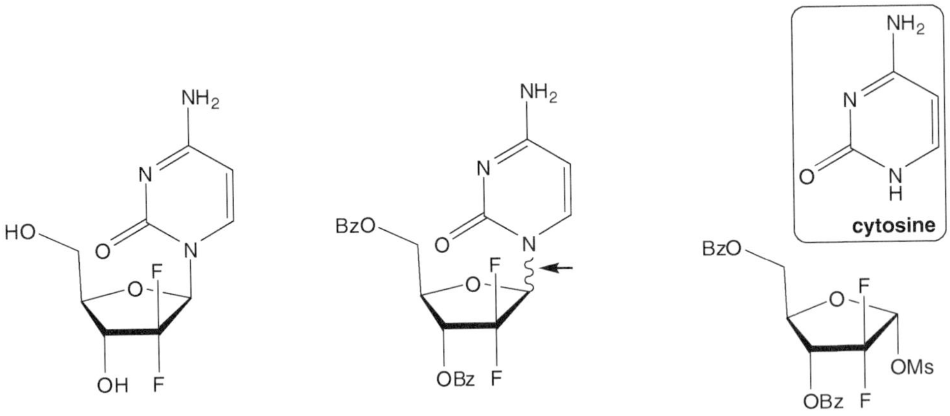

The α-methanesulfonate is crystallized from a mixture of the α- and β-anomers. The methanesulfonate is formed from the alcohol. The alcohol is formed by reduction of the lactone.

Routes to Essential Medicines: A Workbook for Organic Synthesis, First Edition. Peter J. Harrington.

Companion website: www.wiley.com/go/Harrington/routes_essential_medicine

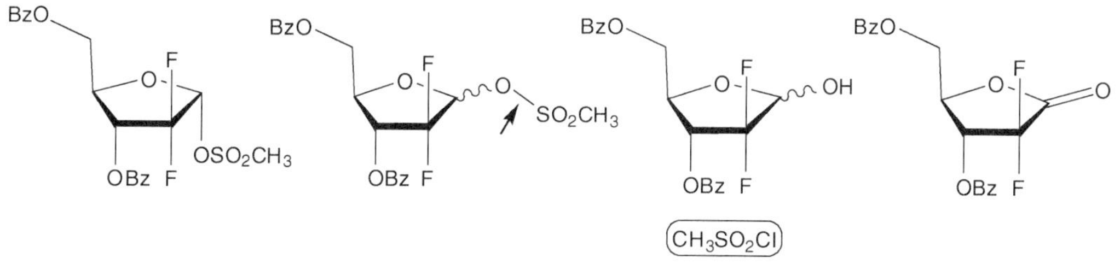

The lactone with the correct (*R*,*R*)-configuration is crystallized from a mixture of the (*R*,*R*)- and (*S*,*R*)-diastereomers. The alcohols at sugar C4 and C5 are released by hydrolysis of the acetonide. The C4 alcohol forms the lactone. The C5 alcohol is then protected as the benzoate ester. An alcohol is formed as a mixture of (*R*,*R*)- and (*R*,*S*)-diastereomers by the addition of an organozinc(II) bromide to D-glyceraldehyde acetonide (**Reformatsky Reaction**). The alcohol is then protected as the benzoate ester. [What is the highest (*R*,*R*) to (*S*,*R*) ratio reported for the mixture of diastereomers? What reaction conditions are associated with the highest ratio?] The organozinc(II) bromide is formed from ethyl bromodifluoroacetate.

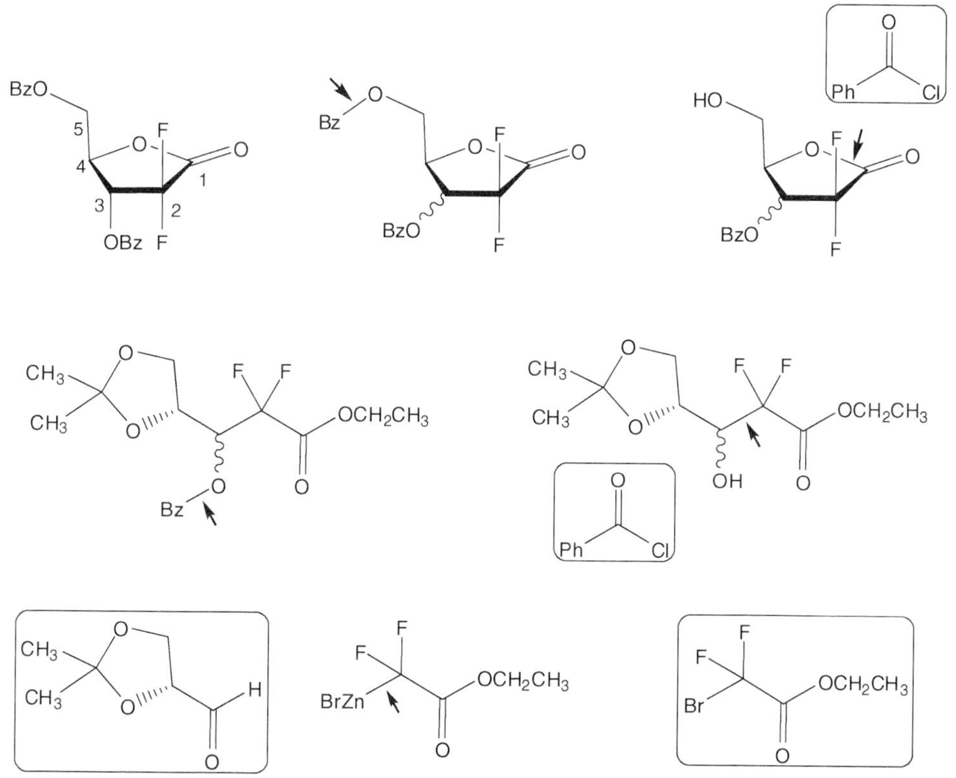

Extended Discussion

Draw the structures of a retrosynthetic analysis of one alternative route to gemcitabine based on fluorination of a carbohydrate derivative.

Gliclazide

Hormones, Other Endocrine Medicines, and Contraceptives/Insulins and Other Medicines Used for Diabetes

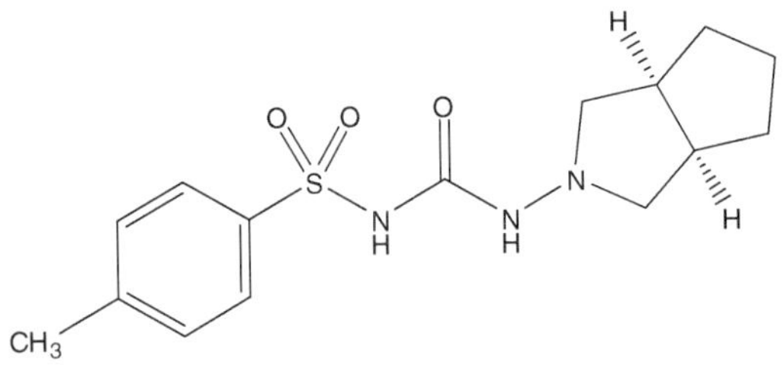

An *N*-arylsulfonylsemicarbazide is often formed by displacement of ammonia from an *N*-arylsulfonylurea by a hydrazine.

Discussion. In the final step, gliclazide is formed by displacement of ammonia from 4-toluenesulfonylurea by the hydrazine. 4-Toluenesulfonylurea is formed from 4-toluenesulfonyl chloride and urea.

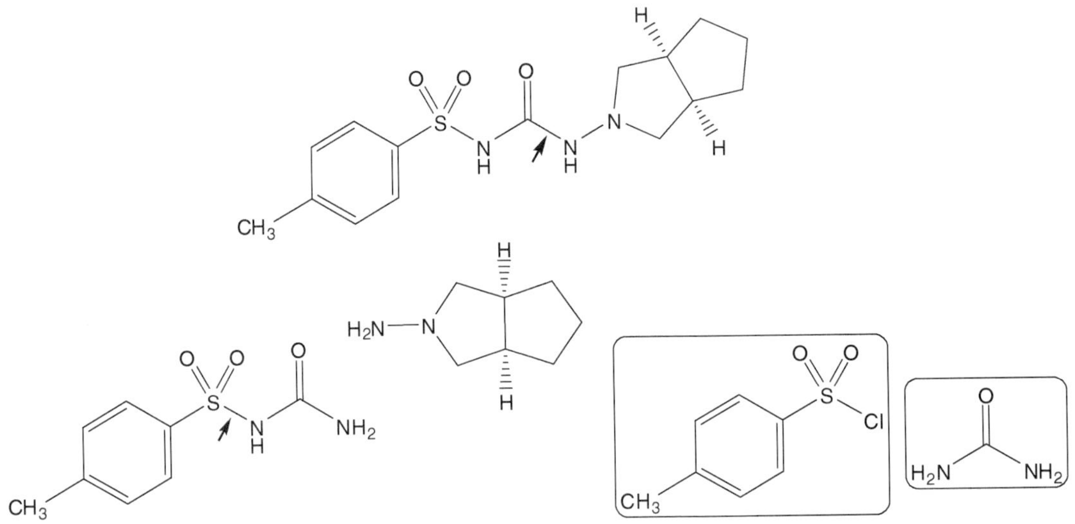

The hydrazine is formed by reduction of the nitrosamine. The nitrosamine is formed by nitrosation of *cis*-3-azabicyclo[3.3.0]octane. The secondary amine of *cis*-3-azabicyclo[3.3.0]octane is formed by reduction of the imide. (List the methods for the imide reduction. Which method is associated with the highest yield?) The imide is formed by ring-opening of the anhydride with ammonia followed by ring-closure and dehydration. The anhydride is formed from crude 1,2-cyclopentanedicarboxylic acid (mixture of *cis*- and *trans*-diastereomers). The crude dicarboxylic acid is formed from ethyl 3-bromo-2-oxocyclohexanecarboxylate (**Favorskii Rearrangement**). The α-bromoketone is formed by bromination of ethyl 2-oxocyclohexanecarboxylate. Ethyl 2-oxocyclohexanecarboxylate is formed from cyclohexanone and diethyl carbonate (mixed **Claisen Condensation**).

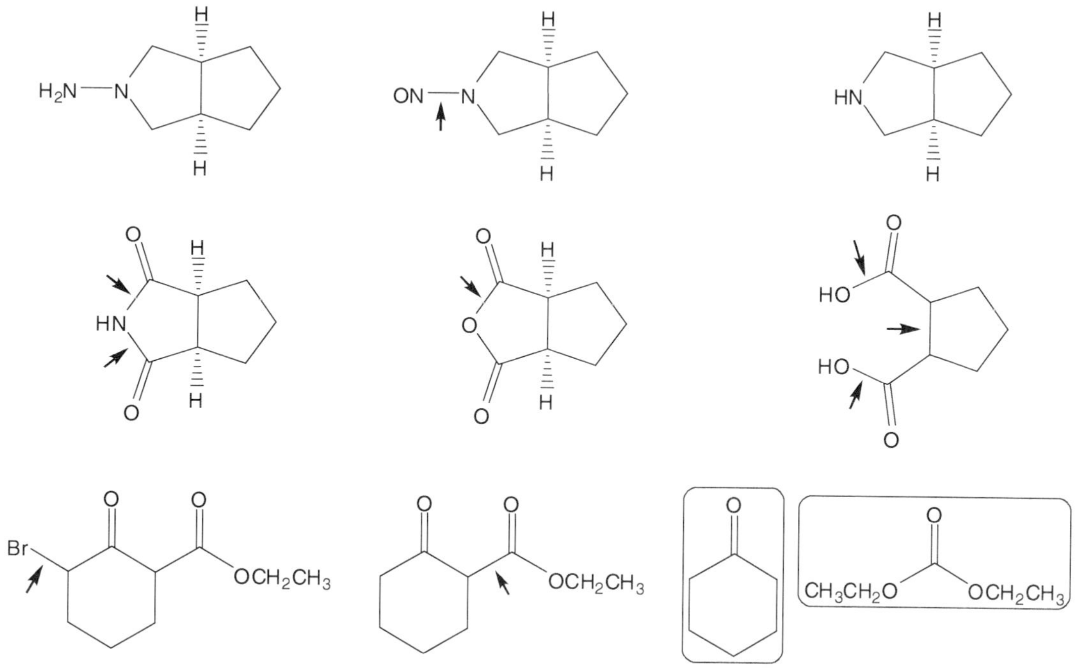

Extended Discussion

Draw the structures of the retrosynthetic analysis of one alternative route to 1,2-cyclopentanedicarboxylic acid. Include the structures of the retrosynthetic analysis of any organic starting material(s) from petrochemical or biochemical raw materials. List the pros and cons for both routes and select one route as the preferred route.

Glutaral

Disinfectants and Antiseptics/Disinfectant

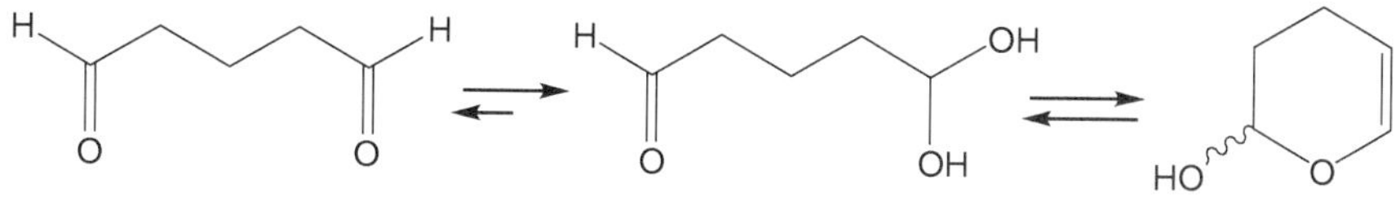

An aldehyde is often formed by oxidation of an alcohol or alkene. The reaction and workup conditions must be selected to minimize over-oxidation to the carboxylic acid or Aldol Condensation of the aldehyde.

Discussion. Glutaral forms a hydrate in aqueous solution. The hydrate is in equilibrium 3,4-dihydro-2*H*-pyran-2-ol. Glutaral hydrate is manufactured by oxidation of cyclopentene or by hydrolysis 3,4-dihydro-2-methoxy-2*H*-pyran.

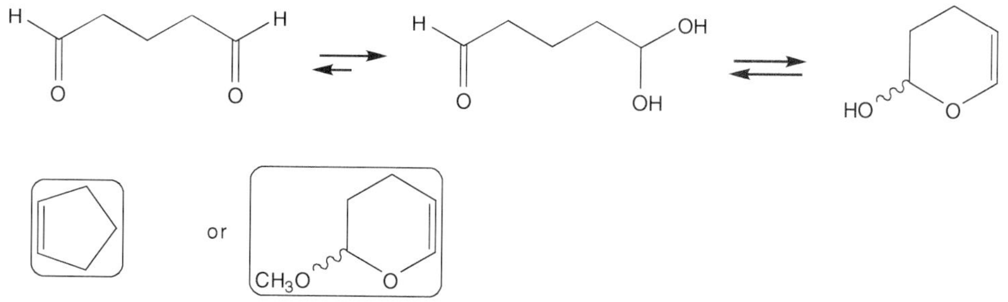

Glyceryl Trinitrate

Cardiovascular Medicines/Antianginal Medicines

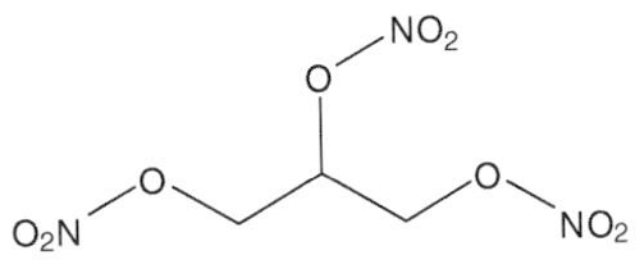

An alkyl nitrate ester is often formed from the alcohol and nitric acid. The alkyl nitrate ester functional group is an explosophore.

Discussion. Glyceryl trinitrate (glycerol trinitrate or nitroglycerin) is formed by the reaction of glycerol with nitric acid. (**NOTE**: Nitroglycerin may explode when subjected to heat, shock, or flame.)

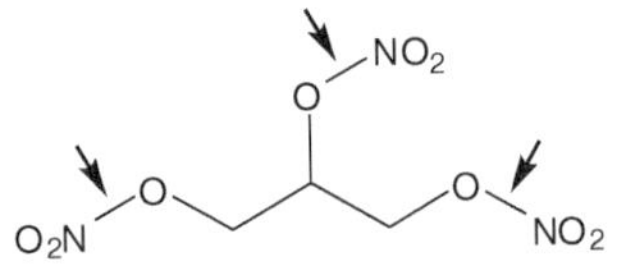

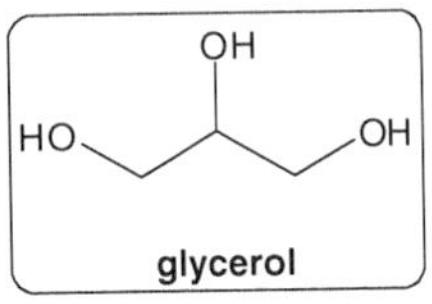

Extended Discussion

Draw the structures and list the names of the functional groups in organic molecules which are explosophores.

H

Haloperidol

Medicines for Pain and Palliative Care/Medicines for Common Symptoms in Palliative Care
Medicines for Mental and Behavioral Disorders/Medicines Used in Psychotic Disorders

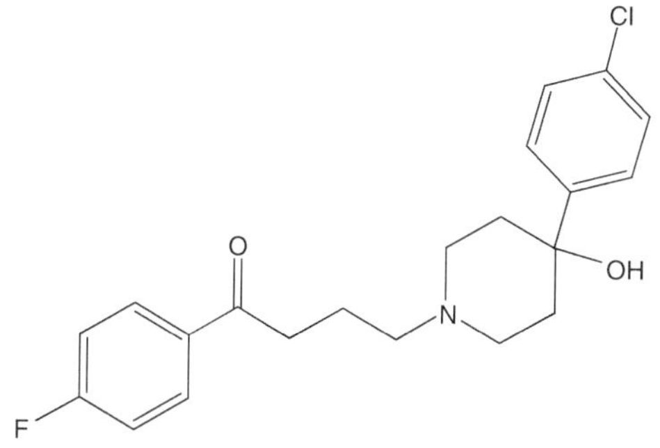

Tertiary amines are ubiquitous in drug structures. A tertiary amine is often formed by alkylation of a secondary amine. The alkylation is most efficient when the amine is used in excess, and the carbon with the leaving group (Cl, Br, I, OTs, OMs) is primary or benzylic.

Discussion. The ketone is released by hydrolysis of the acetal in the final step. The tertiary amine is formed by nucleophilic substitution of chloride by the piperidine. The piperidine is formed by *N*-debenzylation of the 1-benzylpiperidine. The acetal is formed from the ketone and ethylene glycol. The ketone is formed from fluorobenzene and 4-chlorobutanoyl chloride (**Friedel–Crafts Acylation**).

Routes to Essential Medicines: A Workbook for Organic Synthesis, First Edition. Peter J. Harrington.

Companion website: www.wiley.com/go/Harrington/routes_essential_medicine

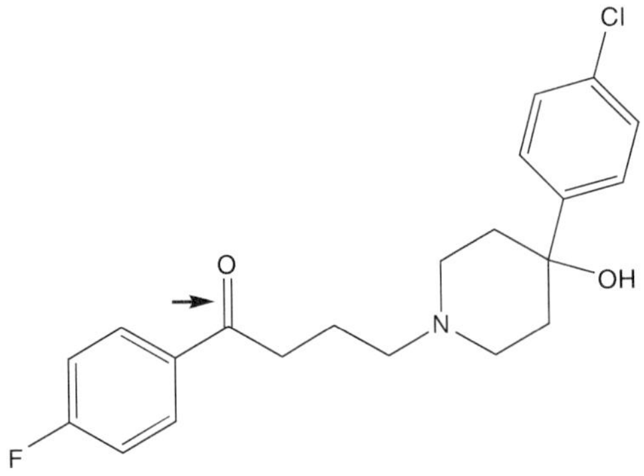

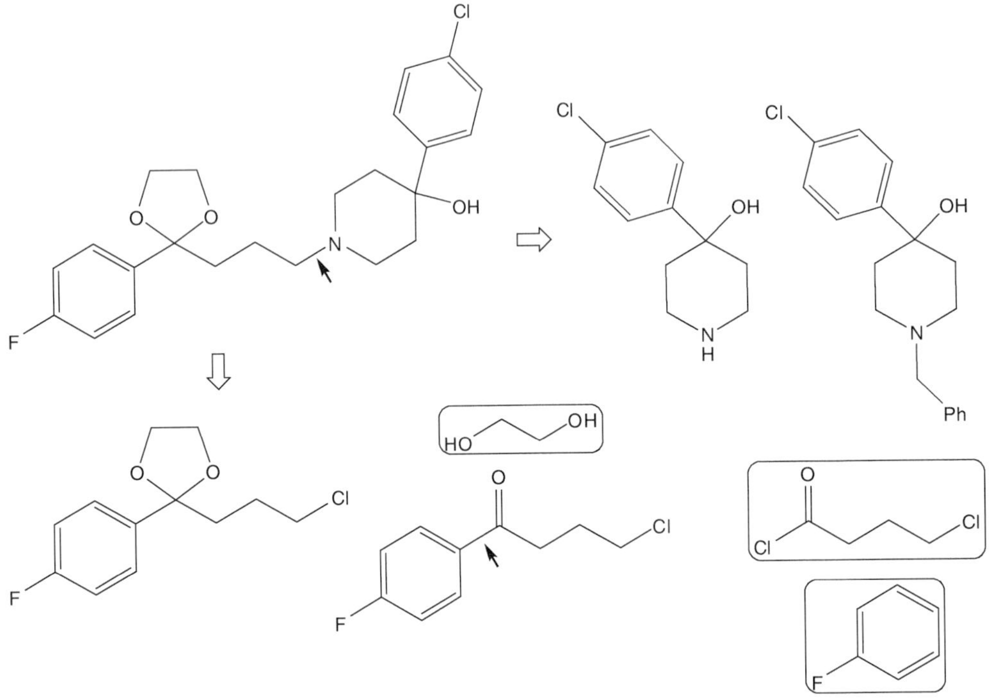

The 1-benzylpiperidin-4-ol is formed by addition of 4-chlorophenylmagnesium bromide to 1-benzyl-4-piperidinone (**Grignard Reaction**). The 4-chlorophenylmagnesium bromide is formed from 1-bromo-4-chlorobenzene.

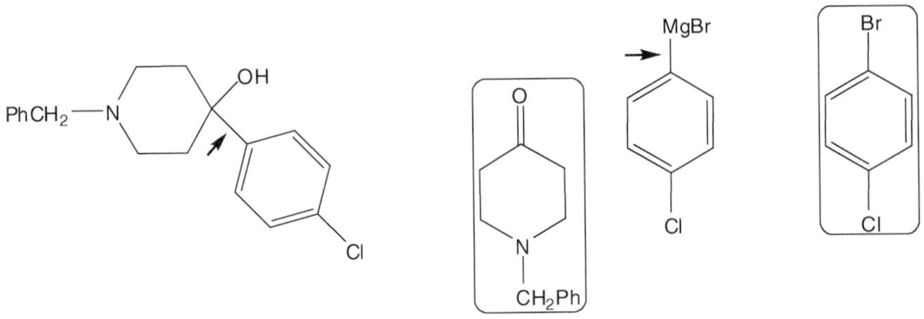

Extended Discussion

4-Chloro-4′-fluorobutyrophenone can also be formed by the reaction of a Grignard reagent with a nitrile. Draw the structures of the retrosynthetic analysis of this alternative route. Include the structures of the retrosynthetic analysis of any organic starting material(s) from petrochemical or biochemical raw materials. List the pros and cons for both routes to 4-chloro-4′-fluoro butyrophenone. Is one route preferred?

Halothane

Anesthetics, Preoperative Medicines and Medical Gases/General Anesthetics and Oxygen/Inhalational Medicines

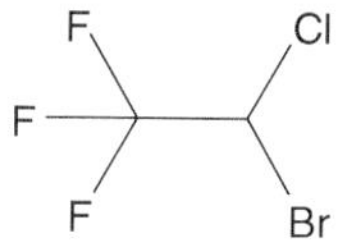

An alkyl fluoride is often formed by the displacement of chloride by fluoride.

Discussion. Halothane is a 1:1 mixture of the (*R*)- and (*S*)-enantiomers. Halothane is formed by bromination of 2-chloro-1,1,1-trifluoroethane. (How can the side product 1,1-dibromo-1-chloro-2,2,2-trifluoroethane be converted to halothane?) 2-Chloro-1,1,1-trifluoroethane (also known as HCFC-133a) is formed from trichloroethylene and hydrogen fluoride.

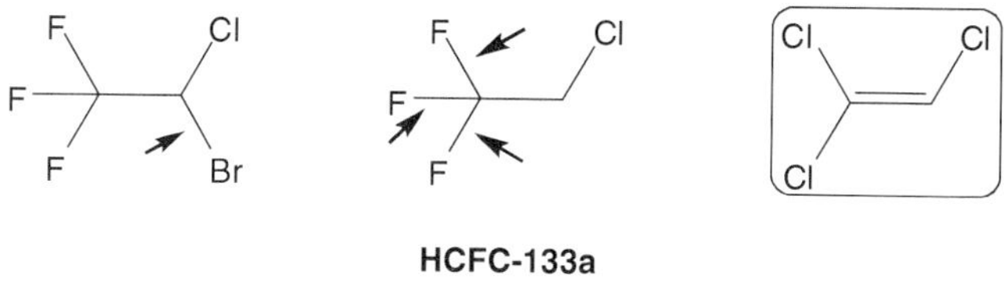

Extended Discussion

Draw the structures of the retrosynthetic analysis of a route to (*S*)-halothane which does not rely on separation of enantiomers by chiral gas chromatography.

Hydralazine

Cardiovascular Medicines/Antihypertensive Medicines

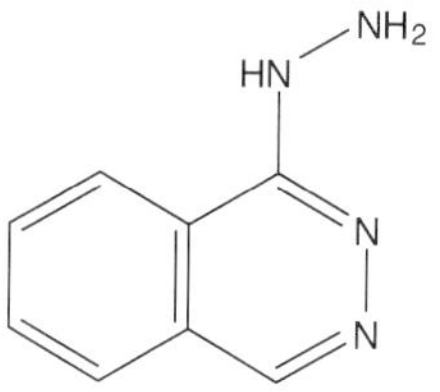

A nitrogen substituent on C1 or C4 of a phthalazine ring is often introduced by displacement of chloride.

Discussion. Hydralazine is formed from 1-chlorophthalazine by chloride displacement by hydrazine. 1-Chlorophthalazine is formed from 1(2*H*)-phthalazinone (phthalazone). Phthalazone is formed by the reaction of phthalaldehydic acid with hydrazine.

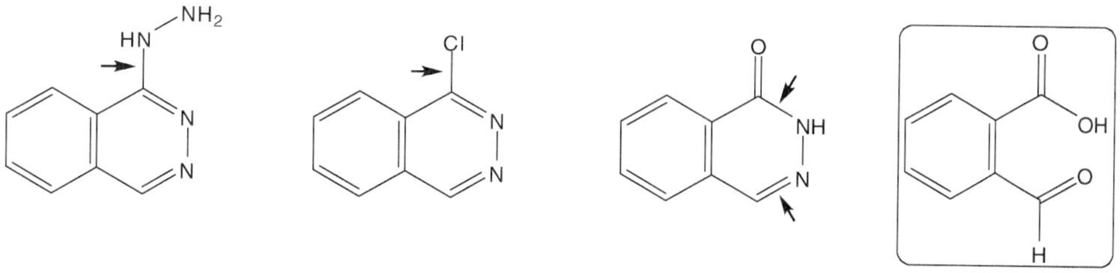

Extended Discussion

Describe the alternative process for formation of 1-chlorophthalazine in a single step from α,α,α,α′,α′-pentachloro-*ortho*-xylene.

and

Draw the structure of a dimer impurity that might form in the final chloride displacement step. Provide details of the procedure for the final step that ensure a low level of this dimer impurity in the isolated hydralazine.

Hydrochlorothiazide

Cardiovascular Medicines/Medicines Used in Heart Failure
Diuretics

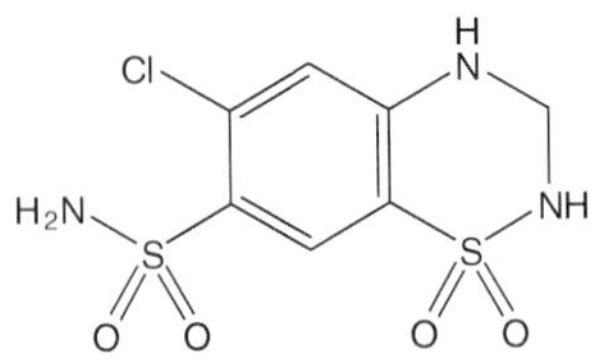

When the target molecule has a benzene ring with multiple substituents, it is likely that one or more of the substituents will be introduced by electrophilic aromatic substitution. Activating/deactivating and directing effects of substituents often determine the order of introduction of ring substituents in the preferred synthetic sequence.

Discussion. The thiadiazine ring of hydrochlorothiazide (HCTZ) is formed in the final step by bridging the aniline and sulfonamide by reaction with formaldehyde. The disulfonamide is formed by reaction of the disulfonyl chloride with ammonia. The disulfonyl chloride is formed by the reaction of 3-chloroaniline with chlorosulfonic acid at elevated temperature. (Since 3-chloroaniline is manufactured from benzene, all substituents on the ring are introduced by electrophilic aromatic substitution.)

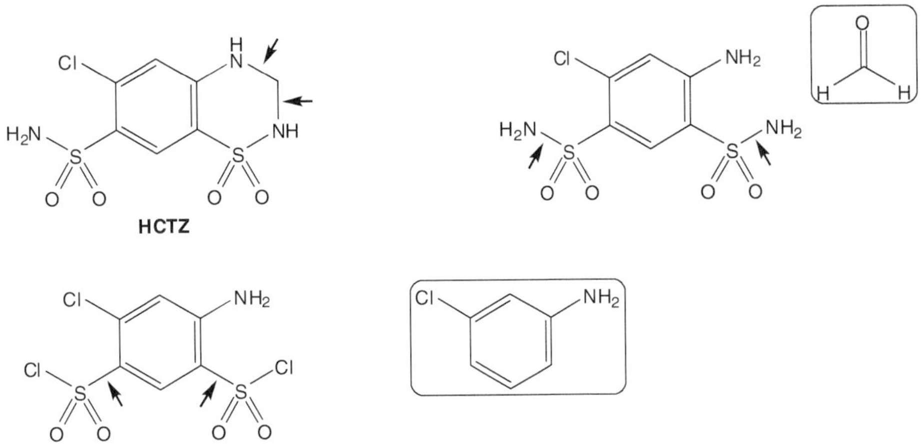

Extended Discussion

Draw the structure of the dimeric impurity (HCTZ-CH_2-HCTZ) formed in the final step. Describe how the workup procedure is designed to reduce the amount of this impurity in the isolated HCTZ.

Hydrocortisone

Antiallergics and Medicines Used in Anaphylaxis
Antineoplastics and Immunosuppressives/Hormones and Antihormones
Dermatological Medicines (Topical)/Anti-inflammatory and Antipruritic Medicines
Gastrointestinal Medicines/Anti-inflammatory Medicines
Hormones, Other Endocrine Medicines and Contraceptives/Adrenal Hormones and Synthetic Substitutes

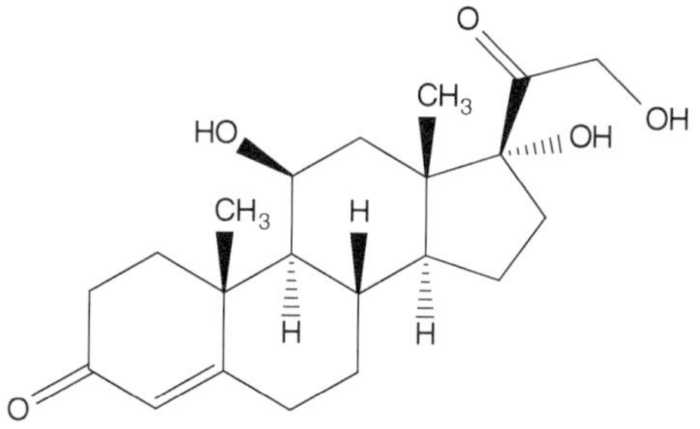

A single-enantiomer molecule with multiple chiral carbons is often formed by modification of a natural product which has most or all of the chiral carbons already in place. An alcohol at C11 of a corticosteroid is often introduced by microbial oxidation.

Discussion. In one preferred route, hydrocortisone (cortisol) is manufactured in seven steps from 16-dehydropregnenolone acetate and in 10 steps from diosgenin. Diosgenin is a phytosteroid sapogenin isolated from the tubers of *Discorea* wild yam.

The alcohol at C11 is introduced in the final step by microbial oxidation of cortexolone 21-acetate. (What is the highest ratio of hydrocortisone to 11-epihydrocortisone in the microbial oxidation? What conditions are associated with the highest ratio? How are hydrocortisone and 11-epihydrocortisone separated?) The α-acetoxyketone is formed by displacement of iodide from the α-iodoketone by potassium acetate. The mixture of the α-iodoketone and the α,α-diiodoketone is formed by iodination of 17α-hydroxyprogesterone. 17α-Hydroxyprogesterone is formed from 16β-bromo-17α-hydroxyprogesterone by hydrogenolysis. The bromohydrin is formed by ring-opening of the 16α,17-epoxide with hydrogen bromide. The 4-ene-3-one of 16α,17-epoxyprogesterone is formed from 16α,17-epoxypregnenolone by oxidation of the C3 alcohol and double bond migration. 16α,17-Epoxypregnenolone is formed from 16-dehydropregnenolone acetate by epoxidation and ester hydrolysis.

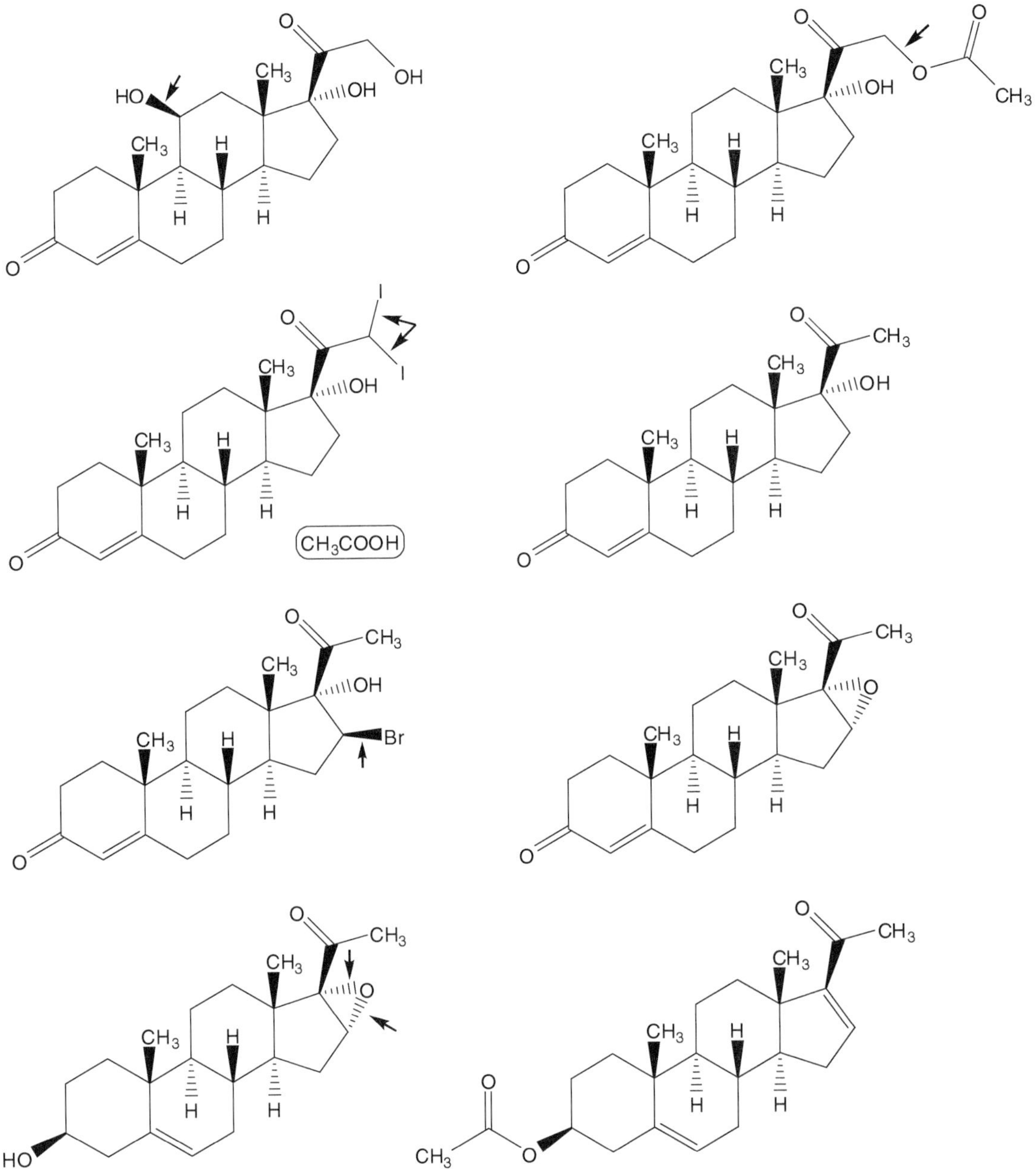

The 16-alkene of 16-dehydropregnenolone acetate is formed from diosone by β-elimination. Diosone is formed by oxidation of the 20(22)-alkene of pseudodiosgenin-3,26-diacetate. Pseudodiosgenin 3,26-diacetate is formed by reaction of diosgenin with acetic anhydride. The three-step synthesis of 16-dehydropregnenolone acetate from diosgenin by acetylation, oxidation, and elimination is known as the **Marker Degradation**.

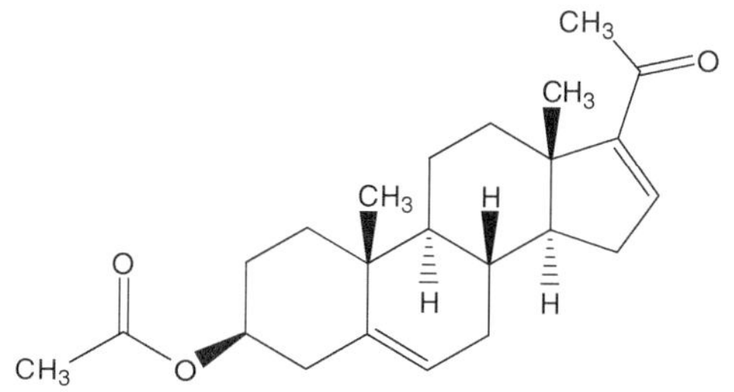

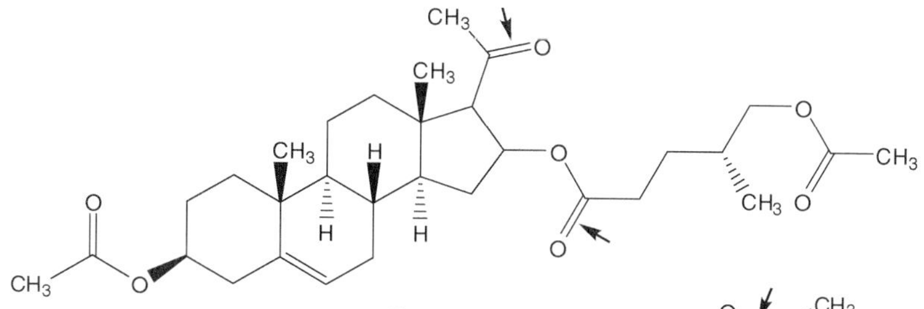

diosone

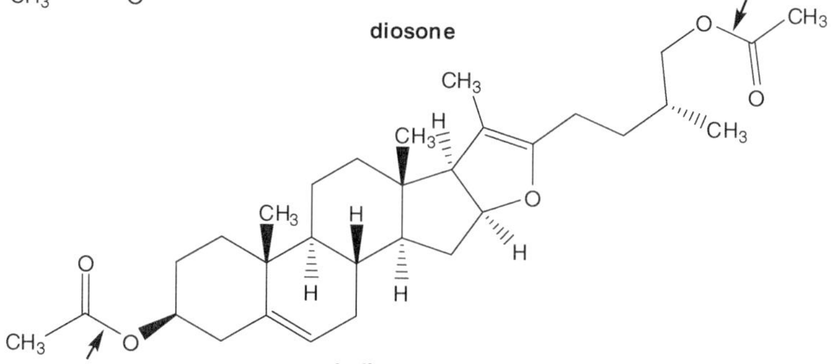

pseudodiosgenin-3,26-diacetate

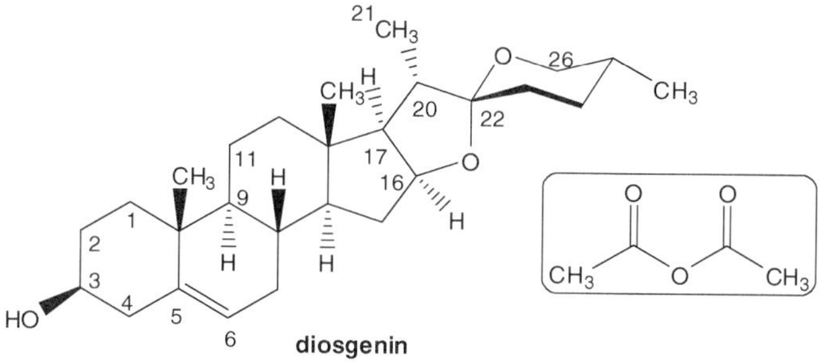

diosgenin

Extended Discussion

Draw the structures of the retrosynthetic analysis of one route to hydrocortisone from 11-epihydrocortisone.

Hydroxycarbamide

Antineoplastics and Immunosuppressives/Cytotoxic and Adjuvant Medicines
Medicines Affecting the Blood/Other Medicines for Haemoglobinopathies

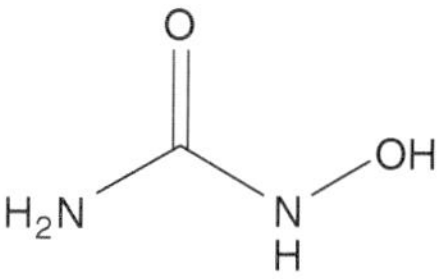

Discussion. Hydroxycarbamide (*N*-hydroxyurea) can be formed by the reaction of hydroxylamine hydrochloride with sodium or potassium cyanate or the reaction of hydroxylamine with methyl or ethyl carbamate (urethane). Hydroxycarbamide decomposes rapidly in aqueous acid. Hydroxycarbamide slowly decomposes and should be stored cold (2–8 °C).

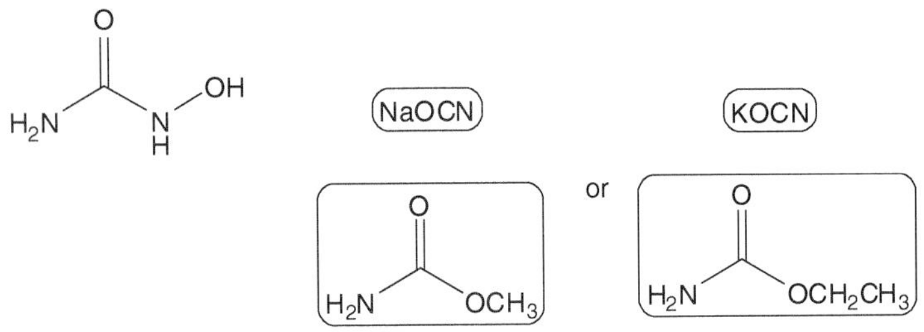

Hydroxychloroquine

Medicines for Diseases of Joints/Disease-modifying Agents Used in Rheumatoid Disorders

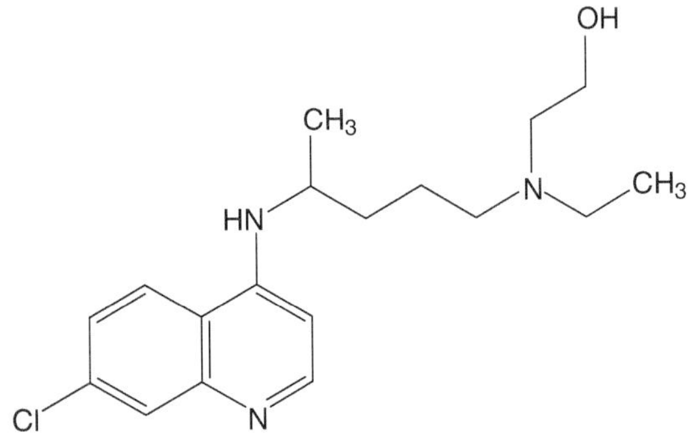

A nitrogen substituent on C2 or C4 of a quinoline ring is often introduced by displacement of chloride. The substitution is facilitated by the ring nitrogen and can be further facilitated by an electron-withdrawing group (NO_2, SO_2R, COOR, CN) on C3.

Discussion. Hydroxychloroquine is a 1:1 mixture of the (*R*)- and (*S*)-enantiomers. Chloride at C4 of 4,7-dichloroquinoline is displaced by the amine in the final step.

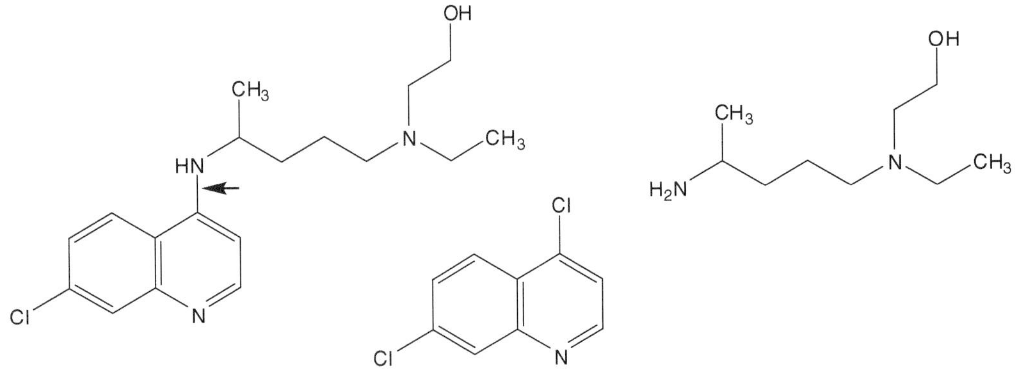

4,7-Dichloroquinoline is formed from 7-chloro-4-hydroxyquinoline (7-chloroquinolin-4-one). The quinolinone is formed by thermolysis/decarboxylation of the quinolin-4-one-3-carboxylic acid. The carboxylic acid is formed by ester hydrolysis. The quinolone ring is formed by intramolecular acylation of the aniline with loss of ethanol. The enamine is formed by reaction of 3-chloroaniline with ethyl ethoxymethylenemalonate. (The four-step sequence from 3-chloroaniline to 7-chloro-4-hydroxyquinoline is an example of the **Gould–Jacobs Reaction**.)

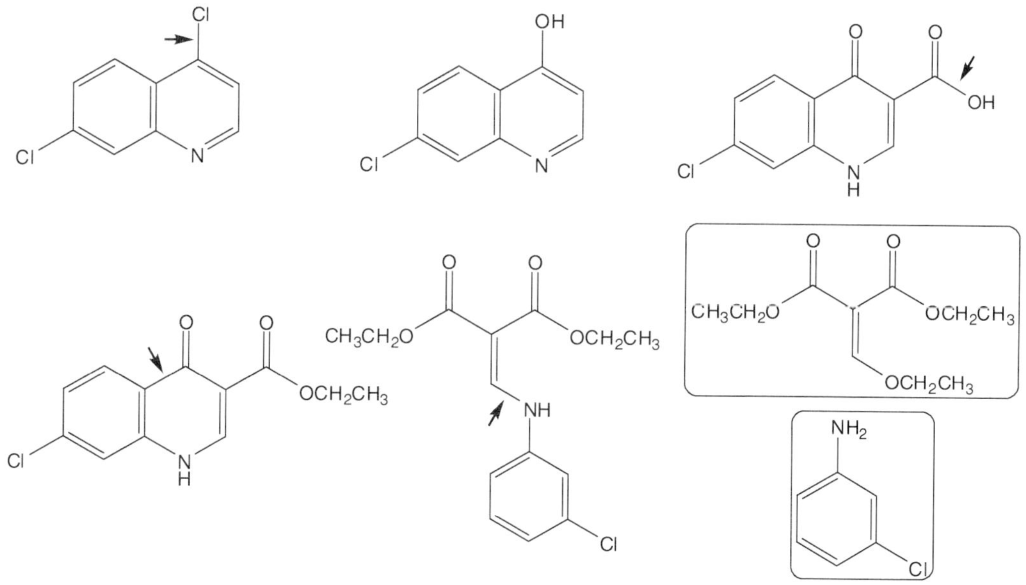

The amine used in the final step is formed by reductive amination of the ketone. 5-[Ethyl(2-hydroxyethyl)amino] pentan-2-one is formed by chloride displacement from 5-chloro-2-pentanone by 2-(ethylamino)ethanol. 5-Chloro-2-pentanone is formed from α-acetyl-γ-butyrolactone.

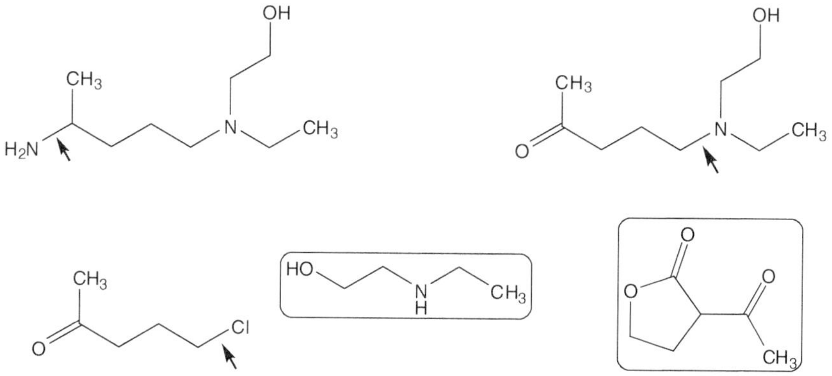

Extended Discussion

In one alternative route to 5-[ethyl(2-hydroxyethyl)amino]pentan-2-one, 4-chloro-2-pentanone is converted to an acetal by reaction with ethylene glycol. The chloride is displaced by 2-(ethylamino)ethanol, and the acetal is then hydrolyzed to regenerate the ketone. List the pros and cons for the two routes to 5-[ethyl(2-hydroxyethyl)amino]pentan-2-one and select one route as the preferred route.

Hyoscine Butylbromide

Medicines for Pain and Palliative Care/Medicines for Other Common Symptoms in Palliative Care

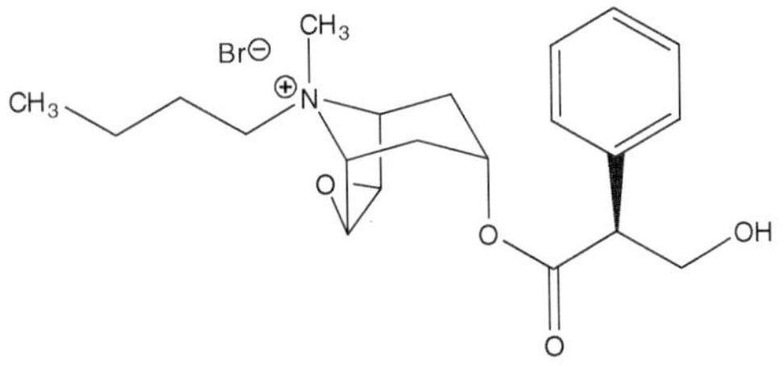

A single-enantiomer molecule with multiple chiral carbons is often formed by modification of a natural product which has most or all of the chiral carbons already in place.

Discussion. The quaternary ammonium salt of hyoscine butylbromide (scopolamine butylbromide) is formed by *N*-alkylation of the tertiary amine of essential medicine hyoscine (scopolamine). Hyoscine is an alkaloid produced by plants of the nightshade family.

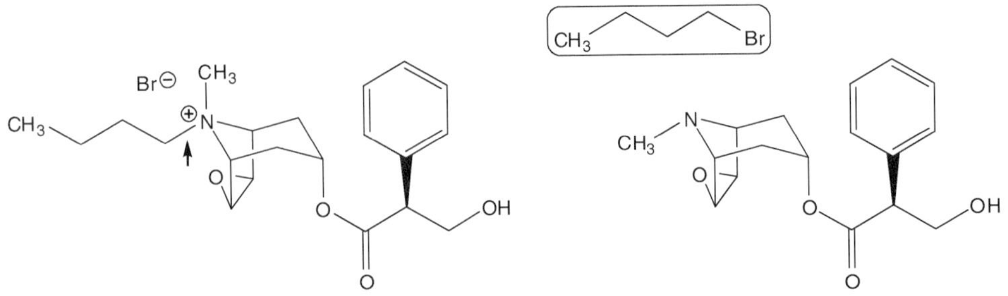

hyoscine

I

Ibuprofen

Medicines for Pain and Palliative Care/Non-opioid and Non-steroidal Anti-inflammatory Medicines
Antimigraine Medicines/For Treatment of Acute Attacks
Specific Medicines for Neonatal Care/Medicines Administered to the Neonate

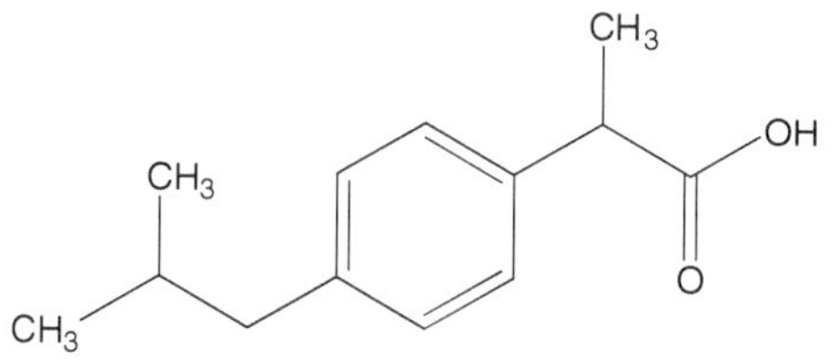

$$\textbf{Atom economy} = \frac{\textbf{mass of target molecule}}{\textbf{mass of all materials used}} \times 100$$

Most reactions produce some chemical waste. The waste could be a side product, a by-product, aqueous waste which must be treated, or solvent that is not recovered. An important consideration in synthetic route design and selection is minimal waste generation or high atom economy.

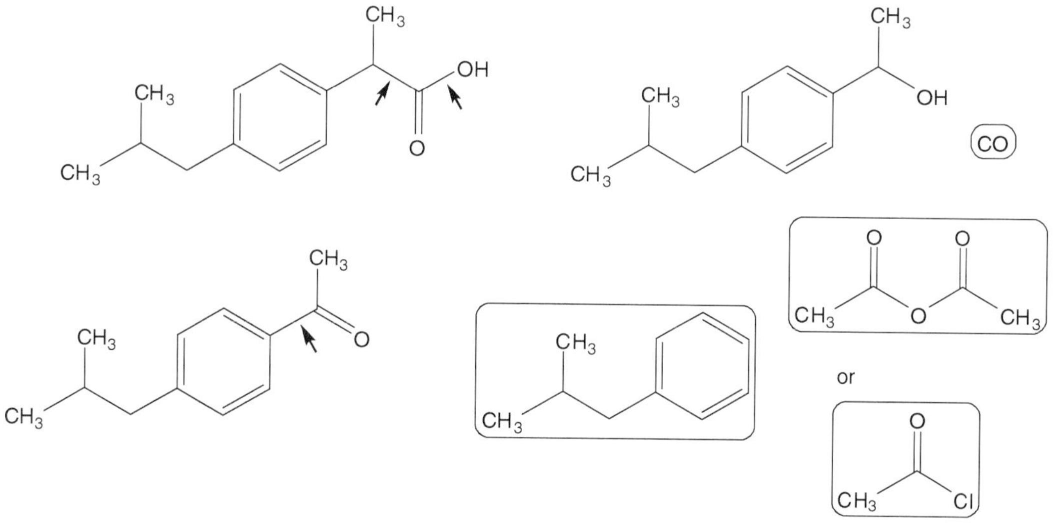

Discussion. The carboxylic acid is formed by carbonylation of the benzylic alcohol. The alcohol is formed by reduction of the ketone. The ketone is formed from isobutylbenzene and acetyl chloride or acetic anhydride (**Friedel–Crafts Acylation**).

Routes to Essential Medicines: A Workbook for Organic Synthesis, First Edition. Peter J. Harrington.

Companion website: www.wiley.com/go/Harrington/routes_essential_medicine

Extended Discussion

Draw the structures of a retrosynthetic analysis of an alternative route from isobutylbenzene to ibuprofen. Evaluate the atom economy for the two routes. Explain all assumptions used in the evaluation.

Ifosfamide

Antineoplastics and Immunosuppressives/Cytotoxic and Adjuvant Medicines

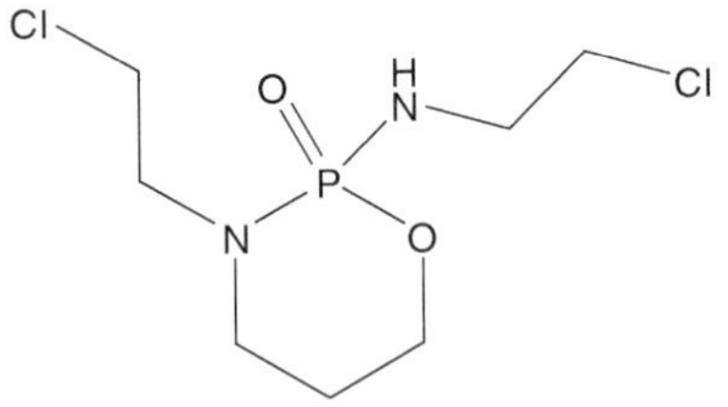

A phosphoric acid ester is often formed from a phosphoryl chloride and an alcohol. A phosphoric acid amide is often formed from a phosphoryl chloride and a primary or secondary amine.

Discussion. Ifosfamide is a 1 : 1 mixture of (*R*)- and (*S*)-enantiomers. In the final step, chlorine attached to phosphorus is displaced by 2-chloroethylamine. 2-Chloroethylamine is released in situ from the hydrochloride salt. 2-Chloro-3-(2-chloroethyl)tetrahydro-2*H*-1,3,2-oxazaphosphorine-2-oxide is formed by displacement of two chlorides from phosphorus oxychloride by the γ-amino alcohol. The γ-amino alcohol is released in situ from the hydrochloride salt. 3-(2-Chloroethyl)amino-1-propanol hydrochloride is formed from 3-(2-hydroxyethyl)amino-1-propanol. (Draw the structure of the major impurity formed in this reaction and present in the crude hydrochloride salt.) The β-amino alcohol 3-(2-hydroxyethyl)amino-1-propanol is formed by the ring-opening of ethylene oxide by 3-amino-1-propanol.

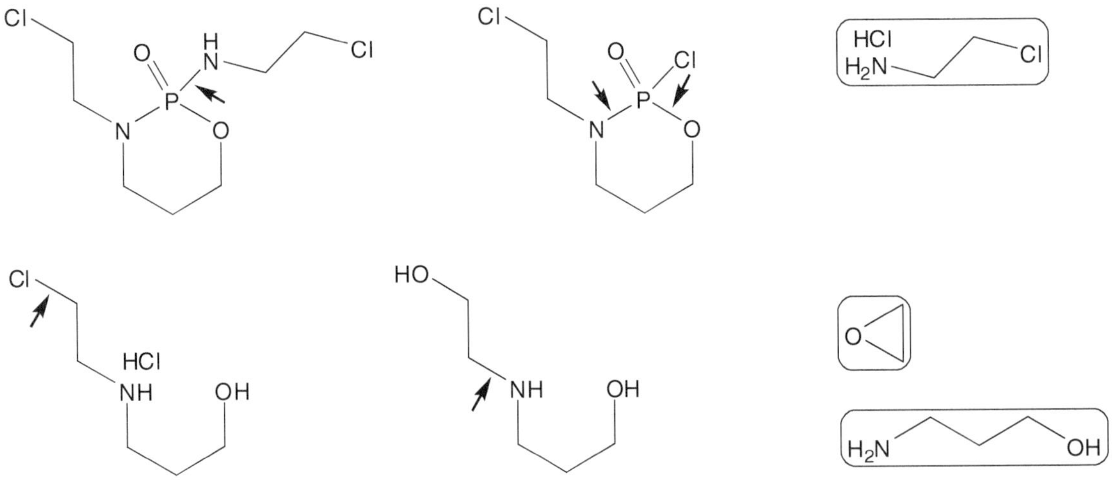

Extended Discussion

2-Chloro-3-(2-chloroethyl)tetrahydro-2*H*-1,3,2-oxazaphosphorine-2-oxide is also formed from phosphorus oxychloride and 3-(1-aziridinyl)-1-propanol. List the pros and cons for both routes and select one route as the preferred route to this intermediate.

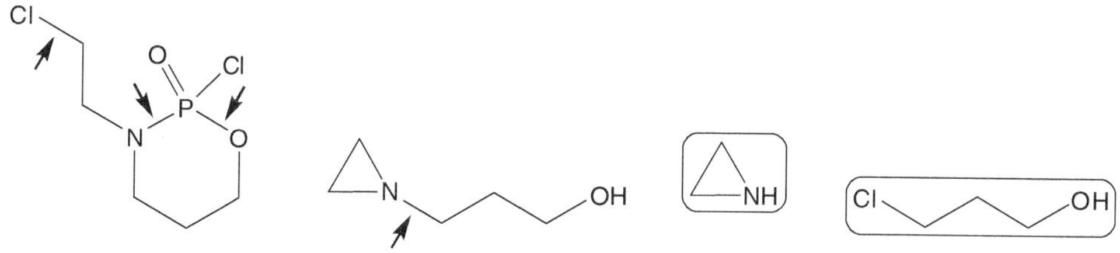

Imatinib

Antineoplastics and Immunosuppressives/Cytotoxic and Adjuvant Medicines

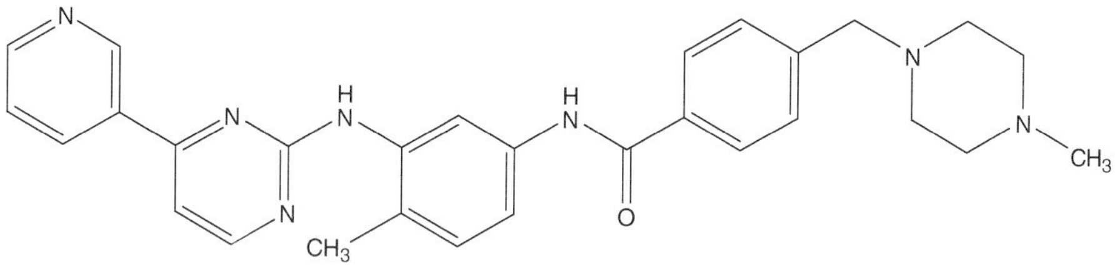

An amide is efficiently formed by reaction of an amine with an acid chloride, anhydride, ester, or carboxylic acid.

Discussion. The amide is formed in the final step from the amine and the carboxylic acid.

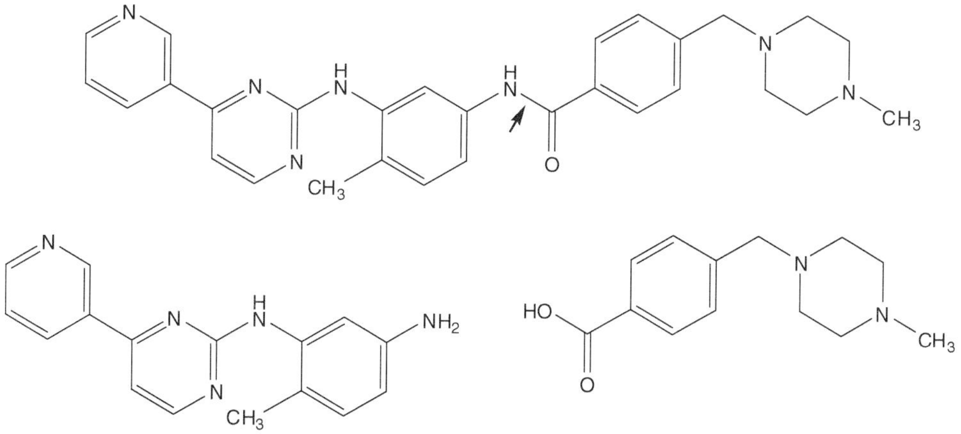

The amine for the final step is formed by reduction of the nitroaromatic. The pyrimidine ring is formed from a guanidine and a β-amino enone (**Pinner Pyrimidine Synthesis**). The guanidine is formed from the aniline and cyanamide. The β-dimethylamino enone is formed by condensation of 3-acetylpyridine and dimethylformamide dimethyl acetal.

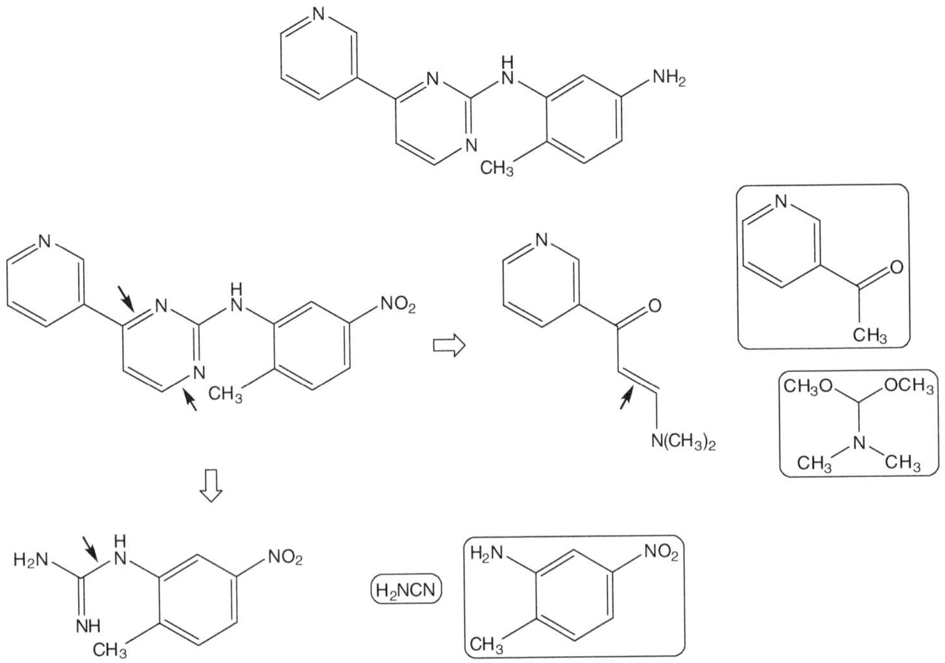

A tertiary amine on the carboxylic acid is formed from 4-formylbenzoic acid and 1-methylpiperazine by reductive amination.

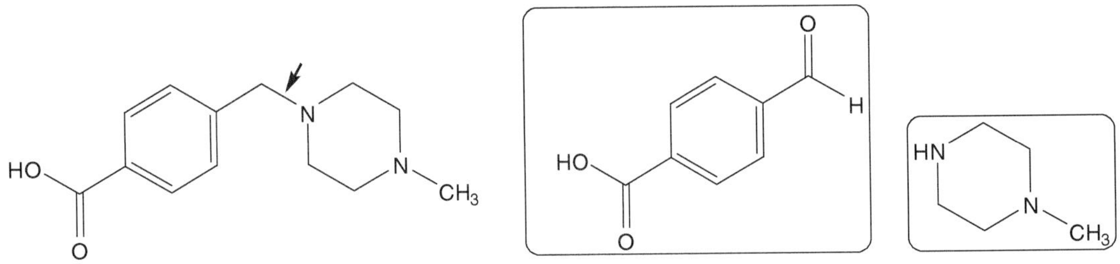

Extended Discussion

Draw the structures of the retrosynthetic analysis of one alternative route to imatinib that does not use cyanamide as a raw material. List the pros and cons for the two routes.

Iohexol

Diagnostic Agents/Radiocontrast Media

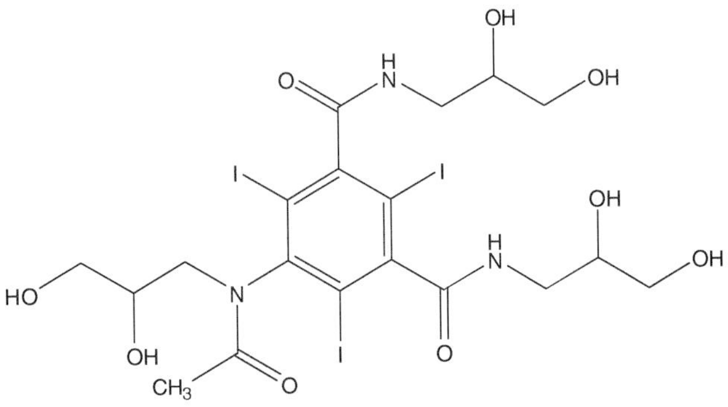

Amides are ubiquitous in drug structures. Amide formation from an amine and acid chloride, anhydride, ester, or acid is often very efficient.

Discussion. Iohexol has three chiral carbons. Iohexol is a mixture of stereoisomers produced from achiral starting materials. The acetanilide is *N*-alkylated by reaction with 3-chloro-1,2-propanediol in the final step. The acetanilide is formed by acetylation of the aniline with acetic anhydride. (**The esters also formed are hydrolyzed.**)

The 5-aminoisophthalic diamide is iodinated at C2, C4, and C6. The 5-aminoisophthalic diamide is formed by reduction of the 5-nitroisophthalic diamide. The 5-nitroisophthalic diamide is formed from the diester, dimethyl 5-nitroisophthalate, and 3-amino-1,2-propanediol.

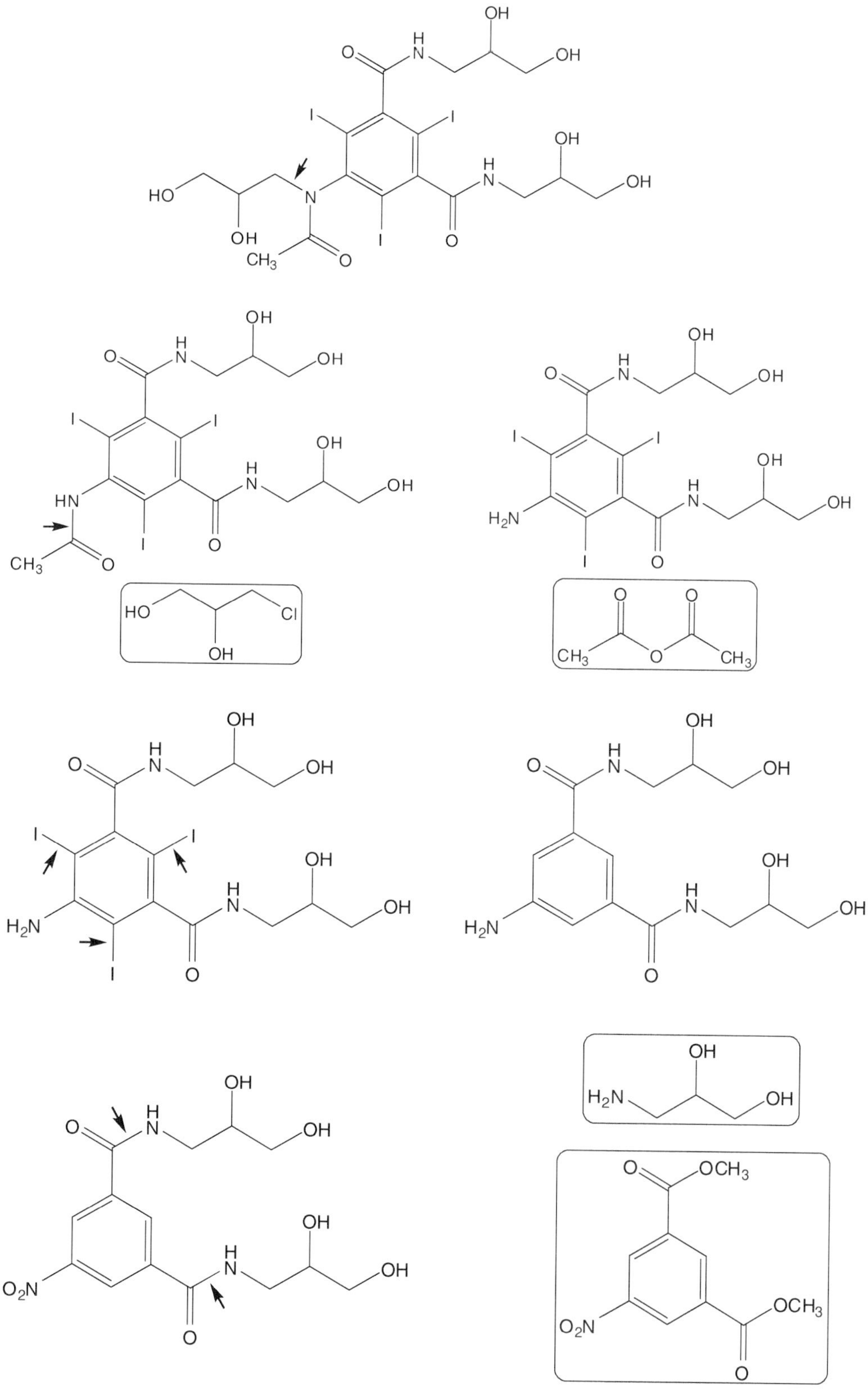

Extended Discussion

Draw a flow diagram for the workup procedure for separation of iohexol from the by-products and side products formed in the final step.

Ipratropium Bromide

Medicines Acting on the Respiratory Tract/Antiasthmatic and Medicines for Chronic Obstructive Pulmonary Disease

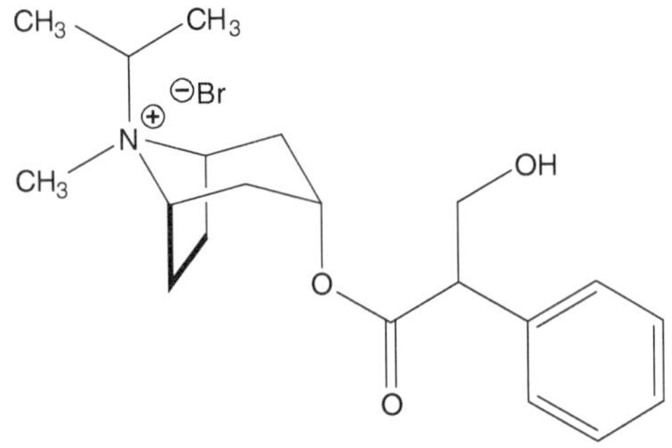

A reaction on or near a rigid bicyclic structure is often highly stereoselective.

Discussion. Ipratropium bromide is a 1 : 1 mixture of the (*R*)- and (*S*)-enantiomers. The primary alcohol is released by acetate ester hydrolysis in the final step. The quaternary ammonium bromide is formed by reaction of the tertiary amine with bromomethane. The ester is formed from the acid chloride and the alcohol, *N*-isopropylnortropine.

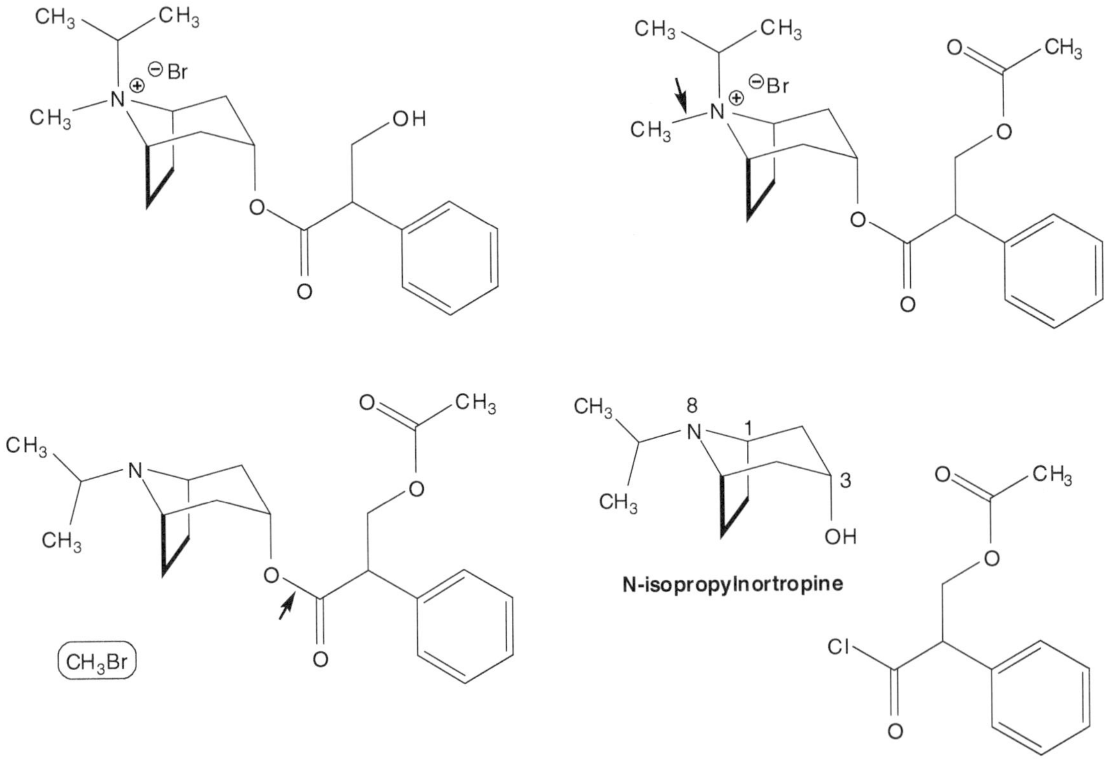

The acid chloride is formed from the carboxylic acid. The acetate ester is formed by the reaction of tropic acid with acetyl chloride. Tropic acid is formed by hydrolysis of the ester. Ethyl 3-hydroxy-2-phenylpropanoate is formed by reduction of ethyl 3-hydroxy-2-phenylacrylate. The 3-hydroxyacrylate is formed by condensation of ethyl phenylacetate with ethyl formate (mixed **Claisen Condensation**).

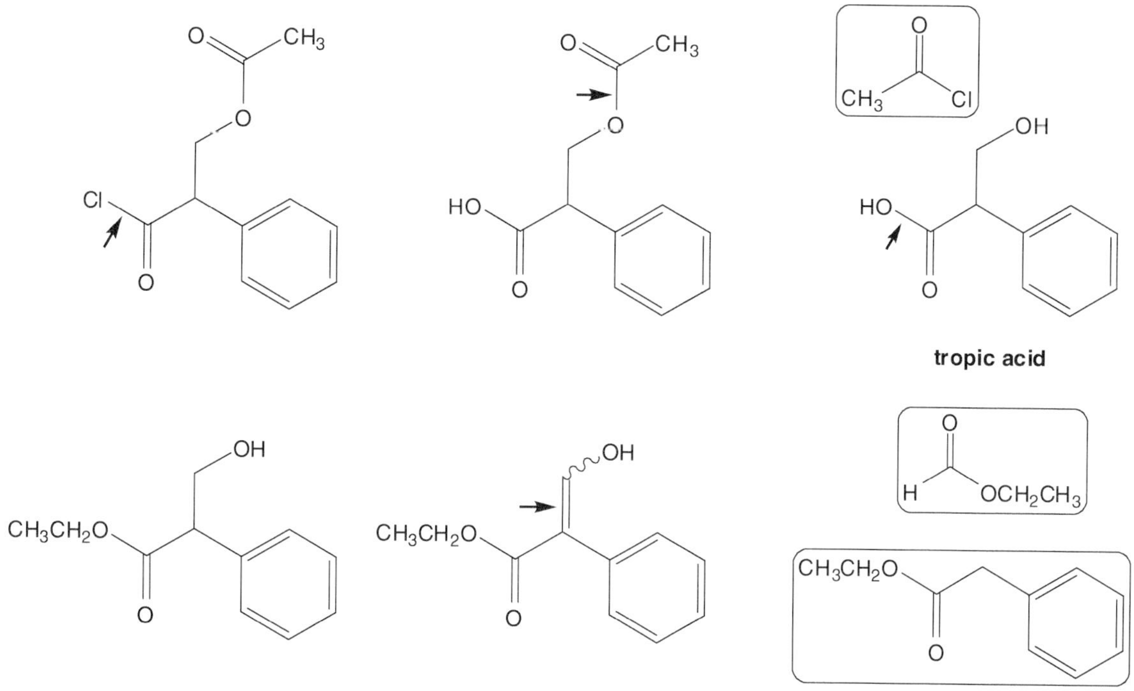

N-Isopropylnortropine is formed by stereoselective reduction of the ketone. *N*-Isopropylnortropinone is efficiently assembled in a single step from isopropylamine, 2,5-dimethoxytetrahydrofuran, and 1,3-acetonedicarboxylic acid (**Robinson–Schopf Reaction**). 1,3-Acetonedicarboxlic acid is formed by oxidation of citric acid. Citric acid is produced by fermentation.

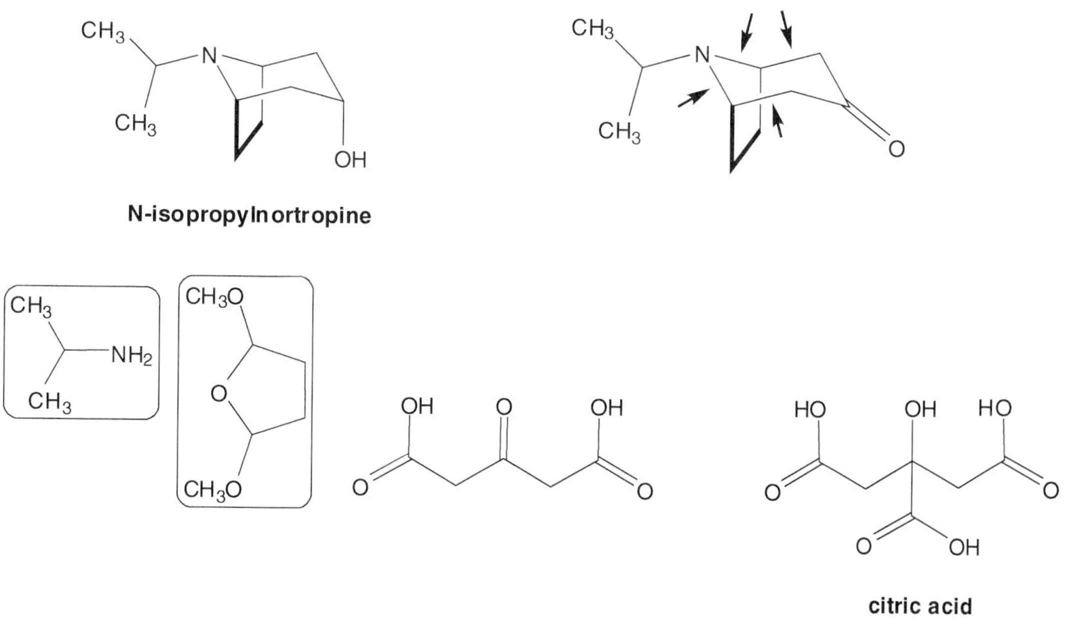

Extended Discussion

Draw the structures of the retrosynthetic analysis of one alternative route to *N*-isopropylnortropine. Include the structures of the retrosynthetic analysis of any organic starting material(s) from petrochemical or biochemical raw materials. List the pros and cons for the two routes to *N*-isopropylnortropine and select one route as the preferred route.

Irinotecan

Antineoplastics and Immunosuppressives/Cytotoxic and Adjuvant Medicines

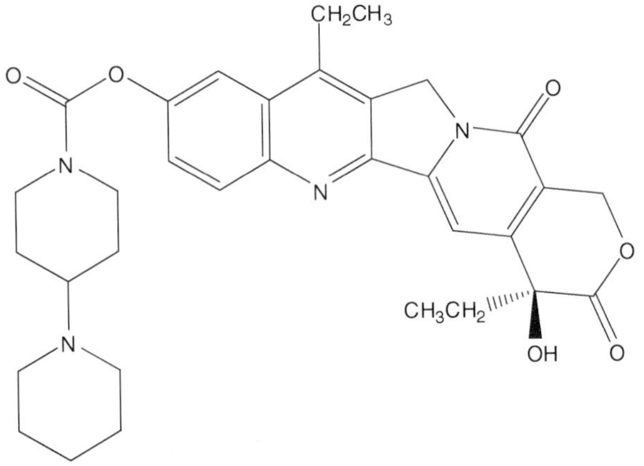

Semisynthesis of a target molecule from a natural product is often based on known reactions or reaction conditions which are uniquely selective when applied to the natural product.

Discussion. Irinotecan is most often manufactured from camptothecin by alkylation at C7 and oxidation at C10. Two routes have the C7 alkylation before the C10 oxidation. In Route A, the carbamate is formed by acylation of 7-ethyl-10-hydroxycamptothecin using the carbamoyl chloride. 7-Ethyl-10-hydroxycamptothecin is formed from 7-ethylcamptothecin-1-oxide by photochemical rearrangement. The *N*-oxide is formed by oxidation of 7-ethylcamptothecin. 7-Ethylcamptothecin is formed from camptothecin and propanal under **Fenton's Reagent** conditions. Camptothecin is an alkaloid first isolated from the deciduous tree *Camptotheca acuminata*.

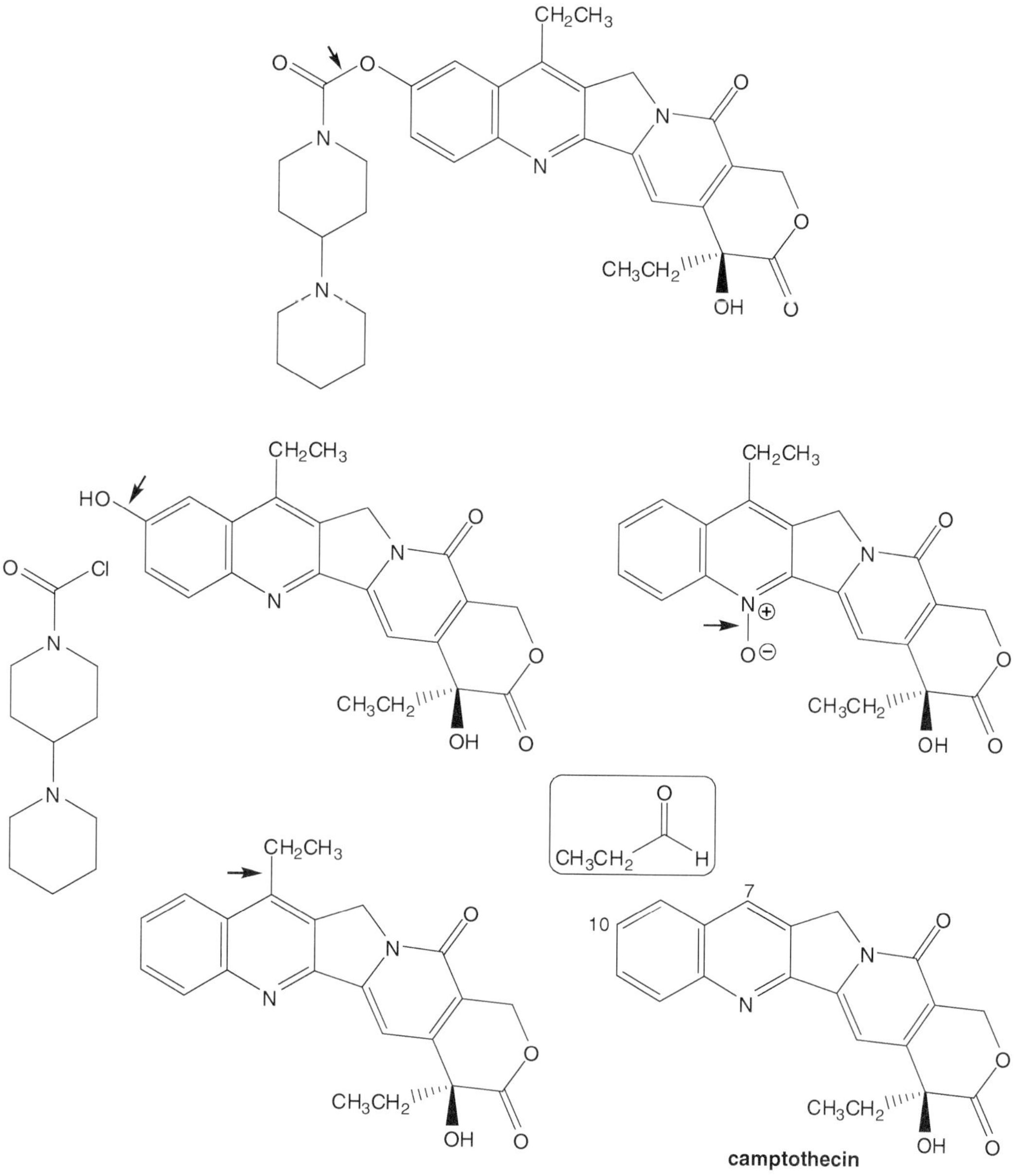

In Route B, the carbamate is again formed by acylation of 7-ethyl-10-hydroxycamptothecin using the carbamoyl chloride. 7-Ethyl-10-hydroxycamptothecin is formed by oxidation of 7-ethyl-1,2,6,7-tetrahydrocamptothecin. The 1,2,6,7-tetrahydrocamptothecin (a mixture of three diastereomers) is formed by hydrogenation of 7-ethylcamptothecin. 7-Ethylcamptothecin is formed from camptothecin and propanal under **Fenton's Reagent** conditions.

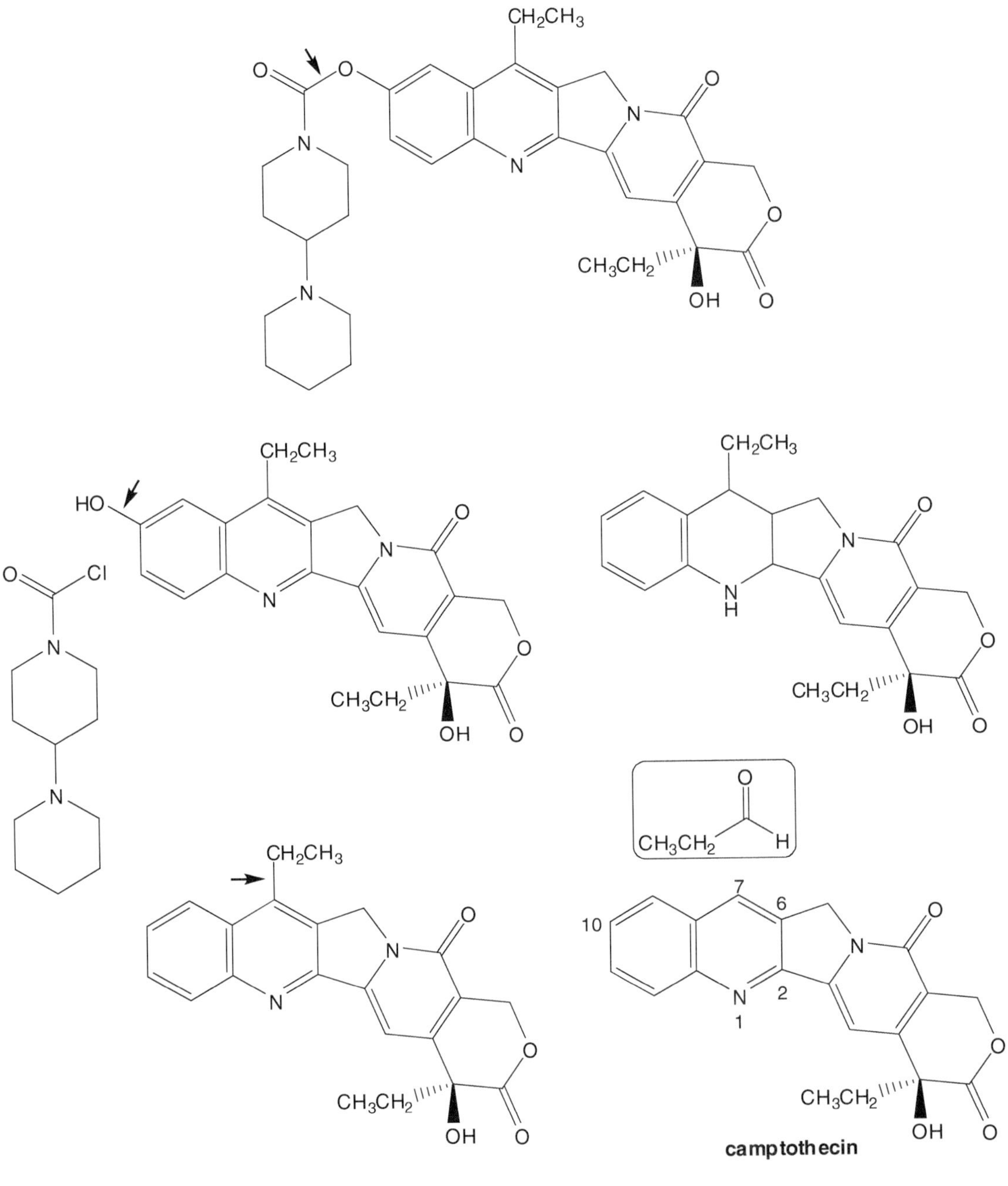

The carbamoyl chloride is formed by reaction of 4-piperidinopiperidine (1,4′-bipiperidine) with phosgene. The secondary amine of 4-piperidinopiperidine is released by hydrogenolysis of the benzyl group of 1-benzyl-4-piperidinopiperidine. 1-Benzyl-4-piperidinopiperidine is formed from 1-benzyl-4-piperidinone and piperidine by reductive amination.

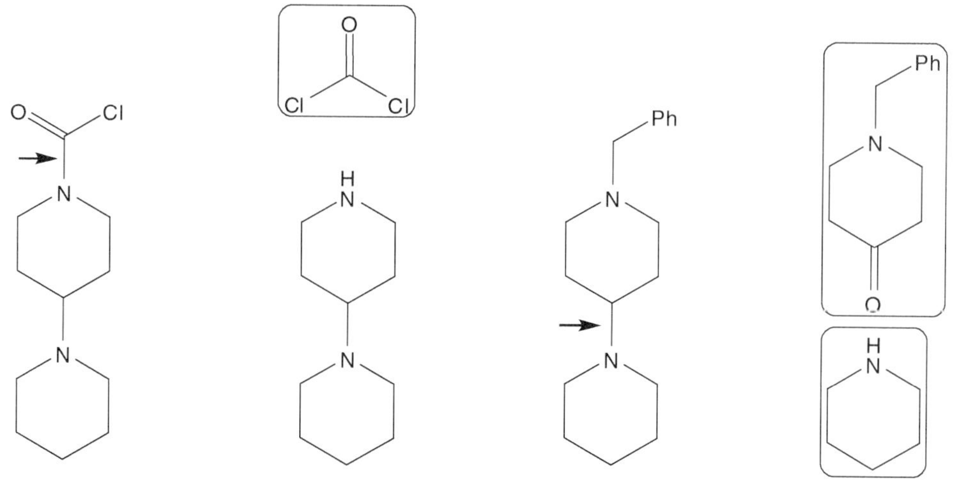

Extended Discussion

Draw the structures of the retrosynthetic analysis of one alternative route (Route C) to irinotecan from camptothecin which has the C10 hydroxylation before the C7 alkylation. List the pros and cons for the three routes. (Include the solubility data for each step.)

Isoflurane

Anesthetics, Preoperative Medicines and Medical Gases/General Anesthetics and Oxygen/Inhalational Medicines

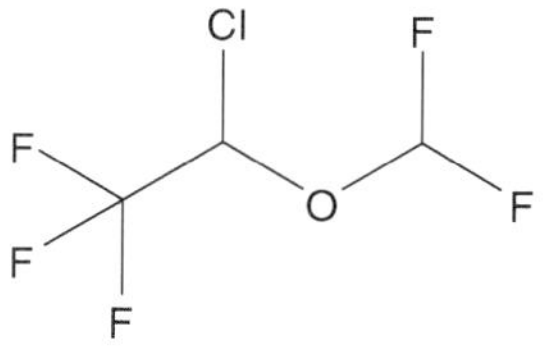

An alkyl fluoride is often formed by the displacement of chloride by fluoride.

Discussion. Isoflurane is a 1 : 1 mixture of the (*R*)- and (*S*)-enantiomers. The α-chloro ether is formed in the final step by chlorination. (The α,α-dichloro ether side product is also converted to isoflurane by reduction. What reducing agents are used?) The difluoromethyl ether is formed by chloride displacement from the dichloromethyl ether (**Swarts Reaction**). The dichloromethyl ether is formed by chlorination of the methyl ether. (How selective is this chlorination?) The methyl ether is formed from 2,2,2-trifluoroethanol and dimethyl sulfate (**Williamson Ether Synthesis**). The alcohol is formed from 2-chloro-1,1,1-trifluoroethane by chloride displacement. 2-Chloro-1,1,1-trifluoroethane (also known as HCFC-133a) is formed from trichloroethylene.

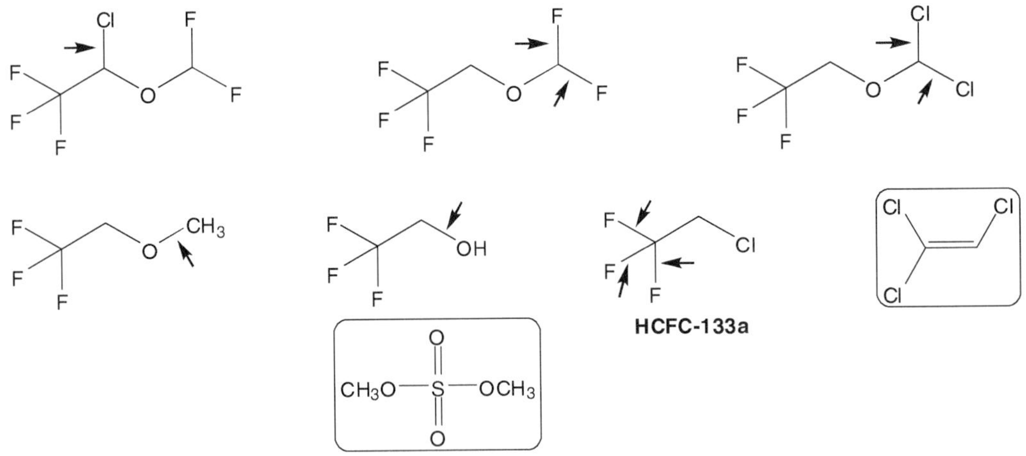

Extended Discussion

Draw the structures of the retrosynthetic analysis of an alternative route to isoflurane from trichloroethylene and chlorodifluoromethane (HCFC-22). List the pros and cons for both routes.

Isoniazid

Anti-infective Medicines/Antibacterials/Antituberculosis Medicines
Anti-infective Medicines/Antiviral Medicines/Antiretrovirals/Medicines for Prevention of HIV-Related Opportunistic Infections

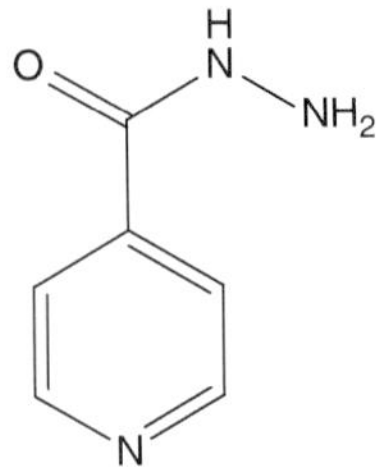

A pyridine with one carbon substituent is often formed from the methylpyridine (picoline).

Discussion. Isoniazid is formed in a single step by reaction of 4-pyridinecarbonitrile (isonicotinonitrile) with hydrazine hydrate.

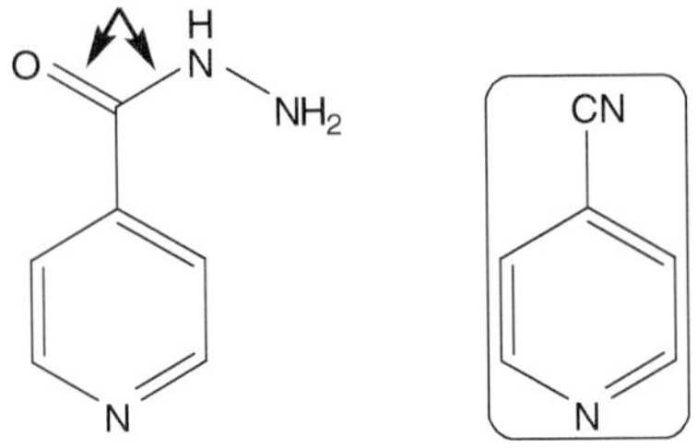

Extended Discussion

Draw the structures of the retrosynthetic analysis of an alternative route to isoniazid from citric acid. List the pros and cons for both routes and select one route as the preferred route.

Isosorbide Dinitrate

Cardiovascular Medicines/Antianginal Medicines

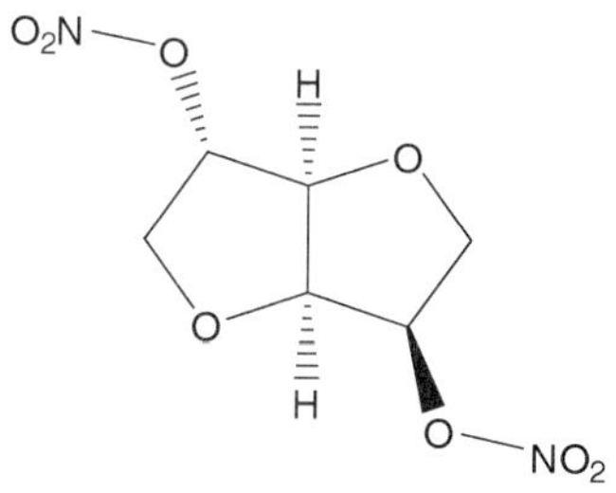

An alkyl nitrate ester is often formed from the alcohol and nitric acid. The alkyl nitrate ester functional group is an explosophore.

Discussion. Isosorbide dinitrate is formed by the reaction of isosorbide with nitric acid. [**NOTE**: The isosorbide dinitrate safety data sheet (SDS) indicates an extreme risk of explosion by fire, friction, or other sources of ignition.]

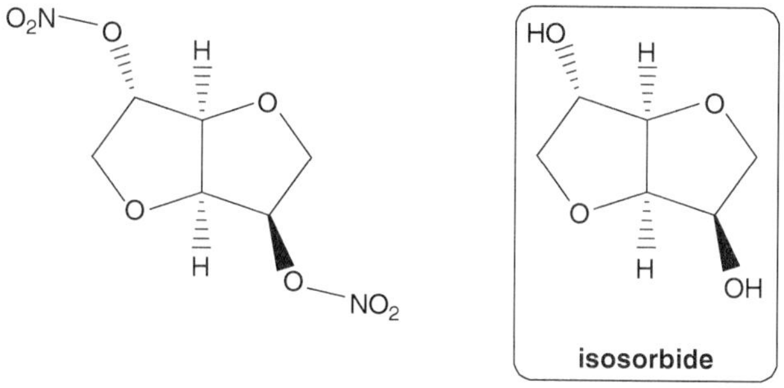

Extended Discussion

Isosorbide mononitrate is also manufactured on metric ton scale. Draw the structures of a retrosynthetic analysis of one route to isosorbide mononitrate.

Itraconazole

Anti-infective Medicines/Antifungal Medicines

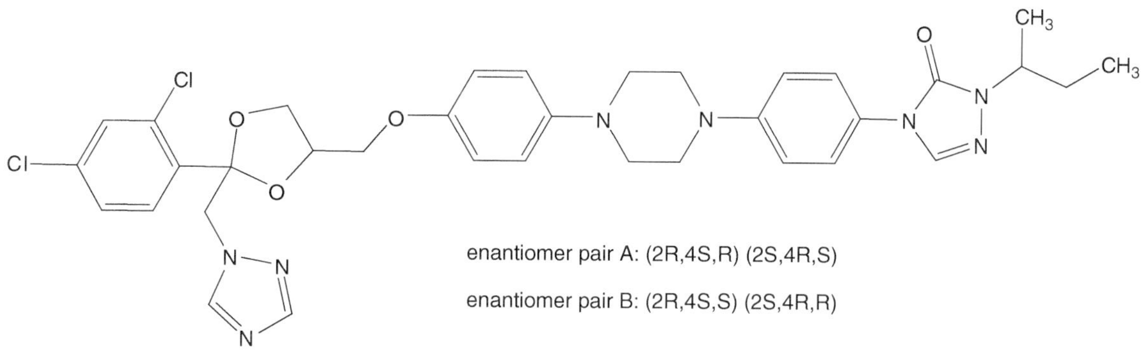

An aldehyde or ketone reacts with an alcohol or diol to form an acetal. The reaction is catalyzed by acid and is especially favorable when it results in formation of a five- or six-membered ring (1,3-dioxolane or 1,3-dioxane).

Discussion. Itraconazole is manufactured as 1 : 1 mixture of two enantiomer pairs. In all four stereoisomers, the dichlorophenyl substituent at C2 and the aryloxymethyl substituent at C4 on the dioxolane ring are *trans*. The ether C-O bond is formed in the final step by displacement of a methanesulfonate by the phenol (**Williamson Ether Synthesis**).

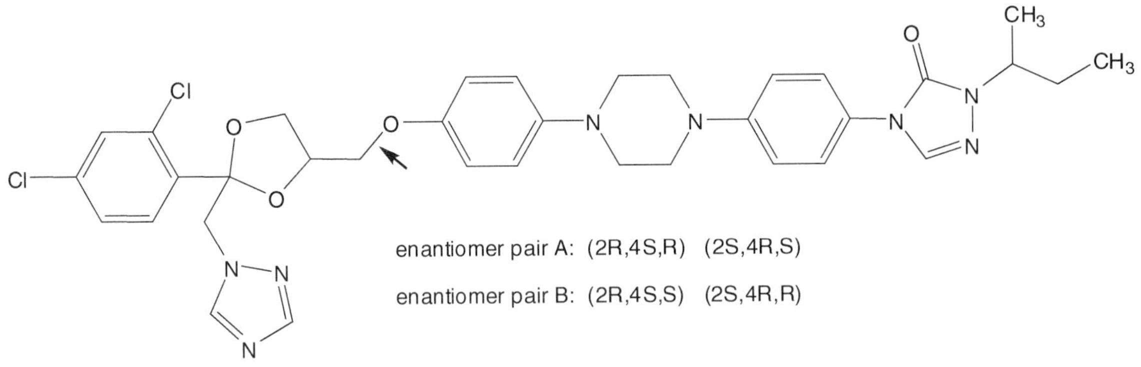

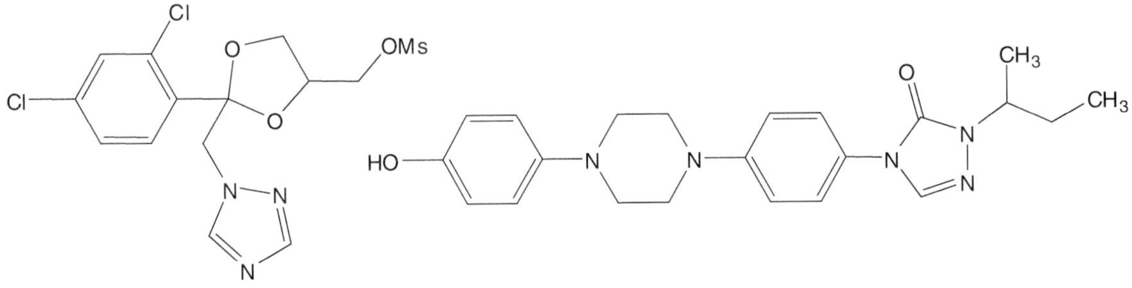

enantiomer pair: (2R,4S) (2S,4R)

enantiomer pair: R S

The methanesulfonate is formed from the alcohol. The alcohol is released by hydrolysis of the benzoate ester. The C-N bond to 1,2,4-triazole is formed by a bromide displacement. (Draw the structure of a side product of the bromide displacement.) The alcohol is protected as a benzoate ester, and the enantiomer pair (2*R*,4*S*) and (2*S*,4*R*) is crystallized from the mixture. The 2-bromomethyl-1,3-dioxolane is formed by bromination of the 2-methyl-1,3-dioxolane. The 1,3-dioxolane is formed from glycerol and 2′,4′-dichloroacetophenone.

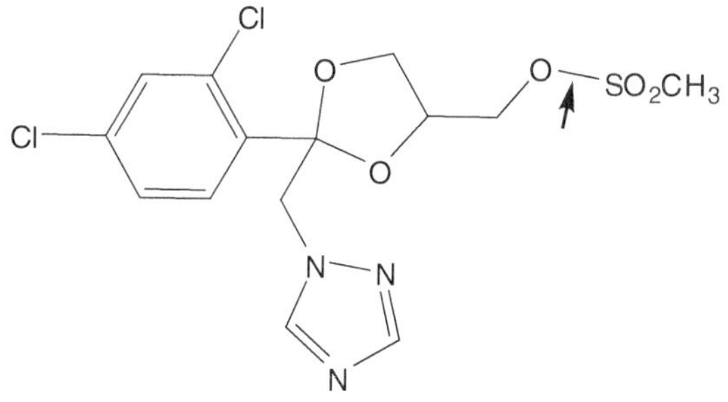

enantiomer pair (2R,4S) (2S,4R)

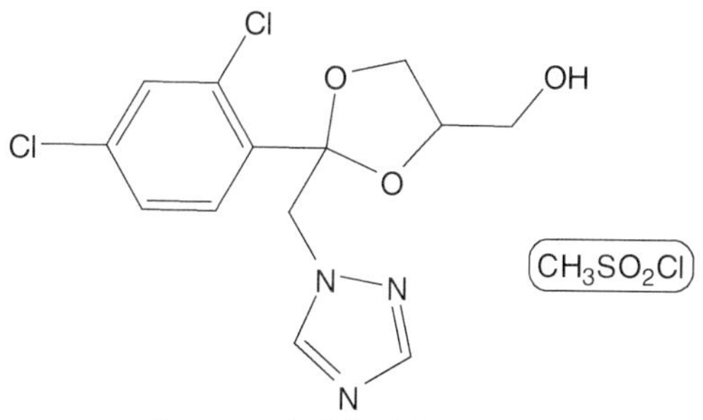

enantiomer pair (2R,4S) (2S,4R)

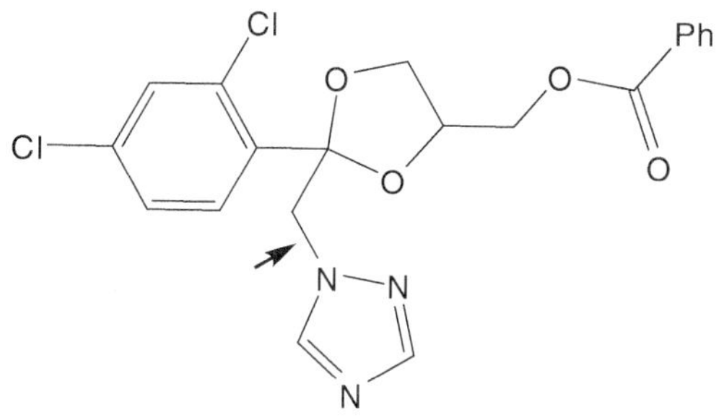

enantiomer pair (2R,4S) (2S,4R)

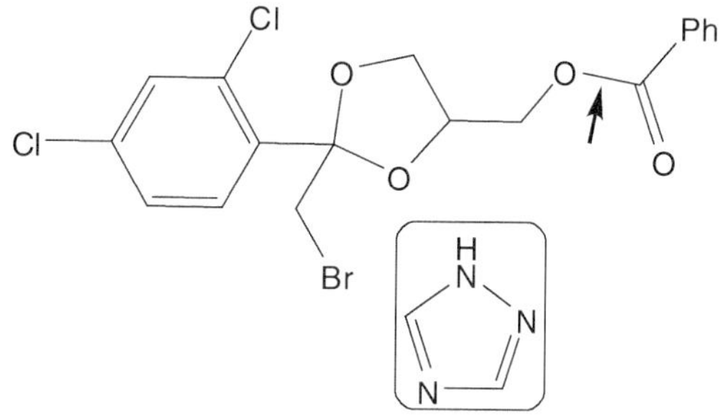

enantiomer pair (2R,4S) (2S,4R)

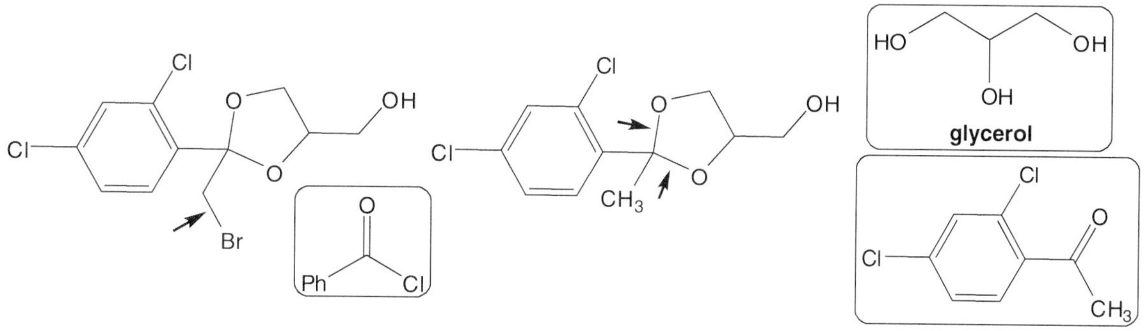

The phenol for the final step is released by cleavage of the methyl ether. The triazolone is alkylated with 2-bromobutane. The triazolone is formed from the semicarbazide and formamidine. The semicarbazide is formed from the phenyl carbamate. The phenyl carbamate is formed from the aniline and phenyl chloroformate. The aniline is formed by reduction of the nitroaromatic.

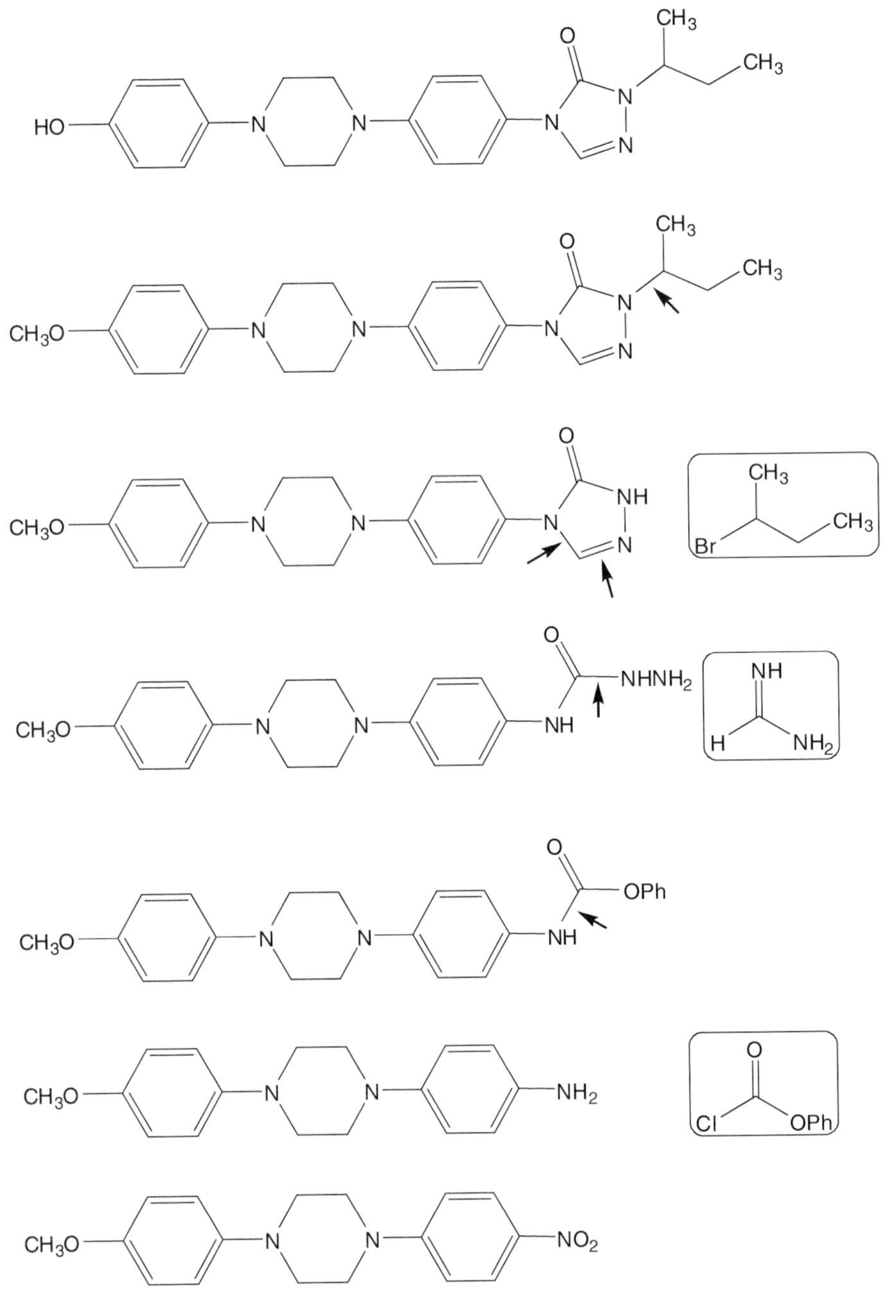

There are many potential routes to 1-(4-methoxyphenyl)-4-(4-nitrophenyl)piperazine. In one analysis, a C-N bond is formed by nucleophilic displacement of chloride from 1-chloro-4-nitrobenzene by 1-(4-methoxyphenyl)piperazine. There are also many potential routes to 1-(4-methoxyphenyl)piperazine. In one analysis, 1-(4-methoxyphenyl)piperazine is formed from *para*-anisidine and bis(2-chloroethyl)amine hydrochloride. Bis(2-chloroethyl)amine hydrochloride could be considered the starting material or could be formed from diethanolamine and carried directly into the piperazine formation.

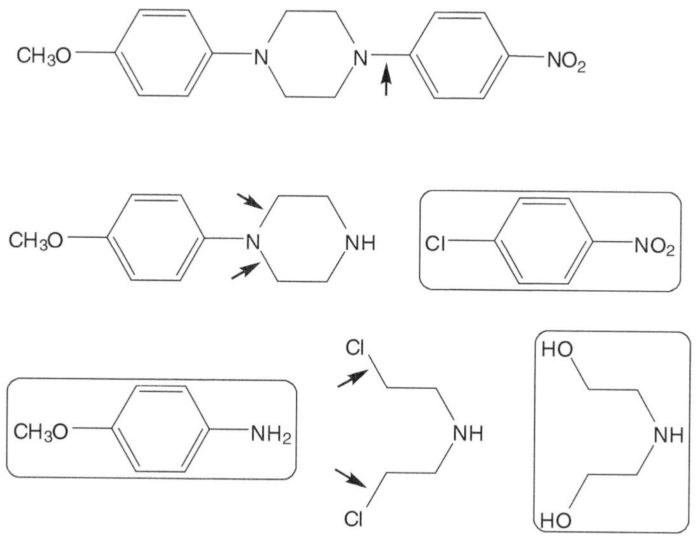

The same starting materials are used in an alternative route to 1-(4-methoxyphenyl)-4-(4-nitrophenyl)piperazine.

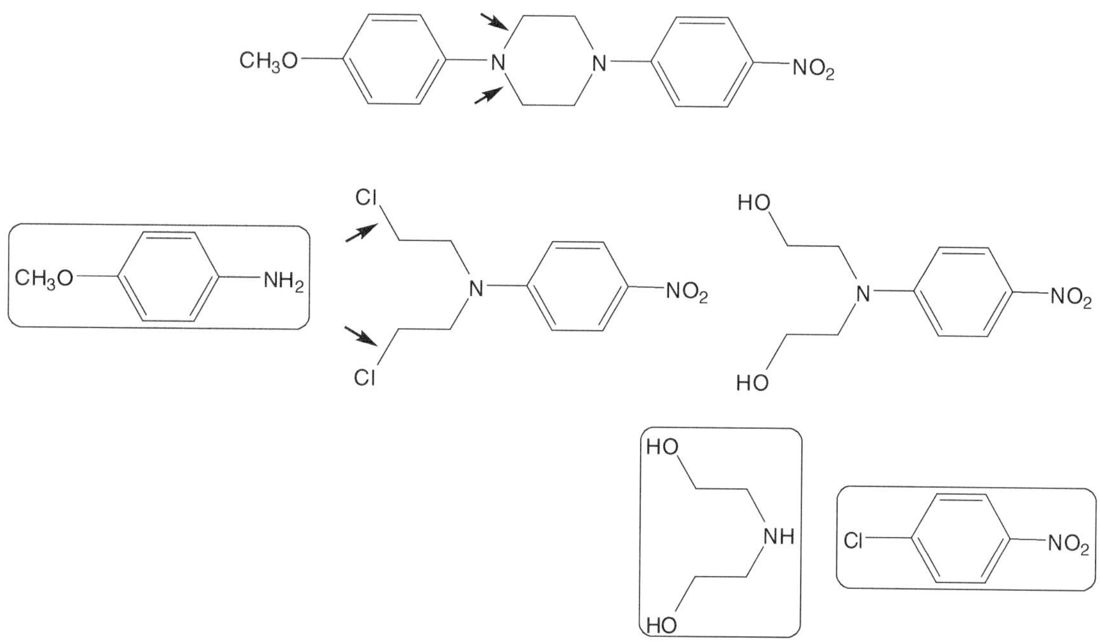

Extended Discussion

Draw the structures of a retrosynthetic analysis of one alternative route to itraconazole from 2′,4′-dichloroacetophenone. Estimate the overall yield of itraconazole from 2′,4′-dichloroacetophenone for both routes.

Ivermectin

Anti-infective Medicines/Antifilarials

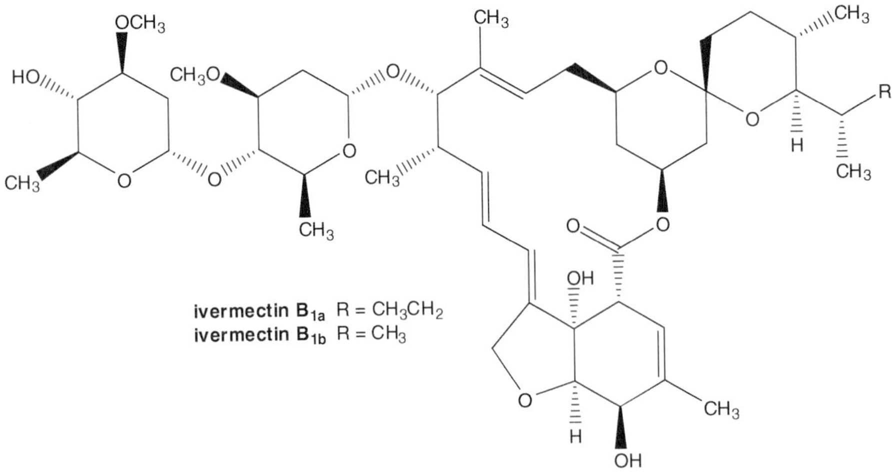

ivermectin B_{1a} R = CH_3CH_2
ivermectin B_{1b} R = CH_3

A single-enantiomer molecule with multiple chiral carbons is often formed by modification of a natural product which has most or all of the chiral carbons already in place.

Discussion. Ivermectin, a mixture of B_{1a} and B_{1b} forms (at least 80% B_{1a} and not more than 20% B_{1b}), is semisynthetic. Ivermectin is formed by selective hydrogenation of the C22-C23 double bond of a mixture of avermectins B_{1a} and B_{1b} (**Wilkinson's catalyst**). Avermectins B_{1a} and B_{1b} are produced by the soil bacterium *Streptomyces avermitilis.*

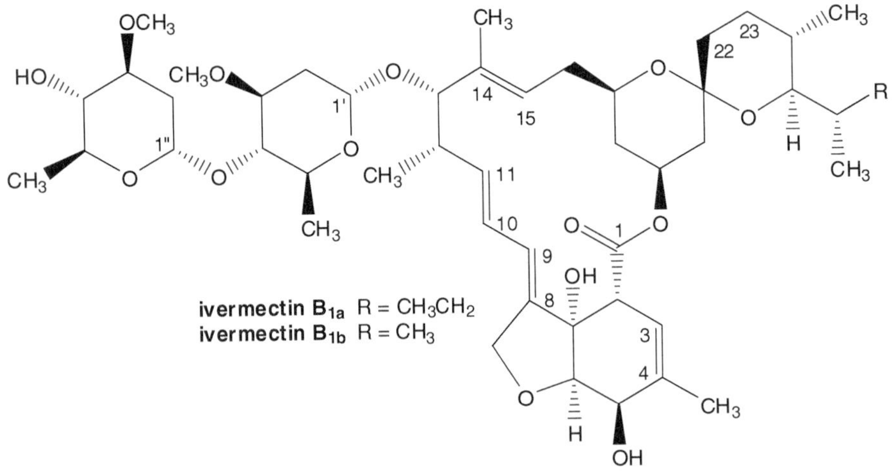

ivermectin B_{1a} R = CH_3CH_2
ivermectin B_{1b} R = CH_3

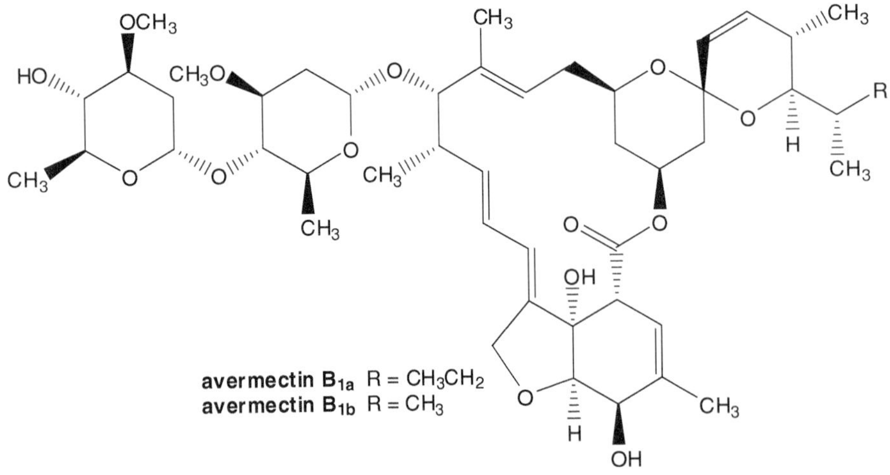

avermectin B_{1a} R = CH_3CH_2
avermectin B_{1b} R = CH_3

Extended Discussion

Draw the structures of three known ivermectin impurities. Is the impurity traced back to an impurity in the starting material or was the impurity formed from avermectin B_{1a} or avermectin B_{1b} in the hydrogenation step?

K

Ketamine

Anesthetics, Preoperative Medicines, and Medical Gases/Injectable Medicines

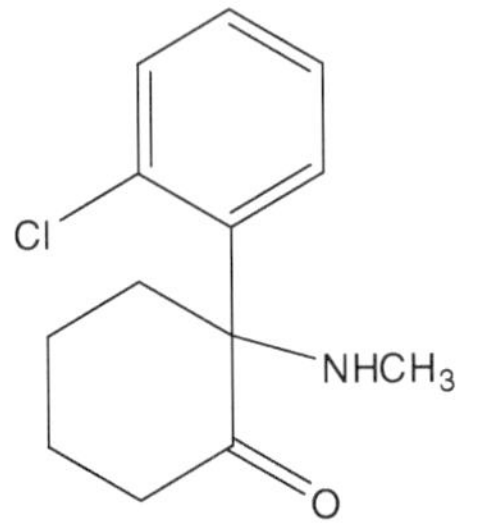

An α-aminoketone is often formed from an α-bromoketone by displacement of bromide by an amine. The reaction is efficient when the α-carbon is primary.

Discussion. Ketamine is a 1:1 mixture of (*R*)- and (*S*)-enantiomers. Ketamine is formed by reaction of the α-bromoketone formed from 2-chlorophenyl cyclopentyl ketone with methylamine. The α-bromoketone is formed by bromination of the ketone. The ketone is formed by the addition of cyclopentylmagnesium bromide to 2-chlorobenzonitrile (**Grignard Reaction**). Cyclopentylmagnesium bromide is formed from bromocyclopentane.

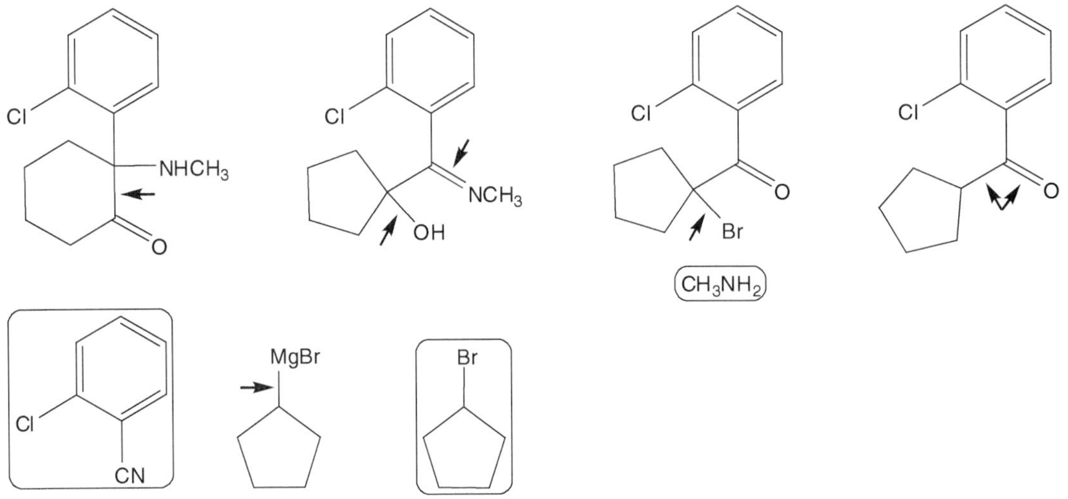

Extended Discussion

(*S*)-Ketamine nasal spray is being evaluated for patients with treatment-resistant depression. Provide details for the synthesis of (*S*)-ketamine by separation of (*R*)- and (*S*)-ketamine and racemization and recycle of (*R*)-ketamine.

Routes to Essential Medicines: A Workbook for Organic Synthesis, First Edition. Peter J. Harrington.

Companion website: www.wiley.com/go/Harrington/routes_essential_medicine

L

Lactulose

Medicines for Pain and Palliative Care/Medicines for Other Common Symptoms in Palliative Care

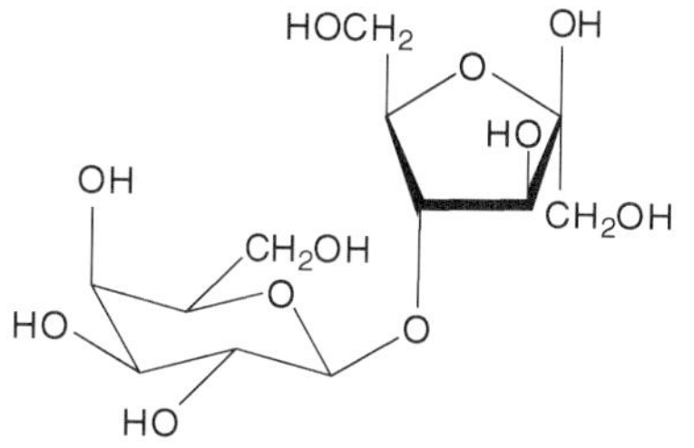

Alcohol adds to an aldehyde or ketone to form a hemiacetal. The addition is reversible: a hemiacetal is in equilibrium with the alcohol and the aldehyde or ketone.

Discussion. Lactulose (4-*O*-β-D-galactopyranosyl-D-fructose) is formed from lactose (4-*O*-β-D-galactopyranosyl-D-glucose) (**Lobry de Bruyn–Alberda van Ekenstein Reaction**). Lactose for manufacture of lactulose is obtained from bovine whey, a byproduct of cheese production.

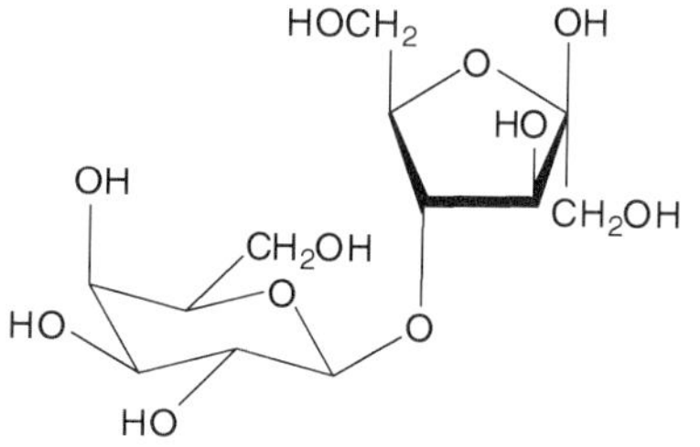

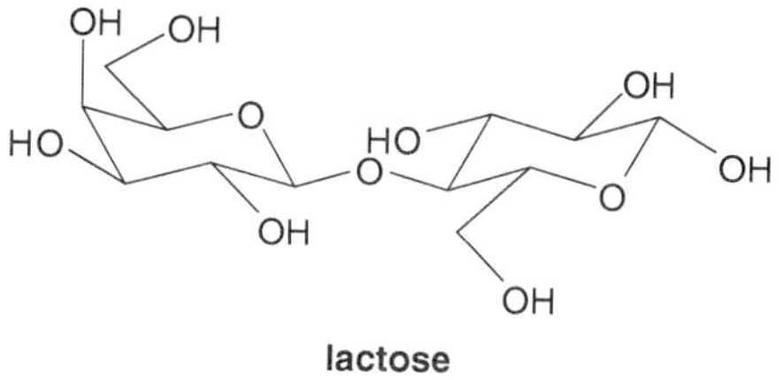

lactose

Extended Discussion

The transformation of lactose into lactulose results in a mixture containing lactulose and smaller amounts of residual lactose, galactose, glucose, tagatose, and epilactose. Draw the structures for tagatose and epilactose. How are tagatose and epilactose formed?

Routes to Essential Medicines: A Workbook for Organic Synthesis, First Edition. Peter J. Harrington.

Companion website: www.wiley.com/go/Harrington/routes_essential_medicine

Lamivudine

Anti-infective Medicines/Antiviral Medicines/Antiretrovirals/Nucleoside–Nucleotide Reverse Transcriptase Inhibitors
Anti-infective Medicines/Antiviral Medicines/Antiretrovirals/Integrase Inhibitors

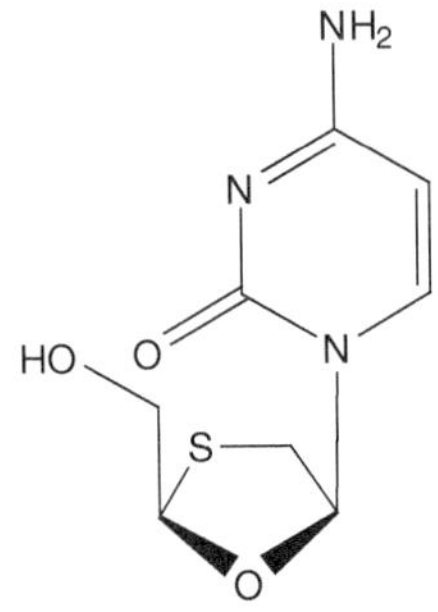

A chiral auxiliary can be used to control the stereochemical outcome of a reaction which forms a nearby chiral carbon. A chiral auxiliary strategy for formation of a chiral carbon is cost-effective when the auxiliary is inexpensive and when the auxiliary can be recovered after the chiral carbon is formed.

Discussion. The primary alcohol of lamivudine is formed by reduction of the (−)-menthyl ester in the final step. (How is (−)-menthol separated from lamivudine?) The bond joining the rings is formed by nucleophilic displacement of an acetate leaving group by N1 of cytosine (**Silyl–Hilbert–Johnson Reaction**). (The (2*R*,5*R*)-product and the (2*R*,5*S*)-diastereomer side product are both formed. What is the highest ratio of product to side product? What reaction conditions are associated with the highest ratio of product to side product?) The acetate is formed from the alcohol. The (5*R*)-hydroxy-1,3-oxathiolane-(2*R*)-carboxylate is formed by the condensation of (1*R*)-(−)-menthyl glyoxalate hydrate with 1,4-dithiane-2,5-diol. Both the (2*R*,5*R*)-alcohol and (2*R*,5*R*)-acetate are crystallized from a mixture of stereoisomers.

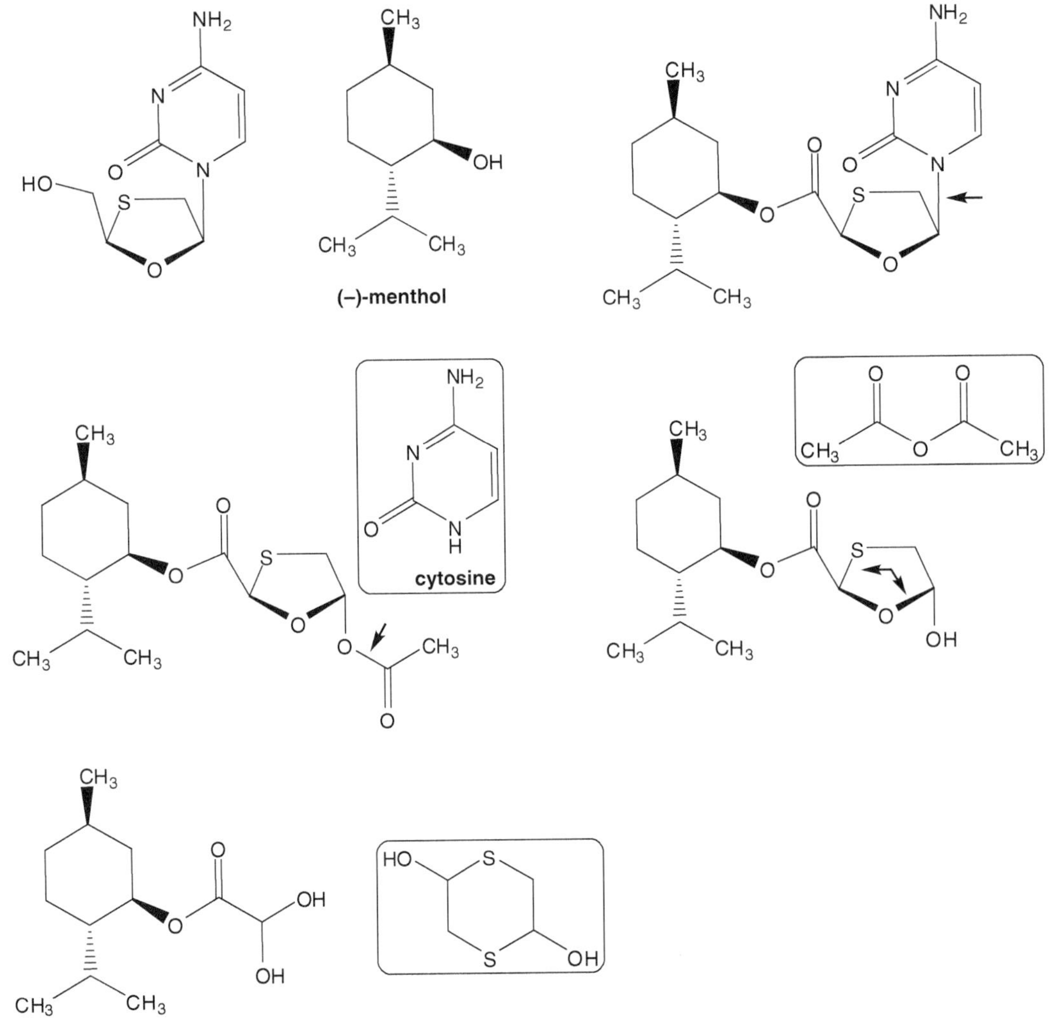

(1*R*)-(−)-Menthyl glyoxalate hydrate is formed by ozololysis of (−)-dimenthyl fumarate. The fumarate diester is formed from fumaric acid and (−)-menthol. (−)-Menthol is produced by synthesis or by isolation from corn mint oil.

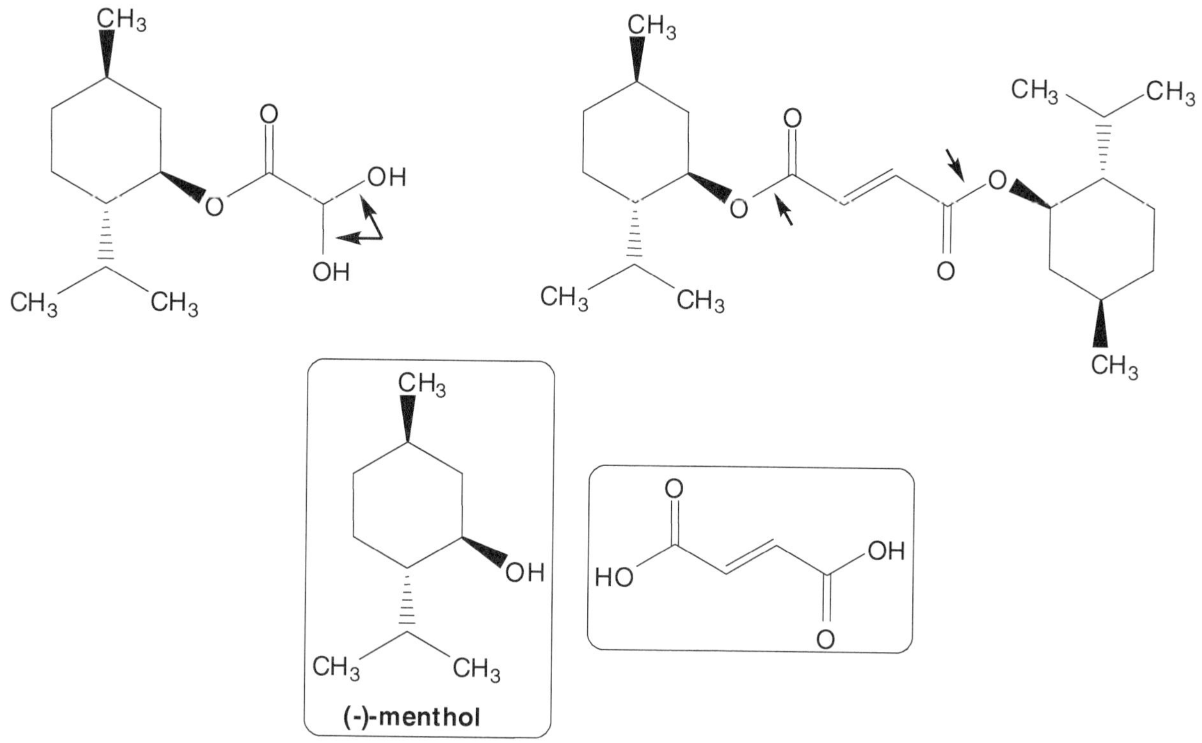

Extended Discussion

Draw the structures of the retrosynthetic analysis of an alternative route to lamivudine using N4-acetylcytosine in place of cytosine. List the pros and cons for both routes and select one route as the preferred route.

Lamotrigine

Anticonvulsants/Antiepileptics

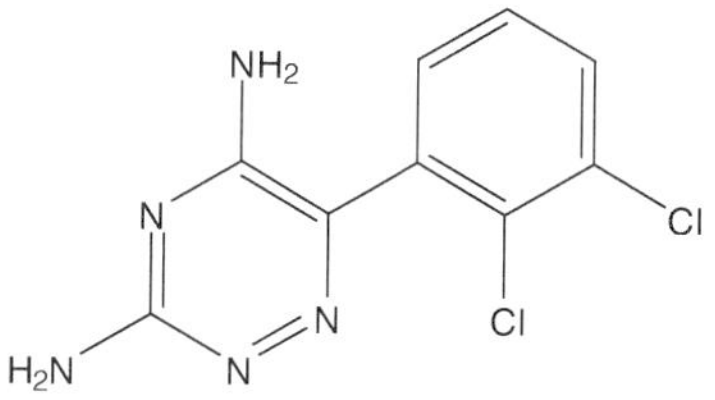

The synthesis of a 1,2,3-trisubstituted benzene by electrophilic aromatic substitution is often inefficient because substitution at a *para* position typically predominates over substitution at an *ortho* position. A 1,2,3-trisubstituted benzene is often formed early in a synthesis when materials are less expensive and the mixture of regioisomers formed is more easily separated.

Discussion. The 1,2,4-triazine ring is formed from the amidinohydrazone in the final step. The amidinohydrazone is formed by addition of aminoguanidine to 2,3-dichlorobenzoyl cyanide. The benzoyl cyanide is formed from the benzoyl chloride and copper cyanide. 2,3-Dichlorobenzoyl chloride is formed by hydrolysis of 2,3-dichlorobenzotrichloride. The benzotrichloride is formed by chlorination of 2,3-dichlorotoluene. 2,3-Dichlorotoluene is formed by chlorination of 2-chlorotoluene. (What is the composition of the mixture produced in the chlorination of 2-chlorotoluene? How is 2,3-dichlorotoluene separated from the mixture?)

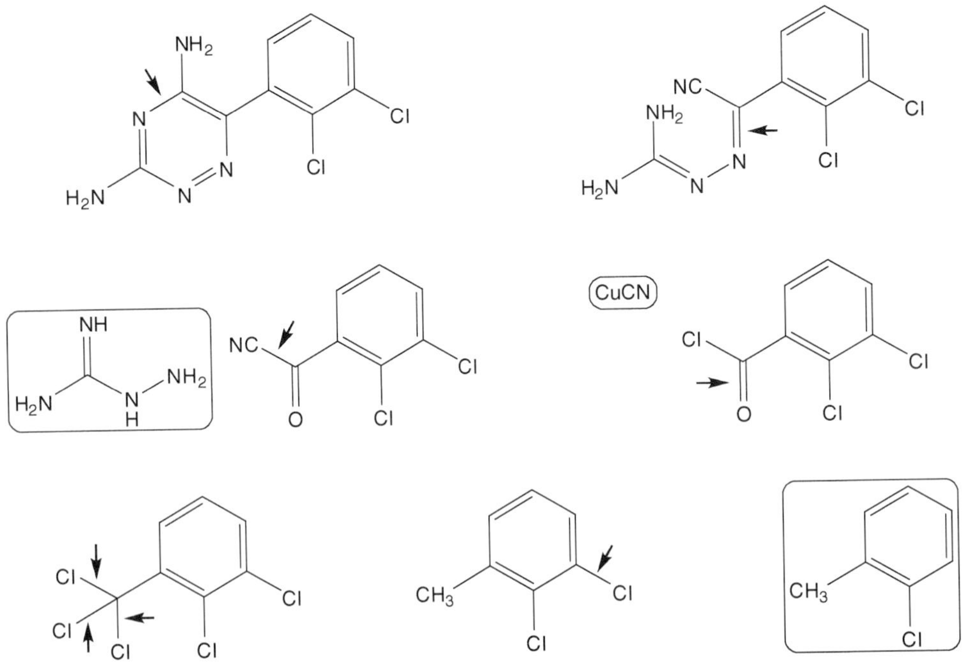

Extended Discussion

Draw the structures of the retrosynthetic analysis of one alternative route to 2,3-dichlorobenzoyl chloride.

Latanoprost

Opthalmological Preparations/Miotics and Antiglaucoma Medicines

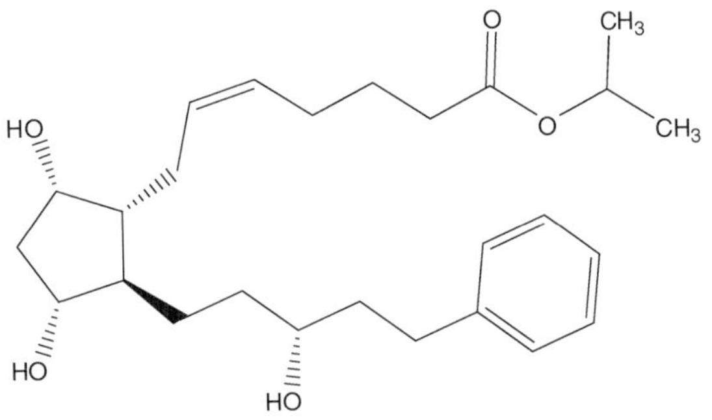

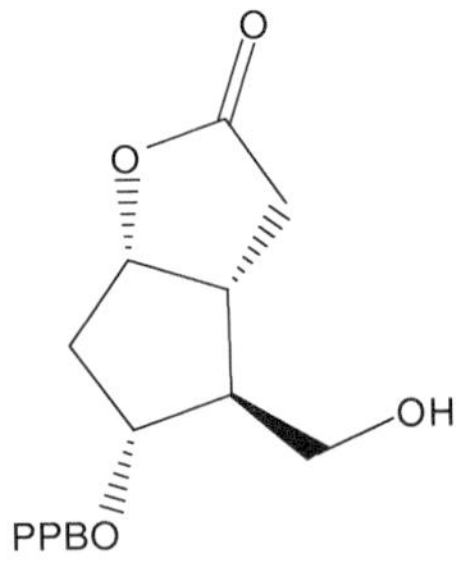

PPB-Coreylactone
PPB = 4-phenylbenzoyl

Prostaglandins are often constructed from a Corey Lactone. Oxidation of the primary alcohol forms an aldehyde which is used to introduce the lower side chain. Reduction of the lactone then forms another aldehyde which is used to introduce the upper side chain.

Discussion. In one preferred route, latanoprost is formed in nine steps from PPB-**Corey Lactone**. In the final step, the ester is formed by alkylation of the carboxylate with 2-iodopropane. The C15 alcohol is released by hydrolysis of the tetrahydropyranyl (THP) ether. The C5–C6 alkene is formed by reaction of the C6 aldehyde with the ylid from (4-carboxybutyl)triphenylphosphonium bromide (**Wittig Reaction**). The aldehyde is formed in situ by ring-opening of the lactol. The lactol is formed by reduction of the lactone. The C11 alcohol is also released during the lactone reduction or by hydrolysis of the 4-phenylbenzoyl (PPB) ester after the reduction.

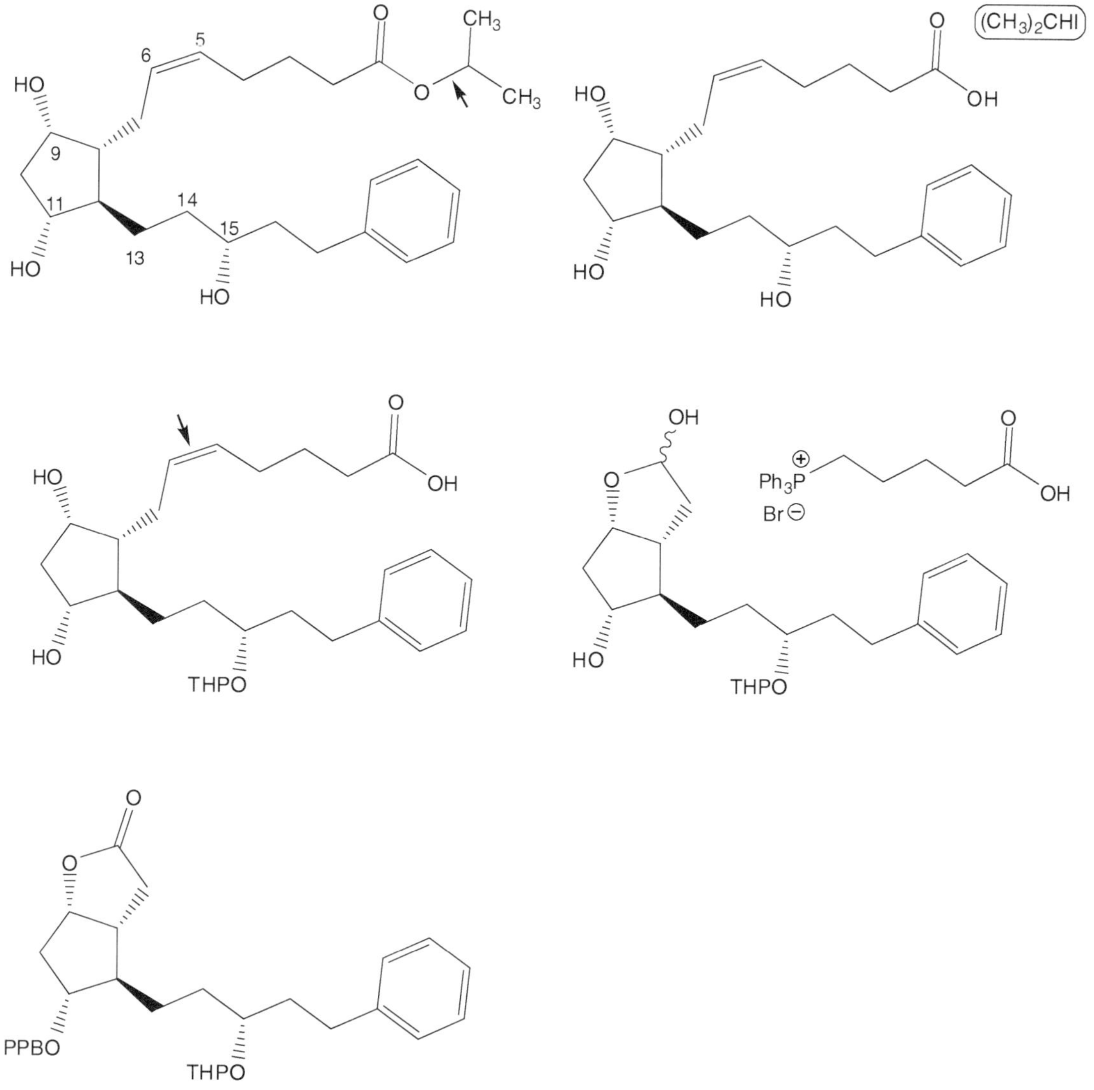

The C13–C14 alkene is reduced. The C15 alcohol is protected as the THP ether. The C15 alcohol is formed by reduction of the ketone. The C13–C14 alkene is formed by reaction of the aldehyde with a phosphonate ester (**Horner–Wadsworth–Emmons Reaction**). The aldehyde is formed by oxidation of the C13 alcohol of PPB-**Corey Lactone**.

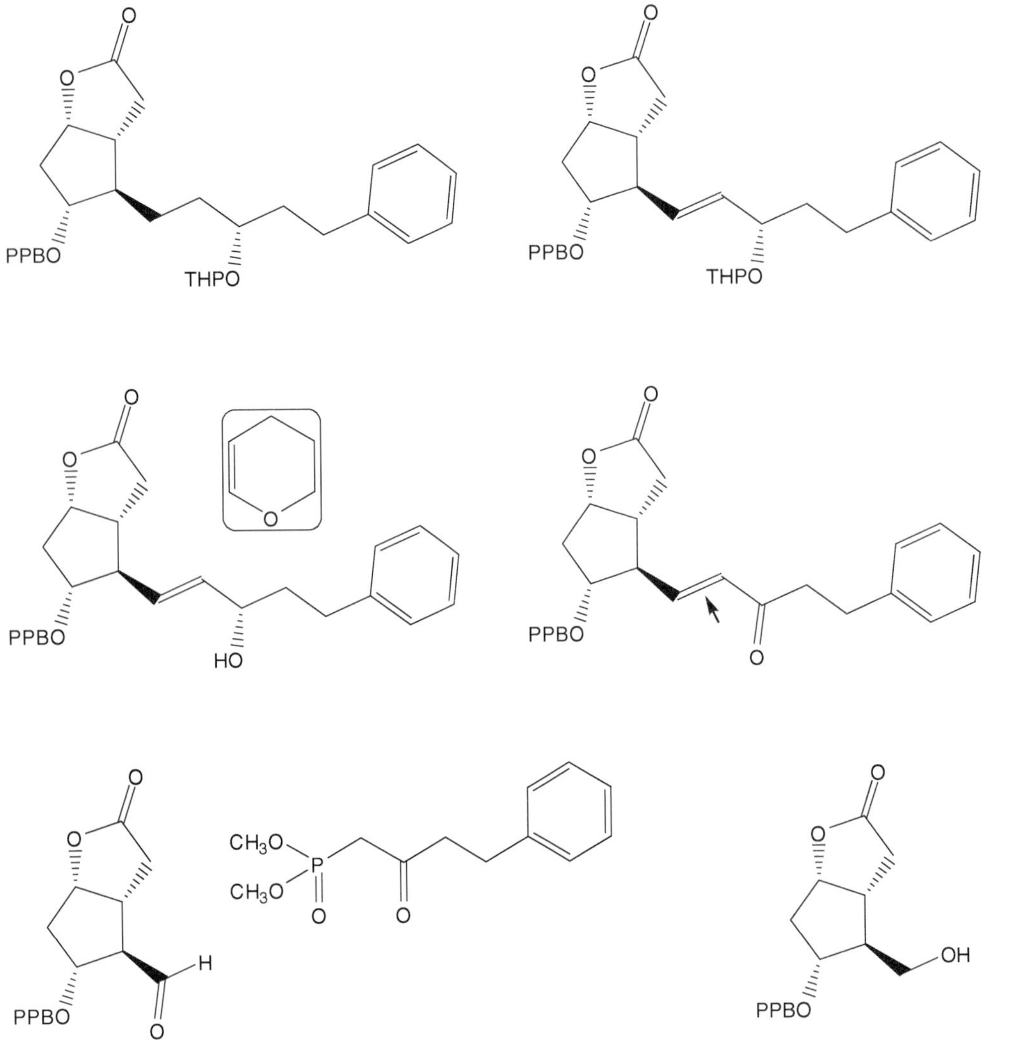

PPB-Corey Lactone

(4-Carboxybutyl)triphenylphosphonium bromide is formed from 5-bromopentanoic acid (5-bromovaleric acid). 5-Bromovaleric acid is formed from δ-valerolactone.

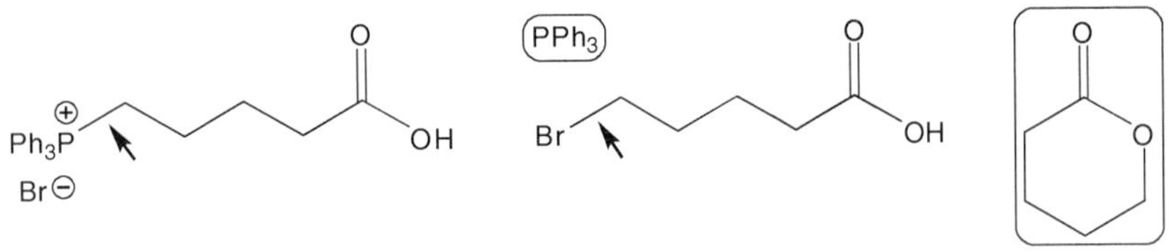

The phosphonate ester, dimethyl (2-oxo-4-phenylbutyl)phosphonate, is formed from the α-iodoketone and trimethyl phosphite (**Michaelis–Arbuzov Reaction**). The α-iodoketone is formed from the α-bromoketone (**Finkelstein Reaction**). The α-bromoketone is formed by bromination of 4-phenyl-2-butanone.

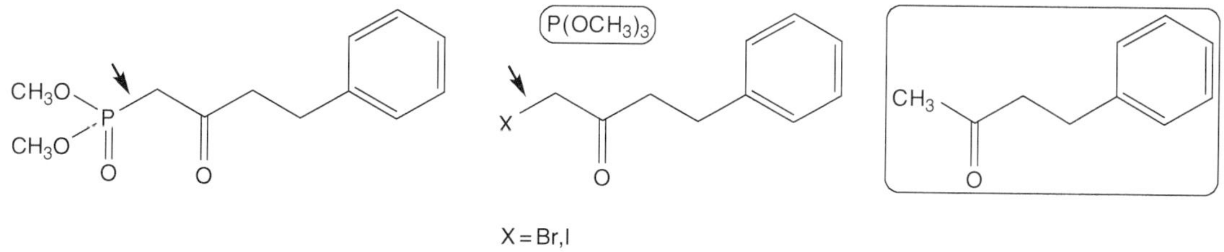

The condensation of dimethyl methylphosphonate with ethyl 3-phenylpropanoate is an alternative route to dimethyl (2-oxo-4-phenylbutyl)phosphonate. (List the pros and cons for both routes to the phosphonate ester. Is one route preferred?)

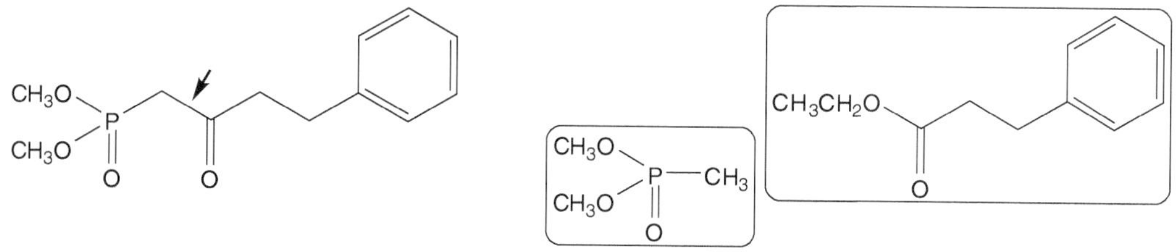

PPB-Corey Lactone is manufactured in ten steps from cyclopentadiene. The C13 alcohol is released by hydrogenolysis of the benzyl ether. The C10 iodide is removed by reduction. The C11 alcohol is protected as an ester by reaction with 4-phenylbenzoyl chloride (PPB-Cl). 4-Phenylbenzoyl chloride is formed from 4-phenylbenzoic acid. The lactone is formed by iodolactonization from the C6 carboxylic acid and C9–C10 alkene. The C6 carboxylic acid and C11 alcohol are formed by hydrolysis of the lactone. The lactone is formed by oxidation of the ketone (**Baeyer–Villiger Oxidation**). The ketone is formed by oxidative decarboxylation of the carboxylic acid. The carboxylic acid is released by ester hydrolysis. (How is the chiral auxiliary D-pantolactone recovered and recycled?) The bicyclo[2.2.1]heptane ring is formed by cycloaddition of the cyclopentadiene and the acrylate ester (**Diels–Alder Cycloaddition**). (Draw the structures of the two diastereoisomers formed by endo-cycloaddition. What is the diastereoselectivity of the cycloaddition?) 5-[(Benzyloxy)methyl]cyclopentadiene is formed by alkylation of cyclopentadiene with benzyl chloromethyl ether. The acrylate ester is formed from the alcohol (D-pantolactone) and acryloyl chloride. Acryloyl chloride is formed from acrylic acid. (Draw the structure of one alternative alcohol used as the chiral auxiliary in the synthesis of a Corey Lactone.)

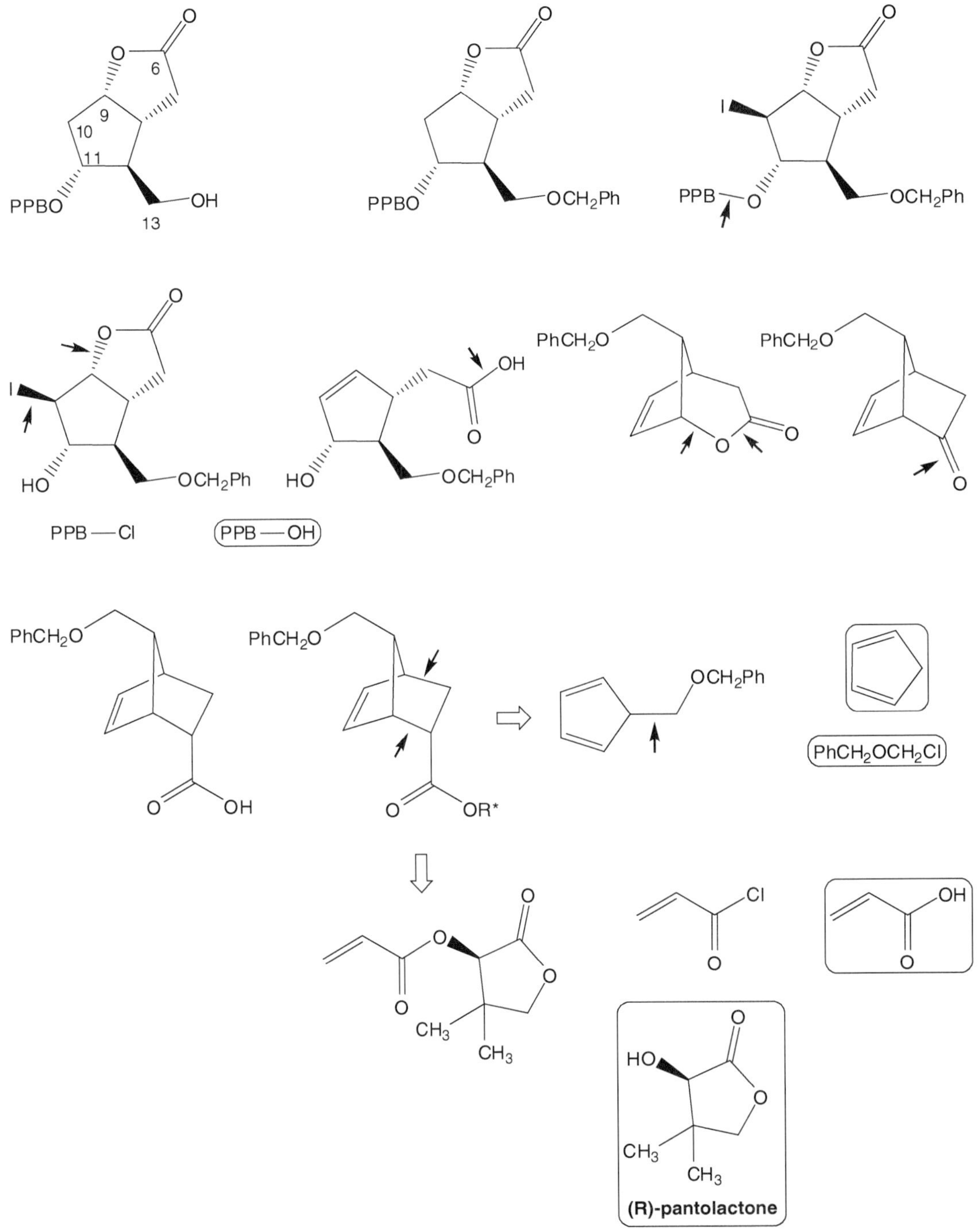

Extended Discussion

Draw the structures of the retrosynthetic analysis of one alternative route to latanoprost that does not use a Corey Lactone. Include the structures of the retrosynthetic analysis of any organic starting material(s) from petrochemical or biochemical raw materials.

Ledipasvir

Anti-infective Medicines/Antiviral Medicines/Antihepatitis Medicines/Medicines for Hepatitis C/Other Antivirals

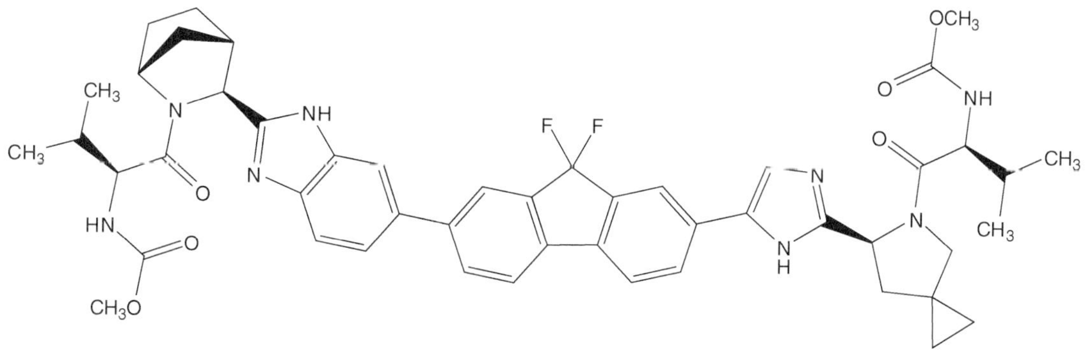

A 2,5-disubstituted imidazole is formed by reaction of an α-acyloxyketone with ammonium acetate. Four C–N bonds of the imidazole ring are formed in the reaction.

Discussion. The amide bonds are formed in the final step by reaction of the amines with *N*-(methoxycarbonyl)-L-valine (Moc-L-valine). The amines are both released by hydrolysis of the *N*-*tert*-butoxycarbonyl (Boc) carbamate.

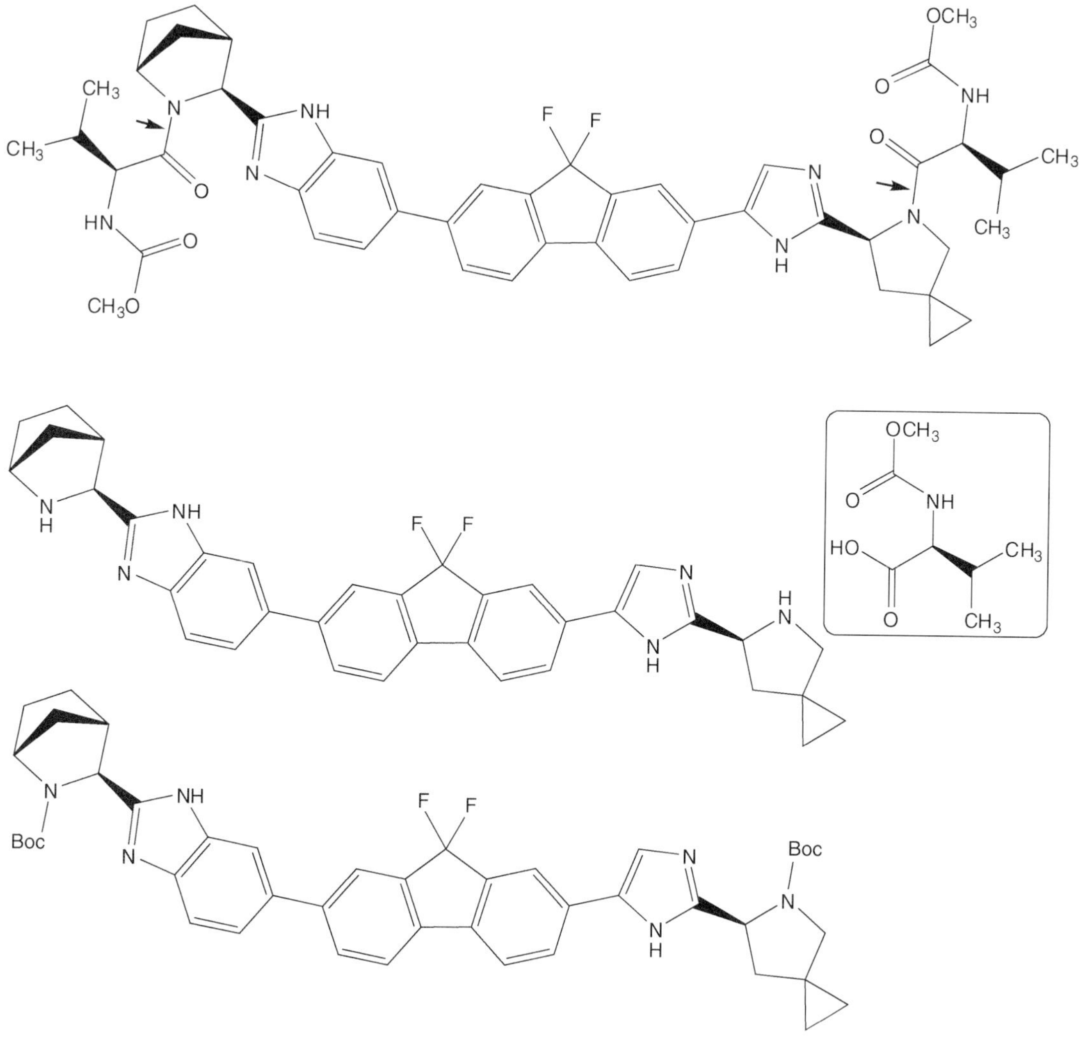

A biaryl C–C bond is formed from the arylboronate ester and the bromofluorene (**Suzuki–Miyaura Coupling**).

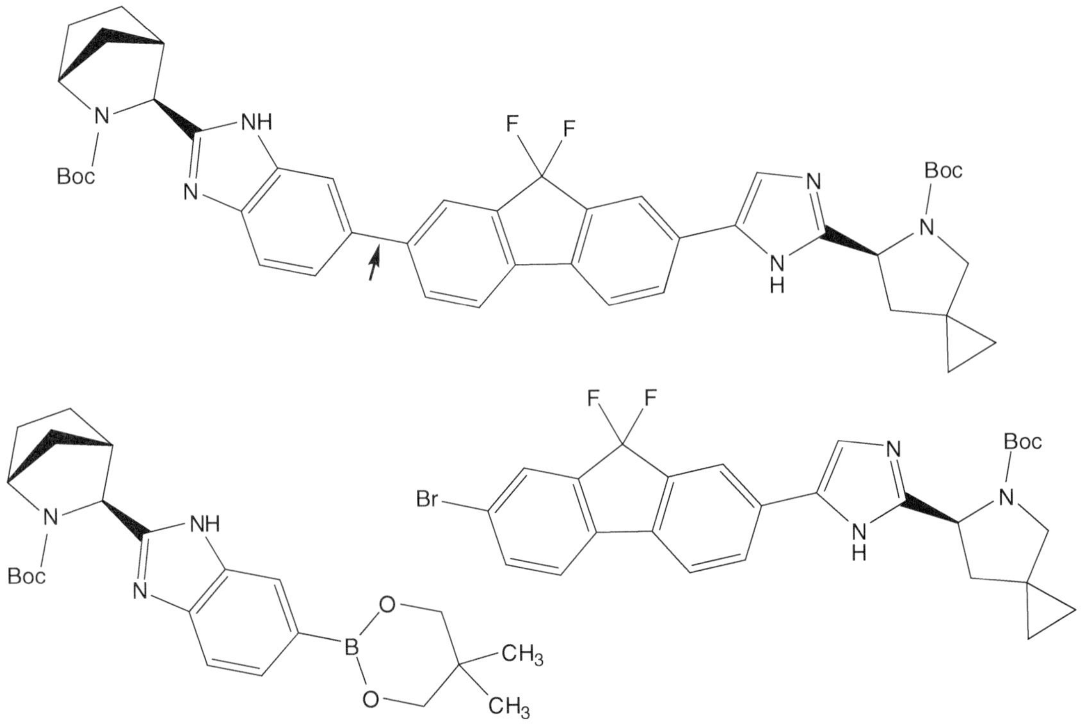

The aryl boronate ester is formed from bis(neopentyl glycolato)diboron and the 6-bromobenzimidazole. The benzimidazole ring is formed by ring-closure of an α-aminoanilide. A mixture of two α-aminoanilides is formed by reaction of 4-bromo-1,2-diaminobenzene with the carboxylic acid. 1,2-diamino-4-bromobenzene is formed by bromination of *ortho*-phenylenediamine.

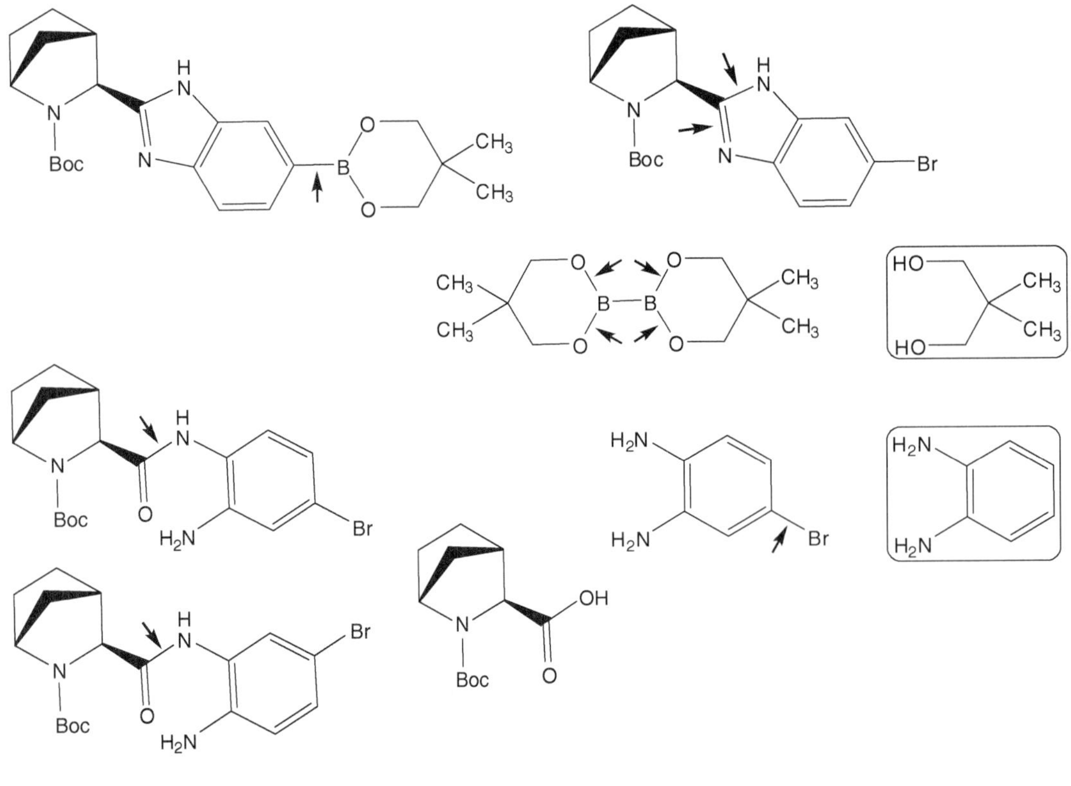

The carboxylic acid is formed by ester hydrolysis. The secondary amine is protected as the Boc-carbamate. The secondary amine is released by hydrogenolysis of the *N*-benzyl tertiary amine and the alkene on the azabicyclo[2.2.1]heptene ring is reduced. Cycloaddition of the imine formed from (*R*)-α-methylbenzylamine and methyl glyoxalate methyl hemiacetal with 1,3-cyclopentadiene forms the azabicyclo[2.2.1]heptene (aza-**Diels–Alder Cycloaddition**).

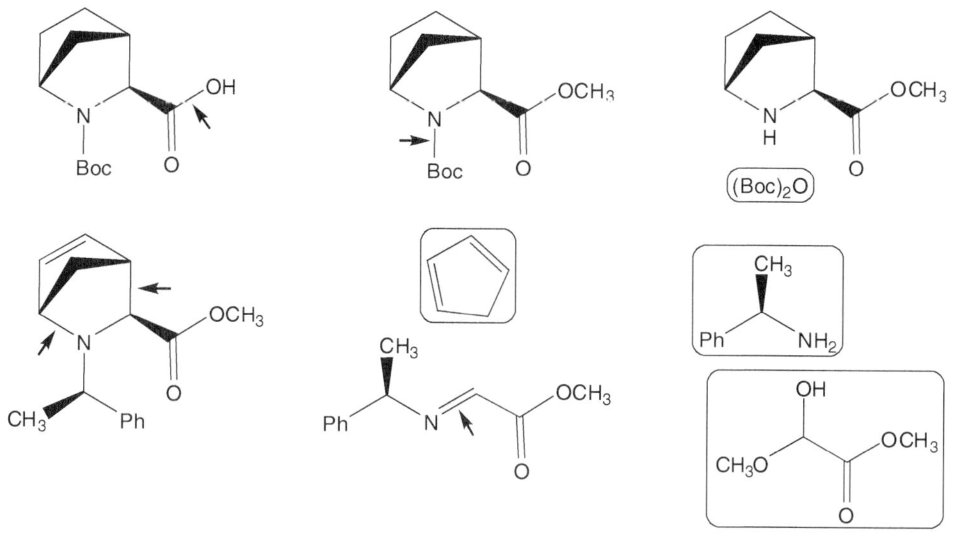

The imidazole ring of the 2-bromofluorene used in the aryl–aryl coupling is formed by reaction of the α-acyloxyketone with ammonium acetate. The α-acyloxyketone is formed from the α-chloroketone by chloride displacement by the carboxylate salt.

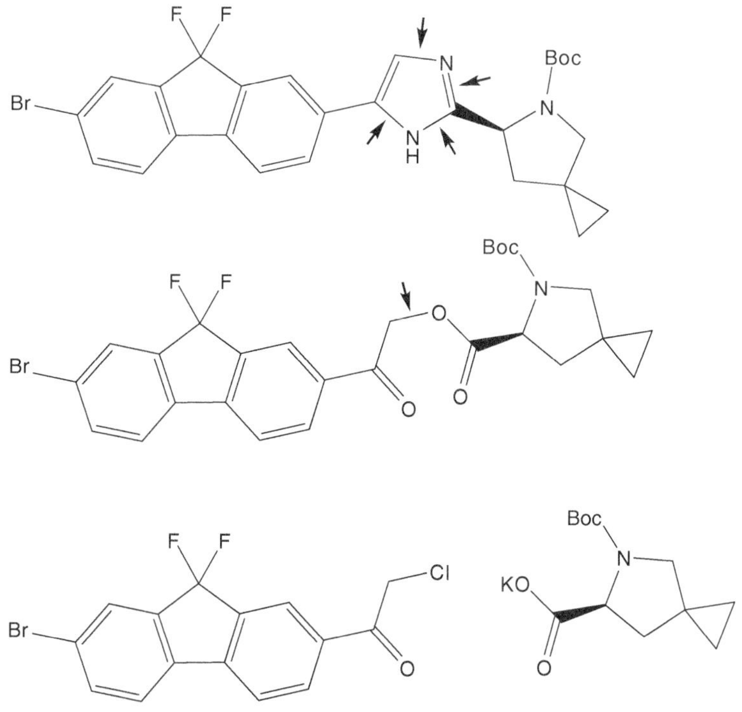

The α-chloroketone is formed by reaction of 2-chloro-*N*-methoxy-*N*-methylacetamide (a **Weinreb Amide**) with a Grignard reagent. The amide is formed from chloroacetyl chloride and *N,O*-dimethylhydroxylamine. The Grignard reagent is formed from the iodofluorene by a metal exchange with isopropylmagnesium chloride. Isopropylmagnesium chloride is formed from 2-chloropropane. 2-Bromo-9,9-difluoro-7-iodofluorene is formed by iodination and then fluorination of 2-bromofluorene.

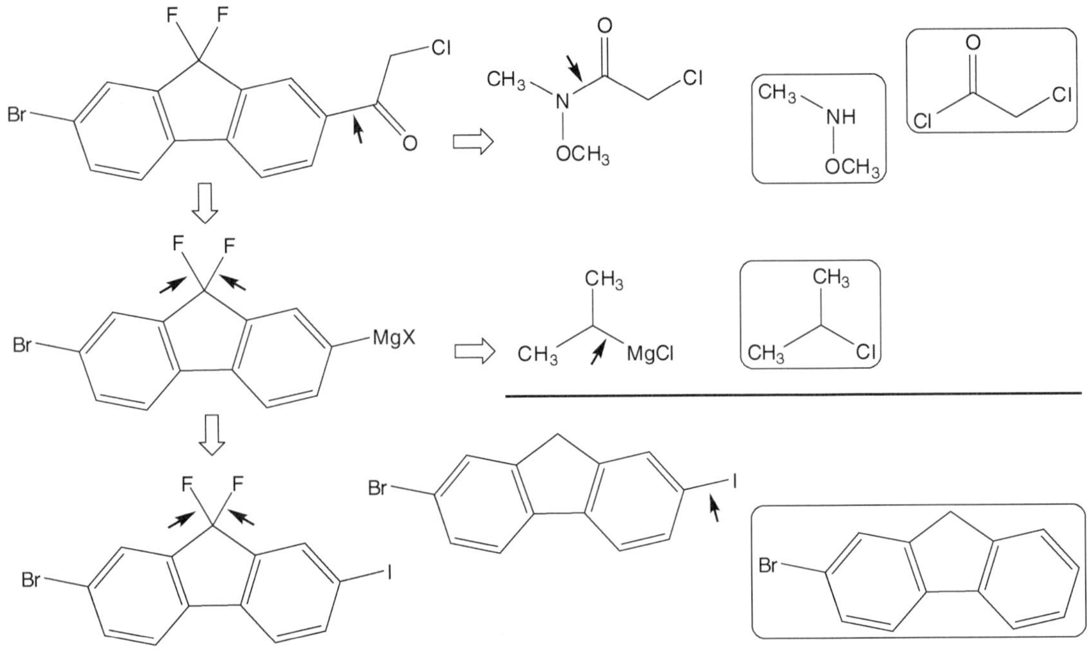

The carboxylate salt is formed from the carboxylic acid, (*S*)-5-(*tert*-butoxycarbonyl)-5-azaspiro[2.4]heptane-6-carboxylic acid. The (*S*)-enantiomer is obtained by resolution. The racemic carboxylic acid is formed by ester hydrolysis. The pyrrolidine ring is formed by N- and C-alkylation of *N*-(*tert*-butoxycarbonyl)glycine ethyl ester. The dialkylating agent (X = leaving group) is formed from 1,1-bis(hydroxymethyl)cyclopropane. (Which leaving group is preferred?)

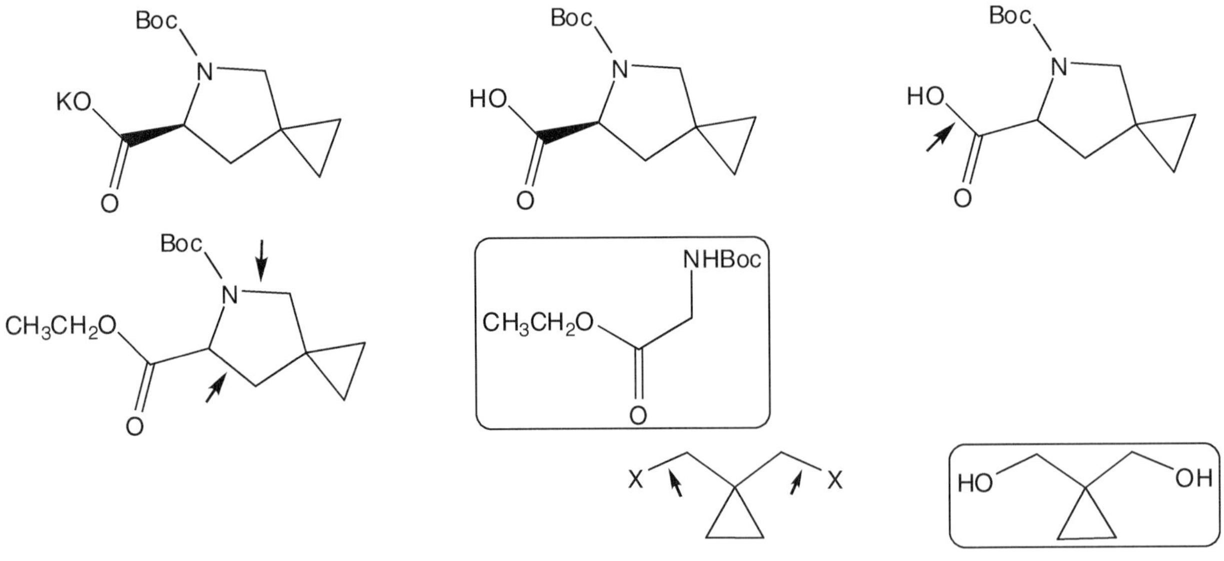

Extended Discussion

Draw the structures of the retrosynthetic analysis of an alternative route to (*S*)-5-(*tert*-butoxycarbonyl)-5-azaspiro[2.4]heptane-6-carboxylic acid that does not utilize 1,1-bis(hydroxymethyl)cyclopropane. Include the structures of the retrosynthetic analysis of any organic starting material(s) from petrochemical or biochemical raw materials. List the pros and cons for both routes and select one route as the preferred route.

Leuprorelin

Hormones and Antihormones

pGlu-His-Trp-Ser-Tyr-D-Leu-Leu-Arg-Pro-$NHCH_2CH_3$

> **A polypeptide is often constructed from the constituent amino acids by forming the amide bonds. The amide bonds are formed via a solid-phase peptide synthesis, a solution-phase peptide synthesis, or a hybrid approach using both solid-phase and solution-phase methods.**

Discussion. To simplify polypeptide synthetic schemes, the amino acids are represented by acronyms and are assumed to be in the L-configuration. Protecting groups used in construction of the polypeptide are also represented by acronyms. Polypeptide amino acid sequences are drawn with the C-terminal amino acid (carboxylic acid) on the right and the N-terminal amino acid (amino group) on the left. (Create a Table of the acronyms used in the analysis. Draw the structure associated with each acronym.)

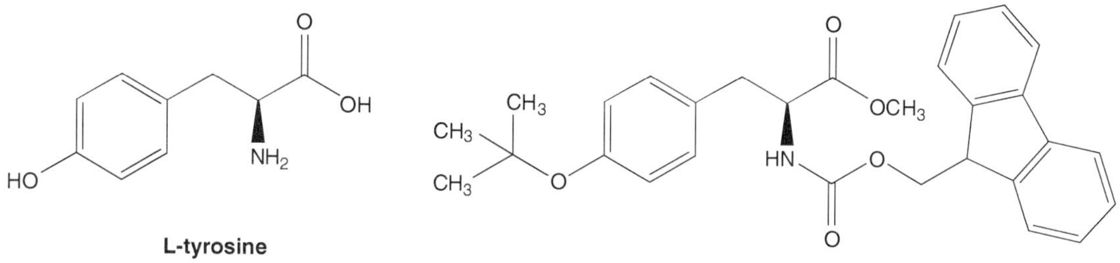

L-tyrosine
H-Tyr-OH
↑ H of α-NH ↑ OH of COOH

Fmoc-Tyr(tBu)-OCH_3
↑ protecting group for α-amino ↑ protecting group off the peptide chain

Construction of leuprorelin by solid-phase peptide synthesis, by solution-phase peptide synthesis, and by the hybrid approach are all described in patents. In a preferred hybrid approach, the final step is a global deprotection of the five functional groups which are off the peptide chain. The final amide bond is formed between the carboxylic acid of a protected pentapeptide and the amine of a protected tetrapeptide.

pGlu-His-Trp-Ser-Tyr-D-Leu-Leu-Arg-Pro-$NHCH_2CH_3$

pGlu-His(Trt)-Trp(Boc)-Ser(tBu)-Tyr(tBu)-D-Leu-Leu-Arg(pbf)-Pro-$NHCH_2CH_3$
↑

pGlu-His(Trt)-Trp(Boc)-Ser(tBu)-Tyr(tBu)-OH   H-D-Leu-Leu-Arg(pbf)-Pro-$NHCH_2CH_3$

The protected tetrapeptide is constructed from two dipeptides by solution-phase peptide synthesis. Dipeptide Fmoc-D-Leu-H-Leu-OH is formed by reaction of the carboxylic acid of Fmoc-D-Leu-OH with the amino group of leucine (H-Leu-OH). Dipeptide Fmoc-Arg(pbf)-Pro-$NHCH_2CH_3$ is formed by reaction of the carboxylic acid of Fmoc-Arg(pbf)-OH with the amine of proline *N*-ethylamide (H-Pro-$NHCH_2CH_3$). The terminal amine of this dipeptide is released by Fmoc-deprotection.

The central amide bond of the tetrapeptide is formed by reaction of the two dipeptides. The terminal amine of the tetrapeptide is released by Fmoc-deprotection.

H-D-Leu-Leu-Arg(pbf)-Pro-$NHCH_2CH_3$

Fmoc-D-Leu-Leu-Arg(pbf)-Pro-$NHCH_2CH_3$

Fmoc-D-Leu-Leu-OH   H-Arg(pbf)-Pro-$NHCH_2CH_3$

Fmoc-D-Leu-OH   H-Leu-OH   Fmoc-Arg(pbf)-Pro-$NHCH_2CH_3$

Fmoc-Arg(pbf)-OH   H-Pro-$NHCH_2CH_3$

The protected pentapeptide is constructed by solid-phase peptide synthesis. Starting with the polymer and the C-terminal amino acid, the peptide chain is constructed right-to-left. The carboxylate of the Fmoc-protected C-terminal amino acid displaces chloride to form an ester link to the polymer. The amino group is released by Fmoc-deprotection. An amide bond is formed by reaction of this amino group with the carboxylic acid of the next Fmoc-protected amino acid. Fmoc-deprotection and amide bond formation are repeated three times to produce the polymer-bound pentapeptide. The carboxylic acid is released when the pentapeptide is released from the resin.

pGlu-His(Trt)-Trp(Boc)-Ser(tBu)-Tyr(tBu)-OH

pGlu-His(Trt)-Trp(Boc)-Ser(tBu)-Tyr(tBu)-O-Polymer

pGlu-OH   H-His(Trt)-Trp(Boc)-Ser(tBu)-Tyr(tBu)-O-Polymer

Fmoc-His(Trt)-Trp(Boc)-Ser(tBu)-Tyr(tBu)-O-Polymer

Fmoc-His(Trt)-OH   H-Trp(Boc)-Ser(tBu)-Tyr(tBu)-O-Polymer

Fmoc-Trp(Boc)-Ser(tBu)-Tyr(tBu)-O-Polymer

Fmoc-Trp(Boc)-OH   H-Ser(tBu)-Tyr(tBu)-O-Polymer

Fmoc-Ser(tBu)-Tyr(tBu)-O-Polymer

Fmoc-Ser(tBu)-OH   H-Tyr(tBu)-O-Polymer

Fmoc-Tyr(tBu)-O-Polymer

Fmoc-Tyr(tBu)-OH   Cl-Polymer

In summary, the nonapeptide backbone is constructed by forming eight amides, four in the solid-phase synthesis and four in solution-phase synthesis. Each amide is formed via an active ester which is formed in situ from the carboxylic acid. Each active ester can be racemized during the amide bond formation. (List reagent(s) used to form active esters used in this peptide synthesis.) Six Fmoc protecting groups are used. (List reagents used to remove Fmoc protecting groups in peptide synthesis.)

Routes to the protected amino acid starting materials are presented. (Draw the structure of each protected amino acid starting material.) Pyroglutamic acid is formed from glutamic acid. D-Leucine can be formed from L-leucine. Leucine, arginine, proline, glutamic acid, histidine, tryptophan, serine, and tyrosine are all produced by fermentation.

Fmoc-D-Leu-OH	H-D-Leu-OH Fmoc-Cl			
Fmoc-Arg(pbf)-OH	H-Arg(pbf)-OH Fmoc-Cl	H-Arg(pbf)-OCH_2CH_3	Boc-Arg(pbf)-OCH_2CH_3	
	Boc-Arg-OCH_2CH_3 pbf-Cl	H-Arg-OCH_2CH_3 $(Boc)_2O$	H-Arg-OH CH_3CH_2OH	
H-Pro-$NHCH_2CH_3$	H-Pro-OH $CH_3CH_2NH_2$			
pGlu-OH	H-Glu-OH			
Fmoc-His(Trt)-OH	H-His(Trt)-OH Fmoc-Cl	H-His-OH Trt-Cl		
Fmoc-Trp(Boc)-OH	H-Trp(Boc)-OH Fmoc-Cl	Z-Trp(Boc)-OCH_2Ph	Z-Trp-OCH_2Ph $(Boc)_2O$	
	H-Trp-OCH_2Ph Z-Cl	H-Trp-OH $PhCH_2OH$		
Fmoc-Ser(tBu)-OH	H-Ser(tBu)-OH Fmoc-Cl	H-Ser(tBu)-OCH_3	H-Ser-OCH_3 $CH_2{=}C(CH_3)_2$	H-Ser-OH CH_3OH
Fmoc-Tyr(tBu)-OH	Fmoc-Tyr(tBu)-OCH_3	Fmoc-Tyr-OCH_3 $CH_2{=}C(CH_3)_2$	H-Tyr-OCH_3 Fmoc-Cl	H-Tyr-OH CH_3OH

Extended Discussion

Draw the structures of the retrosynthetic analysis of 2,2,4,6,7-pentamethyl-2,3-dihydrobenzofuran-5-sulfonyl chloride (pbf–Cl). Include the structures of the retrosynthetic analysis of any organic starting material(s) from petrochemical or biochemical raw materials.

Levamisole

Anti-infective Medicines/Anthelminthics/Intestinal Anthelminthics

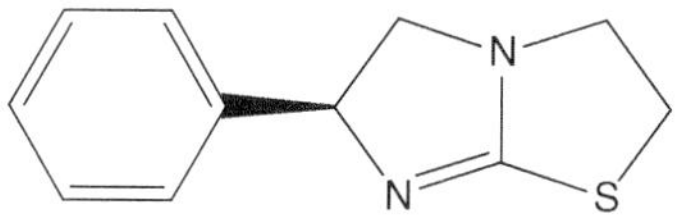

A single-enantiomer molecule is often manufactured by resolution when it can form a salt with a Bronsted acid or Bronsted base and when the "wrong" enantiomer can be racemized and returned to the resolution.

Discussion. Levamisole is produced by resolution of tetramisole. (List the known resolving agents. Select one resolving agent as preferred.) The (*R*)-enantiomer, dexamisole, is converted back to tetramisole and tetramisole is returned to the resolution. (List the reagents and conditions used for the racemization.) The imidazoline ring of tetramisole is formed by nucleophilic displacement of chloride by nitrogen of the 2-iminothiazolidine. The alkyl chloride is formed from the alcohol. The 2-iminothiazolidine hydrochloride is formed by reaction of the aziridine with thiourea. The β-aziridinyl alcohol is formed by the ring-opening of styrene oxide by aziridine. (Review a safety data sheet [SDS] for aziridine. List your concerns regarding the use of aziridine as a starting material.)

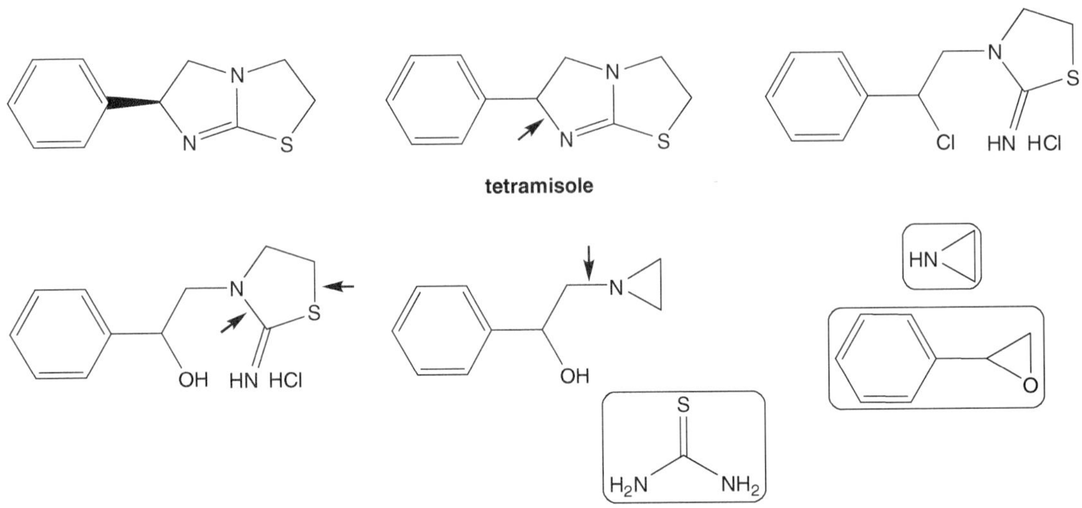

Extended Discussion

Draw the structures of the retrosynthetic analysis of one alternative route to levamisole which does not have aziridine as a starting material.

Levofloxacin

Anti-infective Medicines/Antibacterials/Antituberculosis Medicines

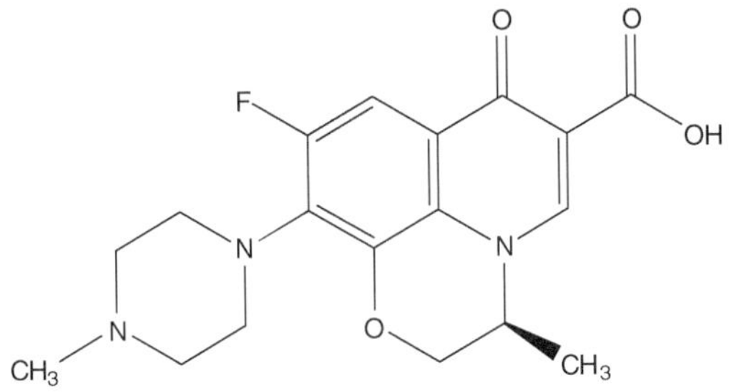

Substitution of a leaving group (F, Cl, NO_2) on an aromatic ring by a nucleophile is often facilitated by an electron-withdrawing group (NO_2, SO_2R, COOR, CN) on an *ortho* or *para* ring carbon. No electron-withdrawing group is required when the displacement results in formation of a five- or six-membered ring.

Discussion. Fluoride at position 7 of the quinolone ring is displaced by 1-methylpiperazine in the final step. The carboxylic acid is released by ester hydrolysis.

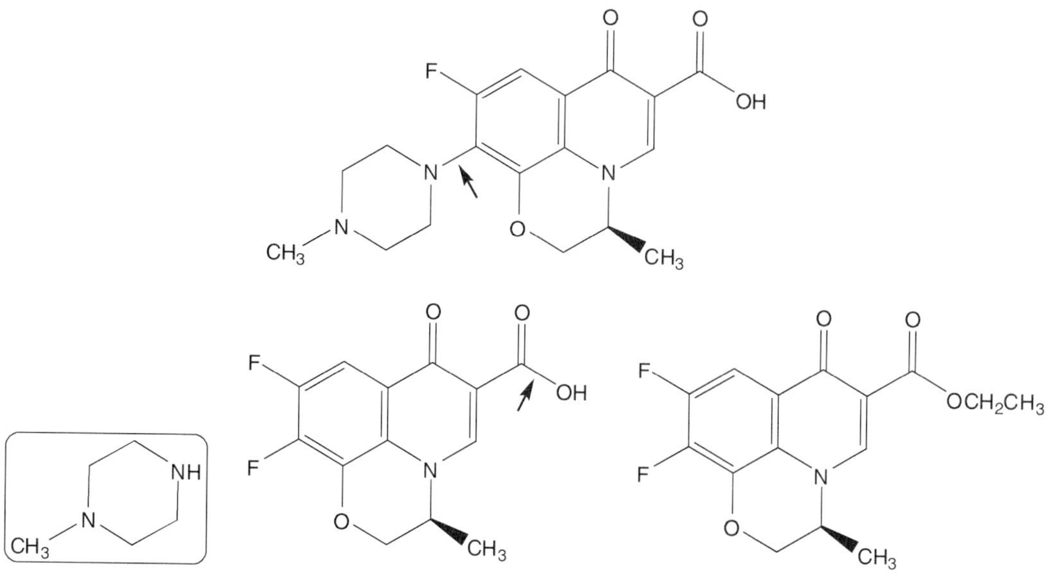

A ring and the ether at quinolone C8 are formed by intramolecular displacement of fluoride by oxygen. The quinolone ring is formed by intramolecular displacement of fluoride by nitrogen. An enamine is formed by reaction of the enol ether with (*S*)-2-amino-1-propanol (L-alaninol).

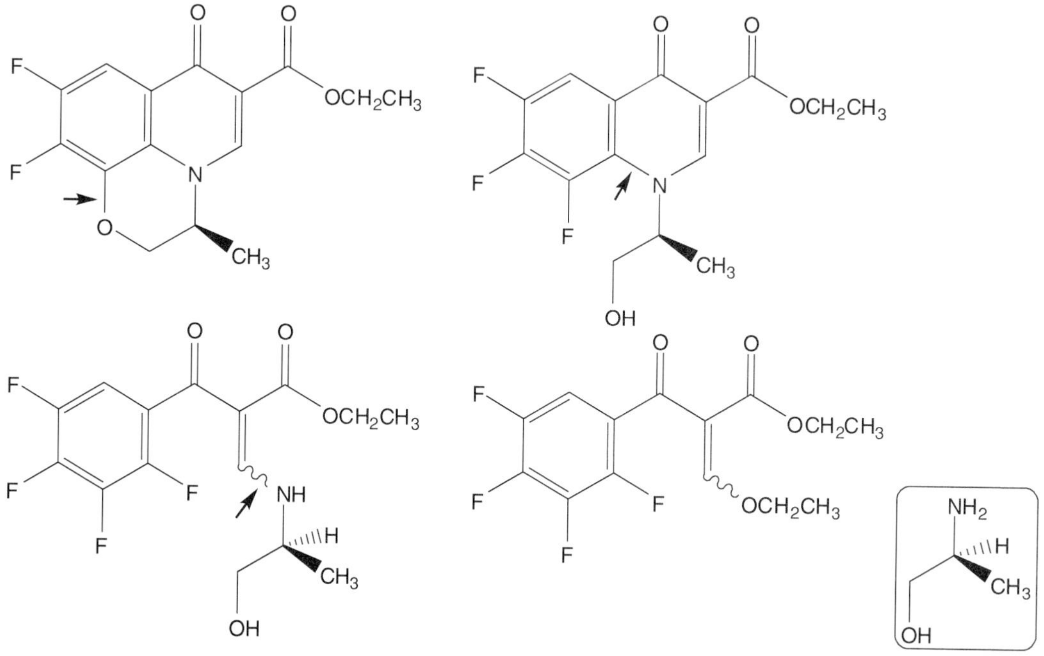

The enol ether is formed by reaction of the β-ketoester with triethyl orthoformate. The β-ketoester is formed from 2,3,4,5-tetrafluorobenzoyl chloride. In one preferred approach, diethyl malonate supplies the added functionality. (The four-step conversion of the 2-fluorobenzoyl chloride to the quinolone is known as the **Grohe–Heitzer Sequence**.) The acid chloride is formed from the carboxylic acid.

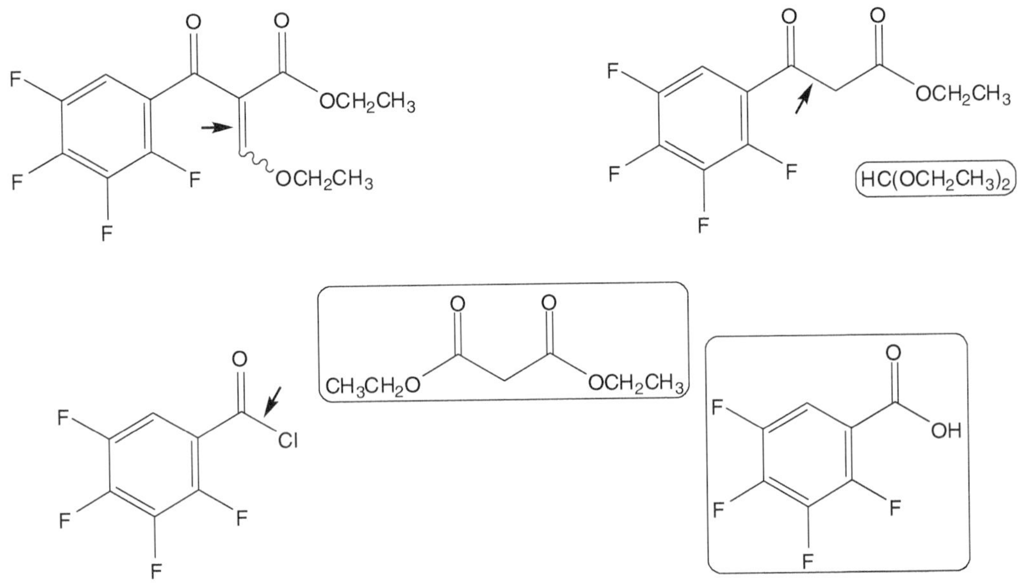

Extended Discussion

List the reagents and conditions used for the reduction of L-alanine and esters of L-alanine to form L-alaninol. List the pros and cons for each method.

Levonorgestrel

Hormones, Other Endocrine Medicines and Contraceptives/Contraceptives/Oral Hormonal Contraceptives
Hormones, Other Endocrine Medicines and Contraceptives/Contraceptives/Intrauterine Devices
Hormones, Other Endocrine Medicines and Contraceptives/Contraceptives/Implantable Contraceptives

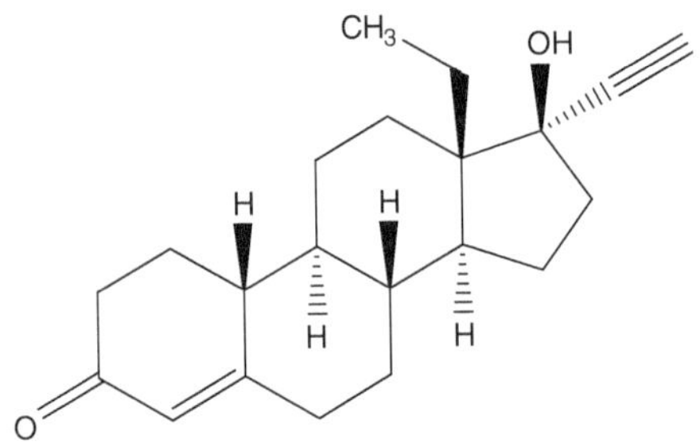

Steroids with an ethyl group at C13 on the gonane structure are often manufactured by total synthesis. The gonane is constructed from 6-methoxy-1-tetralone, vinyl chloride, and 2-ethyl-1,3-cyclopentanedione (Torgov–Smith Synthesis). Chirality at C13 is established by microbial reduction of a 2,2-disubstituted-1,3-cyclopentanedione intermediate.

Discussion. Levonorgestrel is manufactured in six steps from 13-ethyl-3-methoxygona-1,3,5(10),8,14-pentaen-17β-ol acetate and in 11 steps from 6-methoxy-1-tetralone.

The 4-ene-3-one of levonorgestrel is formed in the final step from the 3-methoxy-2,5-diene by hydrolysis of the enol ether and migration of the double bond. The tertiary alcohol at C17 is formed by addition of a metal acetylide to the C17 ketone.

The C17 ketone is formed by oxidation of the C17 secondary alcohol (**Oppenauer Oxidation**). The 3-methoxy-2,5-diene is formed by reduction of the aromatic ring (**Birch Reduction**). The 8,9-alkene is also reduced under the conditions used for reduction of the aromatic ring. The C17 secondary alcohol is released by ester hydrolysis. The C14 alkene of the gona-1,3,5(10),8,14-pentaene is reduced by catalytic hydrogenation.

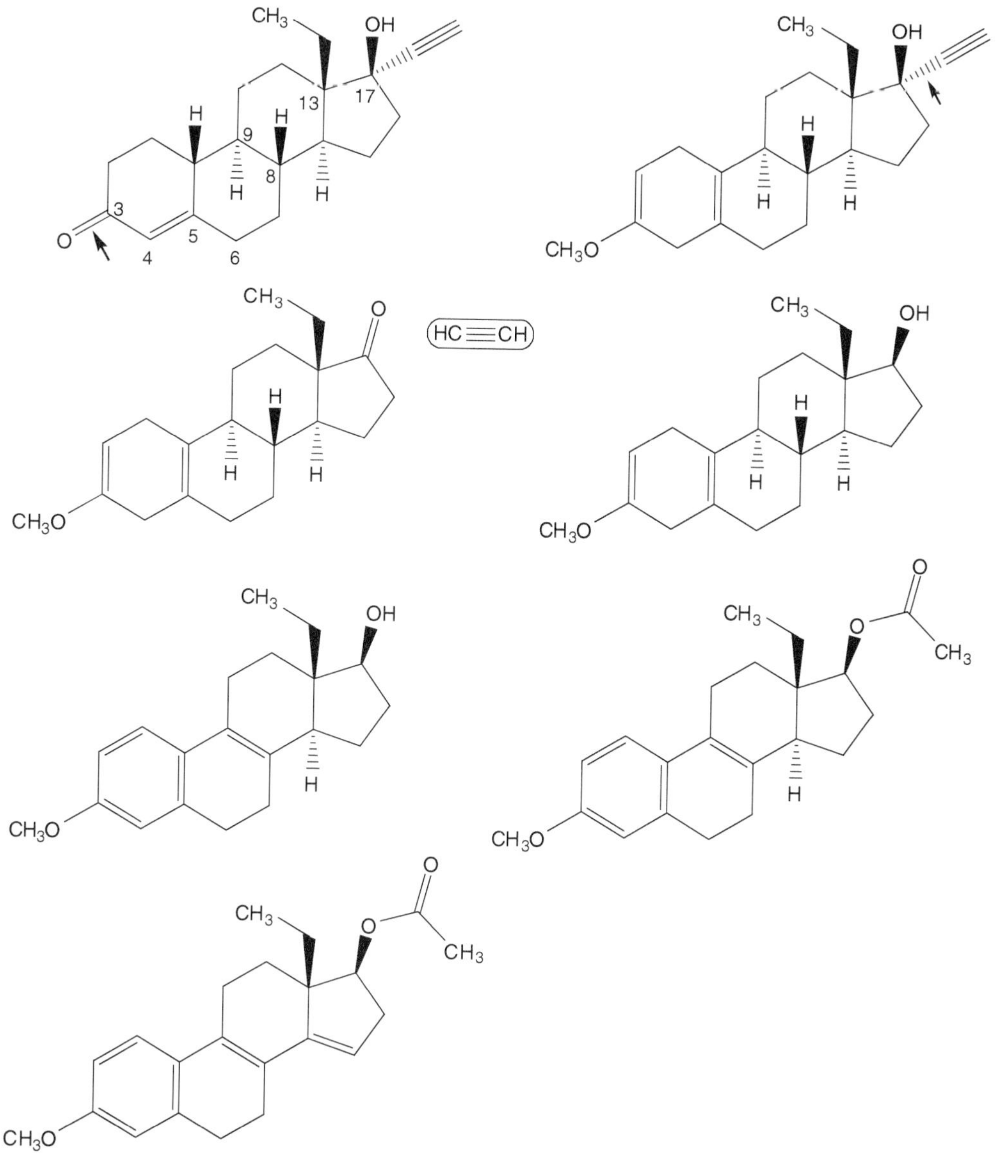

The gona-1,3,5(10),8,14-pentaene is formed from the secosteroid. The C17 acetate of the secosteroid is formed from the C17 alcohol by reaction with acetic anhydride. The chiral carbons at C13 and C17 are formed by microbiological reduction of the secodione. The C12–C13 bond of the secodione is formed by the reaction of 6-methoxy-1-vinyl-1,2,3,4-tetrahydronaph thalen-1-ol with 2-ethyl-1,3-cyclopentanedione. 6-Methoxy-1-vinyl-1,2,3,4-tetrahydronaphthalen-1-ol is formed by addition of vinylmagnesium chloride to 6-methoxy-1-tetralone (**Grignard Reaction**). Vinylmagnesium chloride is formed from vinyl chloride.

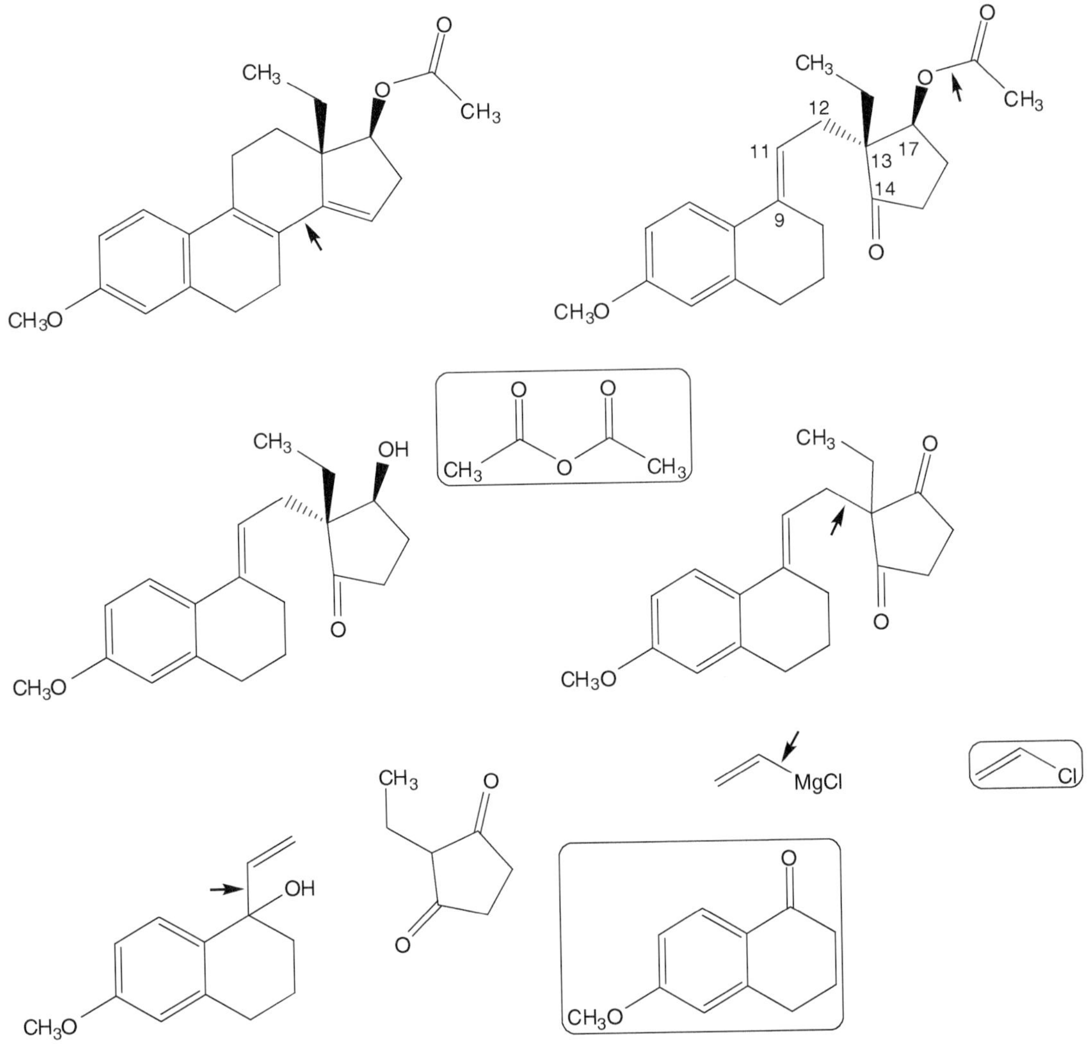

2-Ethyl-1,3-cyclopentanedione is formed from succinic anhydride and butanoic anhydride.

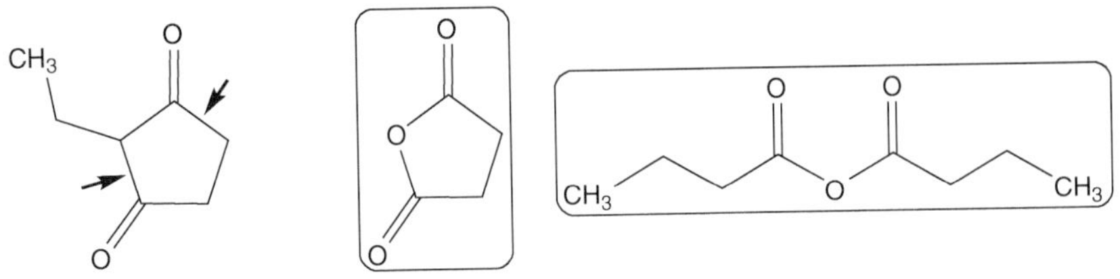

Extended Discussion

Draw the structures of the retrosynthetic analysis of one alternative route to levonorgestrel. Include the structures of the retrosynthetic analysis of any organic starting material(s) from petrochemical or biochemical raw materials. List the pros and cons for both routes and select one route as the preferred route.

Levothyroxine

Hormones, Other Endocrine Medicines and Contraceptives/Thyroid Hormones and Antithyroid Medicines

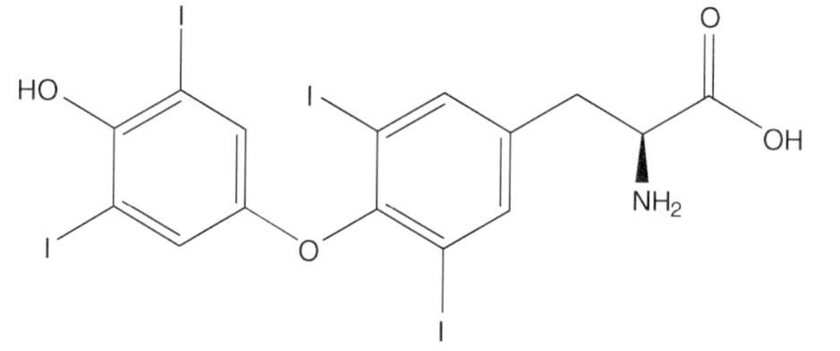

A chiral carbon in a single-enantiomer molecule is often delivered in a starting material.

Discussion. Iodination of the phenol ring of 3,5-diiodothyronine is the final step. The phenol, carboxylic acid, and amine of 3,5-diiodothyronine are released from ether, ester, and amide protecting groups in a single step. The diaryl ether is formed by *O*-arylation of *N*-acetyl-3,5-diiodo-L-tyrosine ethyl ester with a bis(4-methoxyphenyl)iodonium salt.

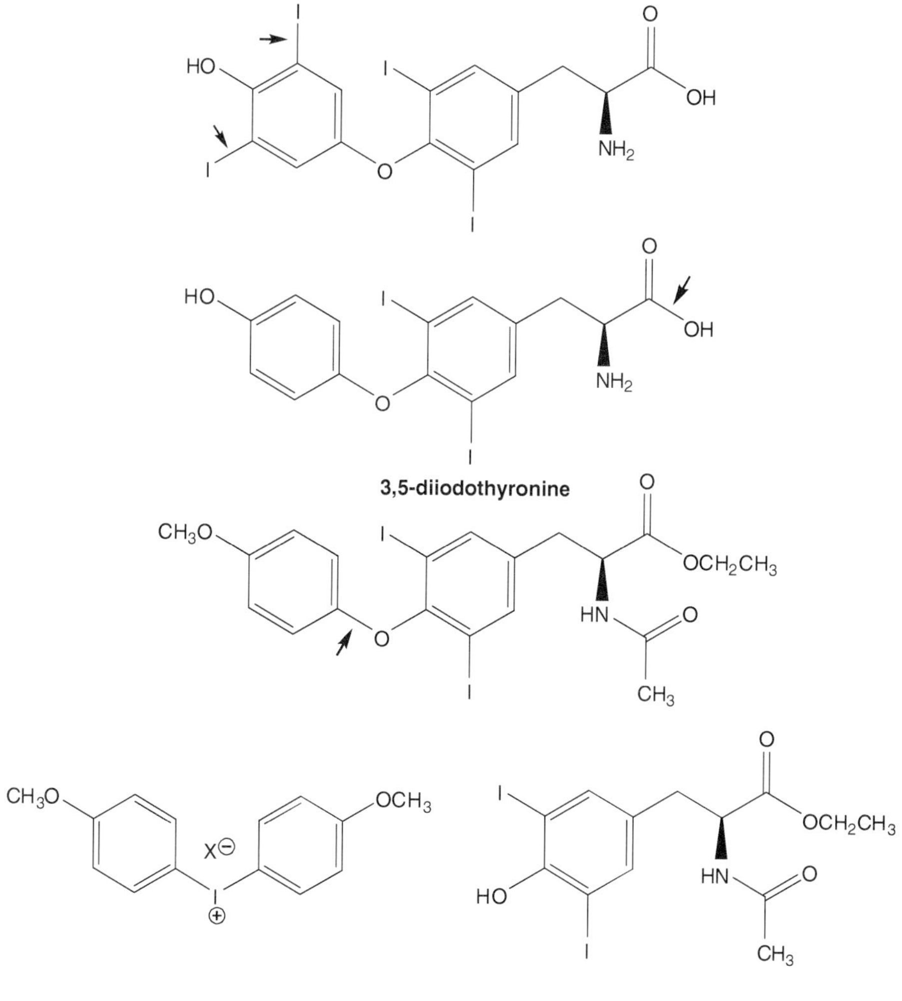

The ester is formed from the carboxylic acid (**Fischer Esterification**). *N*-Acetyl-3,5-diiodo-L-tyrosine is formed by reaction of 3,5-diiodo-L-tyrosine with acetic anhydride. 3,5-Diiodo-L-tyrosine is formed by iodination of L-tyrosine. L-Tyrosine is produced by fermentation.

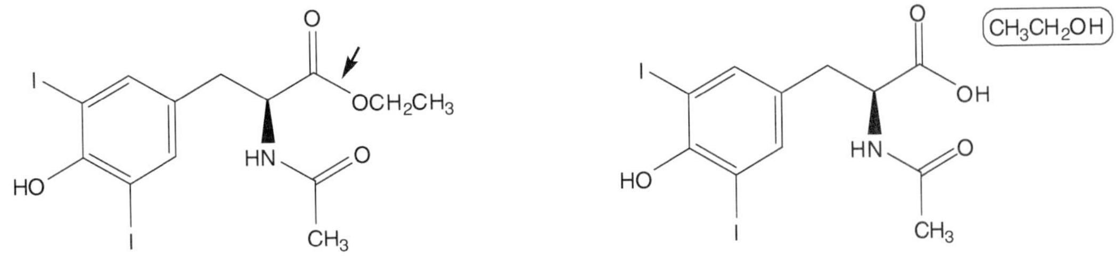

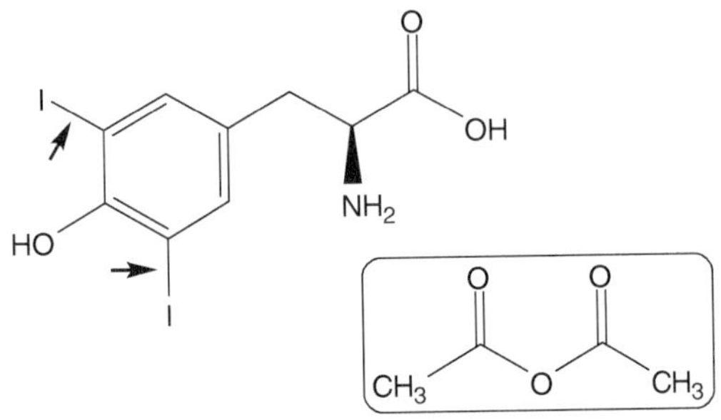

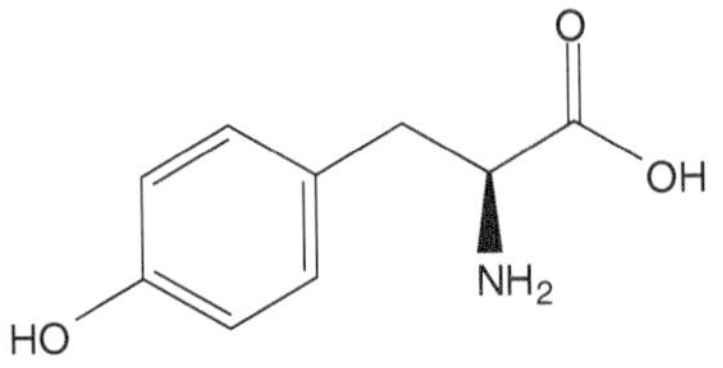

tyrosine

The bis(4-methoxyphenyl)iodonium salt is formed from 4-iodoanisole and anisole. (What is the counterion X and what is the procedure for manufacturing the salt?)

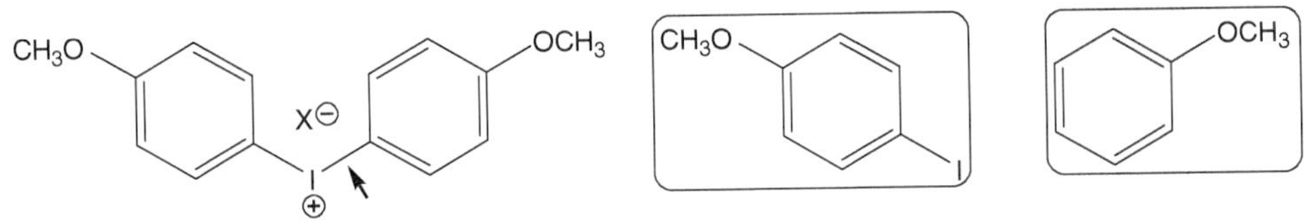

Extended Discussion

Draw the structure for one alternative *O*-arylation reagent used to prepare levothyroxine from 3,5-diiodotyrosine. What reagents and conditions are used for the *O*-arylation step using the alternative reagent? Draw the structures of a retrosynthetic analysis of the reagent. Include the structures of the retrosynthetic analysis of any organic starting material(s) from petrochemical or biochemical raw materials.

Lidocaine

Anesthetics/Preoperative Medicines and Medical Gases/Local Anesthetics
Cardiovascular Medicines/Antiarrhythmic Medicines

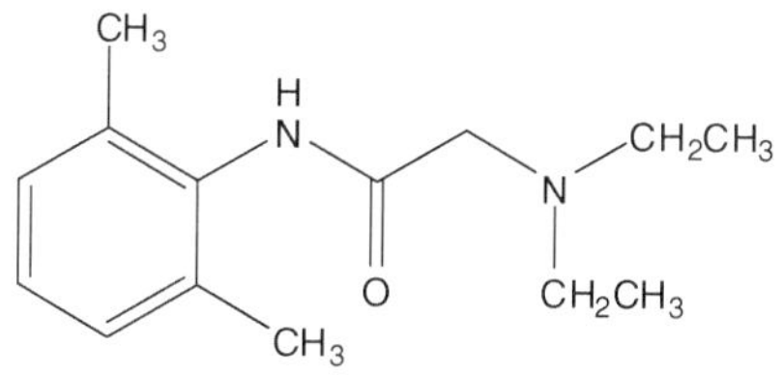

Tertiary amines are ubiquitous in drug structures. A tertiary amine is often formed by alkylation of a secondary amine. The alkylation is most efficient when the amine is used in excess and the carbon with the leaving group (Cl, Br, I, OTs, OMs) is primary or benzylic.

Discussion. Lidocaine is manufactured in just two steps. The tertiary amine is formed in the final step by displacement of chloride from the α-chloroacetamide by *N*,*N*-diethylamine. The α-chloroacetamide is formed by the reaction of 2,6-dimethylaniline with chloroacetyl chloride.

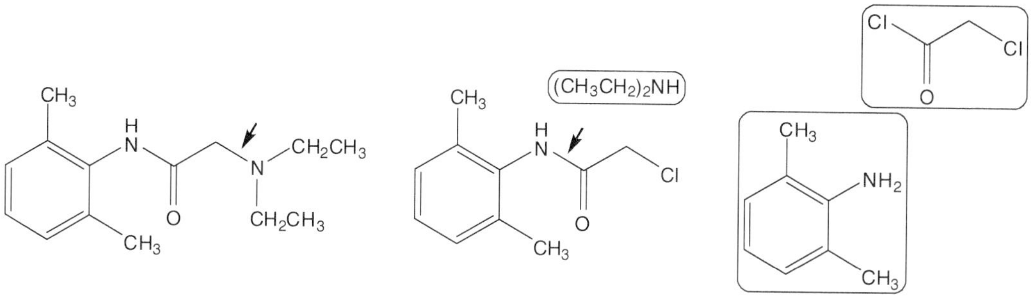

Extended Discussion

Draw the structures of the retrosynthetic analysis of one alternative route to 2,6-dimethylaniline. Discuss the pros and cons for both routes to this starting material and select one route as the preferred route.

Linezolid

Anti-infective Medicines/Antibacterials/Other Antibacterials
Anti-infective Medicines/Antibacterials/Other Antibacterials/Antituberculosis Medicines

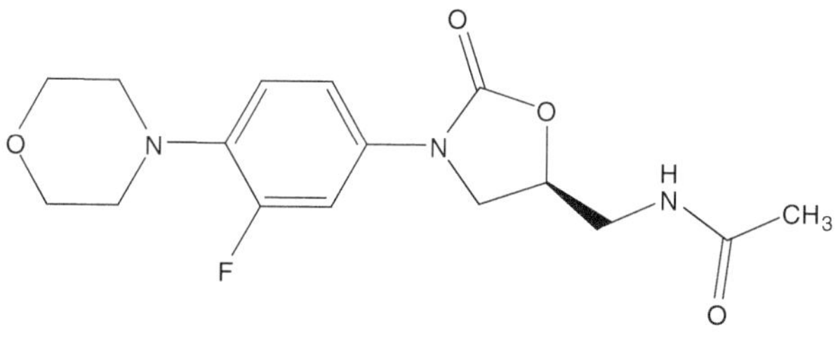

When a single-enantiomer target molecule is manufactured using a resolution, the resolution is likely an early step in the preferred route.

Discussion. In the final step, the acetamide is formed by reaction of the amine with acetic anhydride. The amine is formed by reduction of the azide. The azide is formed by displacement of methanesulfonate. The methanesulfonate is formed from the alcohol. The oxazolidinone is formed by reaction of the benzyl carbamate with (R)-glycidyl butyrate. (Draw structures for the byproduct(s) in this reaction.)

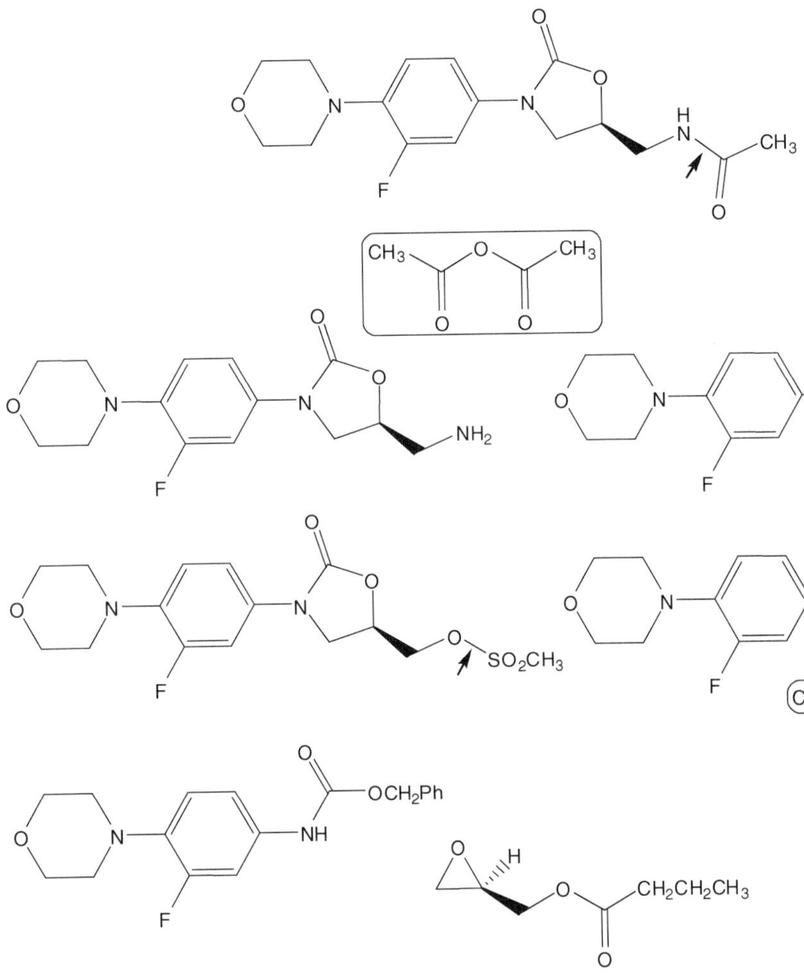

The carbamate is formed from the aniline and benzyl chloroformate. The aniline is formed by reduction of the nitroaromatic. Morpholine is introduced by displacement of a fluoride *para* to the nitro group. 1,2-Difluoro-4-nitrobenzene is formed from 1-chloro-2,4-dinitrobenzene.

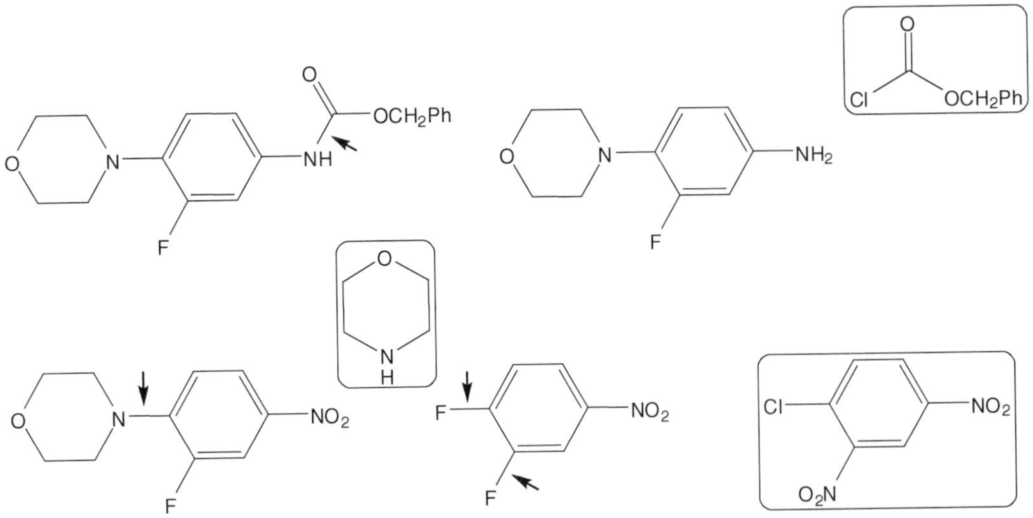

(R)-Glycidyl butyrate is formed by reaction of (*S*)-3-chloro-1,2-propanediol with butanoic anhydride. (*S*)-3-Chloro-1,2-propanediol is formed by hydrolysis of epichlorohydrin.

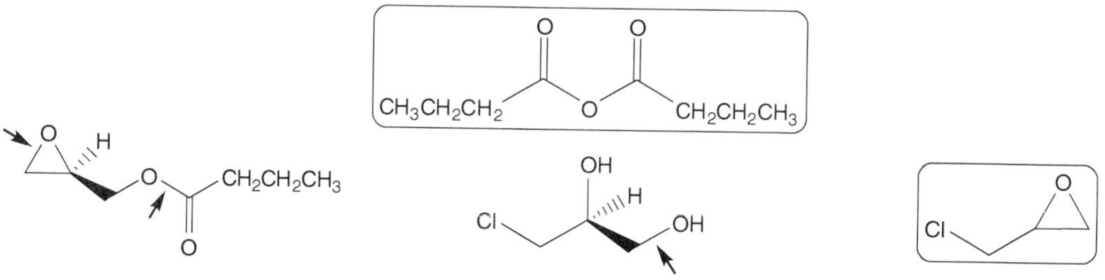

Extended Discussion

Draw the structures of the retrosynthetic analysis of one alternative route to linezolid which does not have an azide intermediate. List the pros and cons for both routes.

Loperamide

Medicines for Pain and Palliative Care/Medicines for Common Symptoms in Palliative Care

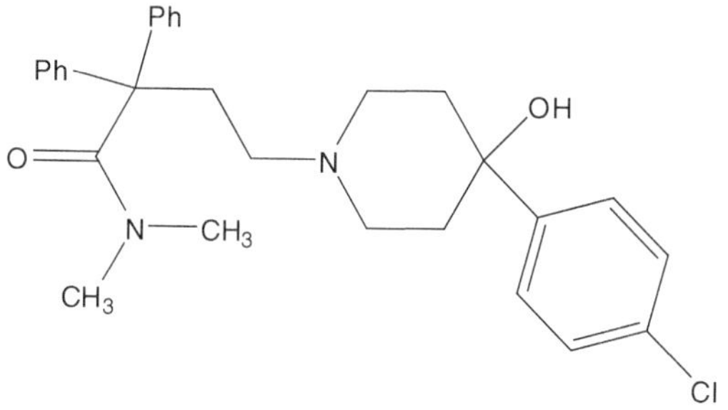

A tertiary alcohol is often formed by addition of a Grignard Reagent RMgX to a ketone.

Discussion. The tertiary amine is formed in the final step by ring-opening of dimethyl(tetrahydro-3,3-diphenyl-2-furylidene)ammonium bromide with the piperidine. 4-(4-Chlorophenyl)piperidin-4-ol is formed by hydrogenolysis of 1-benzyl-4-(4-chlorophenyl)piperidin-4-ol.

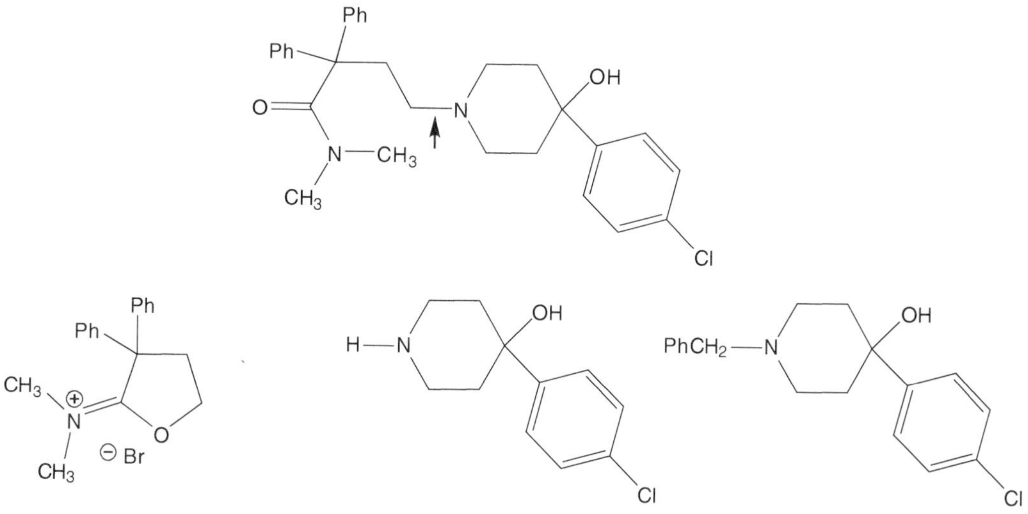

Dimethyl(tetrahydro-3,3-diphenyl-2-furylidene)ammonium bromide is formed by reaction of 4-bromo-2,2-diphenylbutanoyl chloride with *N,N*-dimethylamine. The acid chloride is formed from the carboxylic acid. 4-Bromo-2,2-diphenylbutanoic acid is formed by ring-opening of α,α-diphenyl-γ-butyrolactone. The γ-butyrolactone is formed by the reaction of ethyl diphenylacetate with ethylene oxide.

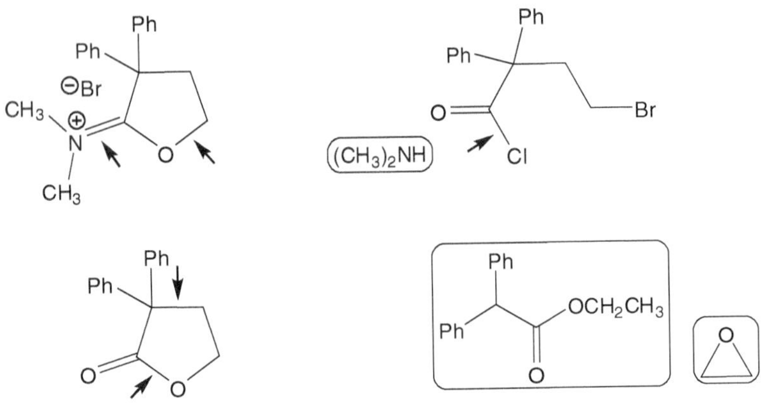

The tertiary alcohol of 1-benzyl-4-(4-chlorophenyl)piperidin-4-ol is formed by addition of 4-chlorophenylmagnesium bromide to 1-benzyl-4-piperidinone (**Grignard Reaction**). 4-Chlorophenylmagnesium bromide is formed from 1-bromo-4-chlorobenzene.

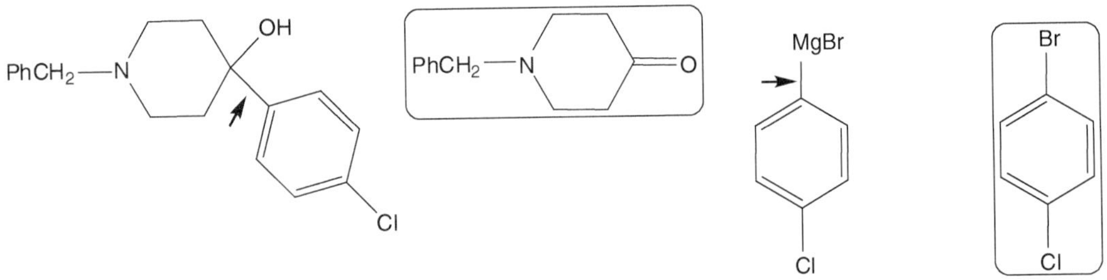

Extended Discussion

Draw structures for side products which might form in the hydrogenolysis to form 4-(4-chlorophenyl)piperidin-4-ol. Describe the reaction conditions selected to minimize the formation of the side products.

Lopinavir

Anti-infective Medicines/Antiviral Medicines/Antiretrovirals/Protease Inhibitor

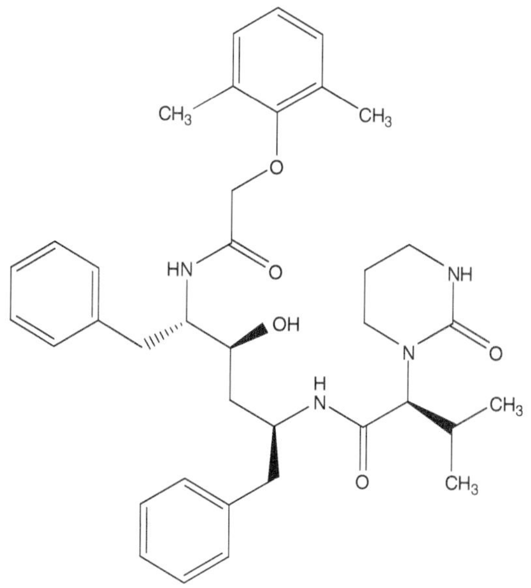

> **Amides, carbamates, and ureas are often efficiently formed from amines: an amide is formed by reaction with an acid chloride, anhydride, ester, or carboxylic acid, a carbamate is formed by reaction with a chloroformate or carbonate, and a urea is formed by reaction with a carbamate.**

Discussion. One amide is formed by reaction of the amine at C2 with the acid chloride in the final step. The amine at C2 is released by hydrogenolysis of the *N*,*N*-dibenzylamine. The other amide is formed by reaction of the amine at C5 with the acid chloride.

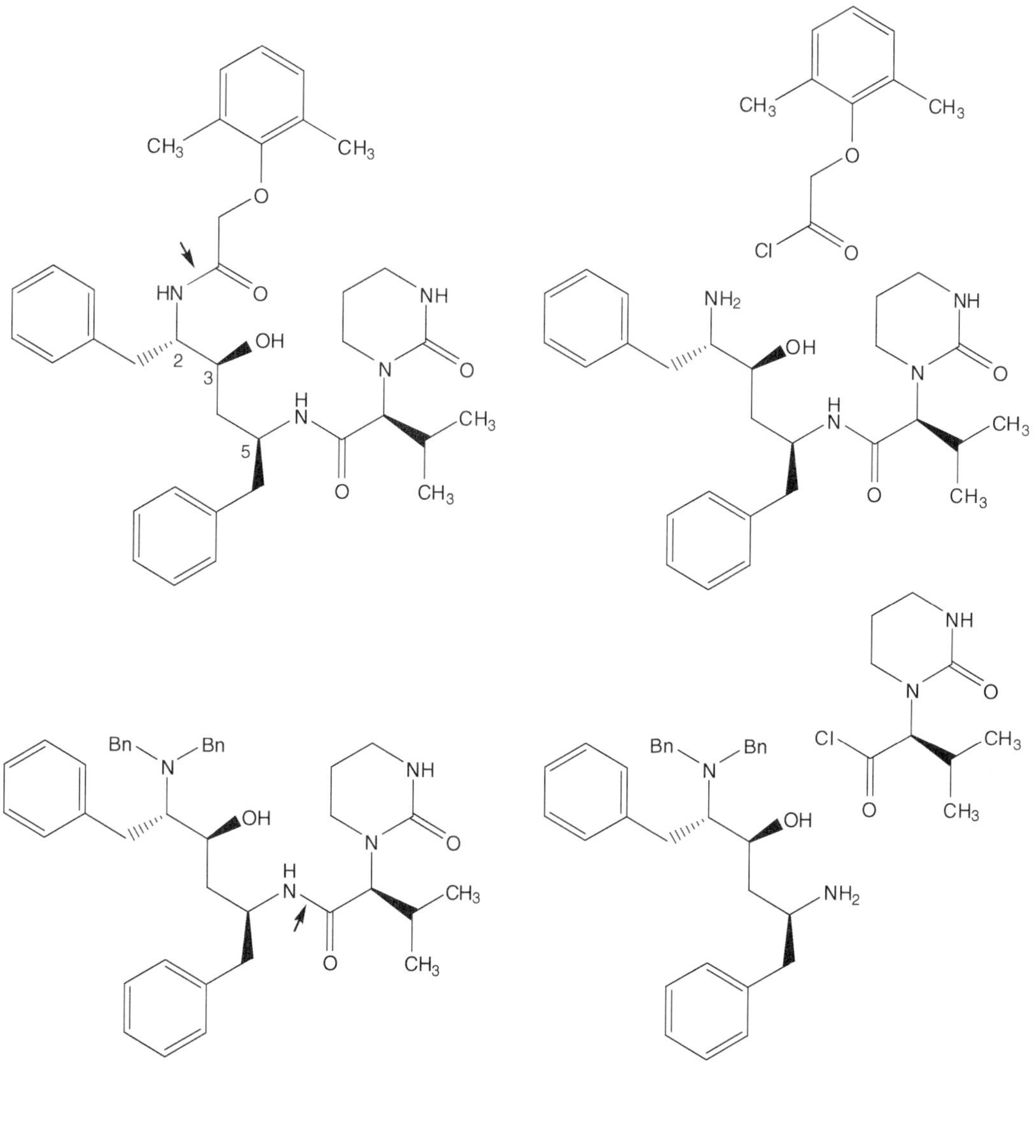

The alcohol of (2*S*,3*S*,5*S*)-5-amino-2-(dibenzylamino)-1,6-diphenylhexan-3-ol is formed by reduction of the C3 ketone. The ketone is formed by reduction of the 4-en-3-one. (Draw structures for three side products which are also formed in this two-reduction sequence.) The 4-en-3-one is formed by addition of benzylmagnesium chloride to the nitrile (**Grignard Reaction**). The nitrile is formed by condensation of acetonitrile and the ester (mixed **Claisen Condensation**). The ester, (*S*)-benzyl 2-(dibenzylamino)-3-phenylpropanoate, is formed by *N*- and *O*-benzylation of L-phenylalanine. L-Phenylalanine is produced by fermentation.

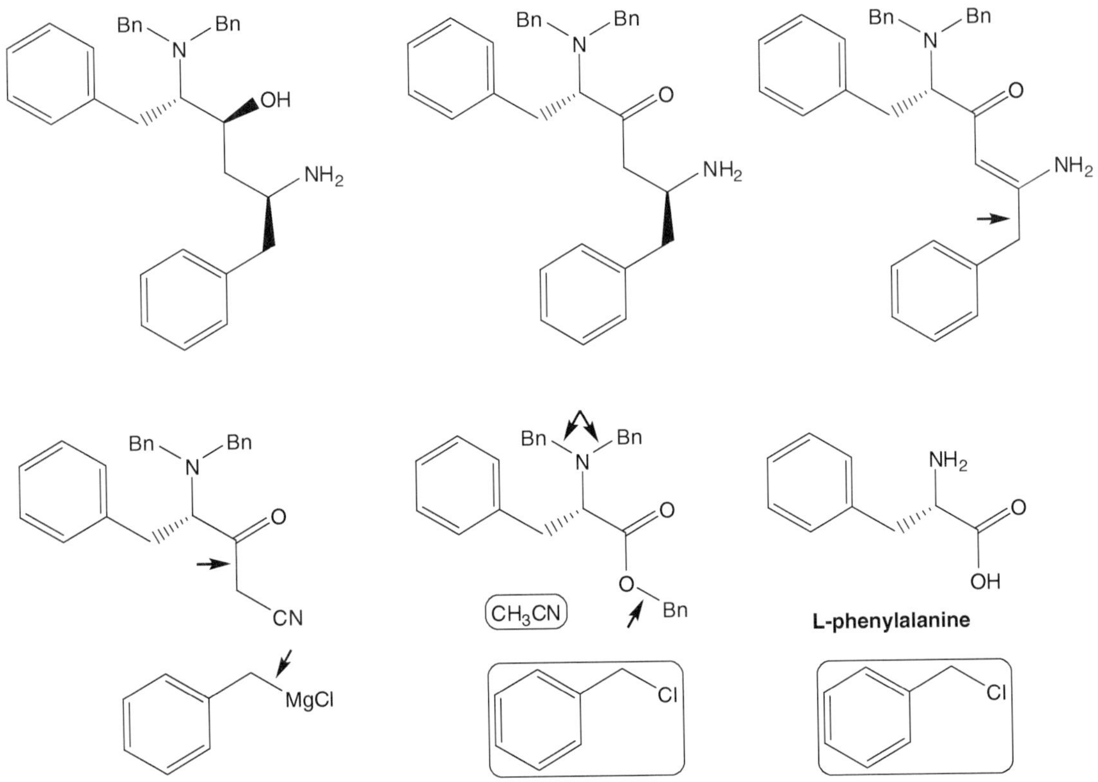

The acid chloride used to form the amide at C2 is formed from the carboxylic acid, (2,6-dimethylphenoxy)acetic acid. The carboxylic acid is formed from 2,6-dimethylphenol (2,6-xylenol) and chloroacetic acid.

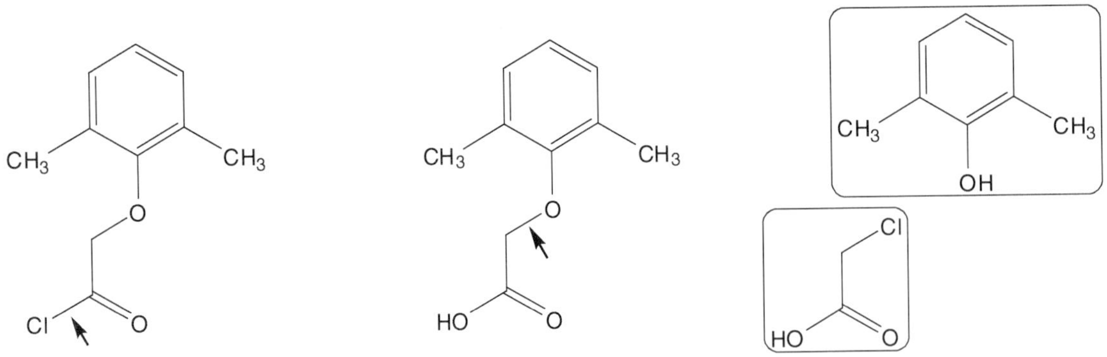

The acid chloride used to form the amide at C5 is formed from the carboxylic acid, (*S*)-3-methyl-2-[2-oxotetrahydropyrimidin-1(2*H*)-yl]-butanoic acid. The displacement of chloride by the urea nitrogen forms the six-membered ring. The urea is formed by reaction of 3-chloropropylamine with the phenyl carbamate. The chloropropylamine is formed from 3-amino-1-propanol. The phenyl carbamate is formed from phenyl chloroformate and L-valine. L-Valine is produced by fermentation.

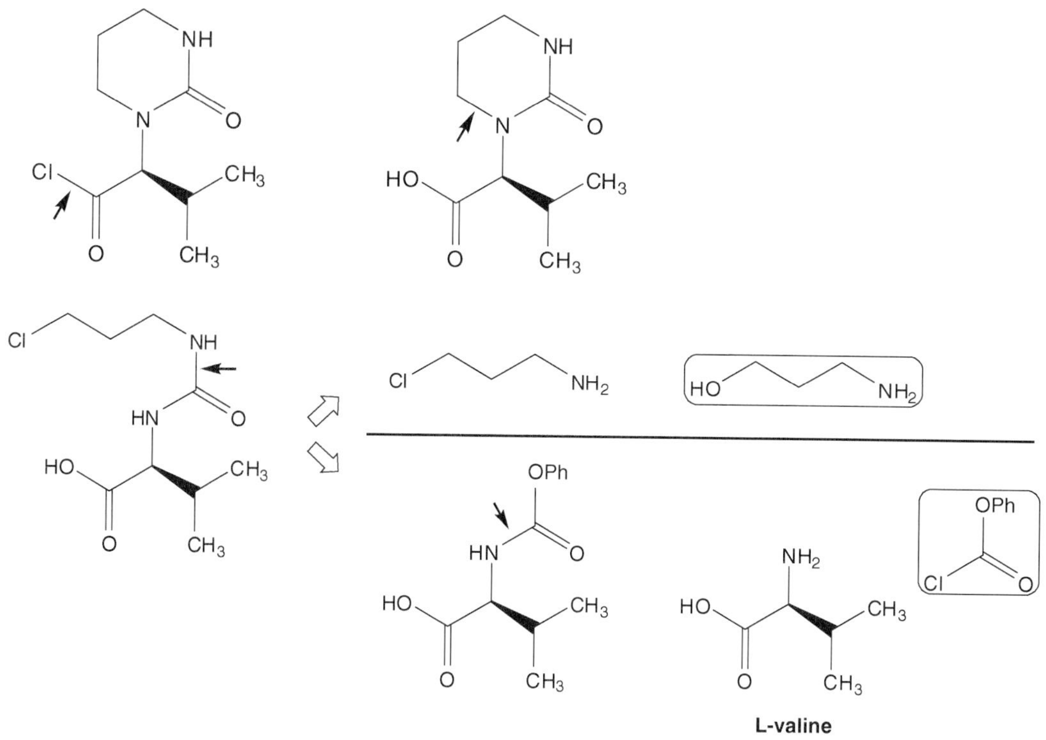

Extended Discussion

Draw the structures of the retrosynthetic analysis of one alternative route to (*S*)-3-methyl-2-[2-oxotetrahydropyrimidin-1(2*H*)-yl]-butanoic acid from L-valine. Include the structures of the retrosynthetic analysis of any organic starting material(s) from petrochemical or biochemical raw materials. List the pros and cons for both routes and select one route as the preferred route.

Loratidine

Antiallergics and Medicines Used in Anaphlaxis

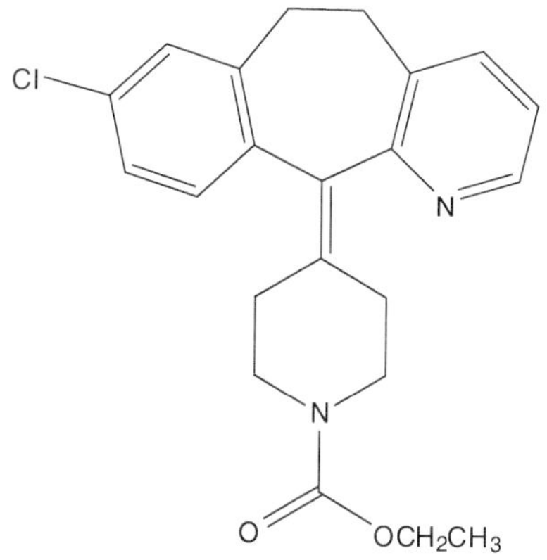

A carbamate is often formed by demethylation of an *N*-methyl tertiary amine.

Discussion. The carbamate is formed by the reaction of the 1-methylpiperidine with ethyl chloroformate.

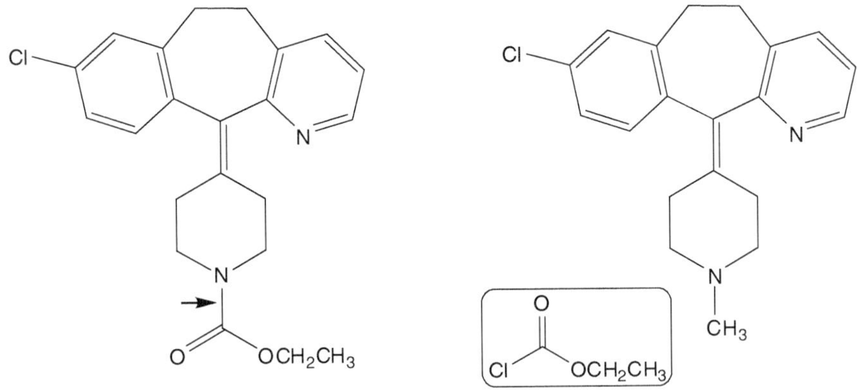

The alkene is formed by dehydration of the tertiary alcohol. The tertiary alcohol and the central ring are formed by intramolecular alkylation of the aromatic ring using the ketone. The ketone is formed by addition of an alkylmagnesium chloride to 3-[2-(3-chlorophenyl)ethyl]-2-pyridinecarbonitrile (**Grignard Reaction**).

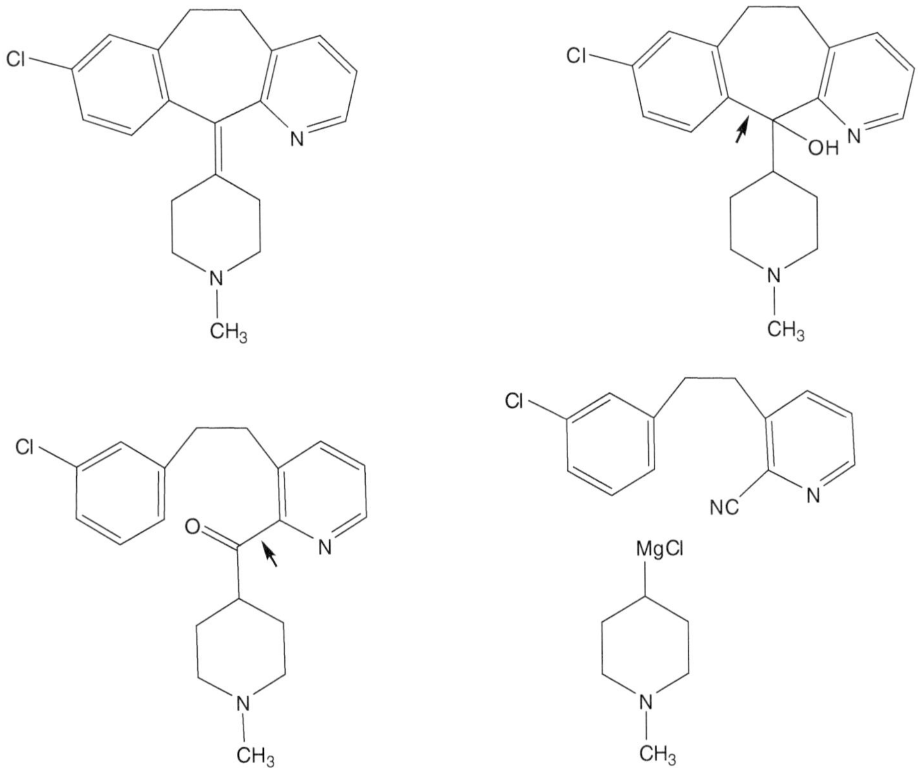

The nitrile is formed from the *N-tert*-butylpyridine-2-carboxamide. The C3 substituent is formed by chloride displacement from 3-chlorobenzyl chloride by the dianion from *N-tert*-butyl-3-methylpyridine-2-carboxamide. *N-tert*-Butyl-3-methylpyridine-2-carboxamide is formed from 3-methylpicolinonitrile (**Ritter Reaction**).

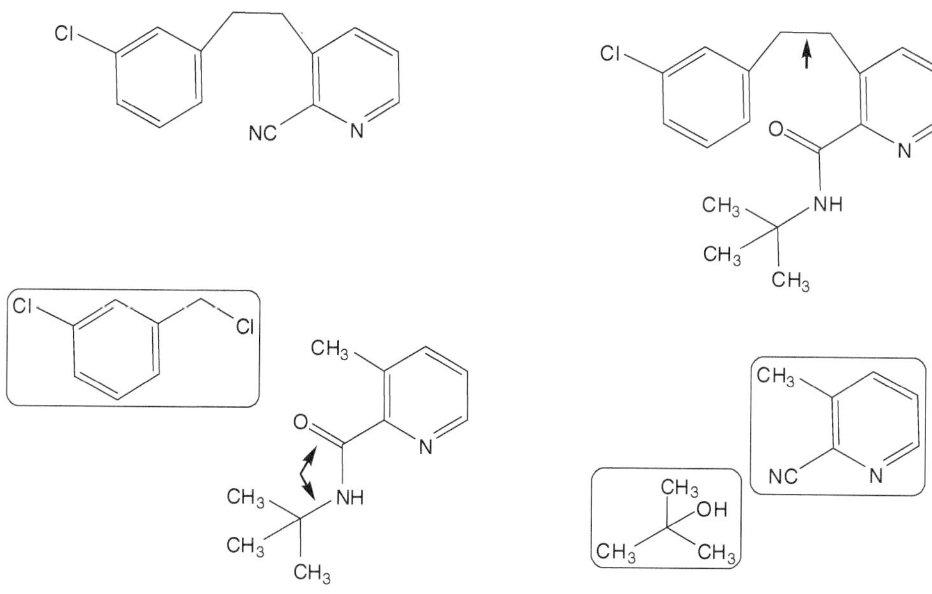

The alkylmagnesium chloride is formed from 4-chloro-1-methylpiperidine. The 4-chloropiperidine is formed from 1-methylpiperidin-4-ol. 1-Methylpiperidin-4-ol is formed by reduction of 1-methyl-4-piperidinone.

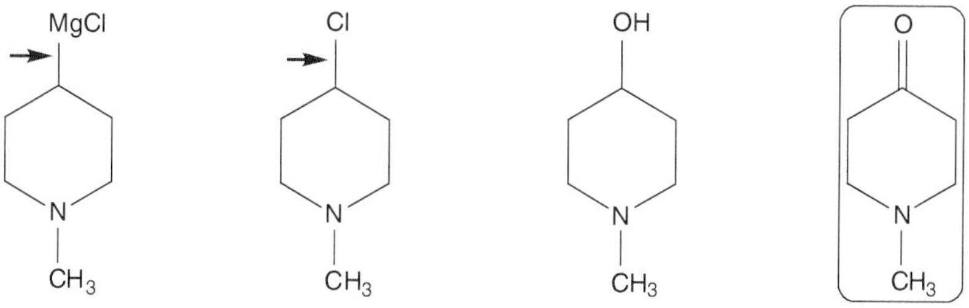

Extended Discussion

Draw the structures of the retrosynthetic analysis of one alternative route to 3-[2-(3-chlorophenyl)ethyl]-2-pyridinecarbonitrile. Include structures of the retrosynthetic analysis of any organic starting material(s) from petrochemical or biochemical raw materials. List the pros and cons for both routes to this intermediate. Is one route preferred?

Lorazepam

Anticonvulsants/Antiepileptics

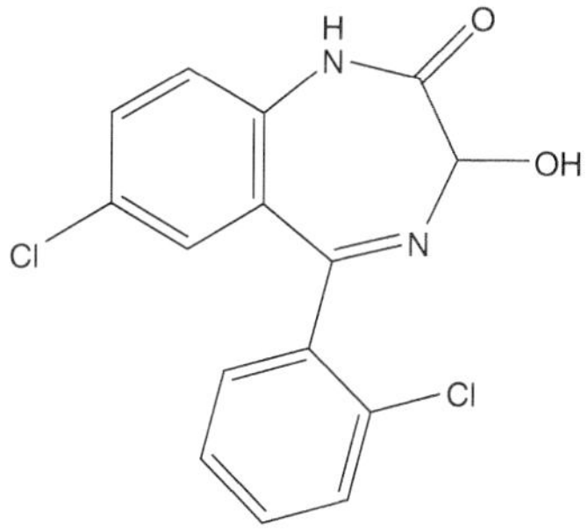

A 5-aryl-1,4-benzodiazepin-2-one is often formed from a 2-aminobenzophenone.

Discussion. Lorazepam is a 1:1 mixture of the (*R*)- and (*S*)-enantiomers. The C3 hydroxyl group is released by hydrolysis of the acetate ester in the final step. The ester is formed by reaction of the N4-oxide with acetic anhydride (**Polonovski Reaction**). The 1,4-benzodiazepin-2-one-4-oxide is formed from 6-chloro-2-chloromethyl-4-(2-chlorophenyl)quinazoline-3-oxide. The quinazoline-3-oxide is formed by reaction of 2-amino-2′,5-dichlorobenzophenone oxime with chloroacetyl chloride. The oxime is formed from 2-amino-2′,5-dichlorobenzophenone.

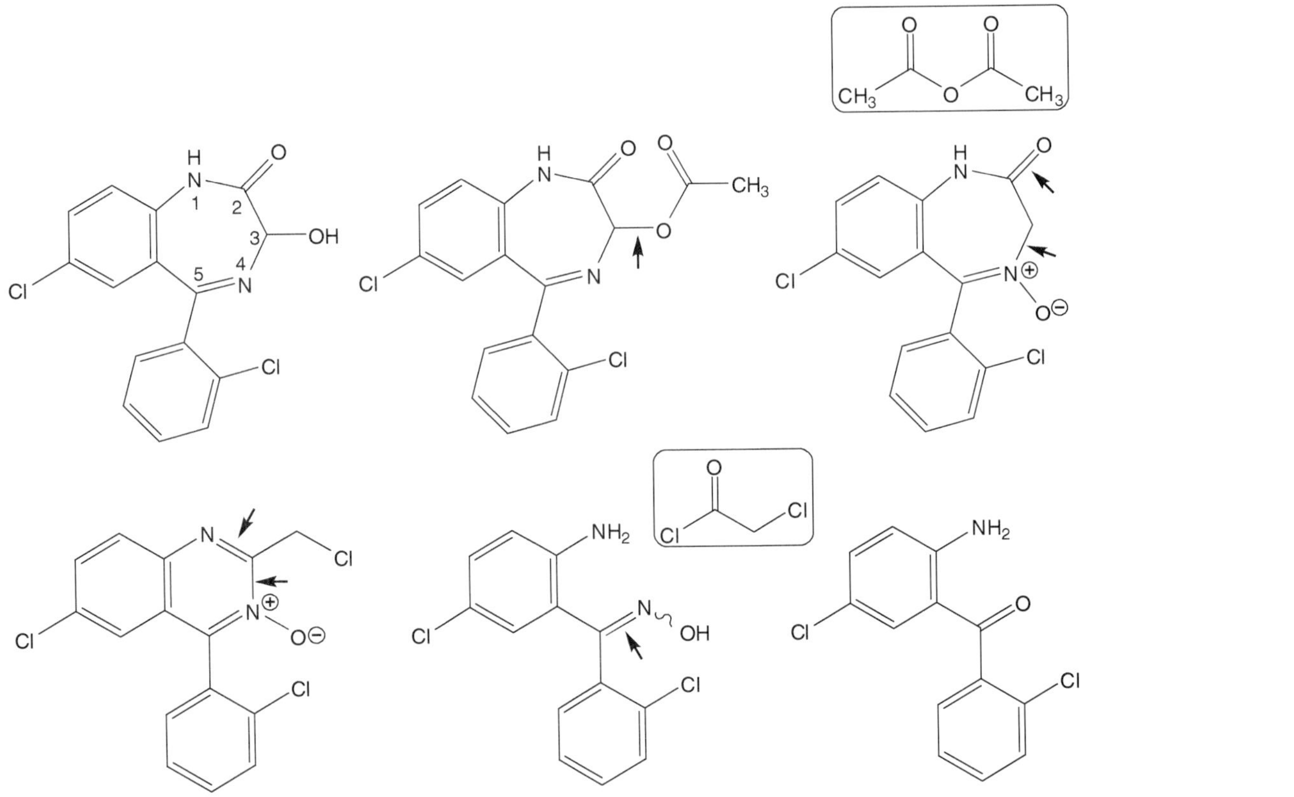

The amine is released by hydrolysis of the acetamide. The benzophenone is formed from 4-chloroacetanilide and 2-chlorobenzoyl chloride (**Friedel–Crafts Acylation**).

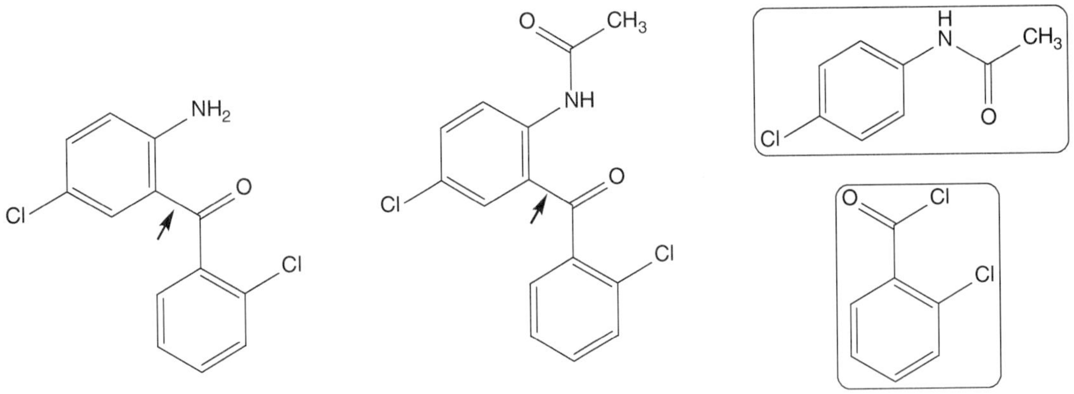

Extended Discussion

Draw the structures of a retrosynthetic analysis of an alternative route to lorazepam which does not proceed via a quinazoline intermediate. List the pros and cons for both routes and select one route as the preferred route.

Losartan Potassium

Cardiovascular Medicines/Antihypertensive Medicines

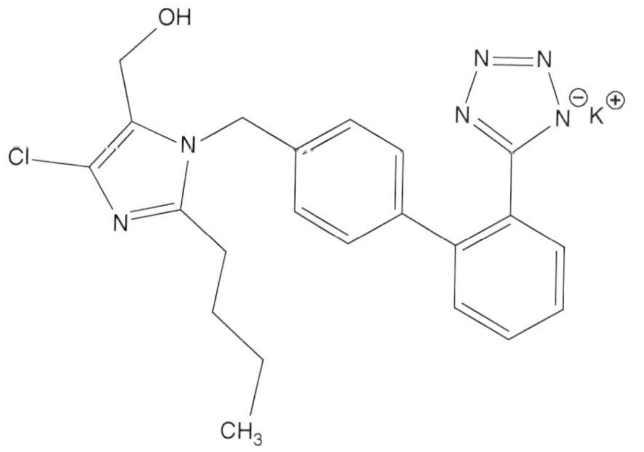

A 5-substituted tetrazole is often formed by reaction of a nitrile with sodium azide-ammonium chloride, sodium azide-triethylammonium chloride, or tributyltin azide.

Discussion. Two routes to losartan will be presented for comparison. In Route A, the tetrazole is delivered in a starting material. In Route B, the tetrazole is formed from a nitrile in the final step.

The tetrazole is released by cleavage of a triphenylmethyl (trityl) protecting group in the final step of Route A. The C–C bond joining the two rings of the biphenyl is formed by coupling a bromoaromatic with an arylboronic acid (**Suzuki–Miyaura Coupling**). The arylboronic acid is formed from the aryllithium and triisopropyl borate. The aryllithium is formed from 5-phenyl-2-trityltetrazole. 5-Phenyl-2-trityltetrazole is formed from 5-phenyltetrazole and trityl chloride.

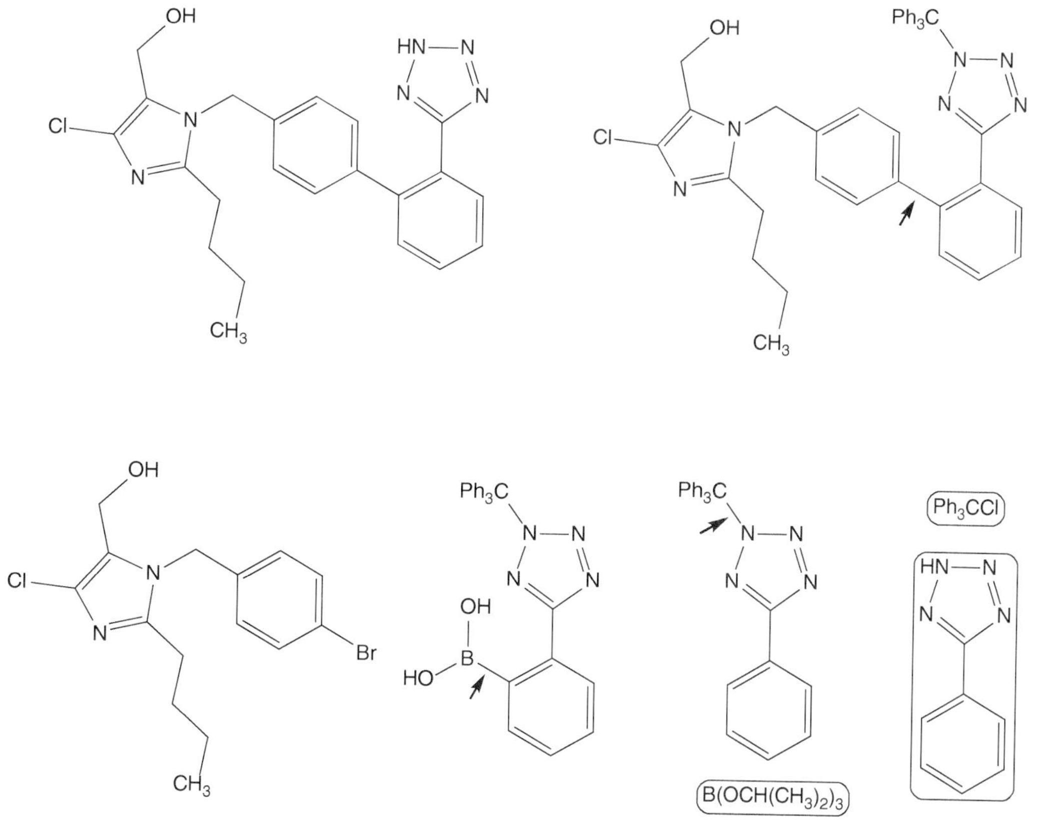

The primary alcohol in the bromoaromatic coupling partner is formed by reduction of the aldehyde. Alkylation of 2-butyl-4-chloro-1*H*-imidazole-5-carboxaldehyde with 4-bromobenzyl bromide is highly selective for imidazole N1.

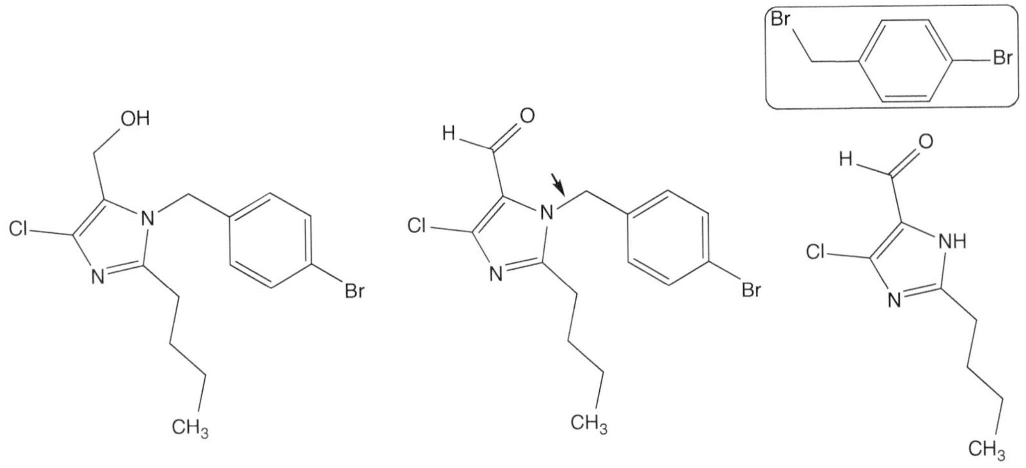

2-Butyl-4-chloro-1*H*-imidazole-5-carboxaldehyde is formed in three steps from pentanenitrile. Reaction of 2-butylimidazolin-4-one with phosphorus oxychloride and *N*,*N*-dimethylformamide converts the amide oxygen at C4 to chlorine and forms the aldehyde at C5 (**Vilsmeier–Haack Reaction**). The imidazolin-4-one is formed from the imidate and glycine. The imidate is formed from pentanenitrile.

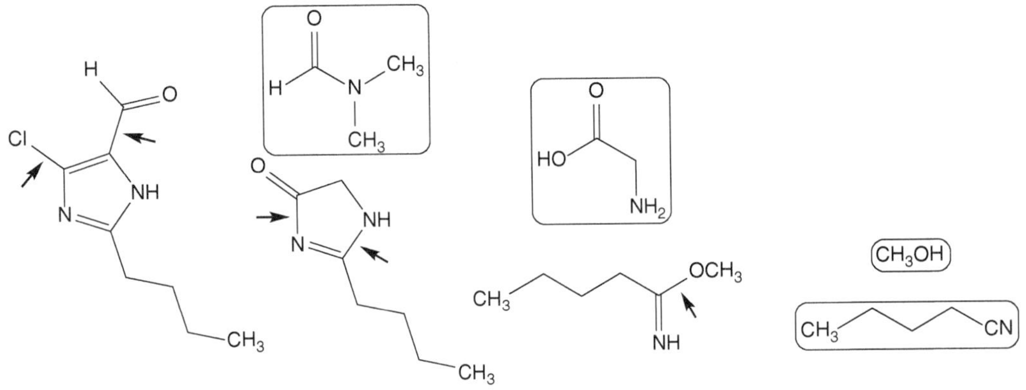

In Route B, the tetrazole is formed from the nitrile in the final step. The primary alcohol is formed by reduction of the aldehyde. Alkylation of 2-butyl-4-chloro-1*H*-imidazole-5-carboxaldehyde with 4′-bromomethyl-2-cyanobiphenyl is highly selective for imidazole N1. The benzyl bromide is formed by bromination of 2-cyano-4′-methylbiphenyl.

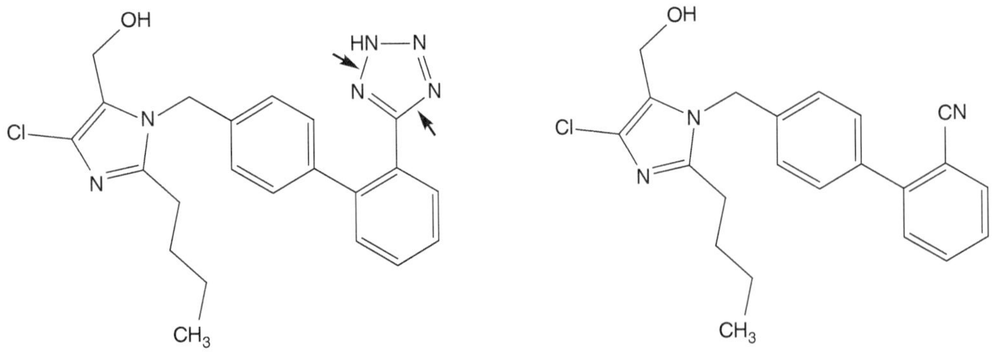

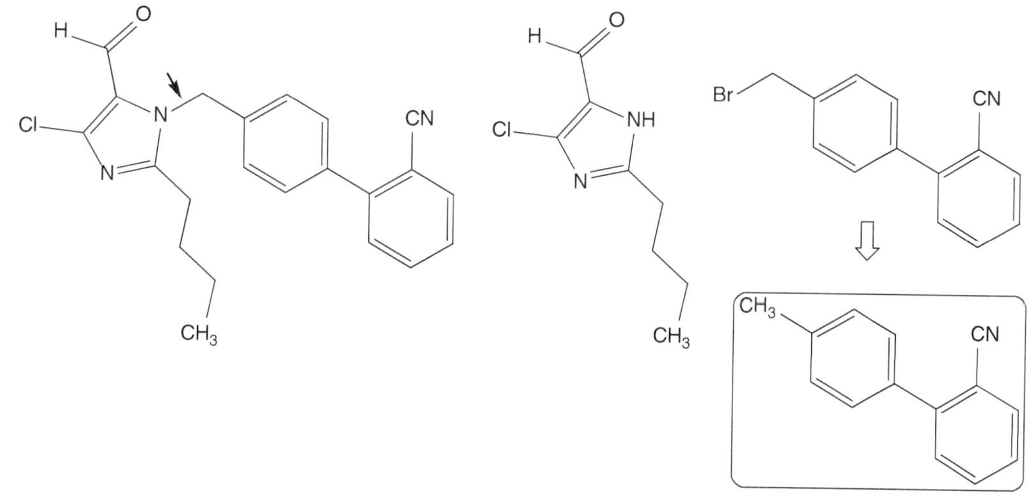

Extended Discussion

Estimate a yield of losartan potassium from 2-butyl-4-chloro-1*H*-imidazole-5-carboxaldehyde for both routes. Show the calculations. List the pros and cons for both routes. Is one route preferred?

Lumefantrine

Anti-infective Medicines/Antiprotozoal Medicines/Antimalarial Medicines/For Curative Treatment

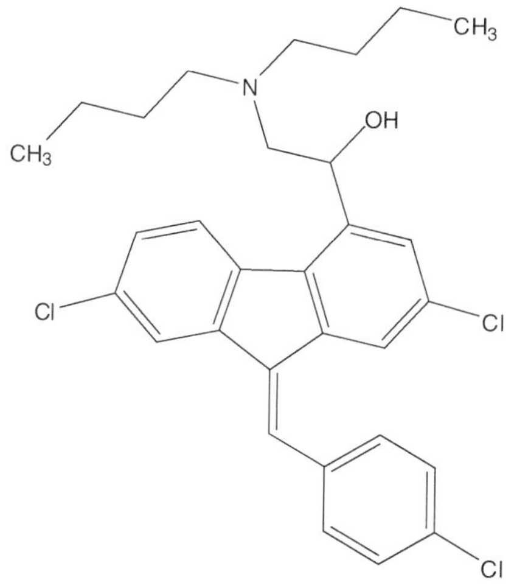

An alkene conjugated to three aromatic rings is often formed by dehydration of an alcohol.

Discussion. Lumefantrine is produced as a 1 : 1 mixture of the (*R*)-and (*S*)-enantiomers of the (*Z*)-alkene. The alkene, conjugated to three aromatic rings, is formed by dehydration of the secondary alcohol. Under the dehydration conditions, the (*E*)- and (*Z*)-alkenes equilibrate and the (*Z*)-alkene crystallizes from the mixture. The secondary alcohol is formed by condensation of the fluorene with 4-chlorobenzaldehyde (**Knoevenagel Condensation**). The β-amino alcohol is formed by ring-opening of an epoxide by *N*,*N*-dibutylamine.

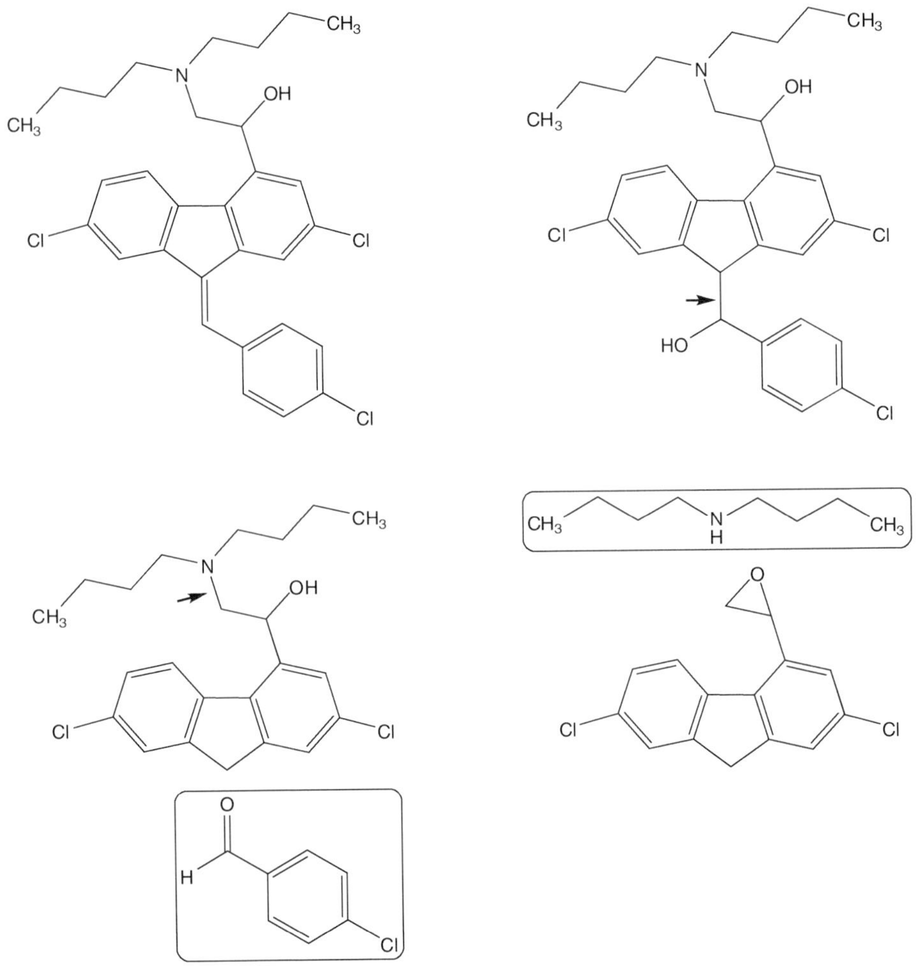

The epoxide is formed from the chlorohydrin. The chlorohydrin is formed by reduction of the α-chloroketone. The α-chloroketone is formed from 2,7-dichlorofluorene and chloroacetyl chloride (**Friedel–Crafts Acylation**). 2,7-Dichlorofluorene is formed by chlorination of fluorene. (Draw the structures for side products which are likely to form in the chlorination of fluorene. List the options for the chlorination reagent(s) and conditions and select one of the options as the preferred option.) Fluorene is obtained from high-temperature coal tar.

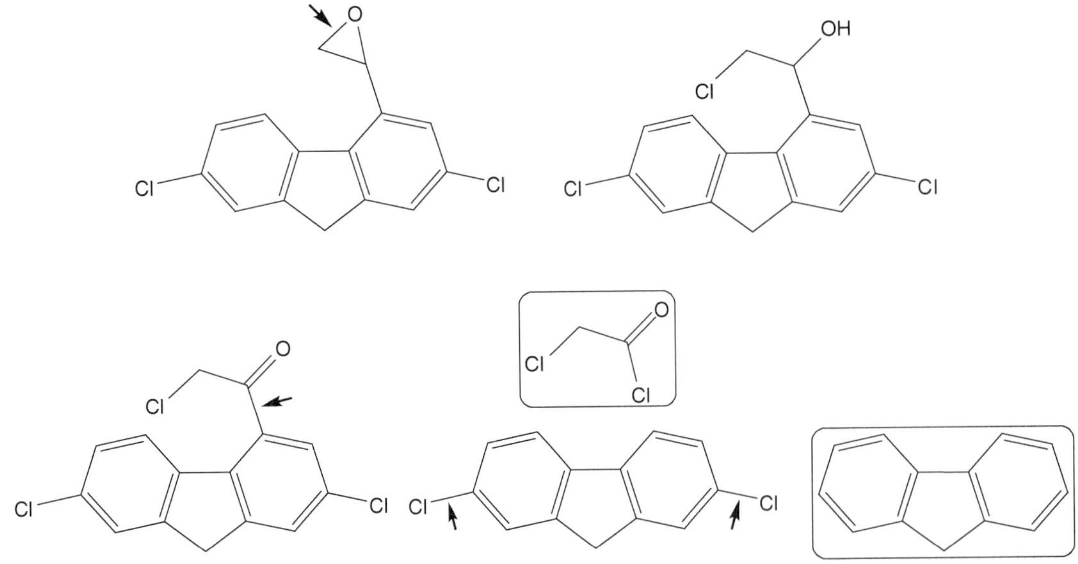

M

Mannitol

Diuretics

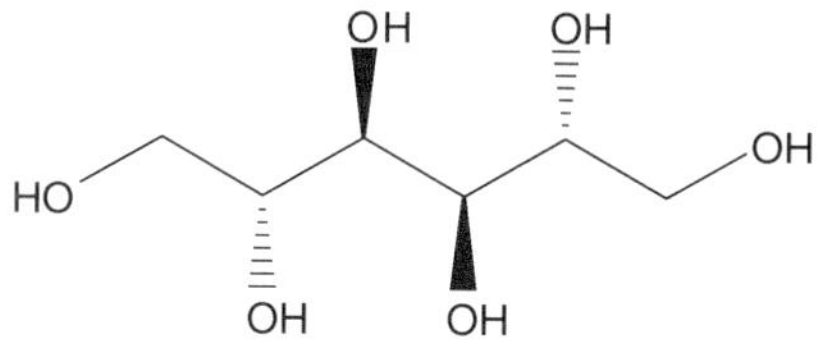

> **A single-enantiomer molecule with multiple chiral carbons is often formed by the modification of a natural product which has most or all of the chiral carbons already in place.**

Discussion. Mannitol (D-mannitol) is separated from a mixture of mannitol and sorbitol (D-sorbitol). The mixture of mannitol and sorbitol is often formed by hydrogenation of a mixture of D-fructose and D-glucose (invert sugar syrup). A mixture with a higher mannitol to sorbitol ratio is formed by hydrogenation of pure D-fructose. (List the catalyst, solvent, temperature, time, pressure, and mannitol: sorbitol ratio for hydrogenations of pure D-fructose.) D-Fructose or D-glucose can also be converted into mixtures of D-mannose and D-glucose or D-mannose, D-fructose, and D-glucose for use in the hydrogenation.

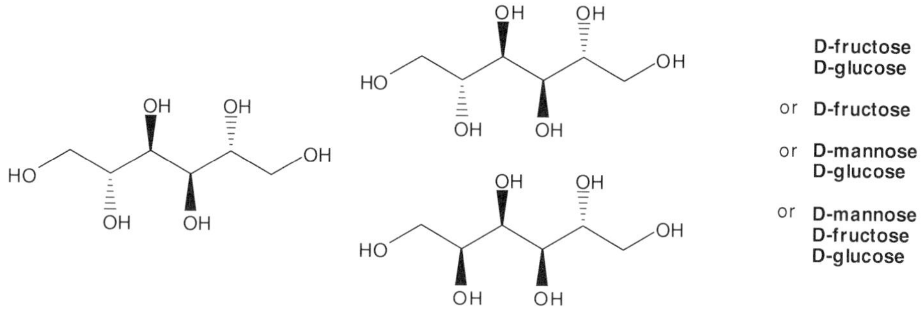

sorbitol

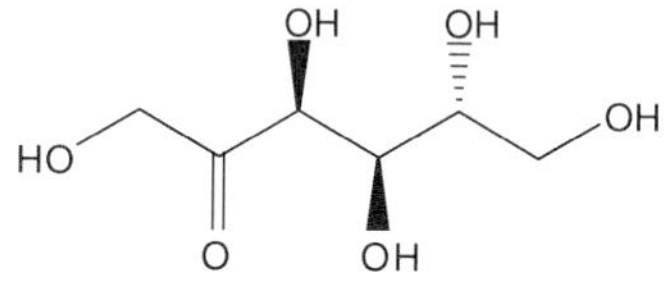

D-fructose

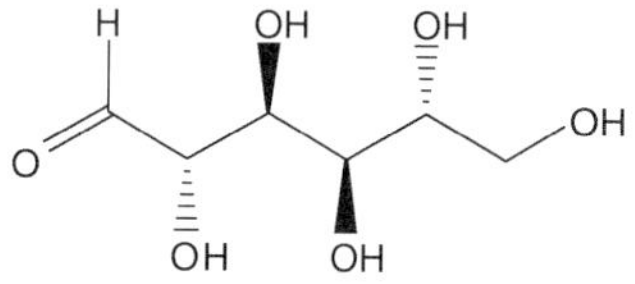

D-mannose

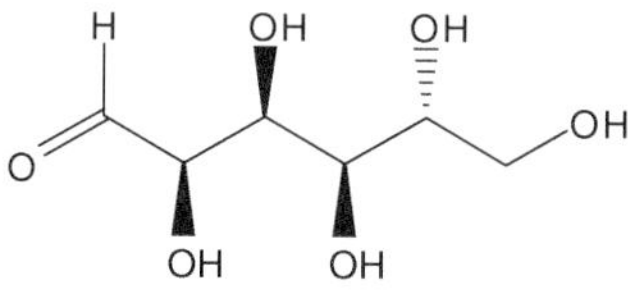

D-glucose

Routes to Essential Medicines: A Workbook for Organic Synthesis, First Edition. Peter J. Harrington.

Companion website: www.wiley.com/go/Harrington/routes_essential_medicine

Mebendazole

Anti-infective Medicines/Intestinal Anthelminthics

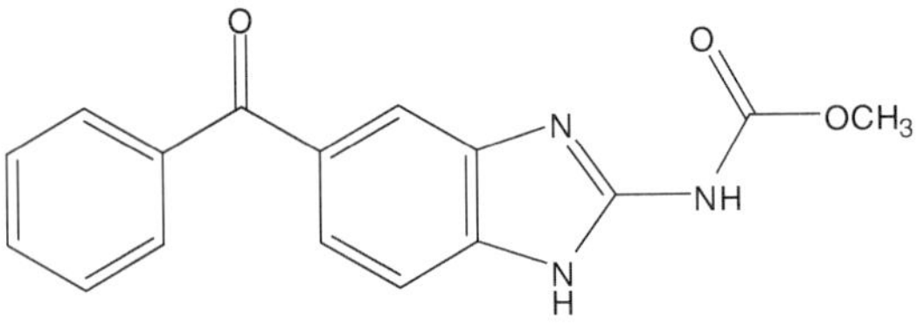

A benzimidazole is often formed from a 1,2-benzenediamine.

Discussion. The benzimidazole ring is formed in the final step by the reaction of 3,4-diaminobenzophenone with *N*-methoxycarbonyl-*S*-methylisothiourea. The isothiourea is formed from *S*-methylisothiouronium hemisulfate and methyl chloroformate.

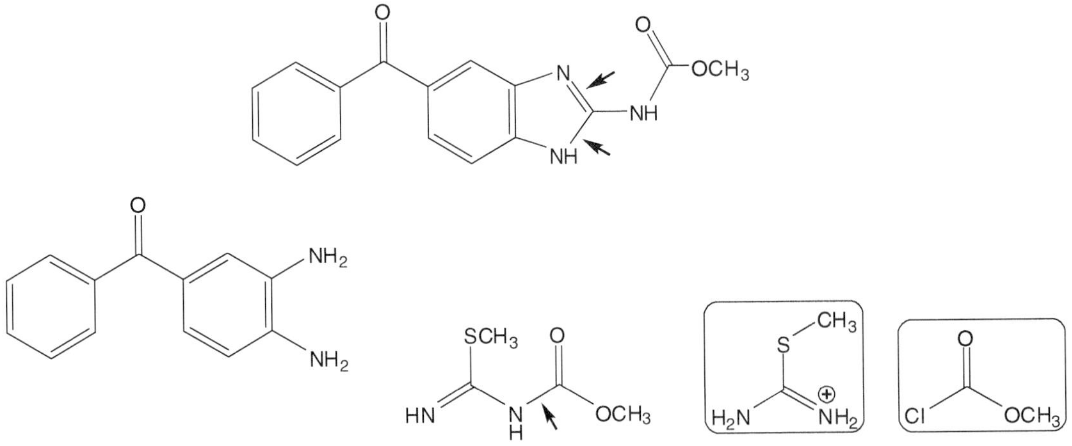

3,4-Diaminobenzophenone is formed by reduction of 4-amino-3-nitrobenzophenone. 4-Amino-3-nitrobenzophenone is formed from 4-chloro-3-nitrobenzophenone by chloride displacement by ammonia. 4-Chloro-3-nitrobenzophenone is formed from 4-chloro-3-nitrobenzoyl chloride and benzene (**Friedel–Crafts Acylation**). The acid chloride is formed from the carboxylic acid.

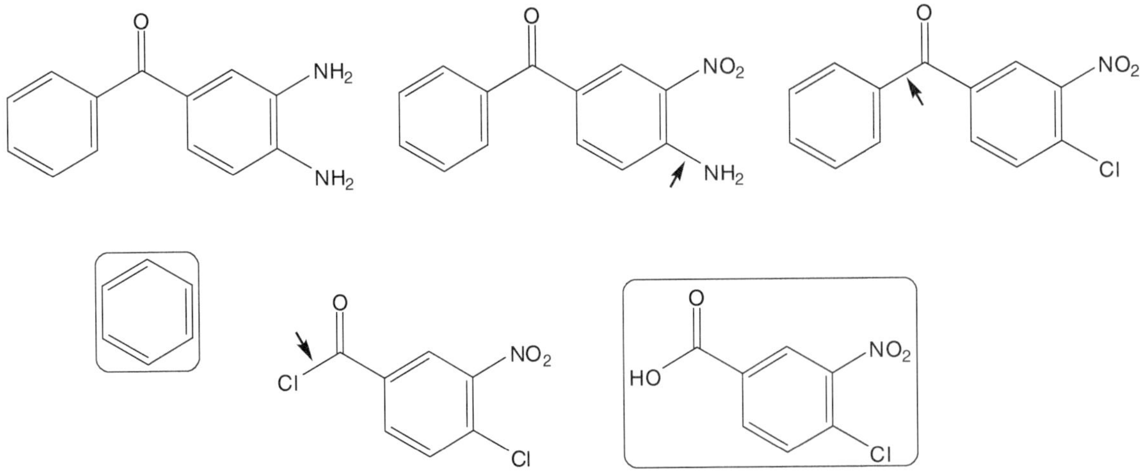

Extended Discussion

Draw the structures of the retrosynthetic analysis of one alternative route to 3,4-diaminobenzophenone. Include the structures of the retrosynthetic analysis of any organic starting material(s) from petrochemical or biochemical raw materials. List the pros and cons for both routes and select one route as the preferred route.

Medroxyprogesterone Acetate

Hormones, Other Endocrine Medicines and Contraceptives/Contraceptives/Injectable Hormonal Contraceptives
Hormones, Other Endocrine Medicines and Contraceptives/Progestogens

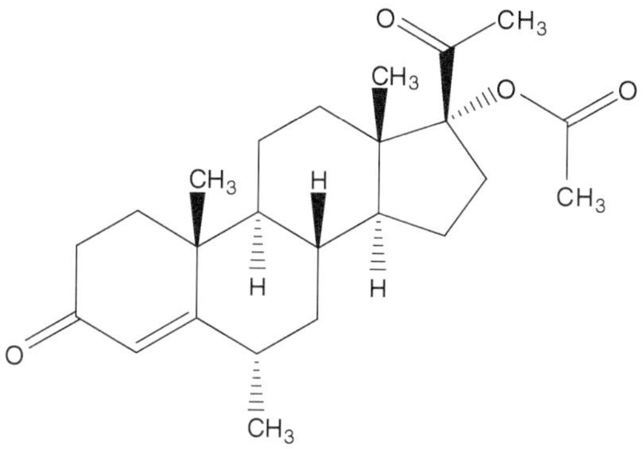

A single-enantiomer molecule with multiple chiral carbons is often formed by modification of a natural product which has most or all of the chiral carbons already in place. A 6α-methyl-4-ene-3-one steroid is often formed by hydrogenation of the 6-methylene-4-ene-3-one.

Discussion. In one preferred route, medroxyprogesterone acetate is manufactured in four steps from 17-α-acetoxyprogesterone, 11 steps from 16-dehydropregnenolone acetate, and 14 steps from diosgenin. Diosgenin is a phytosteroid sapogenin isolated from the tubers of *Discorea* wild yam.

In the final step, the α-methyl at C6 is formed by hydrogenation of the C6 methylene. The C6 methylene is formed by elimination of *N*-methylaniline. The 6-aminomethyl group is formed by reaction of the 3-ethoxy-3,5-diene with formaldehyde and *N*-methylaniline (**Mannich Reaction**). The 3-ethoxy-3,5-diene is formed by reaction of the 4-ene-3-one of 17-α-acetoxyprogesterone with triethyl orthoformate. The 4-ene-3-one of 17-α-acetoxyprogesterone is formed from 17-α-acetoxypregnenolone by oxidation of the secondary alcohol at C3 and migration of the double bond.

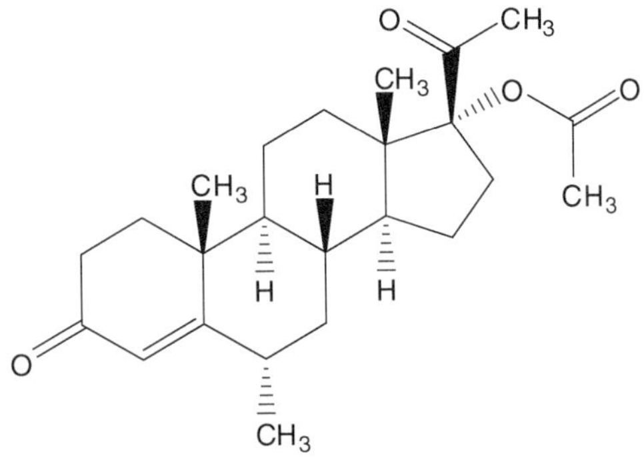

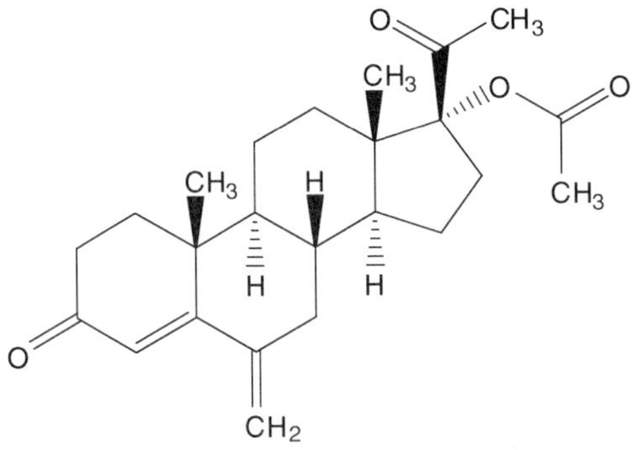

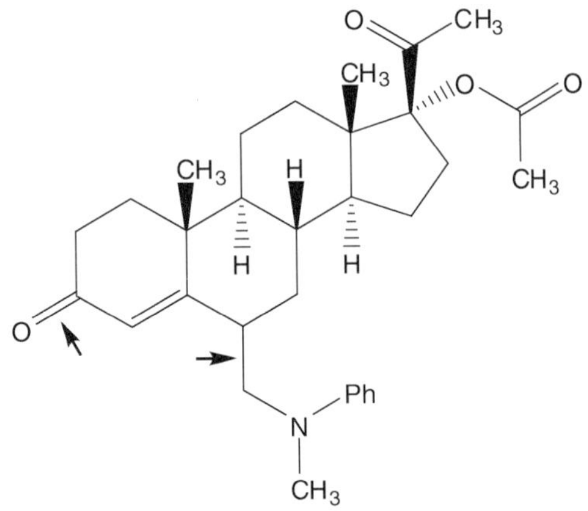

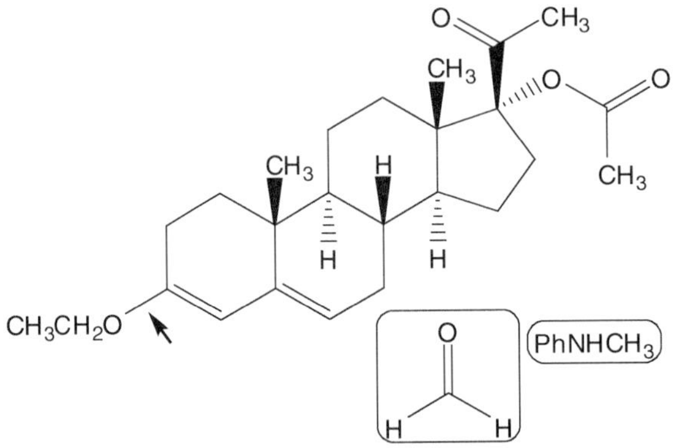

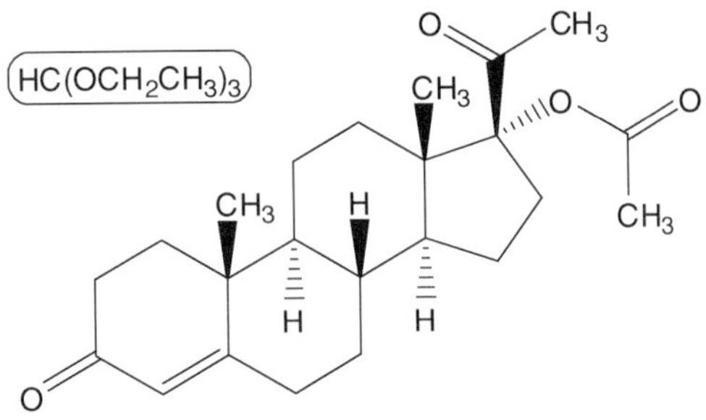

17α-acetoxyprogesterone

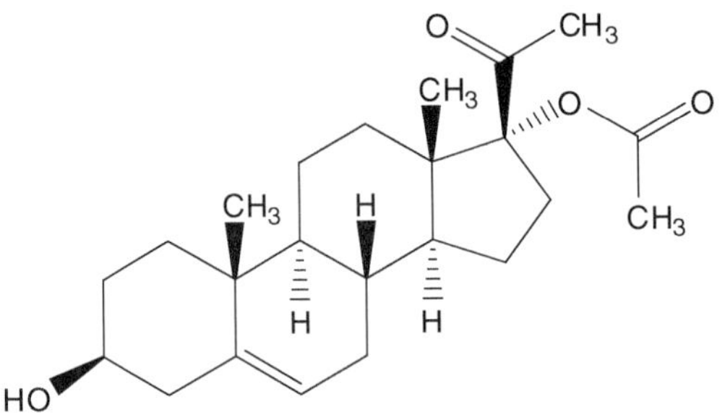

17α-acetoxypregnenolone

The secondary alcohol at C3 of 17α-acetoxypregnenolone is formed by hydrolysis of the acetate. The 3,17-diacetate is formed by reaction of the 3,17-diol with acetic anhydride. A β-bromine at C16 is removed by hydrogenolysis. The bromohydrin is formed by ring-opening of the 16α,17-epoxide. The 16α,17-epoxide is formed from the 16-alkene. The 3β-alcohol is formed by hydrolysis of the acetate ester of 16-dehydropregnenolone acetate under the epoxidation conditions.

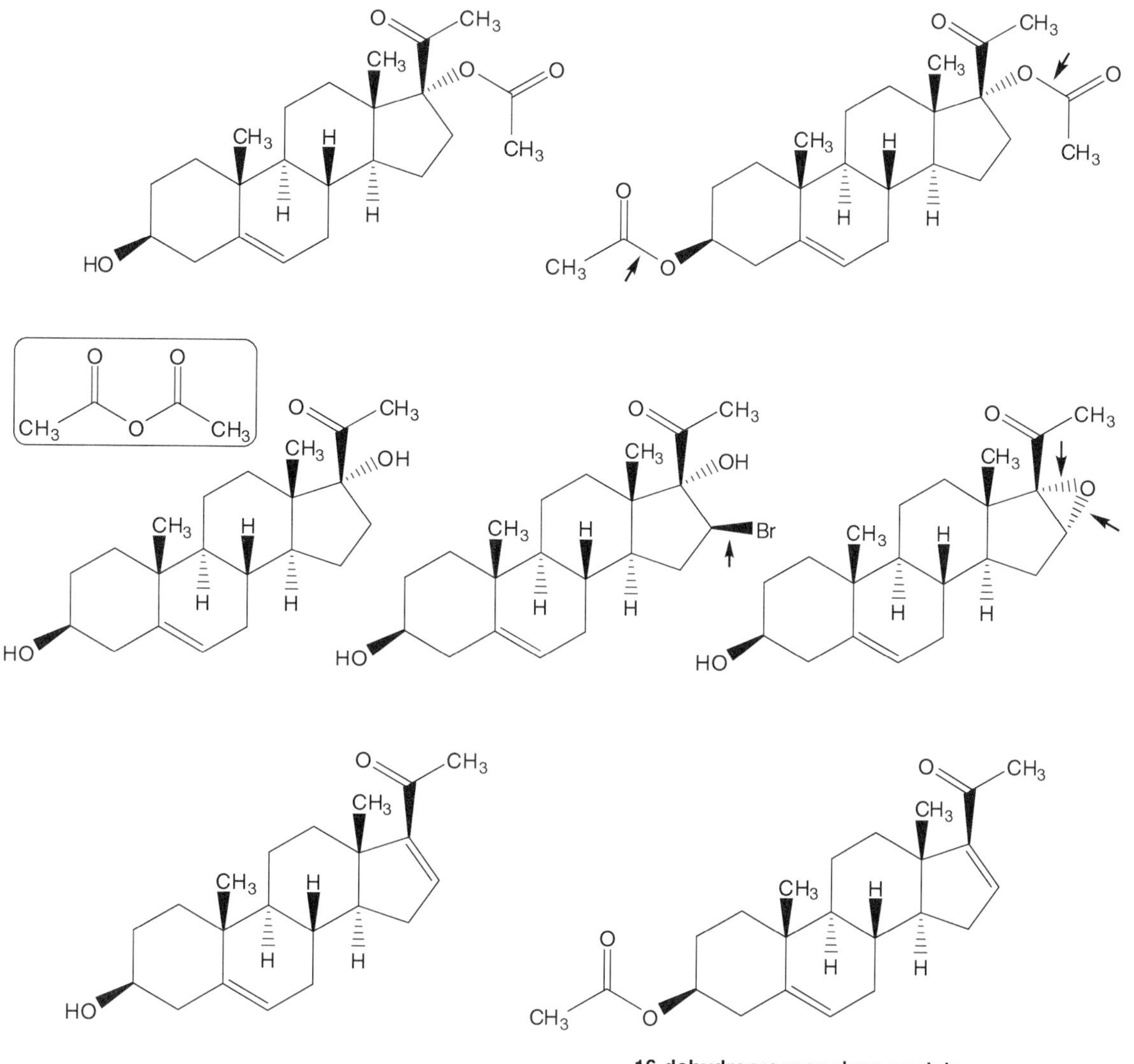

16-dehydropregnenolone acetate

The 16-alkene of 16-dehydropregnenolone acetate is formed from diosone by β-elimination. Diosone is formed by oxidation of the 20(22)-alkene of pseudodiosgenin-3,26-diacetate. Pseudodiosgenin 3,26-diacetate is formed by reaction of diosgenin with acetic anhydride. The three-step synthesis of 16-dehydropregnenolone acetate from diosgenin by acetylation, oxidation, and elimination is known as the **Marker Degradation**.

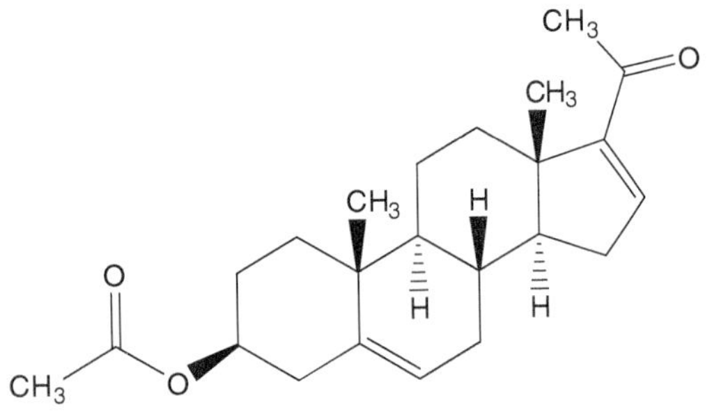

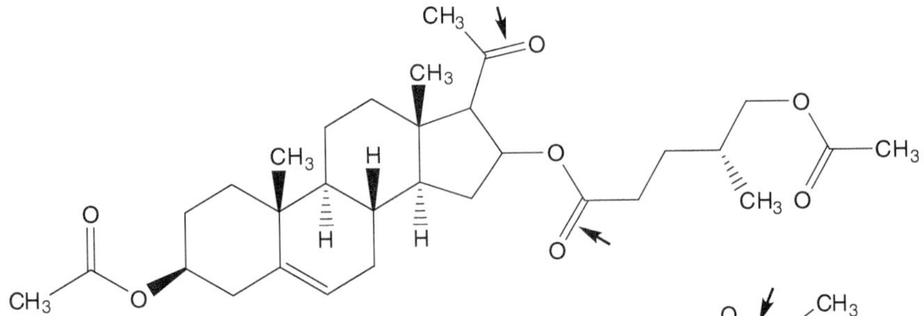

diosone

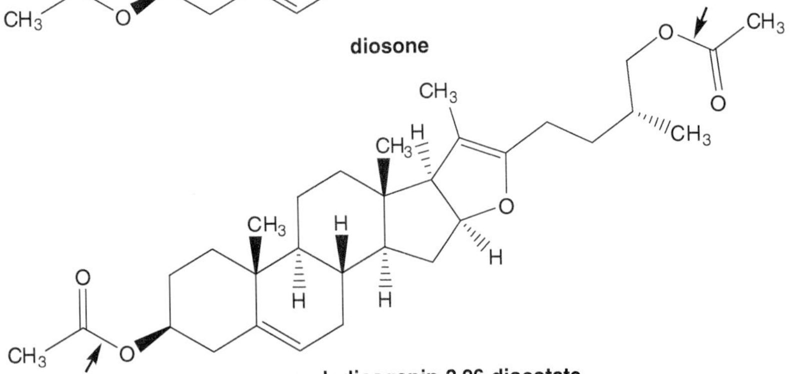

pseudodiosgenin-3,26-diacetate

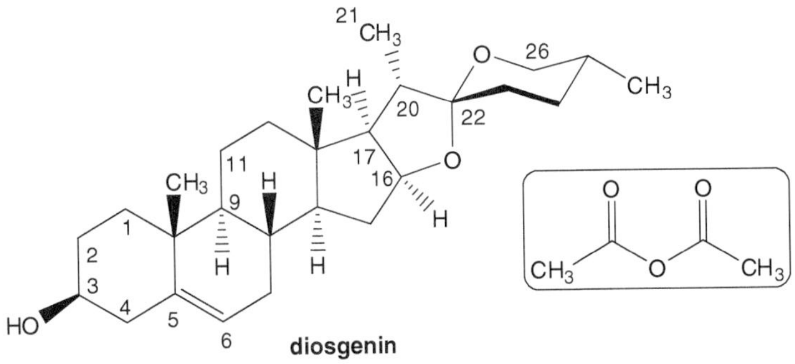

diosgenin

Extended Discussion

Draw the structures of the retrosynthetic analysis of an alternative route to medroxyprogesterone acetate that does not use a Mannich Reaction to introduce the methyl group at C6. List the pros and cons for both routes.

Mefloquine

Anti-infective Medicines/Antiprotozoal Medicines/Antimalarial Medicines/For Prophylaxis

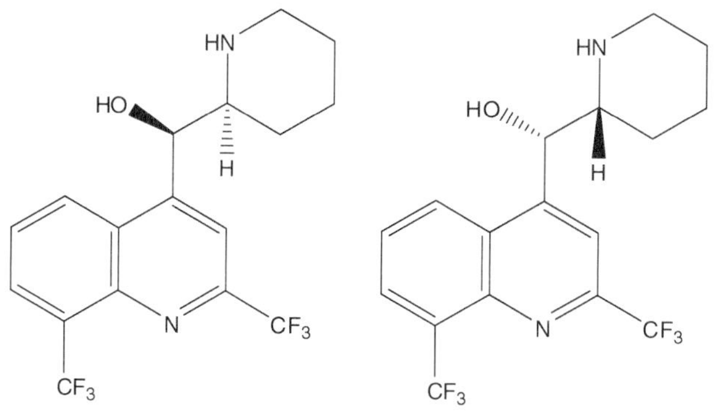

A fluorine-containing substituent on a heterocycle is often delivered in a starting material used to construct the heterocycle.

Discussion. Mefloquine is a 1 : 1 mixture of (11*S*,12*R*)- and (11*R*,12*S*)-enantiomers. The final step is a hydrogenation to form the secondary alcohol from the ketone and the piperidine from the pyridine. The ketone is formed from the cyanohydrin. The cyanohydrin is formed by α-oxidation of the nitrile. The nitrile is formed by displacement of chloride from the 4-chloroquinoline by 2-(2-pyridin-2-yl)acetonitrile. 2-(2-Pyridin-2-yl)acetonitrile is formed from 2-(chloromethyl)pyridine hydrochloride.

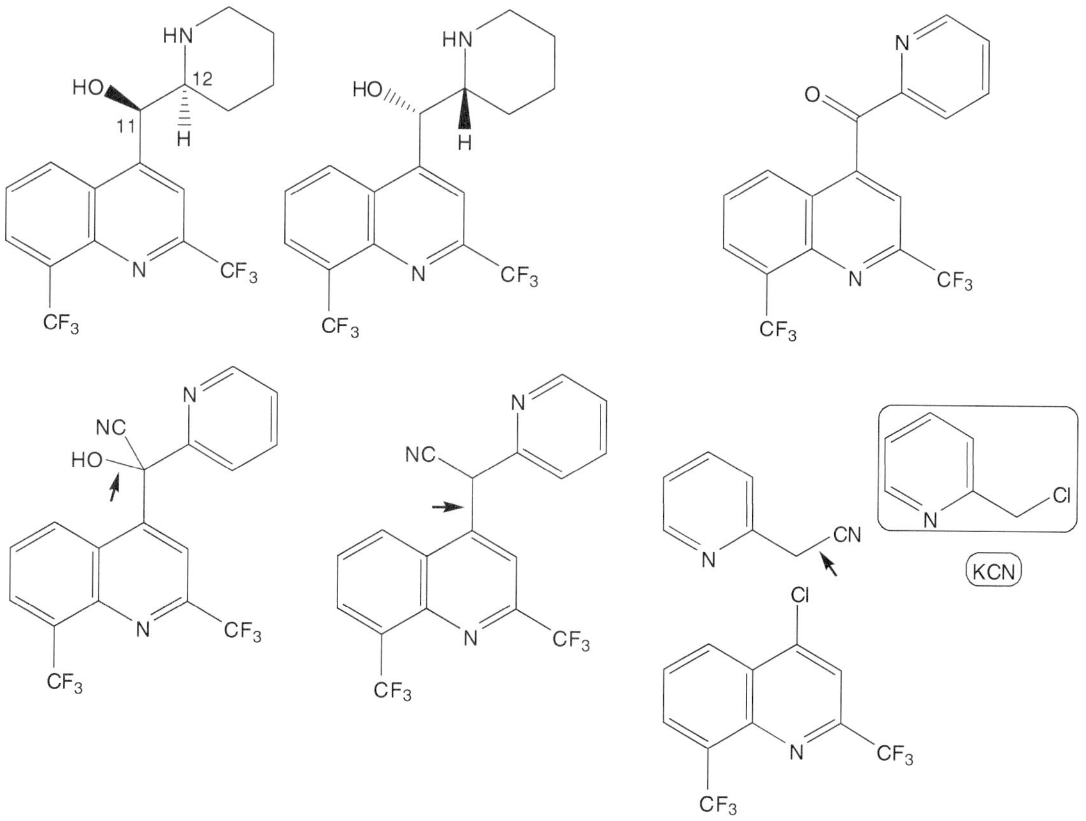

The 4-chloroquinoline is formed from the 4-hydroxyquinoline. The quinoline ring is formed by condensation of ethyl 4,4,4,-trifluoroacetoacetate with 2-trifluoromethylaniline (**Conrad–Limpach Synthesis**). Ethyl 4,4,4,-trifluoroacetoacetate is formed from ethyl trifluoroacetate and ethyl acetate (mixed **Claisen Condensation**). Ethyl trifluoroacetate is formed from trifluoroacetic acid and ethanol (**Fischer Esterification**).

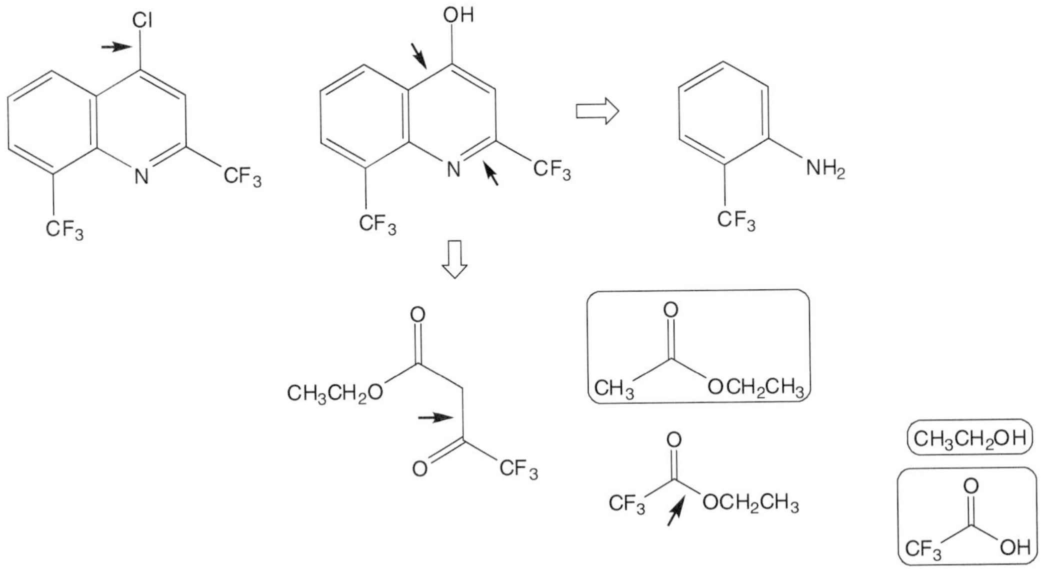

The synthesis of 2-trifluoromethylaniline from α,α,α-trifluorotoluene is accomplished in three steps: chlorination, nitration, and nitro group reduction/reductive dehalogenation.

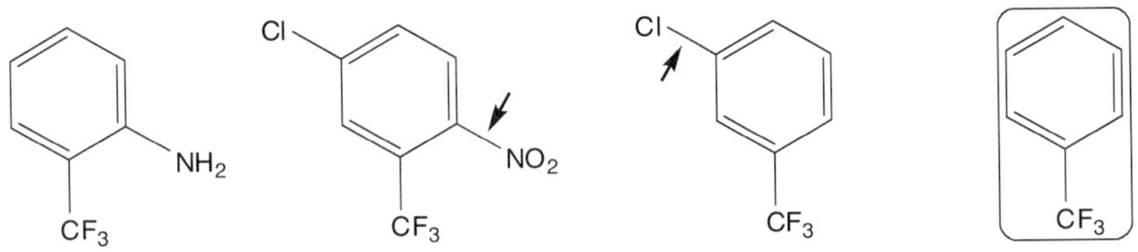

Extended Discussion

Draw the structures of the retrosynthetic analysis of an alternative route to mefloquin which forms the C11 alcohol by ring-opening of an epoxide. Include the structures of the retrosynthetic analysis of any organic starting material(s) from petrochemical or biochemical raw materials. List the pros and cons for both routes.

Meglumine Iotroxate

Diagnostic Agents/Radiocontrast Media

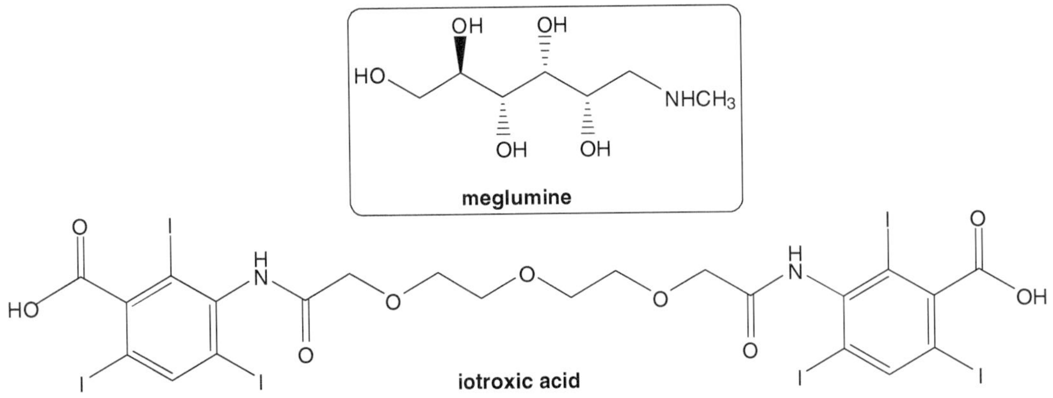

Amides are ubiquitous in drug structures. Amide formation from an amine and acid chloride, anhydride, ester, or carboxylic acid is often very efficient.

Discussion. Meglumine iotroxate is the salt formed from iotroxic acid and meglumine (*N*-methyl-D-glucamine). Iotroxic acid is formed from 3-amino-2,4,6-triiodobenzoic acid and the dicarboxylic acid dichloride. 3-Amino-2,4,6-triiodobenzoic acid is formed by iodination of 3-aminobenzoic acid. The dicarboxylic acid dichloride is formed from the dicarboxylic acid. 3,6,9-Trioxaundecanedioic acid is formed by oxidation of tetraethylene glycol. (List the known options for reagents and conditions for the oxidation of tetraethylene glycol to 3,6,9-trioxaundecanedioic acid. List the pros and cons for each option.)

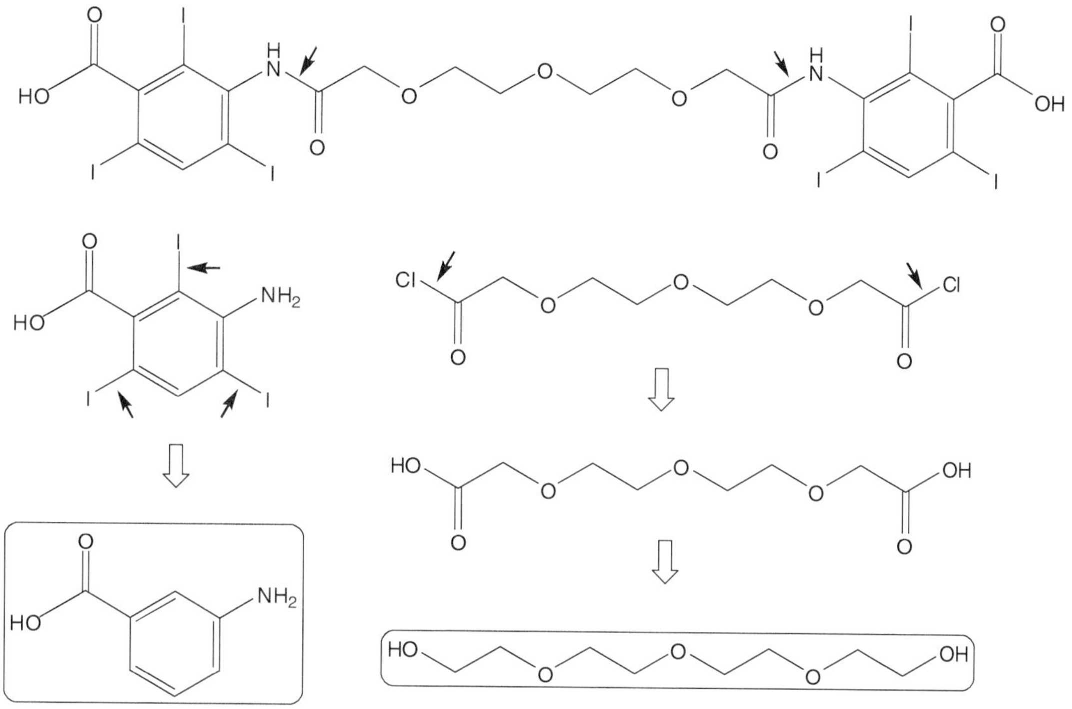

Melarsoprol

Anti-infective Medicines/Antiprotozoal Medicines/Antitrypanosomal Medicines/African Trypanosomiasis

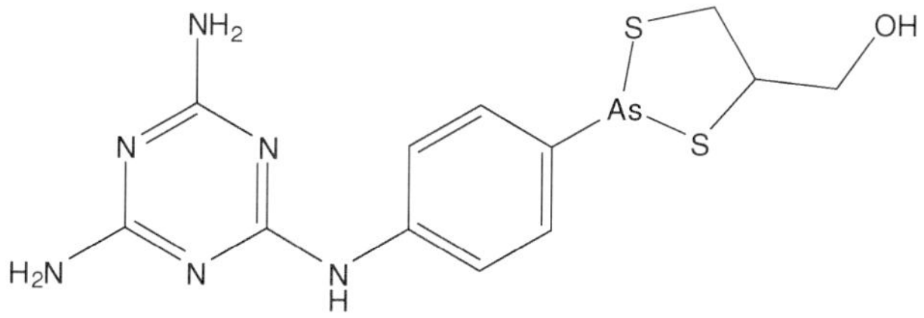

A nitrogen substituent on a ring carbon of 1,3,5-triazine is often introduced by displacement of chloride.

Discussion. Melarsoprol is a mixture of diastereoisomers. Equilibration of the diastereoisomers by slow inversion of the configuration at arsenic has been observed by NMR spectroscopy.

The 1,3,2-dithiarsolane ring of melarsoprol is formed by reaction of melarsen oxide with another essential medicine, dimercaprol. Melarsen oxide is formed by reduction of melarsen acid. Melarsen acid is formed by chloride displacement from 2,4-diamino-6-chloro-1,3,5-triazine by 4-aminophenylarsonic acid (*para*-arsanilic acid). *para*-Arsanilic acid is formed from aniline (**Bechamp Reaction**). 2,4-Diamino-6-chloro-1,3,5-triazine is formed from the trichlorotriazine (cyanuric chloride) by chloride displacement.

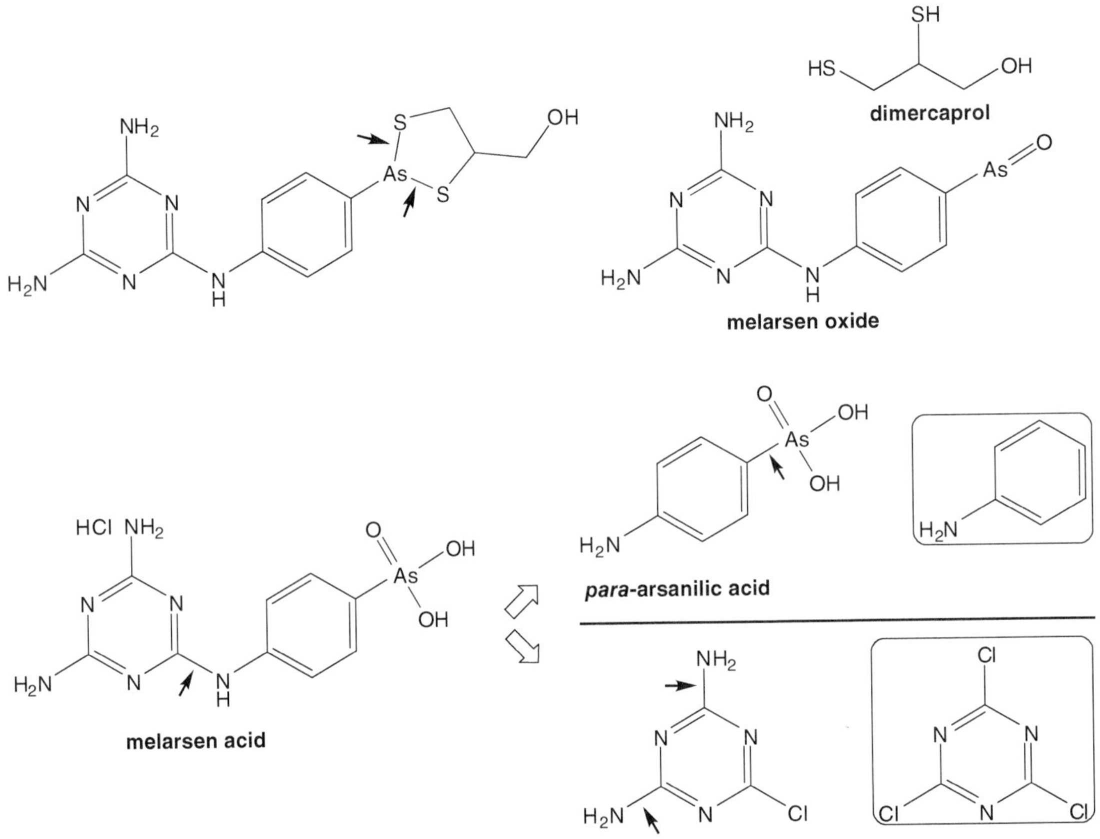

Extended Discussion

Draw the structures of the retrosynthetic analysis of an alternative route to melarsoprol from cyanuric chloride, aniline, and dimercaprol. List the pros and cons for both routes and select one route as the preferred route.

Mercaptopurine

Antineoplastics and Immunosuppressives/Cytotoxic and Adjuvant Medicines

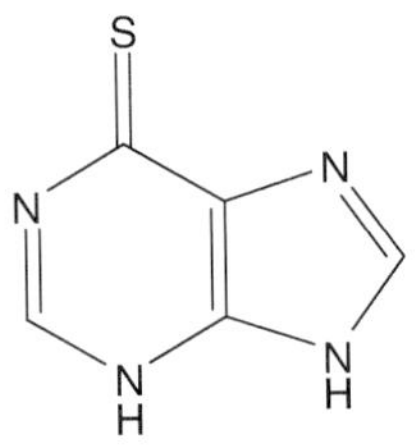

A urea, thiourea, or amidine often provides the carbon at position 2 of a pyrimidine.

Discussion. Sulfur replaces the oxygen of hypoxanthine in the final step. The simplicity of hypoxanthine suggests that functionality must be added, and later removed, to access an efficient route to the purine ring. Two routes to hypoxanthine will be presented.

In Route A, hypoxanthine is formed by hydrogenolysis of 2,8-dichlorohypoxanthine. Oxygen is introduced by displacement of the most reactive chlorine of 2,6,8-trichloropurine. The trichloropurine is formed from uric acid.

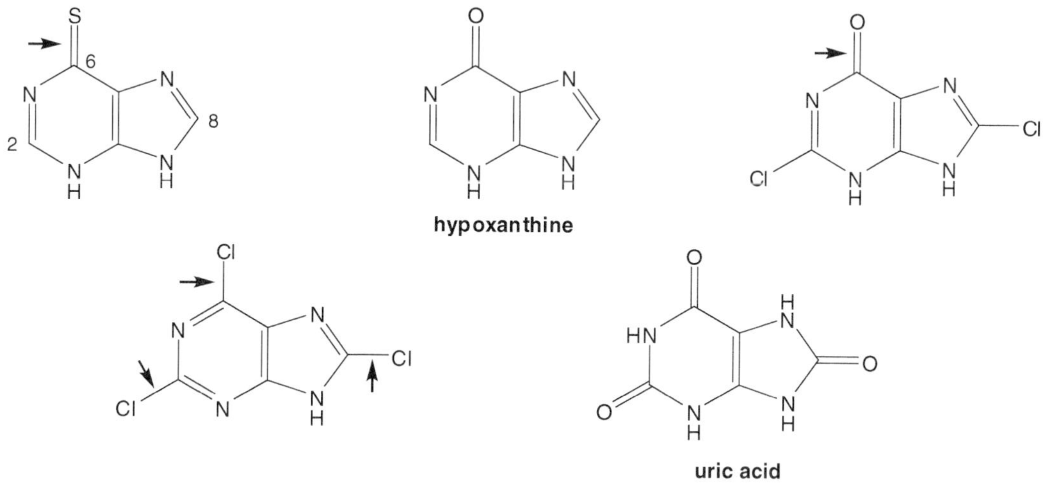

Uric acid is formed by dehydration of pseudouric acid. The urea side chain is formed by reaction of potassium cyanate with the amine of uramil. The amine is formed from the oxime of violuric acid, and the oxime is formed by nitrosation of barbituric acid. Barbituric acid is formed from urea and diethyl malonate.

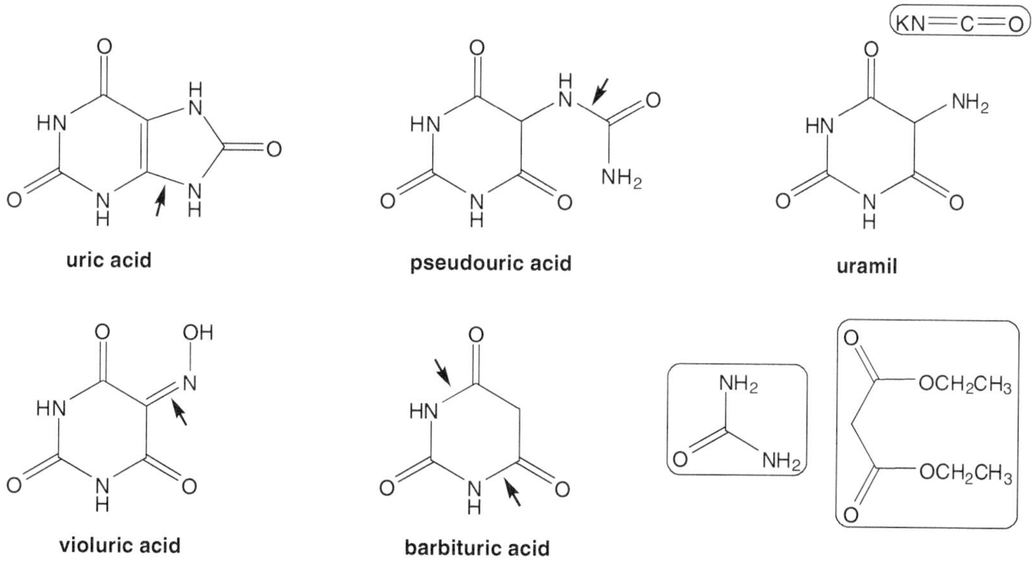

In Route B, hypoxanthine is formed by the reaction of 4,5-diamino-6-hydroxypyrimidine-2-sulfinic acid with formic acid. The sulfinic acid is formed by oxidation of the thiol. The 5-aminopyrimidine is formed by reduction of the 5-nitrosopyrimidine. The 5-nitrosopyrimidine is formed by nitrosation of 4-amino-6-hydroxy-2-mercaptopyrimidine. The pyrimidine ring is formed from thourea and ethyl cyanoacetate.

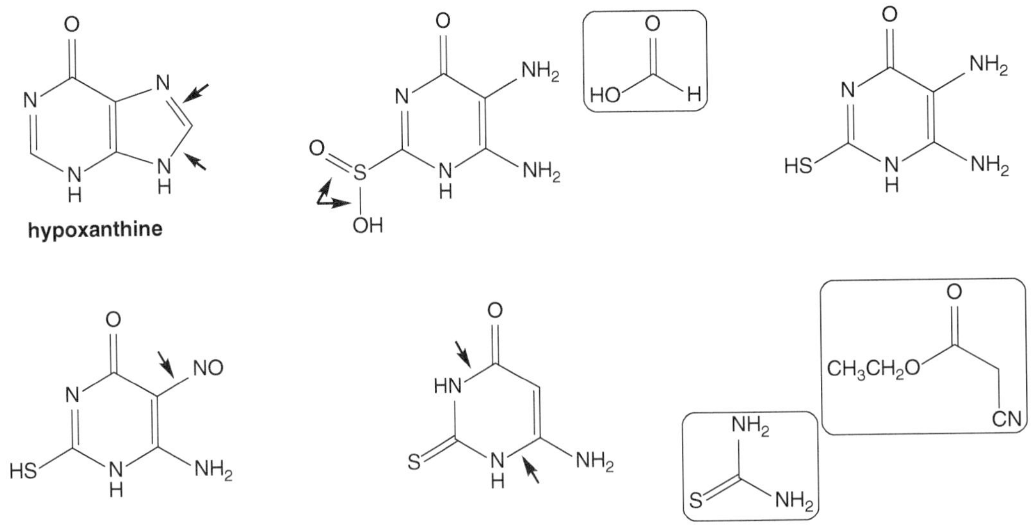

Extended Discussion

List pros and cons for the two routes to hypoxanthine and select one route as the preferred route.

Meropenem

Anti-infective Medicines/Antibacterials/β-Lactam Medicines

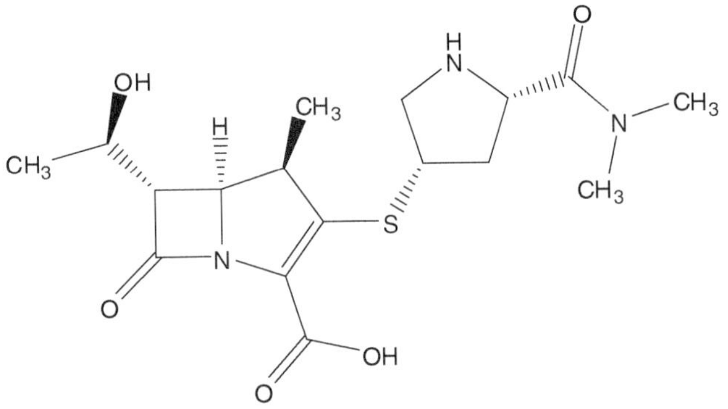

A thioether on the β-carbon of an acrylate ester is often formed by displacement of a leaving group by a thiol.

Discussion. Meropenem is constructed from (4*R*)-hydroxy-L-proline in seven steps and from methyl acetoacetate in 15 steps. The carboxylic acid and amine are released by hydrogenolysis of a *p*-nitrobenzyl (PNB) ester and a *p*-nitrobenzyl carbamate in the final step. The thioether is formed by displacement of diphenyl phosphate by the thiol. The enol phosphate is formed from the ketone. A C-N bond and the five-membered ring are formed by reaction of the azetidinone nitrogen with the α-diazo-β-ketoester. The α-diazo-β-ketoester is formed from the β-ketoester.

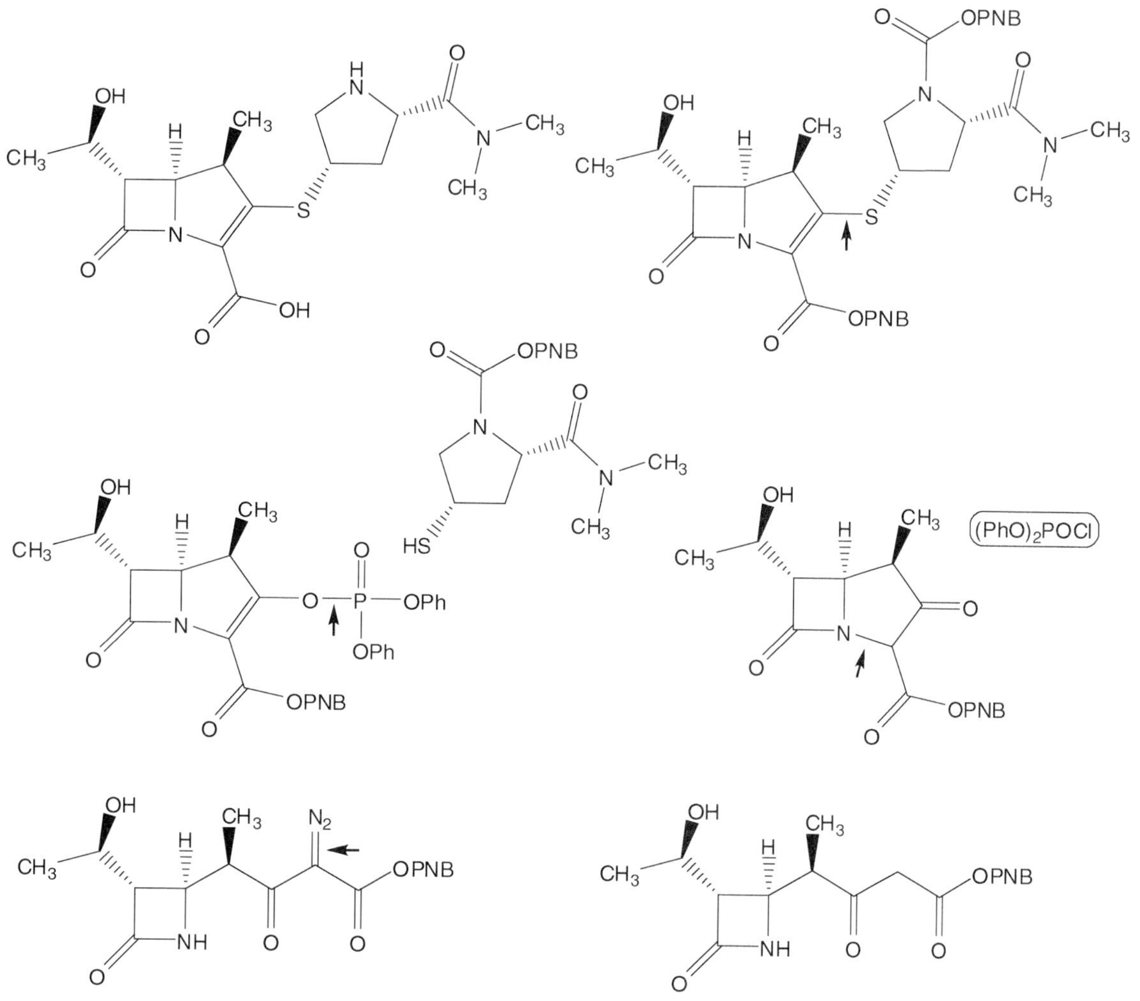

The alcohol is released from the *tert*-butyldimethylsilyl (TBDMS) ether. The β-ketoester is formed by reaction of an active ester, formed in situ from the carboxylic acid, with a magnesium monoalkyl malonate salt. The carboxylic acid is formed by hydrolysis of an imide. The *trans*-3,4-disubstituted azetidinone is formed by displacement of acetate from the 4-acetoxyazetidinone by an organozinc reagent. (The reaction forms two diastereomers, with the methyl group on the β- and α-face. What conditions are associated with the highest β to α ratio?) The organozinc reagent is formed from the α-bromoimide. (Draw the structure of another organometallic reagent used to displace acetate from the 4-acetoxyazetidinone in an alternative synthesis of meropenem.)

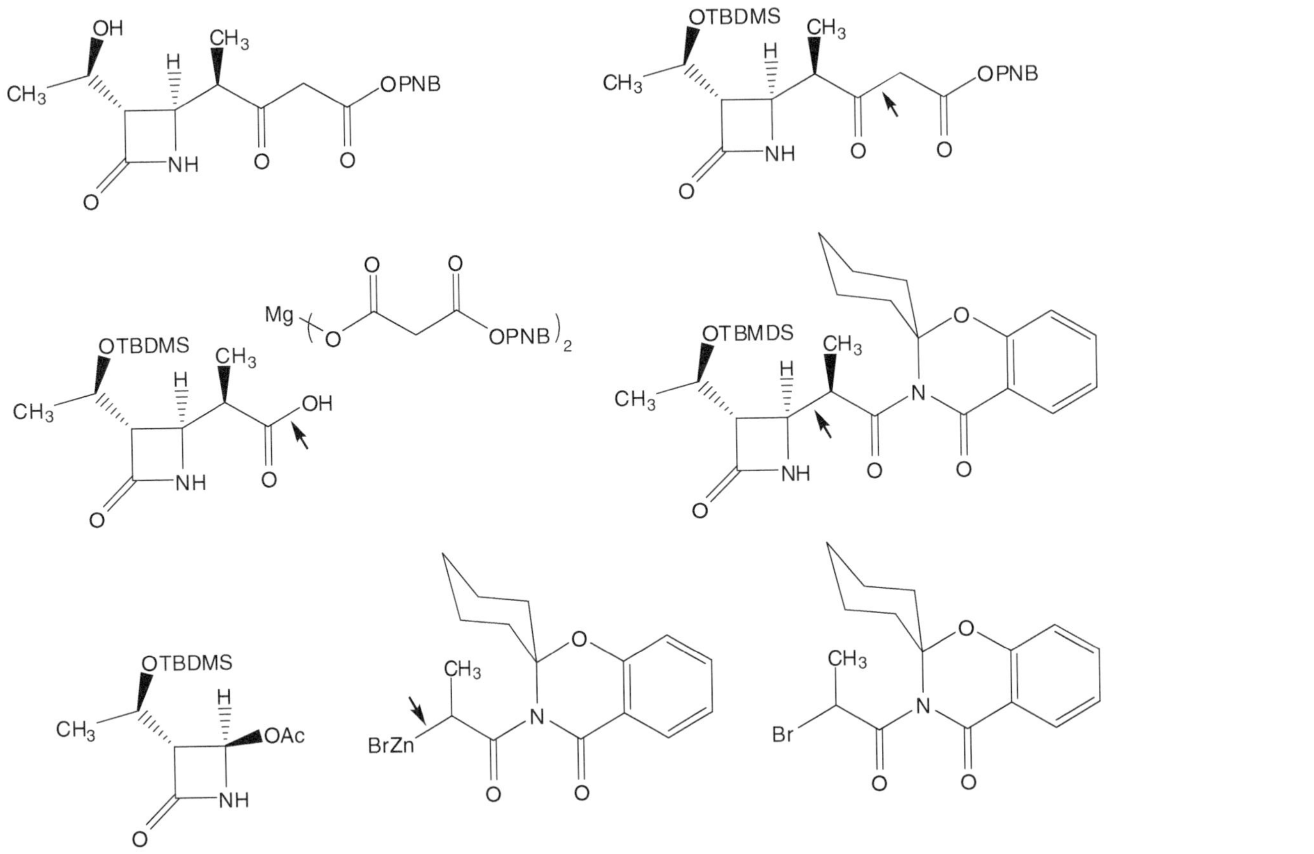

The 4-acetoxyazetidinone is formed by oxidation of the azetidinone. The alcohol is protected as a *tert*-butyldimethylsilyl (TBDMS) ether. The azetidinone ring is formed from the β-aminocarboxylic acid. The amine and carboxylic acid are formed by hydrolysis of an amide and an ester. Two chiral carbons are set by asymmetric hydrogenation of the α-amidoalkyl-β-ketoester using a chiral ruthenium catalyst (**Noyori Asymmetric Hydrogenation**). (What are the highest diastereomeric excess (de) and enantiomeric excess (ee) for the hydrogenation? Draw the structure of the ruthenium catalyst associated with the highest de and ee.) The α-amidoalkyl-β-ketoester is formed by alkylation of the β-ketoester with *N*-chloromethylbenzamide. *N*-Chloromethylbenzamide is formed from *N*-hydroxymethylbenzamide.

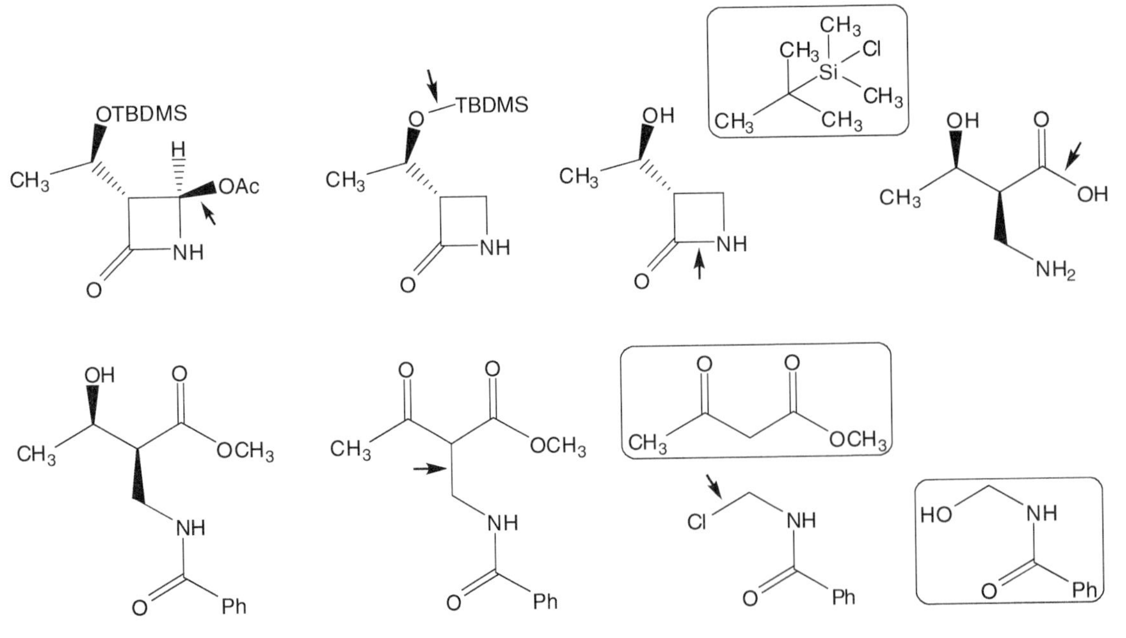

The α-bromoimide is formed from 2-bromopropanoyl bromide and the 1,3-benzoxazine-4(3*H*)-one. Spiro[2*H*-1,3-benzoxazine-2,1′-cyclohexan]-4(3*H*)-one is formed from salicylamide and cyclohexanone.

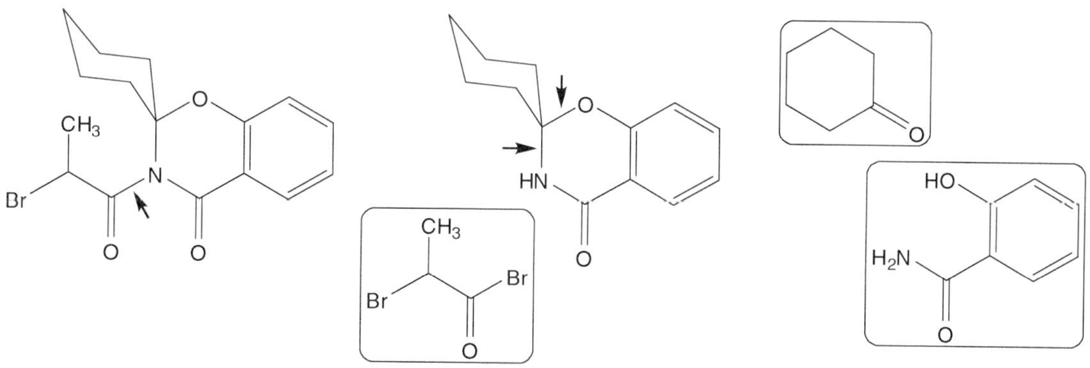

The thiol is formed by hydrolysis of the thioester. The thioester is formed by displacement of methanesulfonate by thioacetic acid. The methanesulfonate is formed from the alcohol. The amide is formed by reaction of an active ester, formed in situ from the carboxylic acid, and dimethylamine. The nitrogen of (4*R*)-hydroxy-L-proline is protected as the PNB carbamate by reaction with 4-nitrobenzyl chloroformate. (4*R*)-Hydroxy-L-proline is produced by fermentation.

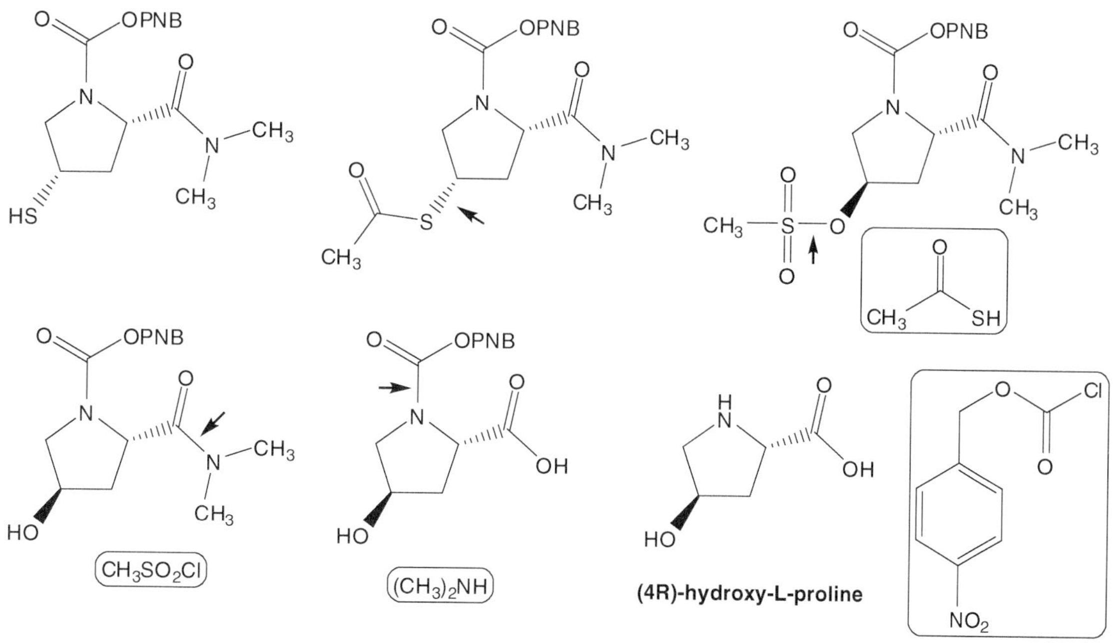

The magnesium carboxylate salt is formed from the monoalkyl malonate. The monoalkyl malonate is formed from malonic acid and 4-nitrobenzyl alcohol.

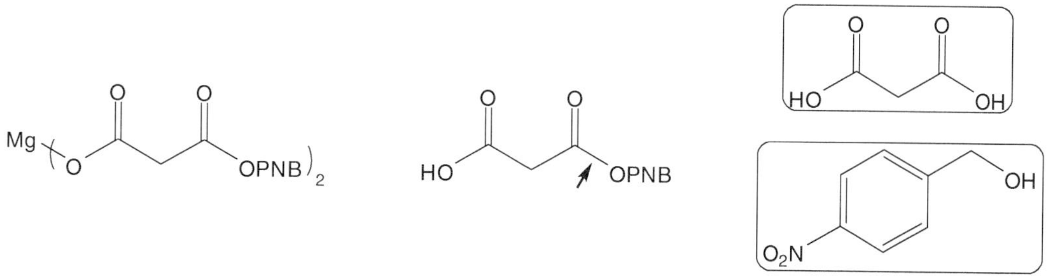

Extended Discussion

Draw the structures of the retrosynthetic analysis of an alternative route to meropenem from an alternative thioester intermediate. Include the structures of the retrosynthetic analysis of any organic starting material(s) from petrochemical or biochemical raw materials. List the pros and cons for both routes.

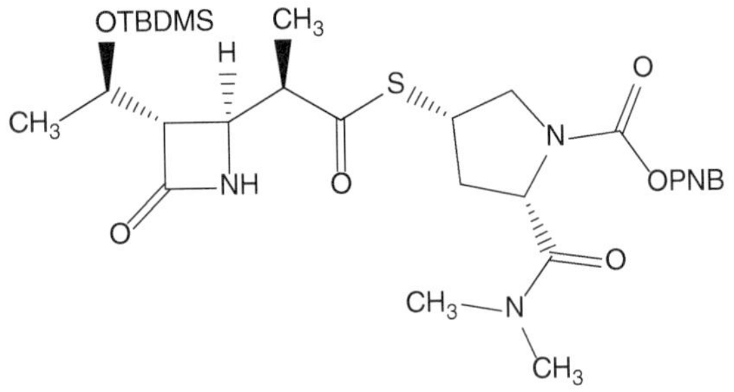

Mesna

Antineoplastics and Immunosuppressives/Cytotoxic and Adjuvant Medicines

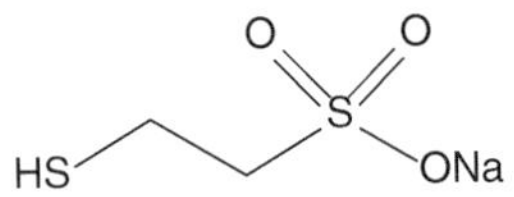

The C-S bond of a thiol is often formed by displacement of a leaving group by sulfur. Reagents providing the sulfur atom include thiourea, potassium ethyl xanthate, sodium trithiocarbonate, and thioacetic acid.

Discussion. Three routes to mesna will be considered. In Route A, mesna is formed from the guanidinium salt by ion exchange. The guanidinium salt is formed from the thiouronium salt. The thiouronium salt is formed by displacement of the bromide of sodium 2-bromoethanesulfonate by thiourea. Sodium 2-bromoethanesulfonate is formed from 1,2-dibromoethane (**Strecker Sulfite Alkylation**).

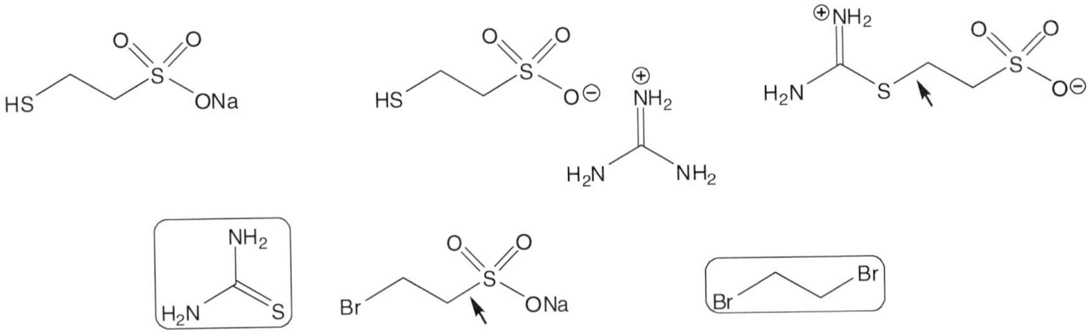

In Route B, the thiol group of mesna is released in the final step by hydrolysis of the thioacetate. The thioacetate is formed by displacement of the bromide of sodium 2-bromoethanesulfonic acid by thioacetic acid.

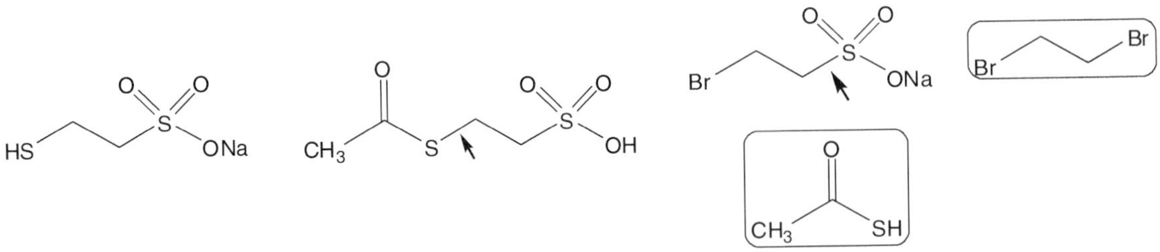

Extended Discussion

Draw the structures of the retrosynthetic analysis of alternative Route C to mesna. Include the structures of the retrosynthetic analysis of any organic starting material(s) from petrochemical or biochemical raw materials. List the pros and cons for the three routes. Is one route preferred?

Metformin

Hormones, Other Endocrine Medicines, and Contraceptives/Insulins and Other Medicines Used for Diabetes

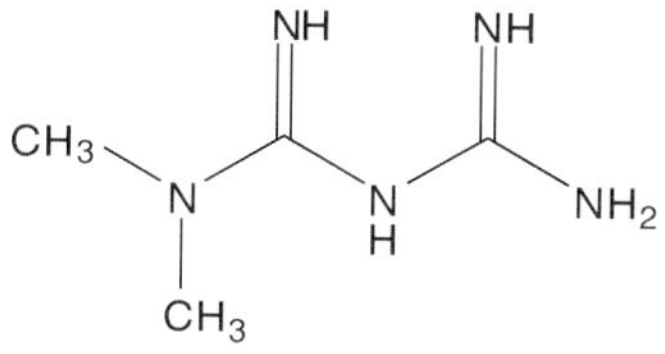

A biguanide is often formed by addition of an amine to the nitrile of cyanoguanidine.

Discussion. Metformin is a 1,1-disubstituted biguanide. The biguanide of metformin is formed in a single step by addition of dimethylamine to the nitrile of cyanoguanidine (dicyandiamide).

Methadone

Medicines for Pain and Palliative Care/Opioid Analgesics
Medicines for Mental and Behavioral Disorders/Medicines for Disorders due to Psychoactive Substance Abuse

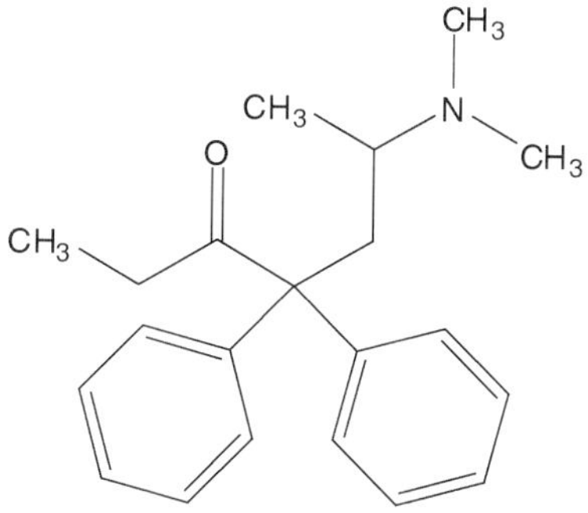

A ketone is often formed by addition of a Grignard reagent to a nitrile.

Discussion. Methadone is a 1:1 mixture of (*R*)- and (*S*)-enantiomers. Methadone is formed in just two steps. The ketone is formed by addition of ethylmagnesium bromide to the nitrile (**Grignard Reaction**). Ethylmagnesium bromide is formed from bromoethane. The nitrile is formed by α-alkylation of diphenylacetonitrile by 2-chloro-*N*,*N*-dimethyl-1-propanamine. (Draw the structure of a side product formed in this reaction. What is the ratio of product to side product? How are the product and side product separated?)

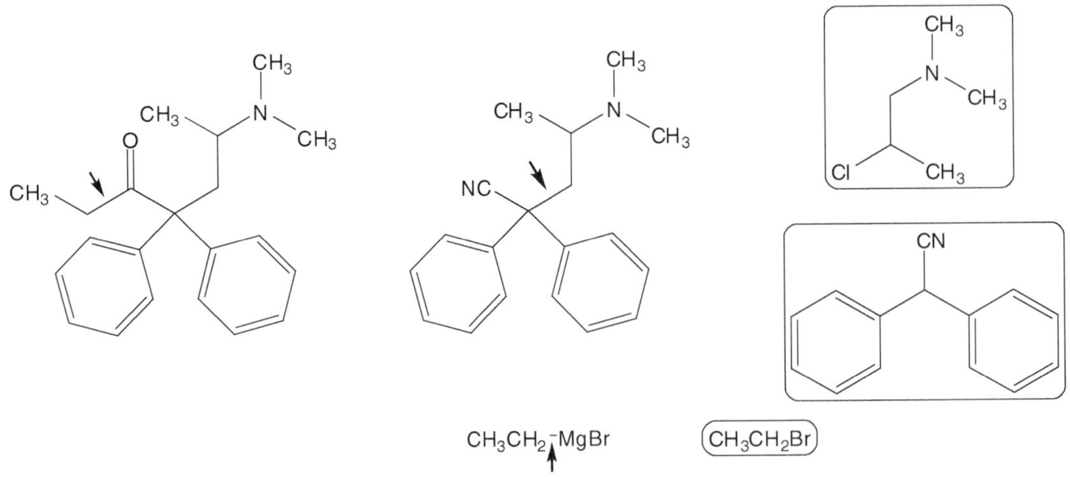

Extended Discussion

One route to diphenylacetonitrile is presented. Draw the structures of the retrosynthetic analysis of one alternative route. Include the structures of the retrosynthetic analysis of any organic starting material(s) from petrochemical or biochemical raw materials. List the pros and cons for both routes and select one route as the preferred route to diphenylacetonitrile.

Methotrexate

Antineoplastics and Immunosuppressives/Cytotoxic and Adjuvant Medicines
Medicines for Diseases of Joints/Disease-modifying Agents Used in Rheumatoid Disorders

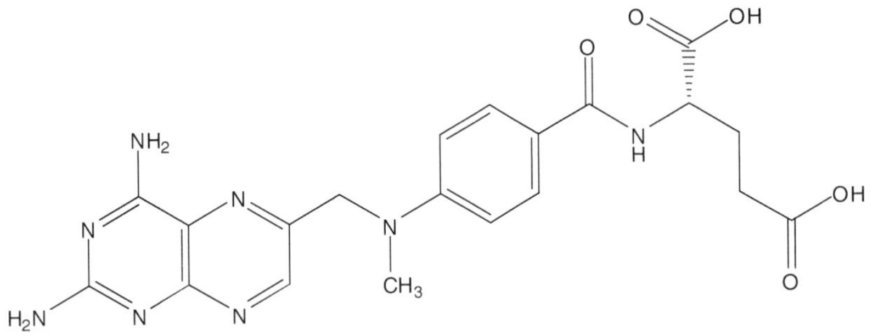

2,4-Diaminopteridines are often formed by condensation of 2,4,5,6-tetraaminopyrimidine with an α-halo- or α-hydroxyaldehyde or ketone.

Discussion. The tertiary amine is formed in the final step by a bromide displacement by a secondary amine. (What are preferred conditions for accomplishing this displacement with minimal racemization?)

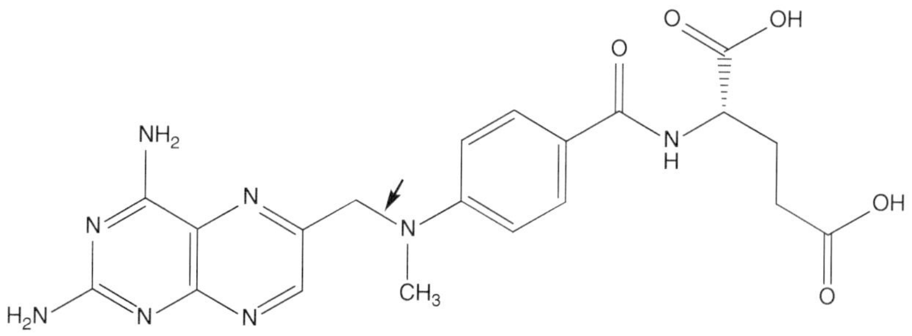

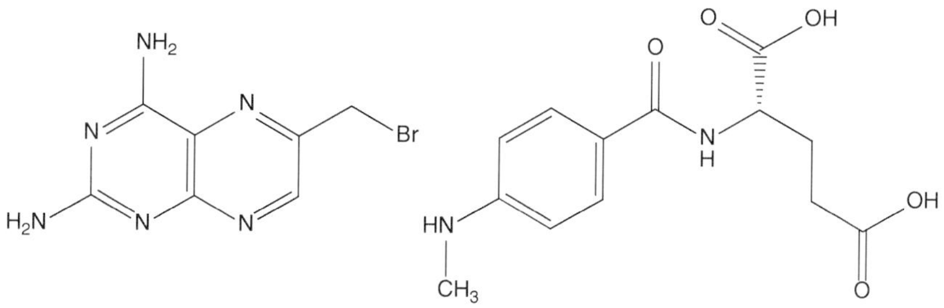

2,4-Diamino-6-bromomethylpteridine is formed from the 6-hydroxymethylpteridine. The 6-hydroxymethylpteridine is formed by condensation of 2,4,5,6-tetraaminopyrimidine with 1,3-dihydroxyacetone. (Draw the structures of two impurities often found in 2,4-diamino-6-hydroxymethylpteridine.) The tetraaminopyrimidine is formed by reduction of 2,4,6-triamino -5-nitrosopyrimidine. The nitrosopyrimidine is formed by nitrosation of 2,4,6-triaminopyrimidine.

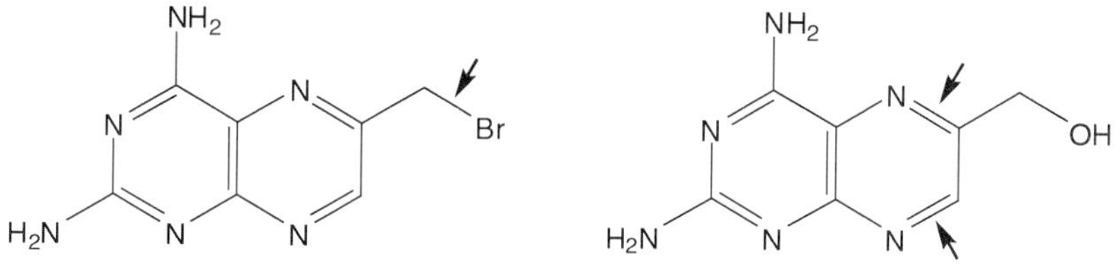

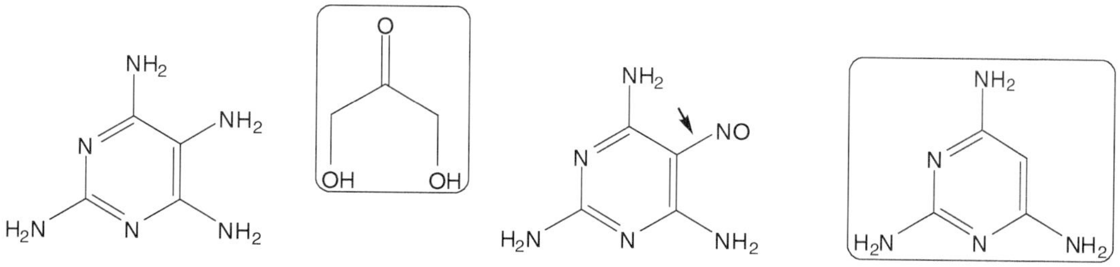

The amine of the 4-(*N*-methylamino)benzamide is released by hydrolysis of the formamide. The benzamide is formed from the acid chloride and L-glutamic acid. The acid chloride is formed from the carboxylic acid. The amine is protected as the formamide by reaction with formic acid. L-Glutamic acid is produced by fermentation.

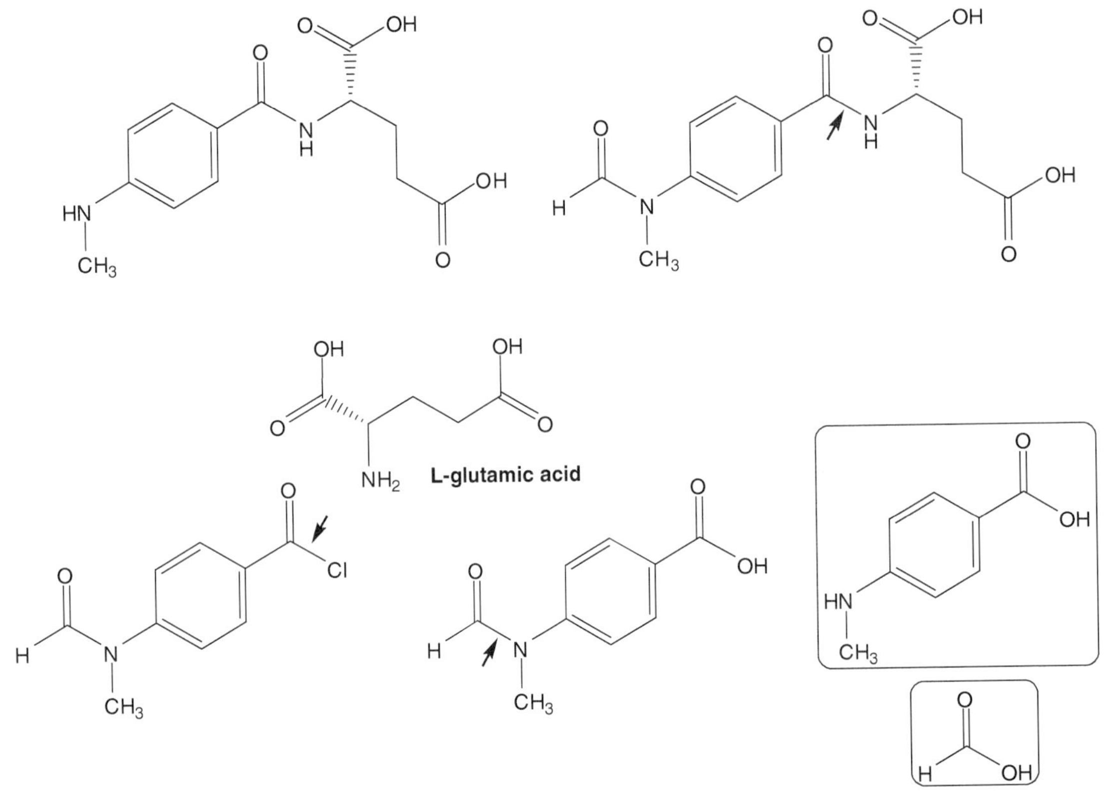

Extended Discussion

Draw the structures of the retrosynthetic analysis of one route to 4-(*N*-methylamino)benzoic acid. Include the structures of the retrosynthetic analysis of any organic starting material(s) from petrochemical or biochemical raw materials.

Methyldopa

Cardiovascular Medicines/Antihypertensive Medicines

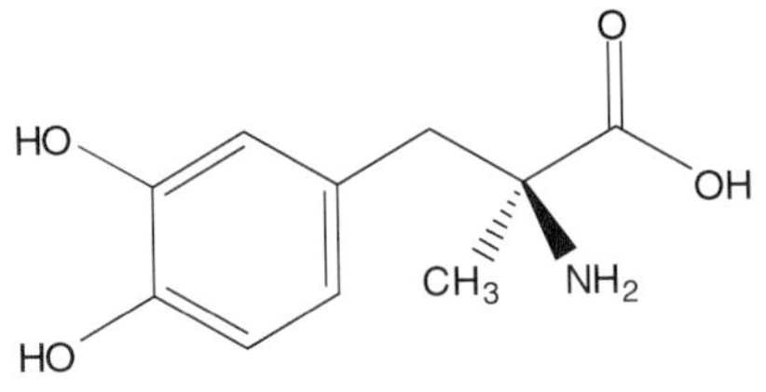

An α-amino acid is often formed by hydrolysis of an α-aminonitrile. An α-aminonitrile is often formed from an aldehyde or ketone.

Discussion. Methyldopa is formed by nitrile hydrolysis and release of the alcohols by *O*-demethylation in the final step. The (*S*)-α-aminonitrile is separated from the racemic mixture by resolution. (The (*R*)-α-aminonitrile is converted to the 1:1 mixture of (*R*)- and (*S*)-enantiomers. What conditions are required for the racemization?) The racemic α-aminonitrile is formed from (3,4-dimethoxyphenyl)acetone and potassium cyanide (**Strecker Synthesis**). (3,4-Dimethoxyphenyl) acetone is formed by hydrolysis of the oxime. The oxime is formed by reduction of the nitroalkene. The nitroalkene is formed from 3,4-dimethoxybenzaldehyde (veratraldehyde) and nitroethane (**Henry Reaction**). Veratraldehyde is formed by *O*-methylation of vanillin.

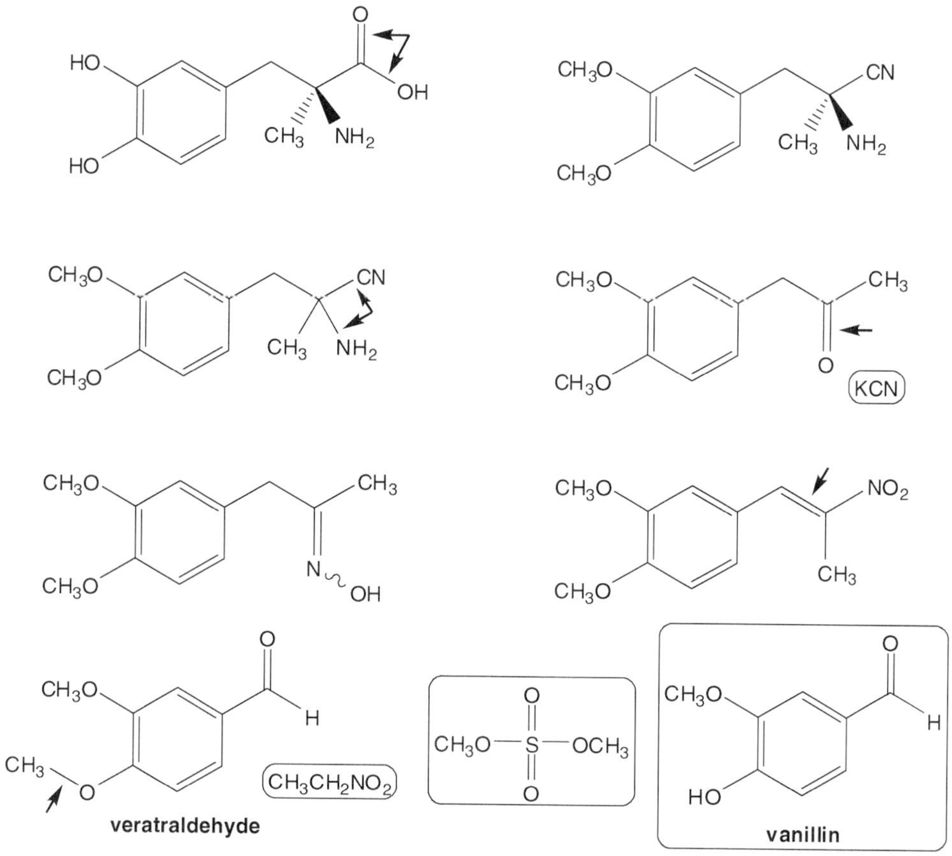

Extended Discussion

Draw the structures of the retrosynthetic analysis of one alternative route to (3,4-dimethoxyphenyl)acetone from veratraldehyde. Include the structures of the retrosynthetic analysis of any organic starting material(s) from petrochemical or biochemical raw materials.

Methylprednisolone

Antineoplastics and Immunosuppressives/Hormones and Antihormones

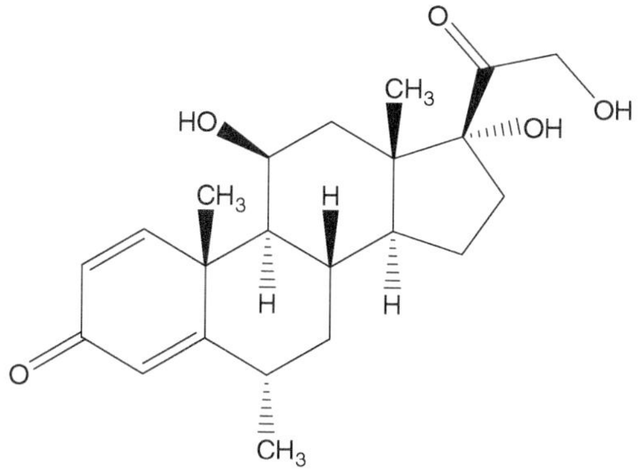

A single-enantiomer molecule with multiple chiral carbons is often formed by modification of a natural product which has most or all of the chiral carbons already in place. A methyl substituent at C6 of a steroid is often introduced by ring-opening of a 5,6-epoxide with a methylmagnesium halide.

Discussion. In one preferred route, methylprednisolone is manufactured in nine steps from 21-deoxycortisone, 15 steps from 16-dehydropregnenolone acetate, and 18 steps from diosgenin. Diosgenin is a phytosteroid sapogenin isolated from the tubers of *Discorea* wild yam.

The alcohol at C21 of methylprednisolone is released by hydrolysis of the acetate ester in the final step. The acetate ester is formed by displacement of iodide from the α-iodoketone. A mixture of the α-iodoketone and α,α-diiodoketone is formed by α-iodination of the methyl ketone. The 1,4-diene-3-one is formed by microbial dehydrogenation of the 4-ene-3-one of medrane. The ketones at C3 and C20 of medrane are released by hydrolysis of the acetals. The 4,5-alkene is formed by elimination from the alcohol at C5 under the acetal hydrolysis conditions.

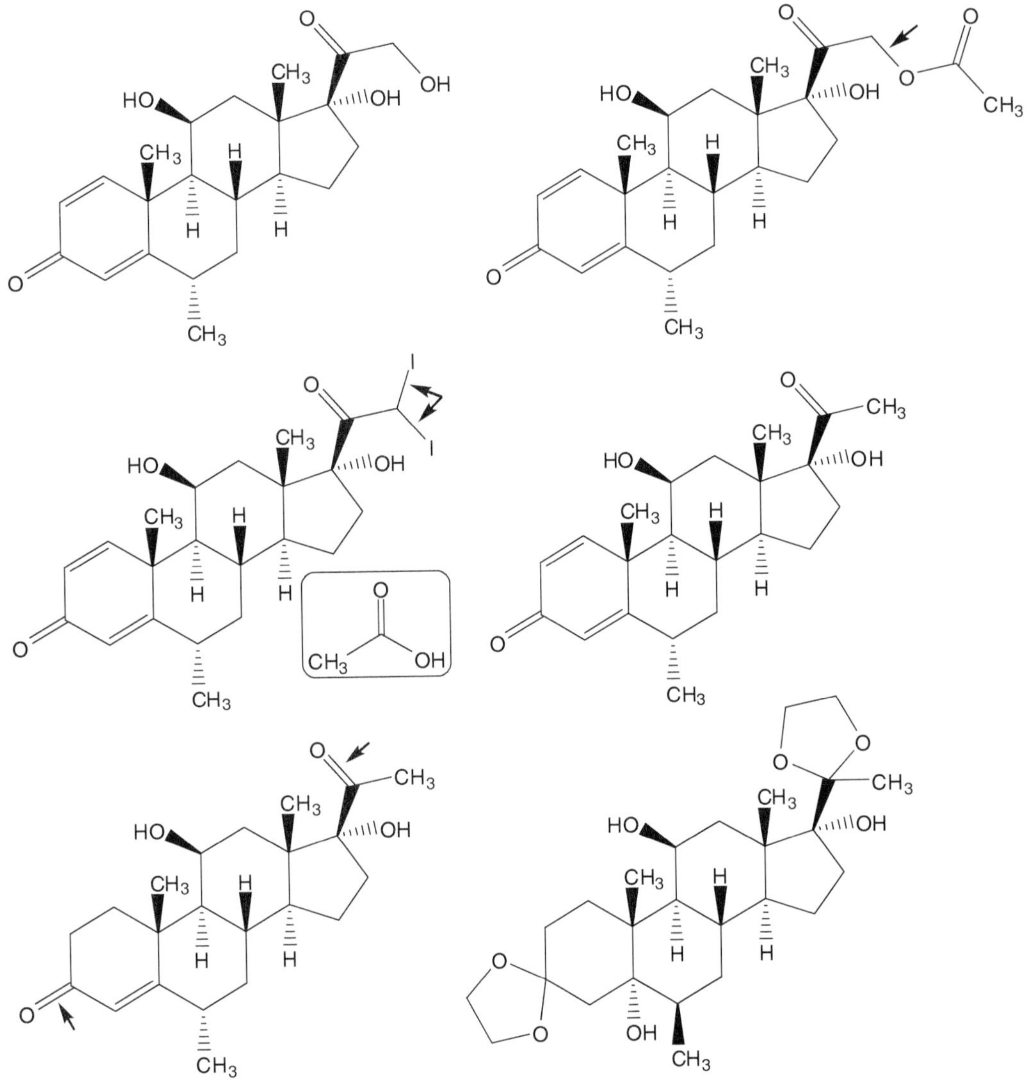

The methyl group at C6 and the alcohol at C5 are formed by ring-opening of the 5α,6-epoxide by methylmagnesium bromide (**Grignard Reaction**). The Grignard reagent is formed from bromomethane. The C11 β-alcohol is formed by reduction of the ketone. The 5α,6-epoxide is formed by epoxidation of the alkene. (What is the ratio of 5α,6-epoxide to 5β,6-epoxide from the epoxidation? How is the 5α,6-epoxide separated from the 5β,6-epoxide? What is the fate of the 5β,6-epoxide?) The acetals are formed by reaction of the ketones at C3 and C20 of 21-deoxycortisone with ethylene glycol.

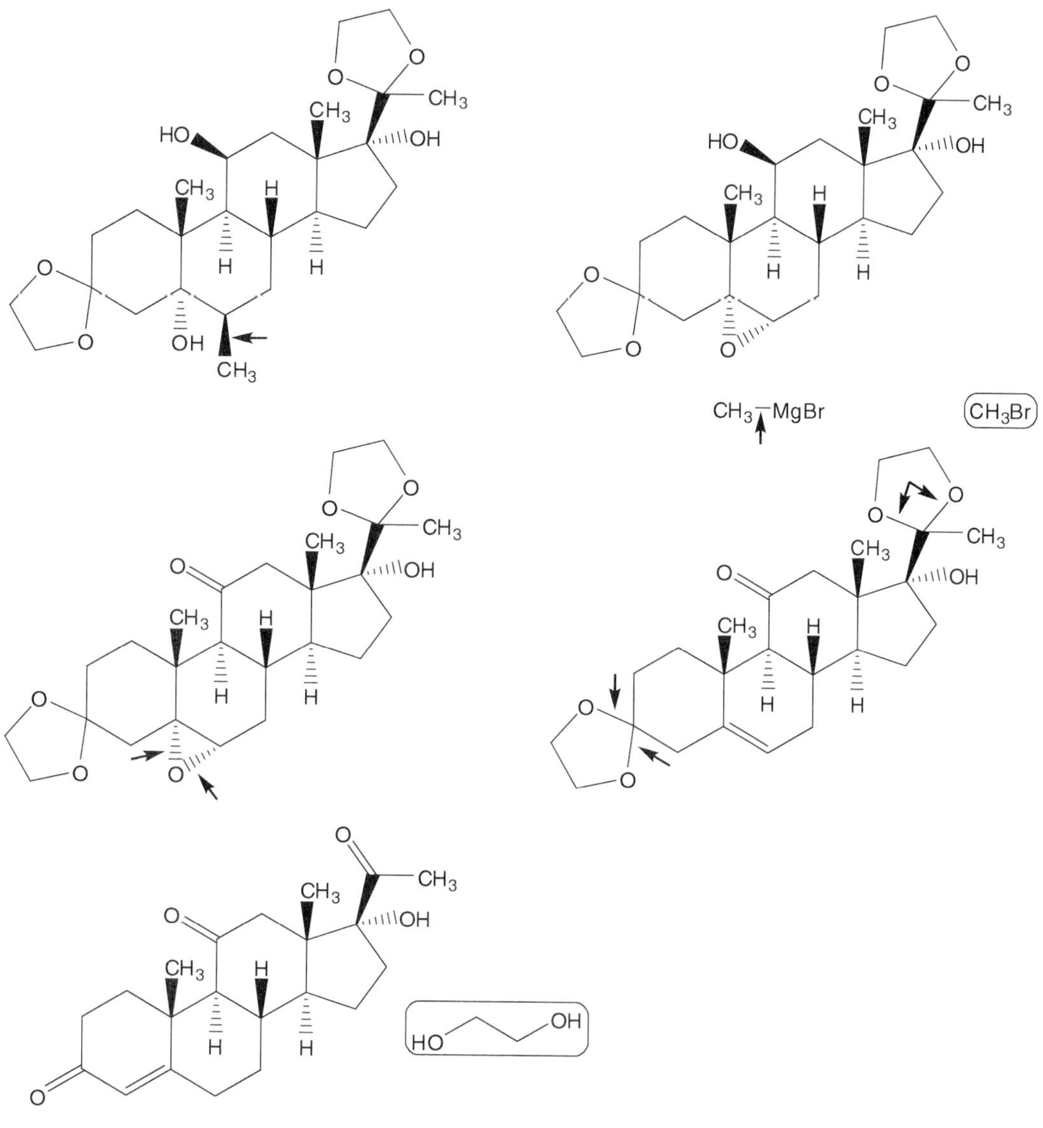

21-deoxycortisone

21-Deoxycortisone is formed by hydrogenolysis of the 16β-bromide. The bromohydrin is formed by ring-opening of the 16α,17-epoxide with hydrogen bromide. The C11 ketone is formed by oxidation of the C11 α-alcohol. The C11 α-alcohol is formed by microbial oxidation. The 4-ene-3-one is formed by oxidation of the alcohol at C3 followed by double bond migration. The 16α,17-epoxide is formed by epoxidation of the 16-alkene of 16-dehydropregnenolone acetate. The alcohol at C3 is formed by hydrolysis of the acetate ester under the epoxidation conditions.

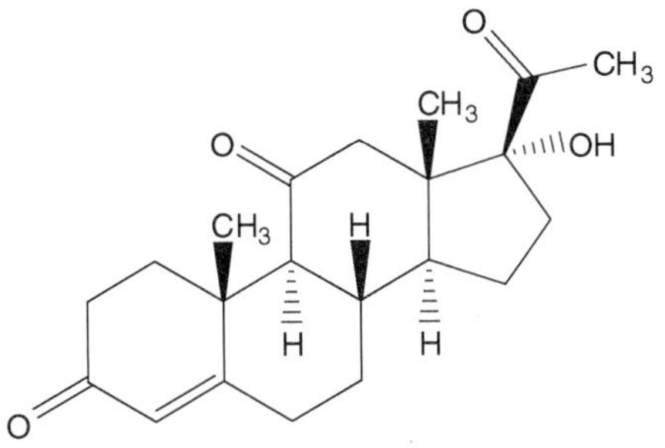

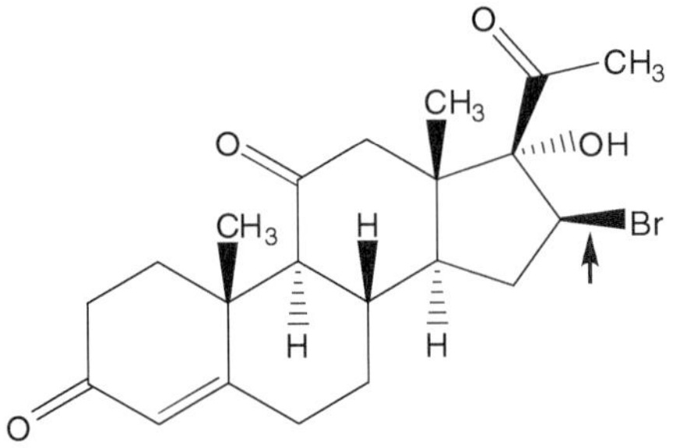

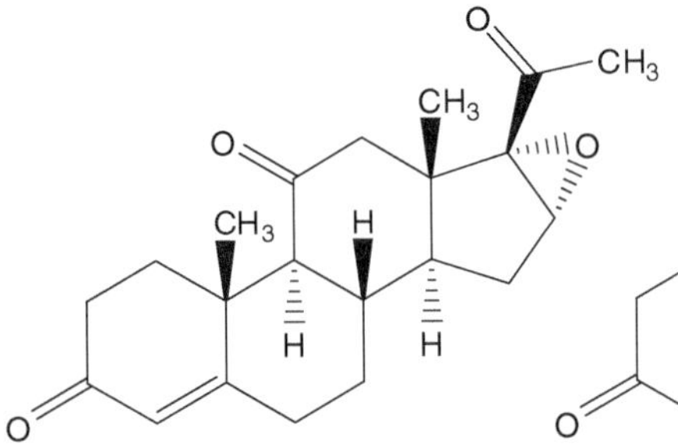

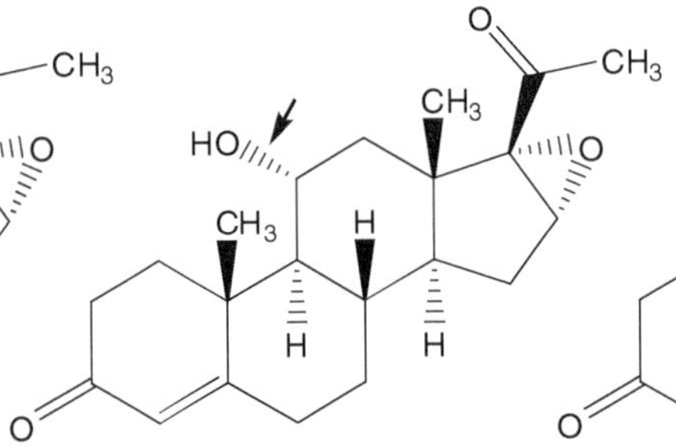

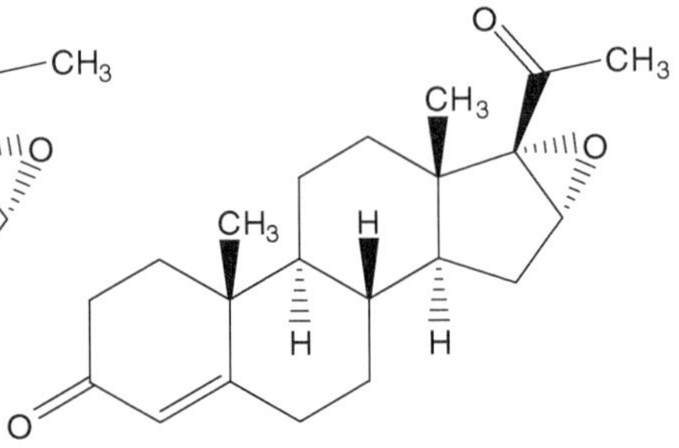

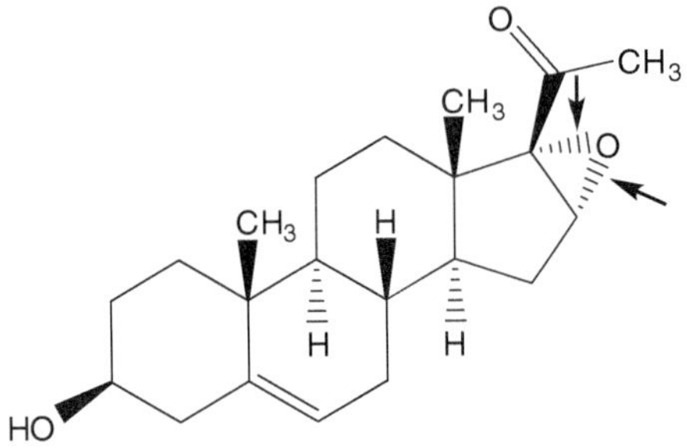

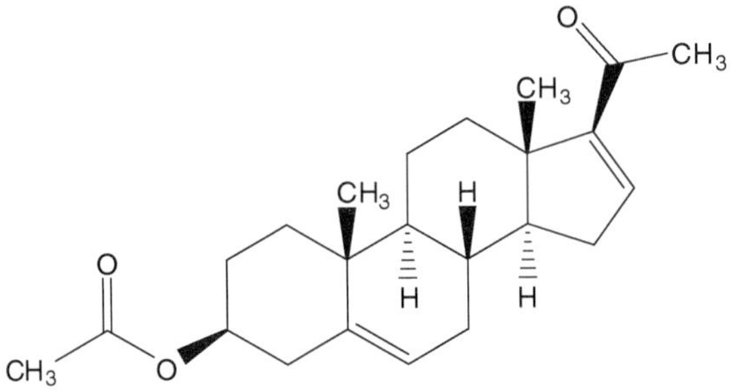

16-dehydropregnenolone acetate

The 16-alkene of 16-dehydropregnenolone acetate is formed from diosone by β-elimination. Diosone is formed by oxidation of the 20(22)-alkene of pseudodiosgenin-3,26-diacetate. Pseudodiosgenin 3,26-diacetate is formed by reaction of diosgenin with acetic anhydride. The three-step synthesis of 16-dehydropregnenolone acetate from diosgenin by acetylation, oxidation, and elimination is known as the **Marker Degradation**.

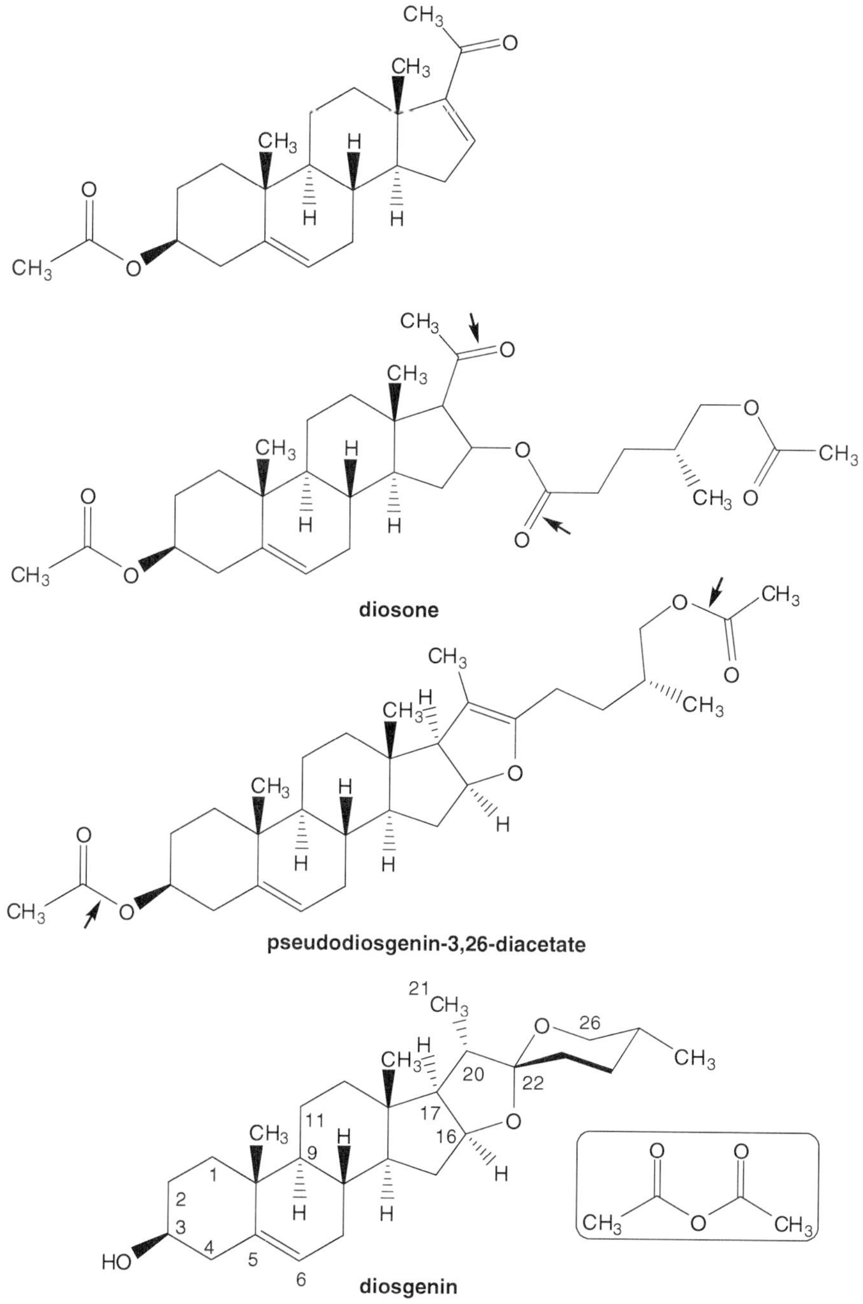

Extended Discussion

Number the steps of the methylprednisolone retrosynthetic analysis. Draw the structures of three methylprednisolone impurities listed in the European Pharmacopeia (EP). Indicate which step in the methylprednisolone retrosynthetic analysis the impurity or a precursor to the impurity is likely to be formed.

Methylthioninium Chloride

Antidotes and Other Substances Used in Poisonings/Specific

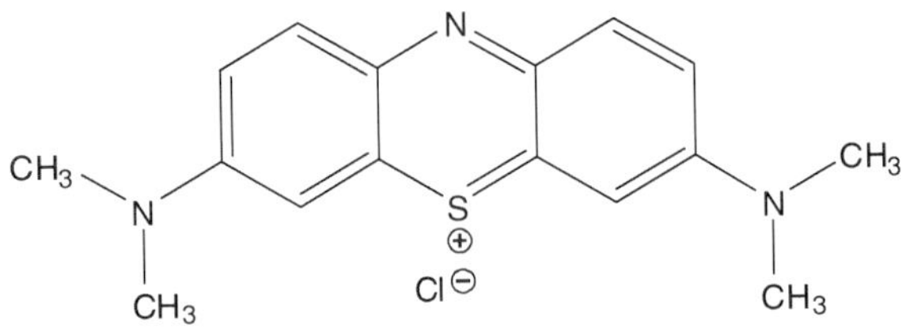

A phenothiazinium ring is often formed from an *ortho*-aminophenylthiosulfonic acid.

Discussion. The phenothiazinium ring of methylthioninium chloride (methylene blue) is formed by reaction of the thiosulfonic acid of Bindschedler's Green with an oxidant. The thiosulfonic acid of Bindschedler's Green is formed by reaction of the thiosulfonic acid of *N*,*N*-dimethyl-*para*-phenylenediamine with *N*,*N*-dimethylaniline and an oxidant. The thiosulfonic acid of *N*,*N*-dimethyl-*para*-phenylenediamine is formed from *N*,*N*-dimethyl-*para*-phenylenediamine. *N*,*N*-dimethyl-*para*-phenylenediamine is formed by reduction of *para*-nitroso-*N*,*N*-dimethylaniline. The nitrosoaniline is formed by nitrosation of *N*,*N*-dimethylaniline.

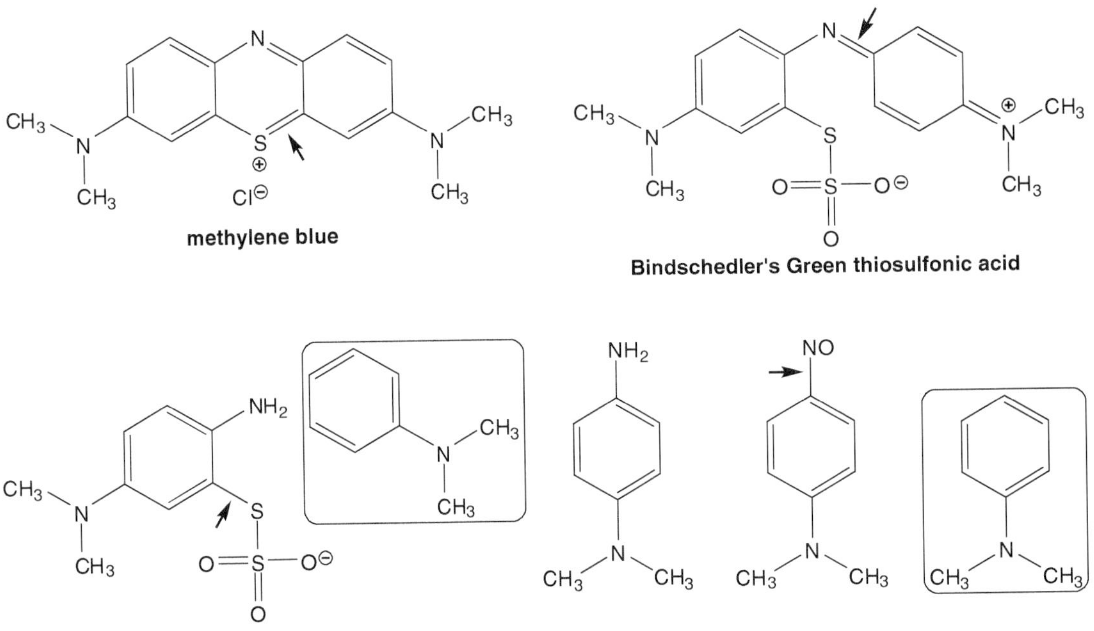

Extended Discussion

Pharmaceutical grade methylthioninium chloride must have very low residual heavy metal content. This specification poses a significant challenge since metal-based oxidants are used in two steps of the five-step synthesis. In the event the process does not consistently deliver methylthioninium chloride with very low residual heavy metal, the metal content can be reduced. Provide the details for the procedure.

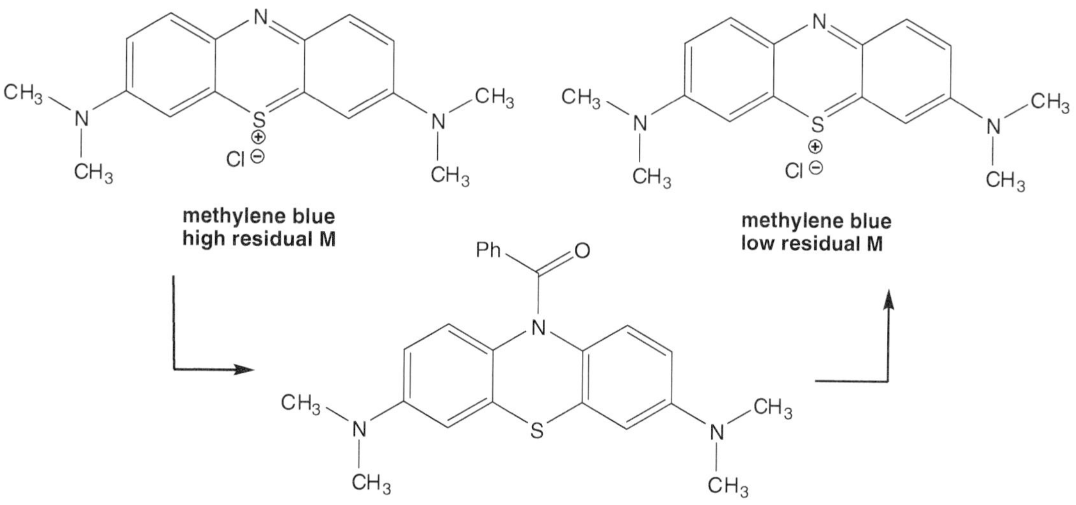

Metoclopramide

Medicines for Pain and Palliative Care/Medicines for Other Common Symptoms in Palliative Care
Gastrointestinal Medicines/Antiemetic Medicines

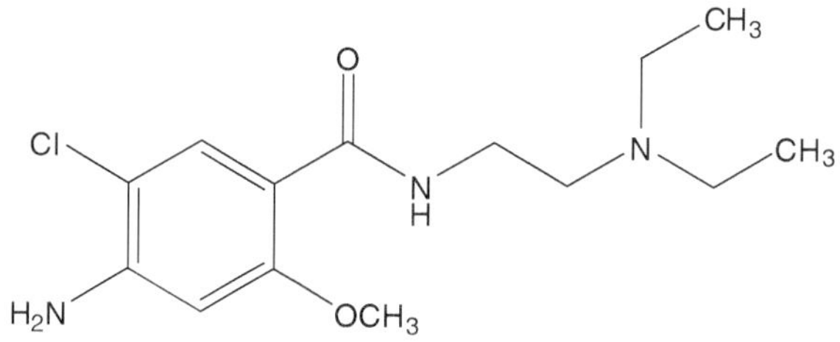

When the target molecule has a benzene ring with multiple substituents, it is likely that one or more of the substituents will be introduced by electrophilic aromatic substitution. Activating/deactivating and directing effects of substituents often determine the order of introduction of ring substituents in the preferred synthetic sequence.

Discussion. The acetamide is hydrolyzed to release the amino group in the final step. The benzamide is formed by reaction of the methyl ester with *N,N*-diethylethylenediamine. The ring is chlorinated at C5. The methyl ether is formed by reaction of the phenol with dimethyl sulfate (**Williamson Ether Synthesis**). The acetamide is formed by reaction of the amine with acetic anhydride. The methyl ester is formed from the carboxylic acid (**Fischer Esterification**). 4-Aminosalicylic acid, another essential medicine, is formed from 3-aminophenol and carbon dioxide (**Kolbe–Schmitt Reaction**).

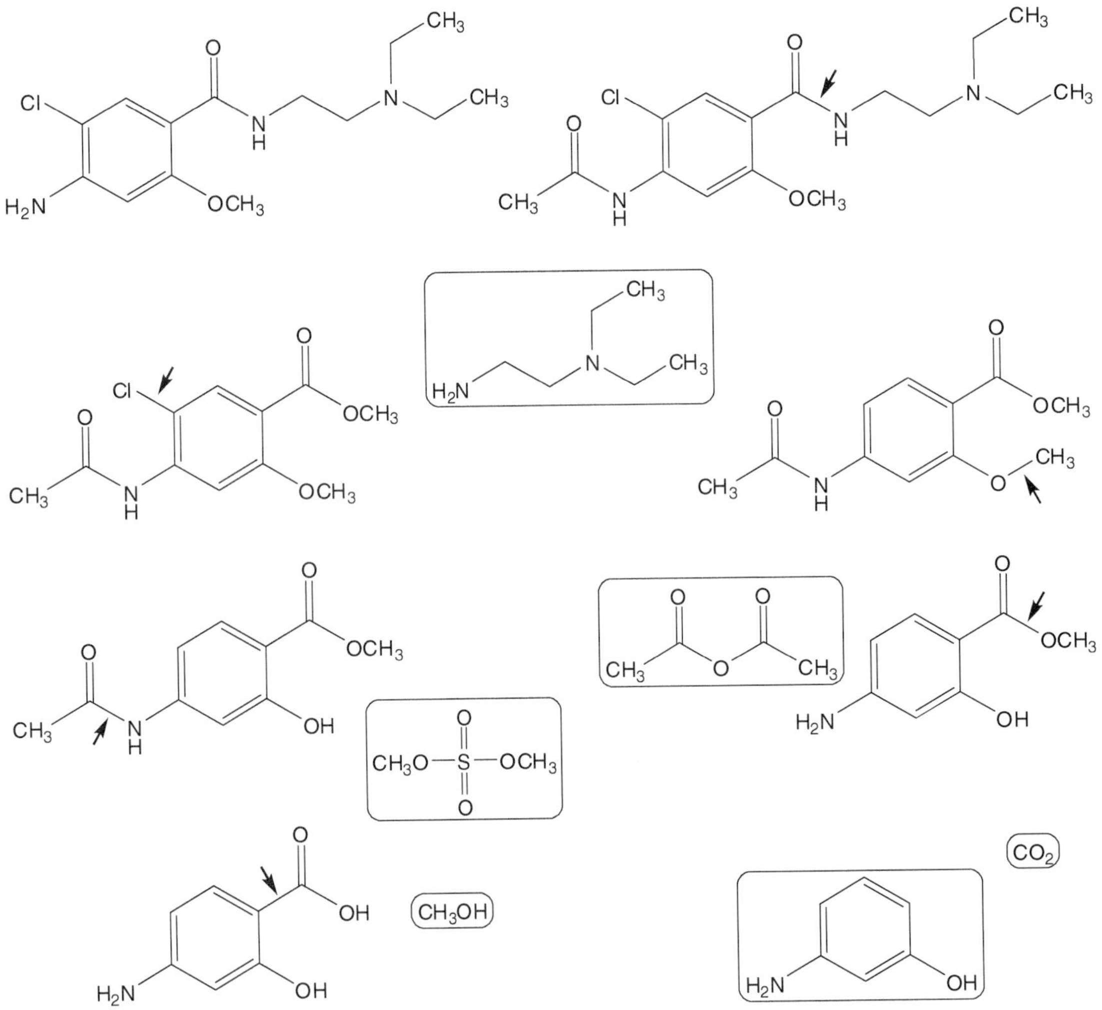

Extended Discussion

Draw the structures of the retrosynthetic analysis for an alternative route to metoclopramide that does not require protection of the amino group as the acetamide. List pros and cons for both routes and select one route as the preferred route.

Metronidazole

Anti-infective Medicines/Antibacterials/Other Antibacterials
Anti-infective Medicines/Antiprotozoal Medicines/Antiamoebic and Antigiardiasis Medicines

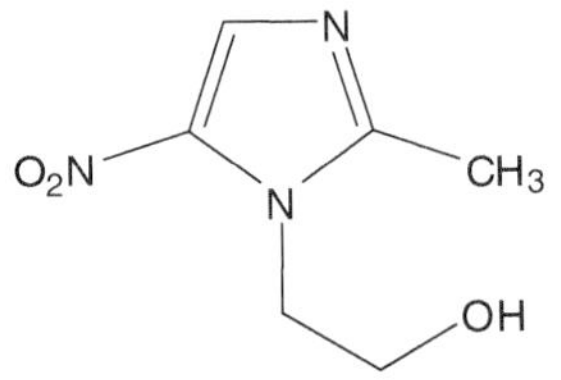

An *N*-alkylimidazole is often formed by alkylation of an imidazole.

Discussion. 2-Methyl-4-nitroimidazole is *N*-alkylated by reaction with ethylene oxide or 2-chloroethanol in the final step. (Which reagent is preferred? Why?) The 2-methyl-4-nitroimidazole is formed by nitration of 2-methylimidazole. (Since 2-methylimidazole is protonated by strong acid, the nitration of 2-methylimidazole using a mixture of nitric acid and sulfuric acid requires elevated temperature. Some decomposition of the imidazole and the formation of nitrogen oxides are observed under these conditions.)

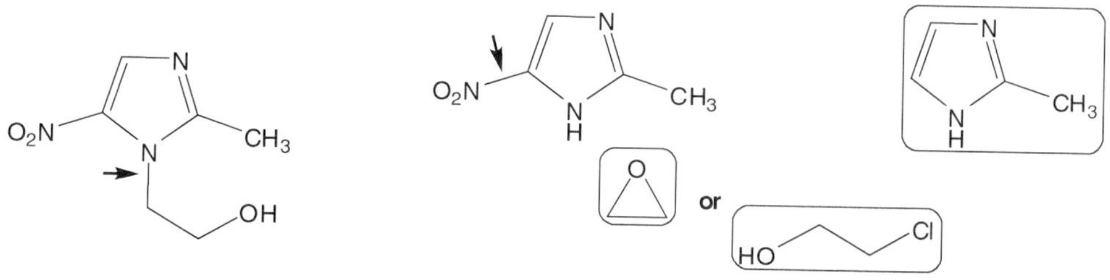

Extended Discussion

Suggest (draw the structure of) a side product which might form in the final step. Is this side product formed?

Miconazole

Dermatological Medicines (Topical)/Antifungal Medicines

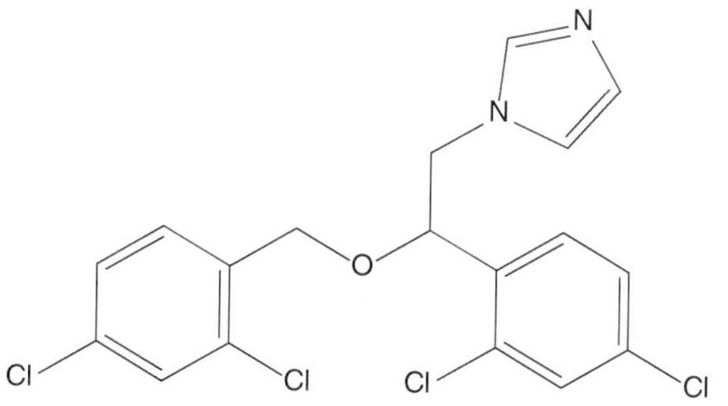

An ether is often formed from an alkyl halide (X = Cl, Br, I) and an alcohol. The C-O bond formation is most efficient when the alkyl halide is primary, allylic, or benzylic.

Discussion. Miconazole is a 1 : 1 mixture of the (*R*)- and (*S*)-enantiomers. The ether is formed in the final step by displacement of the benzylic chloride from 2,4-dichlorobenzyl chloride by the alcohol (**Williamson Ether Synthesis**). The secondary alcohol is formed by reduction of the ketone. A C-N bond is formed by displacement of bromide of the α-bromoketone by imidazole. The α-bromoketone is formed by bromination of 2′,4′-dichloroacetophenone.

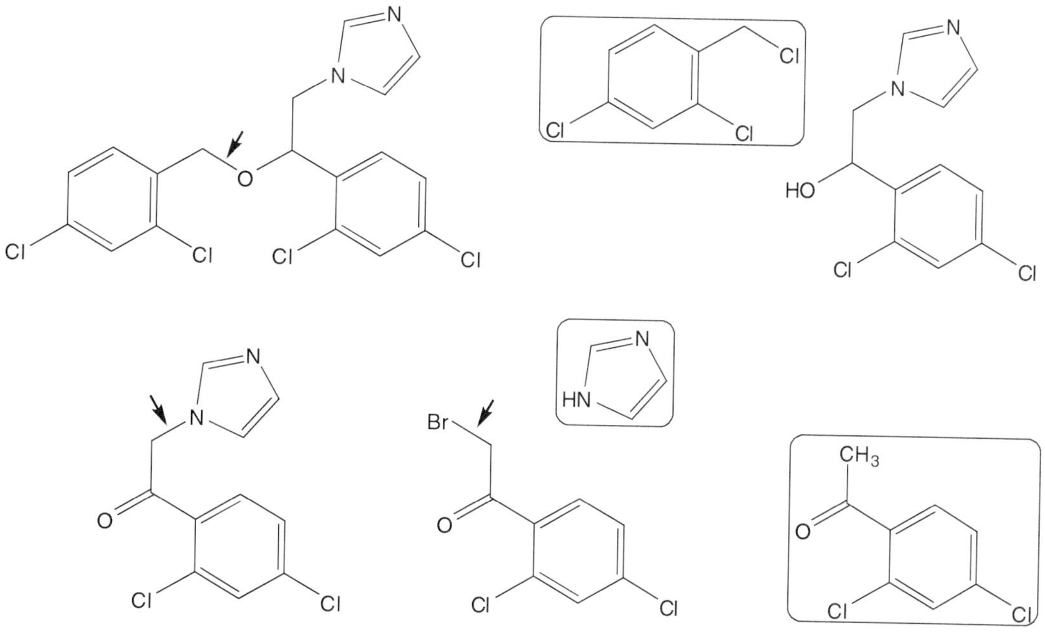

The α-chloroketone, 2,2′,4′-trichloroacetophenone, formed from 1,3-dichlorobenzene and chloroacetyl chloride (**Friedel–Crafts Acylation**), is used in an alternative route to miconazole.

Extended Discussion

List the pros and cons for both routes. Is one route preferred?

Midazolam

Anesthetics, Preoperative Medicines, and Medical Gases/Preoperative Medication and Sedation for Short-term Procedures
Medicines for Pain and Palliative Care/Medicines for Other Common Symptoms in Palliative Care
Anticonvulsants/Antiepileptics

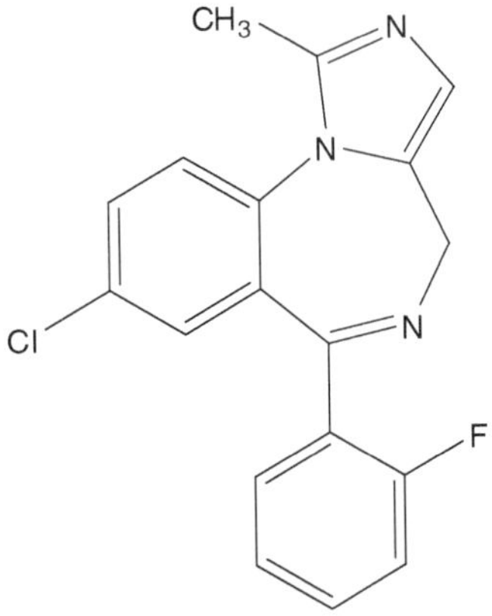

A 5-aryl-1,3-dihydro-2*H*-1,4-benzodiazepin-2-one is often formed in two steps from a 2-aminobenzophenone, chloroacetyl chloride, and ammonia/hexamethy lenetetramine.

Discussion. Midazolam is separated from the mixture of midazolam (major) and isomidazolam (minor) produced by decarboxylation of the imidazolecarboxylic acid. Isomidazolam is also converted to midazolam. (Draw the structure of isomidazolam and provide details on the conversion of isomidazolam to midazolam.) The imidazolecarboxylic acid is formed by hydrolysis of the ester.

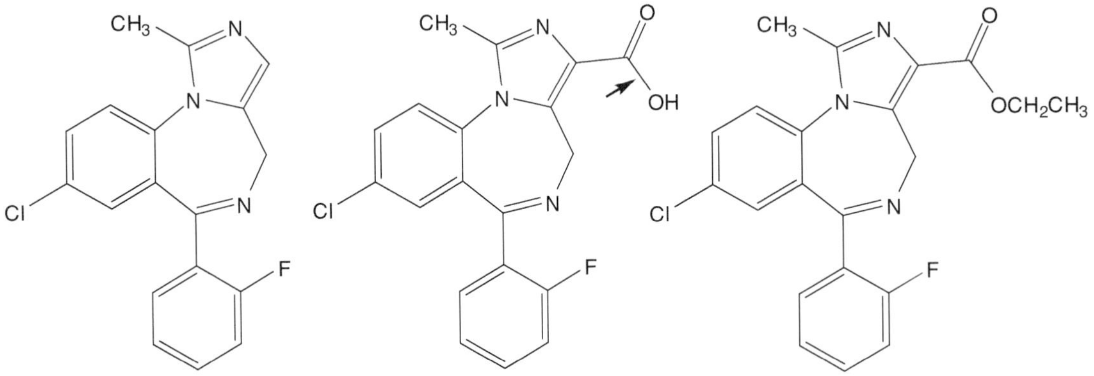

The 2-methylimidazole is formed by cyclization of the amine and the acetamide with elimination of water. The alkene with the amine and acetamide substituents is formed by cleavage of the α-substituted acetamidomalonate. (What are the likely byproducts of this cleavage?) The α-substituted acetamidomalonate is formed by displacement of the leaving group dimorpholinophosphinate by diethyl acetamidomalonate. Reaction conditions are selected so that the three steps (leaving group displacement, cleavage of the α-substituted acetamidomalonate, and cyclization to form the imidazole) are accomplished without isolation of the two intermediates.

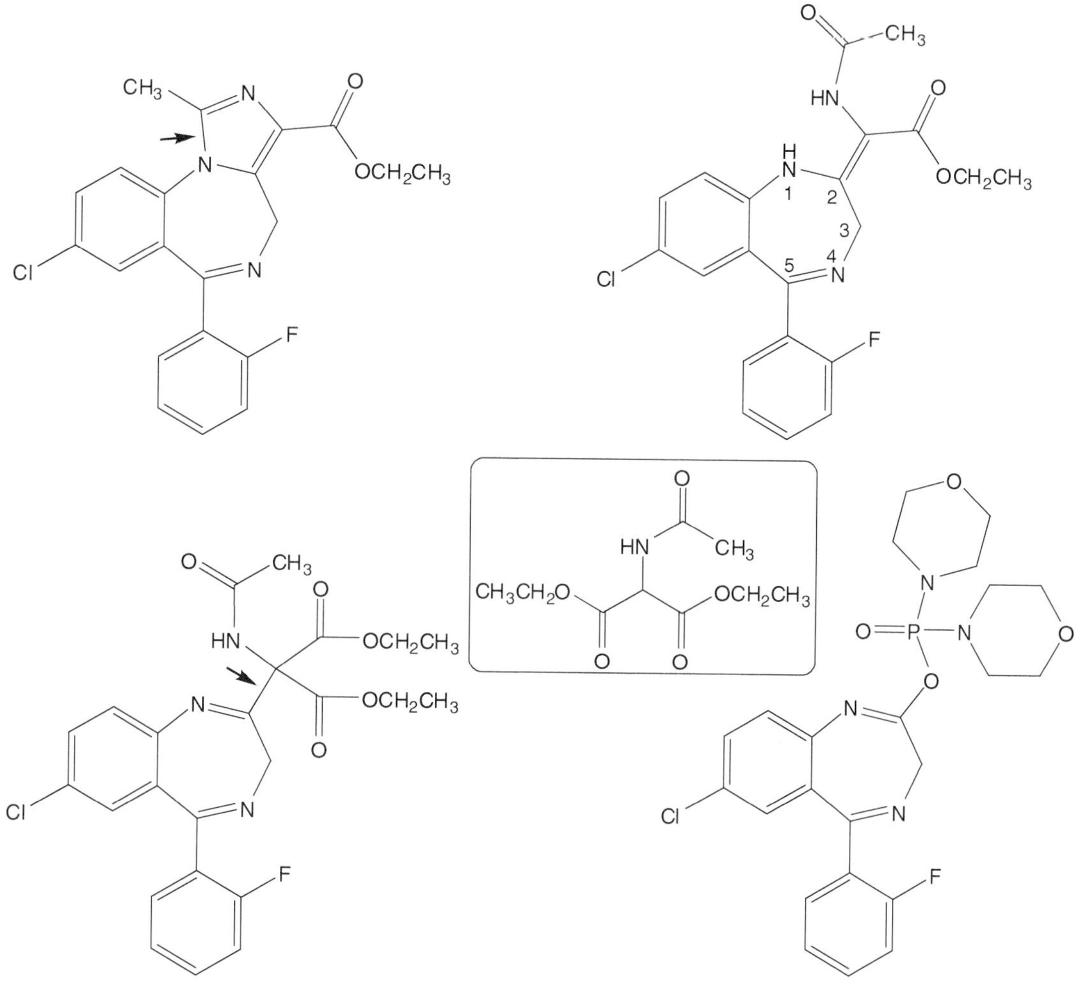

The leaving group is introduced by reaction of dimorpholinophosphinyl chloride with the 1,4-benzodiazepin-2-one. 7-Chloro-5-(2-fluorophenyl)-1,3-dihydro-2*H*-1,4-benzodiazepin-2-one (norflurazepam) is formed after a chloride displacement from the 2-chloroacetamide by ammonia/hexamethylenetetramine. The 2-chloroacetamide is formed from chloroacetyl chloride and 2-amino-4-chloro-2′-fluorobenzophenone. The 2-aminobenzophenone is formed by hydrolysis of the acetamide.

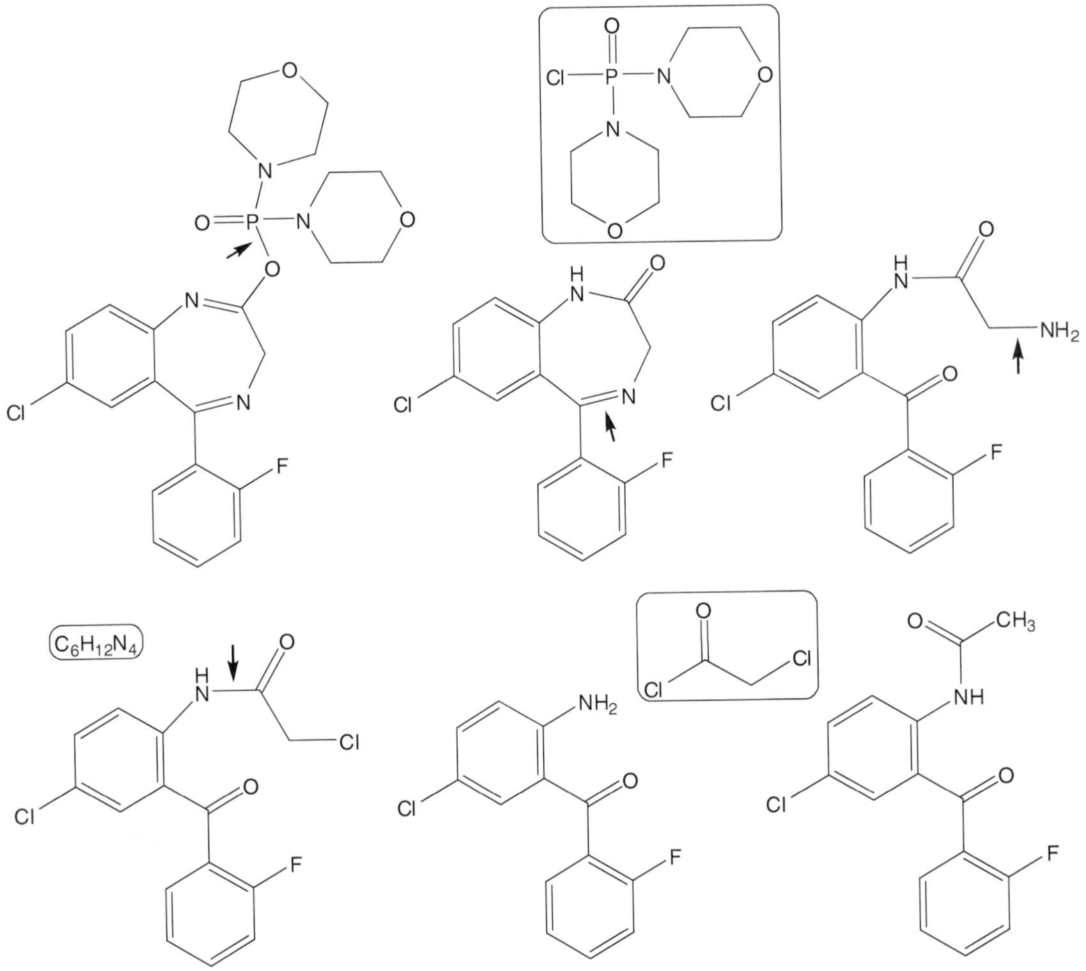

The benzophenone is formed from 4-chloroacetanilide and 2-fluorobenzoyl chloride (**Friedel–Crafts Acylation**). 2-Fluorobenzoyl chloride is formed from the carboxylic acid. 2-Fluorobenzoic acid is formed from anthranilic acid (**Balz–Schiemann Reaction**).

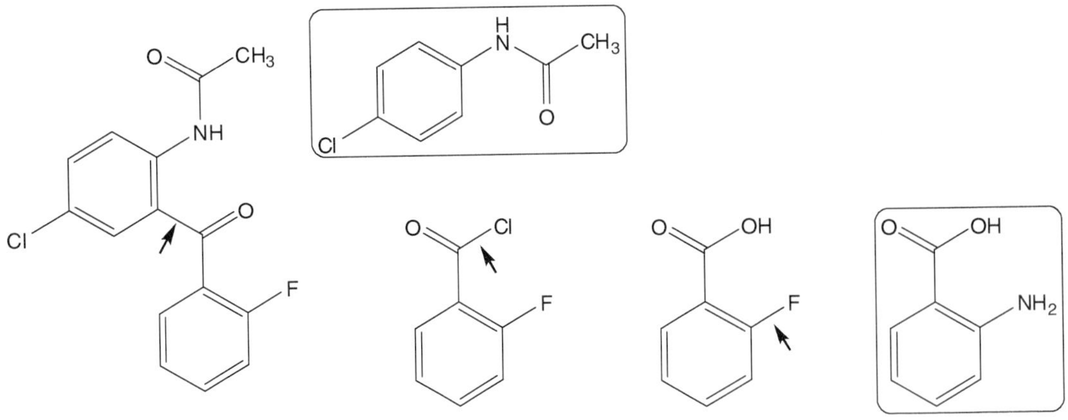

Extended Discussion

Draw the structures of a retrosynthetic analysis of an alternative route to midazolam that does not proceed via decarboxylation of the imidazolecarboxylic acid. List the pros and cons for both routes and select one route as the preferred route.

Mifepristone

Oxytocics and Antioxytocics/Oxytocics

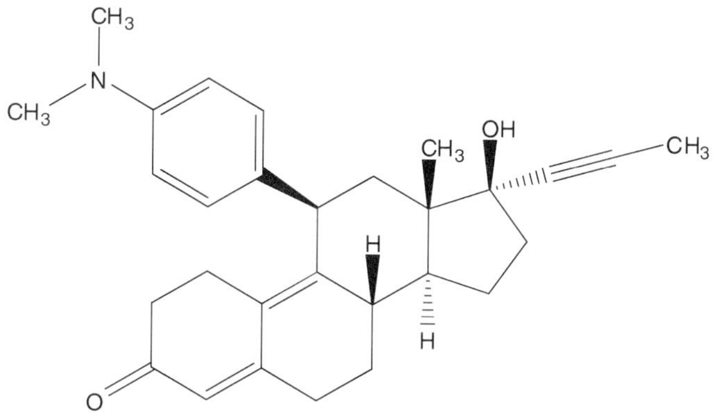

A single-enantiomer molecule with multiple chiral carbons is often formed by modification of a natural product which has most or all of the chiral carbons already in place. Steroids missing the methyl group at position 19 (19-nor-steroids) are often formed from sitolactone. Sitolactone is produced by microbial degradation of phytosterols, including β-sitosterol.

Discussion. Mifepristone (RU-486) is manufactured in four steps from 3,3-ethylenedioxyestra-5(10),9(11)-diene-17-one and in nine steps from β-sitosterol.

The 4,9-diene-3-one is formed in the final step by hydrolysis of the acetal and elimination of water. The reaction of the 9(11)-ene-5α(10)-epoxide with the arylmagnesium halide results in C-C bond formation at C11 and epoxide ring-opening (**Grignard Reaction**). (The 11α-aryl side product is also formed in this reaction. What is the ratio 11β-aryl product to 11α-aryl side product? How is the 11β-aryl product separated from the 11α-aryl side product?) The Grignard reagent is formed from 4-bromo-*N*,*N*-dimethylaniline. The 9(11)-ene-5α(10)-epoxide is formed by epoxidation of the 5(10),9(11)-diene. (The 9(11)-ene-5β(10)-epoxide is also formed in the epoxidation. What is the ratio of the 5α(10)-epoxide to the 5β(10)-epoxide? How is the 5α(10)-epoxide separated from the 5β(10)-epoxide?) The C17 tertiary alcohol is formed by addition a metal acetylide from propyne to the C17 ketone of 3,3-ethylenedioxyestra-5(10),9(11)-diene-17-one.

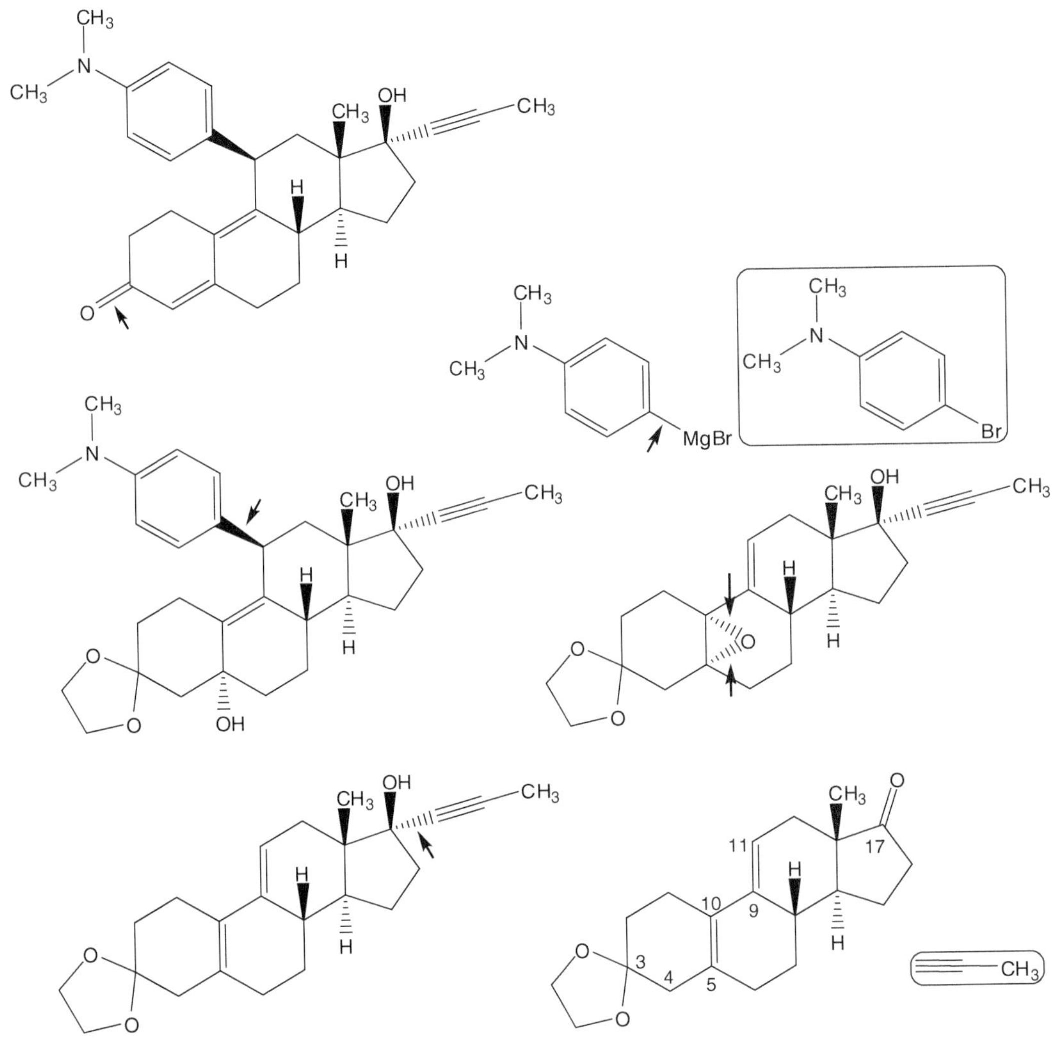

The acetal of 3,3-ethylenedioxyestra-5(10),9(11)-diene-17-one is formed by the reaction of estra-4,9-diene-3,17-dione with ethylene glycol. The 4-ene-3-one and the A ring of estra-4,9-diene-3,17-dione are formed by condensation of the methyl ketone and cyclohexenone of the seco-steroid (**Aldol Condensation**) followed by elimination of water. The methyl ketone of the seco-steroid is released by hydrolysis of the acetal and carried directly into the condensation.

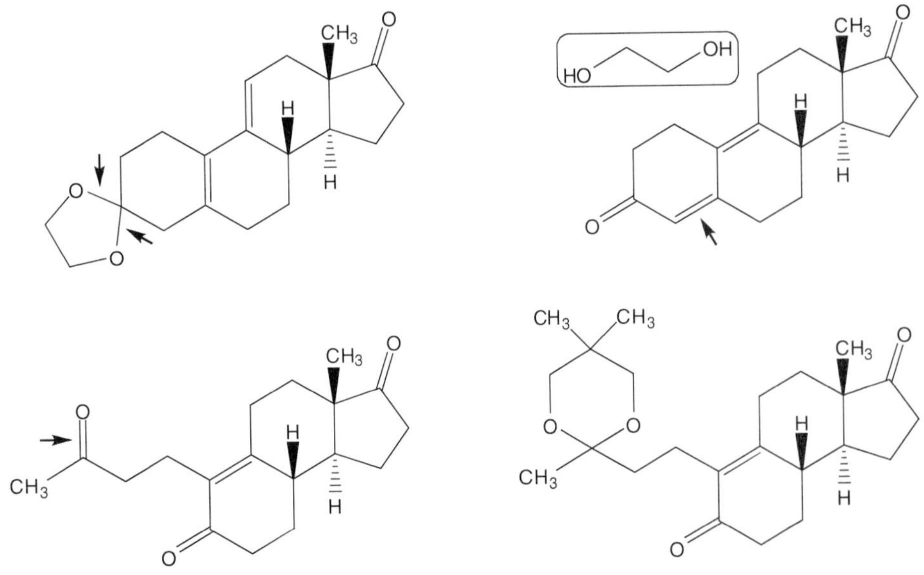

The cyclohexenone and the B ring are formed by condensation of a ketone on a side chain and the cyclohexanone (**Aldol Condensation**) with elimination of water. The cyclohexanone is formed by oxidation of the cyclohexanol and carried directly into the condensation. The ketone on the side chain and the cyclohexanol are both formed by reaction of an alkylmagnesium chloride with the lactone carbonyl of sitolactone (**Grignard Reaction**). Sitolactone is formed by microbial degradation of phytosterols, including β-sitosterol.

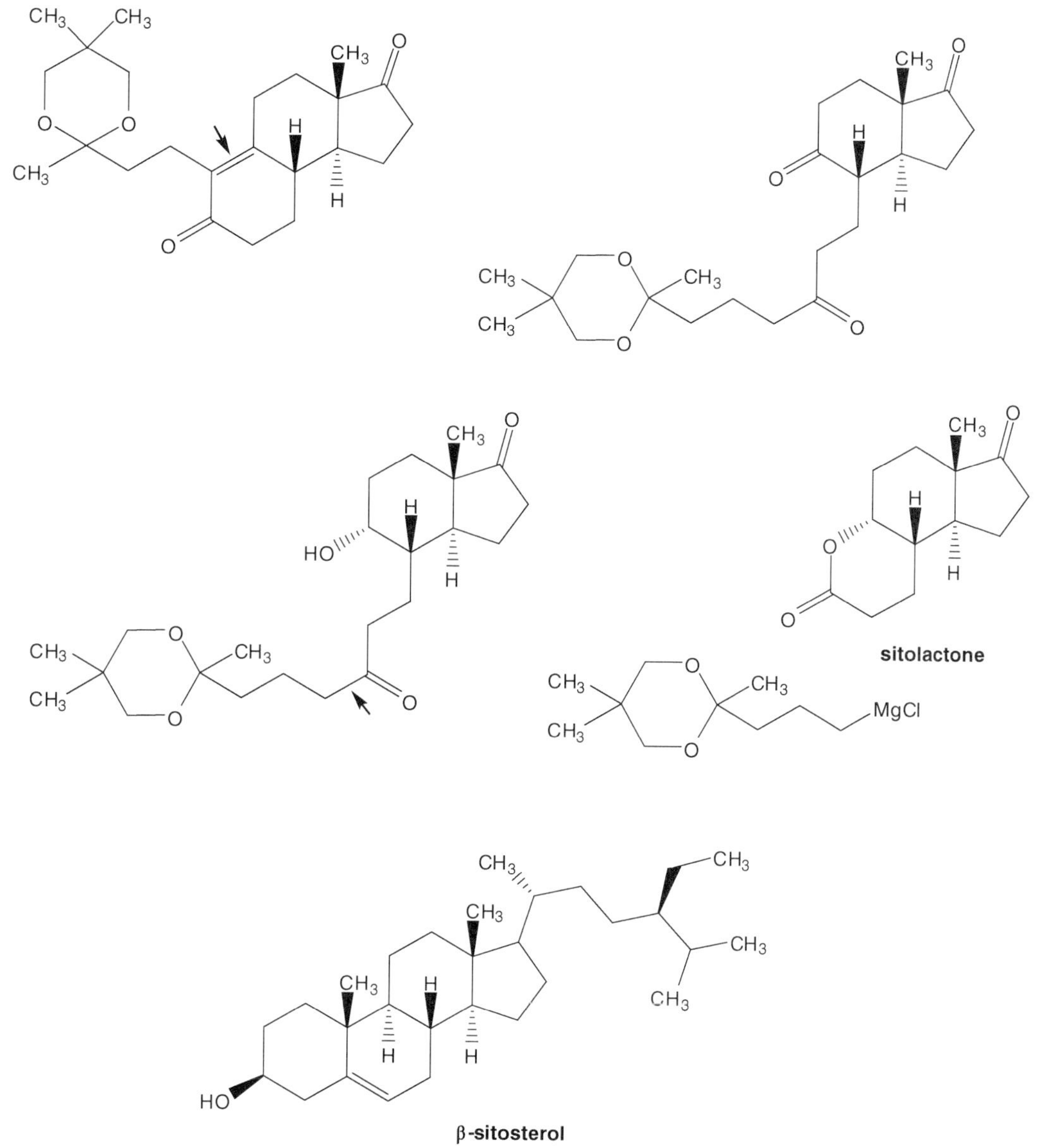

The alkylmagnesium chloride is formed from 2-(3-chloropropyl)-2,5,5,-trimethyl-1,3-dioxane. The 1,3-dioxane/acetal is formed by reaction of 5-chloro-2-pentanone with 2,2-dimethyl-1,3-propanediol (neopentyl glycol).

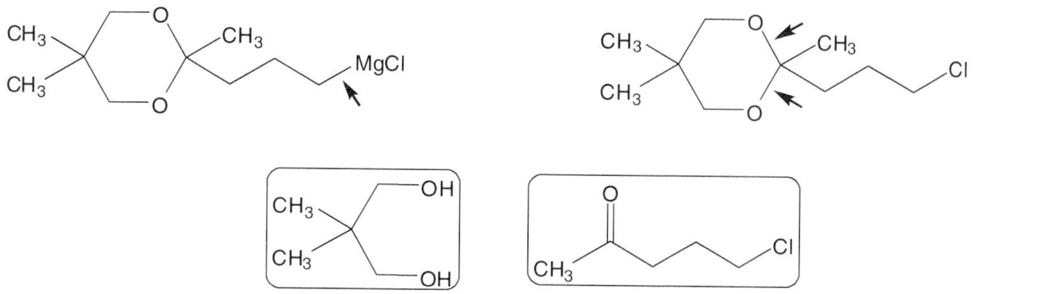

Extended Discussion

Draw structures of the retrosynthetic analysis of an alternative route to the intermediate 3,3-ethylenedioxyestra-5(10),9(11)-diene-17-one from estrone. List the pros and cons for both routes to this intermediate and select one route as the preferred route.

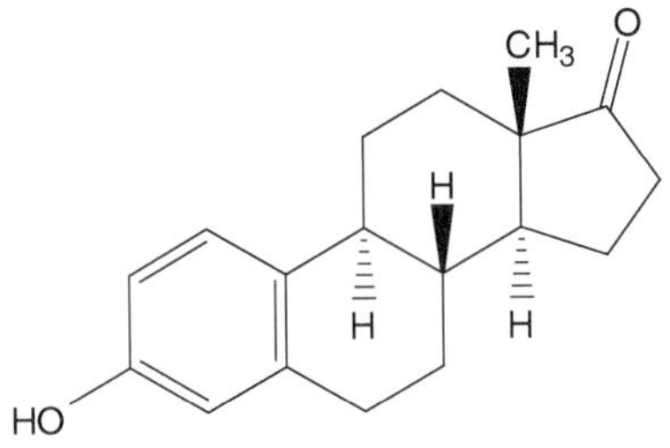

estrone

Miltefosine

Anti-infective Medicines/Antiprotozoal Medicines/Antileishmaniasis Medicines

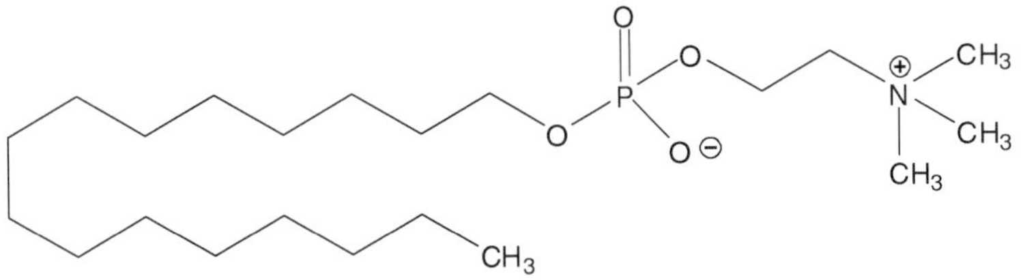

A quaternary ammonium salt is often formed by alkylation of a tertiary amine.

Discussion. Two routes will be presented for comparison. In Route A, the quaternary salt of miltefosine is formed in the final step by alkylation of trimethylamine using a phosphate ester. The phosphate ester is formed by displacement of chloride from 2-chloro-2-oxo-1,3,2-dioxaphospholane (ethylene chlorophosphate) by hexadecane-1-ol (cetyl alcohol). 2-Chloro-2-oxo-1,3,2-dioxaphospholane is formed by oxidation of 2-chloro-1,3,2-dioxaphospholane (ethylene chlorophosphite). 2-Chloro-1,3,2-dioxaphospholane is formed from ethylene glycol.

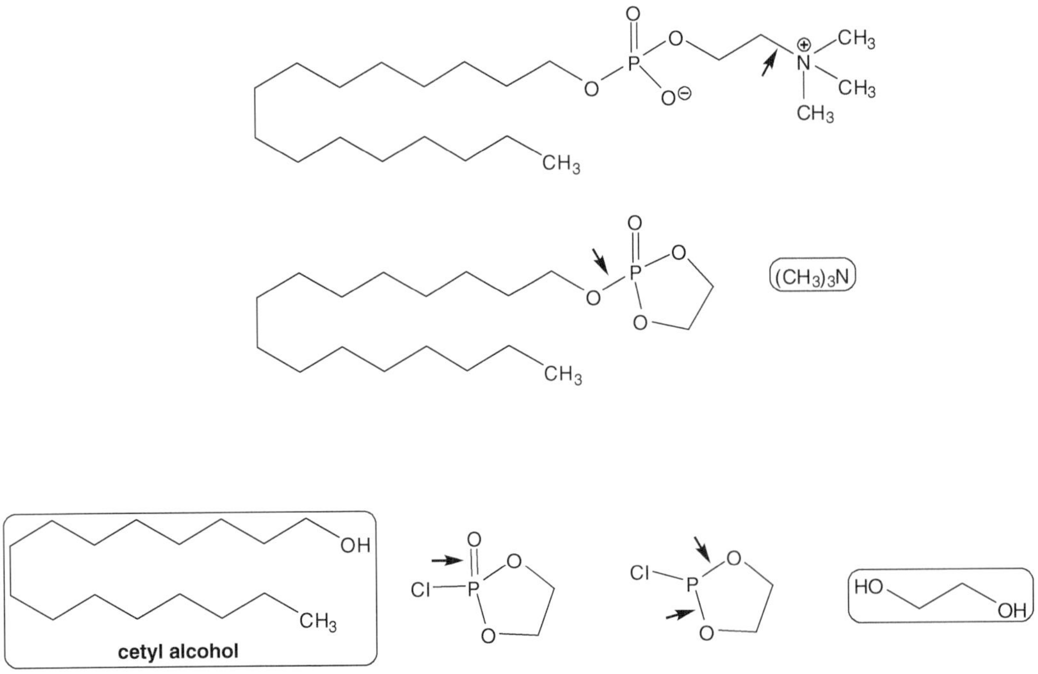

In Route B, the quaternary salt of miltefosine (hexadecylphosphocholine) is formed in the final step by exhaustive methylation of the primary amine (hexadecylphosphoethanolamine) with dimethyl sulfate. The primary amine is formed by hydrolysis of the phosphoric acid amide. The 1,3,2-oxazaphospholane ring is formed by reaction of cetyl dichlorophosphate with ethanolamine. Cetyl dichlorophosphate is formed from cetyl alcohol.

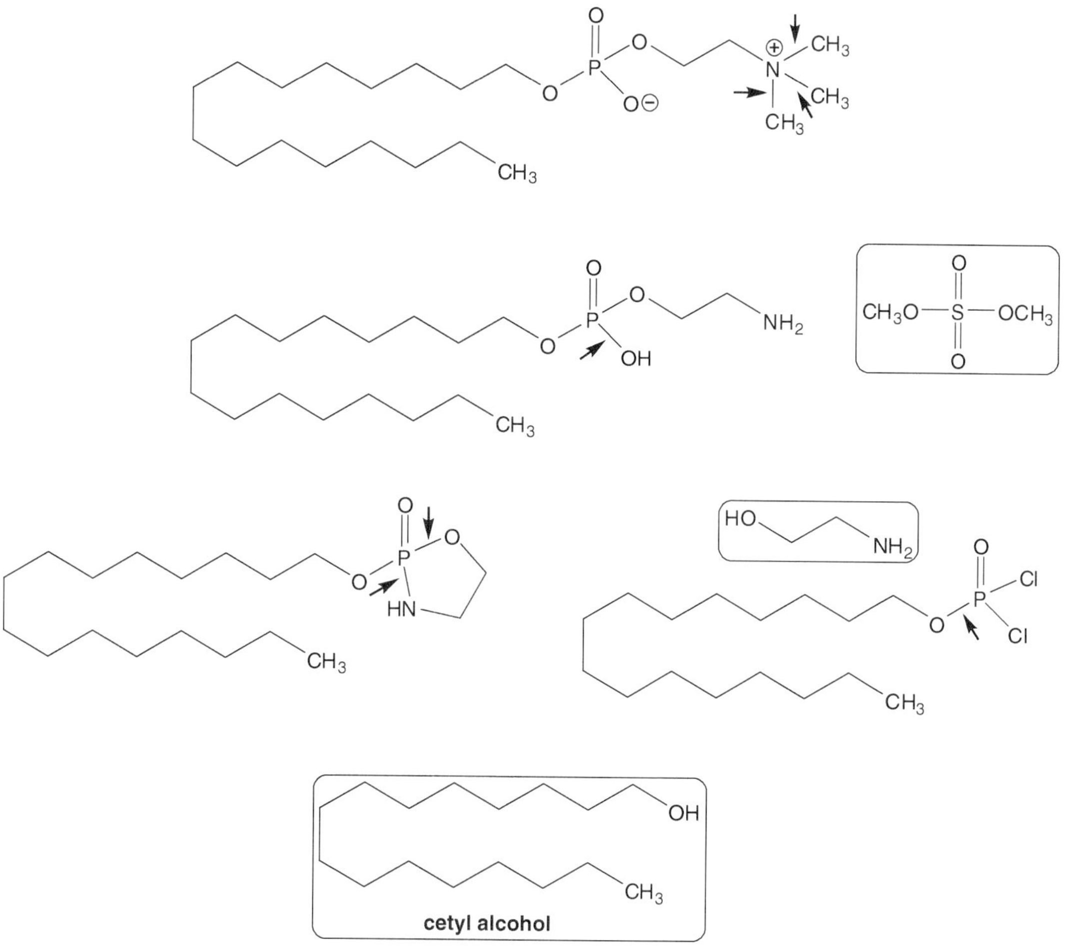

Extended Discussion

List the pros and cons for both routes. Select one route as the preferred route.

Misoprostol

Oxytocics and Antioxytocics/Oxytocics

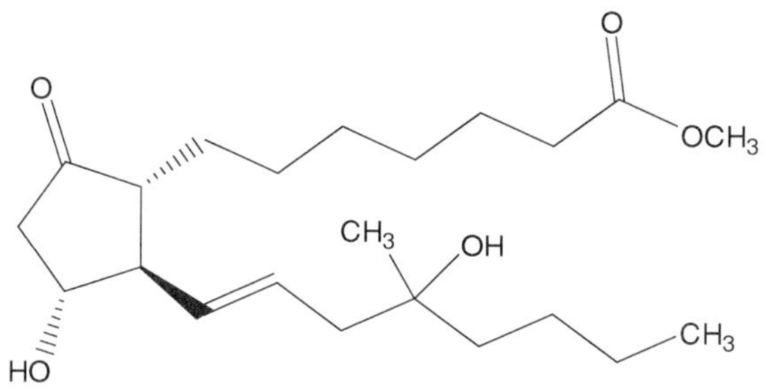

Prostaglandin analogs with the cyclopentanone core of prostaglandin E are often formed by conjugate addition of an alkenylcuprate reagent to a cyclopentenone. An alkoxy trialkylsiloxy group on the α-face of the cyclopentenone directs the conjugate addition at the adjacent carbon to the β-face.

Discussion. Misoprostol is a mixture of (16*R*)- and (16*S*)-diastereomers. Misoprostol is formed in just three steps from (*R*)-norprostol. (*R*)-Norprostol is formed in seven steps from suberic acid (1,8-octanedioic acid).

The C11 and C16 alcohols of misoprostol are released in the final step. (List the protecting groups P[1] and P[2] used for the C11 and C16 alcohols. Select one option for P[1] and one option for P[2] to use in the analysis.) The C12·C13 bond is formed by conjugate addition of an alkenylcuprate to a cyclopentenone. (Draw the structure of one side product which is formed in this reaction.) The alcohol of (*R*)-norprostol is protected. (*R*)-Norprostol is formed by enzyme-mediated hydrolysis of the racemic acetate ester. (*R*)-Norprostol is separated from the (*S*)-acetate and the (*S*)-acetate is racemized and recycled. The racemic acetate ester is formed by reaction of a mixture of the 3-hydroxycyclopent-4-en-1-one and the 4-hydroxycyclopent-2-en-1-one (norprostol) with acetic anhydride.

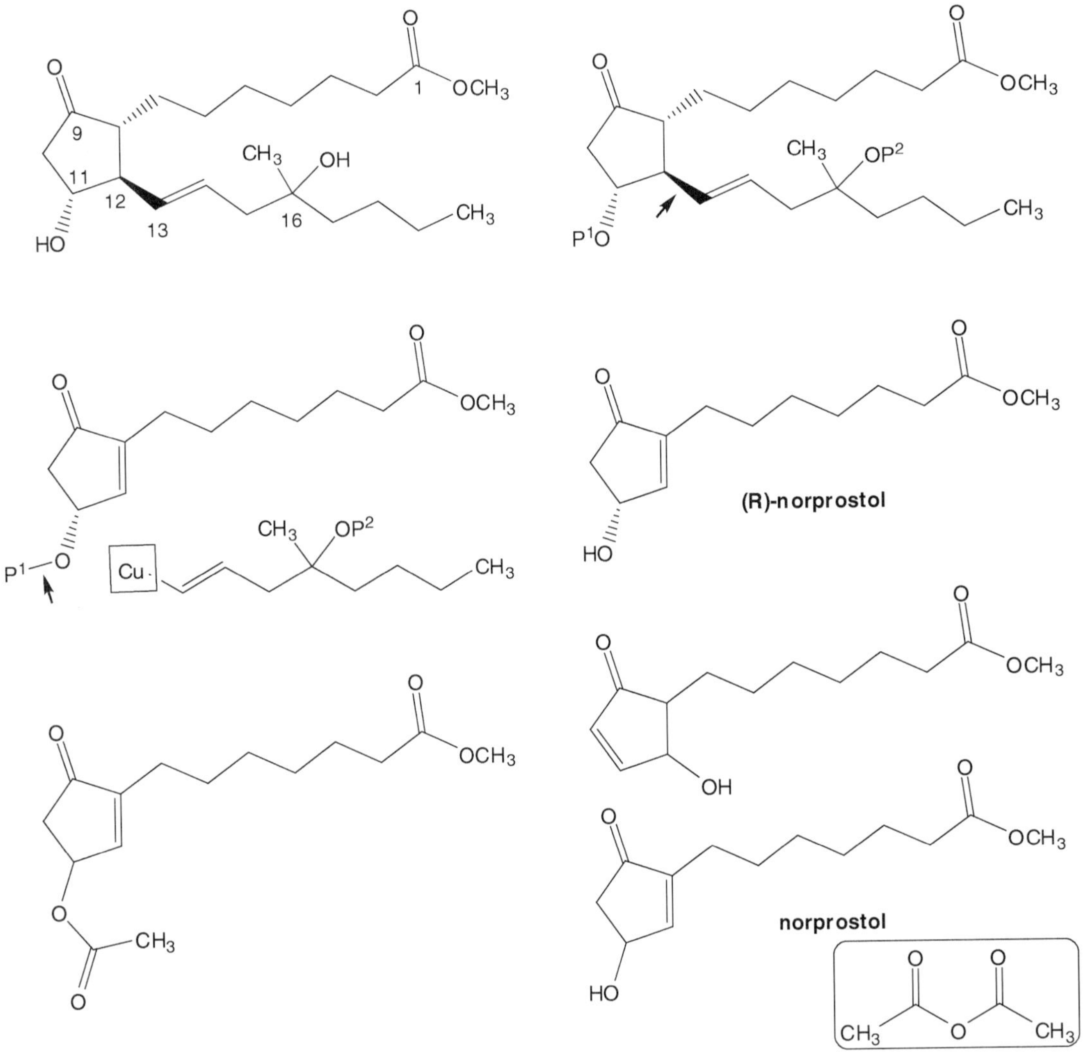

The mixture of the 3-hydroxycyclopent-4-en-1-one and norprostol (3 : 1) is formed by rearrangement of the furfuryl alcohol (**Piancatelli Rearrangement**). The secondary alcohol is formed by reduction of the ketone. The ketone is formed by acylation of furan with an active ester (**Friedel–Crafts Acylation**). The active ester is formed in situ from the carboxylic acid. (List the active esters used in this acylation of furan.) Suberic acid monomethyl ester is formed from suberic acid (**Fischer Esterification**).

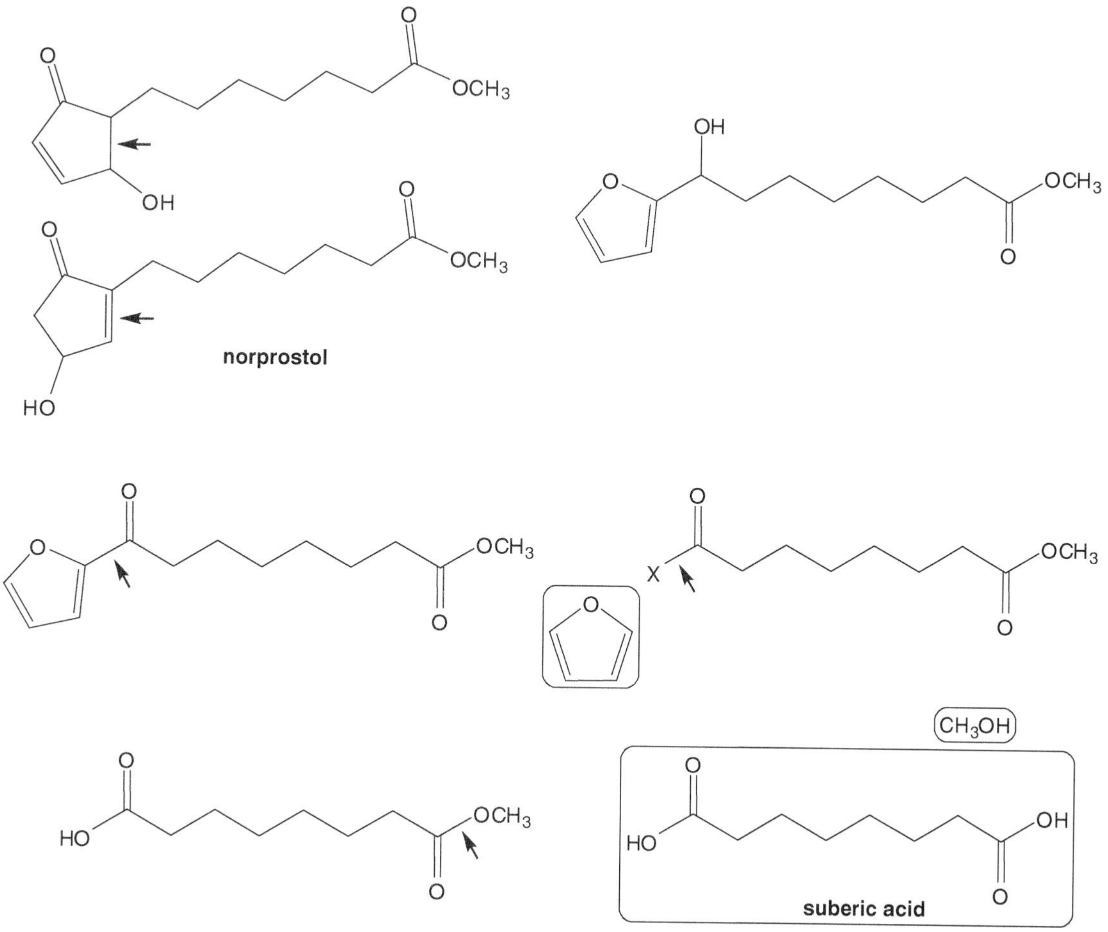

The alkenylcuprate reagent is formed from the alkyne. (Draw the structures for one route to the alkenylcuprate reagent from the alkyne. Draw the structure of the alkenylcuprate showing all the ligands attached to copper.) The alcohol 4-methyl-1-octyn-4-ol is protected. 4-Methyl-1-octyn-4-ol is formed by addition of propargylmagnesium bromide to 2-hexanone (**Grignard Reaction**). Propargylmagnesium bromide is formed from propargyl bromide. Propargyl bromide is formed from propargyl alcohol.

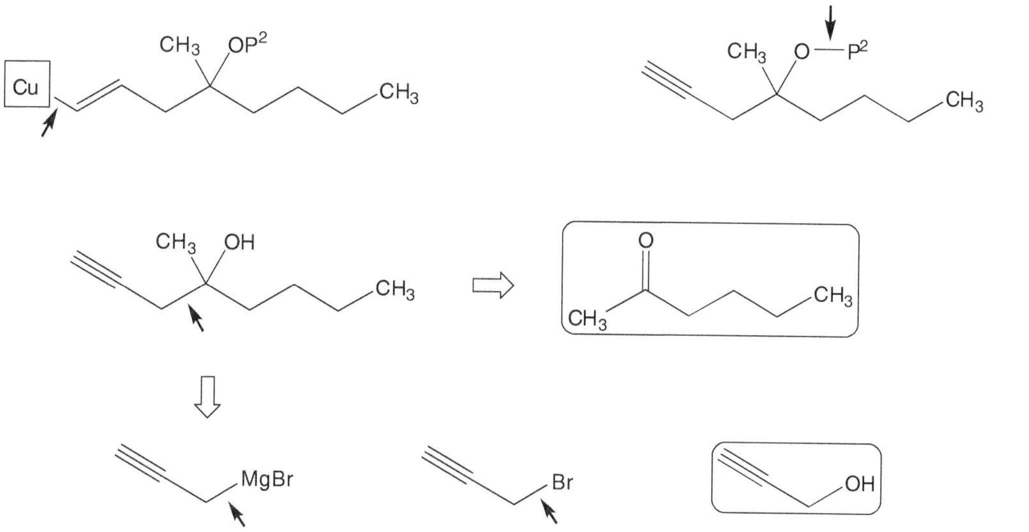

Extended Discussion

Draw the structures of an alternative route to an alkenylcuprate reagent used in the synthesis of misoprostol. Draw the structure of the alkenylcuprate reagent showing all the ligands attached to copper.

Moxifloxacin

Anti-infective Medicines/Antibacterials/Antituberculosis Medicines

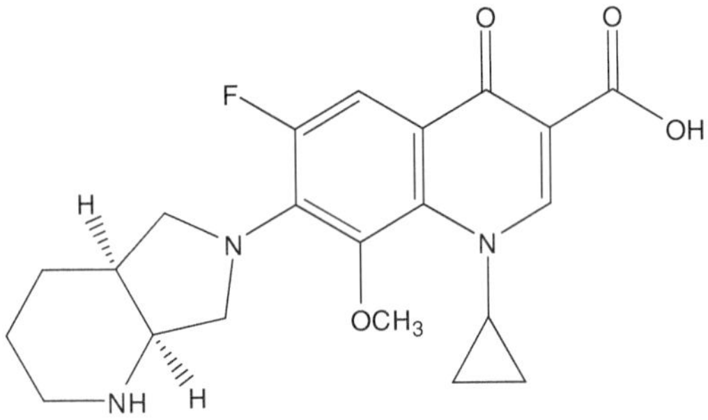

Nucleophilic aromatic substitution is often facilitated by an electron-withdrawing group (NO_2, SO_2R, COOR, CN) on an *ortho* or *para* ring carbon. No electron-withdrawing group is required when the displacement results in formation of a five- or six-membered ring. Leaving groups for nucleophilic aromatic substitution include fluorine, chlorine, and nitro.

Discussion. In the final step, fluoride at C8 is replaced by methoxide. The carboxylic acid is formed by ester hydrolysis. Fluoride at position 7 is replaced by the more nucleophilic nitrogen of the diamine.

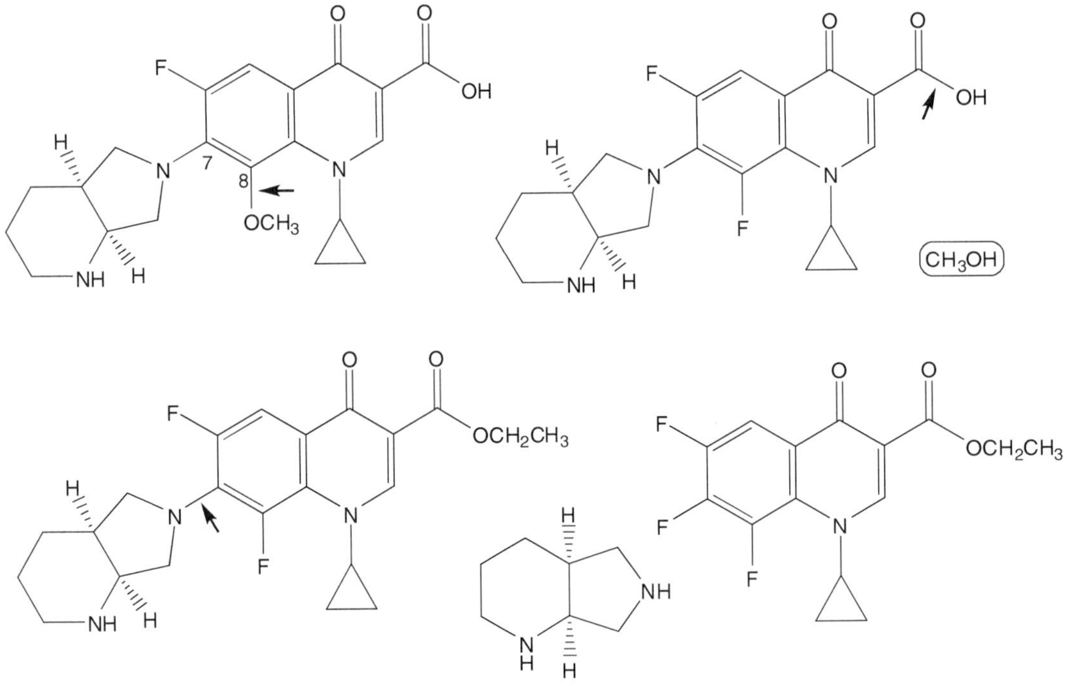

Intramolecular fluoride displacement by nitrogen of the cyclopropanamine forms the quinolone ring. Cyclopropanamine is introduced by displacement of ethanol from the enol ether. The β-ethoxyacrylate is formed by reaction of the β-ketoester with triethyl orthoformate. The β-ketoester is formed from 2,3,4,5-tetrafluorobenzoyl chloride. In one preferred approach, diethyl malonate is the reaction partner. (The four-step conversion of the 2-fluorobenzoyl chloride to the quinolone is known as the **Grohe–Heitzer Sequence**.) The acid chloride is formed from the carboxylic acid.

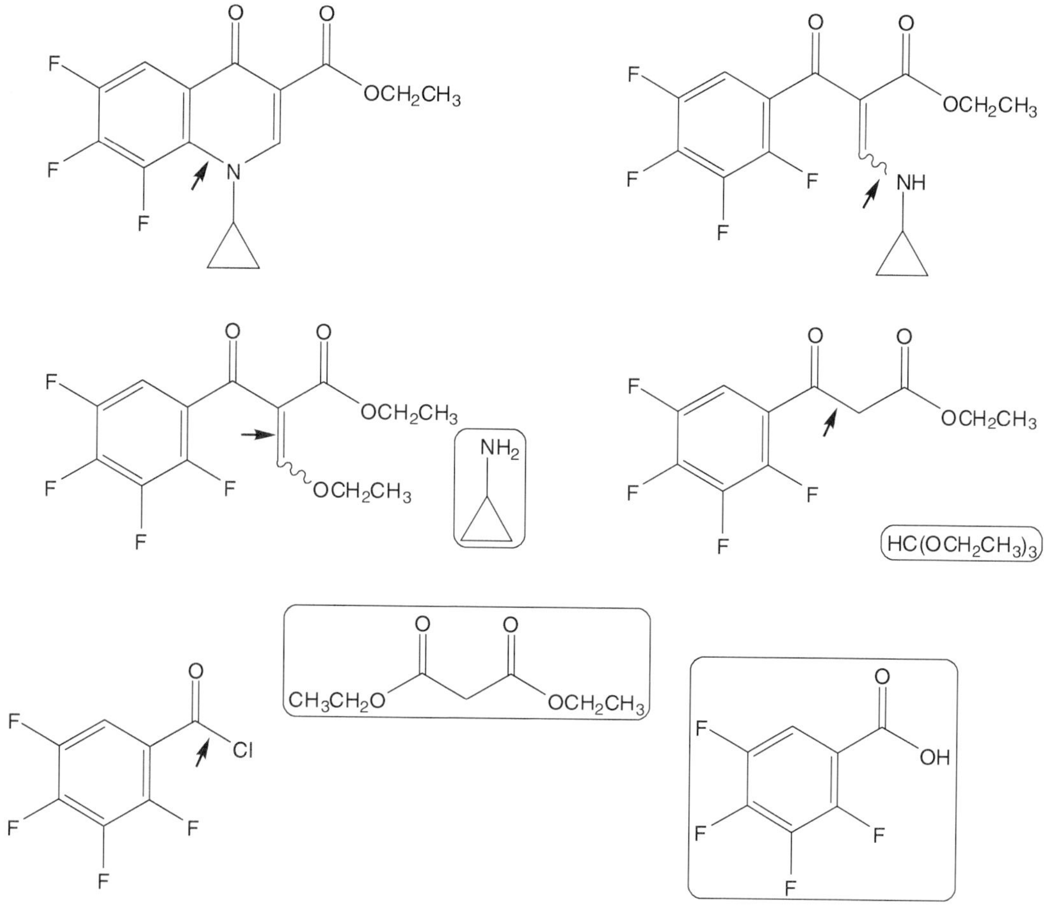

(*S*,*S*)-Octahydro-1*H*-pyrrolo[3,4-*b*]pyridine is formed by debenzylation of the pyrrolidine nitrogen. The (*S*,*S*)- and (*R*,*R*)-enantiomers of 6-benzyloctahydro-1*H*-pyrrolo[3,4-*b*]pyridine are separated by resolution. The pyrrolidine is formed by reduction of the imide. The pyridine ring is reduced. The imide is formed from pyridine-2,3-dicarboxylic acid (quinolinic acid) and benzylamine.

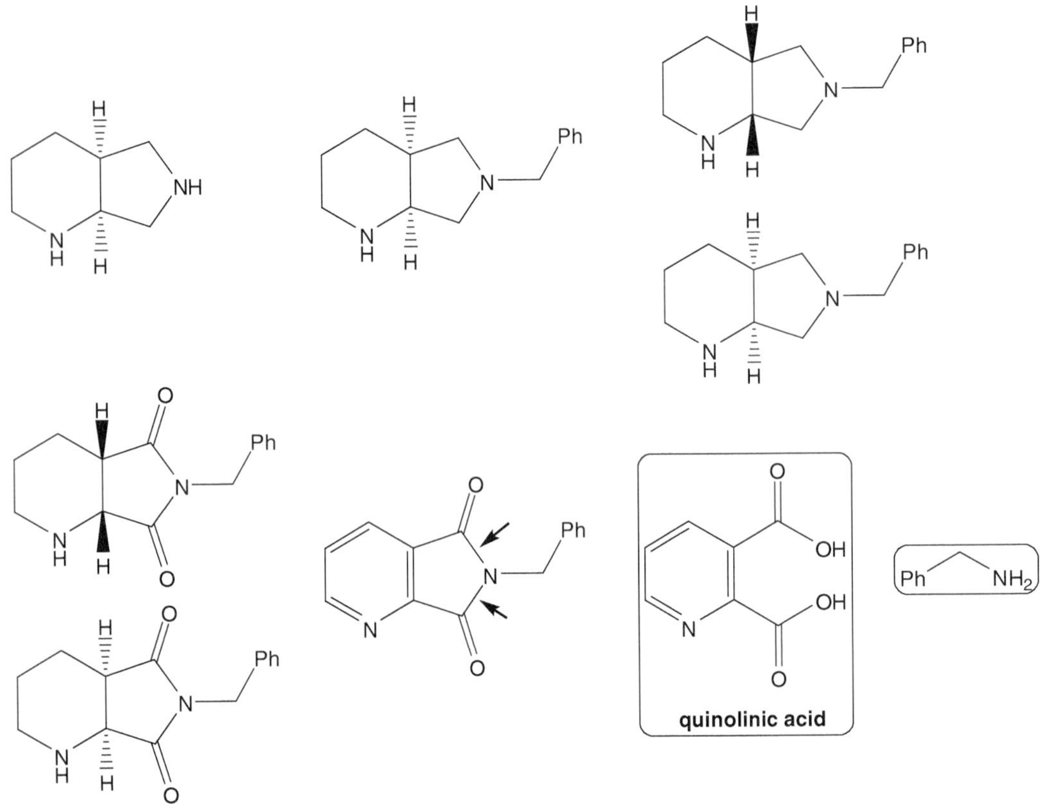

Extended Discussion

An alternative route to moxifloxacin begins with 2,4,5-trifluoro-3-methoxybenzoic acid. Draw the structures of the retrosynthetic analysis of this starting material and then draw the structures of the retrosynthetic analysis of the alternative route. List the pros and cons for the two routes.

N

Naloxone

Antidotes and Other Substances Used in Poisonings/Specific

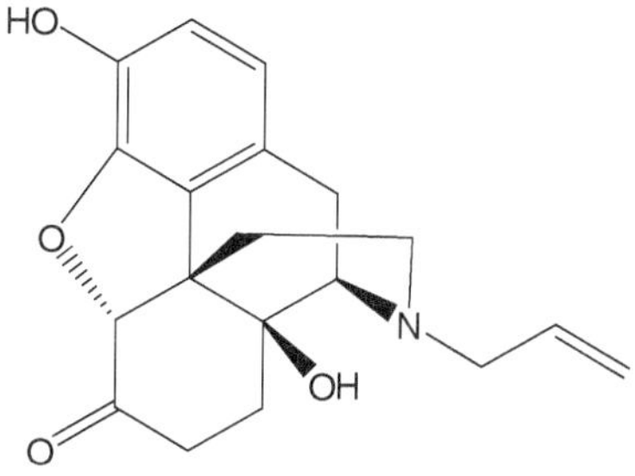

A single-enantiomer molecule with multiple chiral carbons is often formed by the modification of a natural product which has most or all of the chiral carbons already in place.

Discussion. The tertiary amine is formed by bromide displacement from allyl bromide by the secondary amine, noroxymorphone. Noroxymorphone is formed by hydrolysis of acetate esters at C3 and C14 and hydrolysis-decarboxylation of the cyanamide at N17. The cyanamide is formed by *N*-demethylation of the tertiary amine with cyanogen bromide. The diacetate is formed by the reaction of oxymorphone with acetic anhydride. Oxymorphone is formed by *O*-demethylation of oxycodone.

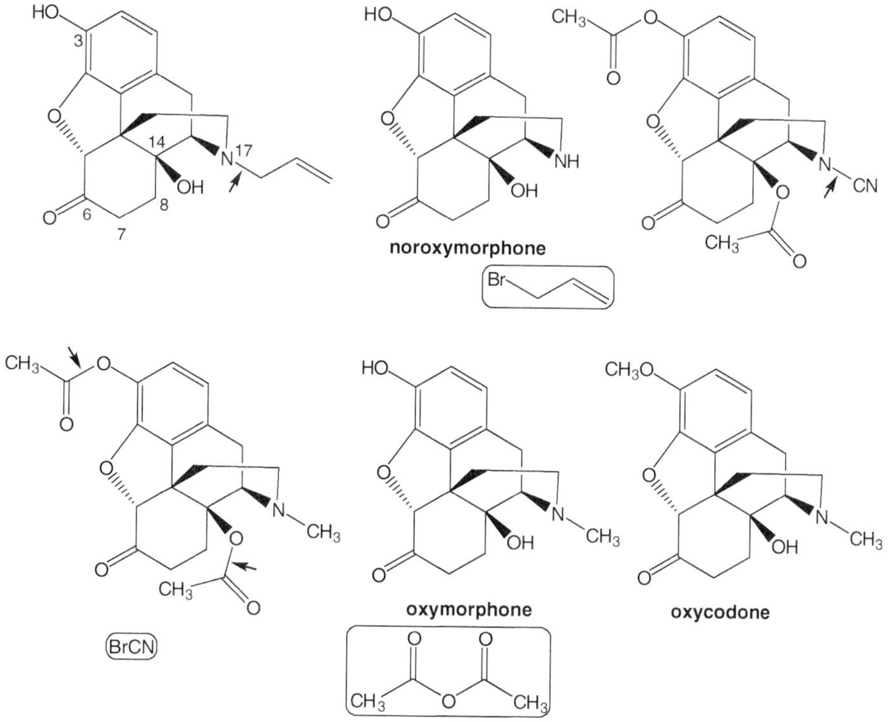

Routes to Essential Medicines: A Workbook for Organic Synthesis, First Edition. Peter J. Harrington.

Companion website: www.wiley.com/go/Harrington/routes_essential_medicine

Oxycodone is manufactured in two steps from the alkaloid thebaine. Oxidation of the thebaine 6,8(14)-diene to the hydroxyenone is followed by reduction of the alkene. Thebaine is the main alkaloid isolated from *Papaver bracteatum* (Persian poppy).

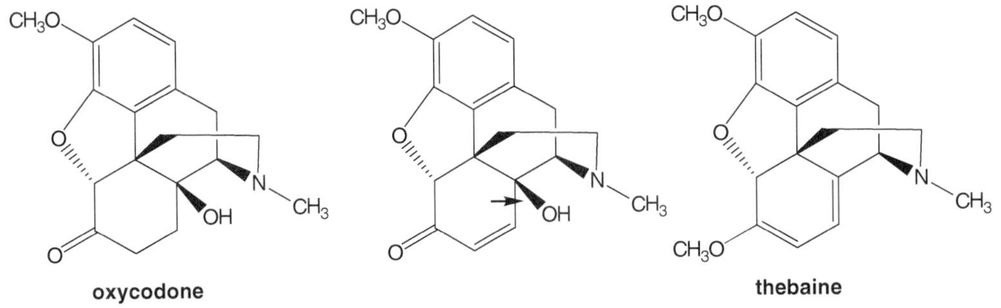

Oxymorphone can also be produced in two steps from the alkaloid oripavine. Oxidation of oripavine 6,8(14)-diene to the hydroxyenone is followed by reduction of the alkene. Oripavine is isolated from the opium poppy *Pavaver somniferum*.

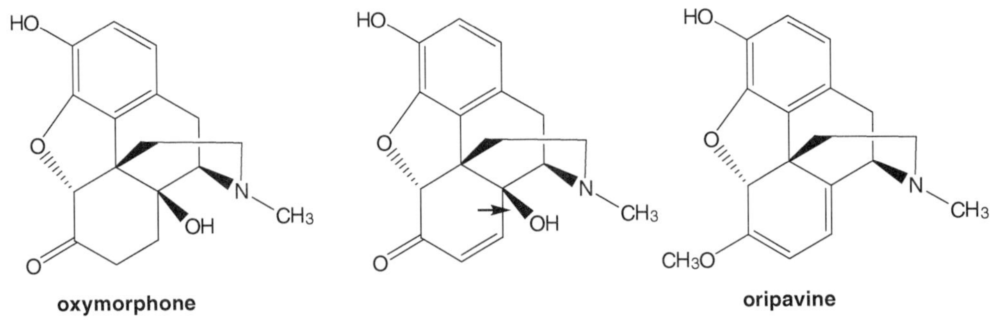

Extended Discussion

Draw the structures of the retrosynthetic analysis of one alternative route from oxymorphone to noroxymorphone. List the pros and cons for both routes and select one route as the preferred route.

Neostigmine Methylsulfate

Muscle-relaxants (Peripherally-acting) and Cholinesterase Inhibitors

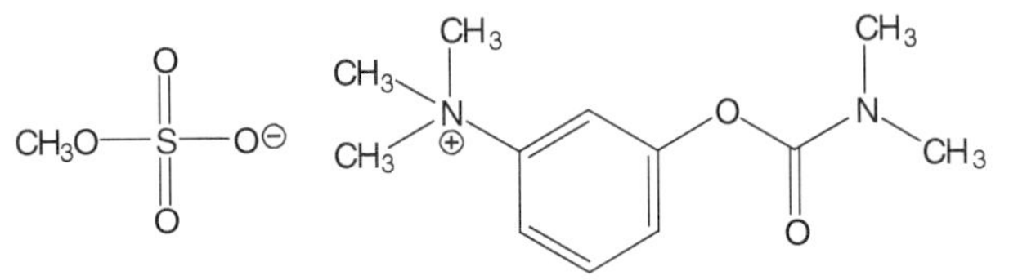

A *meta*-amino phenol is often formed from 1,3-dihydroxybenzene (resorcinol).

Discussion. The quaternary ammonium salt is formed in the final step by methylation of the tertiary amine with dimethyl sulfate. The carbamate is formed from the phenol and dimethylcarbamoyl chloride (dimethylcarbamyl chloride). 3-Dimethylaminophenol is formed from resorcinol and dimethylamine.

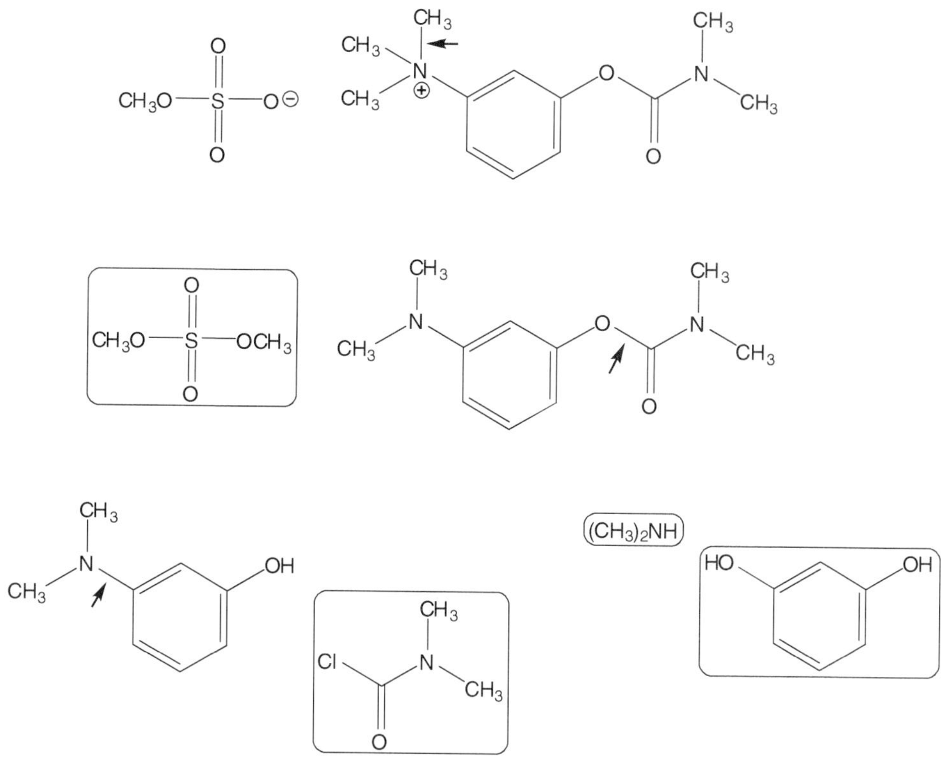

Extended Discussion

Draw the structures of the retrosynthetic analysis of one alternative route to neostigmine methylsulfate that does not utilize dimethylcarbamoyl chloride. Include the structures of the retrosynthetic analysis of any organic starting material(s) from petrochemical or biochemical raw materials. List the pros and cons for both routes. Is one route preferred?

Nevirapine

Anti-infective Medicines/Antiviral Medicines/Antiretrovirals/Non-nucleoside Reverse Transcriptase Inhibitors

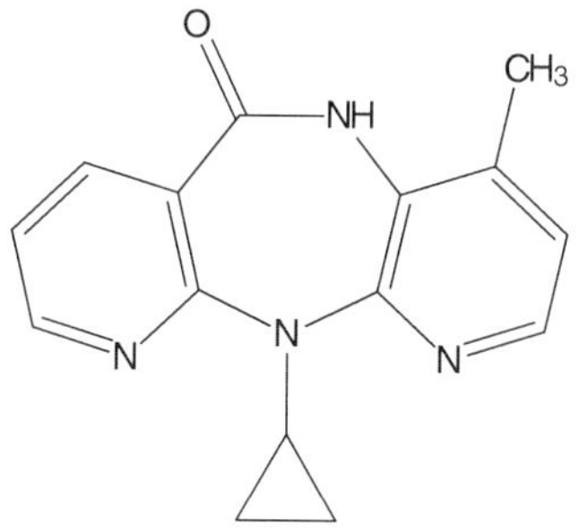

A nitrogen substituent on C2, C4, or C6 of a pyridine ring is often introduced by displacement of chloride by nucleophilic aromatic substitution. The substitution is facilitated by the ring nitrogen and can be further facilitated by an electron-withdrawing group (NO_2, SO_2R, COOR, CN) on pyridine ring C3.

Discussion. Nevirapine is rapidly assembled from two components of similar complexity. The diazepine-5-one ring is formed by displacement of chloride from the 2-chloropyridine by the cyclopropanamine. The amide is formed from the 3-aminopyridine and the pyridine-3-carbonyl chloride.

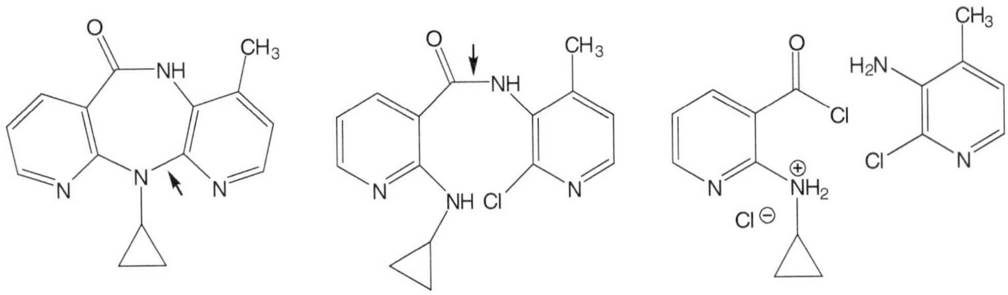

The acid chloride is formed from the carboxylic acid. The carboxylic acid is formed by hydrolysis of the nitrile. 2-Cyclopropyla mino-3-pyridinecarbonitrile is formed by displacement of chloride from 2-chloro-3-pyridinecarbonitrile by cyclopropanamine.

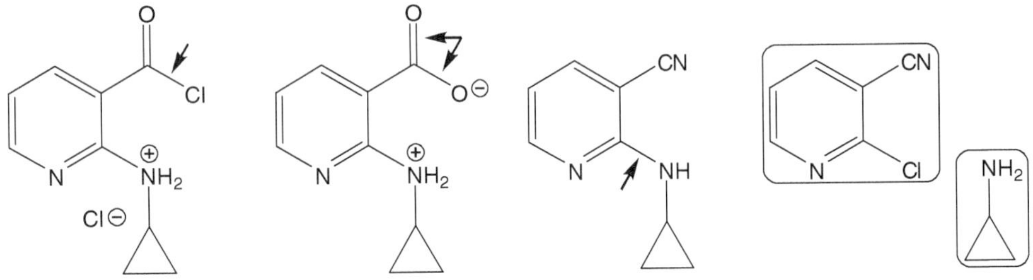

3-Amino-2-chloro-4-methylpyridine is formed from the pyridine-3-carboxamide (**Hofmann Rearrangement**). The 3-carboxamide is formed by hydrolysis of the nitrile. 2-Chloro-4-methylpyridine-3-carbonitrile is formed from the 2-pyridone. The 2-pyridone is formed from the mixture of enol ether and acetal formed by condensation of malononitrile and acetylacetaldehyde dimethyl acetal.

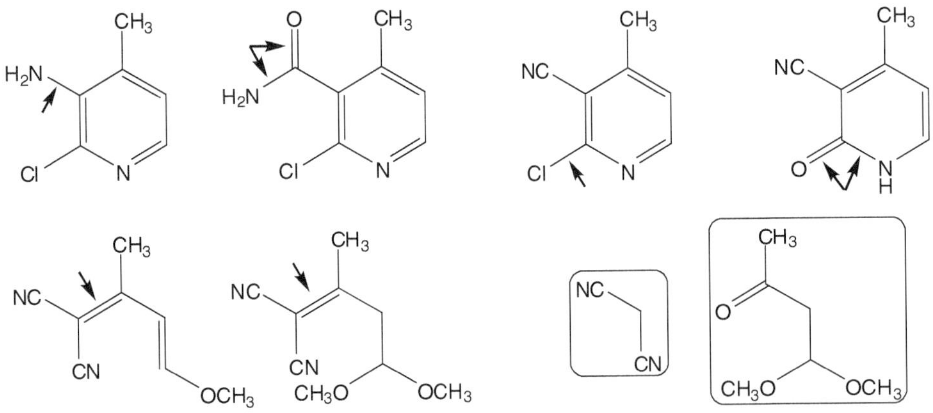

Extended Discussion

Draw the structures of the retrosynthetic analysis of an alternative route to nevirapine from 2-chloropyridine-3-carbonitrile, 3-amino-2-chloro-4-methylpyridine, and cyclopropanamine. List the pros and cons for both routes and select one route as the preferred route.

Niclosamide

Anti-infective Medicines/Intestinal Anthelminthics

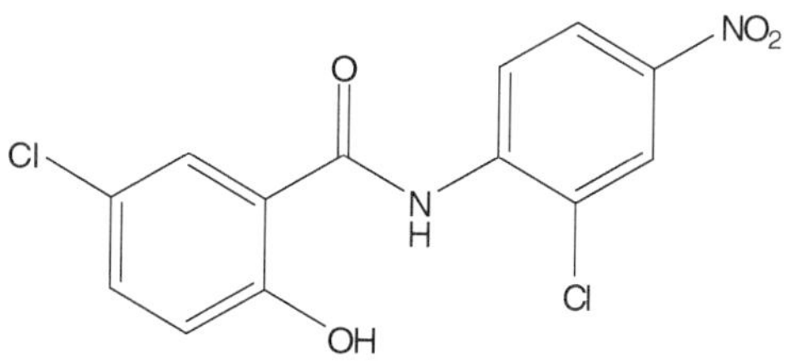

Amides are ubiquitous in drug structures. Amide formation from an amine and acid chloride, anhydride, ester, or carboxylic acid is often very efficient.

Discussion. Niclosamide is formed in a single step from 5-chlorosalicylic acid and 2-chloro-4-nitroaniline.

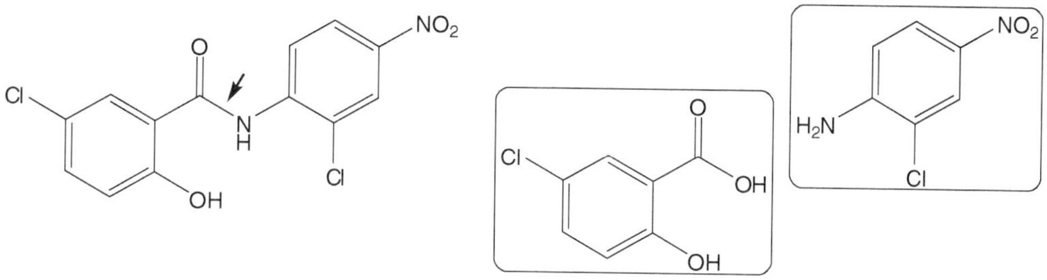

Extended Discussion

Draw the structures of the retrosynthetic analysis of an alternative route to 2-chloro-4-nitroaniline. List the pros and cons for both routes and select one route as the preferred route.

Nicotinamide

Vitamins and Minerals

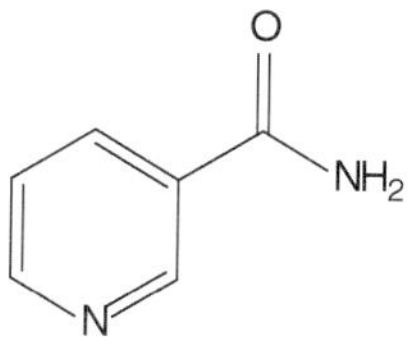

A pyridine with one carbon substituent is often formed from the methylpyridine (picoline).

Discussion. Nicotinamide (niacinamide) is manufactured by nitrile hydratase-mediated hydrolysis of nicotinonitrile.

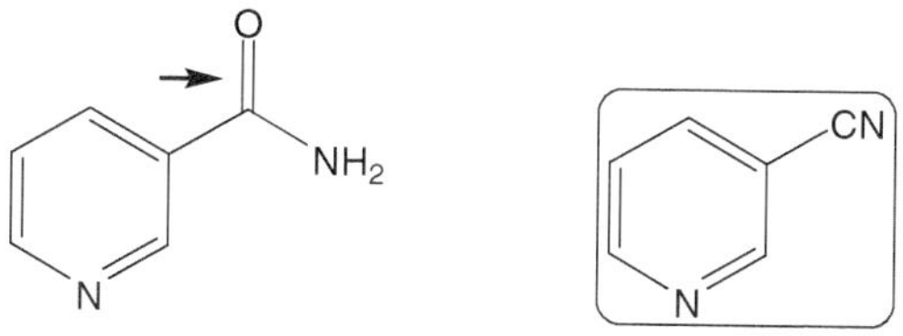

Nifedipine

Oxytocics and Antioxytocics/Antioxytocics (Tocolytics)

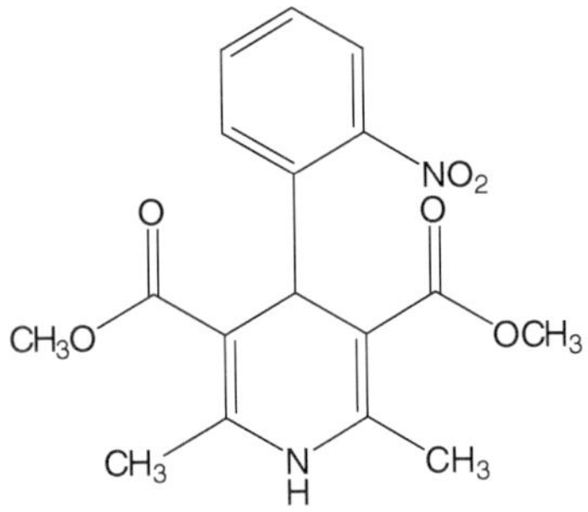

Dihydropyridines are often formed by Hantzsch Synthesis. The key C3–C4 bond of the ring is formed by addition of an enamine to an α,β-unsaturated ketone or ester.

Discussion. Nifedipine is formed by the reaction of methyl 3-amino-2-butenoate (methyl 3-aminocrotonate) with methyl 2-(2-nitrobenzylidene)acetoacetate (**Hantzsch Dihydropyridine Synthesis**). Methyl (2-nitrobenzylidene)acetoacetate is formed from 2-nitrobenzaldehyde and methyl acetoacetate (**Knoevenagel Condensation**).

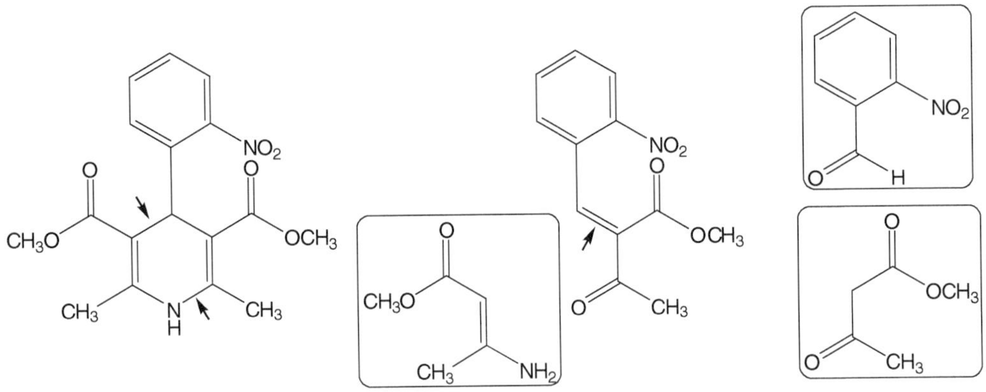

Extended Discussion

The many published methods for synthesis of 2-nitrobenzaldehyde suggest there may be more than one manufacturing process. Draw the structures and the list the key reagents for three routes from 2-nitrotoluene to 2-nitrobenzaldehyde.

Nifurtimox

Anti-infective Medicines/Antiprotozoal Medicines/Antitrypanosomal Medicines/African Trypanosomiasis

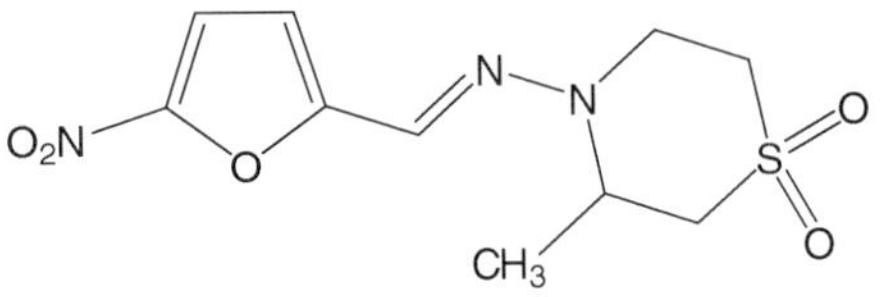

A hydrazone is often formed by condensation of an aldehyde or ketone with a hydrazine.

Discussion. Nifurtimox is a 1 : 1 mixture of the (*R*)- and (*S*)-enantiomers of the (*E*)-hydrazone. The hydrazone is formed by condensation of the aldehyde and the hydrazine in the final step. [Is the (*Z*)-hydrazone also formed in this condensation? If so, how is it separated from nifurtimox?] The aldehyde is formed by in situ hydrolysis of 5-nitro-2-furaldehyde diacetate. The hydrazine, 4-amino-3-methylthiomorpholine-1,1-dioxide is formed from 2-methyl-1,4-oxathiane-4,4-dioxid e. (The hydrazone is likely formed directly from 5-nitro-2-furaldehyde diacetate and the hydrazine. Explain.)

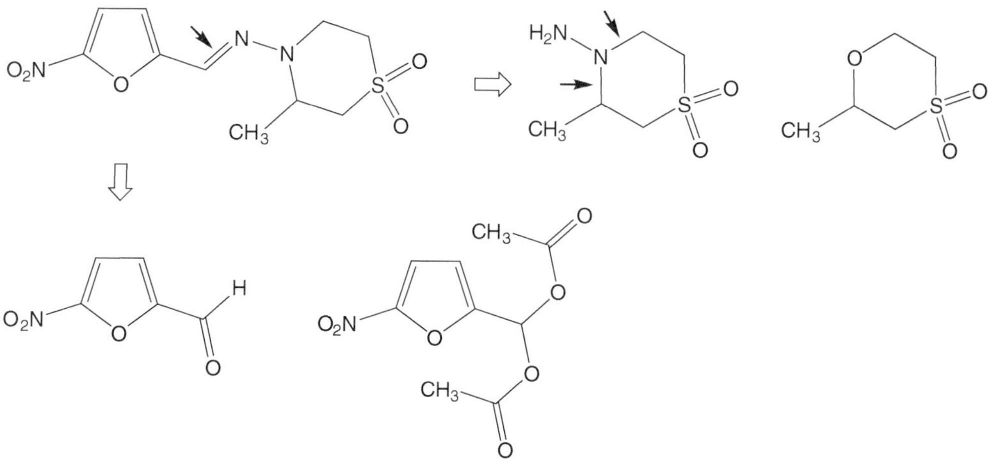

2-Methyl-1,4-oxathiane-4,4-dioxide is formed by oxidation of 2-methyl-1,4-oxathiane. The 1,4-oxathiane is formed from the diol. The diol is formed by ring-opening of propylene oxide with 2-mercaptoethanol.

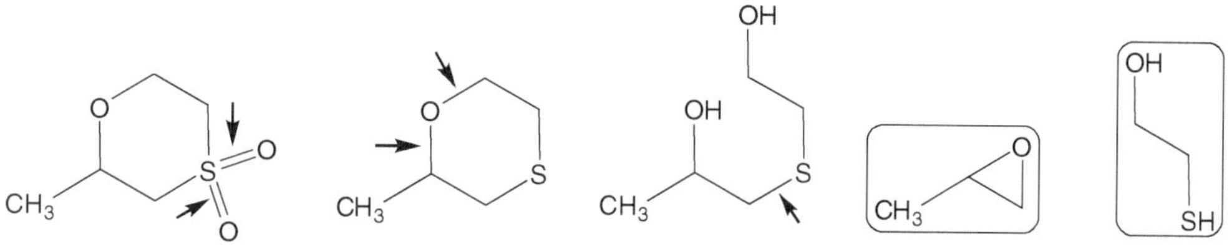

Extended Discussion

Draw the structures of the retrosynthetic analysis of one alternative route to 4-amino-3-methylthiomorpholine-1,1-dioxide. List the pros and cons for both routes to this intermediate and select one route as the preferred route.

Nilotinib

Antineoplastics and Immunosuppressives/Cytotoxic and Adjuvant Medicines

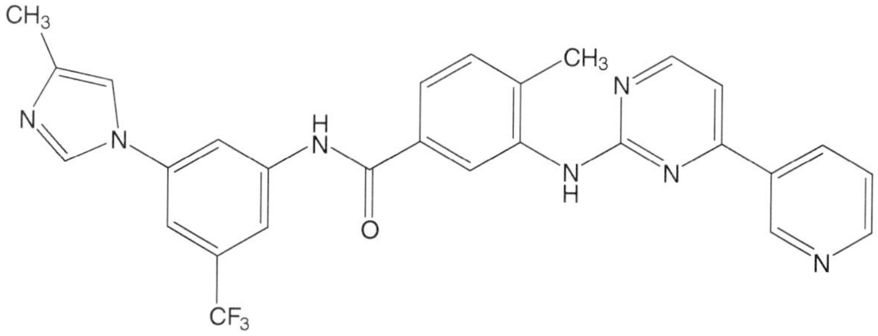

Initial disconnection of a bond near the center of a large molecule is often associated with the most convergent synthesis. Amides are often efficiently formed by reaction of an amine with an acid chloride, anhydride, ester, or carboxylic acid.

Discussion. Nilotinib is rapidly assembled from three components of comparable complexity. The amide C-N bond is formed from the ethyl ester.

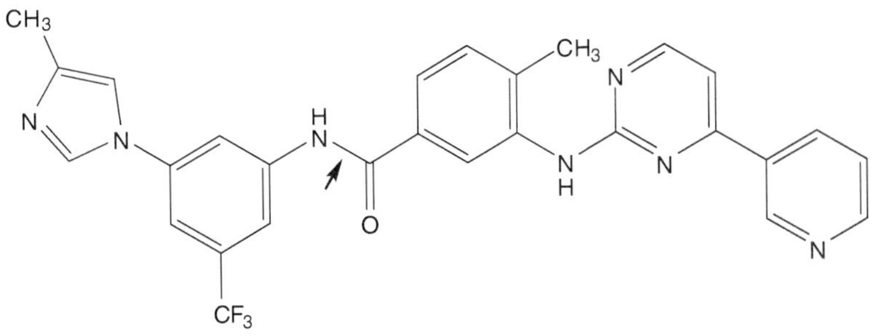

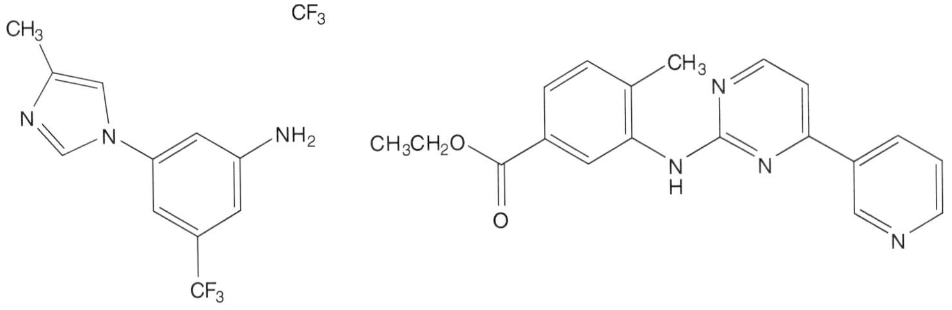

The pyrimidine ring of the ester is formed from a guanidine and a β-amino-α,β-unsaturated ketone (**Pinner Pyrimidine Synthesis**). The β-amino-α,β-unsaturated ketone is formed by the reaction of 3-acetylpyridine with dimethylformamide dimethyl acetal. The guanidine is formed from the amine and cyanamide. Ethyl 3-amino-4-methylbenzoate is formed by reduction of ethyl 4-methyl-3-nitrobenzoate. Ethyl 4-methyl-3-nitrobenzoate is formed by nitration of ethyl *para*-toluate.

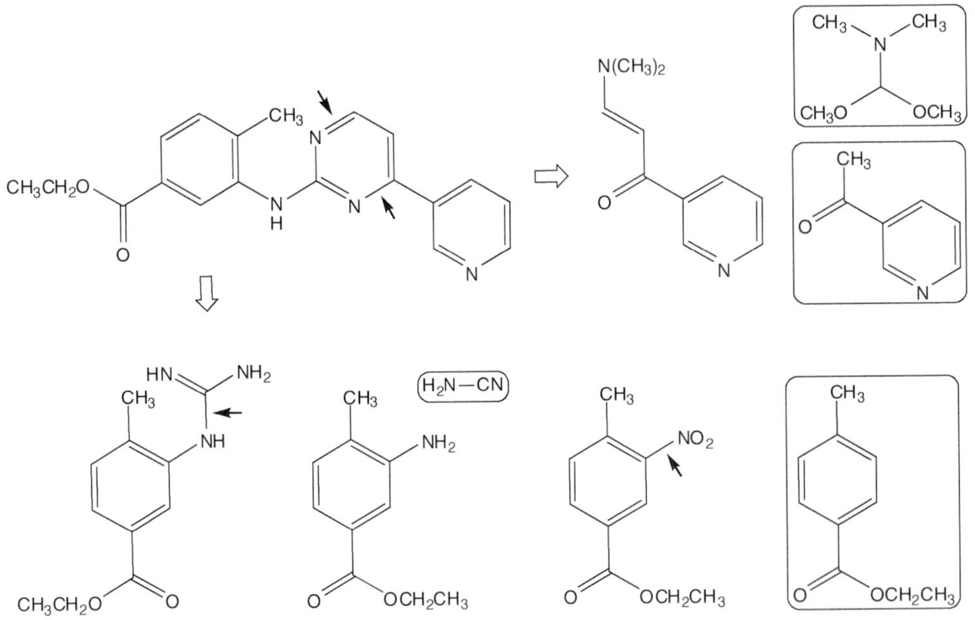

The amine for the final step is formed by displacement of bromide by 4-methylimidazole. [A side product is also formed in this reaction. Draw the structure of the side product. List the options for the reaction conditions (catalyst system, solvent, temperature, time) and the ratio of product to side product and the yield associated with each option.] 3-Bromo-5-(trifluoromethyl)aniline is formed by reduction of 3-bromo-5-nitrobenzotrifluoride. 3-Bromo-5-nitrobenzotrifluoride is formed by bromination of 3-nitrobenzotrifluoride.

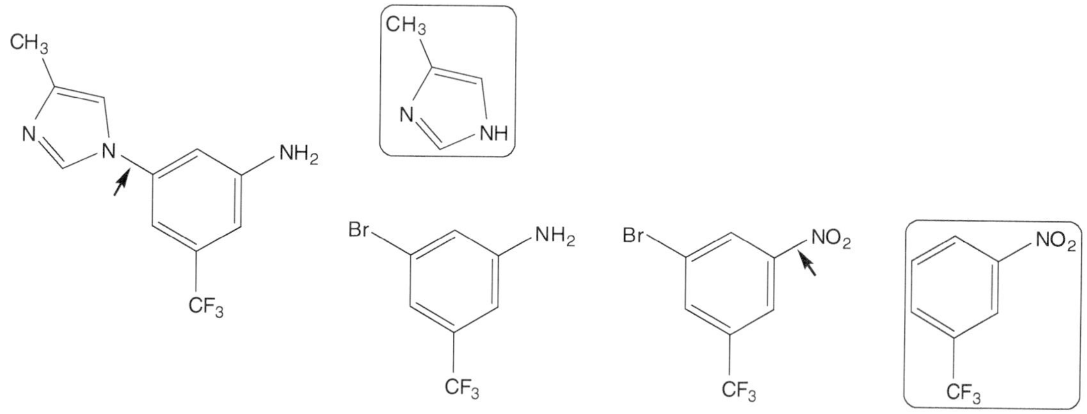

Nitrofurantoin

Anti-infective Medicines/Antibacterials/Other Antibacterials

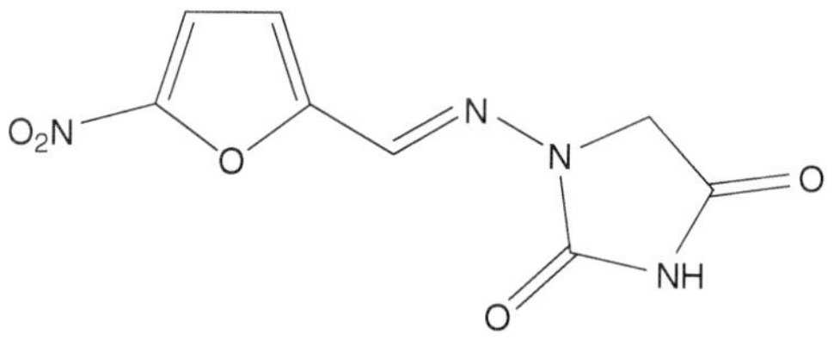

A semicarbazone is often formed by condensation of an aldehyde or ketone with a semicarbazide.

Discussion. Nitrofurantoin is formed by condensation of the aldehyde and 1-aminohydantoin in the final step. [Is the (*Z*)-isomer also formed in this condensation? If so, how is it separated from nitrofurantoin?] The aldehyde is released by in situ hydrolysis of 5-nitro-2-furaldehyde diacetate. 1-Aminohydantoin is released by hydrolysis. (The hydrolysis to form 1-aminohydantoin can be in situ or can be a separate step. Which is preferred?) The hydantoin ring is formed from acetone semicarbazone and ethyl chloroacetate. Acetone semicarbazone is formed from acetone and semicarbazide.

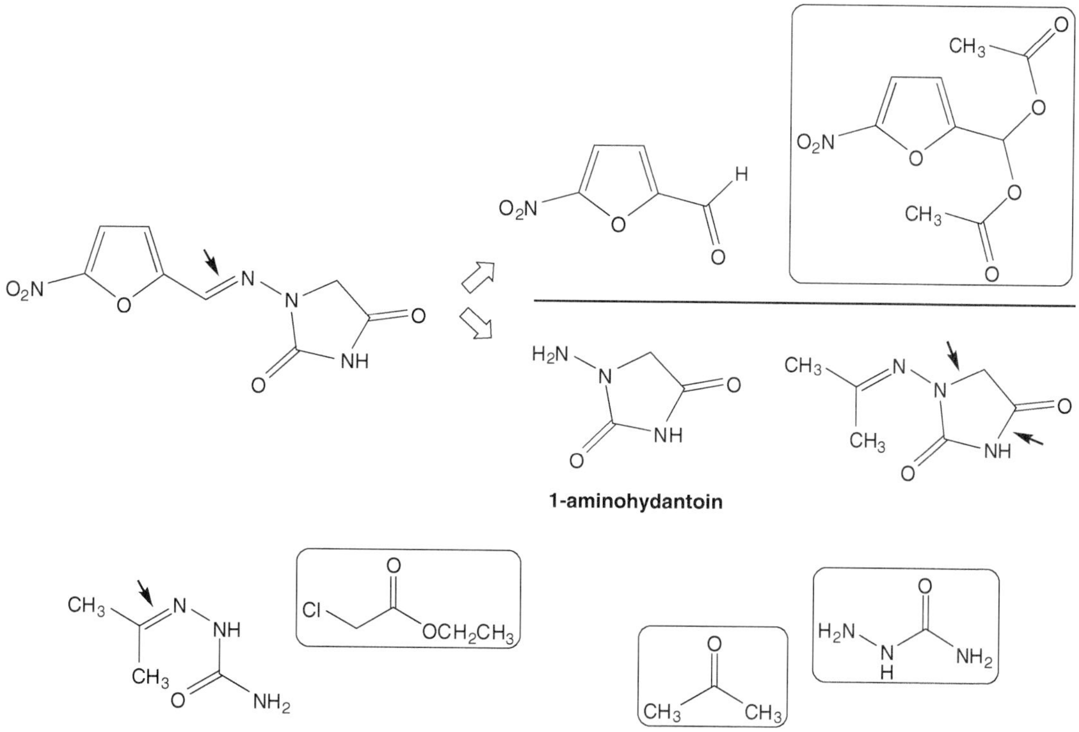

Extended Discussion

1-Aminohydantoin is formed in situ from semicarbazinoacetic acid. Draw the structures of a retrosynthetic analysis of semicarbazinoacetic acid. List the pros and cons for both routes and to nitrofurantoin and select one route as the preferred route.

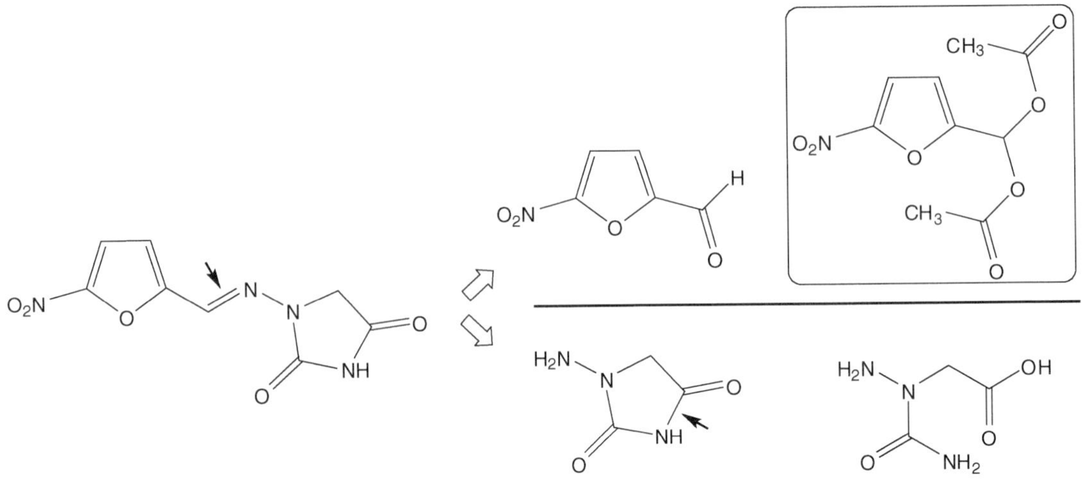

1-aminohydantoin semicarbazinoacetic acid

Norethisterone and Norethisterone Enanate

Hormones, Other Endocrine Medicines, and Contraceptives/Contraceptives/Oral Hormonal Contraceptives
Hormones, Other Endocrine Medicines, and Contraceptives/Contraceptives/Injectable Hormonal Contraceptives

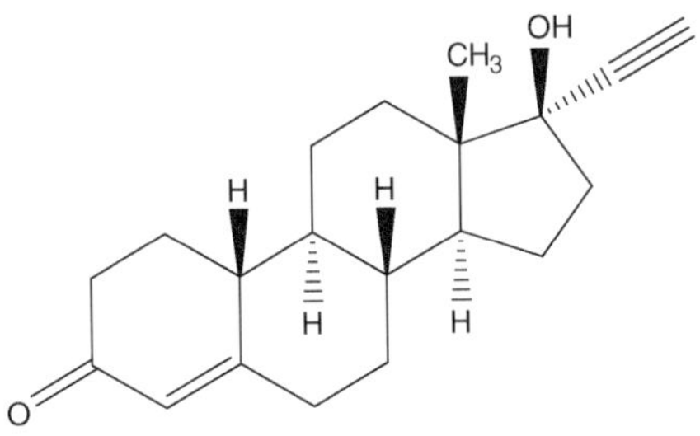

norethisterone

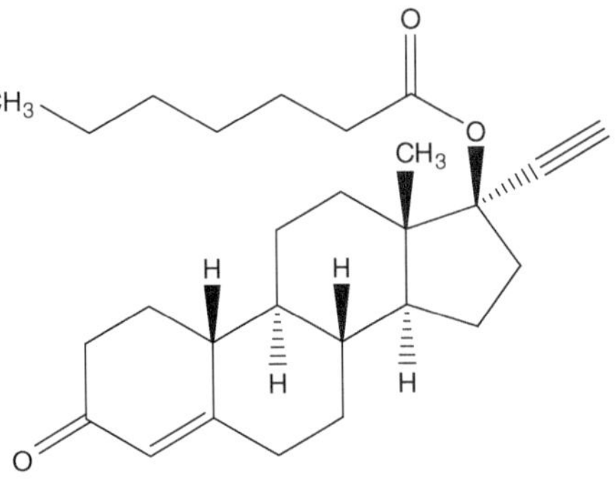

norethisterone enanate

A single-enantiomer molecule with multiple chiral carbons is often formed by modification of a natural product which has most or all of the chiral carbons already in place. A steroid missing the C19 methyl group (a 19-nor-steroid) is often formed by Birch Reduction of a steroid with an aromatic A-ring.

Discussion. Norethisterone is manufactured in seven steps from estrone and in 10 steps from β-sitosterol. Norethisterone enanate is manufactured in 2 steps from norethisterone.

The 4-ene-3-one of norethisterone (norethindrone) is formed from the 3-ethoxy-3,5-diene by hydrolysis of the enol ether followed by double bond migration. The tertiary alcohol at C17 is formed by addition of a metal acetylide to the C17 ketone. The 3-ethoxy-3,5-diene is formed by reaction of estr-4-ene-3,17-dione with triethyl orthoformate. The 4-ene-3-one is formed from the 3-methoxy-2,5-diene by hydrolysis of the enol ether followed by double bond migration. The ketone at C17 is formed by oxidation of the secondary alcohol. The 3-methoxy-2,5-diene and the secondary alcohol at C17 are both formed in the reduction of estrone methyl ether (**Birch Reduction**). Estrone methyl ether is formed by *O*-methylation of estrone (**Williamson Ether Synthesis**). (List the methylating agents used for this *O*-methylation. List the reaction conditions and yield of estrone methyl ether for each methylating agent.)

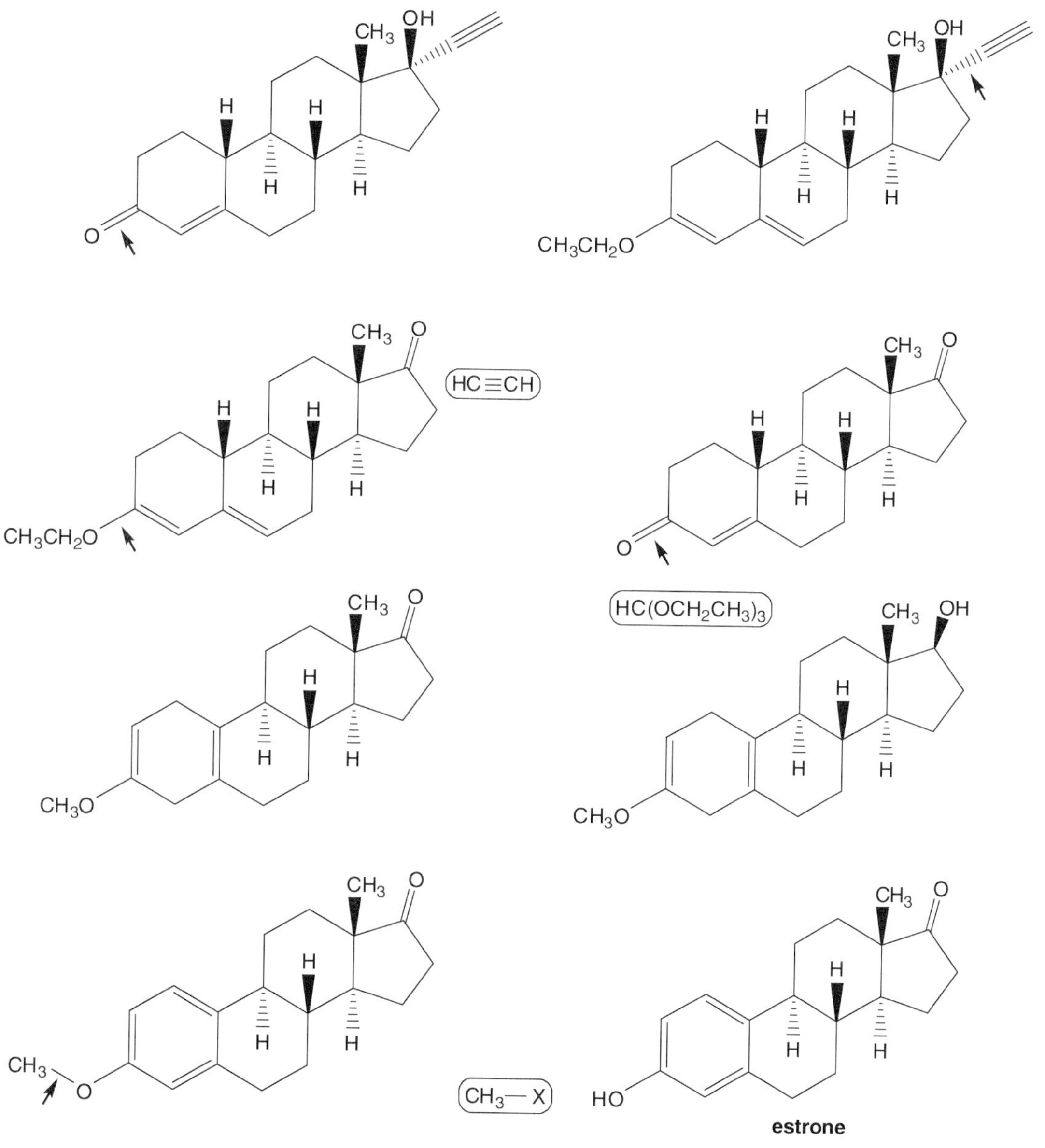

Estrone is formed by A-ring aromatization from androsta-1,4-diene-3,17-dione 17-ethylene glycol ketal (acetal). The acetal is formed from androsta-1,4-diene-3,17-dione (boldione) and ethylene glycol. Androsta-1,4-diene-3,17-dione is formed by microbial oxidation/side chain degradation of phytosterols (plant sterols) including β-sitosterol.

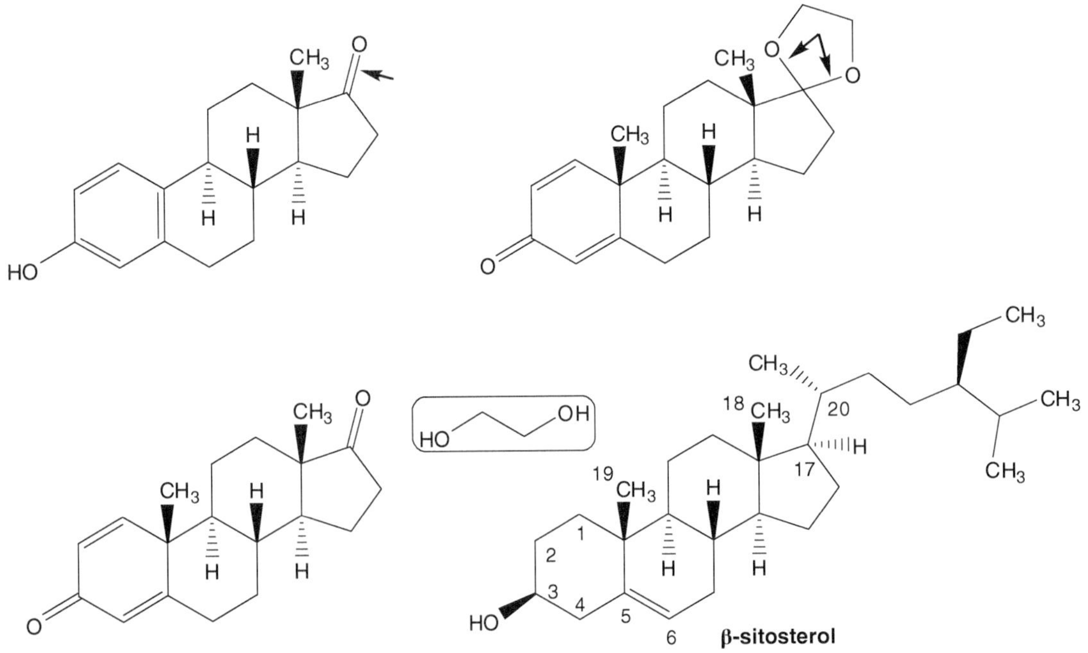

A diester is formed by reaction of norethisterone with heptanoic anhydride (enanthic anhydride). Hydrolysis of the enol ester and double bond migration releases norethisterone enanate.

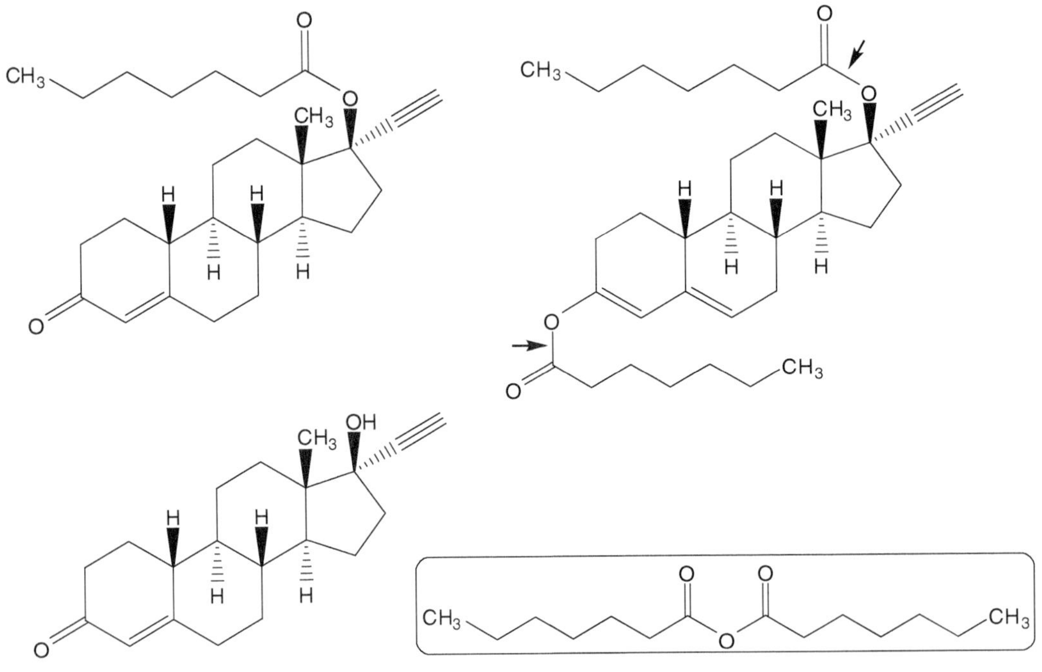

norethisterone

Extended Discussion

Draw the structures of the retrosynthetic analysis of an alternative route to norethisterone from estrone methyl ether that does not have the 3-ethoxy-3,5-diene as an intermediate.

O

Ofloxacin

Ophthalmological Preparations/Anti-infective Agents

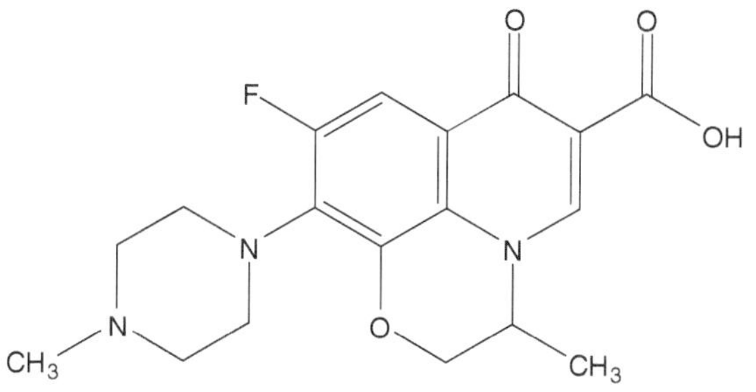

Nucleophilic aromatic substitution is often facilitated by an electron-withdrawing group (NO_2, SO_2R, COOR, CN) on an *ortho* or *para* ring carbon. No electron-withdrawing group is required when the displacement results in the formation of a five- or six-membered ring. Leaving groups for nucleophilic aromatic substitution include fluorine, chlorine, and nitro.

Discussion. Ofloxacin is a 1 : 1 mixture of the (*R*)- and (*S*)-enantiomers. Fluoride at C7 of the quinolone ring is displaced by 1-methylpiperazine in the final step. The carboxylic acid is released by ester hydrolysis.

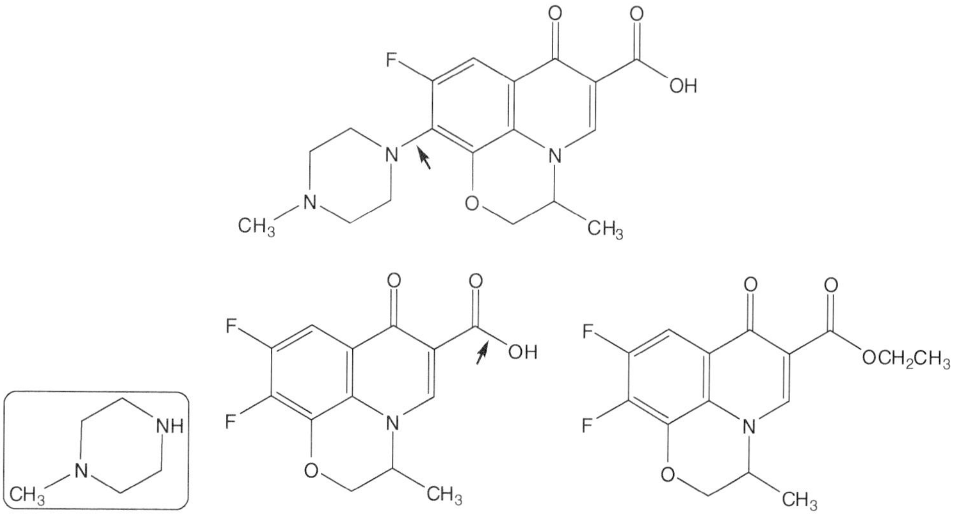

A ring and the C-O bond at quinolone position 8 are formed by intramolecular displacement of fluoride by the alcohol. The quinolone ring is formed by intramolecular displacement of fluoride by the amine. An enamine is formed by displacement of ethanol from the enol ether by 2-amino-1-propanol (DL-alaninol) followed by elimination of ethanol. (Draw the structures of the retrosynthetic analysis of one alternative route to 2-amino-1-propanol.)

Routes to Essential Medicines: A Workbook for Organic Synthesis, First Edition. Peter J. Harrington.

Companion website: www.wiley.com/go/Harrington/routes_essential_medicine

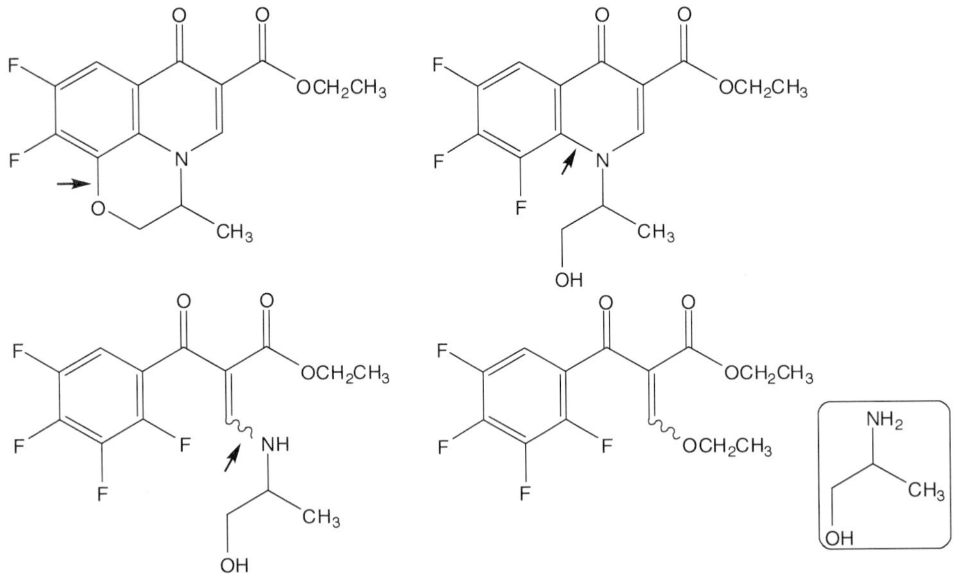

The enol ether is formed by reaction of the β-ketoester with triethyl orthoformate. The β-ketoester is formed from 2,3,4,5-tetrafluorobenzoyl chloride. In one preferred approach, diethyl malonate is the reaction partner. (The four-step conversion of the 2-fluorobenzoyl chloride to the quinolone is known as the **Grohe–Heitzer Sequence**.) The acid chloride is formed from the carboxylic acid.

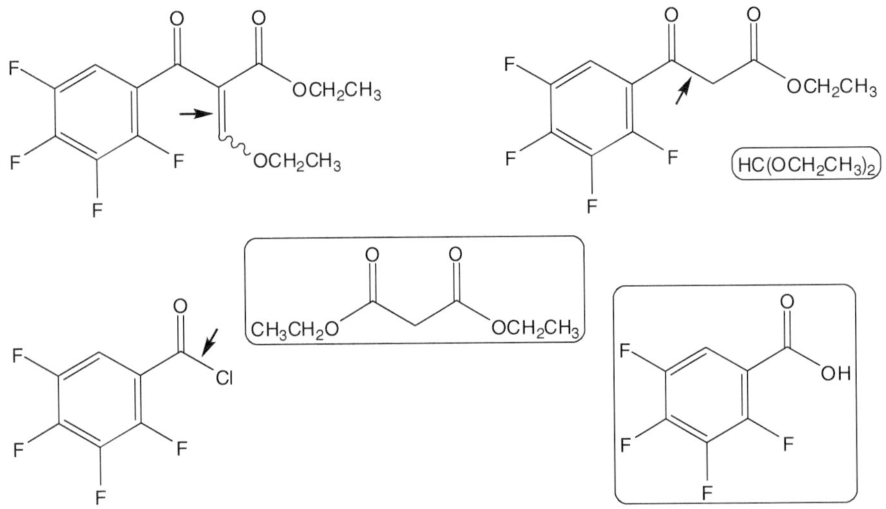

Extended Discussion

Draw the structures of the retrosynthetic analysis of 1,2,3-trifluoro-4-nitrobenzene from petrochemical or biochemical raw materials. Draw the structures of the retrosynthetic analysis of one alternative route to ofloxacin from 1,2,3-trifluoro-4-nitrobenzene. List the pros and cons for the two routes to ofloxacin.

Ombitasvir

Anti-infective Medicines/Antiviral Medicines/Antihepatitis Medicines/Medicines for Hepatitis C/Other Antivirals

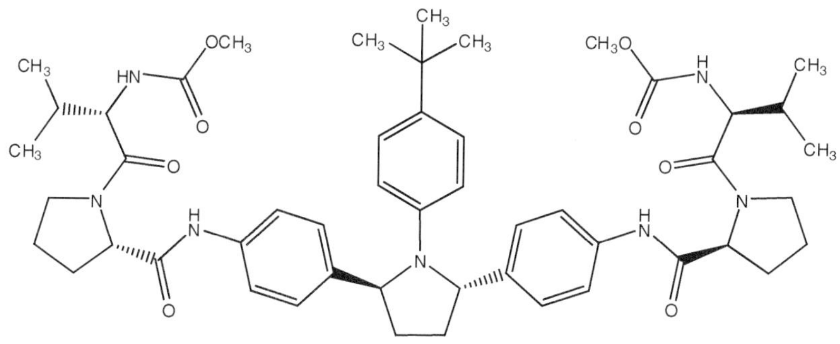

When a target molecule has an axis of symmetry, a preferred route often has intermediates which also have an axis of symmetry.

Discussion. The proline amides are formed in the final step by reaction of the 2,5-bis(4-aminophenyl)pyrrolidine with the carboxylic acid of the Moc-dipeptide *N*-(methoxycarbonyl)-L-valyl-L-proline (Moc-val-pro-OH).

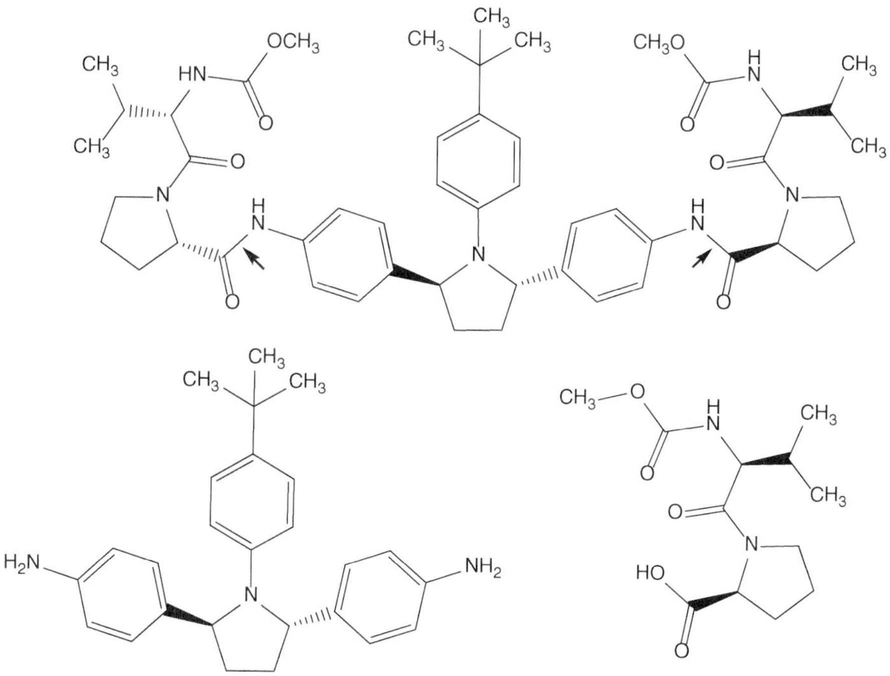

The (2*S*,5*S*)-bis(4-aminophenyl)pyrrolidine is formed by reduction of the (2*S*,5*S*)-bis(4-nitrophenyl)pyrrolidine. The (2*S*,5*S*)-bis(4-nitrophenyl)pyrrolidine is formed by the reaction of 4-*tert*-butylaniline with the (1*R*,4*R*)-dimethanesulfonate. The (1*R*,4*R*)-dimethanesulfonate is formed from the (1*R*,4*R*)-diol. The (1*R*,4*R*)-diol is formed by reduction of the 1,4-dione. 1,4-bis(4-Nitrophenyl)-1,4-butanedione is formed from 2-bromo-4′-nitroacetophenone and 4′-nitroacetophenone.

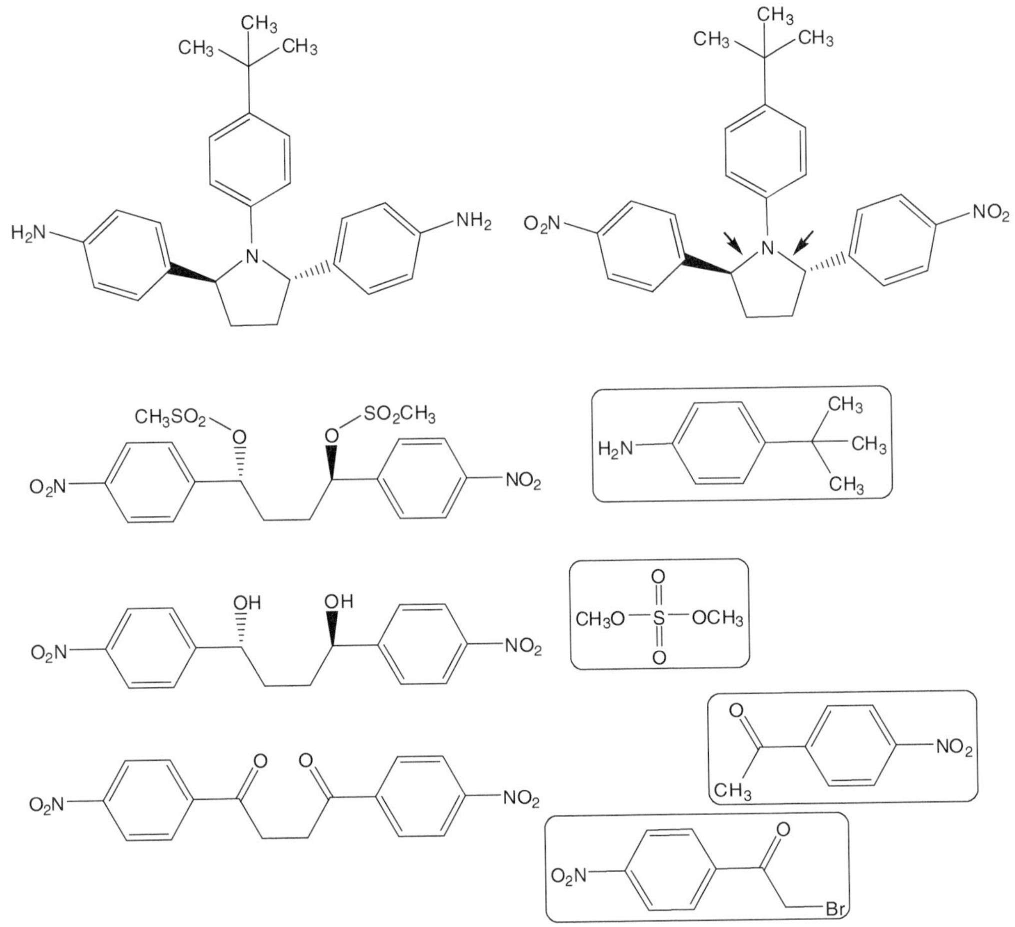

The carboxylic acid of the Moc-dipeptide is released by hydrogenolysis of the benzyl ester (Moc-val-pro-OBn). The Moc-dipeptide amide is formed by reaction of the carboxylic acid of Moc-L-valine (Moc-val-OH) and the amine of L-proline benzyl ester (H-pro-OBn).

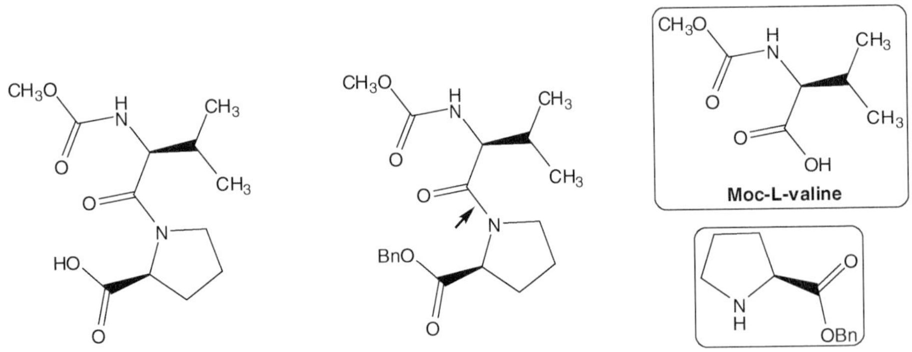

Omeprazole

Gastrointestinal Medicines/Antiulcer Medicines

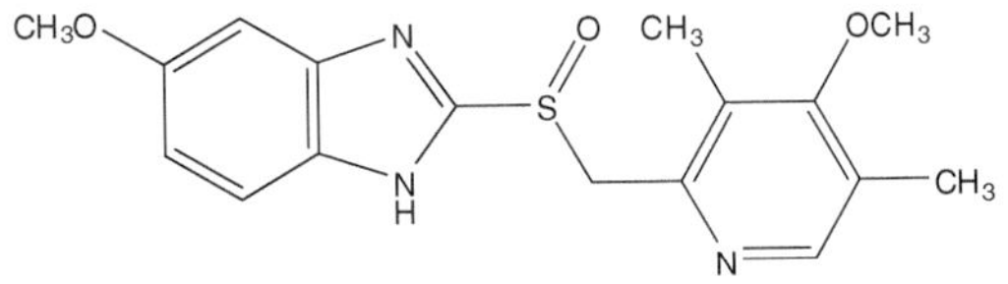

A thioether is often formed by displacement of a leaving group by a thiol.

Discussion. Omeprazole is chiral at sulfur and is manufactured as a 1 : 1 mixture of (*R*)- and (*S*)-enantiomers by oxidation of the thioether to the sulfoxide in the final step. (List the options for oxidizing agents and reaction conditions and draw structures for side products formed for each option.) The thioether is formed by displacement of chloride from 2-chlorom ethyl-4-methoxy-3,5-dimethylpyridine by the thiol, 5-methoxy-2-mercaptobenzimidazole.

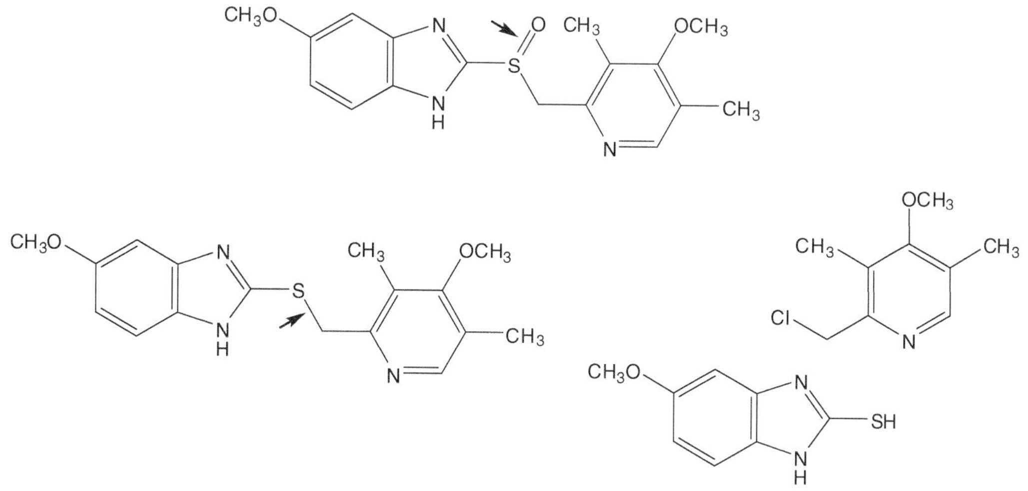

The benzimidazole ring is formed from 4-methoxybenzene-1,2-diamine and potassium ethyl xanthate. The diamine is formed by reduction from 4-methoxy-2-nitroaniline. The nitroaniline is formed by hydrolysis of the nitroacetanilide. 4-Methoxy- 2-nitroacetanilide is formed by nitration of 4-methoxyacetanilide. 4-Methoxyacetanilide is formed from *para*-anisidine and acetic anhydride.

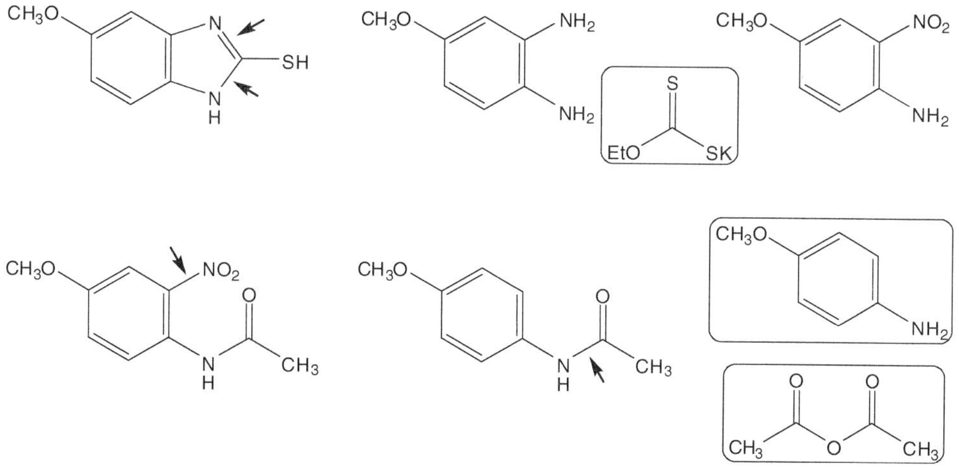

The 2-chloromethylpyridine is formed from the alcohol. The alcohol is formed by hydrolysis of the acetate. The 2-acetoxymethylpyridine is formed by the reaction of 4-methoxy-2,3,5-trimethylpyridine-*N*-oxide with acetic anhydride (**Boekelheide Reaction**). The ether at C4 is formed by displacement of a nitro group by methanol. 2,3,5-Trimethyl-4-nitr opyridine-*N*-oxide is formed by nitration of 2,3,5-trimethylpyridine-*N*-oxide. The pyridine-*N*-oxide is formed by oxidation of 2,3,5-trimethylpyridine (2,3,5-collidine).

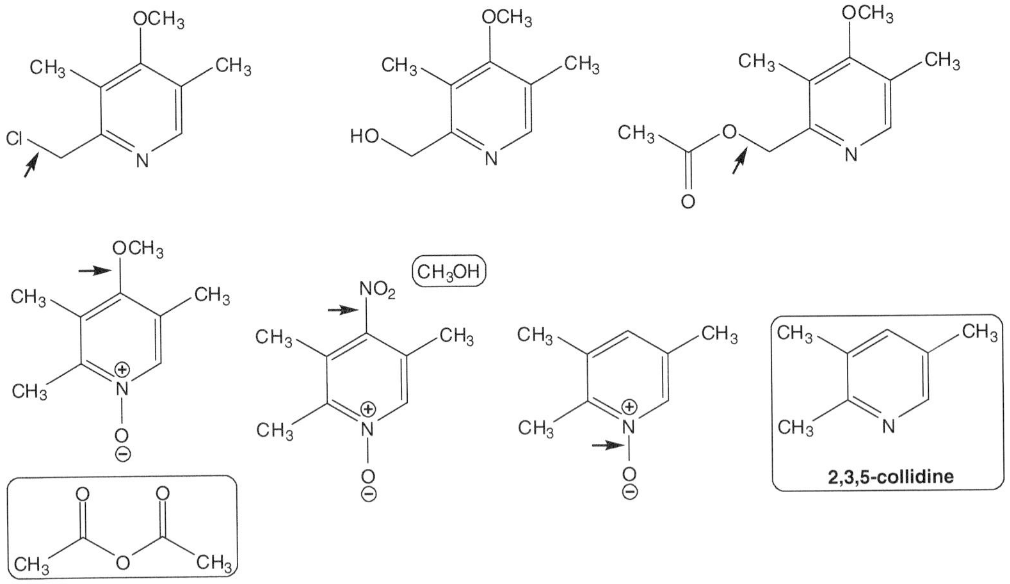

Ondansetron

Medicines for Pain and Palliative Care/Medicines for Other Common Symptoms in Palliative Care
Gastrointestinal Medicines/Antiemetic Medicines

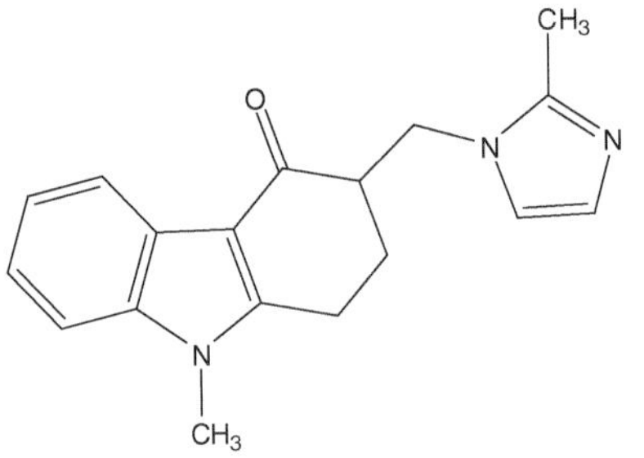

An indole is often formed by Fischer Indole Synthesis from an arylhydrazone. The arylhydrazone is often formed from an arylhydrazine and an aldehyde or ketone.

Discussion. Ondansetron is a 1 : 1 mixture of (*R*)- and (*S*)-enantiomers. Dimethylamine is displaced by 2-methylimidazole in the final step. The dimethylaminomethyl group is formed by the reaction of 9-methyl-4-oxo-1,2,3,4-tetrahydrocarbazole with formaldehyde and dimethylamine (**Mannich Reaction**). The tetrahydrocarbazole is methylated at N9 with dimethyl sulfate. 4-Oxo-1,2,3,4-tetrahydrocarbazole is formed from phenylhydrazine and 1,3-cyclohexanedione (**Fischer Indole Synthesis**).

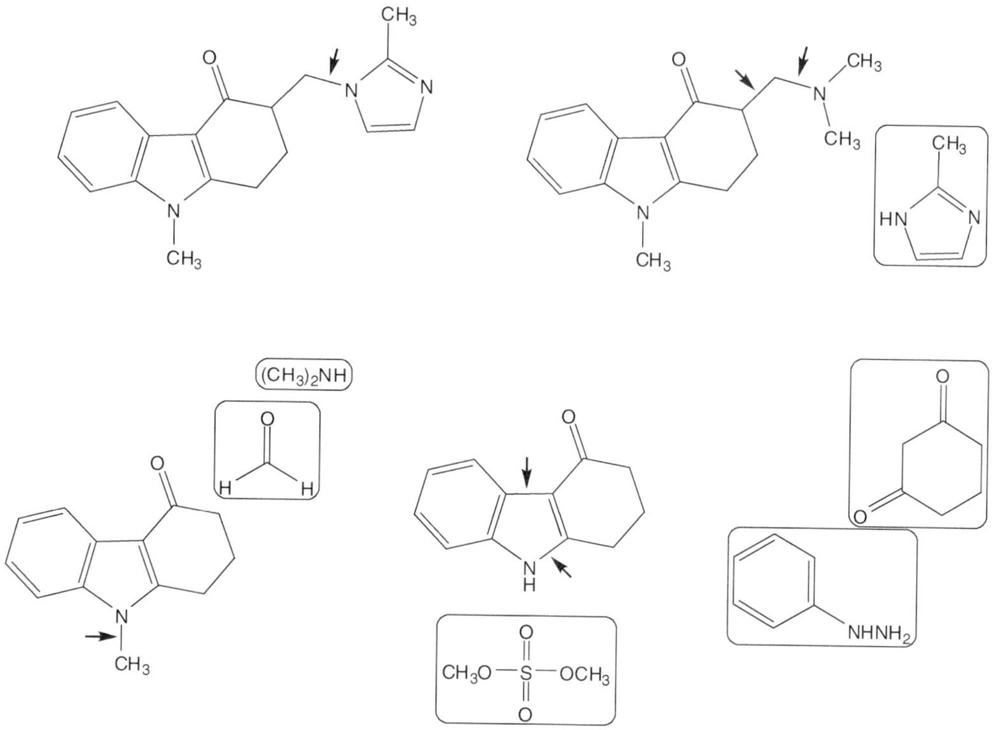

Extended Discussion

Draw the structures of the retrosynthetic analysis of one alternative route to ondansetron that does not utilize a Fischer Indole Synthesis to construct the indole. Include the structures of the retrosynthetic analysis of any organic starting material(s) from petrochemical or biochemical raw materials. List pros and cons for the two routes and select one route as the preferred route.

Oseltamivir

Anti-infective Medicines/Antiviral Medicines/Other Antivirals

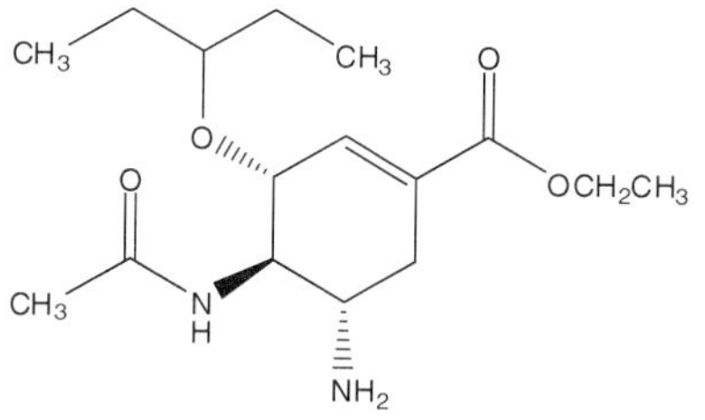

A *trans*-cycloalkane-1,2-diamine is often formed by the ring-opening of an aziridine with an azide or amine.

Discussion. The 5α-primary amine is formed by azide reduction in the final step. The amide is formed by reaction of the 4β-primary amine with acetic anhydride. The 4β-primary amine and the 5α-azide are formed by ring-opening of the 4β,5-aziridine with sodium azide. (What is the regioselectivity of the aziridine ring-opening?) The 4β,5-aziridine is formed from a mixture of two azido alcohols. (What reagent(s) are used for this reaction?) The mixture of azido alcohols is formed by ring-opening of the 4α,5-epoxide with sodium azide. (What is the regioselectivity of the epoxide ring-opening?)

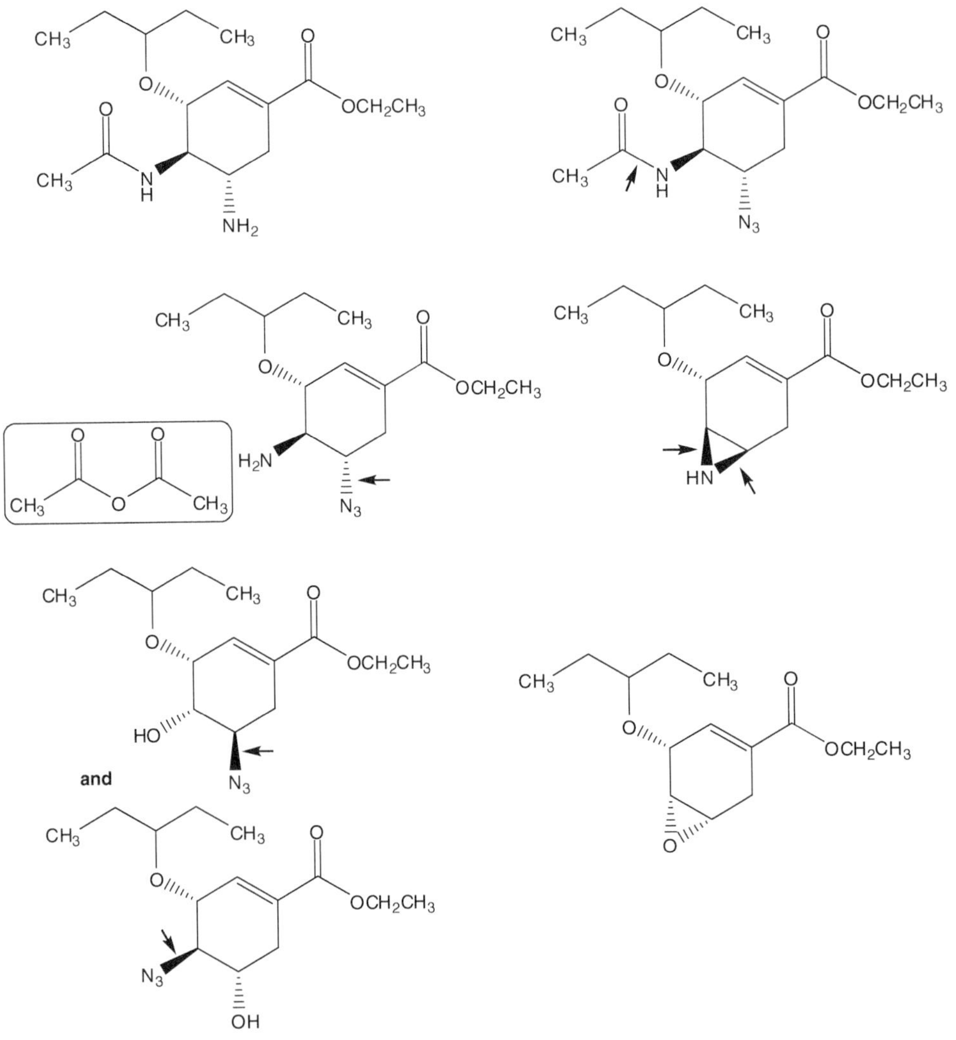

The key 4α,5-epoxide intermediate is formed in six steps from (−)-shikimic acid. The epoxide ring is formed by nucleophilic displacement of the 5β-methanesulfonate by the 4α-alcohol. The 4α-alcohol and the 3α-ether are formed by reductive cleavage of the acetal. (What is the regioselectivity of this reductive cleavage?) The acetal is formed from the acetone acetal and 3-pentanone by acid-catalyzed transacetalization. The 5β-methanesulfonate is formed from the 5β-alcohol. The acetone acetal is formed from ethyl shikimate. (Why is 3-pentanone not used in this step to avoid the transacetalization in a later step?) Ethyl shikimate is formed from (−)-shikimic acid (**Fischer Esterification**). (−)-Shikimic acid is isolated from Chinese star anise.

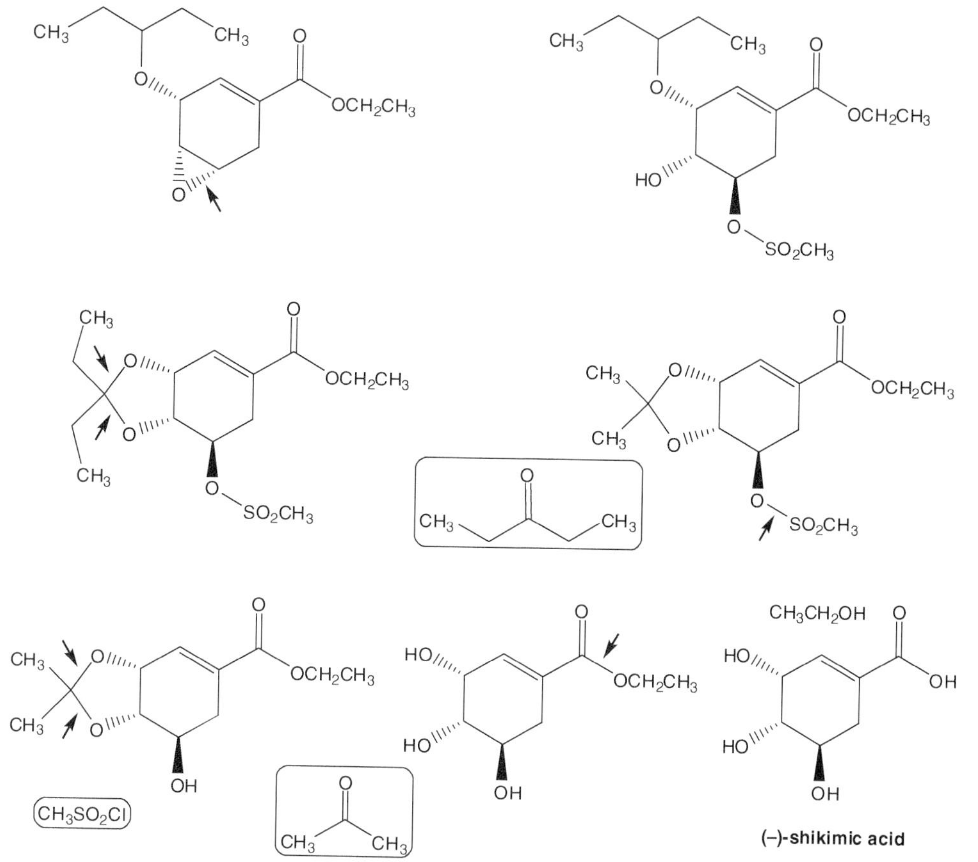

Extended Discussion

Draw the structures of the retrosynthetic analysis of one alternative azide-free route to oseltamivir from (−)-shikimic acid. Include the structures of the retrosynthetic analysis of any organic starting material(s) from petrochemical or biochemical raw materials.

or

Draw the structures of the retrosynthetic analysis of one alternative route to oseltamivir which does not use (−)-shikimic acid as the starting material. Include the structures of the retrosynthetic analysis of any organic starting material(s) from petrochemical or biochemical raw materials.

Oxamniquine

Anti-infective Medicines/Anthelminthics/Antischistosomals and Other Antitrematode Medicines

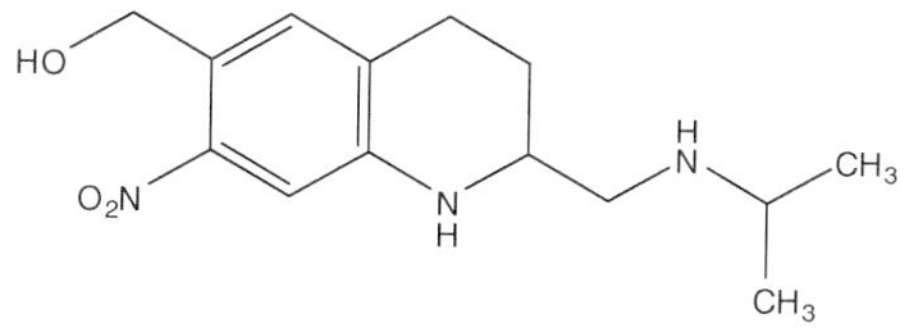

A 1,2,3,4-tetrahydroquinoline is often formed by hydrogenation of a quinoline.

Discussion. Oxamniquine is a 1 : 1 mixture of the (*R*)- and (*S*)-enantiomers. The 1,2,3,4-tetrahydroquinoline ring suggests a late-stage hydrogenation of a quinoline. Since the nitro group would be reduced during a ring hydrogenation, the nitration must be after the ring hydrogenation. The benzylic alcohol would be reduced during a ring hydrogenation and would be oxidized during a nitration so the benzyl alcohol must be formed last.

The alcohol is produced by fermentation in the final step. Nitration of the 6-methyl-1,2,3,4-tetrahydroquinoline results in a mixture of regioisomers. (Why is the 7-nitro isomer the major product?) The tetrahydroquinoline is formed by hydrogenation of the quinoline.

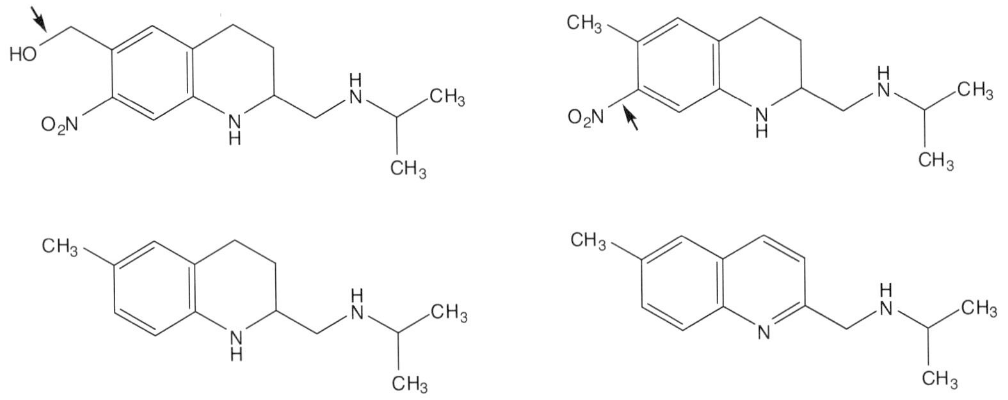

The secondary amine is formed by displacement of chloride by isopropylamine. (What conditions are used to ensure the secondary amine is the major product?) 2-Chloromethyl-6-methylquinoline is formed by chlorination of 2,6-dimethylquinoline. (Why is the chlorination selective for the 2-methyl group?) 2,6-Dimethylquinoline is formed from *para*-toluidine and crotonaldehyde (**Skraup–Doebner-von Miller Synthesis**).

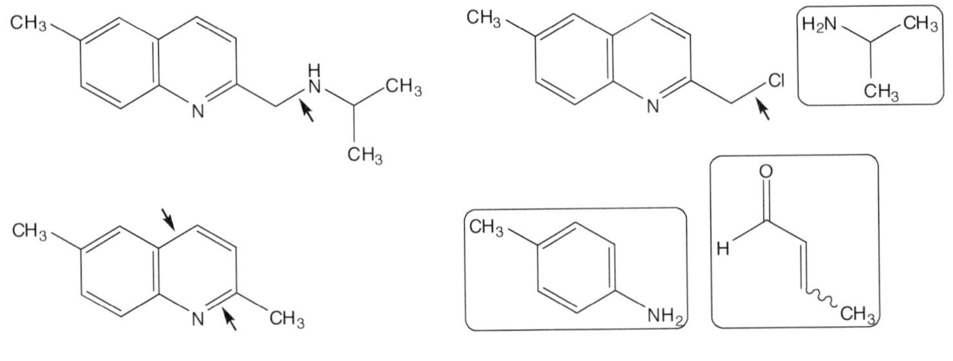

Extended Discussion

Draw the structures of the retrosynthetic analysis of one alternative route to 2,6-dimethylquinoline. Include the structures of the retrosynthetic analysis of any organic starting material(s) from petrochemical or biochemical raw materials. List the pros and cons for both routes to 2,6-dimethylquinoline. Is one route preferred?

Oxytocin

Oxytocics and Antioxytocics/Oxytocics

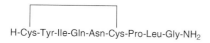

H-Cys-Tyr-Ile-Gln-Asn-Cys-Pro-Leu-Gly-NH_2

A polypeptide is often constructed from the constituent amino acids by forming the amide bonds. The amide bonds are formed via a solid-phase peptide synthesis, a solution-phase peptide synthesis, or a hybrid approach using both solid-phase and solution-phase methods.

Discussion. To simplify polypeptide synthetic schemes, the amino acids are represented by acronyms and are assumed to be in the L-configuration. Protecting groups used in construction of the polypeptide are also represented by acronyms. A disulfide is represented by a bridge connecting the amino acids involved in the bridge. Polypeptide amino acid sequences are drawn with the C-terminal amino acid (carboxylic acid) on the right and the *N*-terminal amino acid (amino group) on the left. (Create a Table of acronyms used in the analysis. Draw the structure associated with each acronym.)

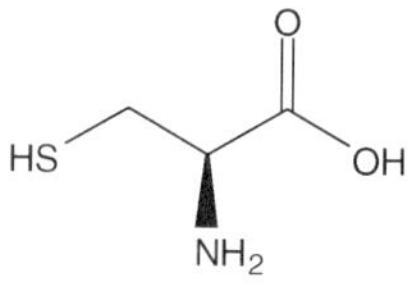

L-cysteine

H-Cys-OH

H of α-NH OH of COOH

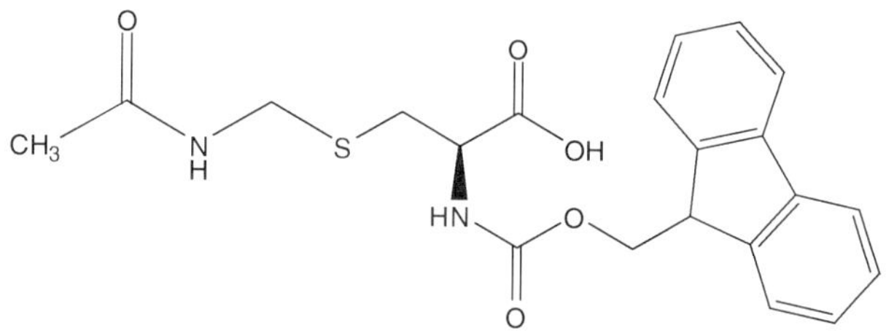

Fmoc-Cys(Acm)-OH

protecting group for α-amino

protecting group off the peptide chain

High-purity oxytocin is constructed by solid-phase peptide synthesis (SPPS). The final step in one SPPS is a global deprotection and release of the amide from a Rink Amide resin. The disulfide is formed on-resin by reaction of the two acetamidomethyl(Acm) thioethers with iodine.

H-Cys-Tyr-Ile-Gln-Asn-Cys-Pro-Leu-Gly-NH_2

↑

Boc-Cys-Tyr(tBu)-Ile-Gln(Trt)-Asn(Trt)-Cys-Pro-Leu-Gly-NH-Resin

Boc-Cys(Acm)-Tyr(tBu)-Ile-Gln(Trt)-Asn(Trt)-Cys(Acm)-Pro-Leu-Gly-NH-Resin

The nonapeptide chain is constructed right-to-left starting with the C-terminal amino acid. The amide is formed by reaction of Fmoc-Gly-OH and a polymer-bound amino group. An amino group is released by Fmoc-deprotection. An amide bond is formed by reaction of this amino group with the carboxylic acid of the next Fmoc-protected amino acid. Fmoc-deprotection and amide bond formation are repeated six times to produce the polymer-bound octaapeptide. The final Fmoc-deprotection is followed by reaction of the amino group with the carboxylic acid of Boc-Cys(Acm)-OH.

Boc-Cys(Acm)-Tyr(tBu)-Ile-Gln(Trt)-Asn(Trt)-Cys(Acm)-Pro-Leu-Gly-NH-Resin

Boc-Cys(Acm)-OH   H-Tyr(tBu)-Ile-Gln(Trt)-Asn(Trt)-Cys(Acm)-Pro-Leu-Gly-NH-Resin

Fmoc-Tyr(tBu)-Ile-Gln(Trt)-Asn(Trt)-Cys(Acm)-Pro-Leu-Gly-NH-Resin

Fmoc-Tyr(tBu)-OH   H-Ile-Gln(Trt)-Asn(Trt)-Cys(Acm)-Pro-Leu-Gly-NH-Resin

Fmoc-Ile-Gln(Trt)-Asn(Trt)-Cys(Acm)-Pro-Leu-Gly-NH-Resin

Fmoc-Ile-OH   H-Gln(Trt)-Asn(Trt)-Cys(Acm)-Pro-Leu-Gly-NH-Resin

Fmoc-Gln(Trt)-Asn(Trt)-Cys(Acm)-Pro-Leu-Gly-NH-Resin

Fmoc-Gln(Trt)-OH   H-Asn(Trt)-Cys(Acm)-Pro-Leu-Gly-NH-Resin

Fmoc-Asn(Trt)-Cys(Acm)-Pro-Leu-Gly-NH-Resin

Fmoc-Asn(Trt)-OH   H-Cys(Acm)-Pro-Leu-Gly-NH-Resin

Fmoc-Cys(Acm)-Pro-Leu-Gly-NH-Resin

Fmoc-Cys(Acm)-OH   H-Pro-Leu-Gly-NH-Resin

Fmoc-Pro-Leu-Gly-NH-Resin

Fmoc-Pro-OH   H-Leu-Gly-NH-Resin

Fmoc-Leu-Gly-NH-Resin

Fmoc-Leu-OH   H-Gly-NH-Resin

Fmoc-Gly-NH-Resin

Fmoc-Gly-OH NH_2 - Resin

In summary, the nonapeptide backbone is constructed by forming eight amides in a solid-phase peptide synthesis. Each amide is formed via an active ester which is formed in situ from the carboxylic acid. Eight of the active esters can be racemized during the amide bond formation (List reagent(s) used to form the active esters in peptide synthesis.) Eight Fmoc protecting groups are used. (List reagents used to remove Fmoc protecting groups in peptide synthesis.)

Routes to the protected amino acid starting materials are presented. (Draw the structure of each protected amino acid starting material.) Cysteine, tyrosine, isoleucine, glutamine, proline, and leucine are produced by fermentation.

Boc-Cys(Acm)-OH	H-Cys(Acm)-OH	H-Cys-OH		
	$(Boc)_2O$	Acm-OH		
Fmoc-Tyr(tBu)-OH	Fmoc-Tyr(tBu)-OCH_3	Fmoc-Tyr-OCH_3	H-Tyr-OCH_3	H-Tyr-OH
		$CH_2{=}C(CH_3)_2$	Fmoc-Cl	CH_3OH
Fmoc-Ile-OH	H-Ile-OH			
	Fmoc-Cl			
Fmoc-Gln(Trt)-OH	H-Gln(Trt)-OH	Z-Gln(Trt)-OH	Z-Gln-OH	H-Gln-OH
	Fmoc-Cl		Trt-Cl	Z-Cl
Fmoc-Asn(Trt)-OH	H-Asn(Trt)-OH	Z-Asn(Trt)-OH	Z-Asn-OH	H-Asn-OH
	Fmoc-Cl		Trt-Cl	Z-Cl
Fmoc-Cys(Acm)-OH	H-Cys(Acm)-OH	H-Cys-OH		
	Fmoc-Cl	Acm-OH		
Fmoc-Pro-OH	H-Pro-OH			
	Fmoc-Cl			
Fmoc-Leu-OH	H-Leu-OH			
	Fmoc-Cl			
Fmoc-Gly-OH	H-Gly-OH			
	Fmoc-Cl			

Extended Discussion

Solution-phase syntheses of oxytocin were developed by du Vigneaud, Stoll, Nobuhara, Velluz, and others. Draw a scheme for the retrosynthetic analysis of one solution-phase synthesis of oxytocin.

P

Paclitaxel

Antineoplastics and Immunosuppressives/Cytotoxic and Adjuvant Medicines

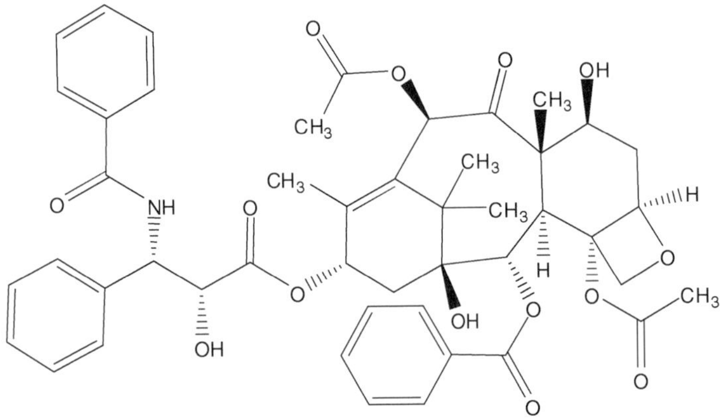

A single-enantiomer molecule with multiple chiral carbons is often formed by the modification of a natural product which has most or all of the chiral carbons already in place. Taxanes are often formed by modification of 10-deacetylbaccatin III. 10-Deacetylbaccatin III is a diterpenoid isolated from the leaves of the European yew Taxus baccata.

Discussion. Paclitaxel is a natural diterpenoid first isolated from the Pacific yew *Taxus brevifola*. Paclitaxel is currently manufactured by plant cell fermentation (PCF) using cells cultured from the needles of the Chinese yew *Taxus chinensis*.

Some paclitaxel is still semisynthetic. In the semisynthesis, the final step is the release of the hydroxyl at C7 from the silyl ether and release of the hydroxyl at C2′ and the amide NH at C3′ from the oxazolidine. The ester at C13 is formed from the carboxylic acid and secondary alcohol.

Routes to Essential Medicines: A Workbook for Organic Synthesis, First Edition. Peter J. Harrington.

Companion website: www.wiley.com/go/Harrington/routes_essential_medicine

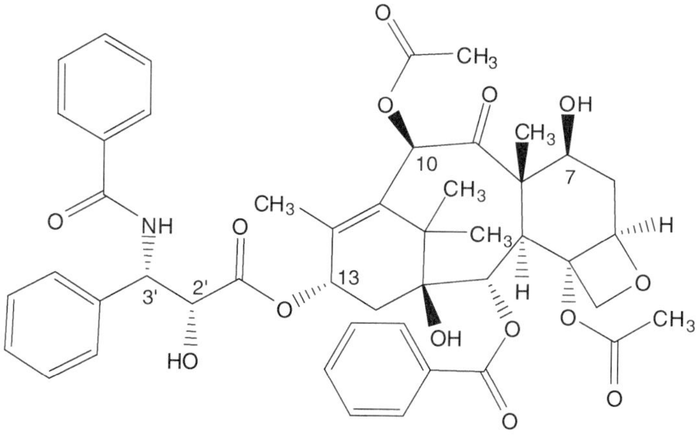

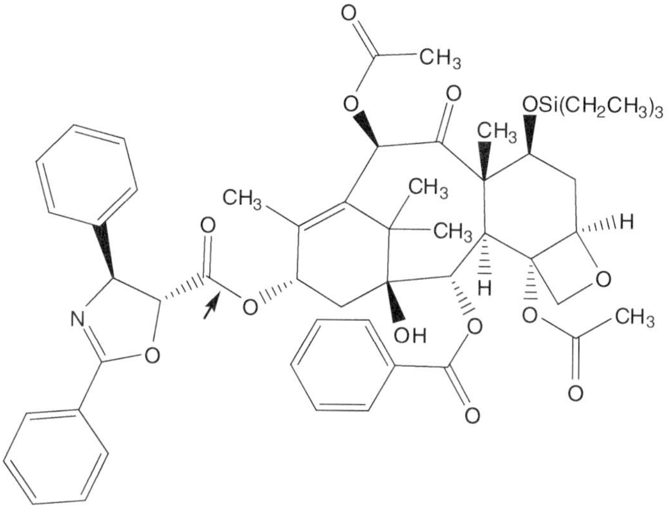

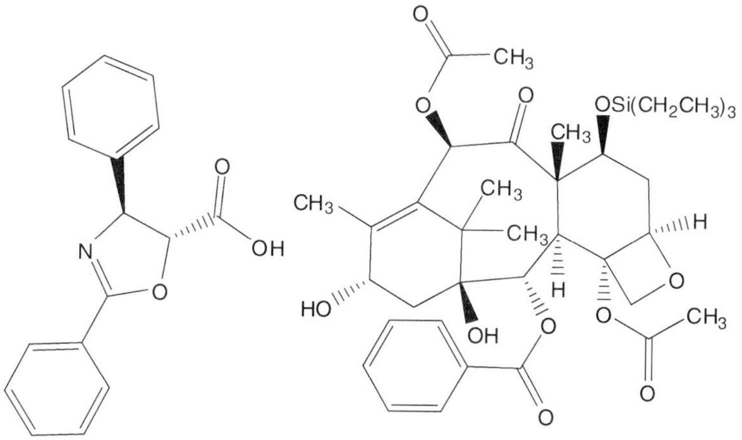

The acetate ester at C10 is formed by the reaction of the secondary alcohol with acetic anhydride. The C7 secondary alcohol of 10-deacetylbaccatin III (10-DAB) is protected as the triethylsilyl ether. 10-Deacetylbaccatin III is a diterpenoid isolated from the leaves of the European yew *Taxus baccata*.

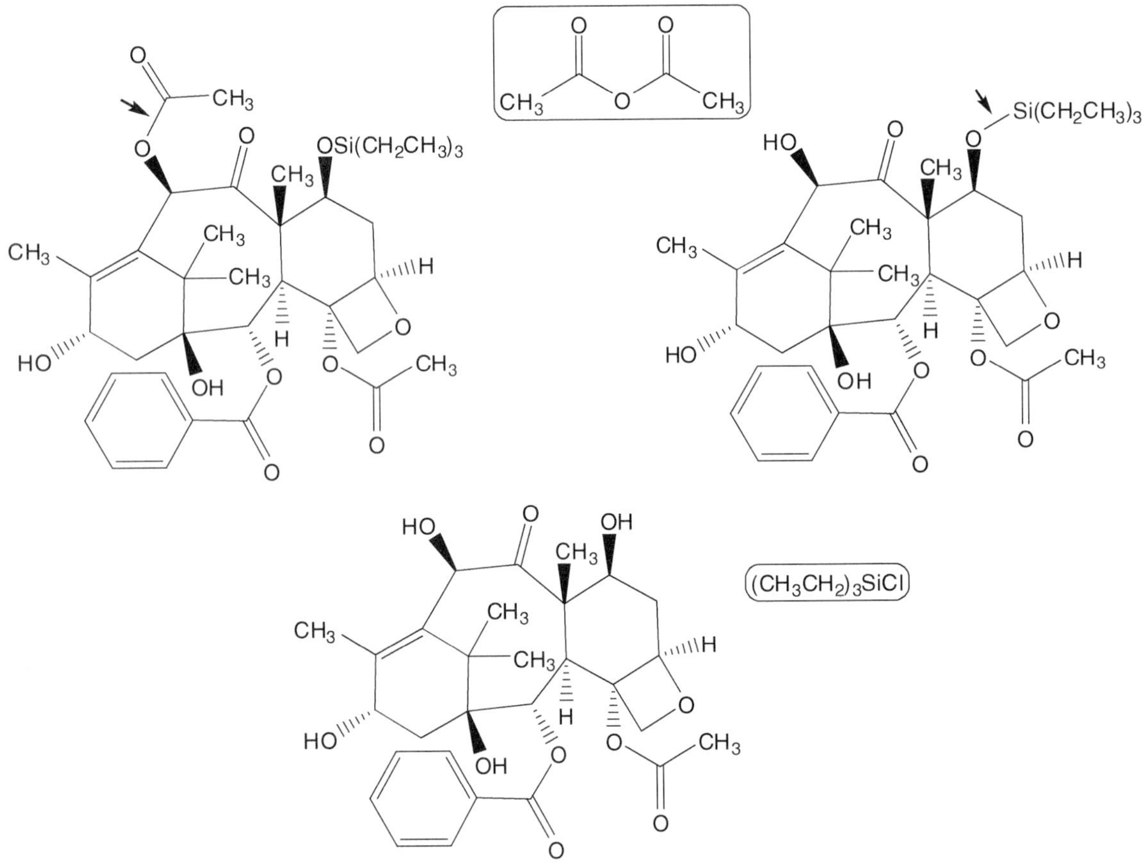

10-deacetylbaccatin III

The carboxylic acid, (4*S*)-*trans*-4,5-dihydro-2,4-diphenyl-5-oxazolecarboxylic acid, is formed by hydrolysis of the methyl ester. The oxazolidine is formed by a reaction of (2*R*,3*S*)-3-phenylisoserine methyl ester with ethyl benzimidate. The benzimidate is formed from benzonitrile.

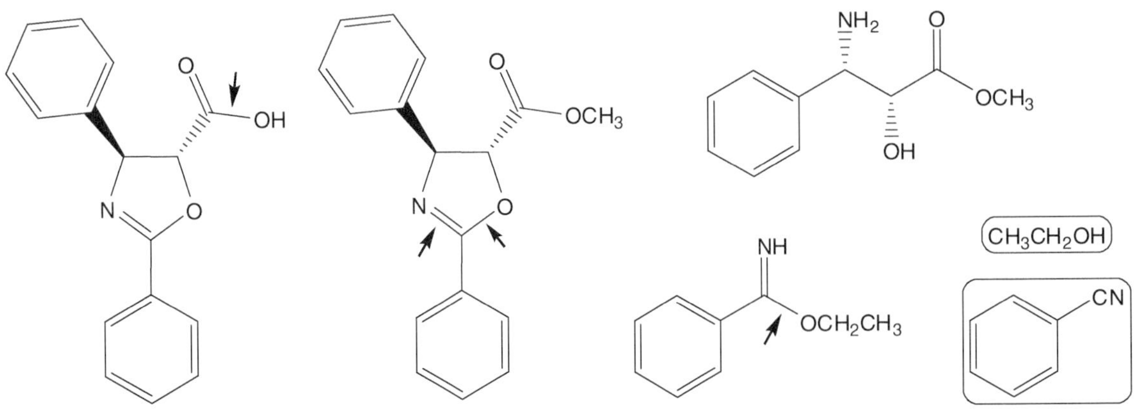

In one preferred route to (2*R*,3*S*)-3-phenylisoserine methyl ester, the ester is formed from the amide. The (2*R*,3*S*)- and (2*S*,3*R*)-3-phenylisoserine amides are separated by resolution. The (2*R*,3*S*)- and (2*S*,3*R*)-amides are formed by reaction of racemic methyl *cis*-3-phenylglycidate with ammonia. Methyl *cis*-3-phenylglycidate is formed from the bromohydrin. Methyl 3-bromo-2-hydroxy-3-phenylpropanoate is formed by the ring-opening of methyl *trans*-3-phenylglycidate by hydrogen bromide. Methyl *trans*-3-phenylglycidate is formed by condensation of benzaldehyde and methyl chloroacetate (**Darzens Reaction**). (Racemic methyl *cis*-3-phenylglycidate is a side product of the Darzens Reaction. How much of the *cis*-side product is present in the crude methyl 3-phenylglycidate? What is the fate of this side product in the next step?)

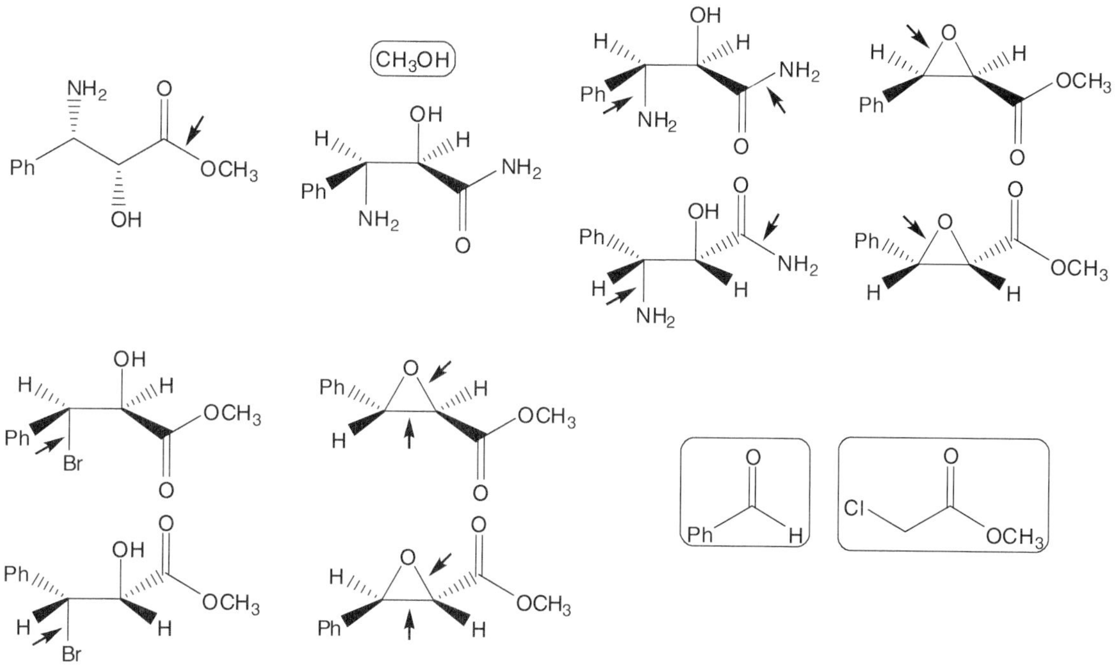

Paracetamol

Medicines for Pain and Palliative Care/Non-opiods and Non-steroidal Anti-inflammatory Medicines
Antimigraine Medicines/For Treatment of Acute Attack

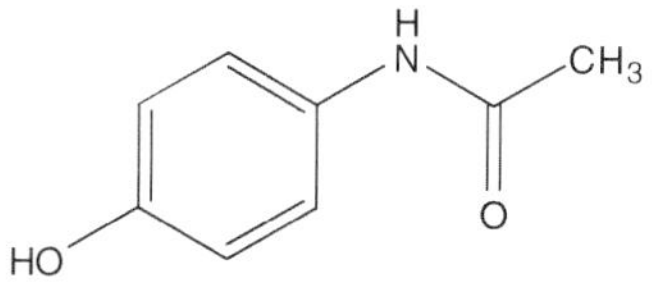

Amides are ubiquitous in drug structures. Amide formation from an amine and acid chloride, anhydride, ester, or carboxylic acid is often very efficient.

Discussion. Paracetamol is manufactured in a single step by acetylation of 4-aminophenol with acetic anhydride.

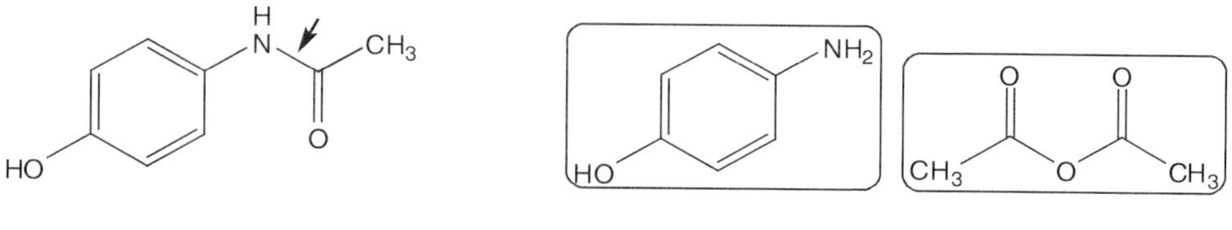

Extended Discussion

Paracetamol can also be formed in a single step from 1,4-hydroquinone and ammonium acetate. List the pros and cons of both routes.

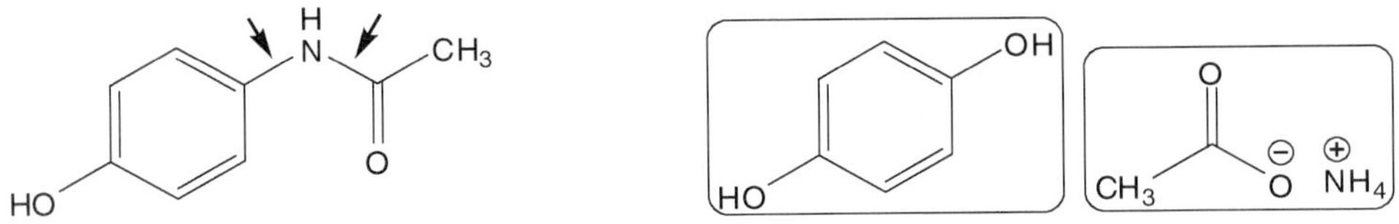

Paritaprevir

Anti-infective Medicines/Antiviral Medicines/Antihepatitis Medicines/Medicines for Hepatitis C/Other Antivirals

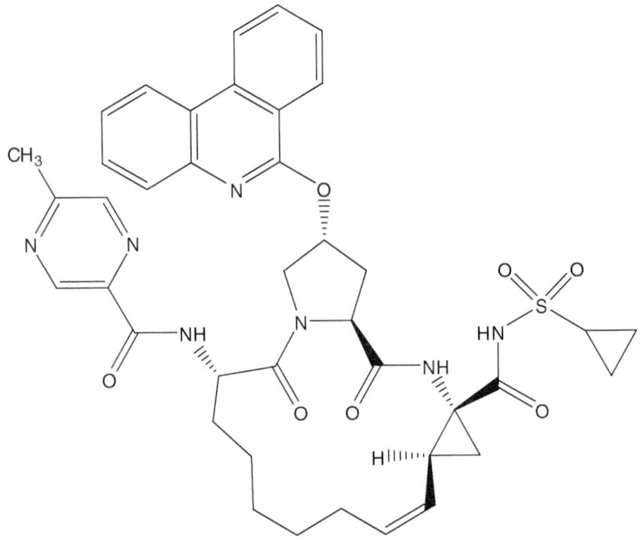

Amides are ubiquitous in drug structures. Amide formation from an amine and acid chloride, anhydride, ester, or carboxylic acid is often very efficient.

Discussion. Some key disconnections reveal the six components of paritaprevir and the modular nature of the synthesis.

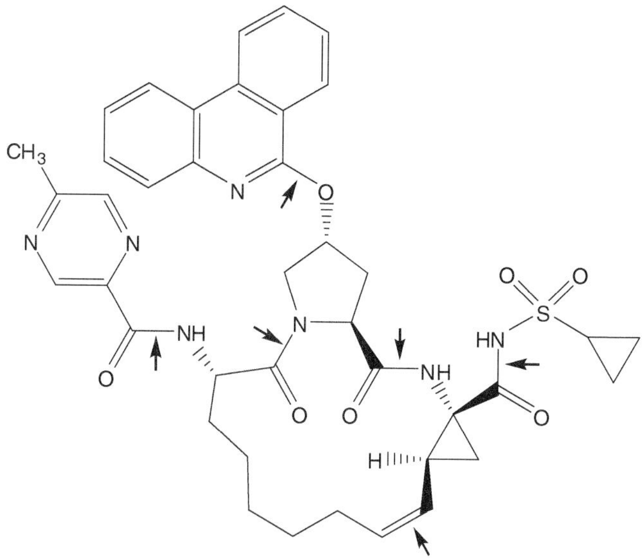

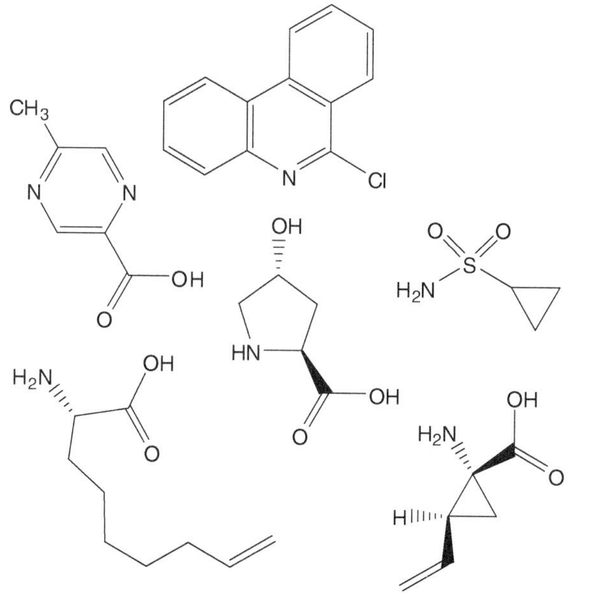

(**NOTE:** The phenanthridine will be represented as Ar until the ether is disconnected.)

The cyclopropanesulfonamide is introduced by *N*-acylation in the final step. The acylating agent is formed from the cyclopropanecarboxylic acid. The carboxylic acid is formed by ester hydrolysis. The *cis*-alkene of the macrocyclic ring is formed by olefin metathesis (**Grubbs Reaction**). The diene for olefin metathesis is assembled when the amide is formed from the carboxylic acid and the pyrrolidine.

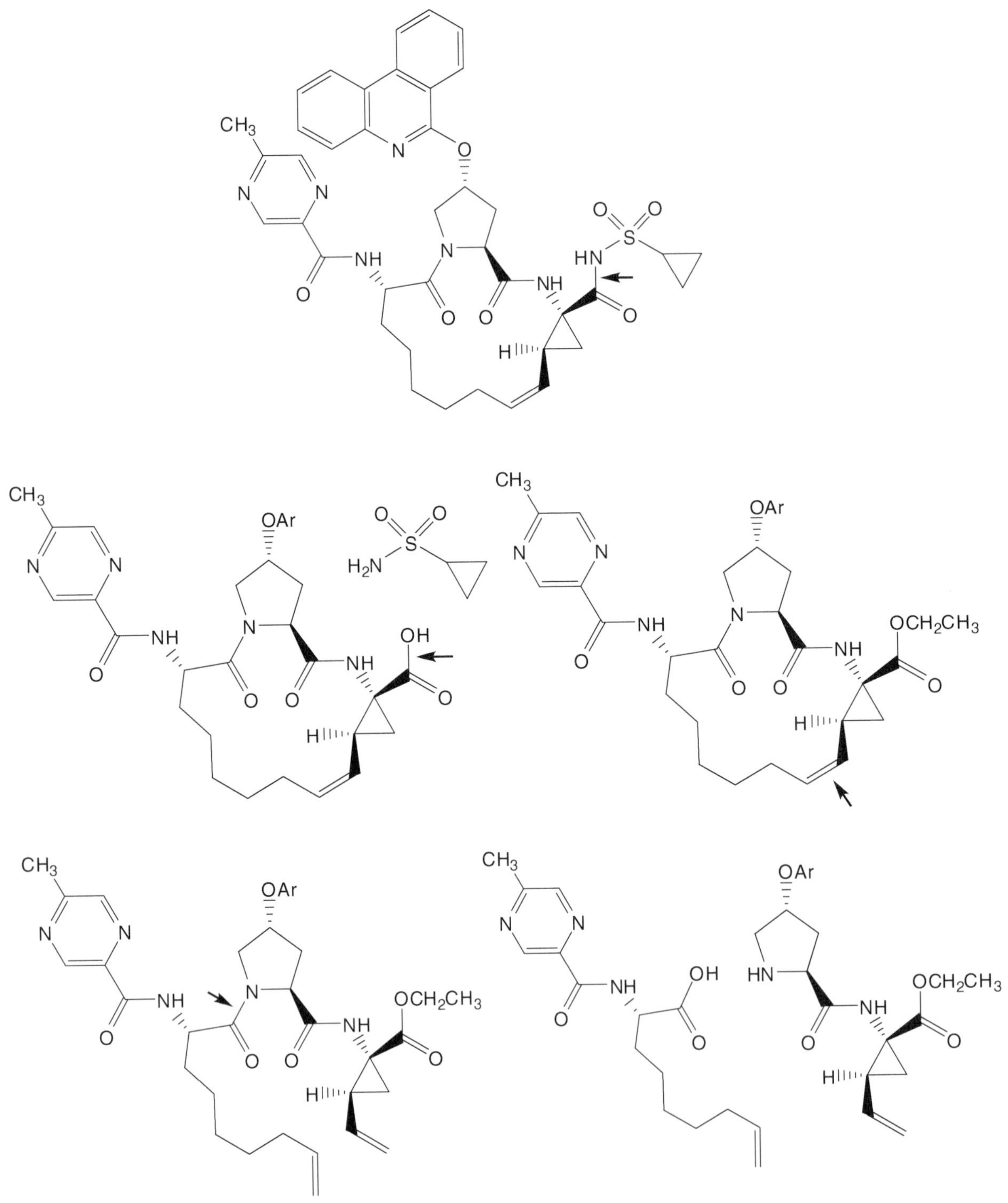

The carboxylic acid component is formed from 5-methylpyrazine-2-carboxylic acid and (*S*)-2-amino-8-nonenoic acid. The (*S*)-α-amino acid is formed by reductive amination of the α-ketoacid. The α-ketoacid is formed by hydrolysis of the α-ketoester. Ethyl 2-oxo-8-nonenoate is formed by the reaction of the alkylmagnesium bromide from 7-bromo-1-heptene with diethyl oxalate (**Grignard Reaction**).

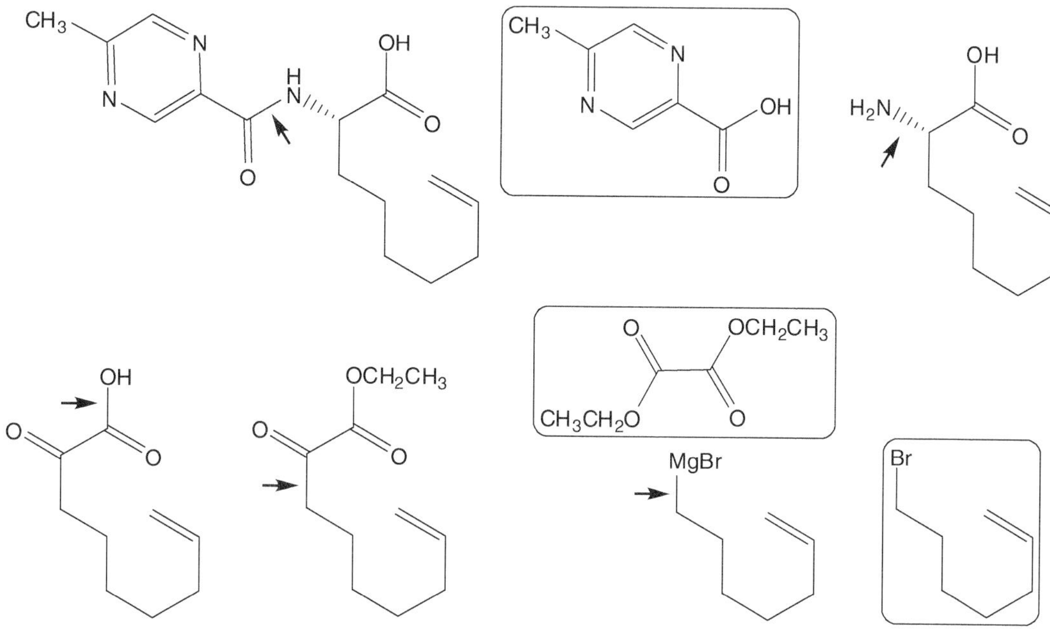

The pyrrolidine is released by hydrolysis of the *tert*-butyl carbamate. The amide bond is formed from the proline carboxylic acid and the cyclopropanamine. The ether C–O bond is formed by nucleophilic displacement of chloride from 6-chlorophenanthridine by the alcohol of *N*-Boc-*trans*-4-hydroxy-L-proline.

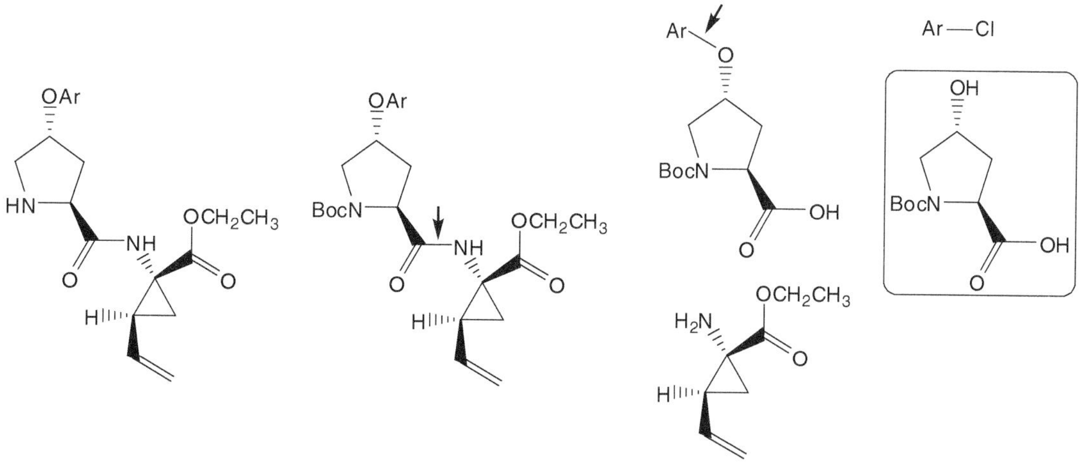

Cleavage of the *tert*-butyl carbamate releases the amine of ethyl (1*R*,2*S*)-1-amino-2-vinylcyclopropanecarboxylate. The carbamate (1*R*,2*S*)-enantiomer is separated from the racemic mixture by selective enzyme-mediated hydrolysis of the (1*S*,2*R*)-ester. The carbamate is formed from the amine. The amine is formed by hydrolysis of the imine. The cyclopropane ring is formed by the reaction of ethyl glycine benzaldehyde imine with *trans*-1,4-dibromo-2-butene. (Draw the structures for four possible stereoisomeric cyclopropane products. Assign (*R*/*S*) configurations to the chiral carbons. This ring-forming reaction is diastereoselective. Identify the desired/isolated products.) The imine is formed from glycine ethyl ester and benzaldehyde.

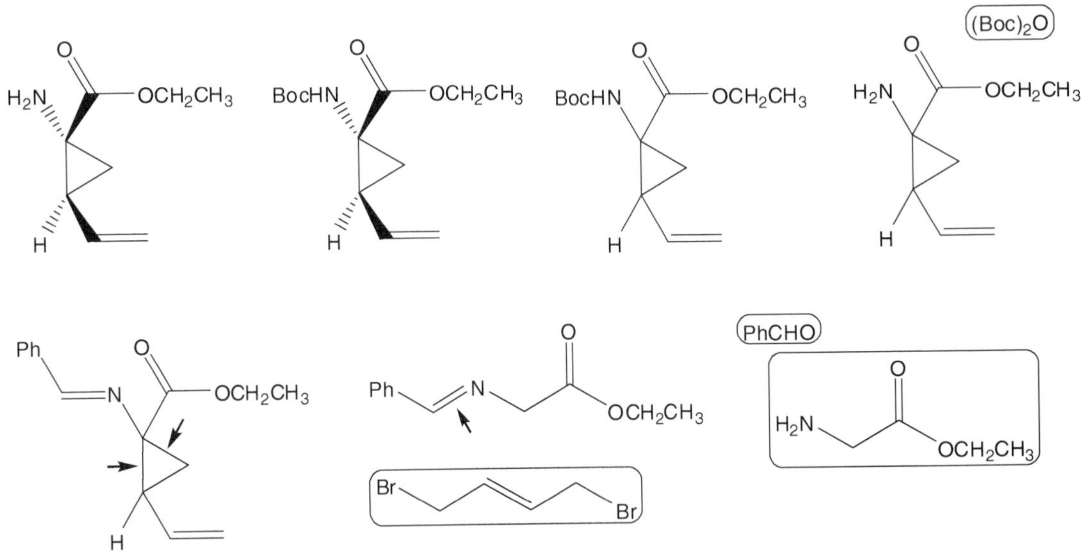

Cyclopropanesulfonamide is formed from *N-tert*-butylcyclopropanesulfonamide. The cyclopropane ring is formed by nucleophilic displacement of chloride by the sulfonamide dianion. The *N-tert*-butylsulfonamide is formed from *tert*-butylamine and the sulfonyl chloride. 3-Chloropropanesulfonyl chloride is formed from 1,3-propanesultone.

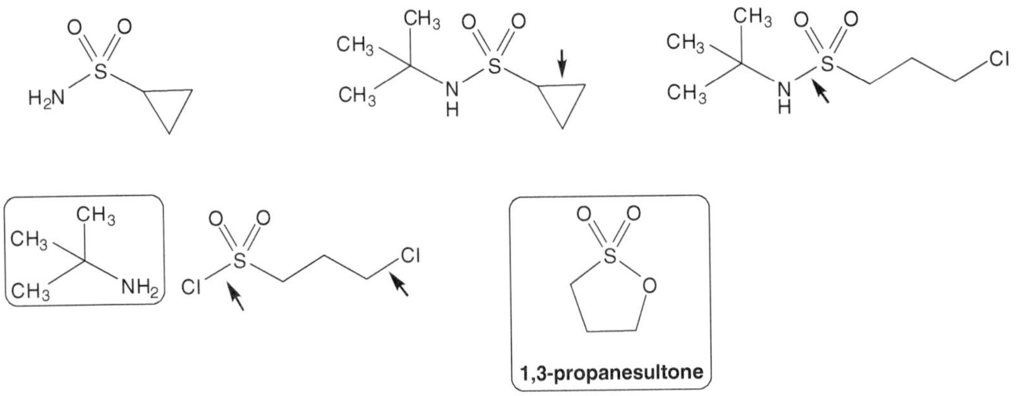

6-Chlorophenanthridine is formed from 6(5*H*)-phenanthridinone. The phenanthridinone can be formed from 9-fluorenone and hydrazoic acid (**Schmidt Reaction**).

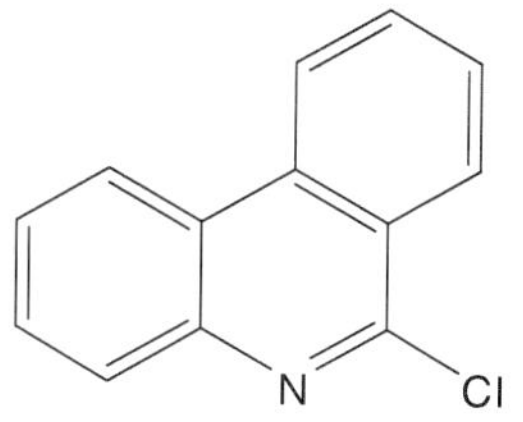

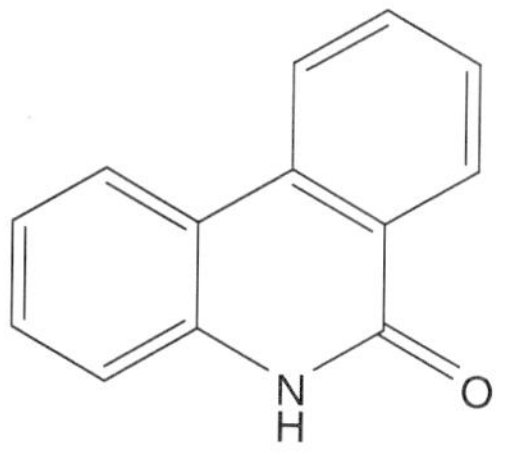

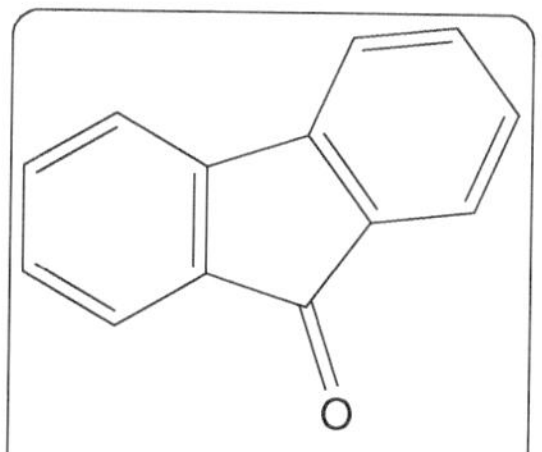

9-fluorenone

Extended Discussion

There are many potential routes to 7-bromo-1-heptene. Draw the structures of a retrosynthetic analysis of one route. Include the structures of the retrosynthetic analysis of any organic starting material(s) from petrochemical or biochemical raw materials.

Penicillamine

Antidotes and Other Substances Used in Poisonings/Specific
Medicines for Diseases of Joints/Disease-modifying Agents Used in Rheumatoid Disorders

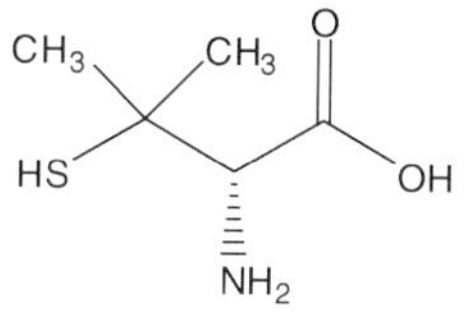

A chiral carbon in a single-enantiomer molecule is often delivered in a starting material.

Discussion. Penicillamine is semisynthetic. Penicillamine is released from a mercury complex by reaction with hydrogen sulfide. The penicillamine mercury complex forms and precipitates when mercuric chloride is added to the mixture produced by hydrolysis of penicillin G (benzylpenicillin). Penicillin G is produced by the fungus *Penicillium chrysogenum*.

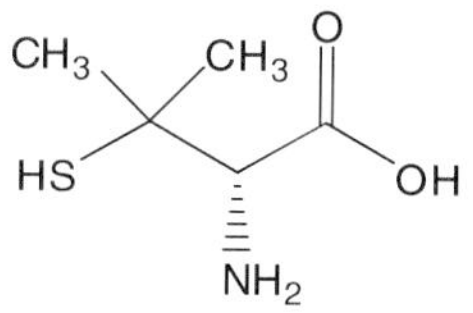

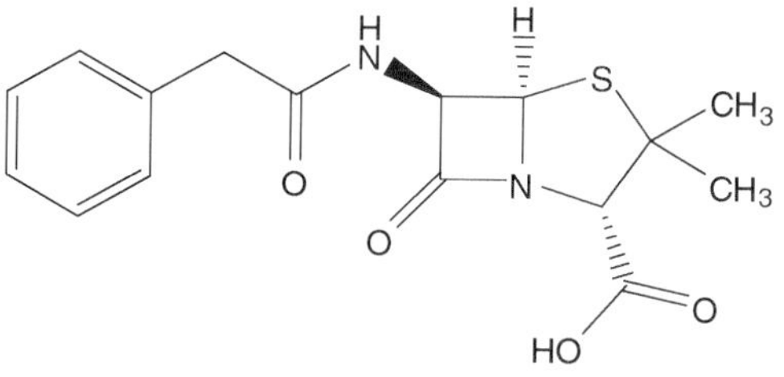

penicillin G (benzylpenicillin)

Extended Discussion

A synthesis of penicillamine usually involves several steps, including a resolution and protection–deprotection of functional groups. Draw the structures of the retrosynthetic analysis of one resolution-based route to penicillamine. (Penicillamine is often used as a chiral auxiliary for asymmetric synthesis. Is there an asymmetric synthesis of penicillamine?)

Pentamidine

Anti-infective Medicines/Antiprotozoal Medicines/Antipneumocystosis and Antitoxoplasmosis Medicines
Anti-infective Medicines/Antiprotozoal Medicines/Antitrypanosomal Medicines/African Trypanosomiasis

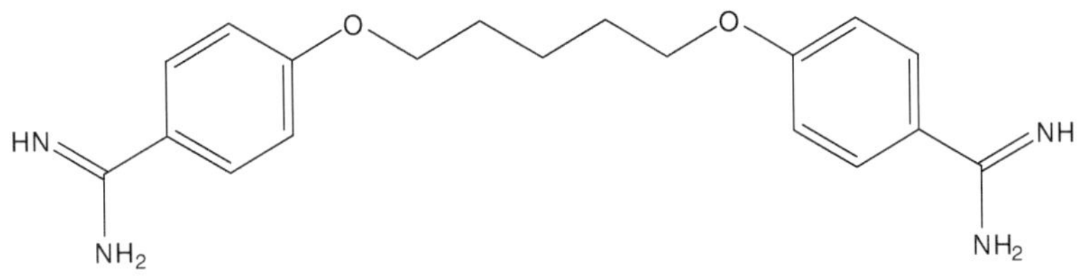

An amidine is often formed by the reaction of an imidate with ammonia or an amine. An imidate is formed by the acid-catalyzed addition of an alcohol to a nitrile.

Discussion. The amidine is formed in the final step by the reaction of the ethyl imidate with ammonia. The imidate is formed by the addition of ethanol to the nitrile. The ether is formed by displacement of a primary bromide by the phenol (**Williamson Ether Synthesis**).

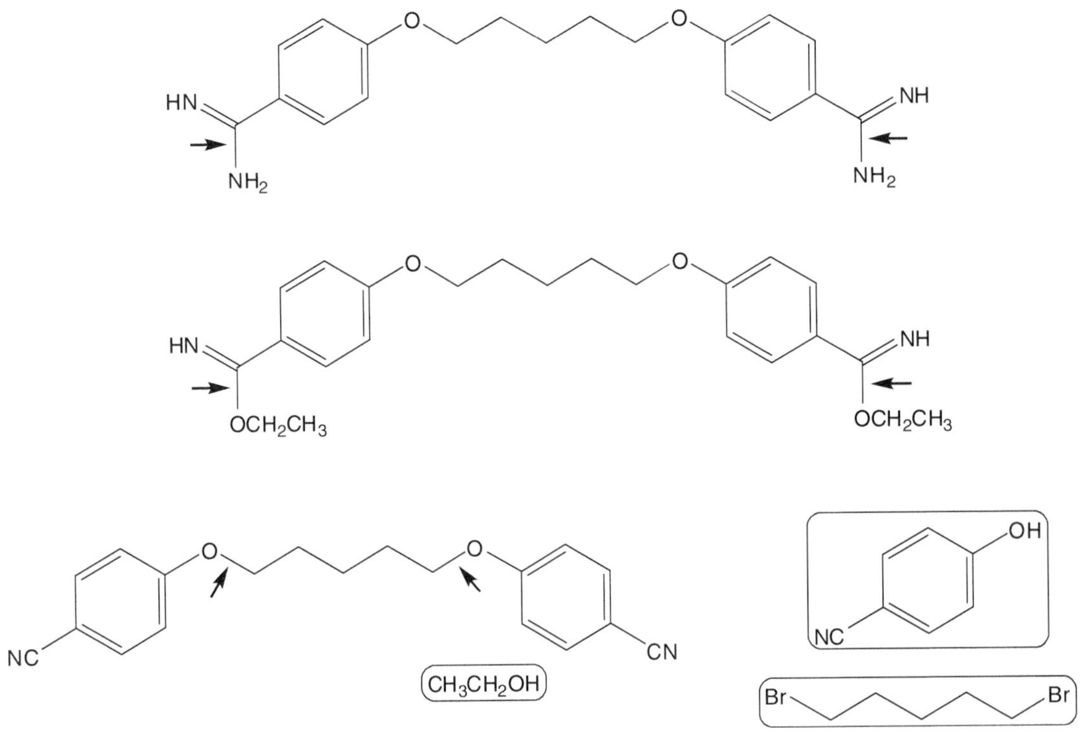

Extended Discussion

Draw the structures of the retrosynthetic analysis of an alternative route to 1,5-dibromopentane from biomass-derived furfural. List the pros and cons of both routes to 1,5-dibromopentane. Is one route preferred?

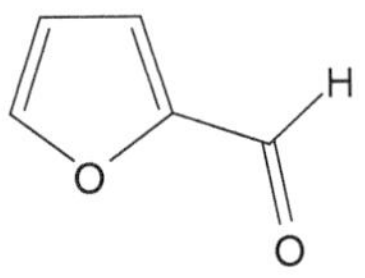

furfural

Permethrin

Dermatologic Medicines (Topical)/Scabicides and Pediculicides

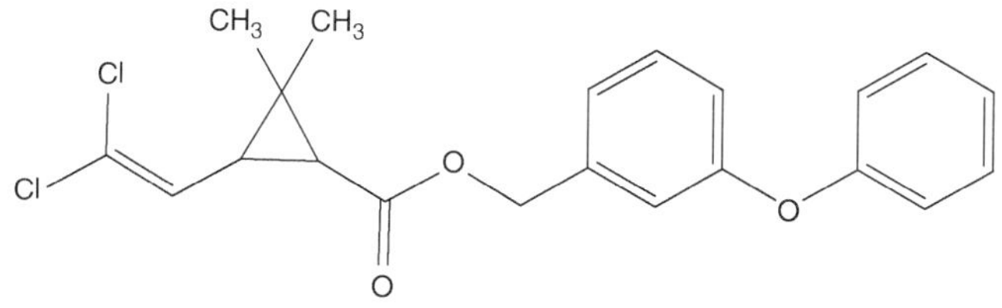

A cyclopropanecarboxylate is often formed from a γ-halocarboxylate (X = Cl, Br).

Discussion. Permethrin is manufactured as a mixture of four stereoisomers. (Draw the structures of the four stereoisomers.) The 3-phenoxybenzyl ester is formed in the final step by transesterification from the methyl ester. The cyclopropanecarboxylate is formed by chloride displacement from the γ-chlorohexenoate. During this reaction, the methyl ester is formed from the ethyl ester by transesterification. (What is the ratio of the *cis*- and *trans*-products?) The γ-chlorohexenoate is formed by the reaction of 1,1,1-trichloro-4-methyl-3-penten-2-ol with triethyl orthoacetate (**Johnson–Claisen Rearrangement**). 1,1,1-Trichloro-4-methyl-3-penten-2-ol is formed by isomerization of 1,1,1-trichloro-4-methyl-4-penten-2-ol. 1,1,1-Trichloro-4-methyl-4-penten-2-ol is formed from chloral and isobutylene (**Prins Reaction**).

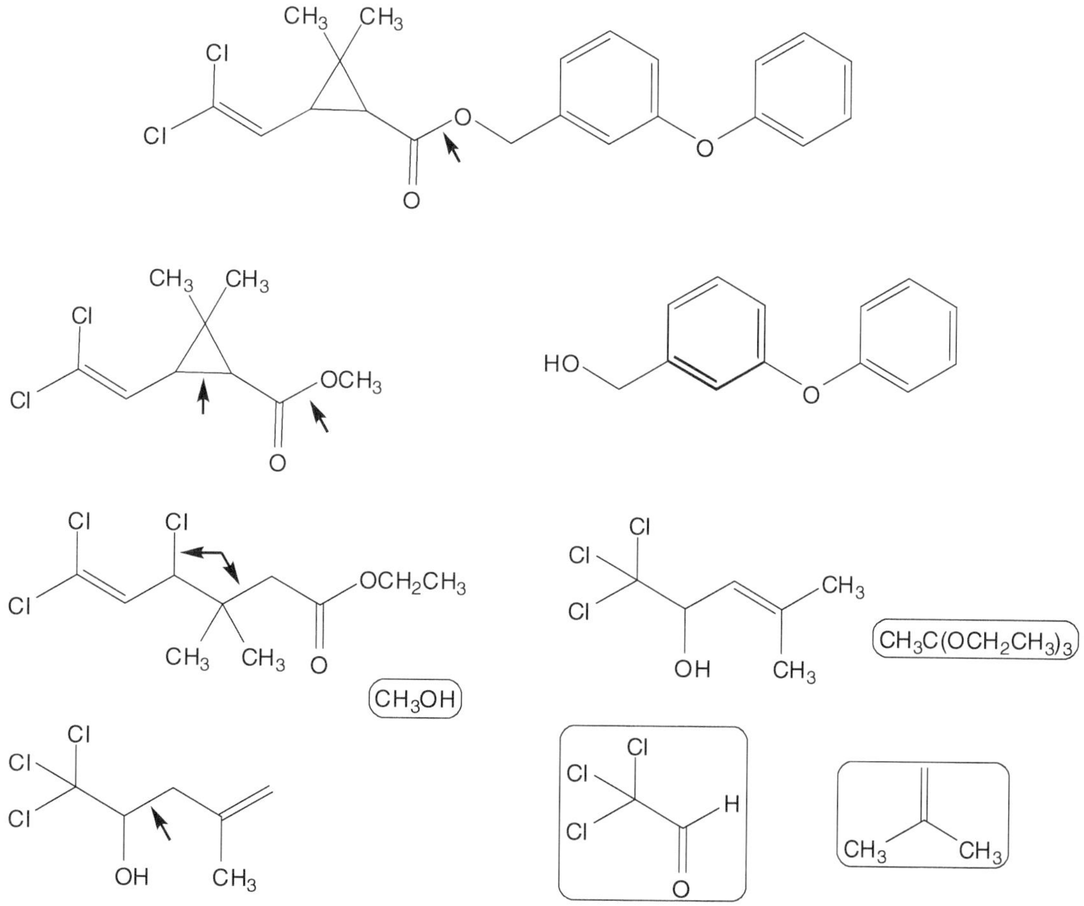

3-Phenoxybenzyl alcohol is formed by reduction of the aldehyde or by hydrolysis of the benzyl chloride. The aldehyde is formed by the oxidation of 3-phenoxytoluene. The benzyl chloride is formed by radical chlorination of 3-phenoxytoluene. 3-Phenoxytoluene is formed by chloride displacement from chlorobenzene by *meta*-cresol (**Ullmann-type Reaction**).

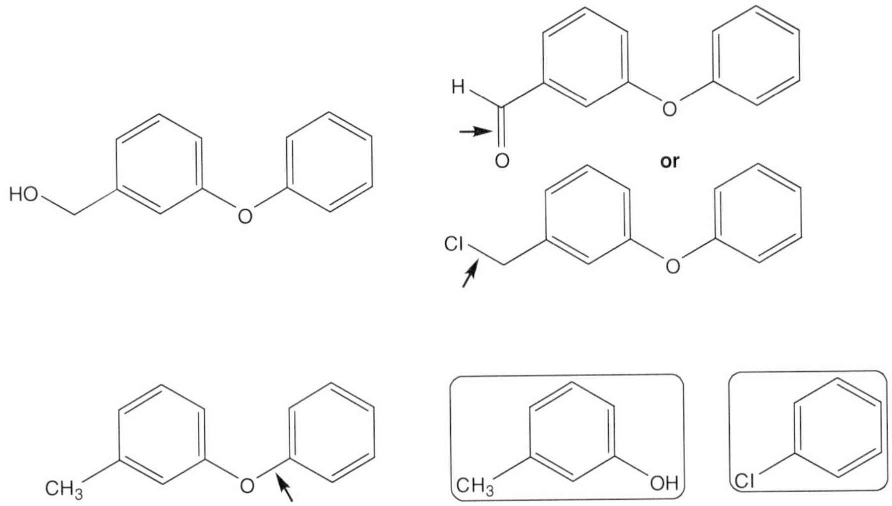

Extended Discussion

Draw the structures of the retrosynthetic analysis of an alternative route to permethrin that uses carbon tetrachloride or bromotrichloromethane as a starting material. List the pros and cons for both routes and select one route as the preferred route.

Phenobarbital

Anticonvulsants/Antiepileptics

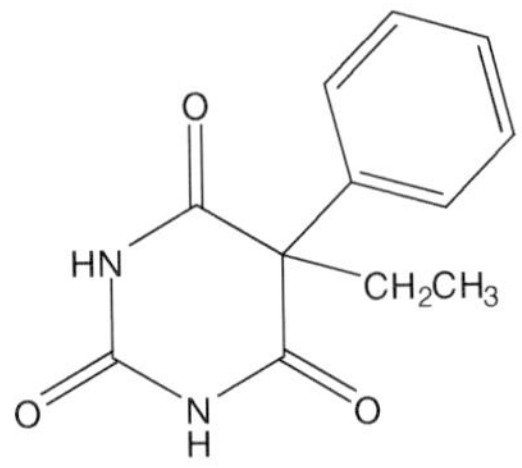

A pyrimidine-4,6-dione is often formed by the reaction of a malonate diester with an amidine, thiourea, or urea.

Discussion. The pyrimidine ring is assembled by the reaction of diethyl 2-ethyl-2-phenylmalonate with urea. The 2,2-disubstituted malonate is formed by alkylation of the 2-phenylmalonate. Diethyl 2-phenylmalonate is formed from ethyl phenylacetate and diethyl carbonate or diethyl oxalate (mixed **Claisen Condensation**).

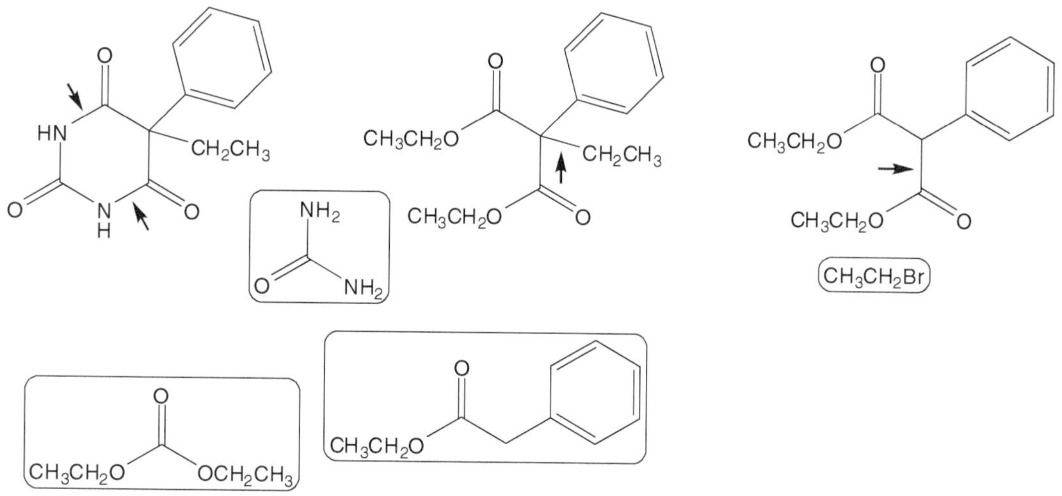

Extended Discussion

Draw the structures of the retrosynthetic analysis of one alternative route to phenobarbital from diethyl malonate or ethyl cyanoacetate.

Phenytoin

Anticonvulsants/Antiepileptics

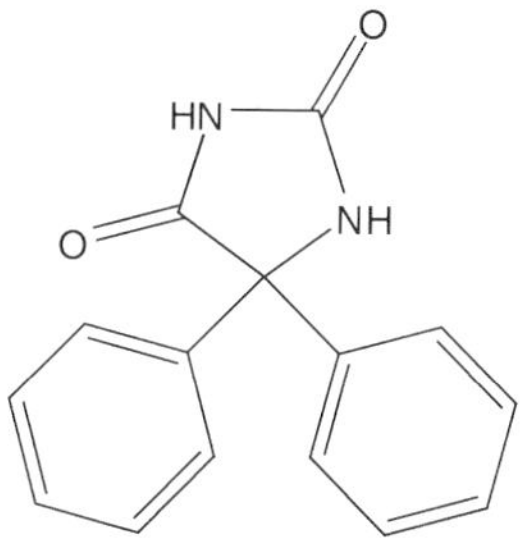

When a target molecule is readily accessible from inexpensive starting materials by more than one route, reaction and workup procedures and equipment requirements become key factors in the selection of a preferred route.

Discussion. Phenytoin is formed in a single step from benzil and urea (**Blitz Synthesis**).

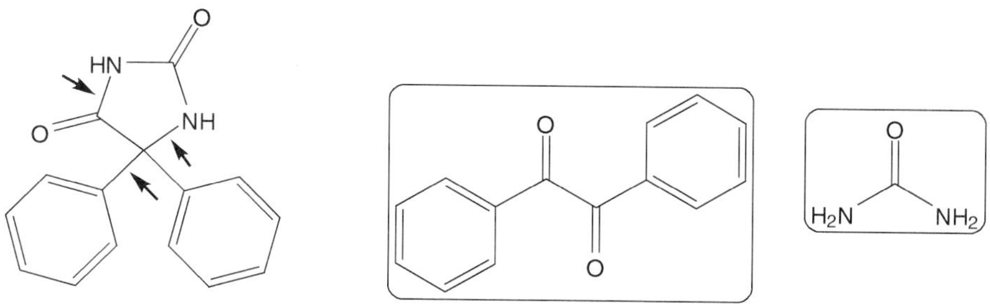

Phenytoin is also formed in a single step from benzophenone, potassium cyanide, and ammonium carbonate (**Bucherer–Bergs Reaction**).

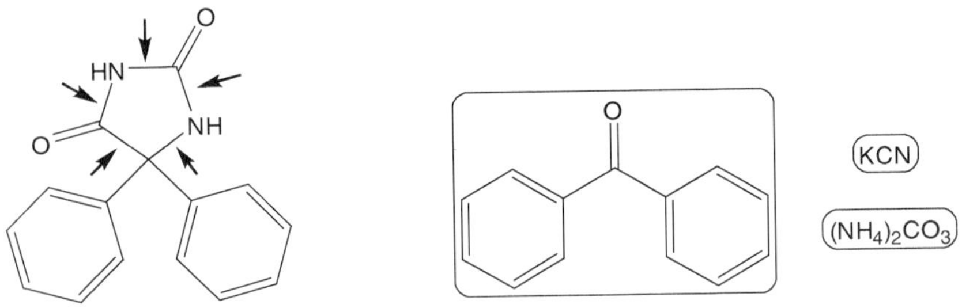

Extended Discussion

List the pros and cons for both routes to phenytoin and select one route as the preferred route.

Piperacillin

Anti-infective Medicines/Antibacterials/Beta-lactam Medicines

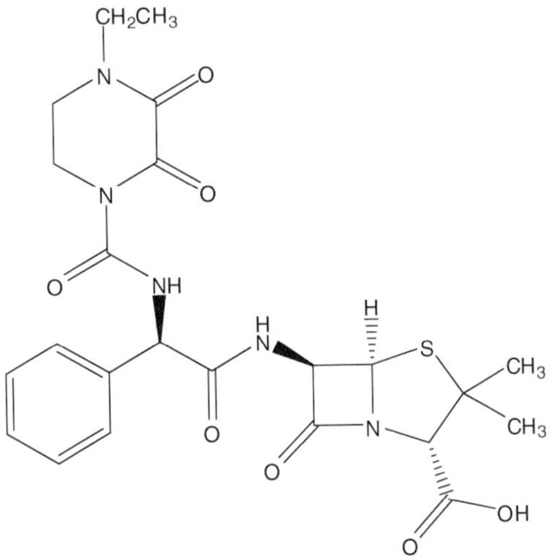

Penicillins are produced by fermentation or are semisynthetic. A semisynthetic penicillin is often formed by acylation of the amine of 6-aminopenicillanic acid (6-APA). 6-APA is produced from penicillin G (benzylpenicillin) by enzyme-mediated hydrolysis of the side-chain amide.

Discussion. Piperacillin is formed in a single step from another essential medicine, ampicillin. The urea is formed in the final step by acylation of the ampicillin amine by a carbamoyl chloride.

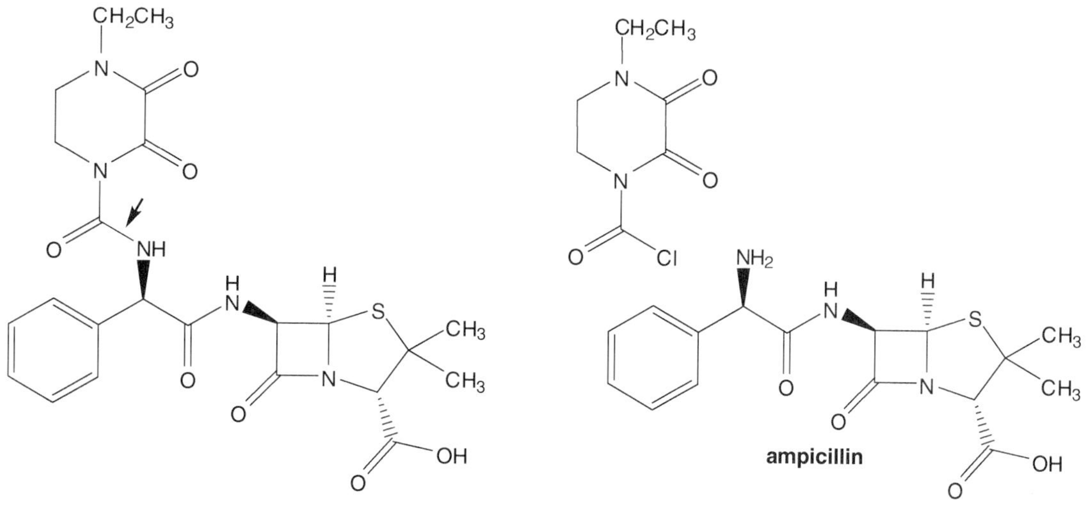

The carbamoyl chloride is formed from 1-ethylpiperazine-2,3-dione and phosgene. The piperazine-2,3-dione is formed from *N*-ethylethylenediamine and diethyl oxalate.

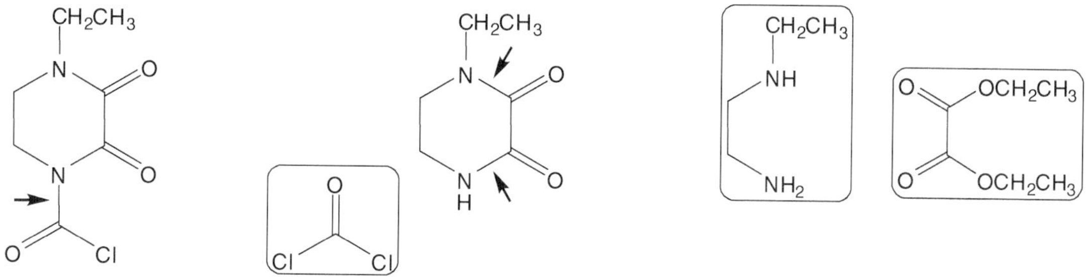

Extended Discussion

Draw the structures of the retrosynthetic analysis of an alternative and more convergent route to piperacillin from 6-aminopenicillanic acid (6-APA). List the pros and cons for both routes.

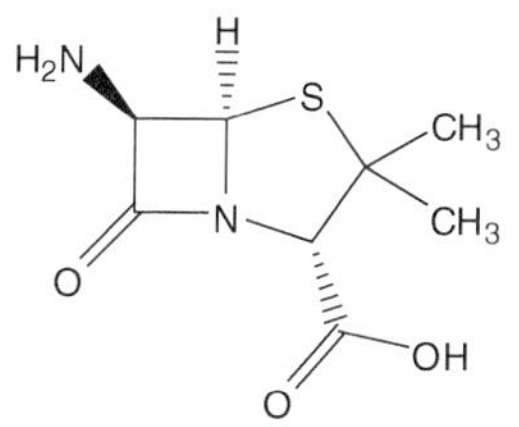

6-aminopenicillanic acid(6-APA)

Piperaquine

Anti-infective Medicines/Antiprotozoal Medicines/Antimalarial Medicines/For Curative Treatment

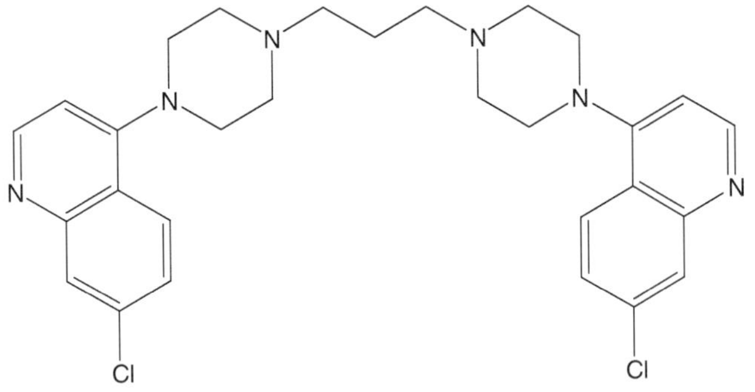

A nitrogen substituent on C2 or C4 of a quinoline ring is often introduced by displacement of chloride by nucleophilic aromatic substitution. The substitution is facilitated by the ring nitrogen and can be further facilitated by an electron-withdrawing group (NO_2, SO_2R, COOR, CN) on C3 of the ring.

Discussion. The symmetry of piperaquine suggests disconnections back to 1,3-dibromopropane and 7-chloro-4-(piperazin-1-yl) quinoline. The 4-(piperazin-1-yl)quinoline is formed by displacement of chloride from 4,7-dichloroquinoline by piperazine.

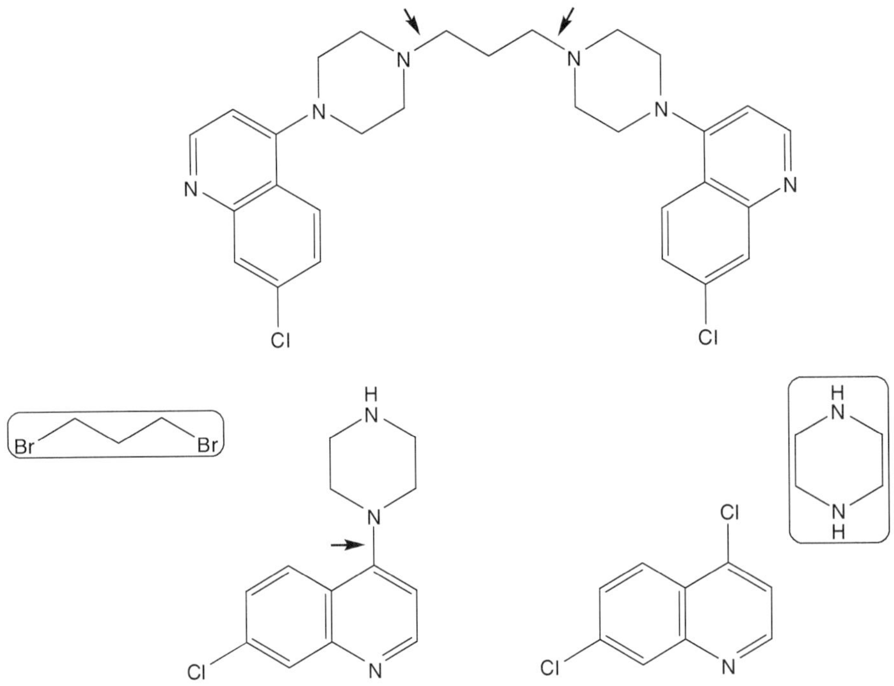

4,7-Dichloroquinoline is formed from 7-chloro-4-hydroxyquinoline (7-chloroquinolin-4-one). The quinolin-4-one is formed by thermolysis/decarboxylation of the quinolin-4-one-3-carboxylic acid. The carboxylic acid is formed by ester hydrolysis. The quinoline ring is formed by acylation of the aromatic ring with loss of ethanol. The enamine is formed by displacement of ethanol from the enol ether of ethyl ethoxymethylenemalonate by 3-chloroaniline. (The four-step sequence from 3-chloroaniline to 7-chloro-4-hydroxyquinoline is an example of the **Gould–Jacobs Reaction**.)

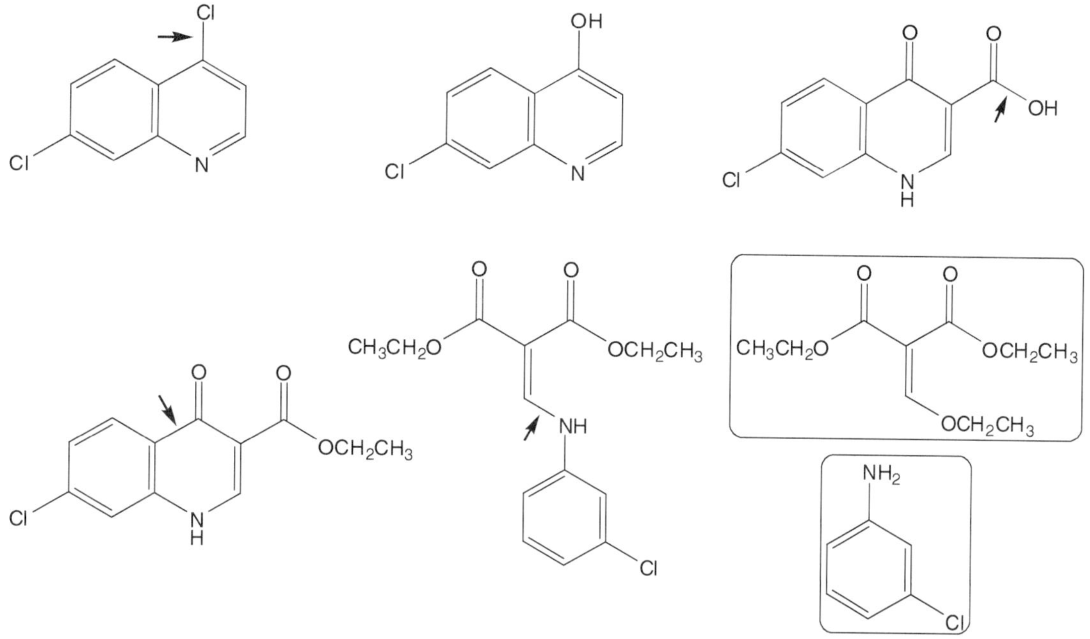

Extended Discussion

Draw the structures for two impurities often found in crude 7-chloro-4-(piperazin-1-yl)quinoline. List the details (equivalents of piperazine, solvent, temperature, time, workup procedure, yield, and purity/impurities) for the procedures used to produce 7-chloro-4-(piperazin-1-yl)quinoline. Select one procedure as the preferred procedure.

Praziquantel

Anti-infective Medicines/Intestinal Anthelminthics
Anti-infective Medicines/Antichistosomal and Other Antitrematode Medicines

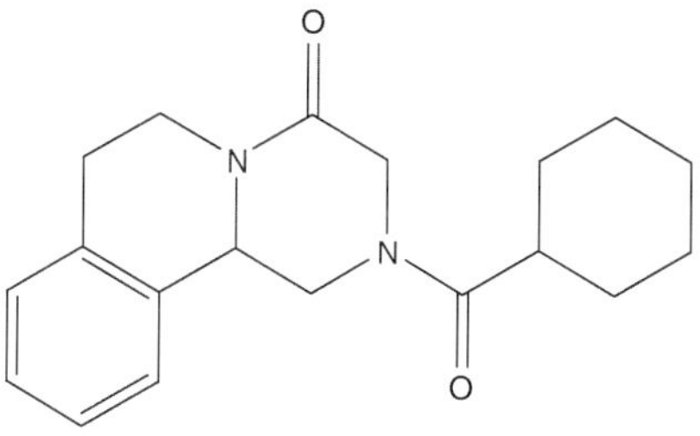

Amides are ubiquitous in drug structures. Amide formation from an amine and acid chloride, anhydride, ester, or carboxylic acid is often very efficient.

Discussion. Praziquantel is produced as a 1 : 1 mixture of the (*R*)- and (*S*)-enantiomers. [Only the (*R*)-enantiomer is active. There is considerable interest in a cost-competitive route to the (*R*)-enantiomer.]

There are two routes to praziquantel. In Route A, the cyclohexanecarboxamide is formed in the final step from the amine and cyclohexanecarbonyl chloride. The amine, 2,3,6,7-tetrahydro-1*H*-pyrazino[2,1-*a*]isoquinolin-4(11b*H*)-one, is released by hydrolysis of the benzamide. The piperazine ring is formed by chloride displacement by the benzamide. The chloroacetamide is formed from chloroacetyl chloride and the secondary amine. In a single step, the primary amine is formed by hydrogenation of the nitrile, the secondary amine is released by benzoyl migration to the primary amine, and the 3,4-double bond in the dihydroisoquinoline is reduced. The 1,2-dihydroisoquinoline is assembled from isoquinoline, benzoyl chloride, and potassium cyanide (**Reissert Reaction**). Isoquinoline is isolated from coal tar.

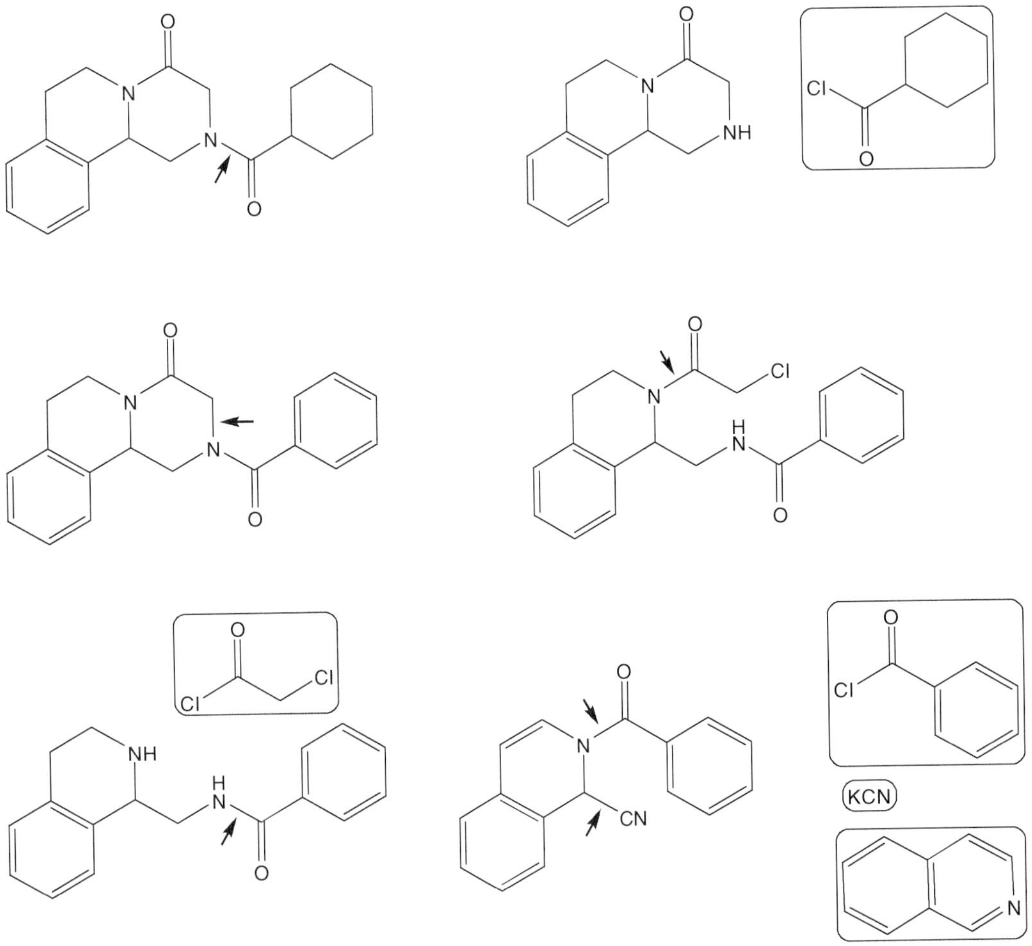

In Route B, the cyclohexanecarboxamide is also formed in the final step from the amine and cyclohexanecarbonyl chloride. The amine, 2,3,6,7-tetrahydro-1*H*-pyrazino[2,1-*a*]isoquinolin-4(11b*H*)-one, is formed from 2-(2,2-dimethoxyethylamino)-*N*-phenethylacetamide (**Pictet–Spengler Reaction**). The 2-aminoacetamide is formed by chloride displacement from the 2-chloroacetamide by aminoacetaldehyde dimethyl acetal. 2-Chloro-*N*-phenethylacetamide is formed from chloroacetyl chloride and phenethylamine. Aminoacetaldehyde dimethyl acetal is formed from chloroacetaldehyde dimethyl acetal.

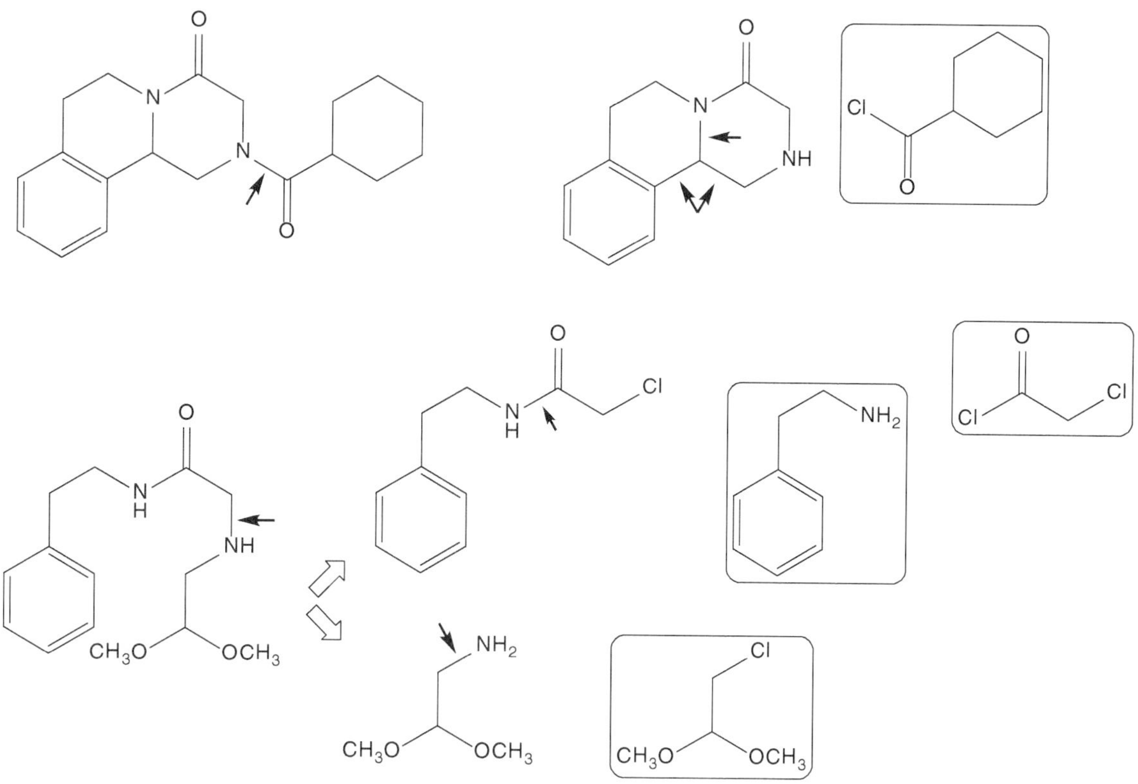

Extended Discussion

Draw the structures of the components used to form praziquantel by a multicomponent **Ugi Reaction**.

Prednisolone

Antiallergics and Medicines Used in Anaphylaxis
Antineoplastics and Immunosuppressives/Hormones and Antihormones
Opthalmological Preparations/Anti-inflammatory Agents

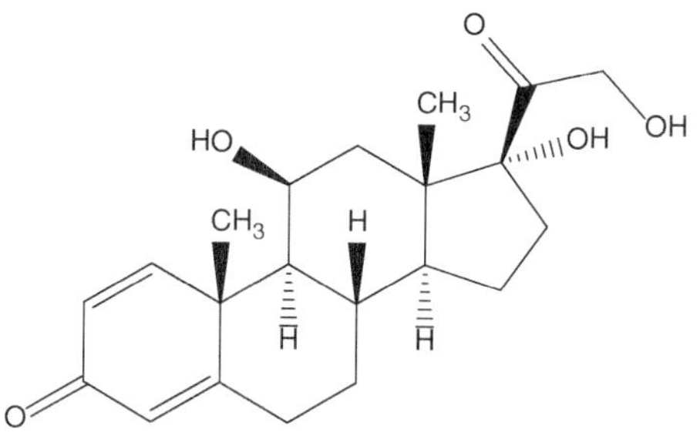

A single-enantiomer molecule with multiple chiral carbons is often formed by the modification of a natural product which has most or all of the chiral carbons already in place. A steroid 1,4-diene-3-one is often formed by microbial dehydrogenation of the 4-ene-3-one.

Discussion. Prednisolone is manufactured by microbial dehydrogenation of another essential medicine, hydrocortisone.

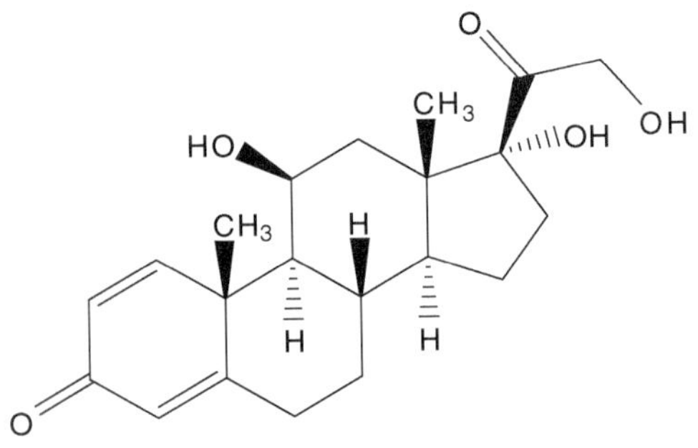

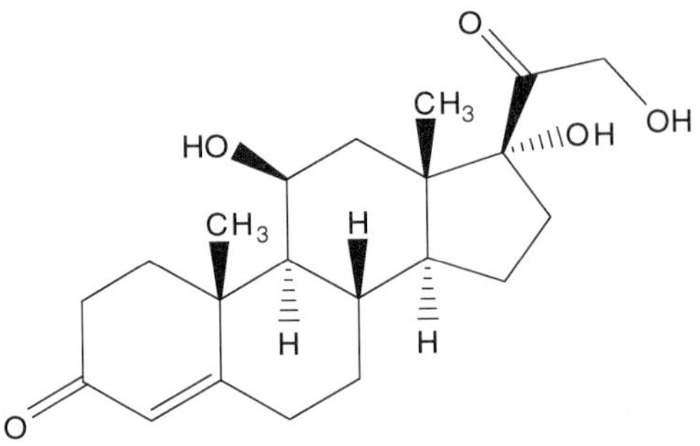

hydrocortisone

Extended Discussion

List the alternative chemical methods used to convert hydrocortisone acetate to prednisolone acetate. List the pros and cons for these methods.

Primaquine

Anti-infective Medicines/Antiprotozoal Medicines/Antimalarial Medicines/For Curative Treatment

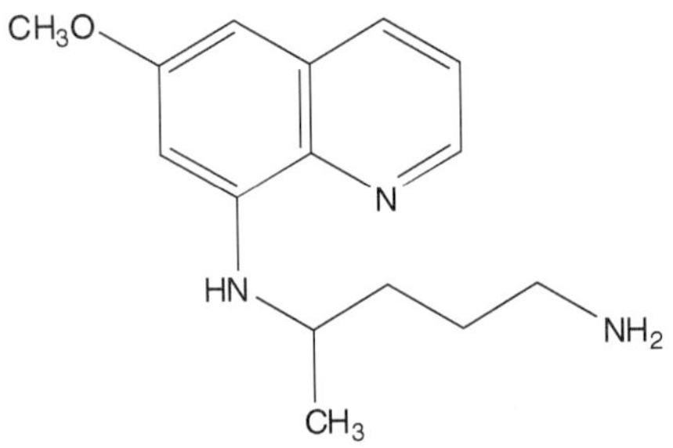

A secondary amine is often formed by reductive amination from an aldehyde or ketone and a primary amine.

Discussion. Primaquine is a mixture of (R)- and (S)-enantiomers. The primary amine of primaquine is released by cleavage of the phthalimide. The secondary amine is formed by reductive amination of *N*-(4-oxopentyl)phthalimide with 8-amino-6-methoxyquinoline.

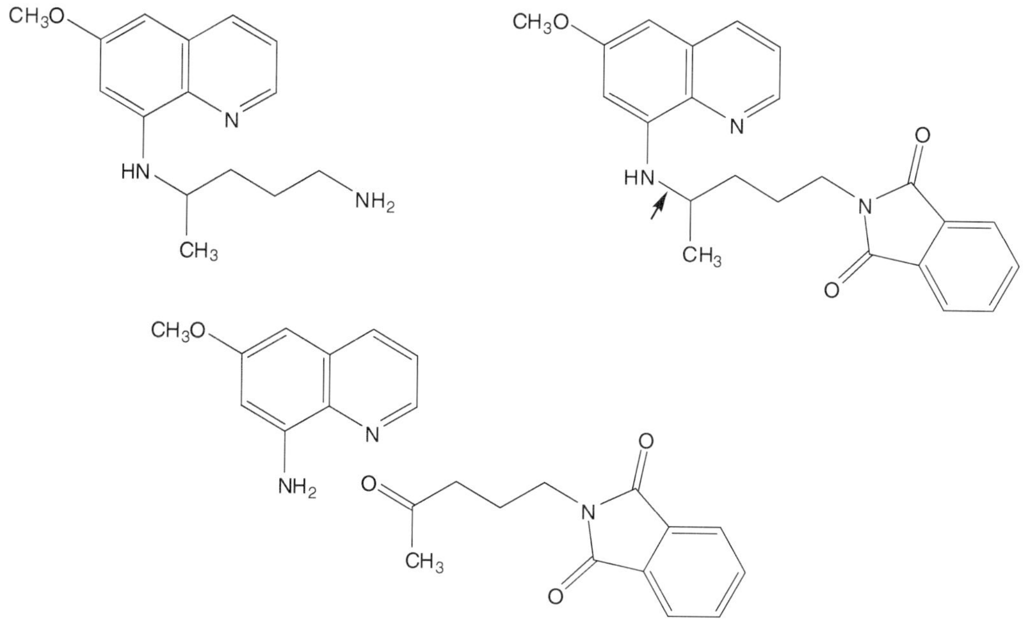

8-Amino-6-methoxyquinoline is formed from the 8-nitroquinoline. 6-Methoxy-8-nitroquinoline is formed from 4-methoxy-2-nitroaniline and glycerol (**Skraup–Doebner–von Miller Synthesis**).

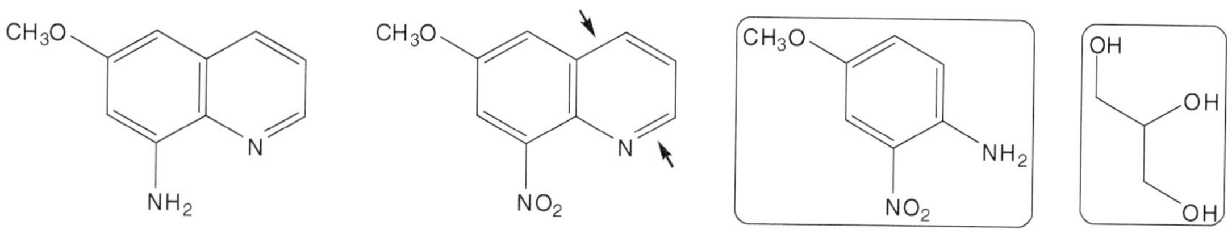

N-(4-Oxopentyl)phthalimide is formed from 5-chloro-2-pentanone by chloride displacement. (The chloride displacement by phthalimide and subsequent cleavage of the phthalimide to release the primary amine is an example of the **Gabriel Synthesis**.) 5-Chloro-2-pentanone is formed from α-acetyl-γ-butyrolactone.

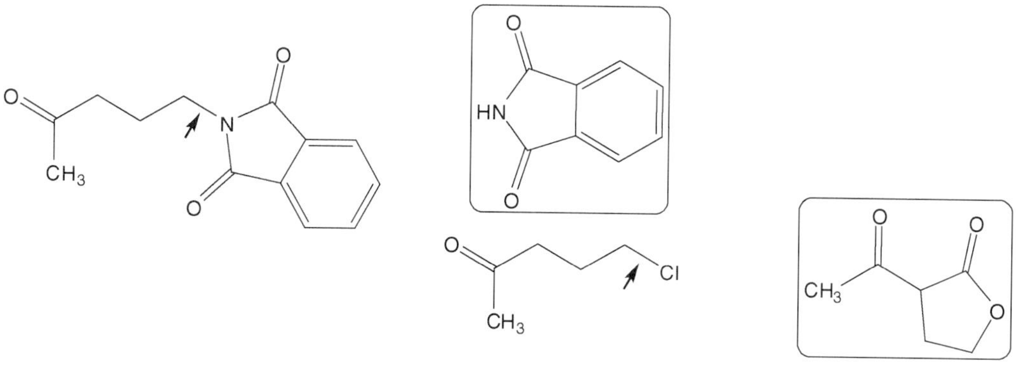

Extended Discussion

Draw the structures of the retrosynthetic analysis of an alternative route to primaquine from 8-amino-6-methoxyquinoline. Include the structures of the retrosynthetic analysis of any organic starting material(s) from petrochemical or biochemical raw materials. List pros and cons for the two routes and select one route as the preferred route.

Procarbazine

Antineoplastics and Immunosuppressives/Cytotoxic and Adjuvant Medicines

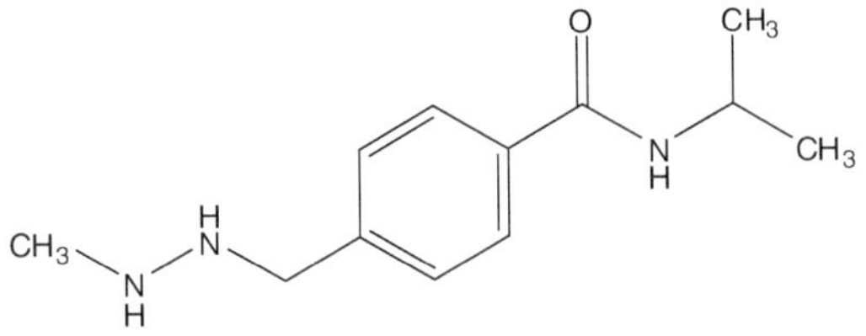

In 1-methylhydrazine, N1 is more nucleophilic than N2.

Discussion. Both nitrogen atoms of the hydrazine are released by cleavage of carbamate protecting groups in the final step. The amide is formed from the acid chloride and isopropylamine. The acid chloride is formed from the carboxylic acid. The carboxylic acid is formed by hydrolysis of the methyl ester.

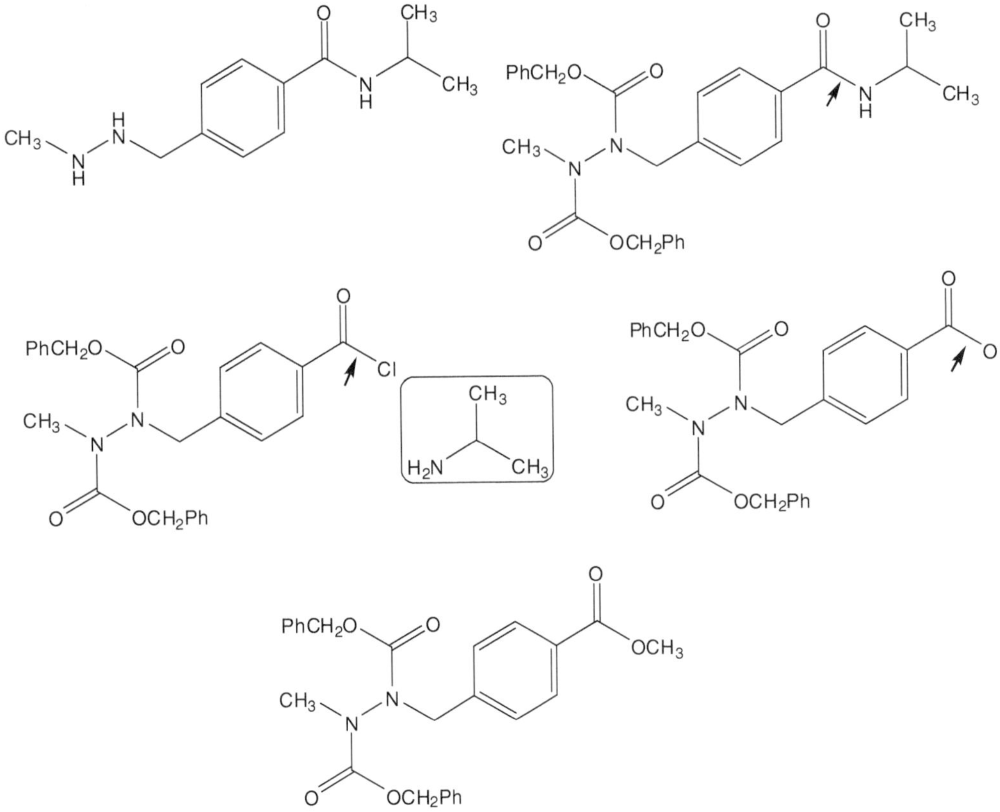

The key C–N bond is formed by bromide displacement by the methylhydrazine dicarboxylate. Methyl 4-bromomethylbenzoate is formed by bromination of methyl *p*-toluate. Dibenzyl 1-methylhydrazine-1,2-dicarboxylate is formed from methylhydrazine and benzyl chloroformate.

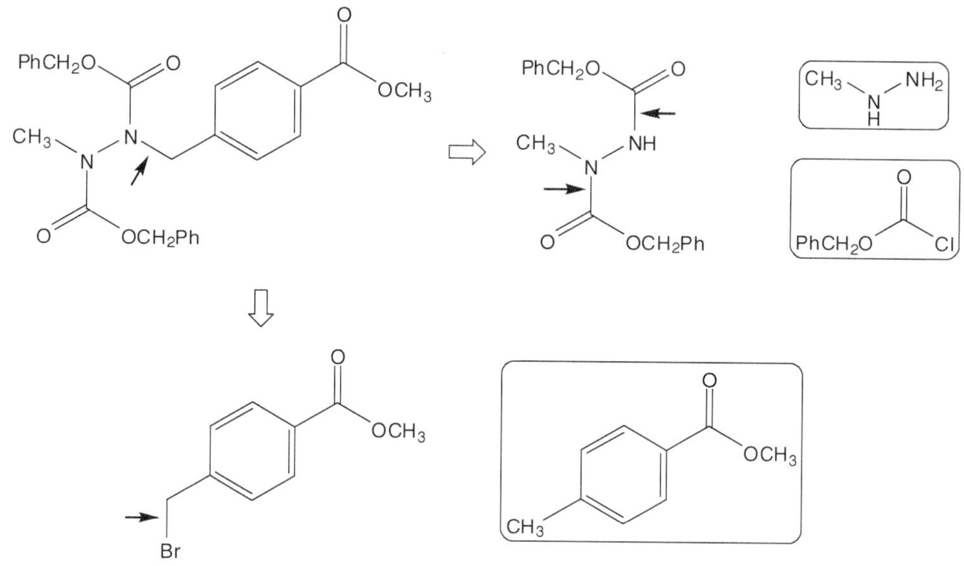

Extended Discussion

Draw the structures of the retrosynthetic analysis of one alternative route to procarbazine. List pros and cons for both routes and select one route as the preferred route.

Progesterone

Hormones, Other Endocrine Medicines and Contraceptives/Contraceptives/Intravaginal Contraceptives

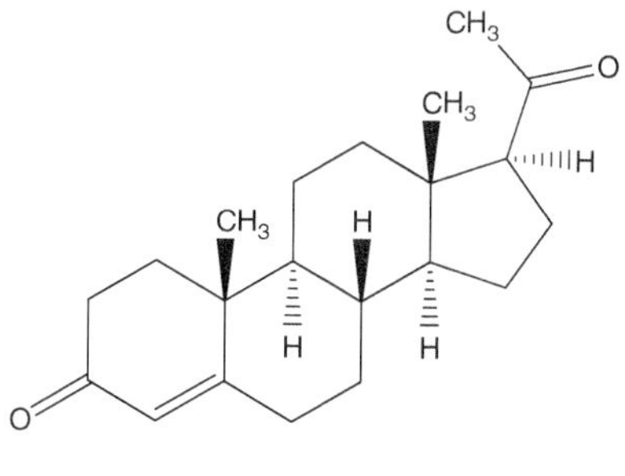

A single-enantiomer molecule with multiple chiral carbons is often formed by the modification of a natural product which has most or all of the chiral carbons already in place.

Discussion. Progesterone is manufactured in three steps from 16-dehydropregnenolone acetate and in six steps from diosgenin. Diosgenin is a phytosteroid sapogenin isolated from the tubers of *Discorea* wild yam.

In the final step, progesterone is formed from pregnenolone by oxidation of the secondary alcohol at C3 and migration of the double bond. Pregnenolone is formed by hydrolysis of pregnenolone acetate. Pregnenolone acetate is formed by hydrogenation of 16-dehydropregnenolone acetate.

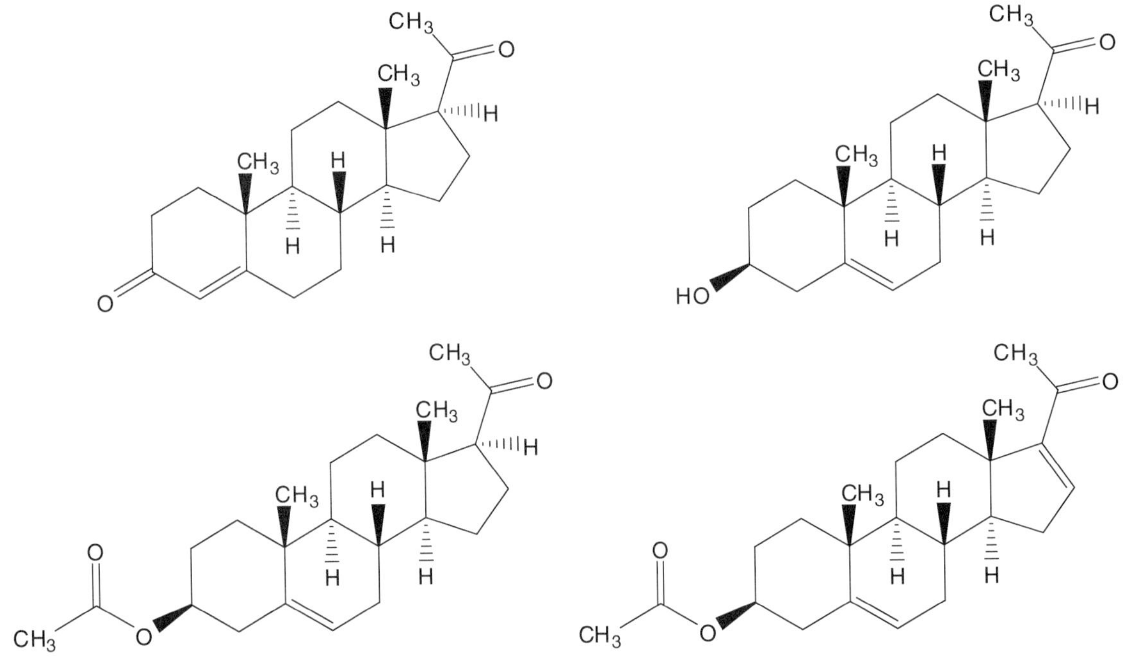

16-dehydropregnenolone acetate

The 16-alkene of 16-dehydropregnenolone acetate is formed from diosone by β-elimination. Diosone is formed by oxidation of the 20(22)-alkene of pseudodiosgenin-3,26-diacetate. Pseudodiosgenin 3,26-diacetate is formed by the reaction of diosgenin with acetic anhydride. The three-step synthesis of 16-dehydropregnenolone acetate from diosgenin by acetylation, oxidation, and elimination is known as the **Marker Degradation**.

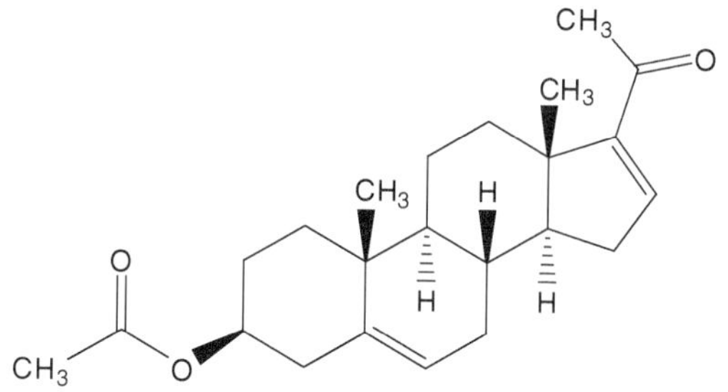

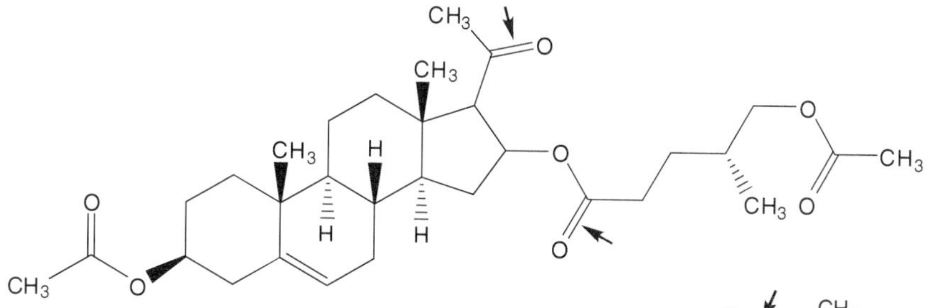

diosone

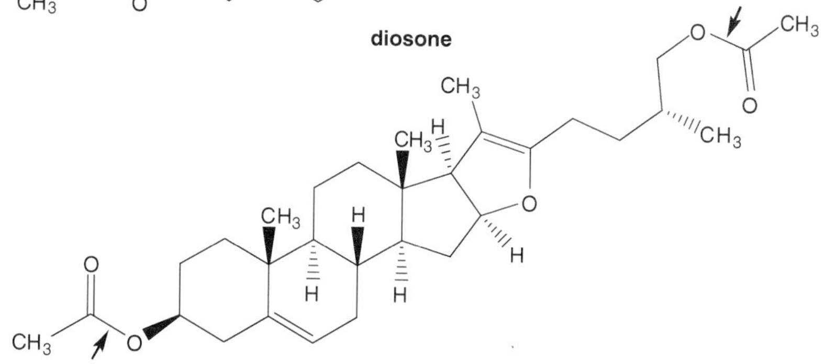

pseudodiosgenin-3,26-diacetate

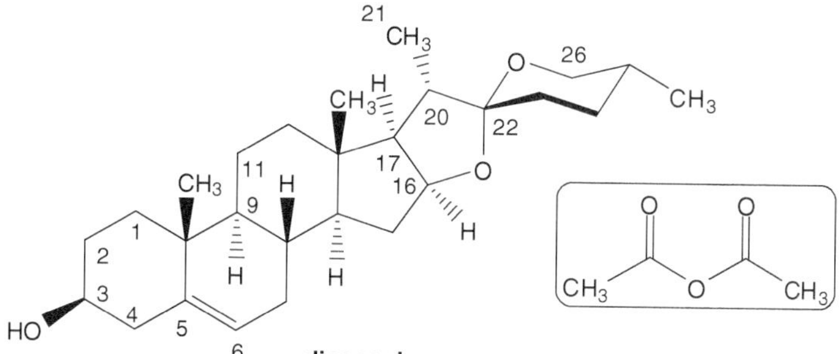

diosgenin

Extended Discussion

Draw structures of the retrosynthetic analysis of an alternative route to progesterone from stigmasterol. List the pros and cons for both routes.

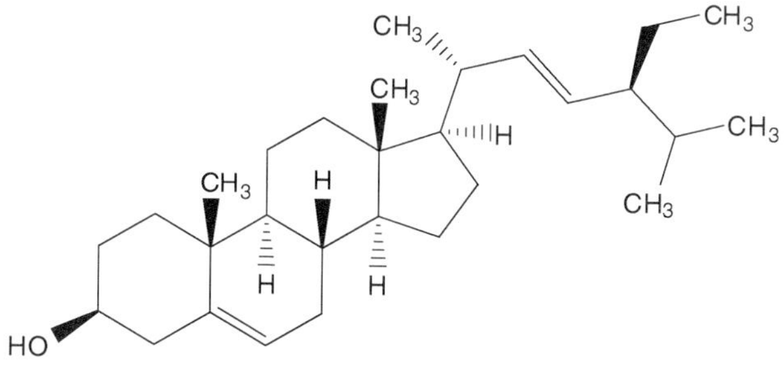

stigmasterol

Proguanil

Anti-infective Medicines/Antiprotozoal Medicines/Antimalarial Medicines/For Prophylaxis

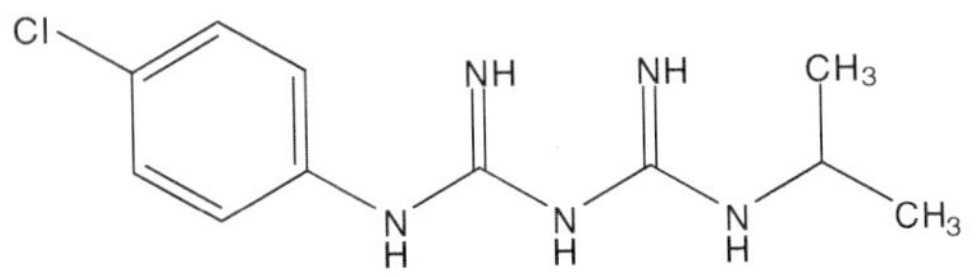

A biguanide is formed by addition of an amine to a cyanoguanidine.

Discussion. The biguanide of proguanil is formed from 4-chloroaniline, isopropylamine, and dicyanamide. Dicyanamide is available as the sodium salt.

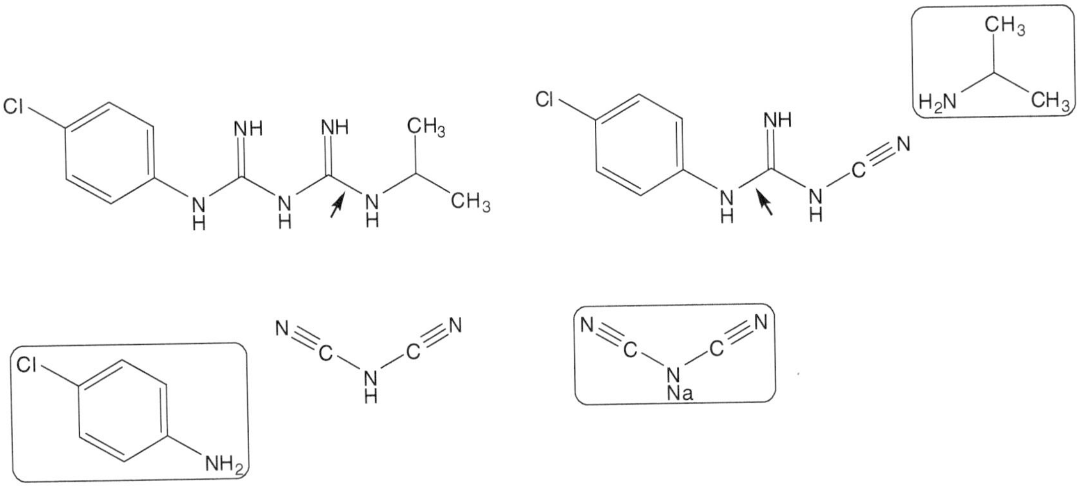

Extended Discussion

Draw the structures of the retrosynthetic analysis of one alternative route to proguanil from the same starting materials. List the pros and cons for both routes. Is one route preferred?

Propofol

Anesthetics, Preoperative Medicines, and Medical Gases/Injectable Medicines

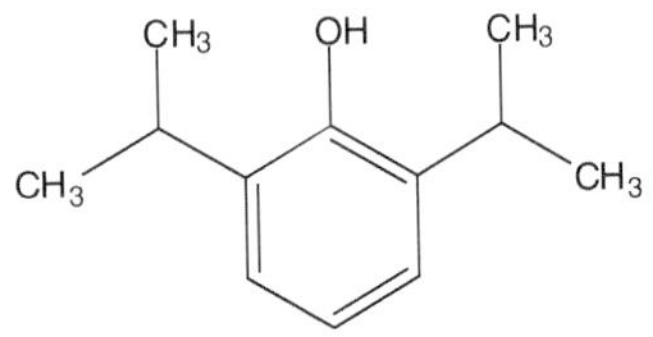

Every target molecule in this workbook must meet high purity specifications. When the starting materials are inexpensive and the route is just one or two steps, the target molecule purification procedure is often a major contributor to the cost of the drug.

Discussion. Disconnection of the isopropyl substituents suggests phenol and propene or isopropanol are the starting materials. Catalysts for the alkylation reaction include mineral acids and acidic zeolites. The search for the catalyst and reaction conditions that result in both high phenol conversion and high propofol selectivity continues to this day. Since propofol is a liquid (melting point 18 °C) with a high boiling point (242 °C), the procedure for conversion of the crude product mixture into high-purity propofol is an important contributor to the manufacturing cost.

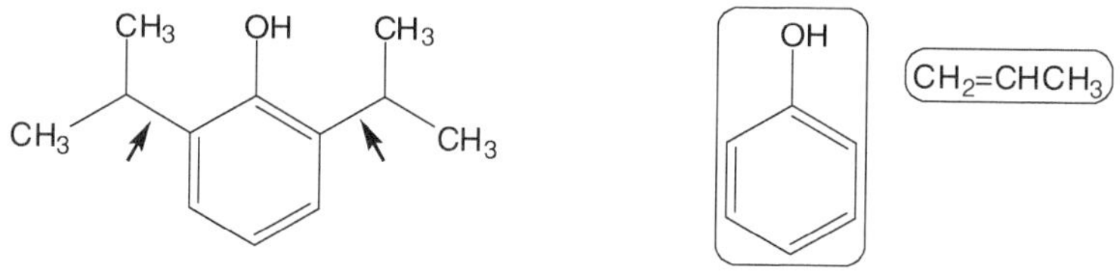

In an alternative route, the final step is the decarboxylation of the 4-hydroxy-3,5-diisopropylbenzoic acid.

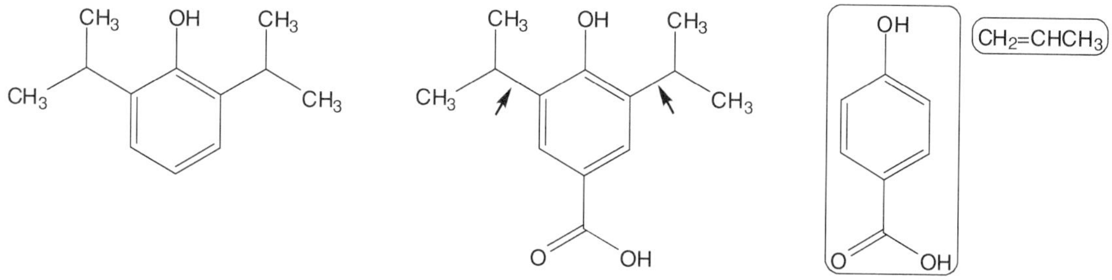

Extended Discussion

Each target molecule has a list of known impurities. Draw the structures for propofol impurities **A** through **P** from the European Pharmacopeia. Identify seven of these impurities which are most likely to be easily separated from propofol by distillation.

Propranolol

Antimigraine Medicines/For Prophylaxis

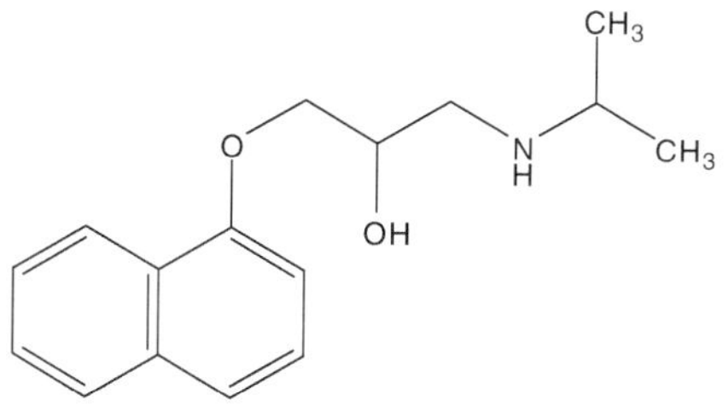

A β-amino alcohol is often formed by ring-opening of an epoxide by an amine.

Discussion. Propranolol is a 1 : 1 mixture of the (*R*)- and (*S*)-enantiomers. The (*S*)-enantiomer is active. The β-amino alcohol is formed by ring-opening of the epoxide of glycidyl 1-naphthyl ether by isopropylamine. Glycidyl 1-naphthyl ether is formed by chloride displacement from epichlorohydrin by 1-naphthol (**Williamson Ether Synthesis**).

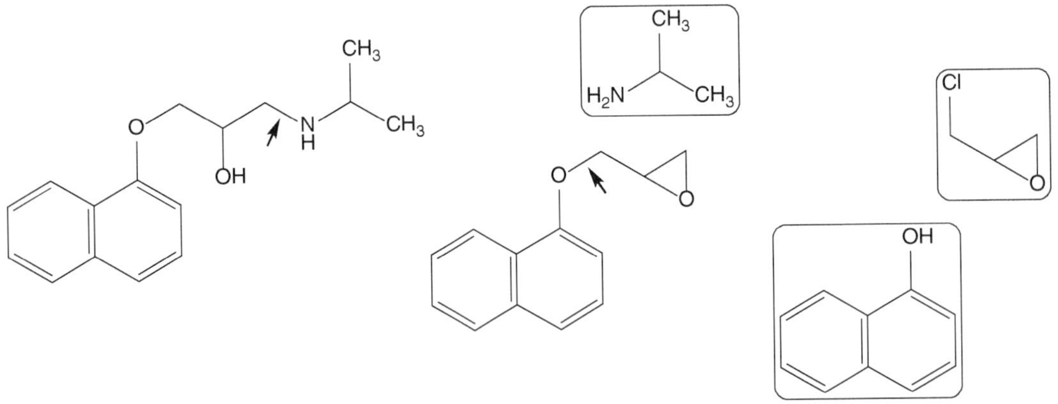

Extended Discussion

Review the *Safety Data Sheet* for epichlorohydrin. List the precautions to take before, during, and after the reaction of epichlorohydrin with 1-naphthol.

Propylthiouracil

Hormones, Other Endocrine Medicines and Contraceptives/Thyroid Hormones, and Antithyroid Medicines

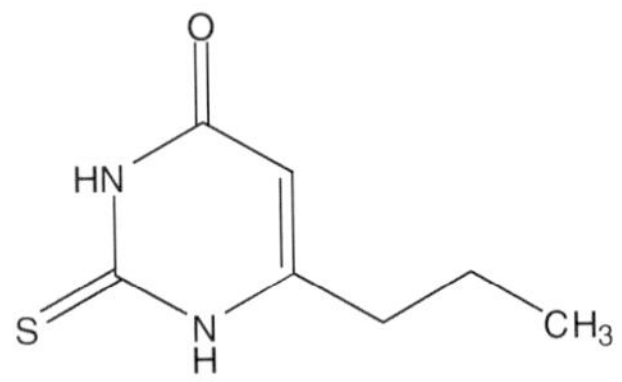

A thiouracil is often formed by the reaction of a β-ketoester with thiourea.

Discussion. The ring is formed by condensation of thiourea with ethyl 3-oxohexanoate. Ethyl 3-oxohexanoate is formed by the reaction of the acyl Meldrum's acid with ethanol. Meldrum's acid is acylated with butanoyl chloride.

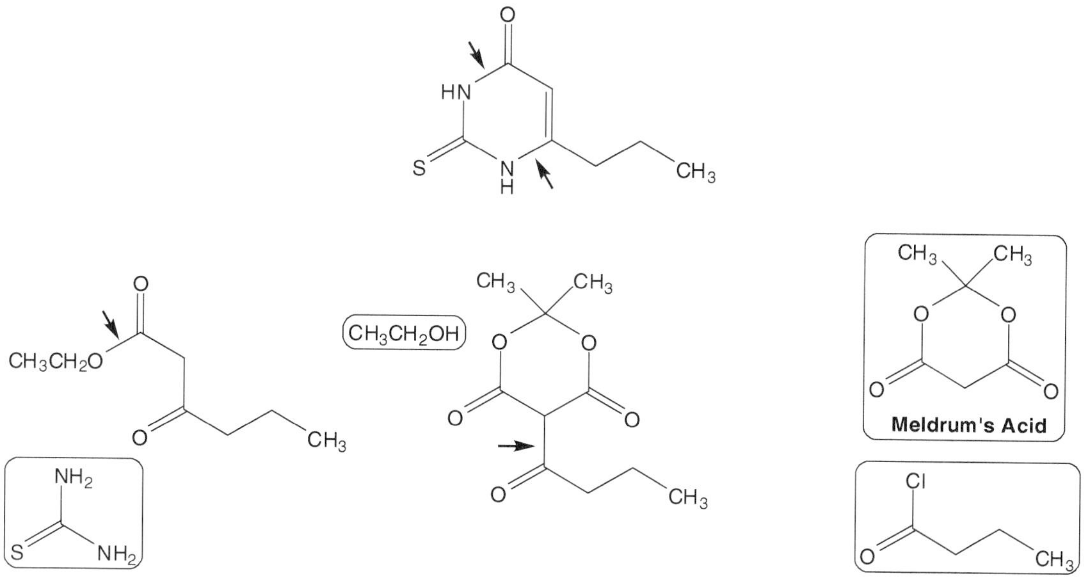

Extended Discussion

Draw the structures of a retrosynthetic analysis of an alternative route to ethyl 3-oxohexanoate. Include the structures of the retrosynthetic analysis of any organic starting material(s) from petrochemical or biochemical raw materials. List the pros and cons for both routes and select one route as the preferred route.

Prostaglandin E_1

Specific Medicines for Neonatal Care/Medicines Administered to the Neonate

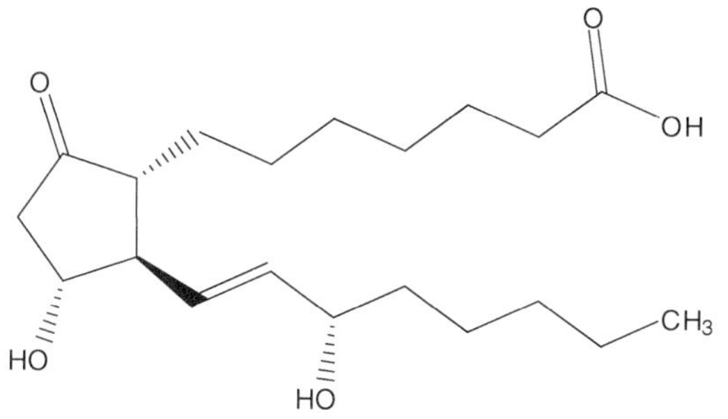

Prostaglandin analogs with the cyclopentanone core of prostaglandin E are often formed by conjugate addition of an alkenylcuprate reagent to a cyclopent-2-en-1-one. A 4-alkoxy or 4-trialkylsiloxy group on the α-face of the cyclopent-2-en-1-one directs the conjugate addition at C3 to the β-face.

Discussion. Prostaglandin E_1 (alprostadil) is formed in just three steps from (*R*)-norprostol. (*R*)-Norprostol is formed in seven steps from suberic acid (1,8-octanedioic acid).

The C11 and C15 alcohols and the carboxylic acid are all released in the final step. (List the protecting groups P^1 and P^2 used in this route to prostaglandin E1. Select one option for P^1 and one option for P^2 to use in the analysis.) The C–C bond joining C3 of the ring and the side chain is formed by conjugate addition of an alkenylcuprate to the cyclopent-2-en-1-one of an alcohol-protected (*R*)-norprostol. The alcohol of (*R*)-norprostol is protected. (*R*)-Norprostol is formed by enzyme-mediated hydrolysis of the racemic acetate ester. The racemic acetate ester is formed by the reaction of a mixture of norprostol and the isomeric 5-substituted 4-hydroxycyclopent-2-en-1-one with acetic anhydride. (*R*)-Norprostol is then separated from the (*S*)-acetate and the (*S*)-acetate is racemized and recycled.

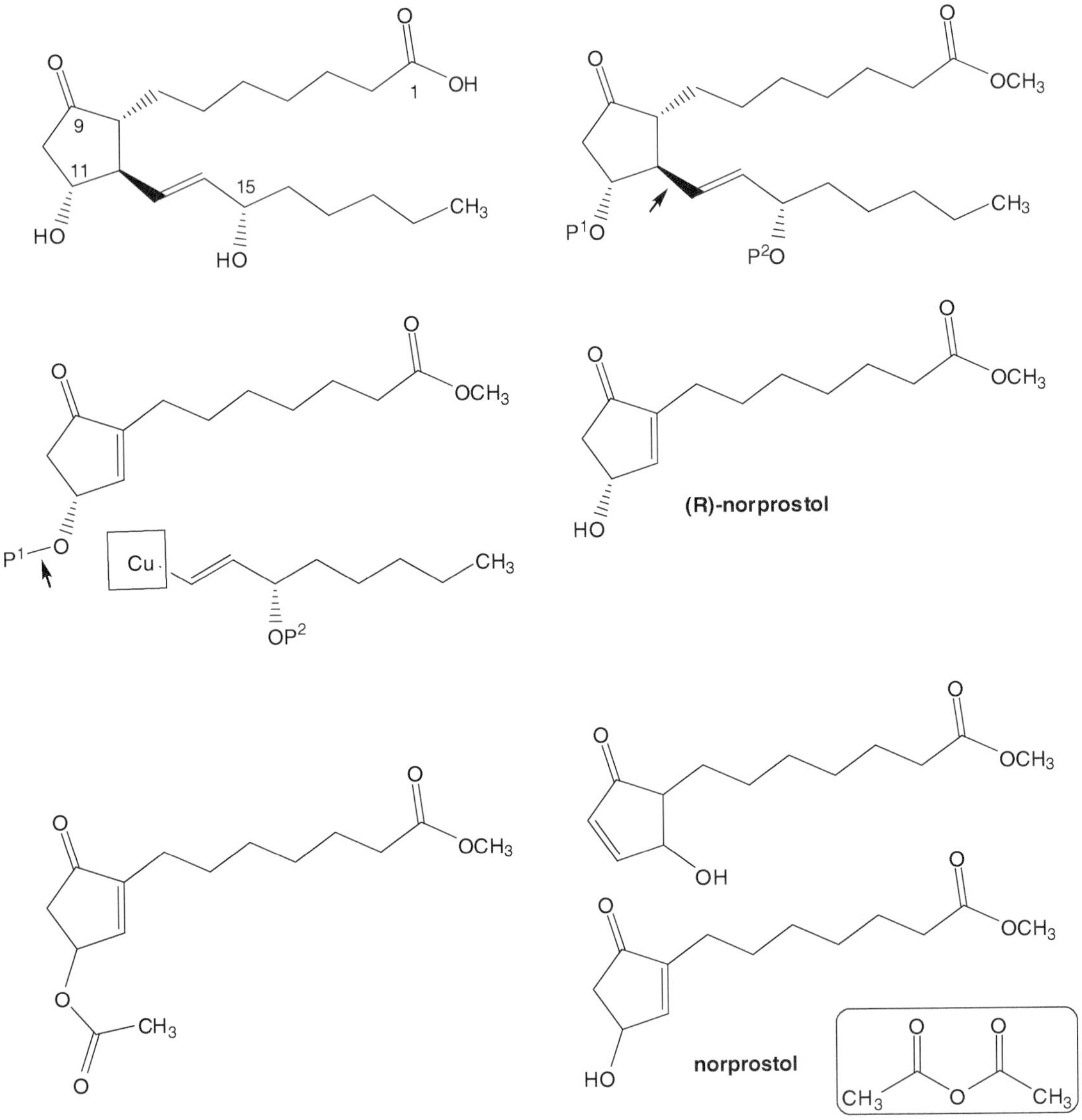

The mixture of norprostol and the isomer (1 : 3) is formed by the rearrangement of methyl 8-(2-furanyl)-8-hydroxyoctanoate (**Piancatelli Rearrangement**). The secondary alcohol is formed by the reduction of the ketone. Methyl 8-(2-furanyl)-2-oxooctanoate is formed by acylation of furan with an active ester (**Friedel–Crafts Acylation**). The active ester is formed in situ from the carboxylic acid. Suberic acid monomethyl ester is formed from suberic acid (**Fischer Esterification**).

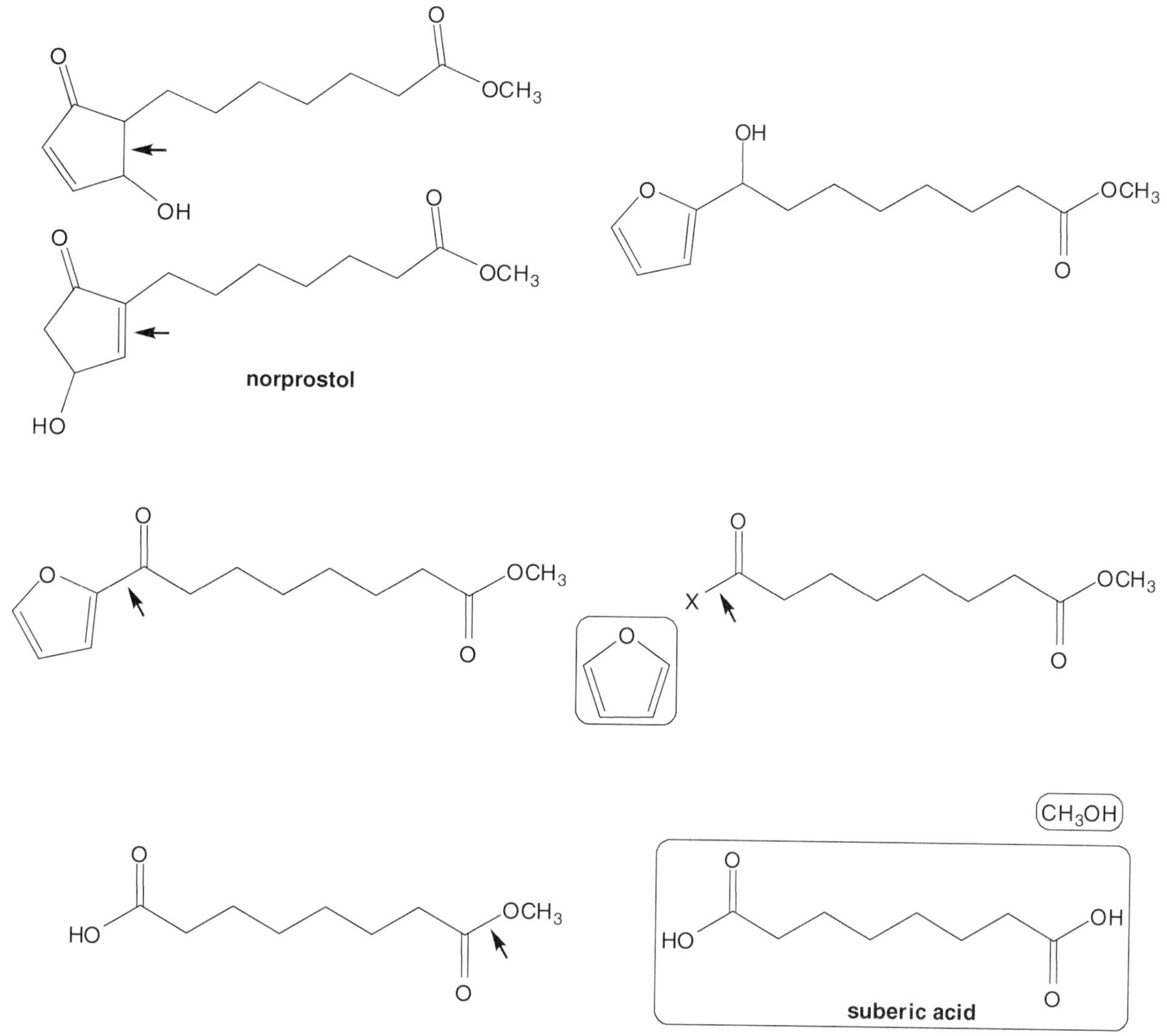

The alkenylcuprate reagent is formed from the alkyne. (Draw the structures for one route to the alkenylcuprate reagent from the alkyne. Draw the structure of the alkenylcuprate showing all the ligands attached to copper.) The alcohol is protected. (*S*)-1-Ocytn-3-ol is formed from racemic 1-octyn-3-yl acetate. (*S*)-1-Octyn-3-ol is formed by enzyme-mediated hydrolysis of the (*S*)-1-octyn-3-yl acetate. [(*S*)-1-Octyn-3-ol and (*R*)-1-octyn-3-yl acetate are separated and the (*R*)-1-octyn-3-yl acetate is racemized and recycled.] Racemic 1-octyn-3-yl acetate is formed by the reaction of 1-octyn-3-ol with acetic anhydride.1-Octyn-3-ol is formed from acetylene and hexanal.

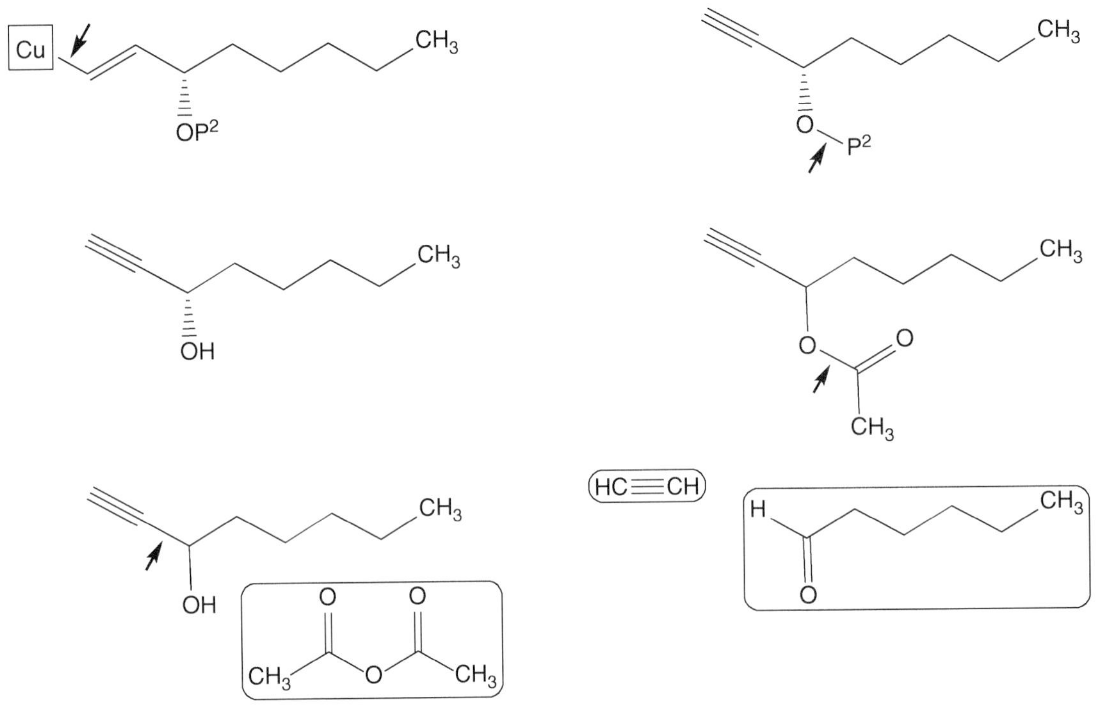

Extended Discussion

Draw the structures of the retrosynthetic analysis of an alternative route to norprostol. Include the structures of the retrosynthetic analysis of any organic starting material(s) from petrochemical or biochemical raw materials.

Prostaglandin E_2

Specific Medicines for Neonatal Care/Medicines Administered to the Neonate

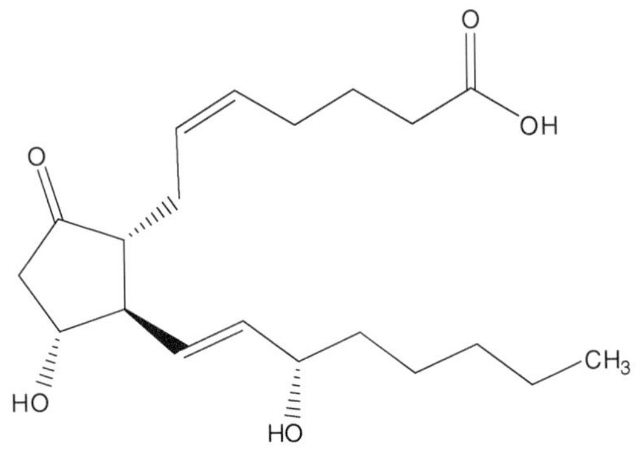

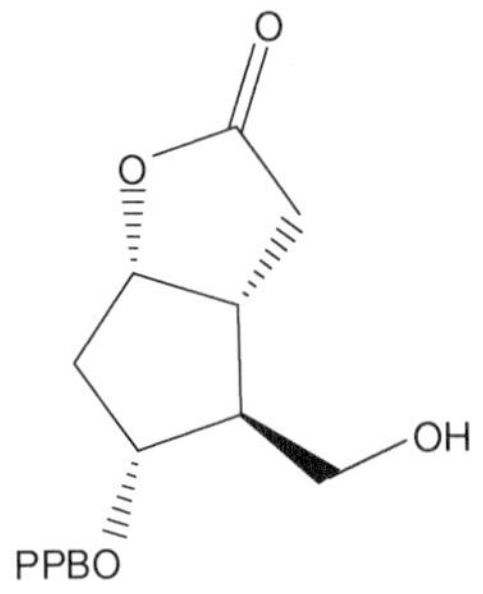

PPB-Corey lactone

PPB = 4-phenylbenzoyl

Prostaglandins are often constructed from a Corey Lactone. Oxidation of the primary alcohol of the Corey lactone forms an aldehyde which is used to introduce the lower side chain. Reduction of the lactone then forms another aldehyde which is used to introduce the upper side chain.

Discussion. In one preferred route, prostaglandin E_2 (dinoprostone) is formed in nine steps from PPB-**Corey Lactone**. In the final step, the C11 and C15 alcohols are released by hydrolysis of tetrahydropyranyl (THP) ethers. The C9 ketone is formed by oxidation of the alcohol. The C5–C6 alkene is formed by the reaction of the aldehyde with the ylid from (4-carboxybutyl)triphenylphosphonium bromide (**Wittig Reaction**). The aldehyde is formed in situ by ring-opening of the lactol. The lactol is formed by the reduction of the lactone. The THP ethers are formed from the C11 and C15 alcohols.

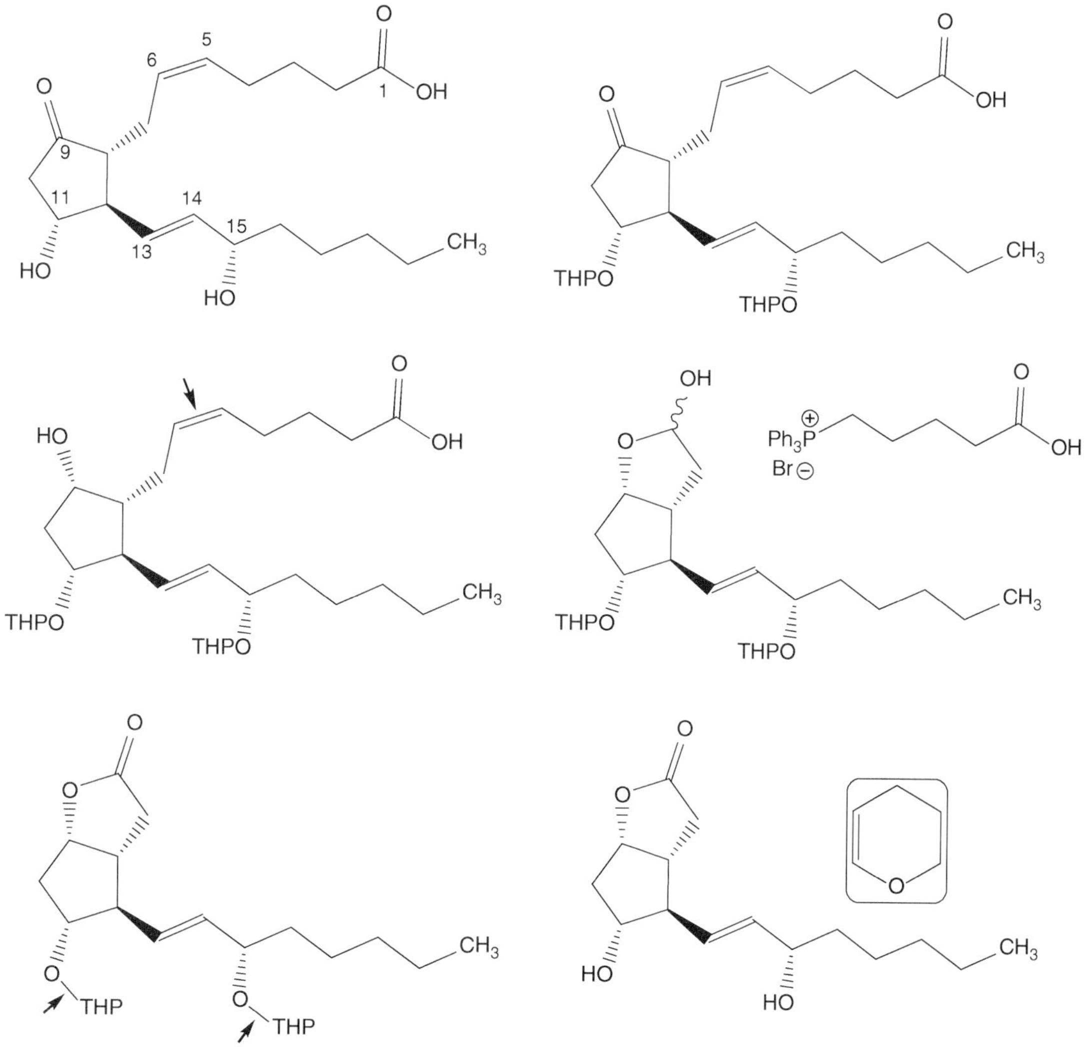

The C11 alcohol is released by ester hydrolysis. The C15 alcohol is formed by the reduction of the ketone. The C13–C14 alkene is formed by the reaction of the aldehyde with a phosphonate ester (**Horner–Wadsworth–Emmons Reaction**). The aldehyde is formed by oxidation of the primary alcohol of PPB-**Corey Lactone**.

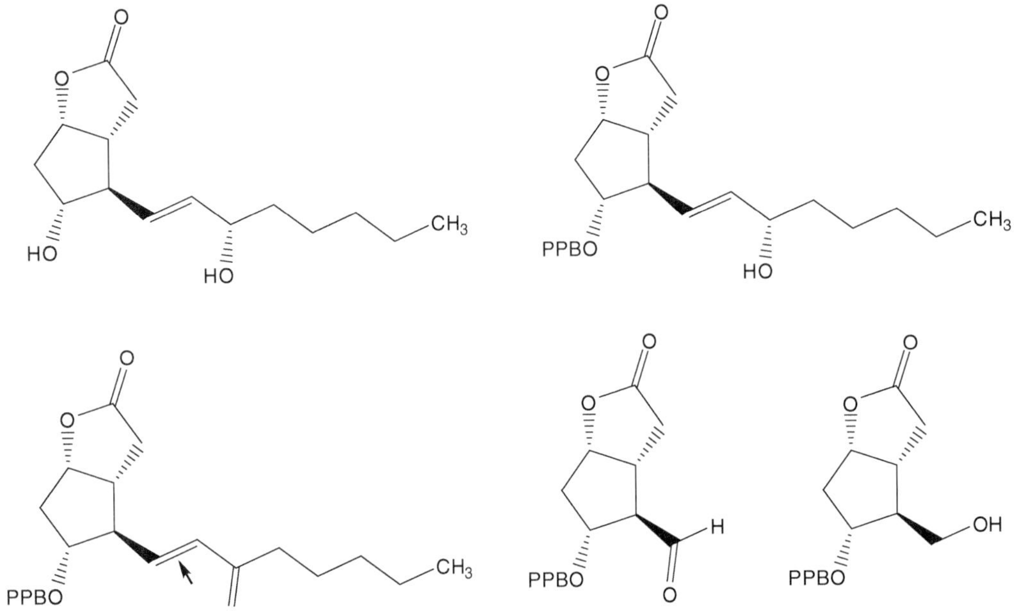

Corey Lactone

PPB = 4-phenylbenzoyl

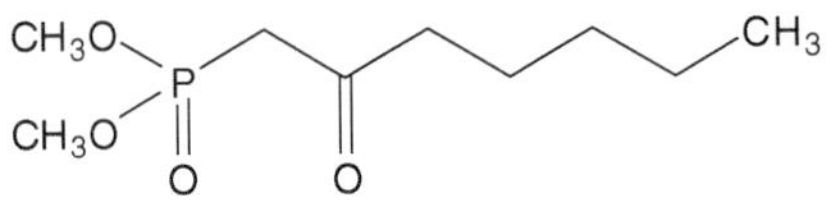

(4-Carboxybutyl)triphenylphosphonium bromide is formed from 5-bromopentanoic acid (5-bromovaleric acid). 5-Bromovareric acid is formed from δ-valerolactone.

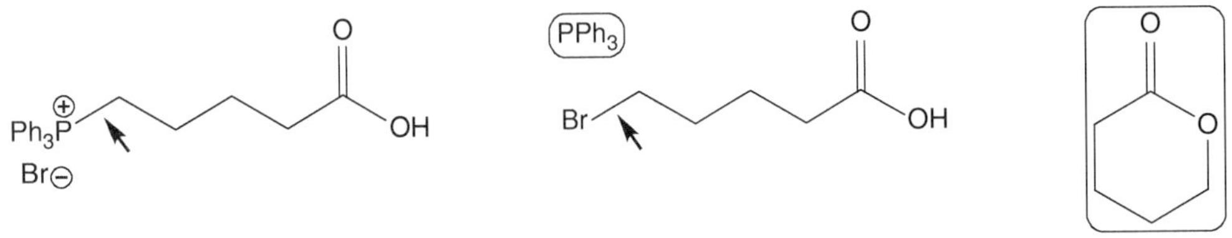

Dimethyl 2-oxoheptylphosphonate is formed by condensation of dimethyl methylphosphonate with ethyl hexanoate (ethyl caproate).

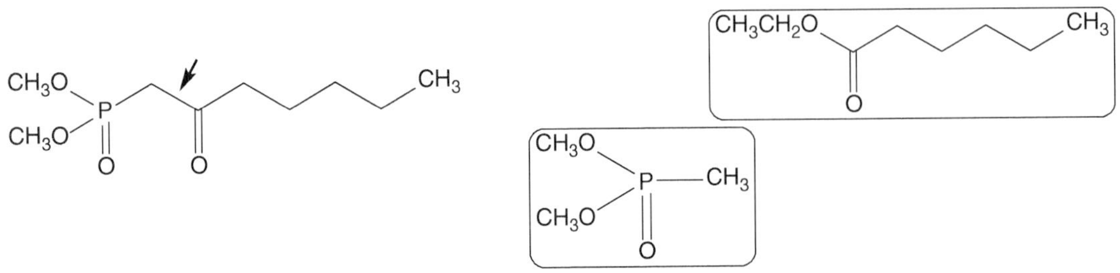

The Corey Lactone is formed in 10 steps from cyclopentadiene. The primary alcohol of the Corey Lactone is released by hydrogenolysis of the benzyl ether. The C10 iodide is removed by reduction. The secondary alcohol is protected as an ester by reaction with 4-phenylbenzoyl chloride (PPB-Cl). The lactone is formed by iodolactonization from the carboxylic acid and alkene. The carboxylic acid and secondary alcohol are formed by hydrolysis of the 2-oxabicyclo[3.2.1]oct-6-en-3-one. The 2-oxabicyclo[3.2.1]oct-6-en-3-one is formed by oxidation of the bicyclo[2.2.1]hept-5-en-2-one (**Baeyer–Villiger Oxidation**). The ketone is formed by oxidative decarboxylation of the carboxylic acid. The carboxylic acid is released by ester hydrolysis. (How is the chiral auxiliary D-pantolactone recovered and recycled?) The bicyclo[2.2.1]heptane is formed by cycloaddition of the 1,3-cyclopentadiene and the acrylate ester (**Diels–Alder Cycloaddition**). (Draw the structures of the two diastereoisomers formed by endo-cycloaddition. What is the diastereoselectivity of the cycloaddition?) 5-(Benzyloxy)methyl-1,3-cyclopentadiene is formed by alkylation of 1,3-cyclopentadiene with benzyl chloromethyl ether. The acrylate ester is formed from the alcohol (D-pantolactone) and acryloyl chloride. The acid chloride is formed from acrylic acid. (Draw the structure of one alternative alcohol used as the chiral auxiliary in the synthesis of the Corey Lactone.)

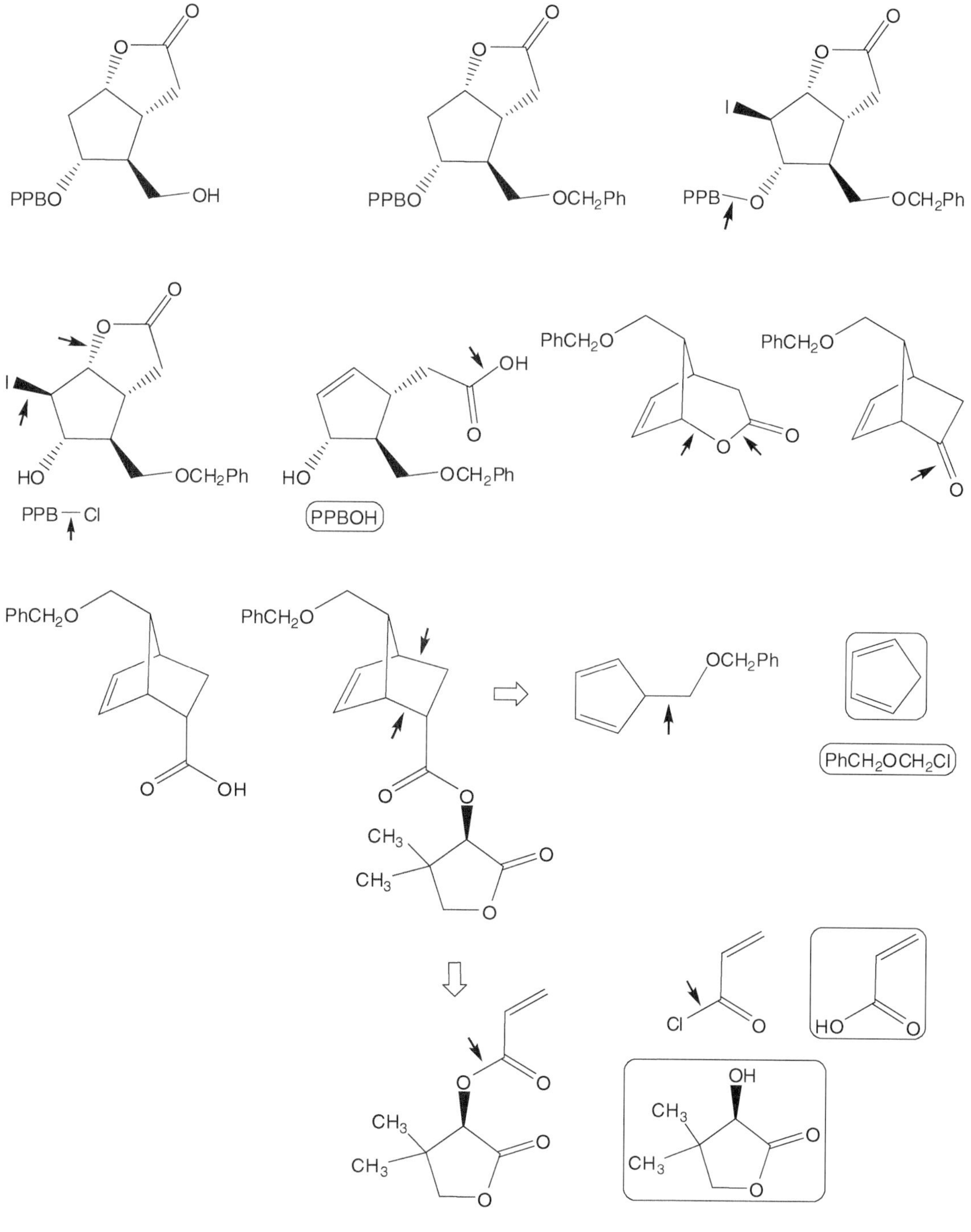

Extended Discussion

Draw the structures of the retrosynthetic analysis of one route to prostaglandin E2 that does not use a Corey Lactone. Include the structures of the retrosynthetic analysis of any organic starting material(s) from petrochemical or biochemical raw materials.

Pyrantel

Anti-infective Medicines/Anthelminthics/Intestinal Anthelminthics

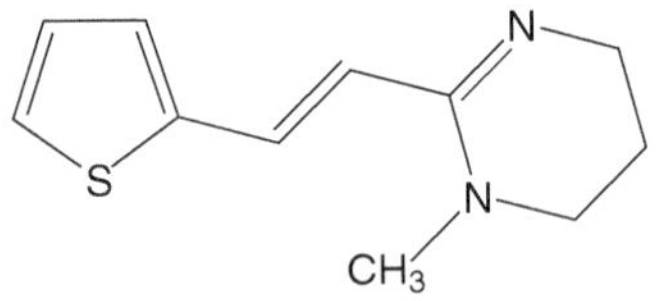

A thiophene ring is often delivered in a starting material. A 2-substituted thiophene is often formed by electrophilic aromatic substitution of thiophene.

Discussion. The alkene is formed by condensation of 2-thiophenecarboxaldehyde with 1,2-dimethyl-1,4,5,6-tetrahydropy rimidine. The aldehyde is formed from thiophene, *N*,*N*-dimethylformamide, and phosgene (**Vilsmeier–Haack Reaction**). 1,2-Dimethyl-1,4,5,6-tetrahydropyrimidine is formed by the reaction of *N*-methyl-1,3-propanediamine with ethyl acetoacetate. (What byproduct(s) are formed in this reaction?)

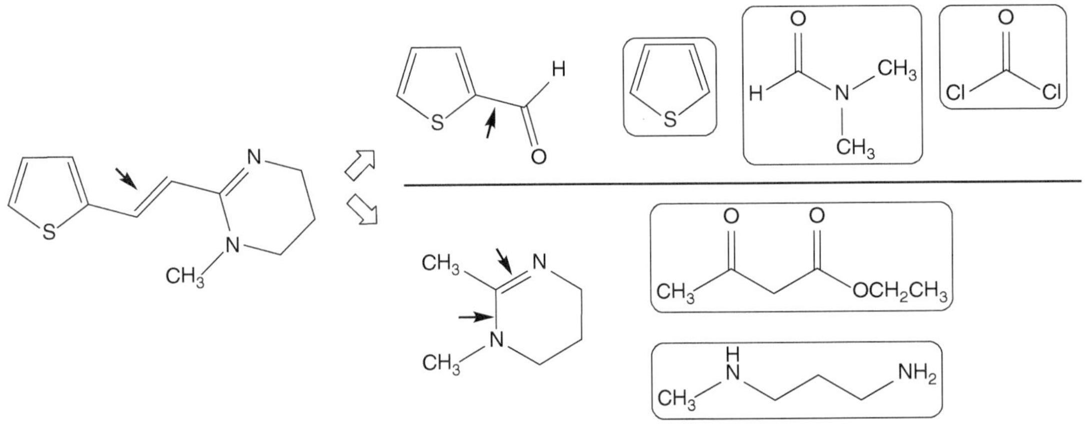

Extended Discussion

Draw the structure of one alternative reaction partner used to convert *N*-methyl-1,3-propanediamine to 1,2-dimethyl-1,4,5 ,6-tetrahydropyrimidine. What byproduct(s) are formed in the tetrahydropyrimidine formation using this alternative partner?

Pyrazinamide

Anti-infective Medicines/Antibacterials/Antituberculosis Medicines

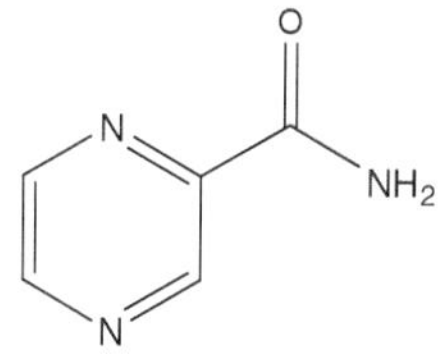

A primary amide is often formed by hydrolysis of a nitrile.

Discussion. Pyrazinamide is produced in a single step by hydrolysis of 2-pyrazinecarbonitrile.

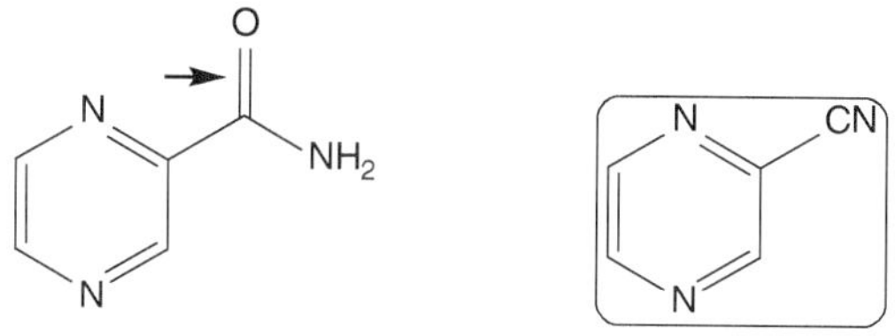

Extended Discussion

Draw the structures of the retrosynthetic analysis of an alternative multi-step route to pyrazinamide from *ortho*-phenylenediamine and glyoxal.

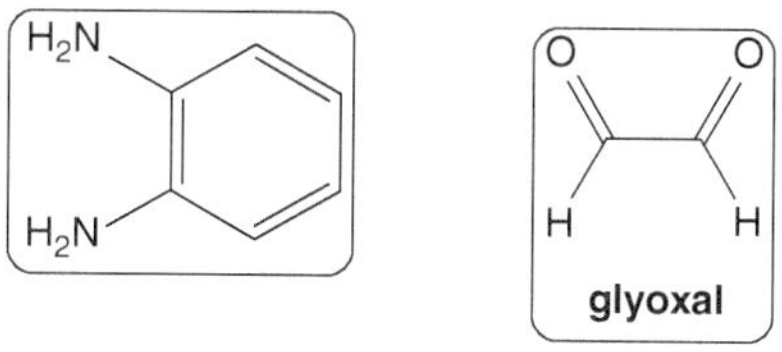

Pyridoxine

Vitamins and Minerals

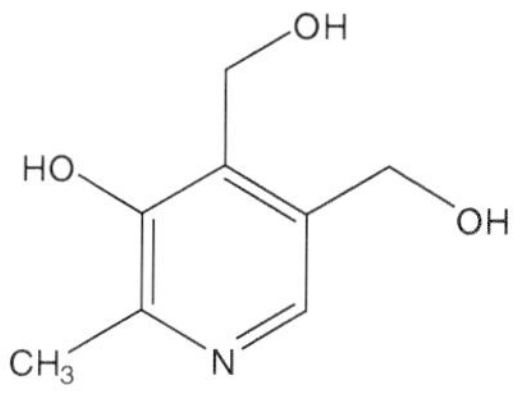

A 3-hydroxypyridine is often formed by construction of the pyridine ring by [4+2]-cycloaddition using an oxazole as the diene.

Discussion. In one preferred approach, the side-chain alcohols are released by hydrolysis of the acetal in the final step. The six-membered ring of the pyridine is formed by [4+2]-cycloaddition of an oxazole diene and alkene dienophile (**Diels–Alder Reaction**). The 3-hydroxypyridine is then formed by loss of ethanol from the cycloadduct. A preferred alkene dienophile is formed from *cis*-2-butene-1,4-diol and butanal.

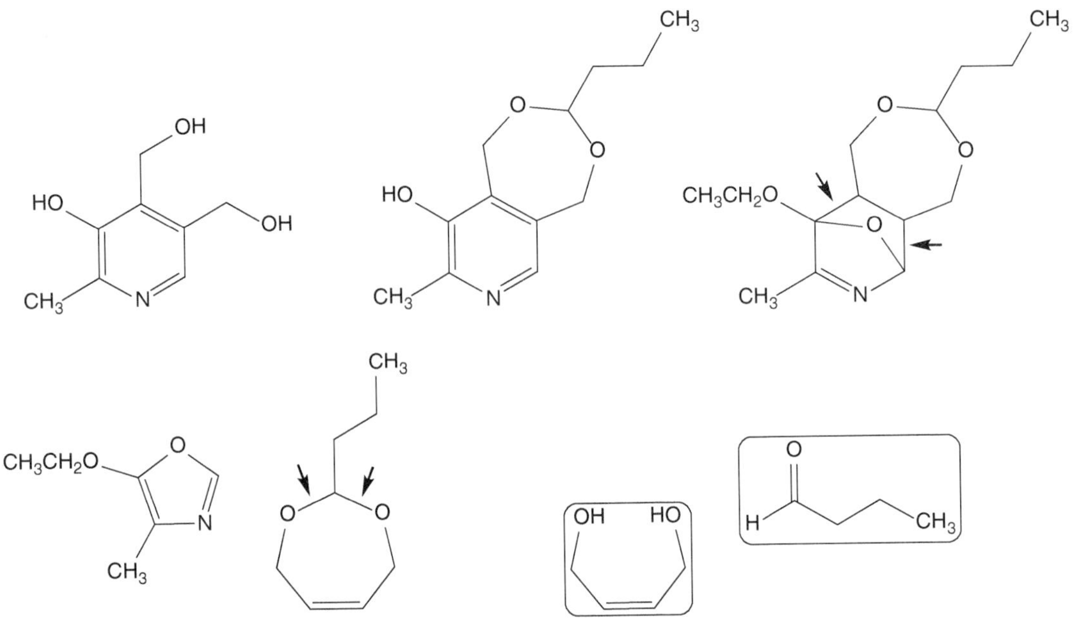

5-Ethoxy-4-methyl-1,3-oxazole, is formed by decarboxylation of the oxazole-2-carboxylic acid. The carboxylic acid is formed by hydrolysis of the ester. The oxazole ring is formed by cyclodehydration of ethyl *N*-ethoxalylalaninate. Ethyl *N*-ethoxalylalaninate is formed from alanine ethyl ester and diethyl oxalate.

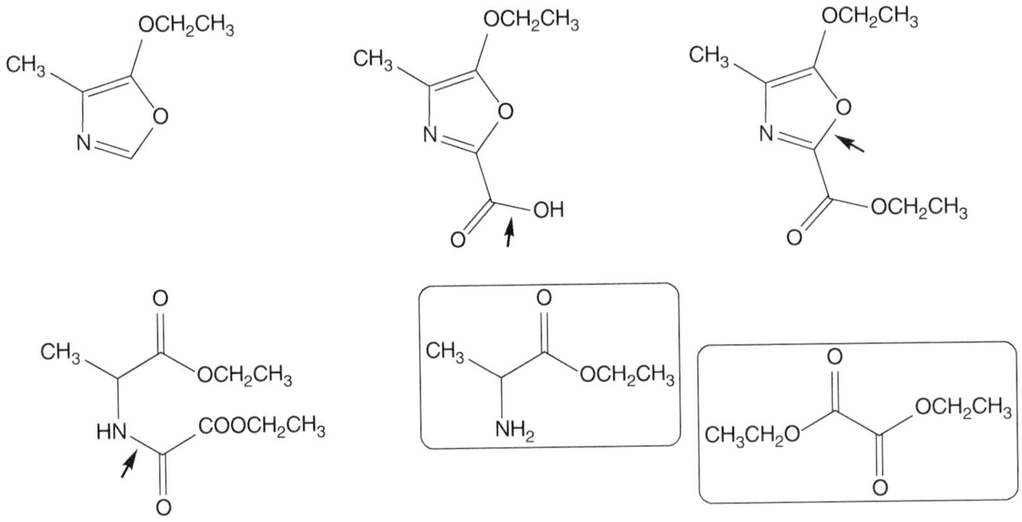

Extended Discussion

Draw the structures of a retrosynthetic analysis of one alternative route to a 5-alkoxy-4-methyloxazole. Discuss the pros and cons for both routes to the 5-alkoxy-4-methyloxazole and select one route as the preferred route.

Pyridostigmine Bromide

Muscle-relaxants (Peripherally-acting) and Cholinesterase Inhibitors

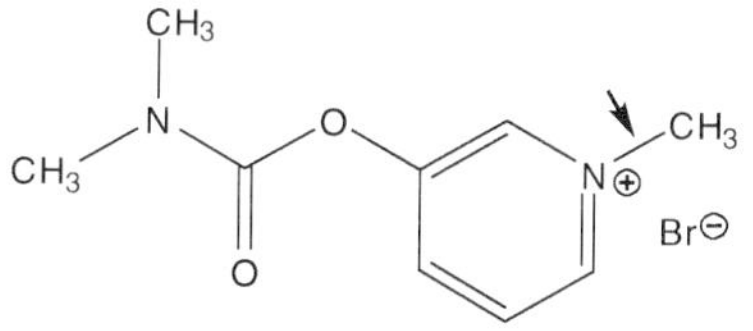

A phenol is often formed from an arylsulfonic acid by alkali fusion.

Discussion. Pyridostigmine is manufactured in just two steps. The *N*-methylpyridinium salt is formed by *N*-alkylation of the pyridine with bromomethane. The pyridine-3-yloxycarbamate is formed from dimethylcarbamoyl chloride and 3-hydroxypyridine.

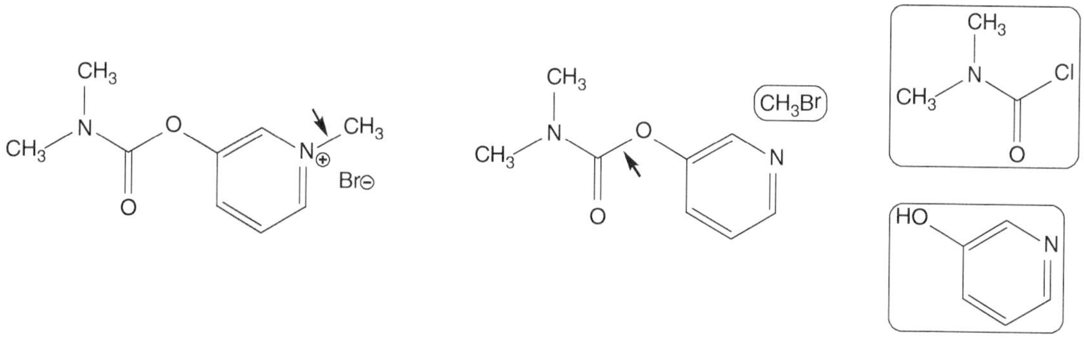

Extended Discussion

3-Hydroxypyridine is formed from pyridine-3-sulfonic acid by alkali fusion. Draw the structures of the retrosynthetic analysis of one alternative route to 3-hydroxypyridine from petrochemical or biochemical raw materials.

Pyrimethamine

Anti-infective Medicines/Antiprotozoal Medicines/Antipneumocytosis and Antitoxoplasmosis Medicines

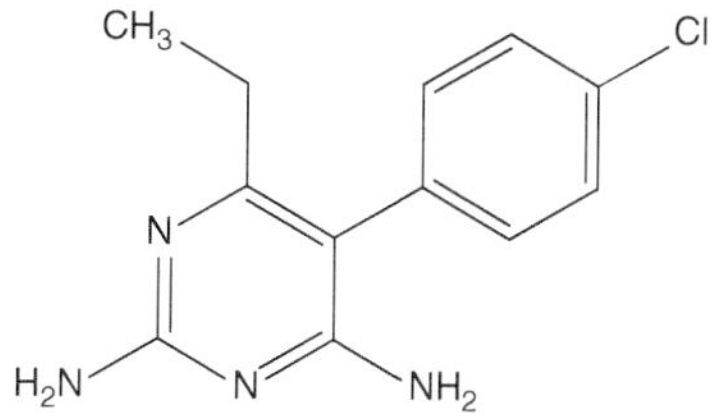

A pyrimidine ring with an amino group at C4 or C6 is often formed by ring construction from a 3-oxoalkanenitrile and an amidine, thiourea, or urea.

Discussion. The pyrimidine ring is formed in the final step by the reaction of the 3,3-ethylenedioxy-2-pentanenitrile with guanidine. The acetal is formed from the ketone and ethylene glycol. 2-(4-Chlorophenyl)-3-oxopentanenitrile is formed by condensation of (4-chlorophenyl)acetonitrile with ethyl propanoate.

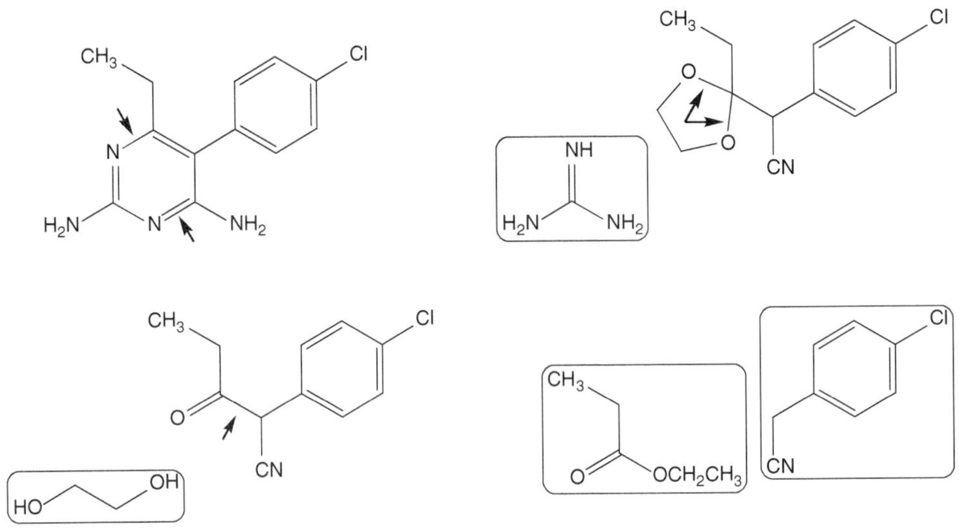

Extended Discussion

The C–C bond joining the rings is formed by a **Suzuki–Miyaura Coupling** in an alternative route to pyrimethamine. Draw the structures of the retrosynthetic analysis of this alternative route. Include the structures of the retrosynthetic analysis of any organic starting material(s) from petrochemical or biochemical raw materials.

Pyronaridine

Anti-infective Medicines/Antiprotozoal Medicines/Antimalarial Medicines/For Curative Treatment

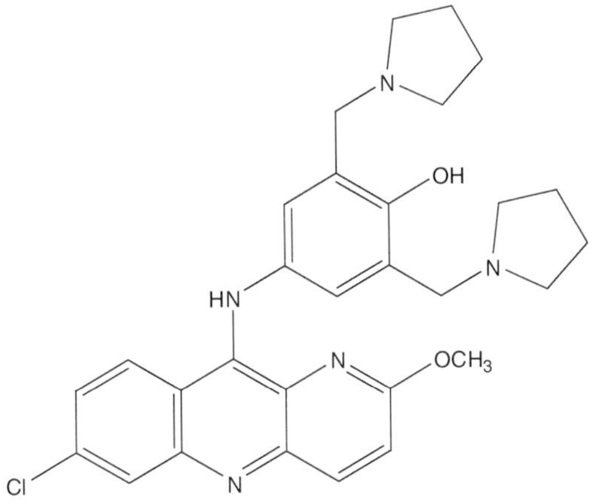

A nitrogen substituent on C2, C4, or C6 of a pyridine ring is often introduced by displacement of chloride by nucleophilic aromatic substitution. The substitution is facilitated by the ring nitrogen and can be further facilitated by an electron-withdrawing group (NO_2, SO_2R, COOR, CN) on C3.

Discussion. The C10 chloride of 7,10-dichloro-2-methoxybenzo[*b*][1,5]naphthyridine is displaced by 4-amino-2,6-bis(pyrrolidin-1-ylmethyl)phenol in the final step.

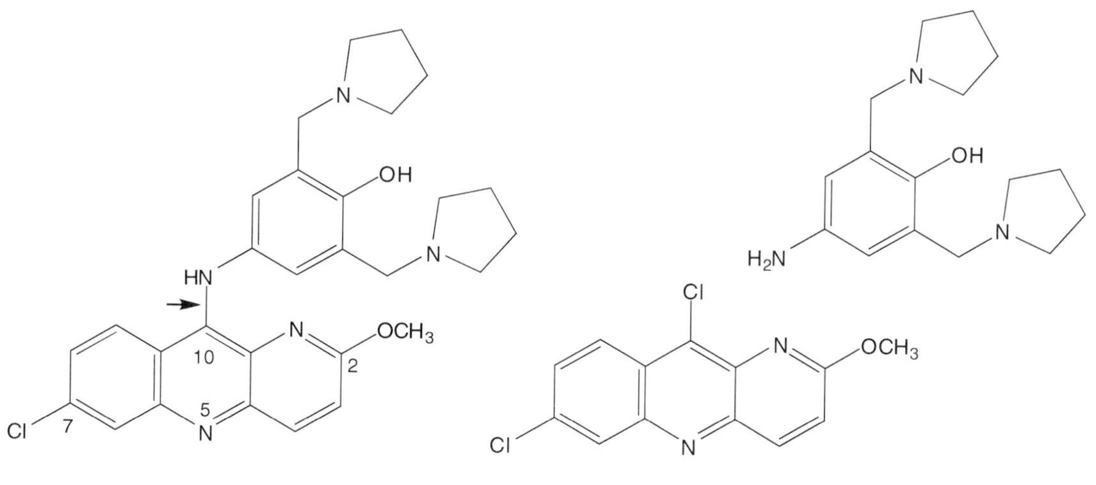

7,10-Dichloro-2-methoxybenzo[*b*][1,5]naphthyridine is formed from the benzo[*b*][1,5]naphthyridine-10(5*H*)-one. The benzo[*b*][1,5]naphthyridine-10(5*H*)-one is formed by intramolecular acylation at C6 of the 5-arylamino-2-methoxypyridine. The 5-arylamino-2-methoxypyridine is formed by copper-catalyzed displacement of chloride from 2,4-dichlorobenzoic acid by 5-amino-2-methoxypyridine (**Ullmann–Goldberg Reaction**). 5-Amino-2-methoxypyridine is formed by reduction of the nitropyridine. 2-Methoxy-5-nitropyridine is formed by chloride displacement from 2-chloro-5-nitropyridine by methanol.

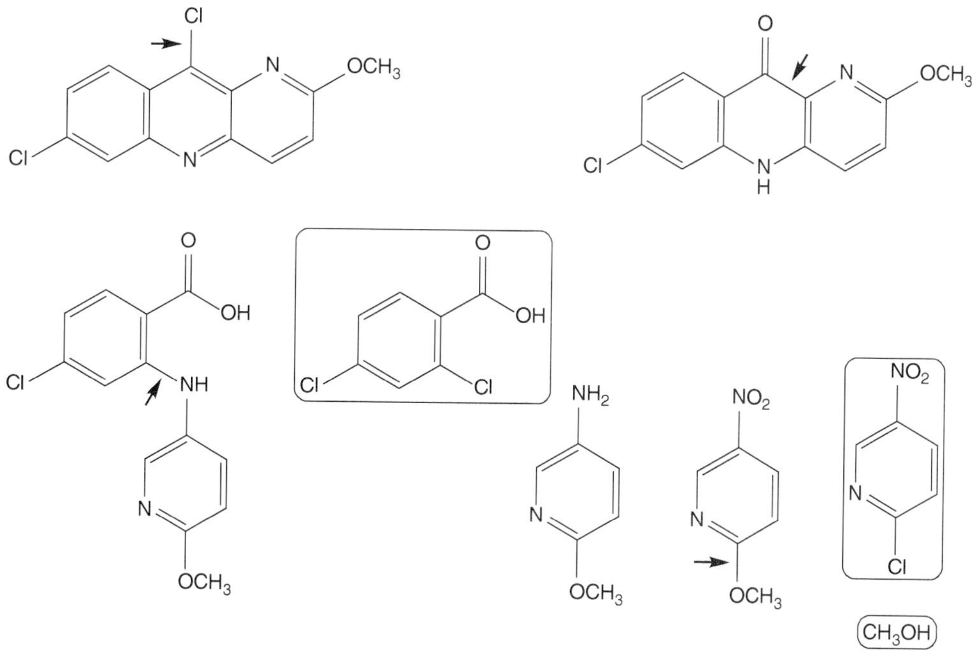

4-Amino-2,6-bis(pyrrolidin-1-ylmethyl)phenol is formed by reduction of the 4-nitrophenol. The pyrrolidin-1-ylmethyl substituents are introduced by the reaction of 4-nitrophenol with paraformaldehyde and pyrrolidine (**Mannich Reaction**).

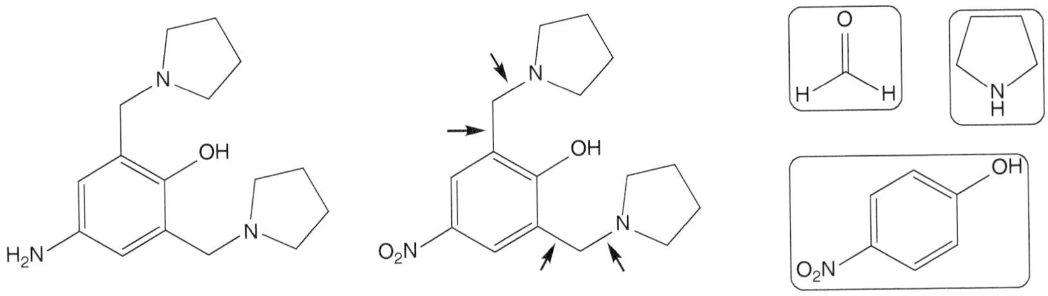

Extended Discussion

The **Mannich Reaction** is the final step in an alternative route to pyronaridine. List pros and cons for the two routes and select one route as the preferred route.

R

Raltegravir

Anti-infective Medicines/Antiviral Medicines/Antiretrovirals/Integrase Inhibitors

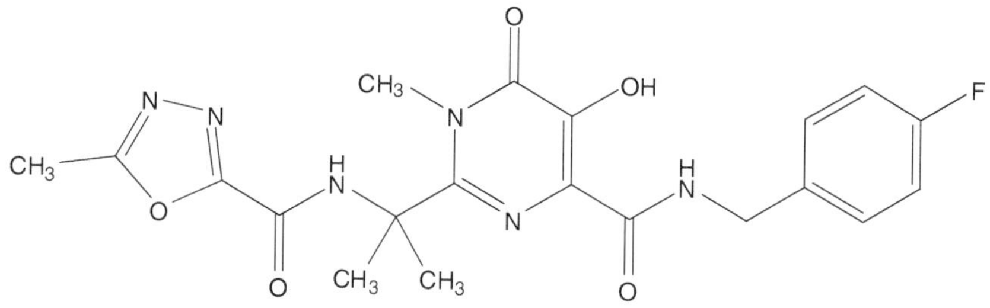

When one component of a target molecule is more expensive than the others, the preferred process often has that more expensive component introduced late in the synthesis.

Discussion. The hydroxyl group at C5 of the pyrimidine is released in the final step by hydrolysis of the pivalate ester. An amide is formed by the reaction of the amine with 5-methyl-1,3,4-oxadiazole-2-carbonyl chloride. The amine is released by cleavage of the benzyl carbamate. The 5-hydroxyl group is protected as a pivalate ester.

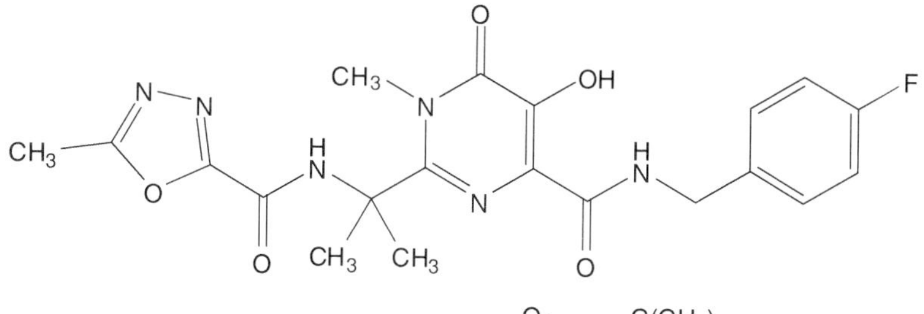

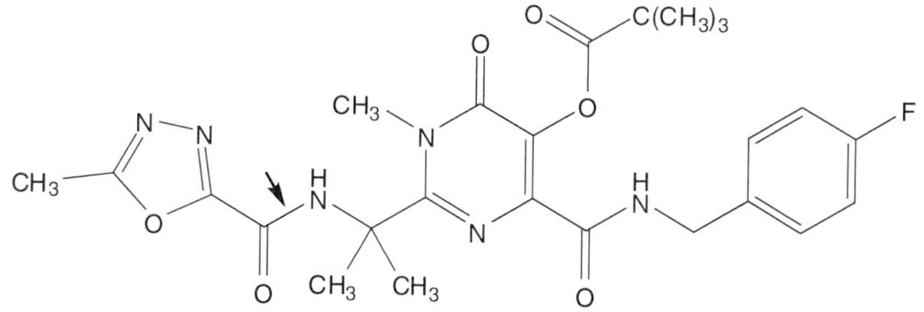

Routes to Essential Medicines: A Workbook for Organic Synthesis, First Edition. Peter J. Harrington.

Companion website: www.wiley.com/go/Harrington/routes_essential_medicine

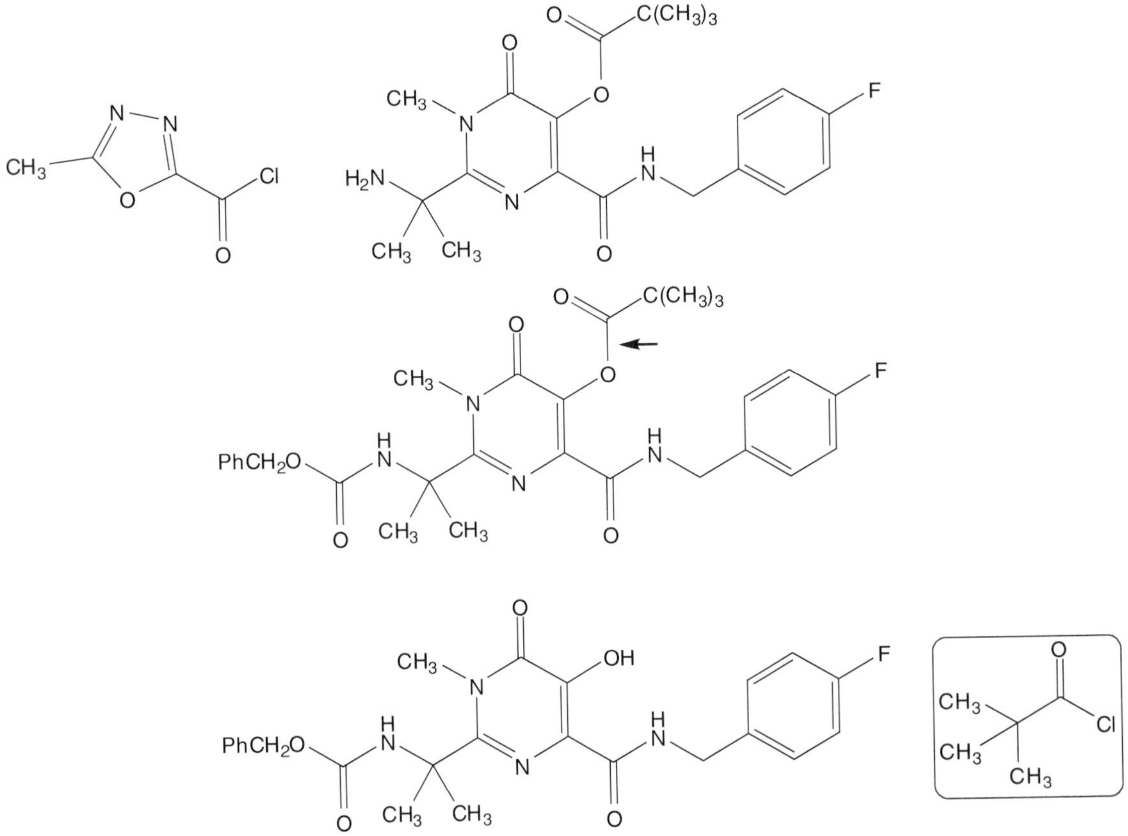

Methylation of the 5,6-dihydroxypyrimidine-4-carboxamide with trimethylsulfoxonium iodide is selective for N1. The pyrimidine-4-carboxamide is formed by the reaction of the methyl ester with 4-fluorobenzylamine.

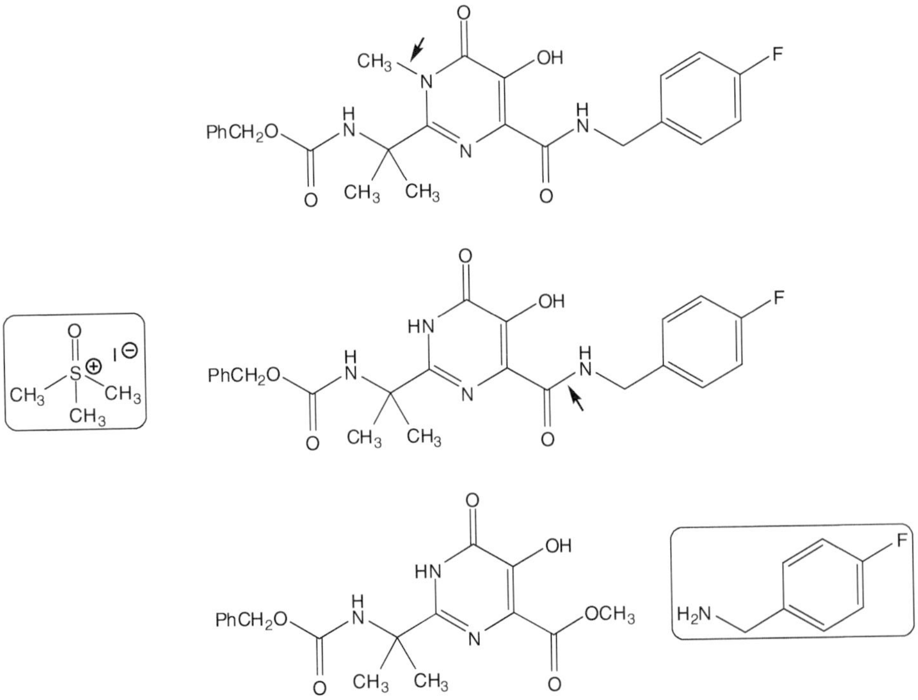

The 5,6-dihydroxypyrimidine-4-carboxylate is formed by the addition of the amidoxime to dimethyl acetylenedicarboxylate followed by rearrangement of the adduct on heating. The amidoxime is formed from the nitrile. The amine is protected as the benzyl carbamate. 2-Amino-2-methylpropanenitrile is formed from acetone, cyanide, and ammonia (**Strecker Synthesis**).

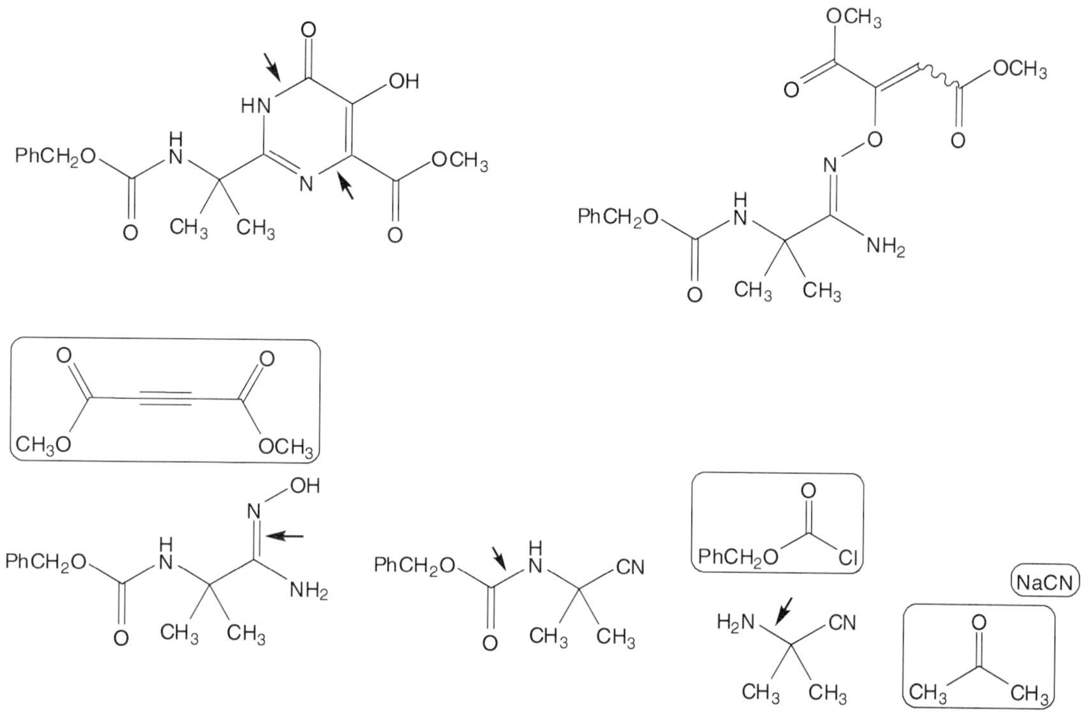

5-Methyl-1,3,4-oxadiazole-2-carbonyl chloride is formed from the carboxylate salt. The carboxylate salt is formed by hydrolysis of the ester. The 1,3,4-oxadiazole ring is assembled in one step from 5-methyl-1*H*-tetrazole and ethyl oxalyl chloride (ethyl chlorooxoacetate).

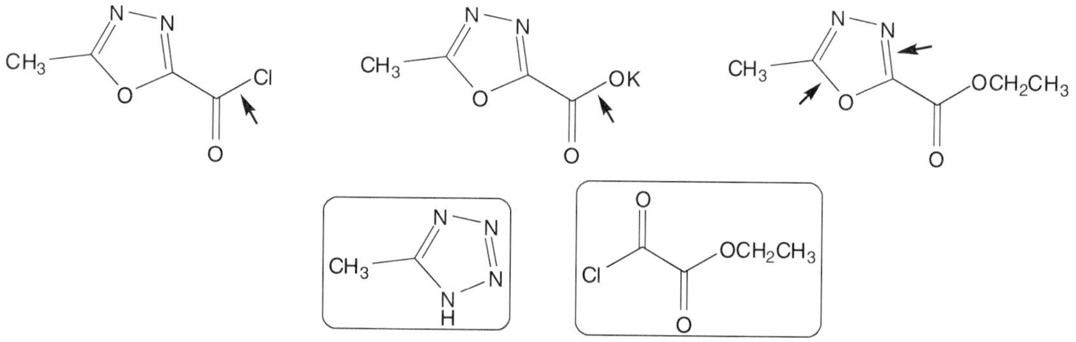

Extended Discussion

Draw the structures for the retrosynthetic analysis for one alternative route to ethyl 5-methyl-1,3,4-oxadiazole-2-carboxylate. Would this key intermediate be less expensive if produced by the alternative route? List assumptions used to answer the question.

Ranitidine

Gastrointestinal Medicines/Antiulcer Medicines

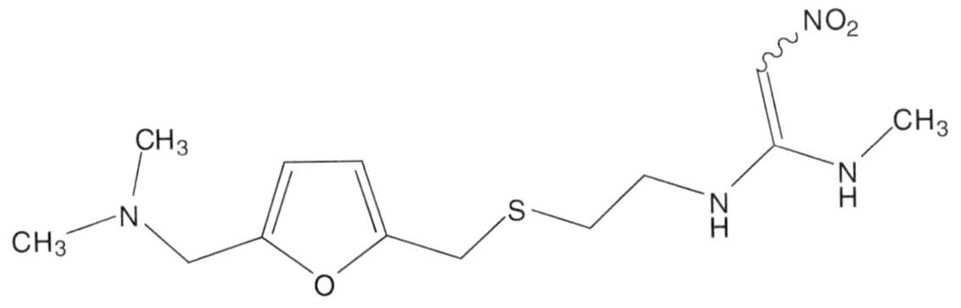

A β-nitroenamine is often formed by displacement of a leaving group from the β-carbon of a nitroalkene by an amine.

Discussion. The *S*-alkylcysteamine displaces methanethiol (methyl mercaptan) from *N*-methyl-1-(methylthio)-2-nitroethenamine in the final step. The *S*-alkylcysteamine is formed from the furfuryl alcohol and cysteamine. 5-(Dimethylaminomethyl)furfuryl alcohol is formed by the reaction of furfuryl alcohol with dimethylamine and formaldehyde (**Mannich Reaction**). Cysteamine is formed by hydrolysis of 2-thiazoline-2-thiol. (Poisonous hydrogen sulfide is a byproduct of this hydrolysis. Draw the structures of one alternative route to cysteamine from thiazoline-2-thiol which does not produce hydrogen sulfide as a byproduct.)

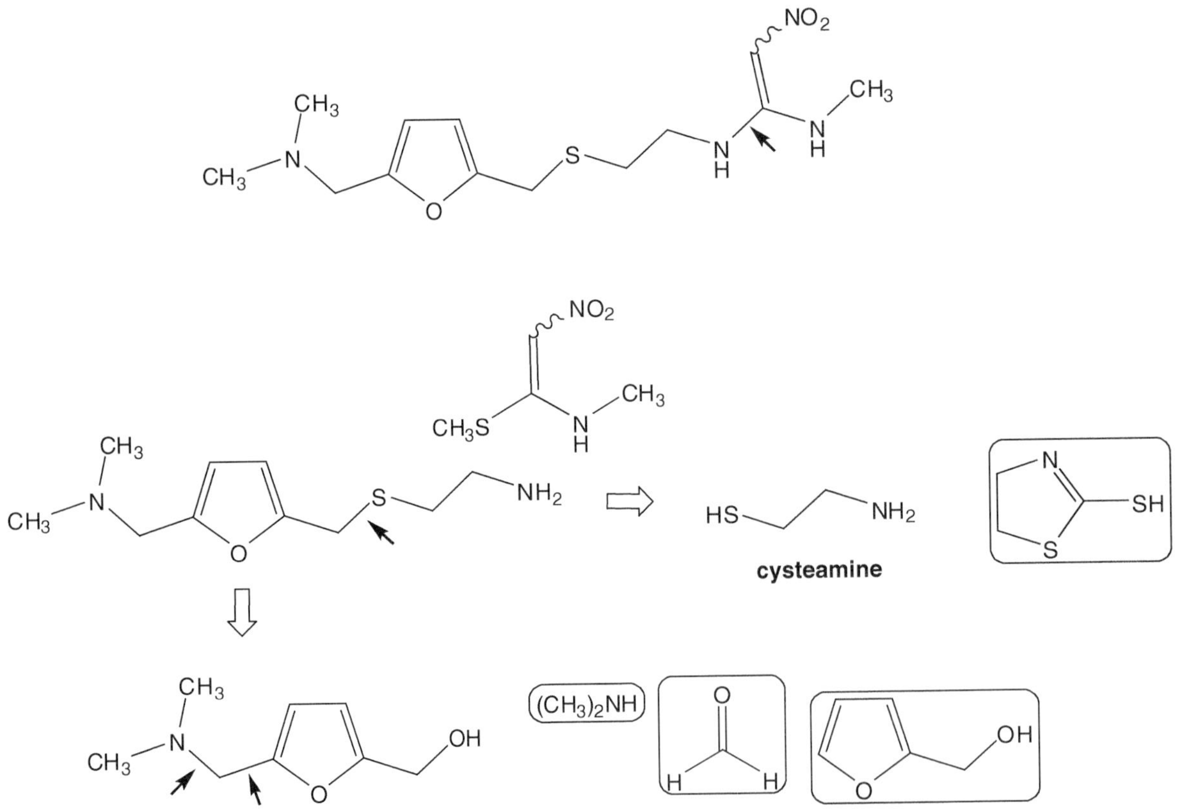

N-Methyl-1-(methylthio)-2-nitroethenamine is formed by the addition of nitromethane to methyl isothiocyanate followed by *S*-methylation.

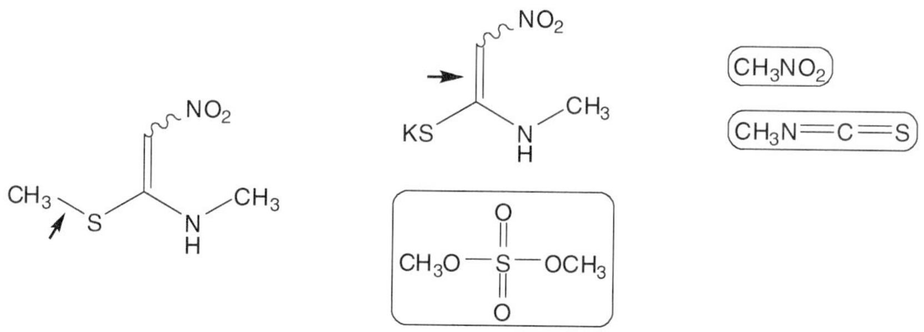

Extended Discussion

Draw the structures of a retrosynthetic analysis of one alternative route to *N*-methyl-1-(methylthio)-2-nitroethenamine. List the pros and cons for both routes to this intermediate and select one route as the preferred route.

Retinoic Acid (All-*trans*)

Antineoplastics and Immunosuppressives/Cytotoxic and Adjuvant Medicines

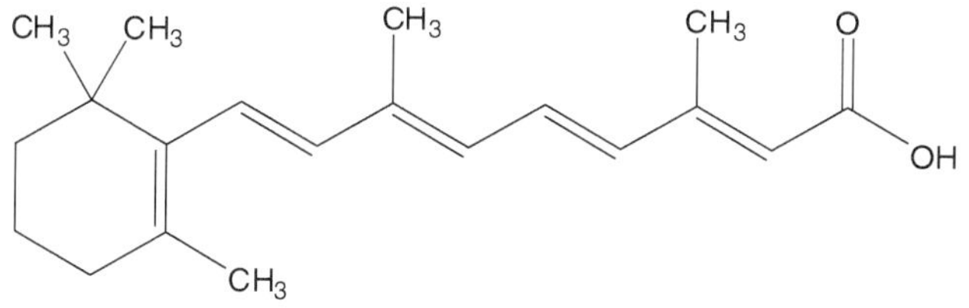

An alkene is often formed by the reaction of an aldehyde or ketone with a phosphorus ylid (Wittig Reaction).

Discussion. The carboxylic acid of all-*trans* retinoic acid (Tretinoin) is released by ester hydrolysis in the final step. The C11-C12 double bond is formed from a phosphonium salt and an aldehyde (**Wittig Reaction**). (What is the *E*/*Z* ratio of the C11-C12 bond in the product mixture?) The phosphonium salt is formed by the reaction of vinyl-β-ionol with triphenylphosphine. (Draw the structures of four diene diastereomers formed in this reaction.) Vinyl-β-ionol is formed by the addition of vinylmagnesium bromide to β-ionone (**Grignard Reaction**). Vinylmagnesium bromide is formed from vinyl bromide.

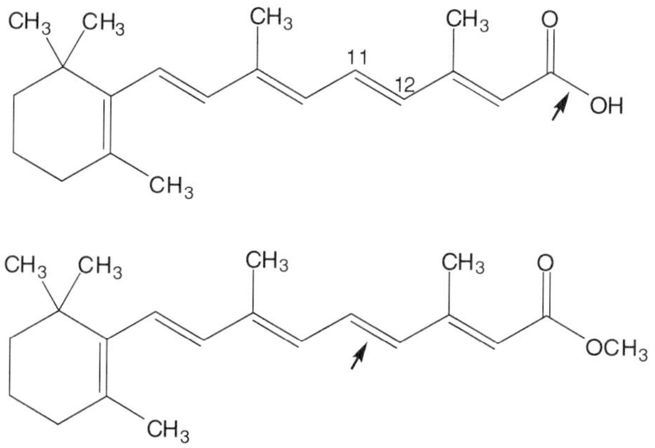

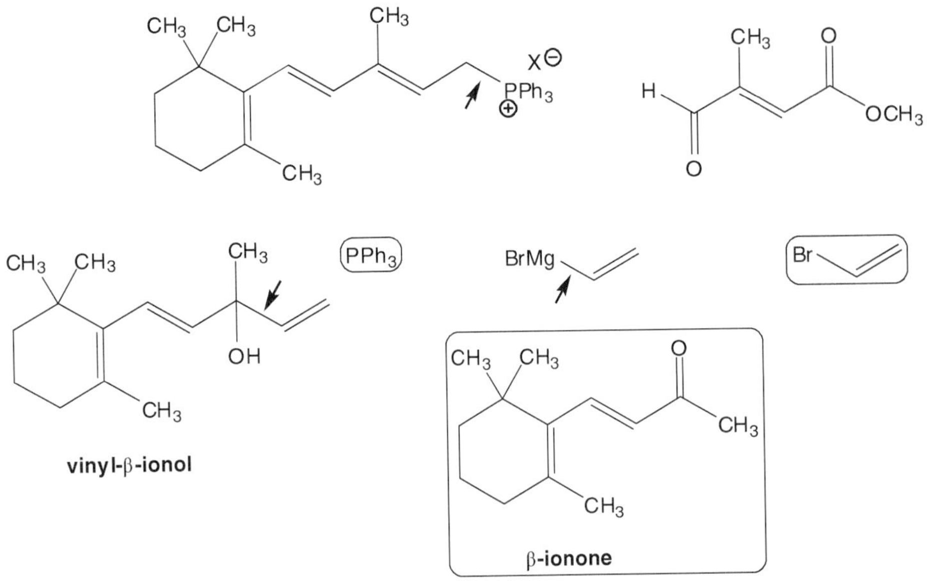

Methyl (*E*)-3-methyl-4-oxo-2-butenoate, the aldehyde for the Wittig reaction, is formed by condensation of methyl glyoxalate and propanal. (What is the *E*/*Z* ratio in the product mixture?)

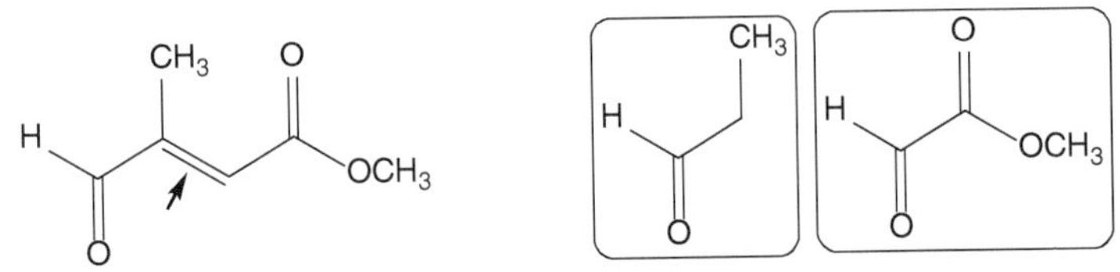

Retinol (Vitamin A$_1$)

Vitamins and Minerals

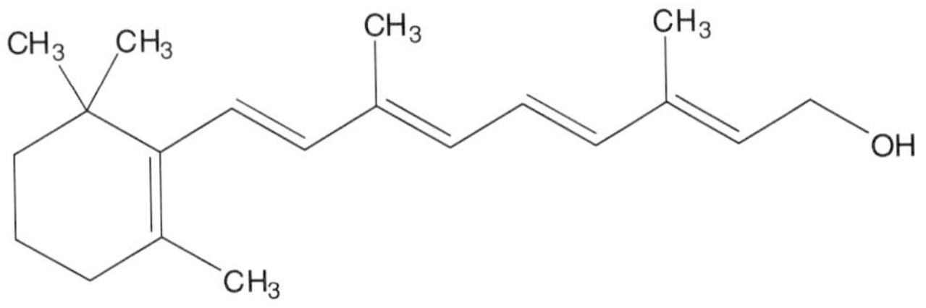

An alkene is often formed by dehydration of an alcohol. The dehydration is facilitated when the alkene formed is stabilized by conjugation. If the alkene can form (*E*)- and (*Z*)-diastereomers, both are formed in the dehydration. The more stable alkene is the major product when the diastereomers can interconvert under dehydration conditions.

Discussion. In the final step of the retinol synthesis associated with Hoffmann-La Roche and now DSM, the alcohol is released by hydrolysis of the acetate ester. The all-(*E*) conjugated double bond system forms by dehydration of secondary alcohol at C10 of the side chain. The acetate ester is formed by acetylation of the C15 primary alcohol. The C11-C12 alkene is formed by hydrogenation of an alkyne using Lindlar catalyst. The secondary alcohol at C10 is formed by addition of the alkynylmagnesium halide to the aldehyde 2-methyl-4-(2,6,6-trimethyl-1-cyclohexen-1-yl)-2-butenal (**Grignard Reaction**).

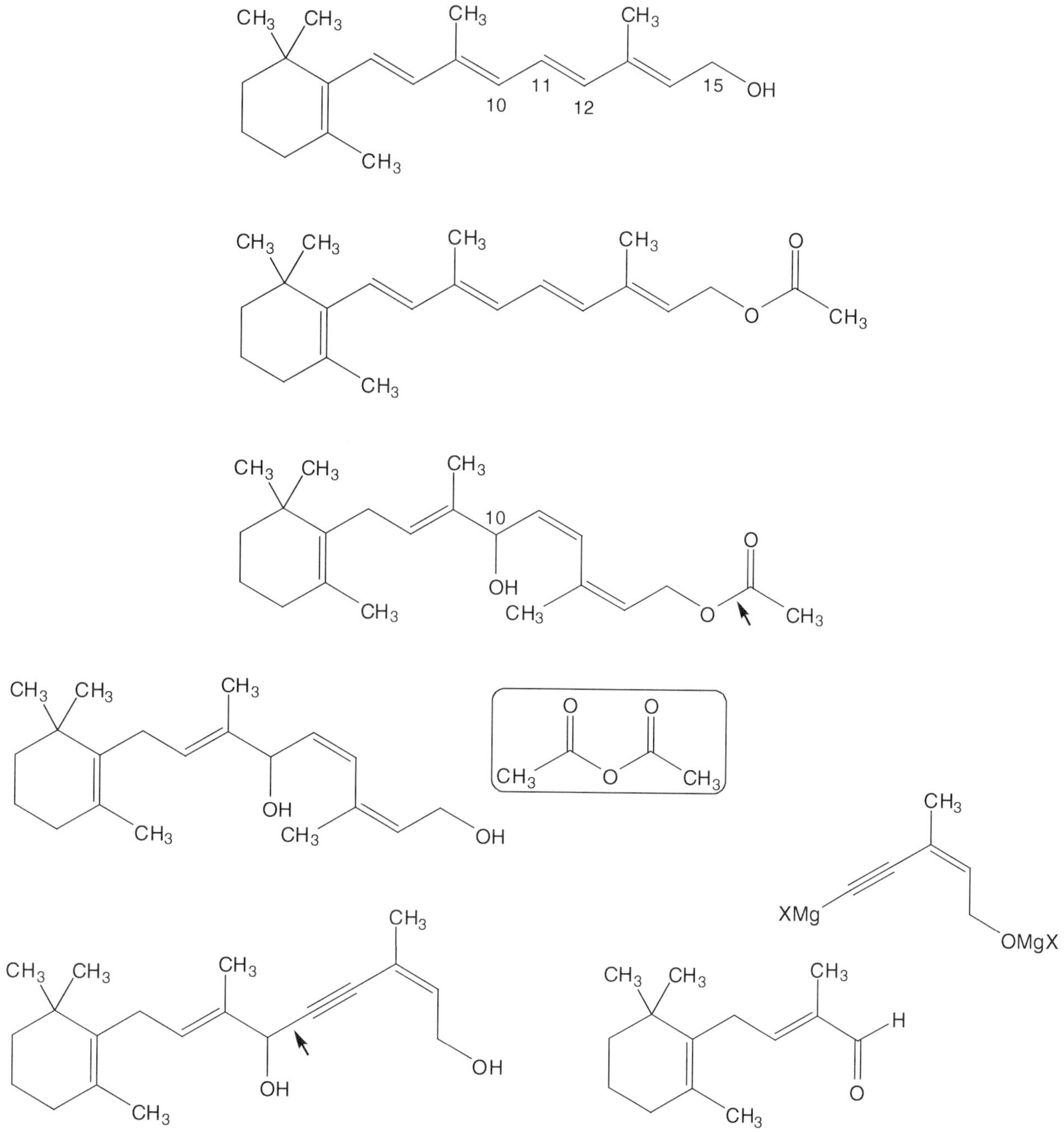

The alkynylmagnesium halide is formed by reaction of (*Z*)-3-methylpent-2-en-4-yn-1-ol with ethylmagnesium bromide. Ethylmagnesium bromide is formed from bromoethane. (*Z*)-3-Methylpent-2-en-4-yn-1-ol is formed by rearrangement of 3-methylpent-1-en-4-yn-3-ol. This tertiary alcohol is formed by addition of a metal acetylide to methyl vinyl ketone.

2-Methyl-4-(2,6,6-trimethyl-1-cyclohexen-1-yl)-2-butenal is formed from β-ionone in two steps. Addition of ethyl chloroacetate to the ketone forms the glycidic acid ester (**Darzens Reaction**). The aldehyde is formed by hydrolysis of the ester followed by decarboxylation.

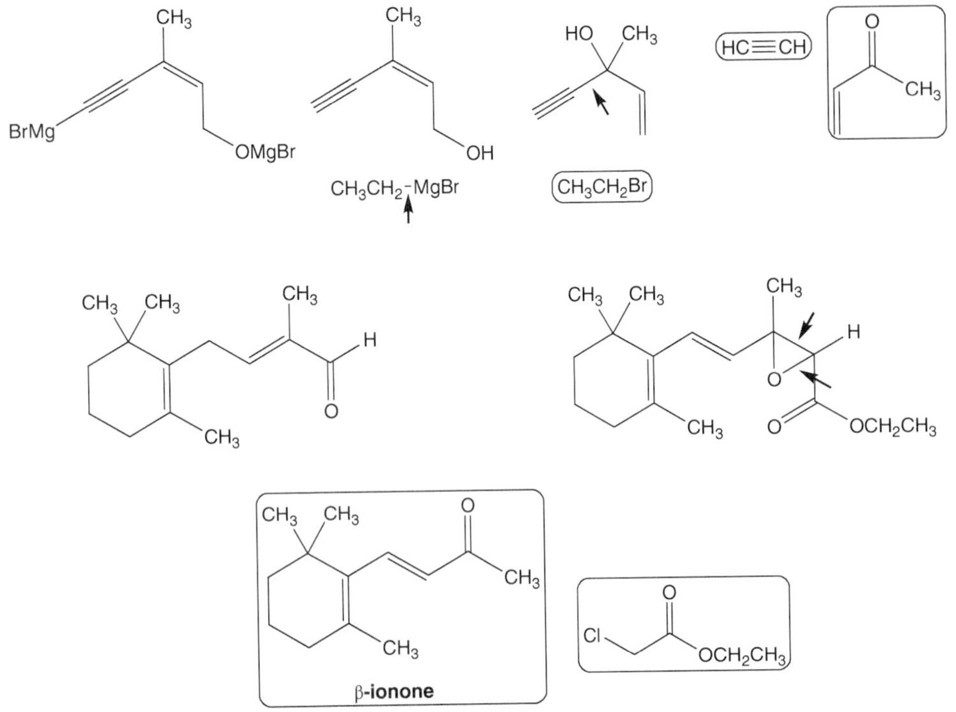

Extended Discussion

The alternative BASF route to retinol utilizes the same Wittig reagent that is used to produce retinoic acid. The aldehyde partner required for the Wittig reaction leading to retinol is (*E*)-4-acetoxy-2-methyl-2-butenal. Draw the structures of a retrosynthetic analysis of one route to this aldehyde. Include the structures of the retrosynthetic analysis of any organic starting material(s) from petrochemical or biochemical raw materials.

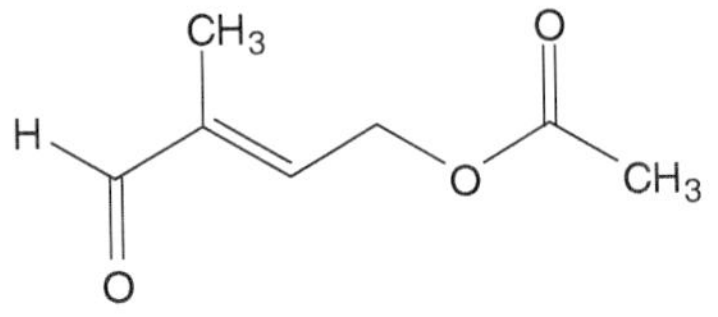

(E)-4-acetoxy-2-methyl-2-butenal

Ribavirin

Anti-infective Medicines/Antiviral Medicines/Other Antivirals
Anti-infective Medicines/Antiviral Medicines/Antihepatitis Medicines/Medicines for Hepatitis C/Other Antivirals

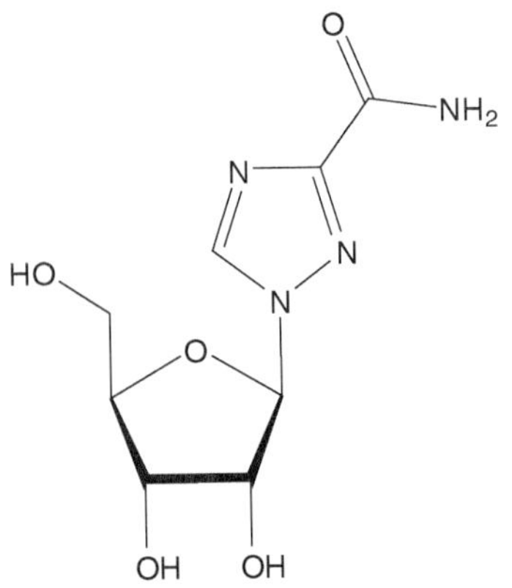

A nucleoside analog is often formed by displacement of a leaving group on the sugar by nitrogen of the heterocycle. The displacement usually results in a mixture of two products, with the heterocycle on the top face (β) or the bottom face (α) of the sugar. Factors which influence the β-product to α-product ratio include the heterocycle, the sugar, the leaving group, and the reaction conditions.

Discussion. The ribavirin carboxamide is formed from the methyl ester in the final step. The three alcohols on the sugar are released by transesterification. The C-N bond is formed by displacement of the C1 acetate of β-D-ribofuranose 1,2,3,5-tetraacetate by methyl 1,2,4-triazole-3-carboxylate.

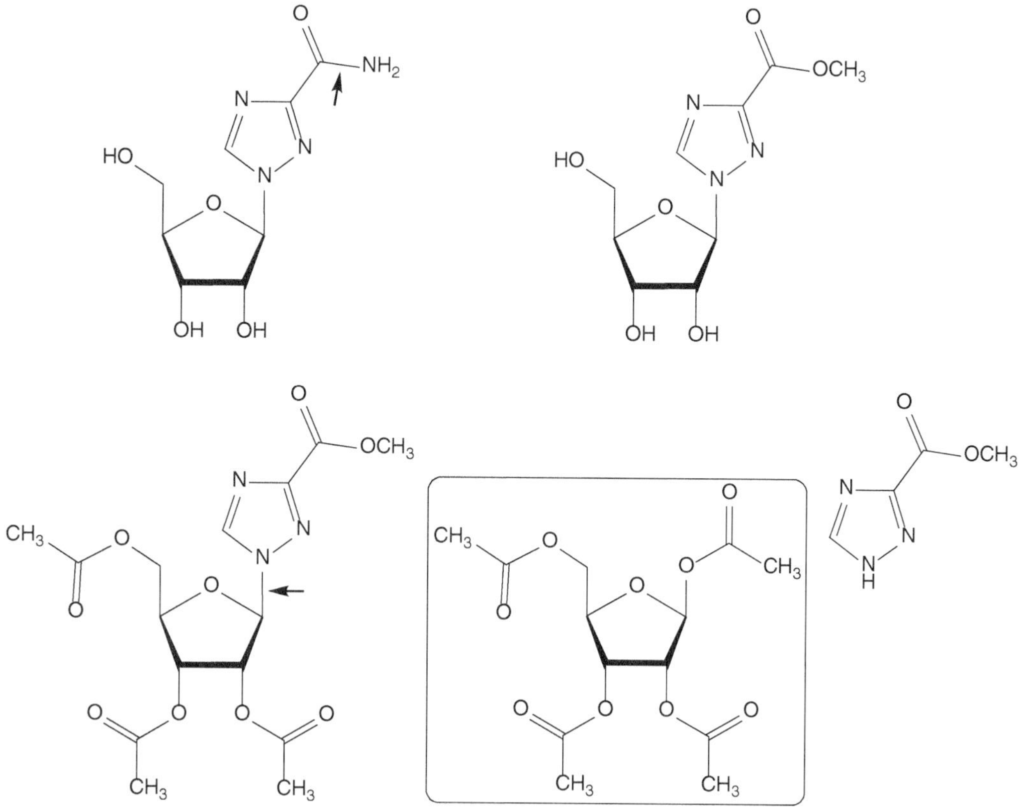

Methyl 1,2,4-triazole-3-carboxylate is formed from the carboxylic acid (**Fischer Esterification**). 1,2,4-Triazole-3-carboxylic acid is formed from 5-amino-1,2,4-triazole-3-carboxylic acid via the diazonium salt. 5-Amino-1,2,4-triazole-3-carboxylic acid is formed by condensation of aminoguanidine bicarbonate with oxalic acid.

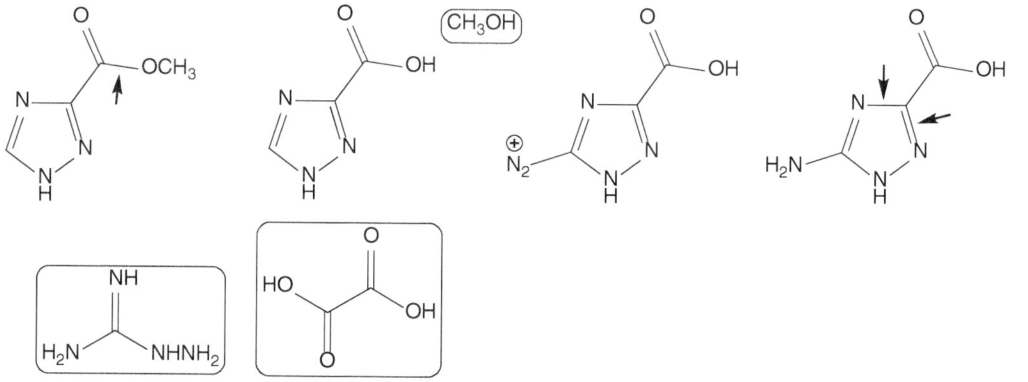

Extended Discussion

Draw the structures of a retrosynthetic analysis of one alternative route to methyl 1,2,4-triazole-3-carboxylate. Include the structures of the retrosynthetic analysis of any organic starting material(s) from petrochemical or biochemical raw materials. List the pros and cons for both routes to methyl 1,2,4-triazole-3-carboxylate and select one route as the preferred route.

Rifabutin

Anti-infective Medicines/Antibacterials/Antituberculosis Medicines

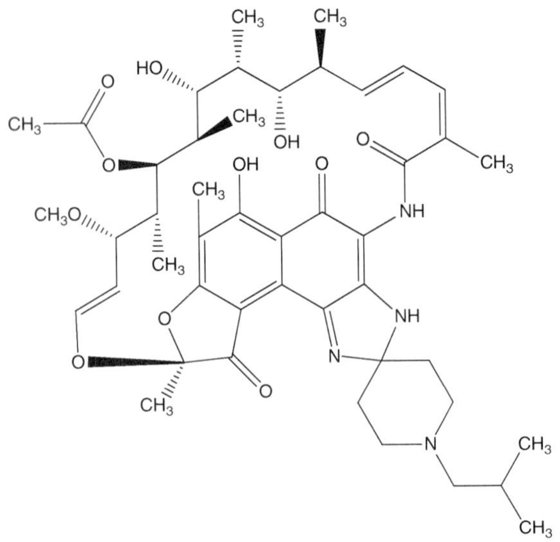

A single-enantiomer molecule with multiple chiral carbons is often formed by modification of a natural product which has most or all of the chiral carbons already in place. The rifamycins are often formed by modification of rifamycin S. Rifamycin S is formed from the fermentation product rifamycin B.

Discussion. Rifabutin is formed by reaction of 3-amino-4-deoxy-4-iminorifamycin S with 1-isobutyl-4-piperidinone. 3-Amino-4-deoxy-4-iminorifamycin S is formed from 3-aminorifamycin S. 3-Aminorifamycin S is formed by bromide displacement from 3-bromorifamycin S. 3-Bromorifamycin S is formed by bromination of rifamycin S.

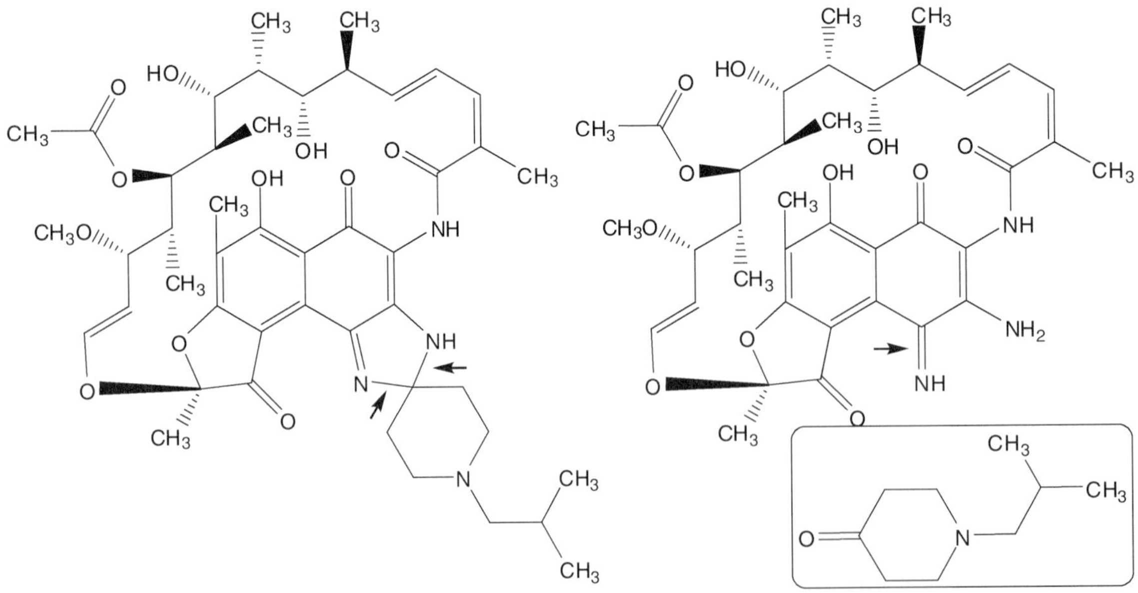

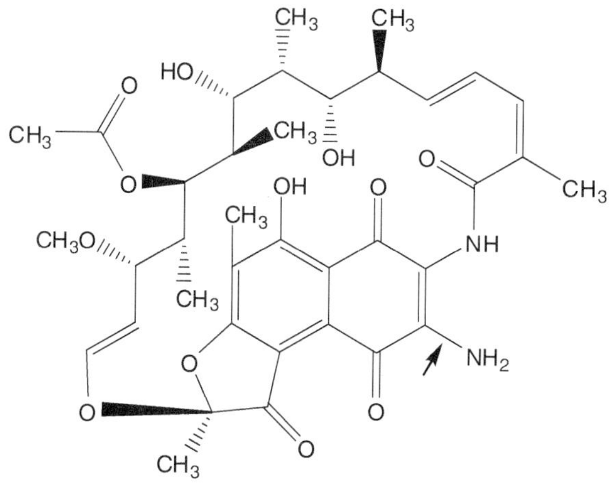

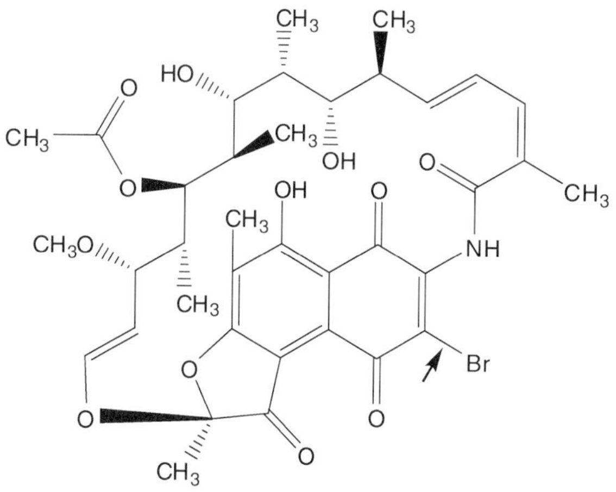

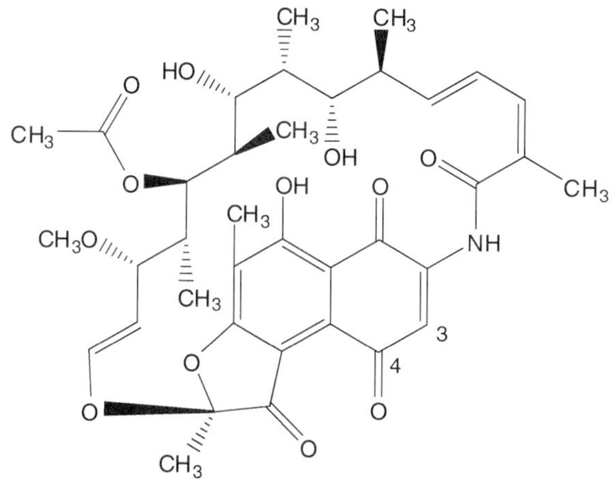

rifamycin S

Rifamycin S is formed by hydrolysis of rifamycin O. Rifamycin O is formed by oxidation of rifamycin B. Rifamycin B is produced by fermentation.

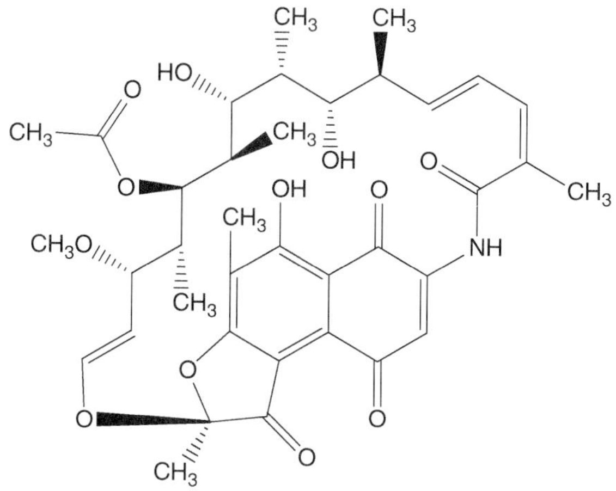

rifamycin S

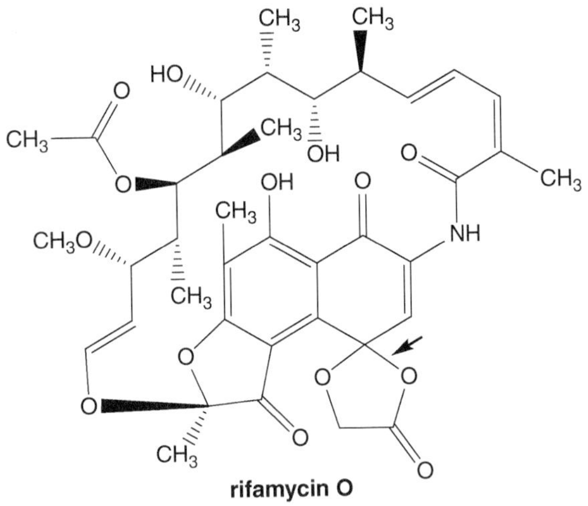

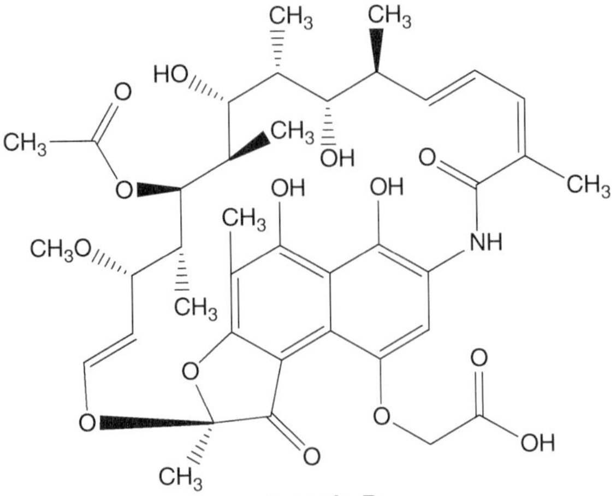

Extended Discussion

Draw the structures of the retrosynthetic analysis of an alternative route to 3-aminorifamycin S. List the pros and cons for both routes. Is one route preferred?

Rifampicin

Anti-infective Medicines/Antibacterials/Antituberculosis Medicines

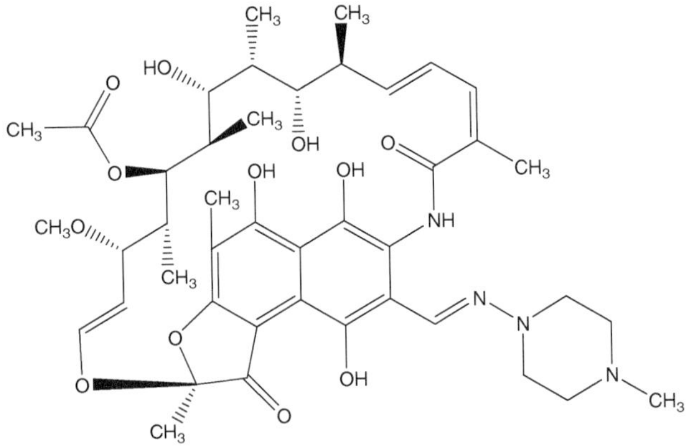

A single-enantiomer molecule with multiple chiral carbons is often formed by modification of a natural product which has most or all of the chiral carbons already in place. The rifamycins are often formed by modification of rifamycin S. Rifamycin S is formed from the fermentation product rifamycin B.

Discussion. The rifampicin hydrazone is formed by reaction of rifamycin oxazine with 1-amino-4-methylpiperazine. Rifamycin oxazine is formed by reaction of rifamycin S with 1,3,5-tri-*tert*-butylhexahydro-1,3,5-triazine and paraformaldehyde.

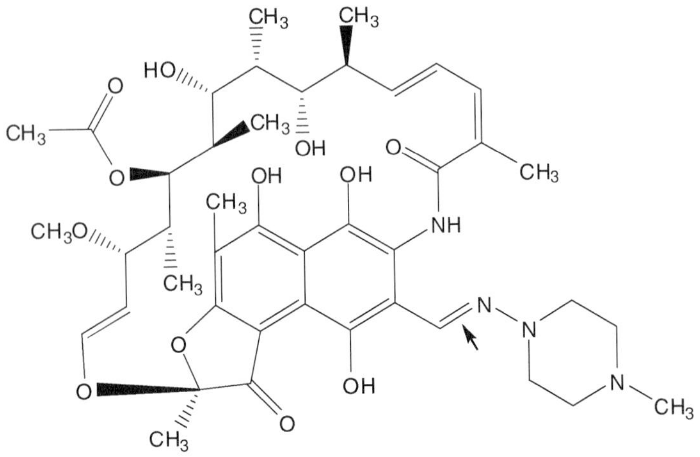

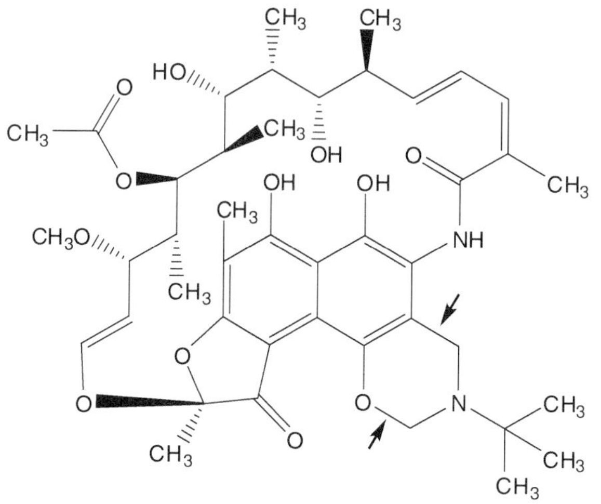

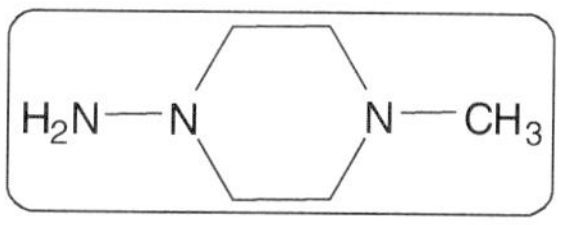

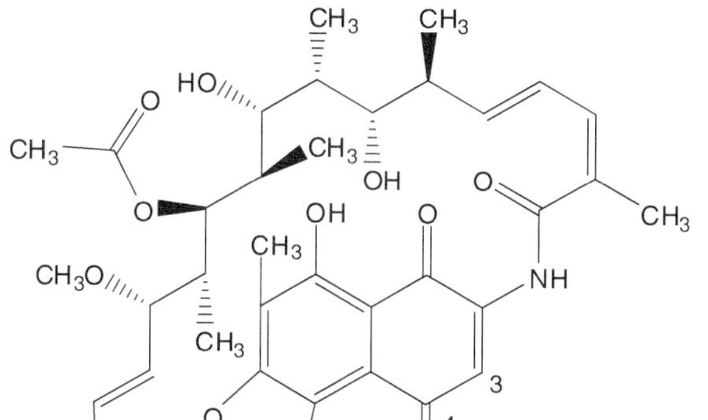

rifamycin S

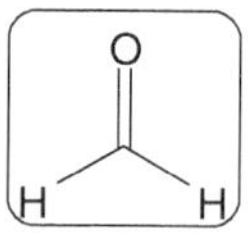

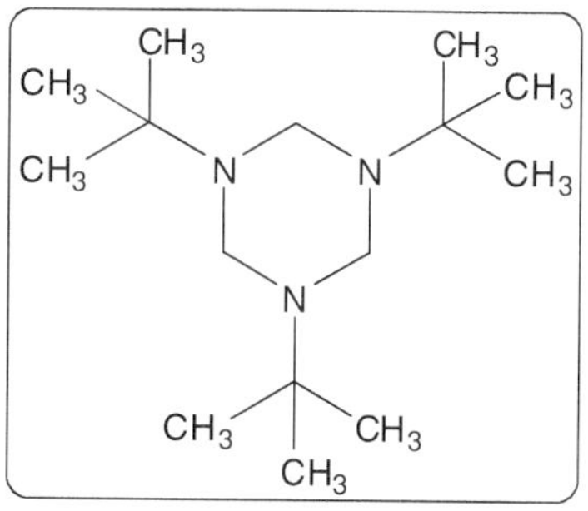

Rifamycin S is formed by hydrolysis of rifamycin O. Rifamycin O is formed by oxidation of rifamycin B. Rifamycin B is produced by fermentation.

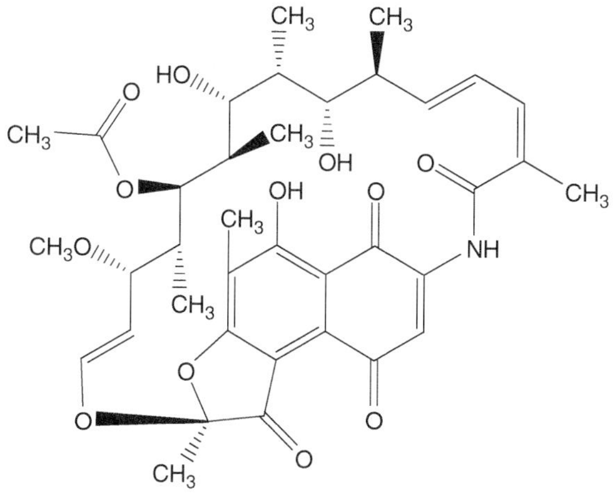

rifamycin S

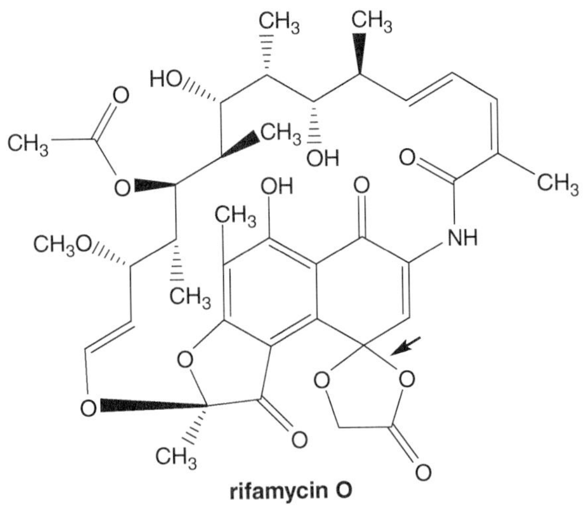

rifamycin O

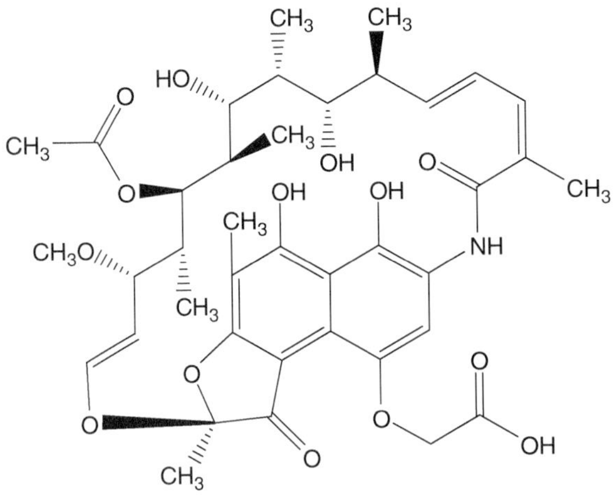

rifamycin B

Rifapentine

Anti-infective Medicines/Antibacterials/Antituberculosis Medicines

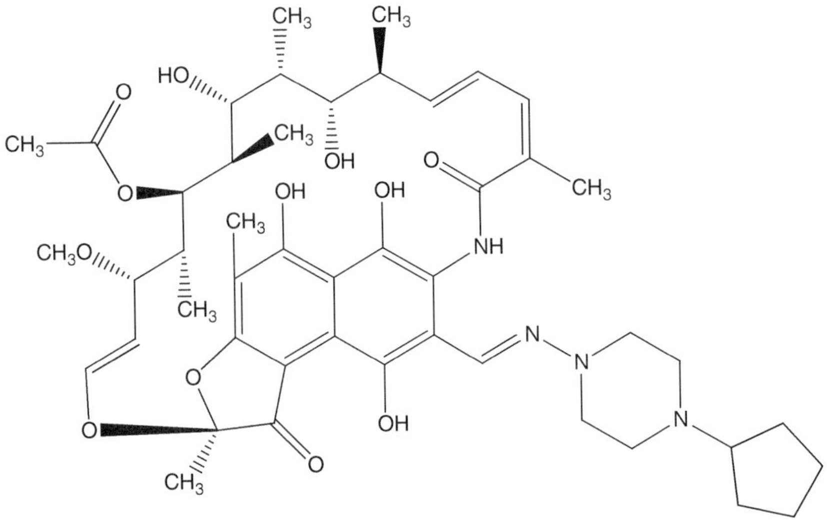

A single-enantiomer molecule with multiple chiral carbons is often formed by modification of a natural product which has most or all of the chiral carbons already in place. The rifamycins are often formed by modification of rifamycin S. Rifamycin S is formed from the fermentation product rifamycin B.

Discussion. The rifapentine hydrazone is formed by the reaction of rifamycin oxazine with 1-amino-4-cyclopentylpipera zine. Rifamycin oxazine is formed by the reaction of rifamycin S with 1,3,5-tri-*tert*-butylhexahydro-1,3,5-triazine and paraformaldehyde.

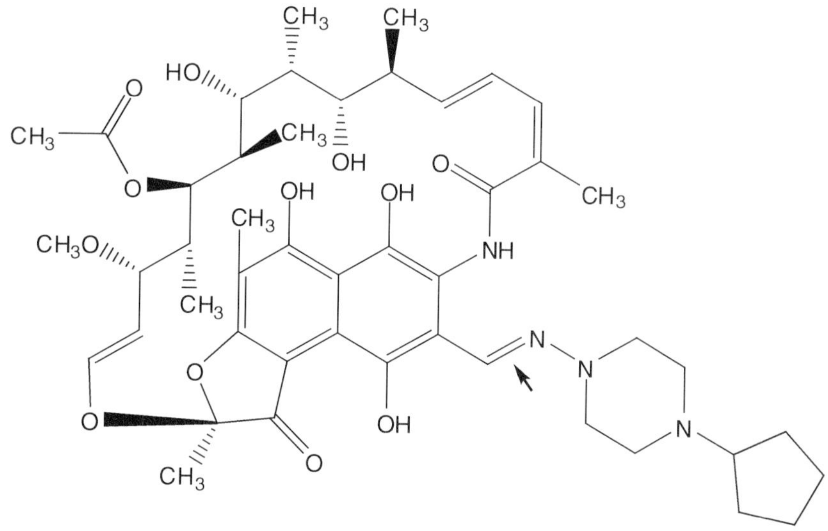

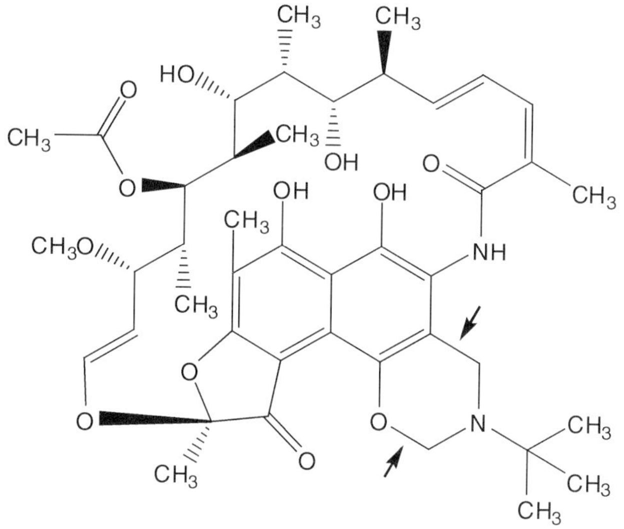

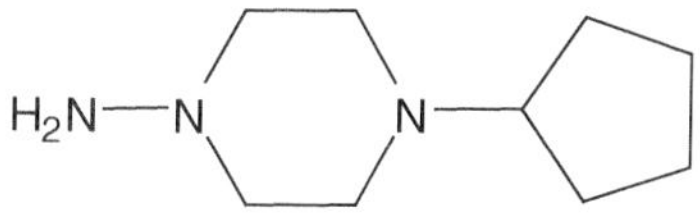

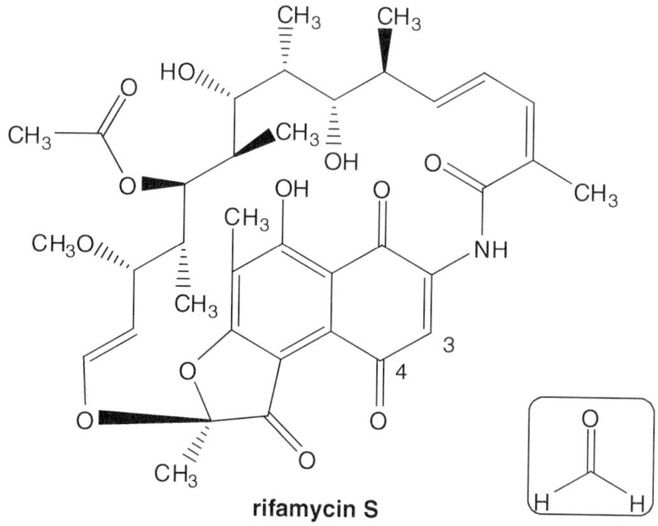

rifamycin S

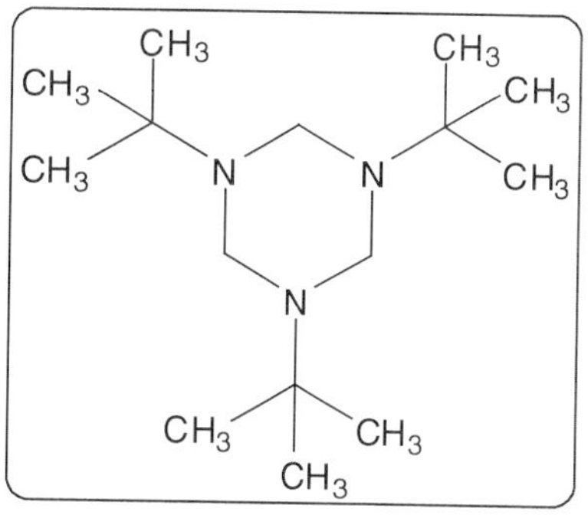

Rifamycin S is formed by hydrolysis of rifamycin O. Rifamycin O is formed by oxidation of rifamycin B. Rifamycin B is produced by fermentation.

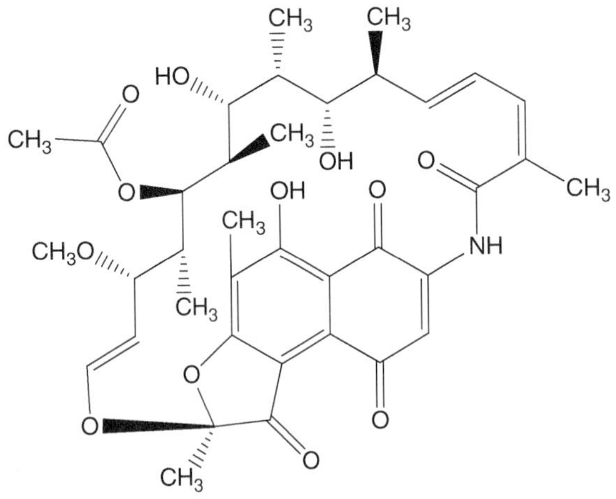

rifamycin S

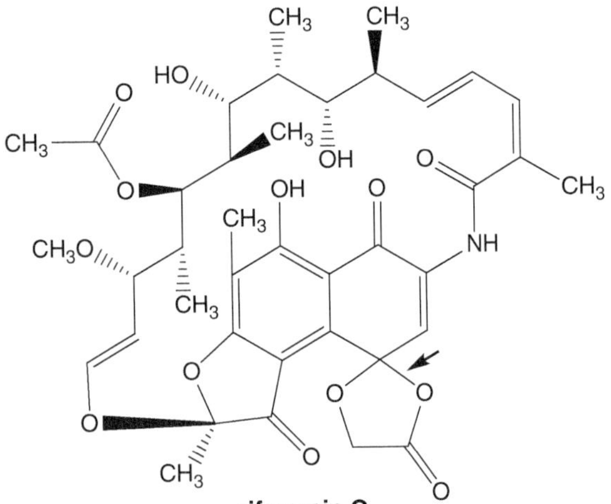

rifamycin O

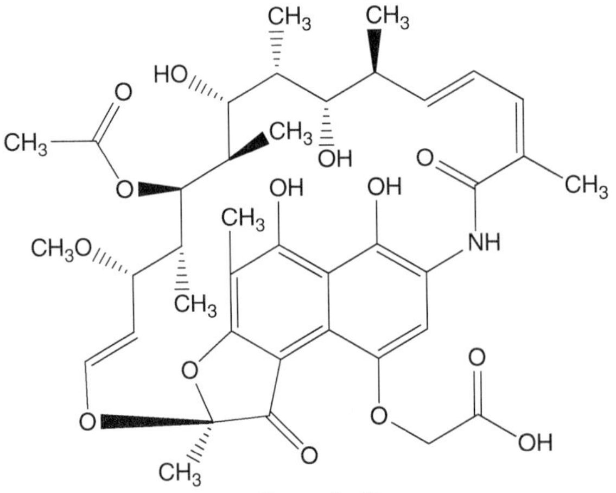

rifamycin B

1-Amino-4-cyclopentylpiperazine is formed by reduction of the nitrosopiperazine. The nitrosopiperazine is formed by nitrosation of 1-cyclopentylpiperazine. 1-Cyclopentylpiperazine is formed from cyclopentanone and piperazine by reductive amination.

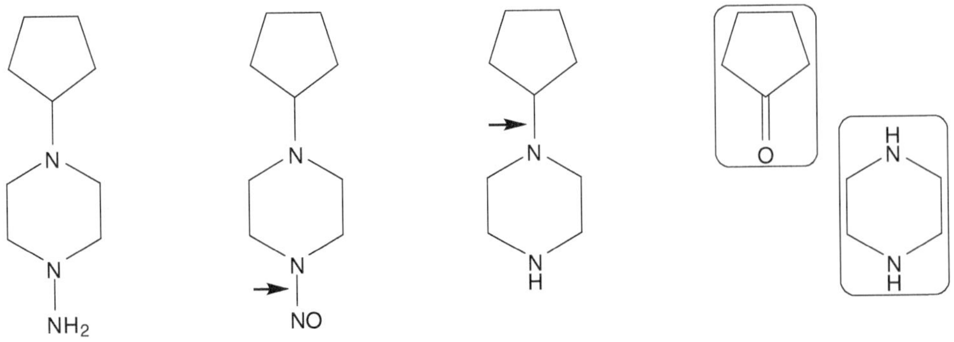

Extended Discussion

Draw the structures of the retrosynthetic analysis of one alternative route to 1-cyclopentylpiperazine. Include the structures of the retrosynthetic analysis of any organic starting material(s) from petrochemical or biochemical raw materials. List the pros and cons for both routes to 1-cyclopentylpiperazine.

Risperidone

Medicines for Mental and Behavioral Disorders/Medicines Used in Psychotic Disorders

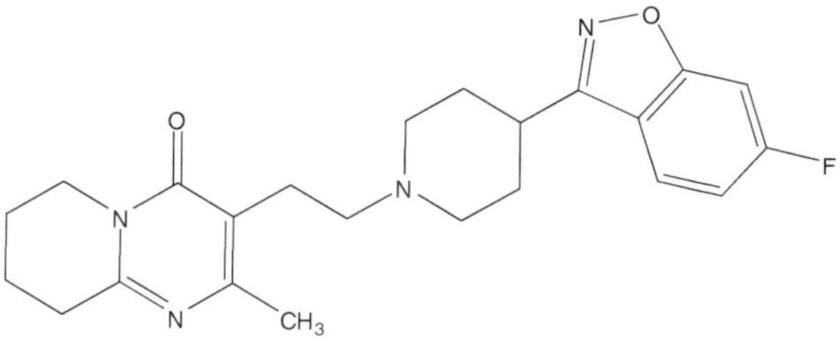

Tertiary amines are ubiquitous in drug structures. A tertiary amine is often formed by alkylation of a secondary amine. The alkylation is most efficient when the amine is used in excess and the carbon with the leaving group (Cl, Br, I, OTs, OMs) is primary or benzylic.

Discussion. The tertiary amine is formed in the final step by chloride displacement by the piperidine. The 6,7,8,9-tetrahy dro-4*H*-pyrido[1,2-*a*]pyrimidin-4-one is formed by reduction of the 4*H*-pyrido[1,2-*a*]pyrimidin-4-one. 3-(2-Chloroethyl)-2-methyl-4*H*-pyrido[1,2-*a*]pyrimidin-4-one is formed from 2-aminopyridine and α-acetyl-γ-butyrolactone (**Pinner Pyrimidine Synthesis**).

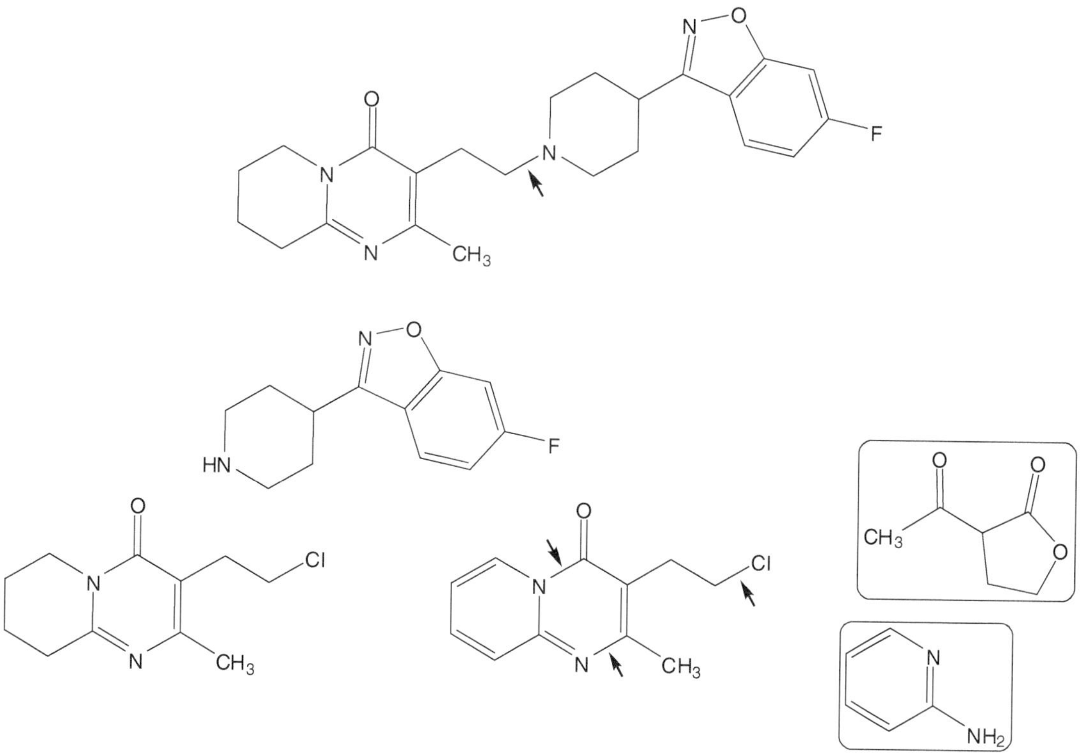

The isoxazole ring of 6-fluoro-3-(4-piperidinyl)benzo[*d*]isoxazole is formed by intramolecular displacement of fluoride by the oxime. The oxime is formed from the ketone. The piperidine is released by amide hydrolysis. The ketone is formed from 1,3-difluorobenzene and the acid chloride (**Friedel Crafts Acylation**). The acid chloride is formed from the carboxylic acid. 1-Acetylpiperidine-4-carboxylic acid is formed by the reaction of piperidine-4-carboxylic acid (isonipecotic acid) with acetic anhydride.

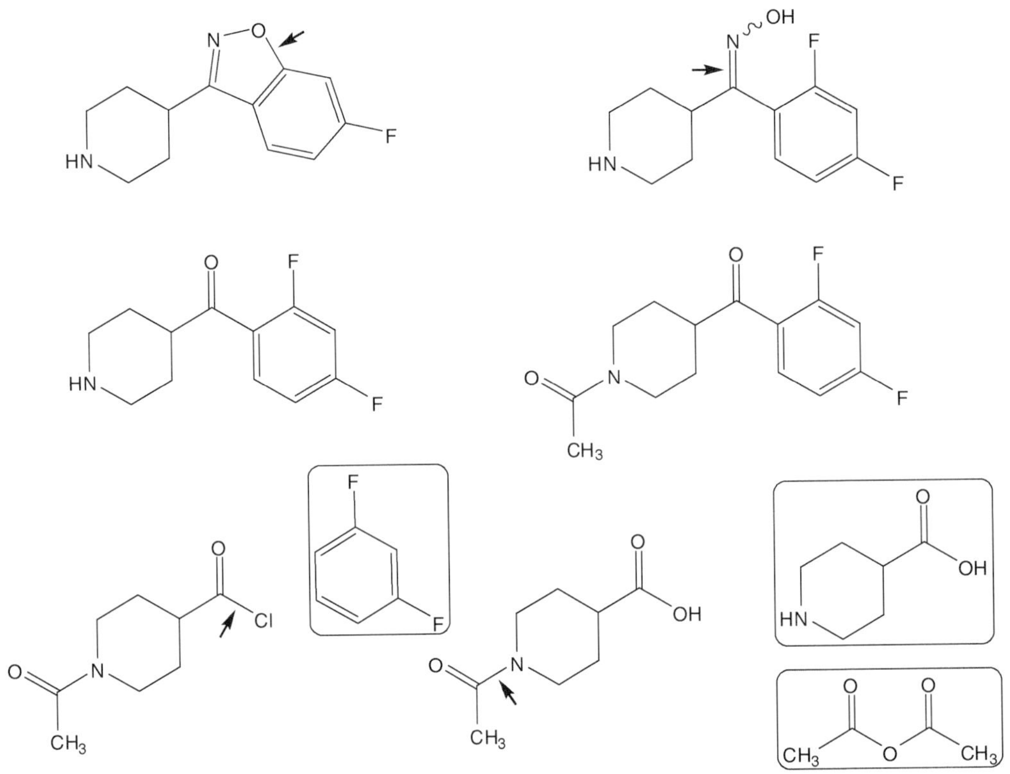

Extended Discussion

Draw structures of a retrosynthetic analysis of one route to 1,3-difluorobenzene from petrochemical or biochemical raw materials.

Ritonavir

Anti-infective Medicines/Antiviral Medicines/Antiretrovirals/Protease Inhibitors

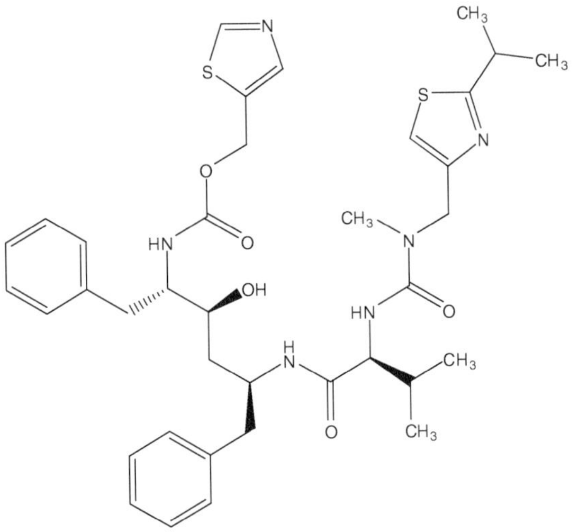

Amides, carbamates, and ureas are often efficiently formed from amines: an amide is formed by acylation with an acid chloride, anhydride, ester or carboxylic acid, a carbamate is formed by acylation with a chloroformate or carbonate, and a urea is formed by acylation with a carbamate.

Discussion. Disconnection of the C-N bonds of the amide, carbamate, and urea reveals the key components of ritonavir. The two primary amines of (2*S*,3*S*,5*S*)-2,5-diamino-1,6-diphenylhexan-3-ol suggest the *N*-protection–deprotection strategy will be an important feature of an efficient ritonavir synthesis.

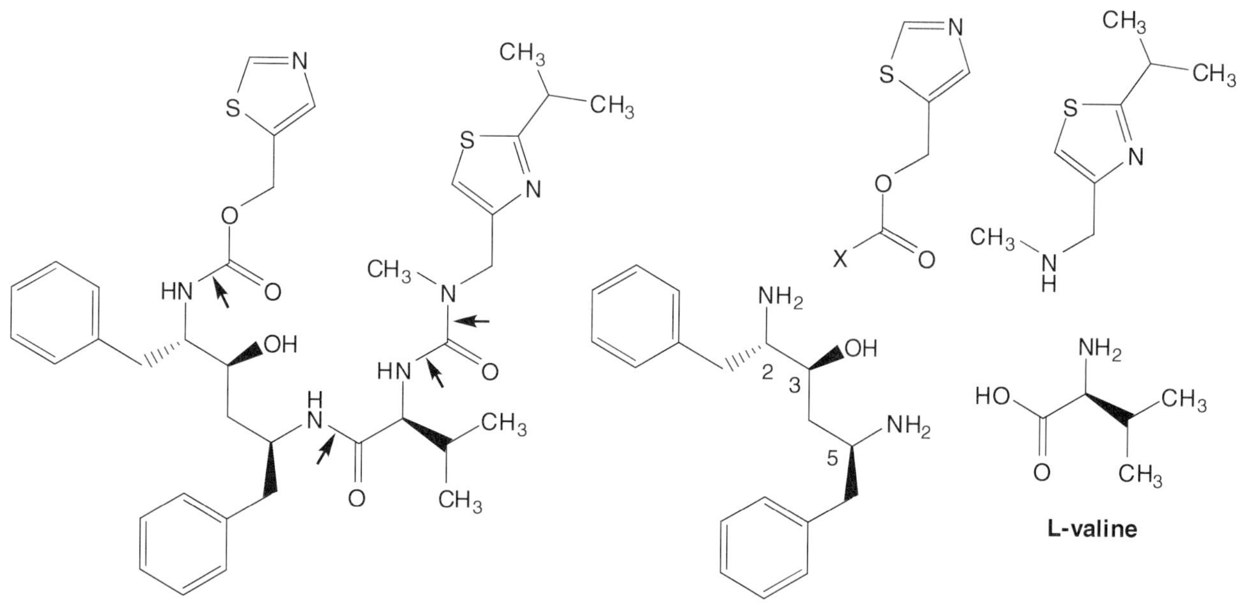

The amide bond at C5 is formed from the carboxylic acid and amine in the final step.

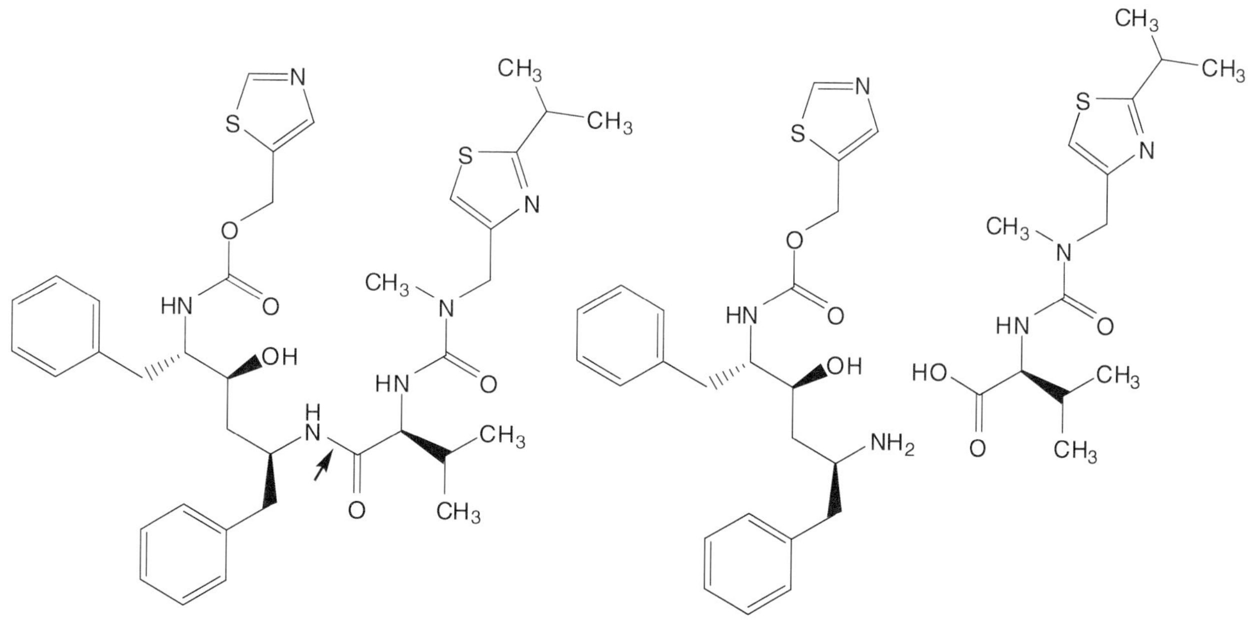

The carboxylic acid is formed by hydrolysis of the methyl ester. The urea is formed by the reaction of the 4-nitrophenyl carbamate with the secondary amine. The carbamate is formed from L-valine methyl ester and 4-nitrophenyl chloroformate.

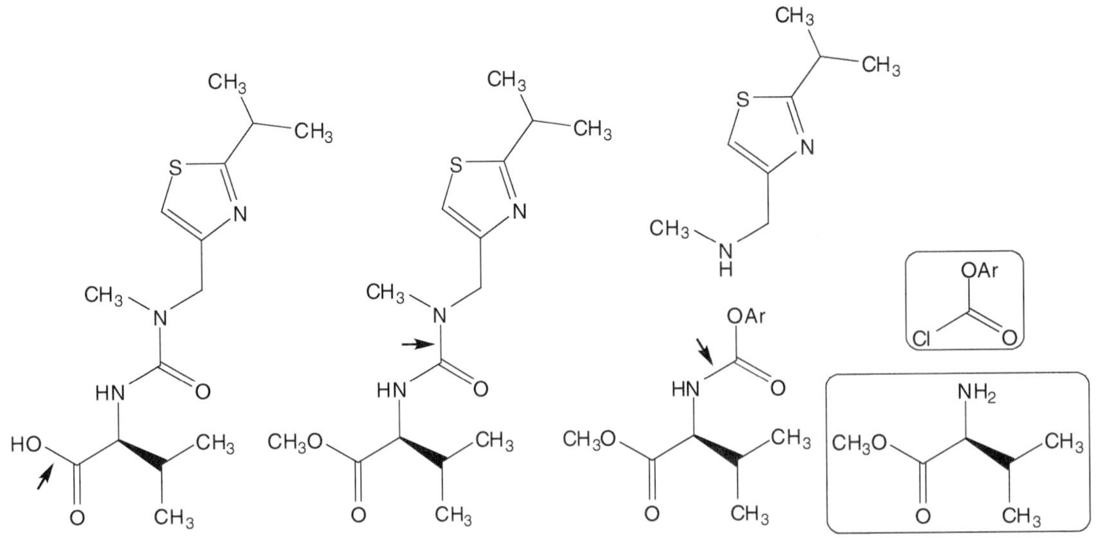

The secondary amine, 2-isopropyl-4-(methylaminomethyl)isothiazole, is formed by chloride displacement from 4-(chloromethyl)-2-isopropylthiazole by methylamine. The thiazole is formed by the condensation of thioisobutyramide and 1,3-dichloroacetone. The thioamide is formed from isobutyramide.

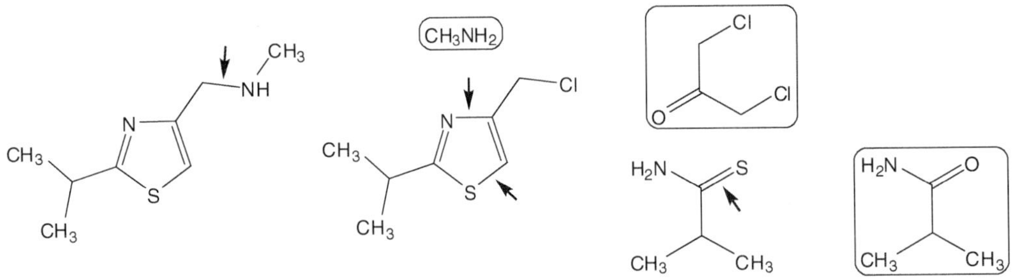

The amine for the ritonavir final step is released by hydrolysis of the *tert*-butyl carbamate (Boc) at C5. The carbamate at C2 is formed by the reaction of a 4-nitrophenyl carbonate with the amine at C2.

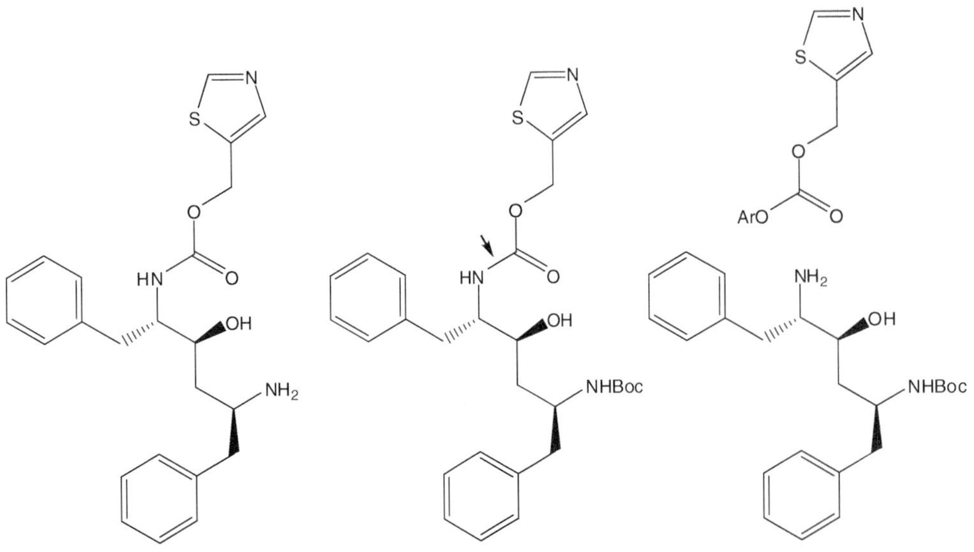

The carbonate is formed from 5-(hydroxymethyl)thiazole and 4-nitrophenyl chloroformate. 5-(Hydroxymethyl)thiazole is formed by hydrogenolysis of the 2-chlorothiazole. 2-Chloro-5-(hydroxymethyl)thiazole is formed by hydrolysis of 2-chloro-5-chloromethylthiazole.

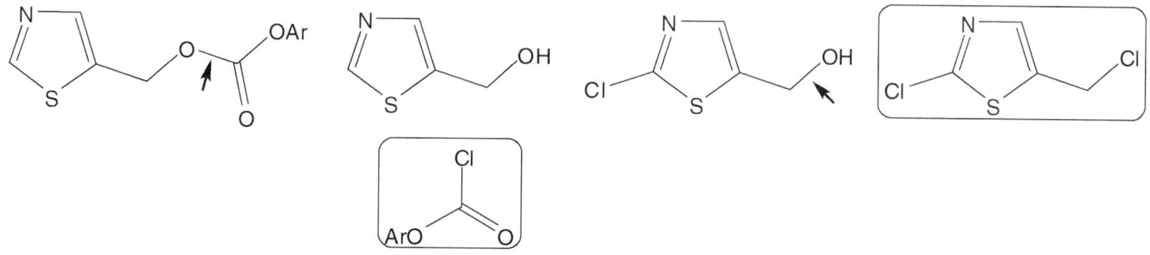

The amine at C2 of the ritonavir scaffold is released from the tertiary amine by hydrogenolysis. The primary amine at C5 is protected as the *tert*-butyl carbamate (Boc). The alcohol of (2*S*,3*S*,5*S*)-5-amino-2-(dibenzylamino)-1,6-diphenylhexan-3-ol is formed by reduction of the ketone and the ketone is formed by reduction of the enone. (Name three diastereomers which are also formed in this two-reduction sequence. Identify the downstream process intermediate which is upgraded to remove diastereomer impurities.) The enone is formed by addition of benzylmagnesium chloride to the nitrile (**Grignard Reaction**). Benzylmagnesium chloride is formed from benzyl chloride. The nitrile is formed by condensation of acetonitrile with the benzyl ester (mixed **Claisen Condensation**). The ester, (*S*)-benzyl 2-(dibenzylamino)-3-phenylpropanoate, is formed by *N*- and *O*-alkylation of L-phenylalanine with benzyl chloride. L-Phenylalanine is produced by fermentation.

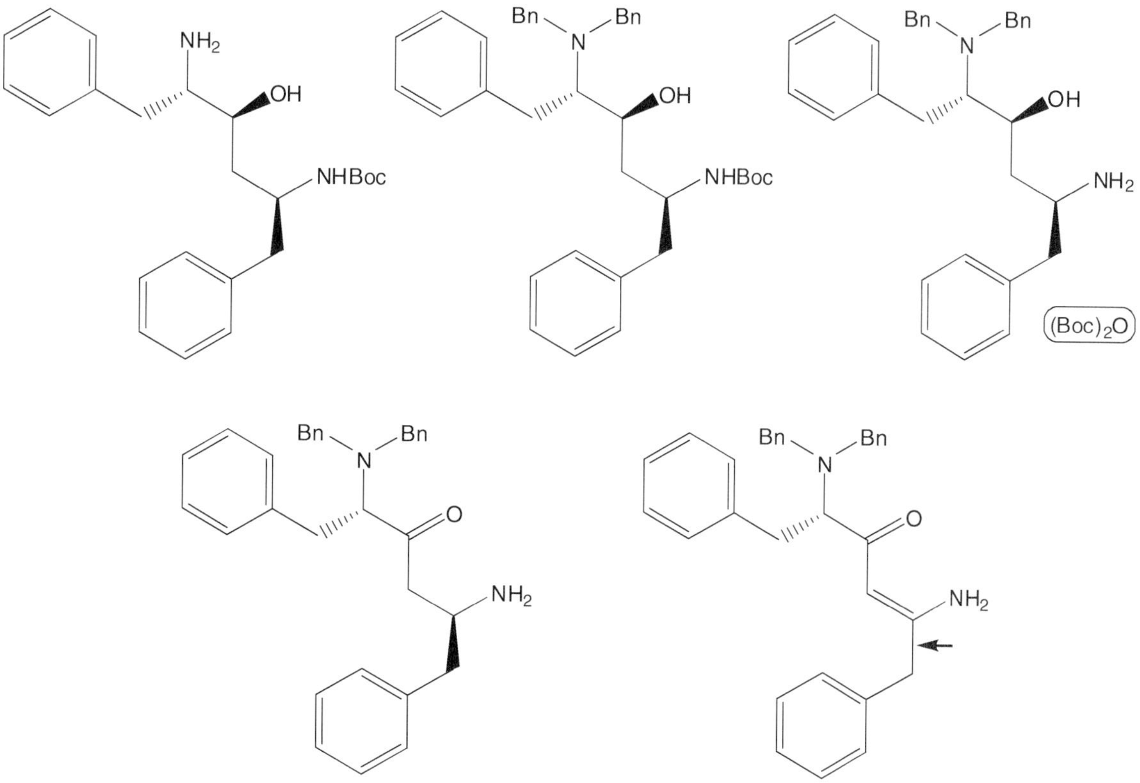

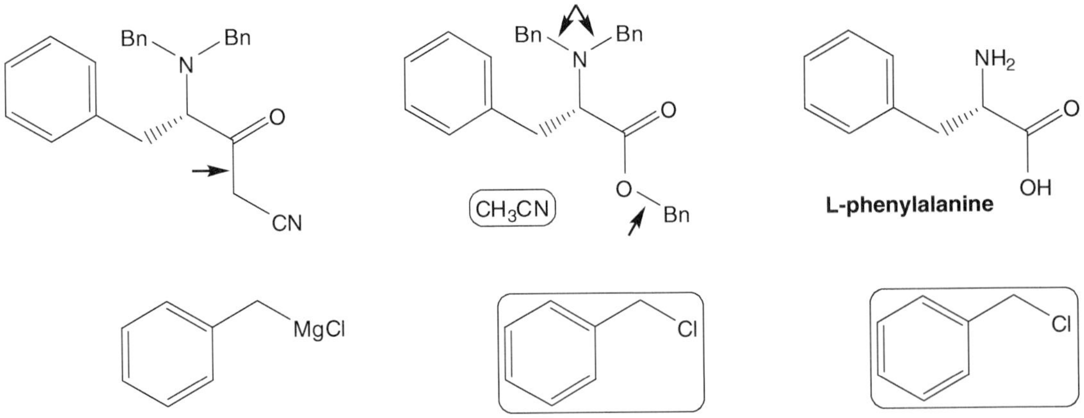

Extended Discussion

Draw the structures of the retrosynthetic analysis of one alternative route to ritonavir from (2*S*,3*S*,5*S*)-5-amino-2-(dibenzylamino)-1,6-diphenylhexan-3-ol. List the pros and cons for both routes and select one route as the preferred route.

S

Salbutamol

Medicines Acting on the Respiratory Tract/Antiasthmatic and Medicines for Chronic Obstructive Pulmonary Disease

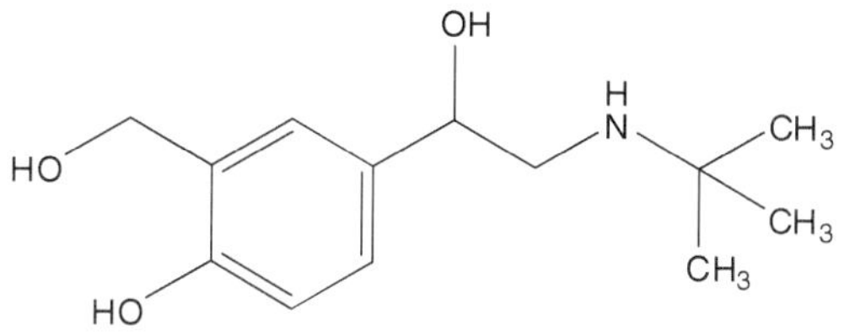

A benzylic alcohol is often formed by the reduction of an aldehyde or ketone.

Discussion. Salbutamol is a 1 : 1 mixture of the (*R*)- and (*S*)-enantiomers. In the final step, the primary alcohol and phenol are released by hydrolysis of the 4H-1,3-benzodioxin. The secondary amine is formed by hydrogenolysis of the tertiary amine. The secondary alcohol is formed by the reduction of the ketone. The tertiary amine is formed by bromide displacement from 2-bromo-1-(2,2-dimethyl-4H-1,3-benzodioxin-6-yl)ethanone by *N*-benzyl-*tert*-butylamine.

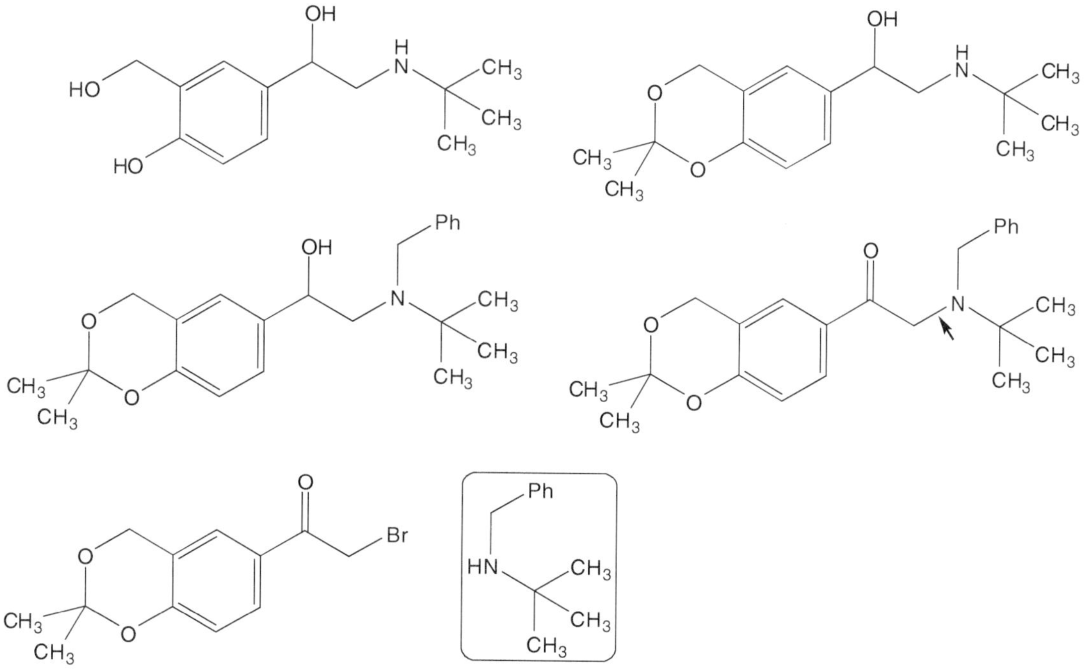

Routes to Essential Medicines: A Workbook for Organic Synthesis, First Edition. Peter J. Harrington.

Companion website: www.wiley.com/go/Harrington/routes_essential_medicine

The 4H-1,3-benzodioxin is formed by the reaction of the diol with 2-methoxypropene. The primary alcohol of 2-bromo-4′-hydroxy-3′-(hydroxymethyl)acetophenone is formed by the reduction of the aldehyde. The acetophenone is formed from salicylaldehyde and bromoacetyl bromide (**Friedel–Crafts Acylation**).

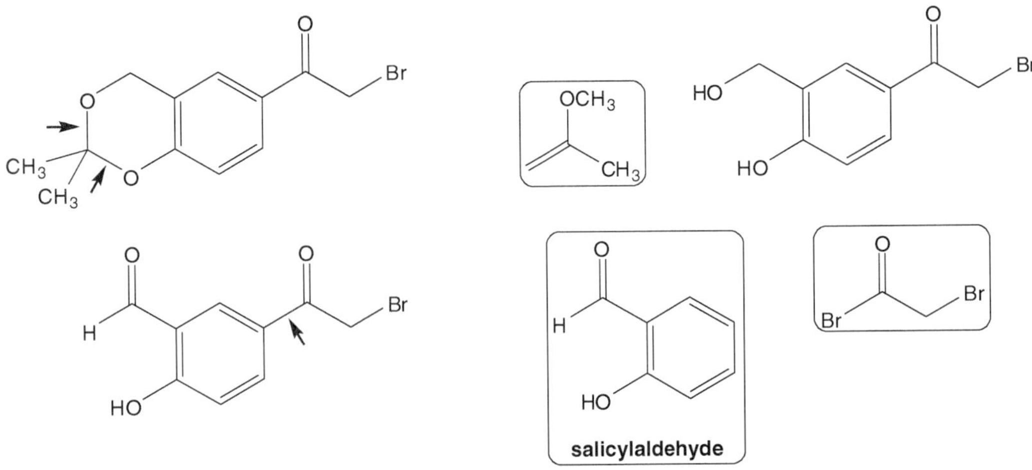

Extended Discussion

In one alternative route to salbutamol, the β–amino alcohol is formed by ring-opening an epoxide with *tert*-butylamine. The epoxide is formed from the bromohydrin. The bromohydrin is formed by the reduction of the α-bromoketone. List the pros and cons for both routes. Which route is preferred?

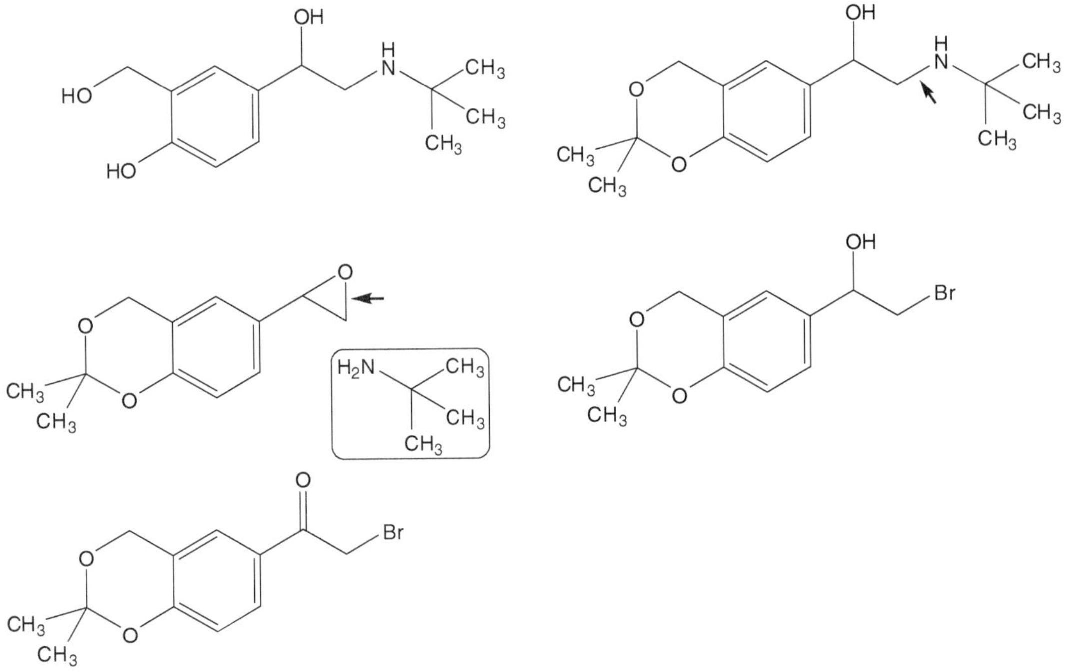

Simeprevir

Anti-infective Medicines/Antiviral Medicines/Antihepatitis Medicines/Medicines for Hepatitis C/Protease Inhibitors

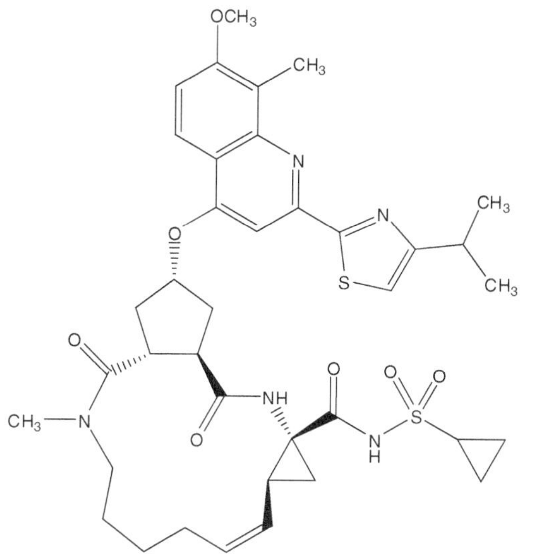

Amides are often efficiently formed by the reaction of an amine with an acid chloride, anhydride, ester, or carboxylic acid.

Discussion. Five disconnections reveal the five components of simeprevir and the modular nature of the synthesis.

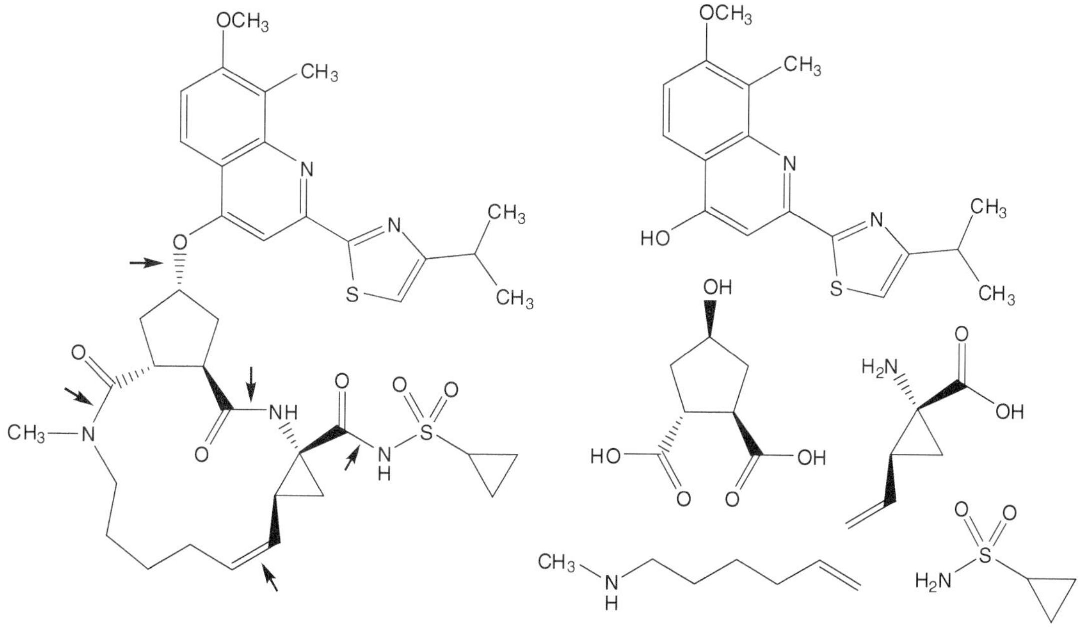

(**NOTE:** The heterocyclic ether will be represented as OAr until the ether is disconnected.)

The *N*-acylsulfonamide is formed in the final step by the reaction of an acylating agent with cyclopropanesulfonamide. The acylating agent is formed in situ from the cyclopropanecarboxylic acid. (Draw the structure of the acylating agent.) The cyclopropanecarboxylic acid is formed by ester hydrolysis. The *cis*-alkene of the macrocyclic ring is formed by olefin metathesis (**Grubbs Reaction**).

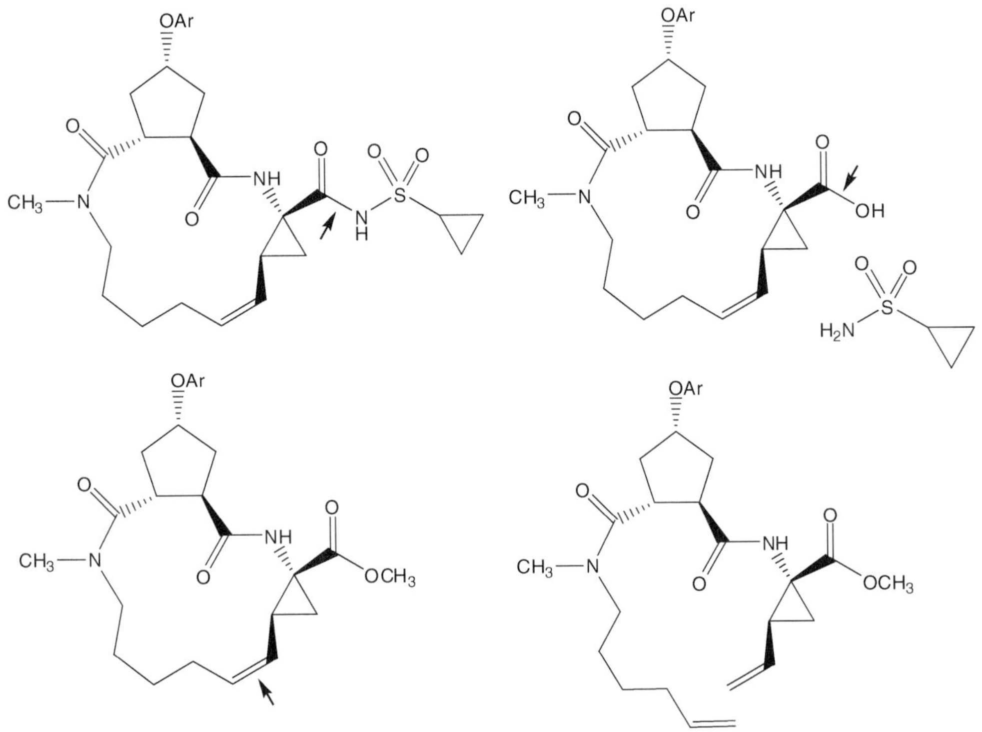

One amide is formed by the reaction of the amine, methyl (1*R*,2*S*)-1-amino-2-vinylcyclopropane-1-carboxylate, with the carboxylic acid. The carboxylic acid is formed by hydrolysis of the methyl ester. The C-O bond of the ether is formed with inversion of the stereochemistry from the 4-hydroxyquinoline and the secondary alcohol (**Mitsunobu Reaction**).

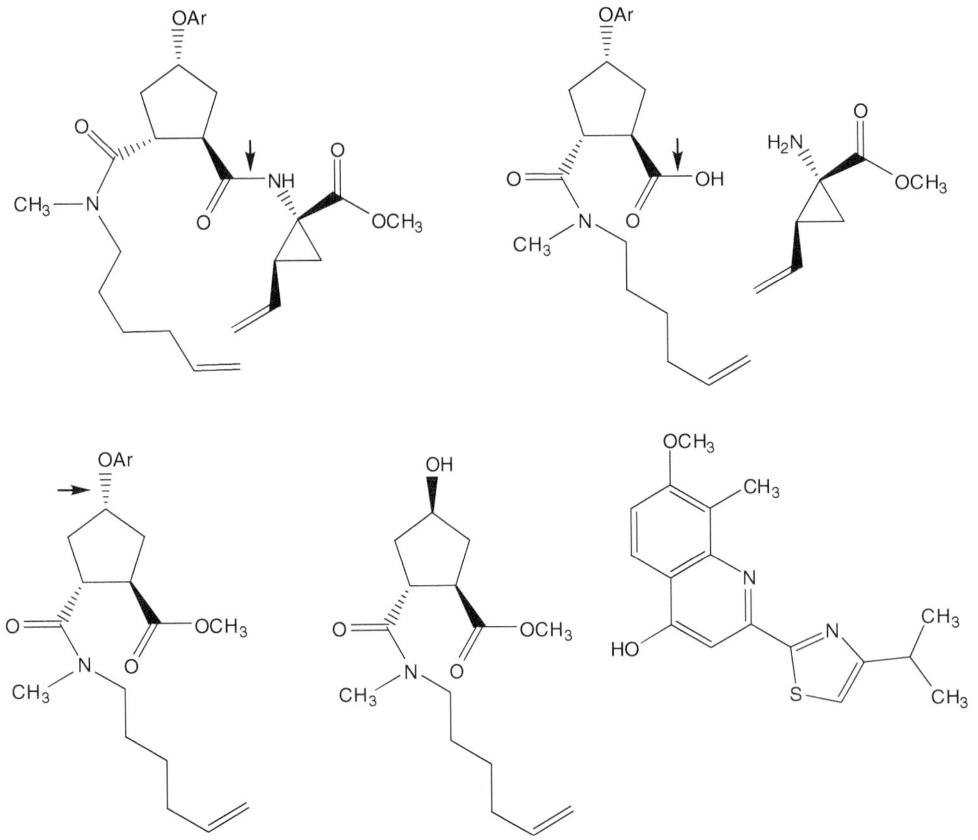

The secondary alcohol and methyl ester are both formed by ring-opening the lactone with methanol. The amide is formed by the reaction of N-methyl-5-hexen-1-amine with (*R,R*)-3-oxo-2-oxabicyclo[2.2.1]heptan-5-carboxylic acid.

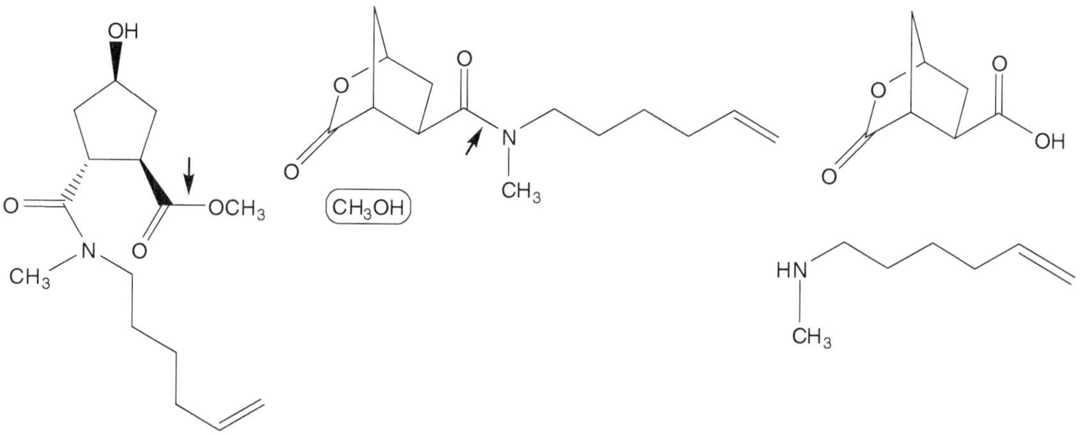

The (*R,R*)-lactone is produced by resolution. The racemic lactone is formed by cyclization of 4-hydroxy-*trans*-1,2-cyclop entanedicarboxylic acid. The secondary alcohol is formed by the reduction of the ketone. The diacid is formed by hydrolysis of the diester. The ketone is formed by cyclization of the 1,6-diacid and decarboxylation of the resulting β-ketoacid. The 1,6-diacid is formed by oxidative cleavage of the cyclohexene. The cyclohexene is formed by cycloaddition of 1,3-butadiene and dimethyl fumarate (**Diels-Alder Reaction**). 1,3-Butadiene is formed by thermolysis of 3-sulfolene.

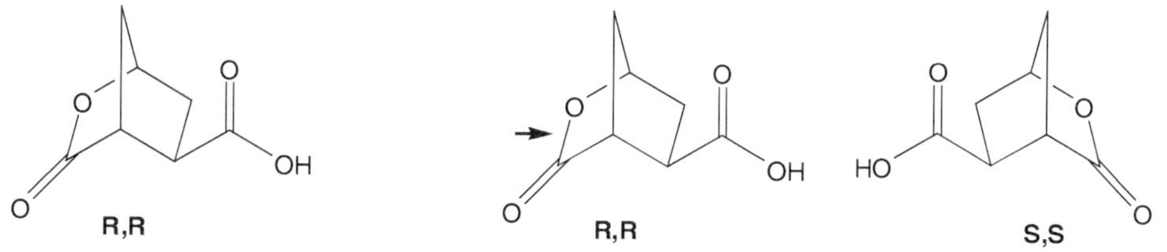

Following only the R,R structures:

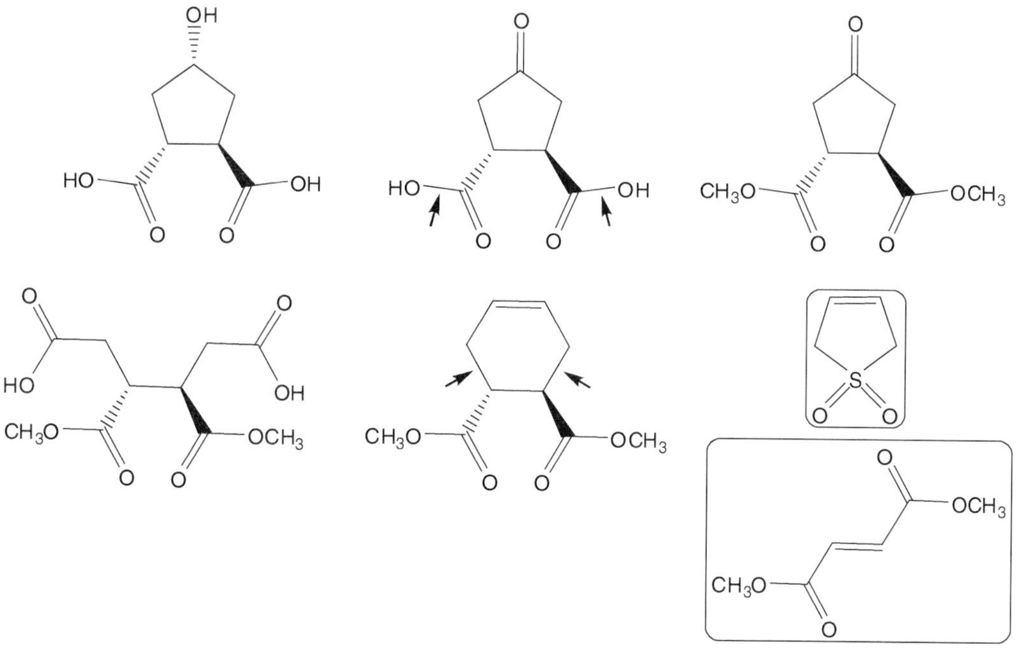

Cyclopropanesulfonamide is formed from *N*-*tert*-butylcyclopropanesulfonamide. The cyclopropane ring is formed by nucleophilic displacement of chloride by the sulfonamide dianion. The *N*-*tert*-butylsulfonamide is formed from *tert*-butylamine and the sulfonyl chloride. 3-Chloropropanesulfonyl chloride is formed from 1,3-propanesultone.

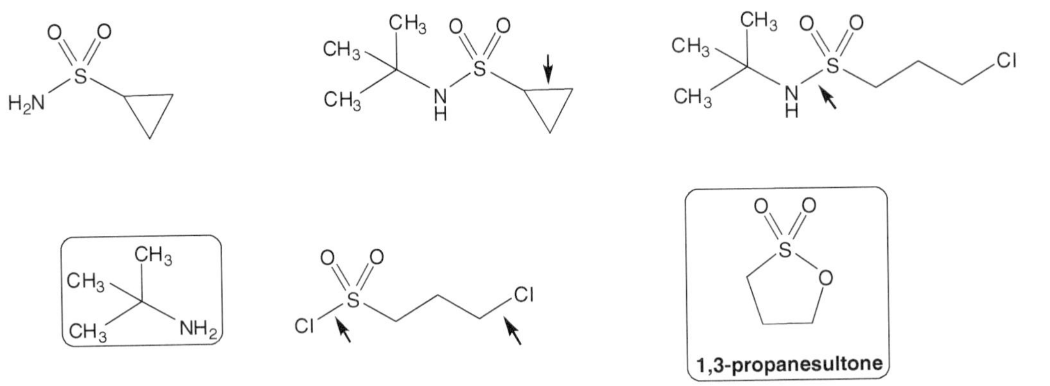

Cleavage of the *tert*-butyl carbamate releases key component methyl (1*R*,2*S*)-1-amino-2-vinylcyclopropanecarboxylate. The (1*R*,2*S*)-enantiomer of the carbamate is separated from the racemic mixture by a selective enzyme-mediated hydrolysis of the (1*S*,2*R*)-ester. The carbamate is formed from the amine. The amine is formed by hydrolysis of the benzaldehyde imine. The cyclopropane ring is formed by the reaction of methyl glycine benzaldehyde imine with *trans*-1,4-dibromo-2-butene. (Draw the structures for four possible stereoisomeric cyclopropane products. Assign (R/S) configurations to the chiral carbons. This ring-forming reaction is diastereoselective. Identify the desired/isolated products.) The imine is formed from methyl glycine and benzaldehyde.

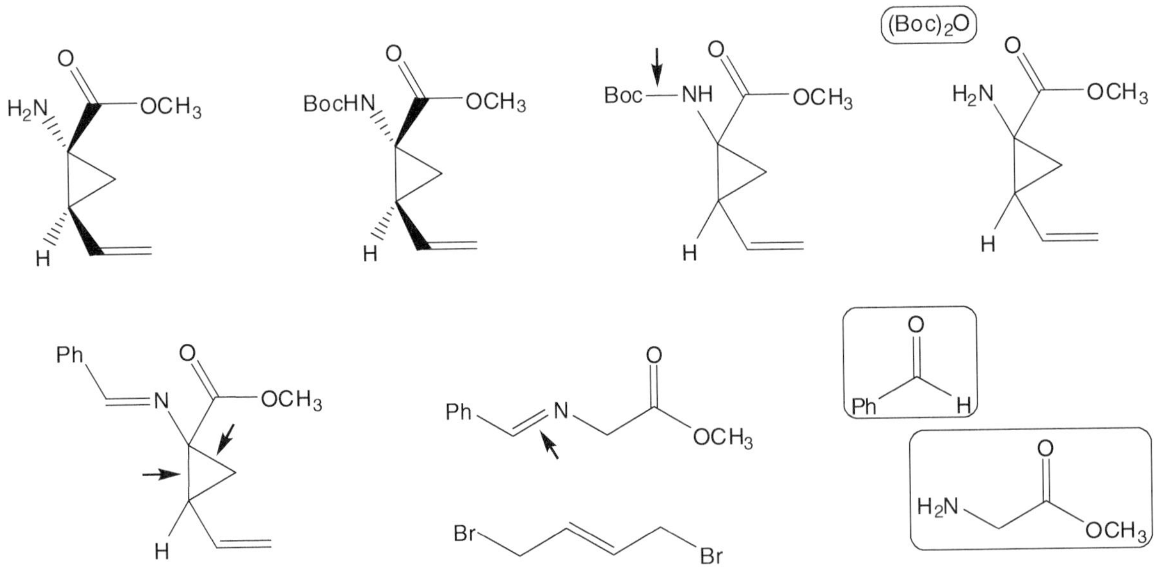

N-Methyl-5-hexen-1-amine is formed by hydrolysis of the trifluoroacetamide. The trifluoroacetamide is formed by bromide displacement from 6-bromo-1-hexene by *N*-methyltrifluoroacetamide. 6-Bromo-1-hexene is formed by hydrogen bromide elimination from 1,6-dibromohexane. *N*-Methyltrifluoroacetamide is formed from ethyl trifluoroacetate and methylamine.

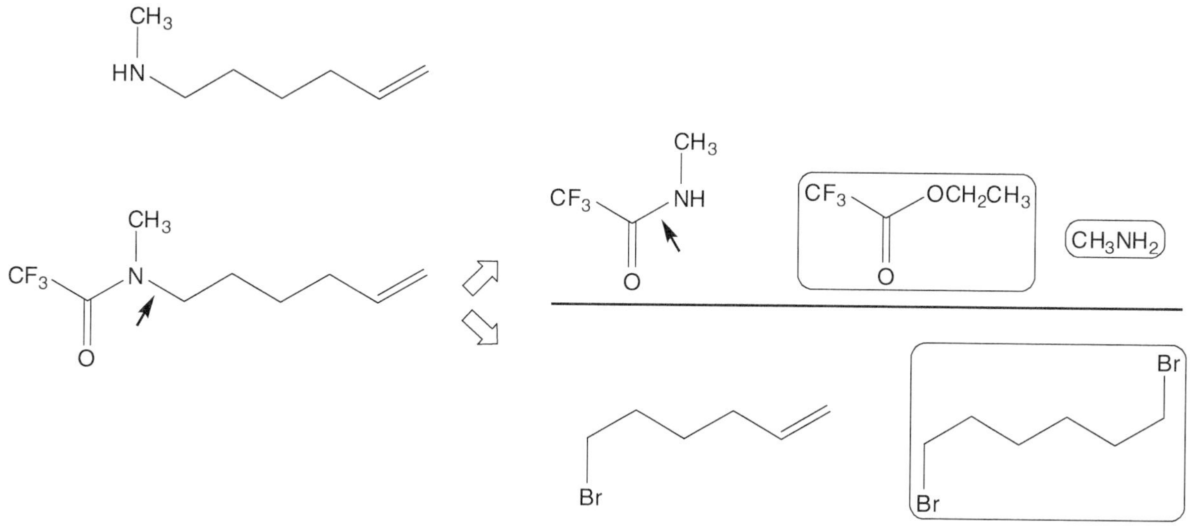

The 4-hydroxyquinoline (quinolin-4-one) ring is formed by condensation of an acetophenone with an amide and elimination of water. The amide is formed from the amine and the acid chloride.

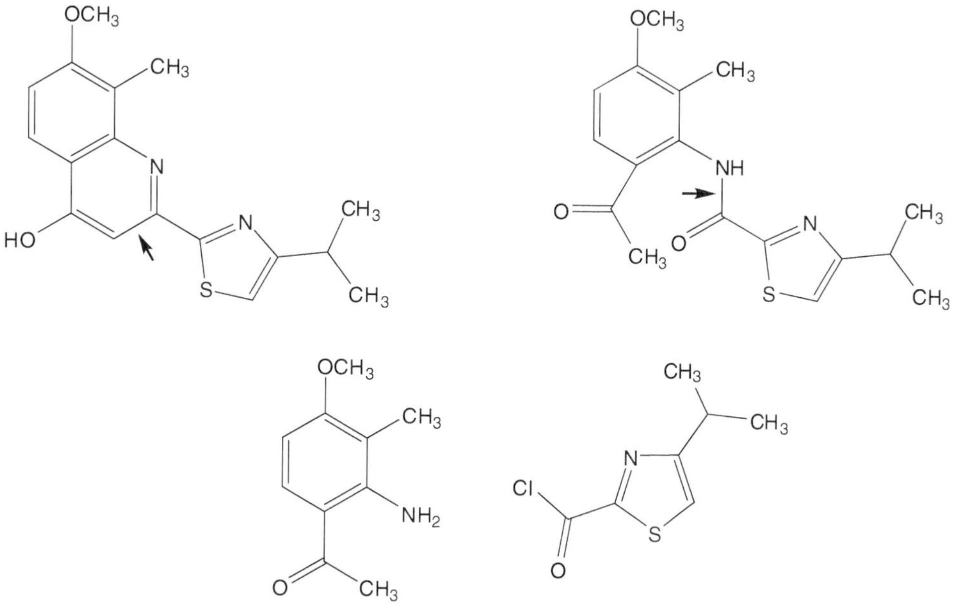

2-Amino-4-methoxy-3-methylacetophenone is formed by the reaction of 3-methoxy-2-methylaniline with acetonitrile (**Sugasawa Reaction**). The aniline is formed from the benzamide (**Hofmann Rearrangement**). The benzamide is formed from 3-methoxy-2-methylbenzoic acid.

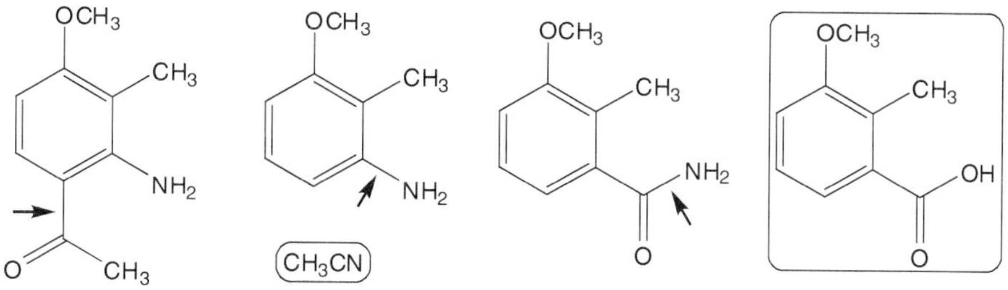

The thiazole-2-carbonyl chloride is formed from the carboxylic acid. The carboxylic acid is formed by ester hydrolysis. Ethyl 4-isopropylthiazole-2-carboxylate is formed by condensation of the α–haloketone (X = Cl or Br) with ethyl thiooxamate. The α–haloketone is formed from isopropyl methyl ketone. Ethyl thiooxamate is formed from ethyl cyanoformate. Ethyl cyanoformate is formed by the reaction of ethyl chloroformate with cyanide.

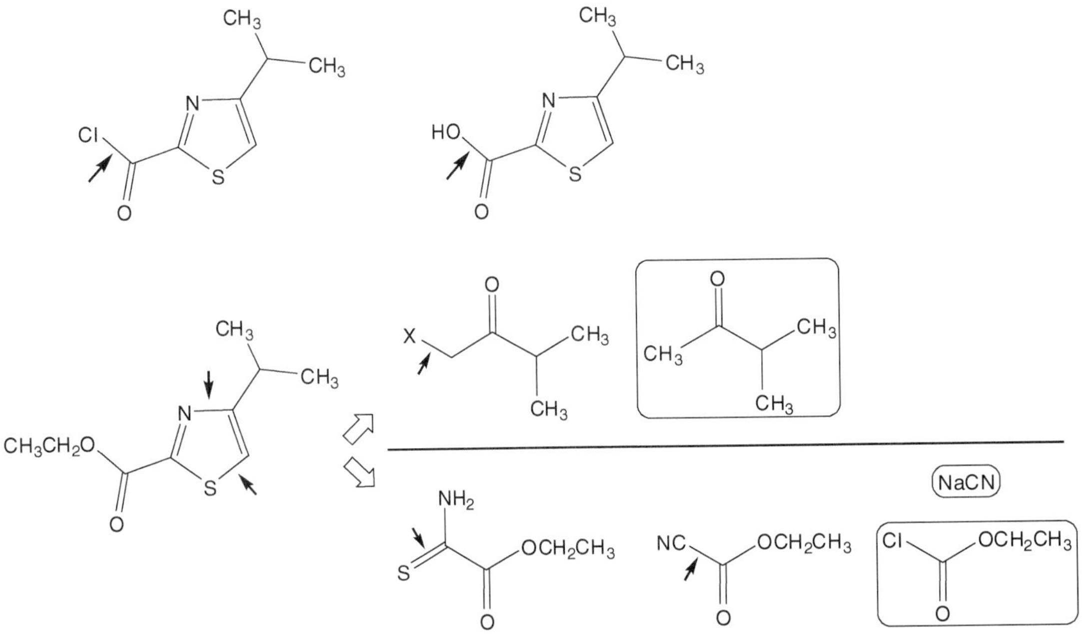

Extended Discussion

Draw the structures of a retrosynthetic analysis of an alternative route to a (1*R*,2*S*)-1-amino-2-vinylcyclopropane-1-carboxylate which utilizes a **Curtius Rearrangement** or **Hofmann Rearrangement**. Include the structures of the retrosynthetic analysis of any organic starting material(s) from petrochemical or biochemical raw materials.

Simvastatin

Cardiovascular Medicines/Lipid-Lowering Agents

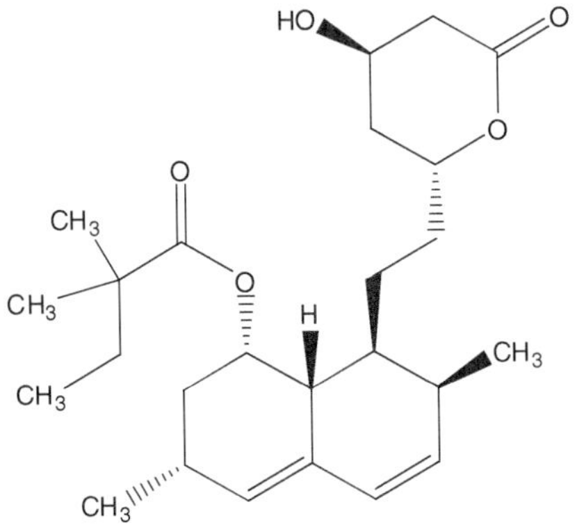

A single-enantiomer molecule with multiple chiral carbons is often formed by modification of a natural product that has most or all of the chiral carbons already in place.

Discussion. Simvastatin is manufactured from lovastatin, which is produced by the filamentous fungus *Aspergillis terreus*. While lovastatin and simvastatin differ by just a methyl group on the ester at C8, a protection-deprotection strategy is required to address the reactivity of the secondary alcohol and the lactone.

The simvastatin lactone is formed from the C11 alcohol and ammonium carboxylate salt in the final step. The highly crystalline ammonium carboxylate salt is formed from the carboxylic acid. In a single operation, the carboxylic acid is formed by amide hydrolysis and the C11 and C13 alcohols are released from the silyl ethers. (What reagents and conditions are associated with minimal hydrolysis of the ester at C8?) The α,α-dimethyl ester is formed by alkylation of the α-methyl ester using iodomethane. (What base is used to form the ester enolate? How many equivalents of base are used?) The C11 and C13 alcohols are protected as silyl ethers. The carboxamide is formed by opening the lovastatin lactone with butylamine.

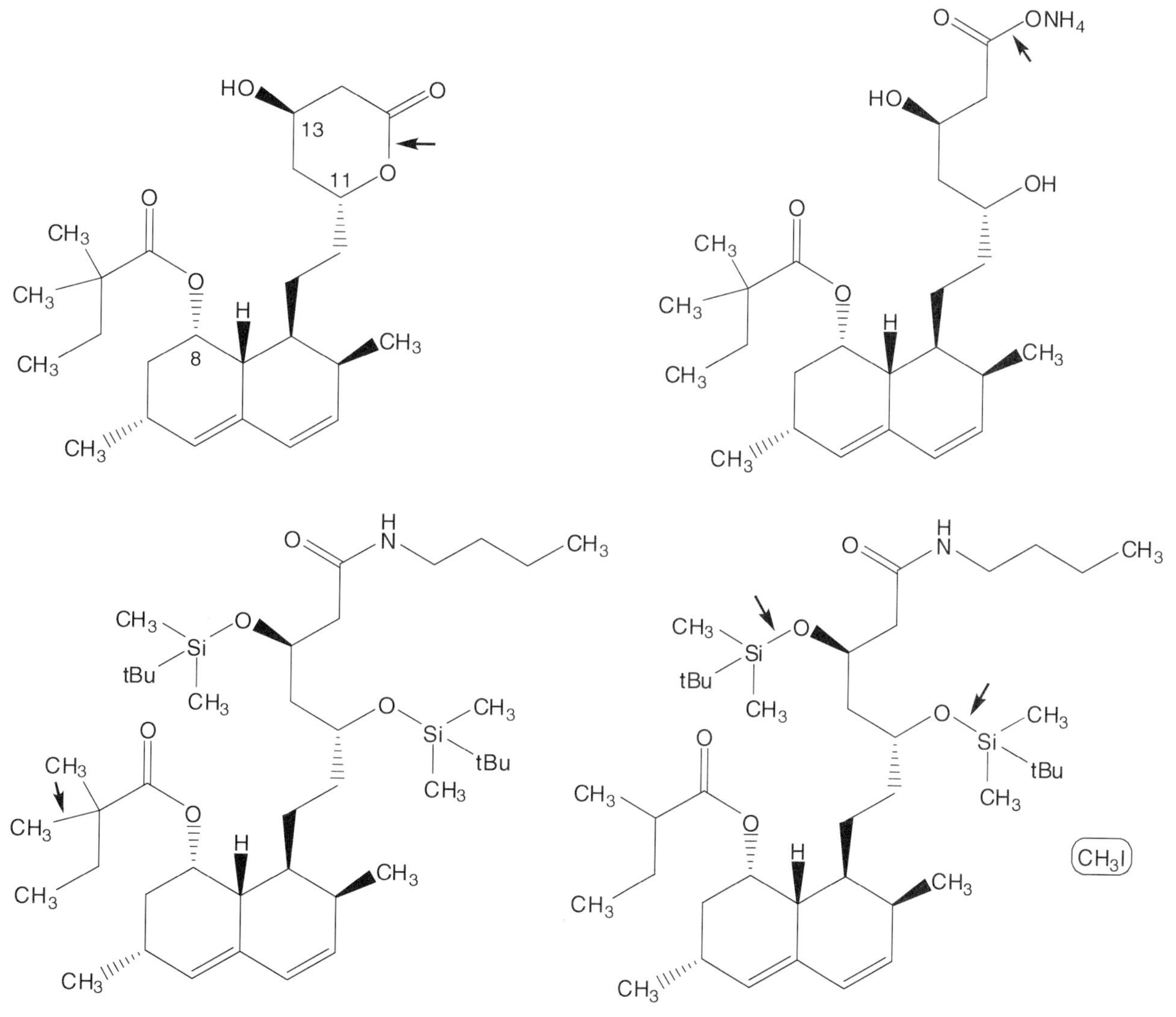

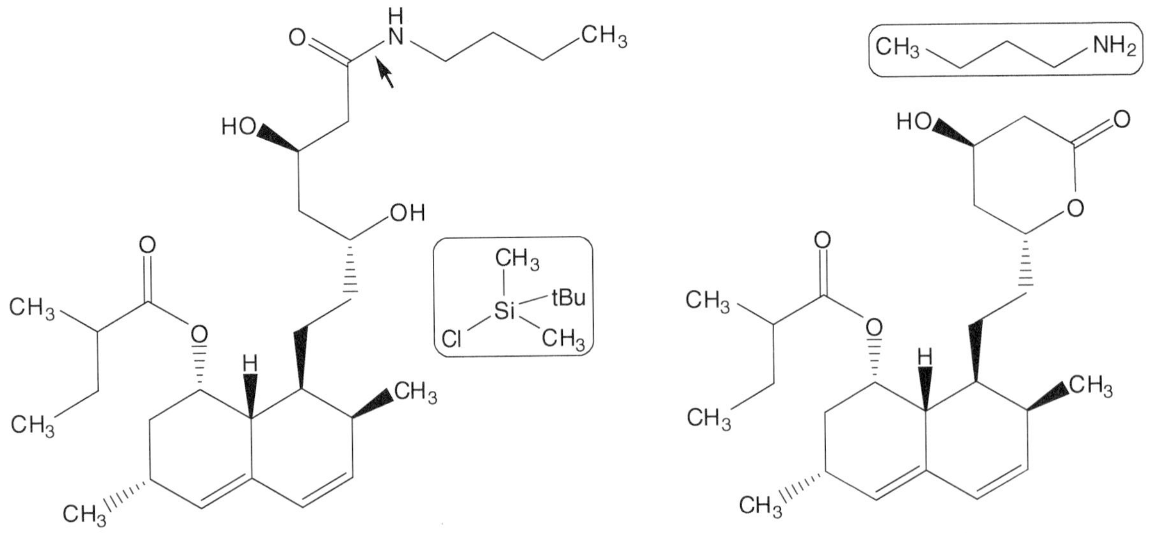

lovastatin

Extended Discussion

Other amines and alcohol protecting groups (or no alcohol protecting groups) are used in alternative routes to simvastatin. Draw the structures of the retrosynthetic analysis of one of these alternative routes. List the pros and cons for both routes and select one route as the preferred route.

Sodium Calcium Edetate

Antidotes and Other Substances Used in Poisonings/Specific

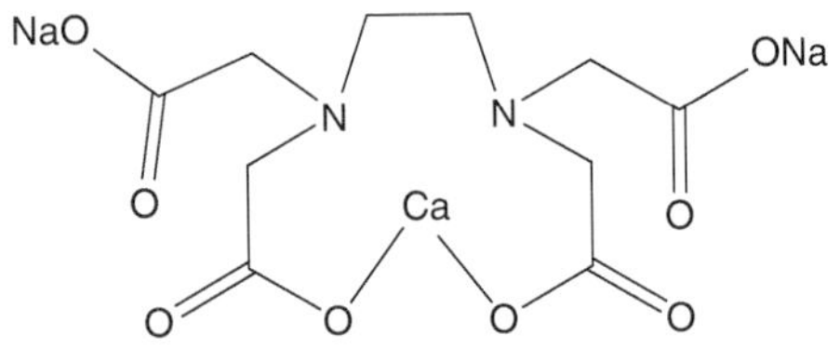

An α-aminocarboxylic acid is often formed by hydrolysis of an α-aminonitrile.

Discussion. Sodium calcium edetate is formed from disodium edetate (disodium ethylenediaminetetraacetate, disodium EDTA). Disodium EDTA is formed from ethylenediaminetetraacetic acid (EDTA). EDTA is precipitated from an aqueous solution of tetrasodium EDTA by the addition of acid. The aqueous solution of tetrasodium EDTA is formed by a reaction of ethylenediamine, formaldehyde, and sodium cyanide in aqueous hydroxide.

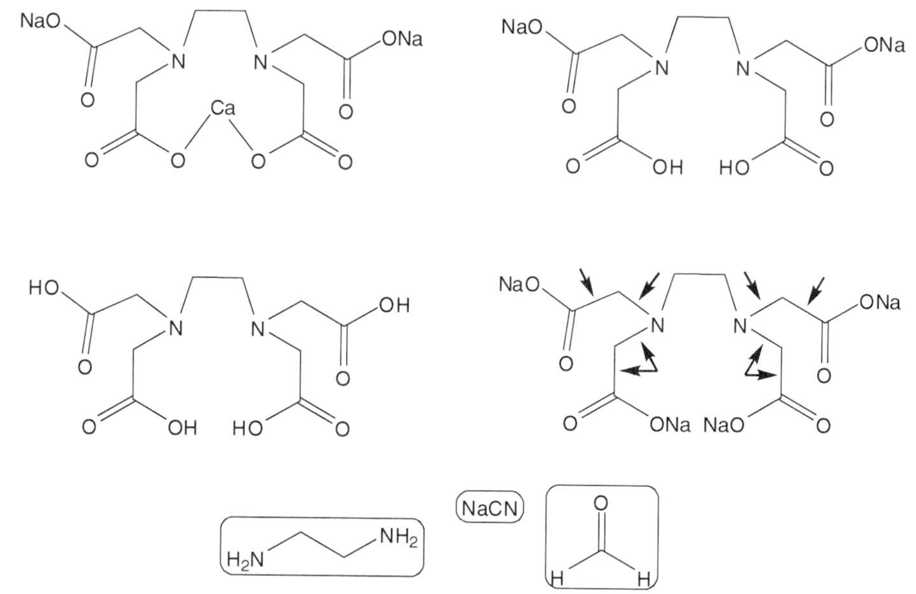

Extended Discussion

Why is ethylenediamine tetraacetic acid (EDTA) isolated?

Sofosbuvir

Anti-infective Medicines/Antiviral Medicines/Antihepatitis Medicines/Medicines for Hepatitis C/Nucleotide Polymerase Inhibitors
Anti-infective Medicines/Antiviral Medicines/Antihepatitis Medicines/Medicines for Hepatitis C/Other Antivirals

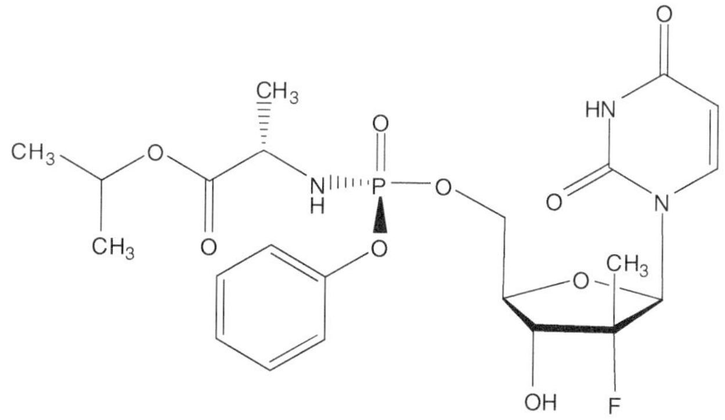

A nucleotide analog is often formed from the nucleoside analog (the phosphate is often introduced last). A nucleoside analog is often formed by displacement of a leaving group on the sugar by nitrogen of the heterocycle. The displacement usually results in a mixture of two products, with the heterocycle on the top face (β) or the bottom face (α) of the sugar. Factors that influence the β-product to α-product ratio include the heterocycle, the sugar, the leaving group, and the reaction conditions.

Discussion. In the final step, pentafluorophenol is displaced with inversion of configuration at phosphorus from the (S,S_P)-pentafluorophenyl phenyl phosphoramidate by the 5′-alcohol of the nucleoside. The 3′- and 5′-alcohols of the nucleoside are released by benzoate ester (Bz) hydrolysis. The uridine is formed by hydrolysis of the N4-benzoylcytosine.

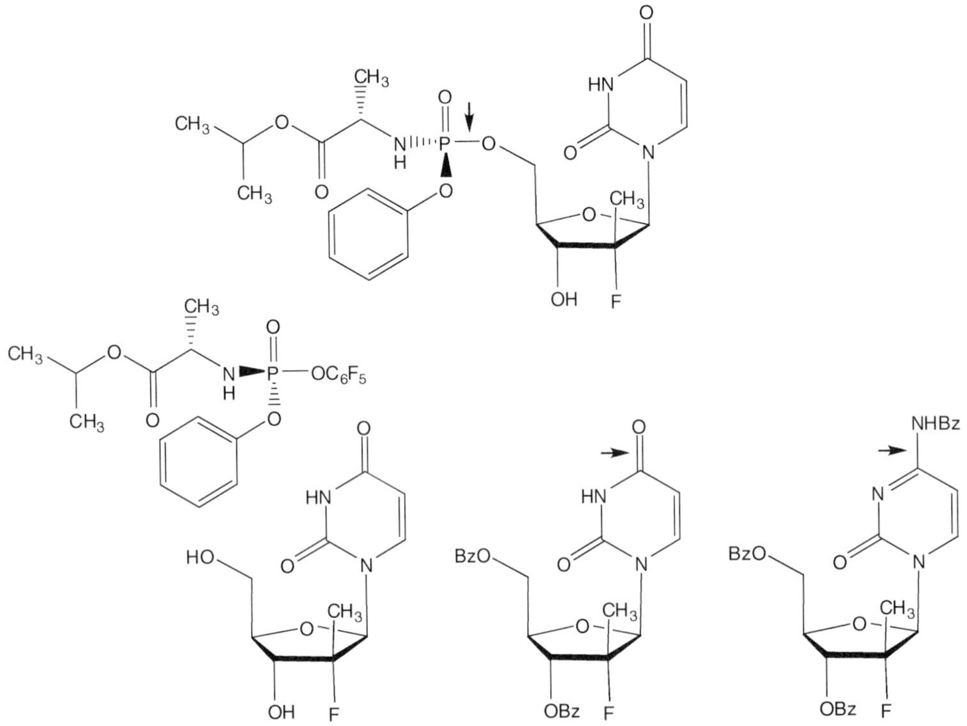

The (S,S_p)-pentafluorophenyl phenyl phosphoramidate is produced from the mixture of (S,S_p)- and (S,R_p)-diastereomers. The phosphoramidate diastereomers are formed by displacement of chloride from the phosphoramidochloridate by pentafluorophenol. The mixture of phosphoramidochloridate diastereomers is formed by displacement of chloride from phenyl phosphorodichloridate (phenyl dichlorophosphate) by L-alanine isopropyl ester.

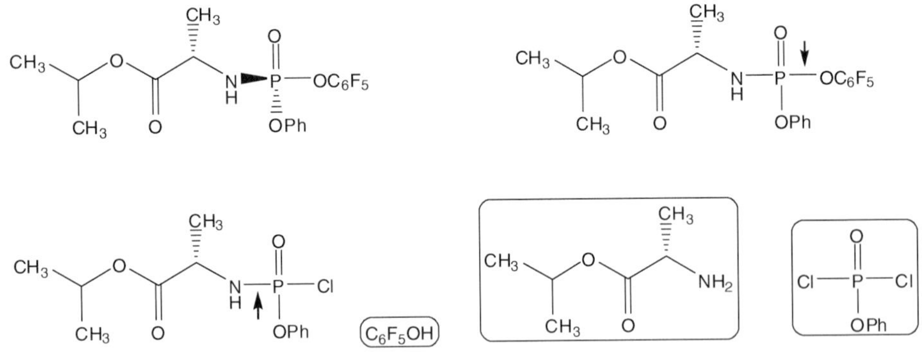

The nucleoside is formed by displacement of an acetate leaving group on the sugar by nitrogen of the O-TMS derivative of N4-benzoylcytosine. (What is the highest reported ratio of β-product to α-product? What conditions are associated with the highest ratio? How is the β-product separated from the product mixture?) *N*-Benzoylcytosine is formed from cytosine. The sugar acetate is formed by the reaction of the hemiacetal with acetic anhydride. The hemiacetal is formed by the reduction of the γ-lactone.

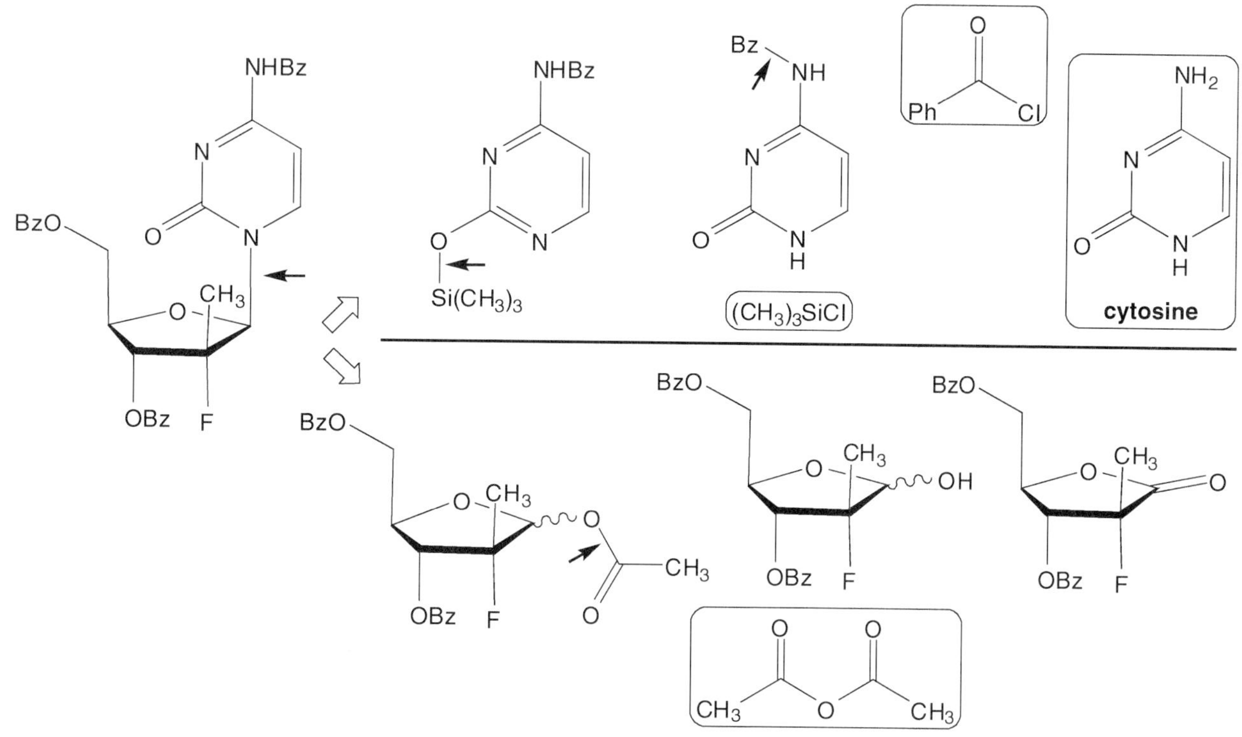

The 3′- and 5′-alcohols of the γ-lactone are protected as benzoate esters. The γ-lactone is formed by acetal hydrolysis followed by transesterification. Inversion of configuration is observed in the nucleophilic displacement of a cyclic sulfonate ester by fluoride. The cyclic sulfonate ester is formed by oxidation of the cyclic sulfite ester. The cyclic sulfite ester is formed from the diol. The diol is formed by *syn*-dihydroxylation of the (E)-alkene. The (E)-alkene is formed from D-glyceraldehyde acetonide and (ethoxcarbonylmethyl)triphenylphosphonium bromide (**Wittig Reaction**).

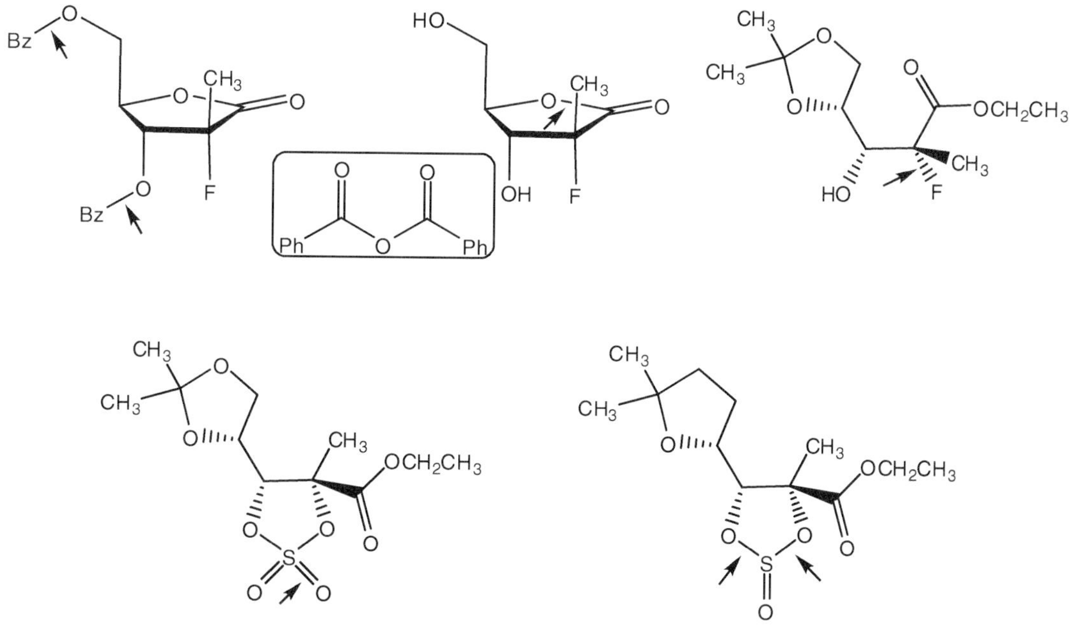

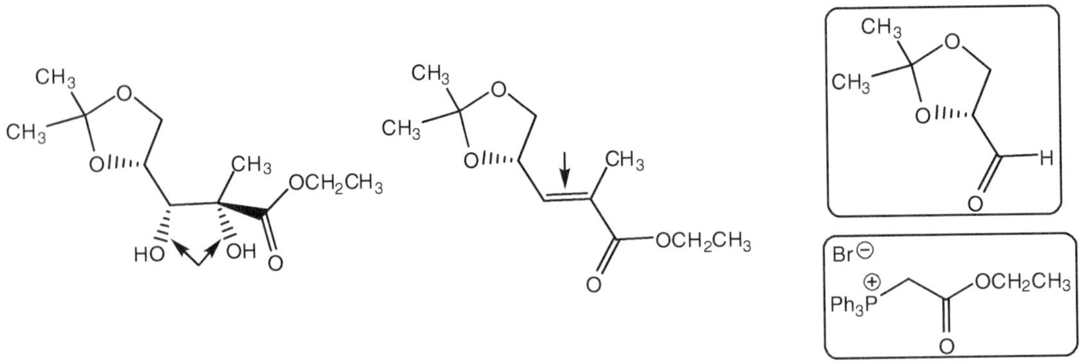

Extended Discussion

Draw the structures of the retrosynthetic analysis of one alternative route to sofosbuvir using other protecting groups on the 3′- and 5′-alcohols and/or a different leaving group on the sugar. What is the ratio of β–product to α–product for the nucleoside formation by the alternative route?

Spironolactone

Cardiovascular Medicines/Medicines Used in Heart Failure
Diuretics

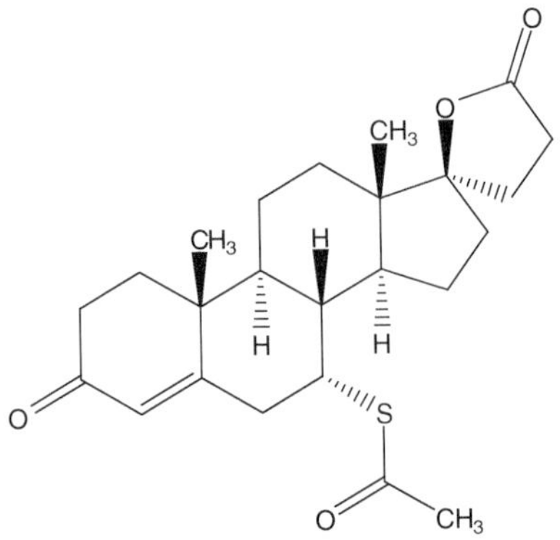

A single-enantiomer molecule with multiple chiral carbons is often formed by the modification of a natural product that has most or all of the chiral carbons already in place. A substituent at C7 of an androstane steroid is often introduced by conjugate addition to a 4,6-diene-3-one. The 4,6-diene-3-one is often formed by oxidation of the 4-ene-3-one.

Discussion. In one preferred route, spironolactone is formed in seven steps from androst-4-ene-3,17-dione and in eight steps from a phytosterol.

Thioacetic acid is added to the 6,7-alkene of canrenone in the final step. The 4,6-diene-3-one of canrenone is formed from the 3-acetoxy-3,5-diene. The 3-acetoxy-3,5-diene is formed by reaction of spirolactone with acetic anhydride. The γ-hydroxypropanoate needed to form the lactone is formed by hydrogenation of the γ–hydroxypropiolate. The addition of acid then results in spirolactone by formation of the γ-lactone, release of the C3 ketone from the acetal, and double bond migration. The γ–hydroxypropiolate is formed by carboxylation of the alkyne.

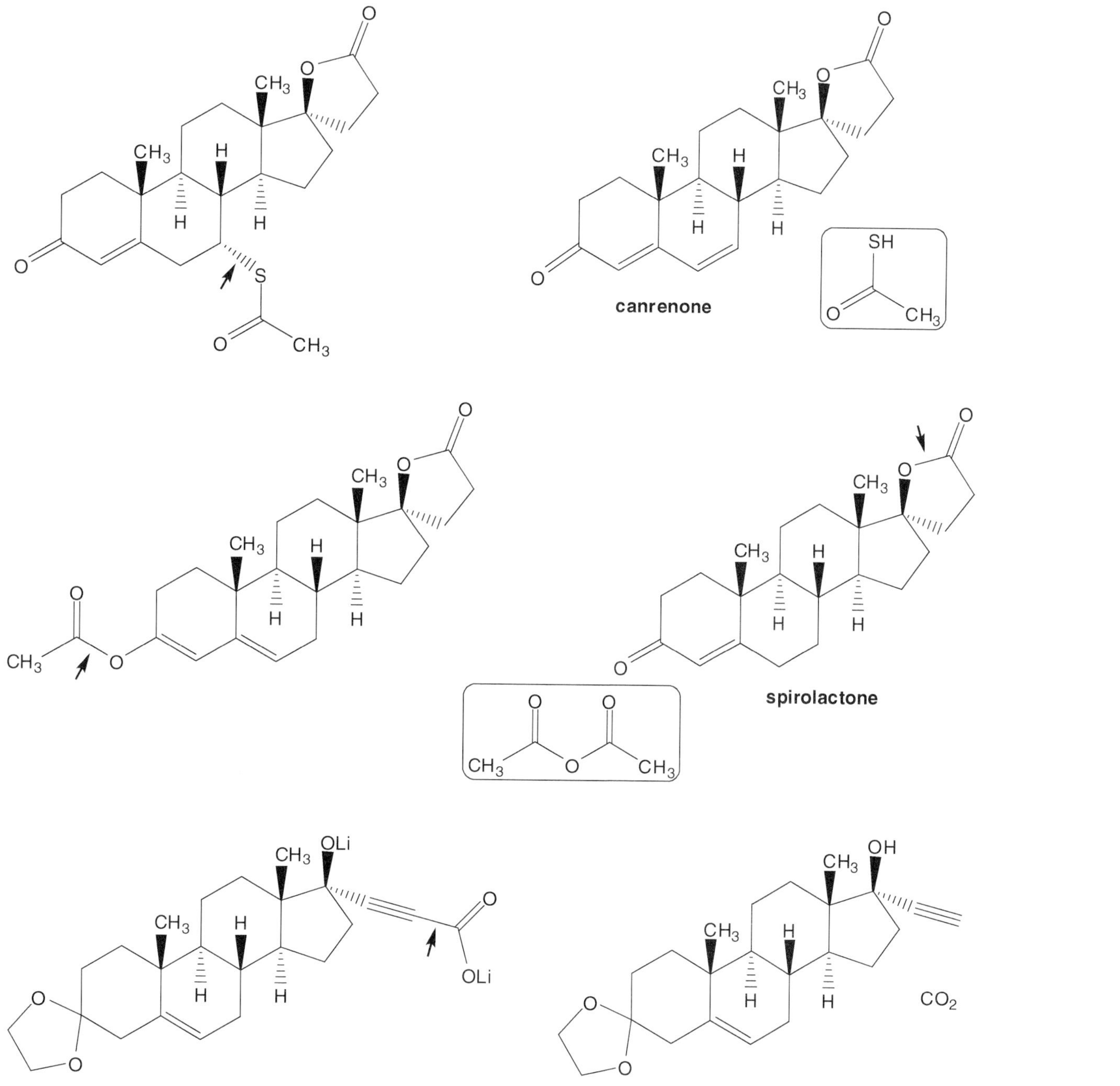

The 4,5-alkene migrates as the C3 ketone of ethisterone is protected as the acetal. Ethisterone is formed by the addition of acetylene to the C17 ketone of androst-4-ene-3,17-dione. Androst-4-ene-3,17-dione is formed by microbial oxidation/side chain degradation of phytosterols (plant sterols) including stigmasterol.

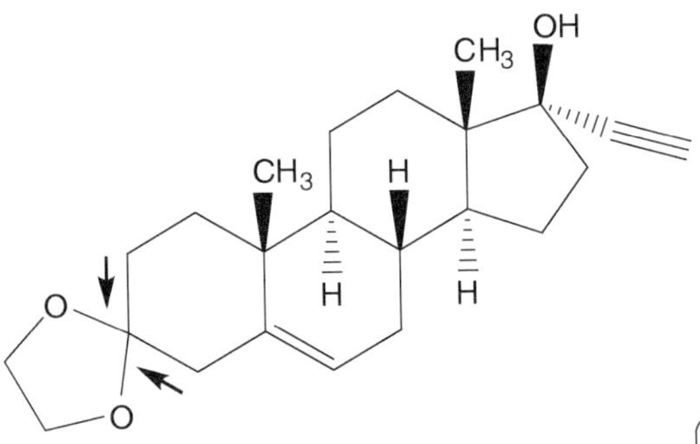

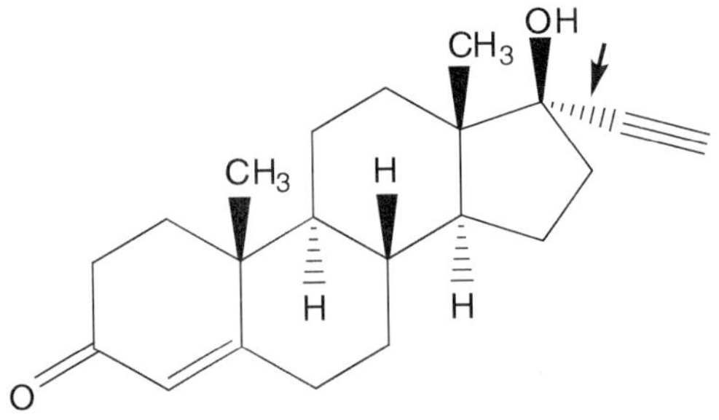

ethisterone

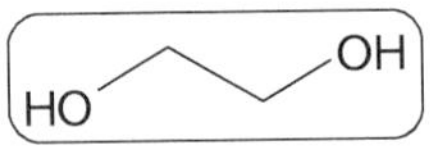

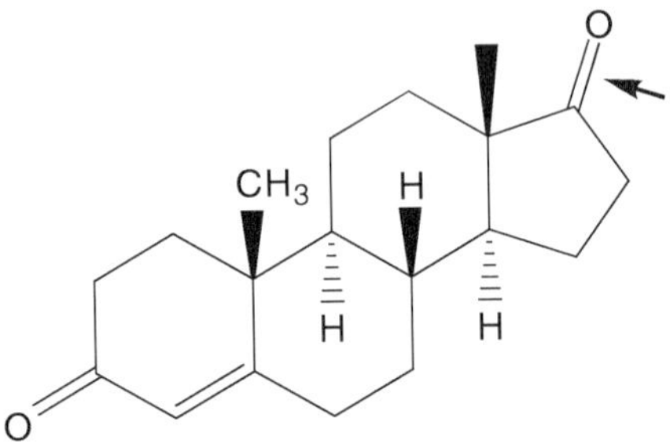

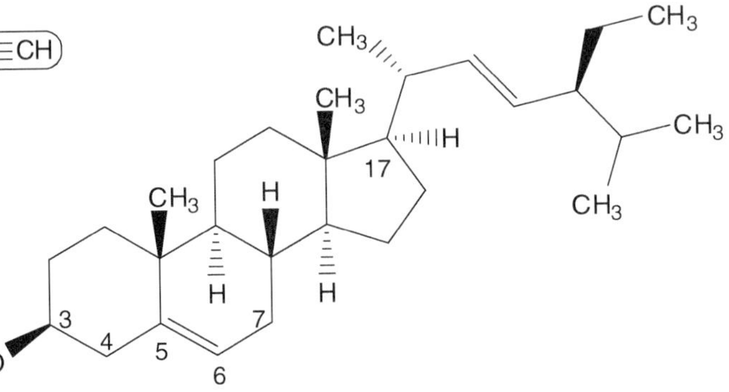

androst-4-ene-3,17-dione

stigmasterol

Extended Discussion

Draw the structures of the retrosynthetic analysis of one alternative route to spironolactone from androst-4-ene-3,17-dione that does not have a hydrogenation step. Include the structures of the retrosynthetic analysis of any organic starting material(s) from petrochemical or biochemical raw materials. List the pros and cons for both routes and select one route as the preferred route.

Succimer

Antidotes and Other Substances Used in Poisonings/Specific

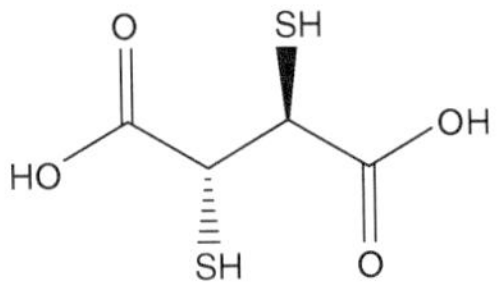

A β-mercaptocarboxylic acid is often formed by conjugate addition of hydrogen sulfide or a hydrogen sulfide equivalent to an α,β-unsaturated carboxylic acid.

Discussion. Succimer is optically inactive (*R,S*)-2,3-dimercaptosuccinic acid. A mixture of the (*R,R*)- and (*S,S*)-enantiomers and the (*R,S*)-*meso* compound is formed by double addition of thiosulfate across the triple bond of acetylenedicarboxylic acid. The (*R,R*)- and (*S,S*)-enantiomers in the mixture are converted to the much less soluble (*R,S*)-*meso* succimer during thiosulfonic acid hydrolysis.

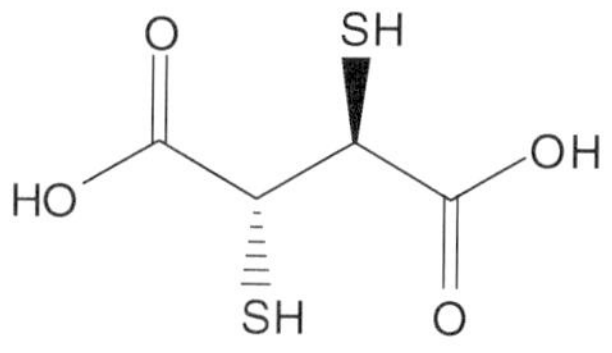

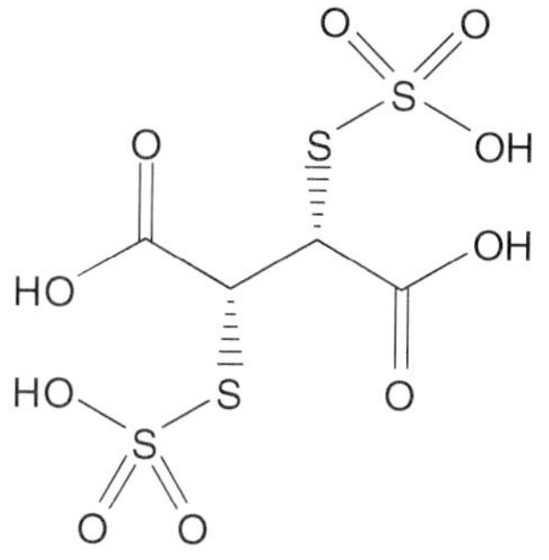

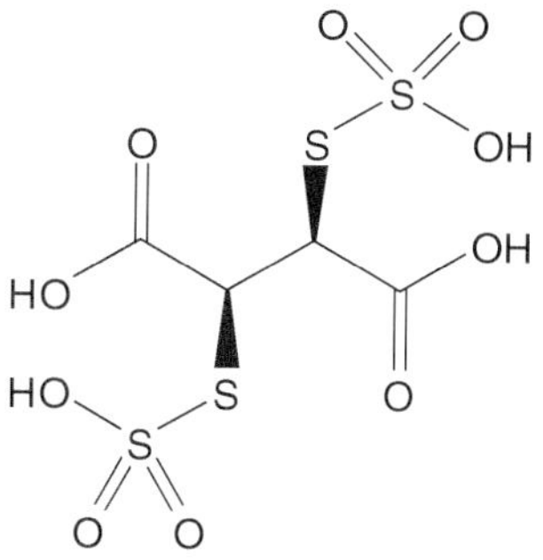

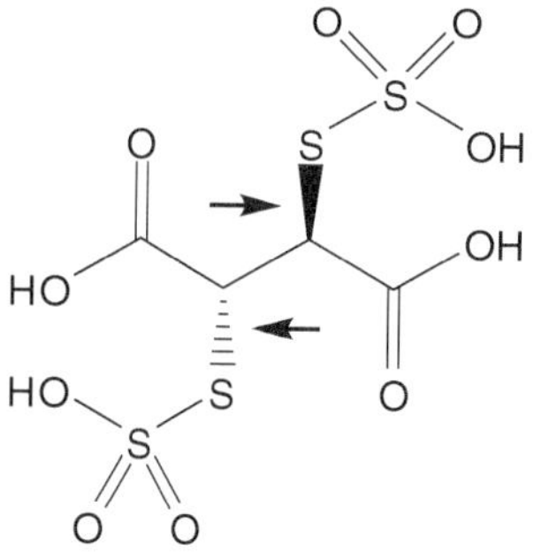

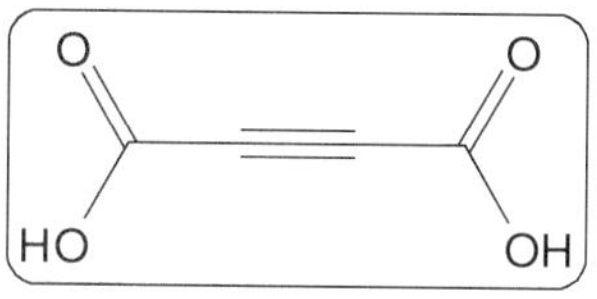

Extended Discussion

Hydrogen sulfide is a poisonous, flammable, and corrosive gas. List alternative sulfur-containing reagents used to make thiols.

Sulfadiazine and Silver Sulfadiazine

Anti-infective Medicines/Antiprotozoal Medicines/Antipneumocystosis and Antitoxoplasmosis Medicines
Dermatological Medicines(topical)/Anti-infective Medicines

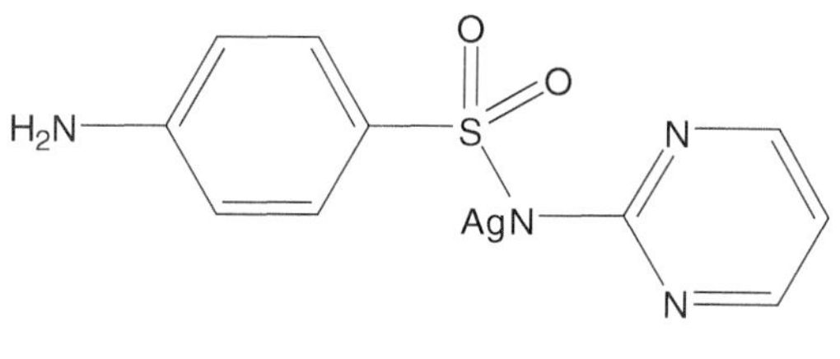

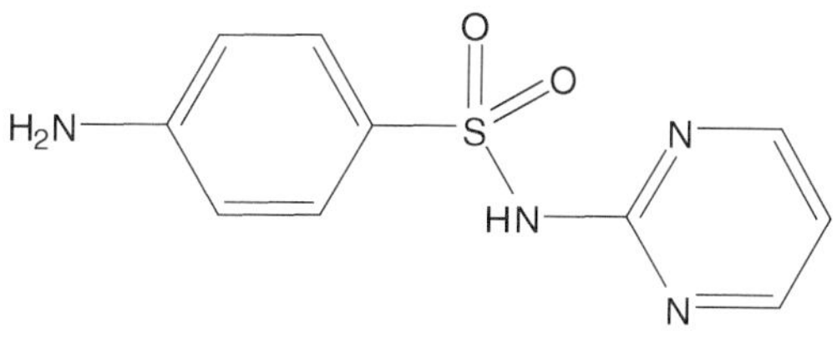

A sulfonamide is often formed by the reaction of a sulfonyl chloride with an amine.

Discussion. Silver sulfadiazine is formed from sulfadiazine. The amino group of sulfadiazine is released in the final step by hydrolysis of the acetamide. The sulfonamine is formed by a reaction of 4-acetylaminobenzenesulfonyl chloride (*N*-acetylsulfanilyl chloride) with 2-aminopyrimidine.

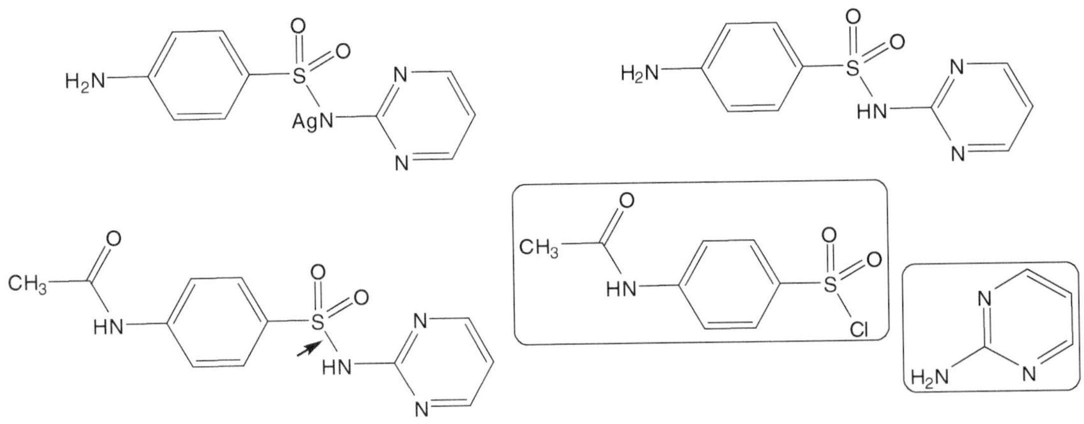

Extended Discussion

Draw the structures of the retrosynthetic analysis of an alternative route to sulfadiazine from 4-nitrobenzenesulfonyl chloride (nosyl chloride). Include the structures of the retrosynthetic analysis of nosyl chloride from petrochemical or biochemical raw materials. List the pros and cons for both routes and select one route as the preferred route.

Sulfadoxine

Anti-infective Medicines/Antiprotozoal Medicines/Antimalarial Medicines/For Curative Treatment

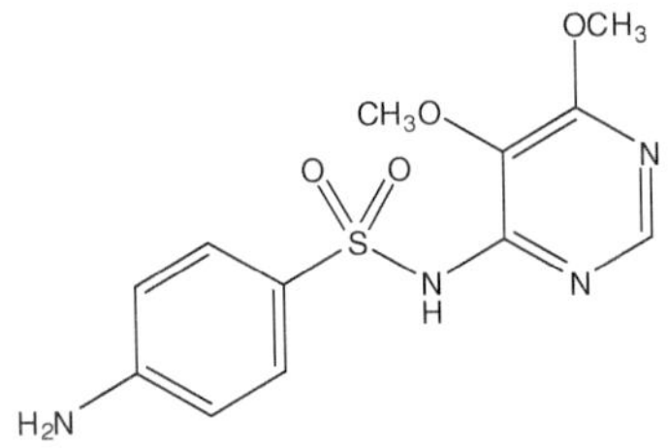

A sulfonamide is often formed by reaction of a sulfonyl chloride with an amine.

Discussion. The amino group of sulfadoxine is released in the final step by hydrolysis of the acetamide. The sulfonamine is formed by the reaction of the aminopyrimidine with 4-acetylaminobenzenesulfonyl chloride (*N*-acetylsulfanilyl chloride).

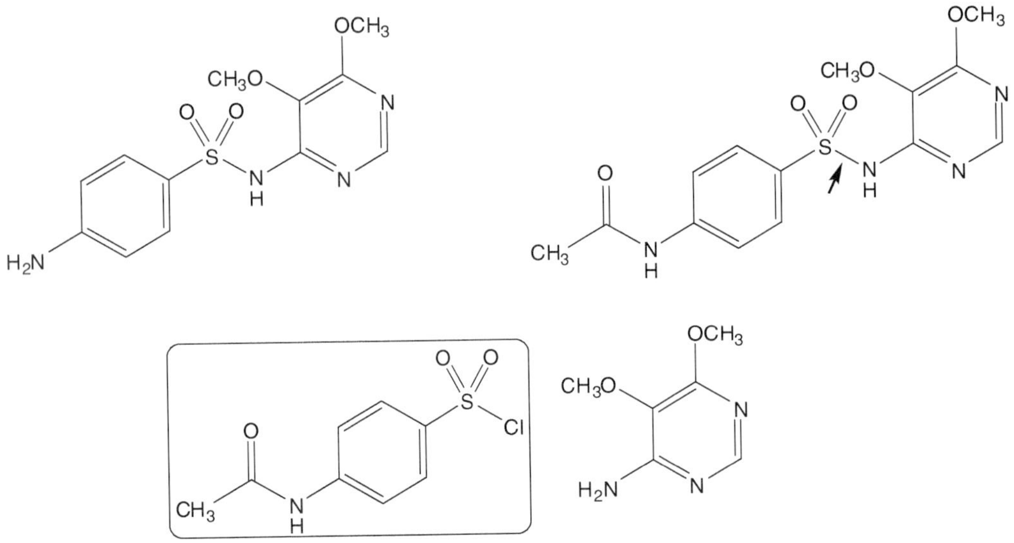

4-Amino-5,6-dimethoxypyrimidine is formed from 4,6-dichloro-5-methoxypyrimidine by displacement of one chloride by ammonia then displacement of the remaining chloride by methanol. The 4,6-dichloropyrimidine is formed from the pyrimidine-4,6-dione. 5-Methoxy-4,6-pyrimidinedione is formed by the reaction of dimethyl methoxymalonate with formamidine acetate (**Pinner Pyrimidine Synthesis**). Dimethyl methoxymalonate is formed in two steps from methyl methoxyacetate and dimethyl oxalate (mixed **Claisen Condensation**).

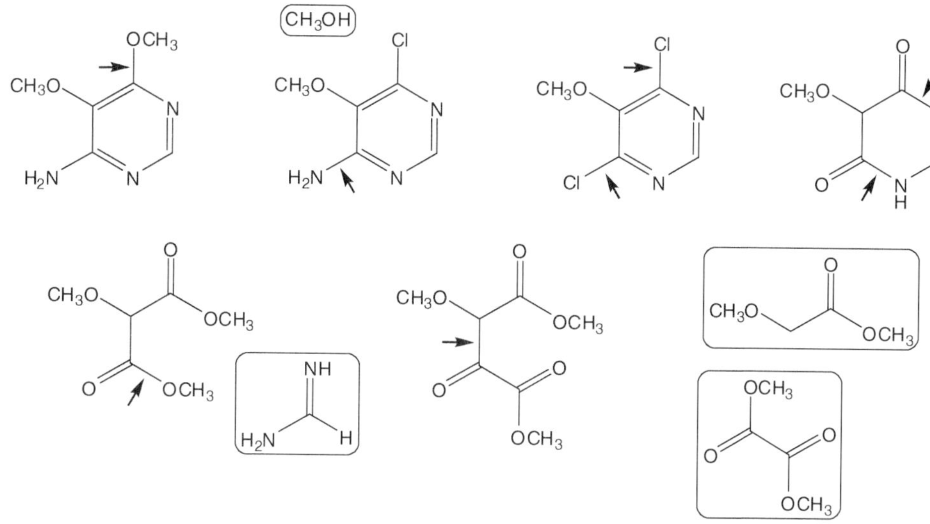

Extended Discussion

Draw the structures of a retrosynthetic analysis of an alternative route to sulfadoxine from 4-aminobenzenesulfonamide (sulfanilamide). Include the structures of the retrosynthetic analysis of sulfanilamide from petrochemical or biochemical raw materials. List the pros and cons for both routes. Is one route preferred?

Sulfamethoxazole

Anti-infective Medicines/Antibacterials/Other Antibacterials
Anti-infective Medicines/Antiprotozoal Medicines/Antipneumocystosis and Antitoxoplasmosis Medicines

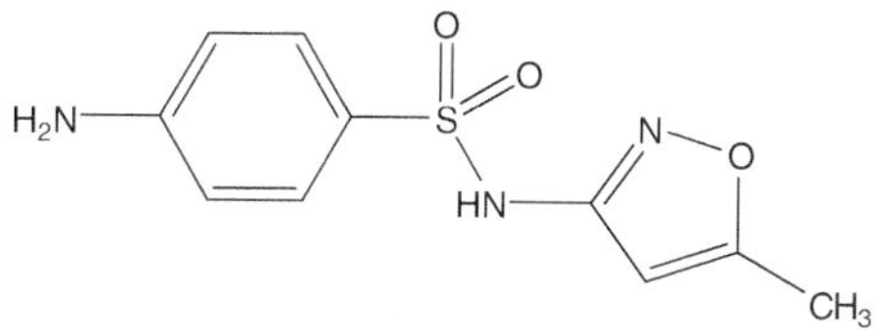

A 3-aminoisoxazole is often formed by Hofmann Rearrangement of a carboxamide.

Discussion. The amino group of sulfamethoxazole is released in the final step by hydrolysis of the acetamide. The sulfonamide is formed by the reaction of 3-amino-5-methylisoxazole with 4-acetylaminobenzenesulfonyl chloride (*N*-acetylsulfanilyl chloride).

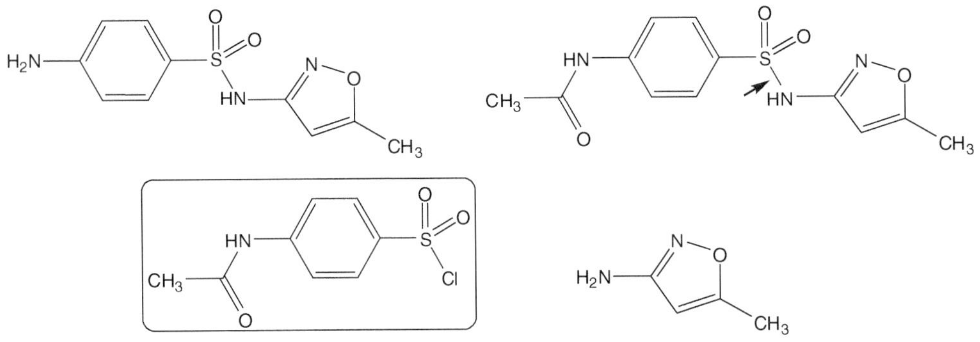

3-Amino-5-methylisoxazole is formed by hydrolysis of the ethyl carbamate. The ethyl carbamate is formed from the amide (**Hofmann Rearrangement**). The amide is formed from the ester. Ethyl 5-methylisoxazole-3-carboxylate is formed from ethyl 2,4-dioxopentanoate (ethyl acetopyruvate). Ethyl acetopyruvate is formed by the condensation of acetone with diethyl oxalate (mixed **Claisen Condensation**).

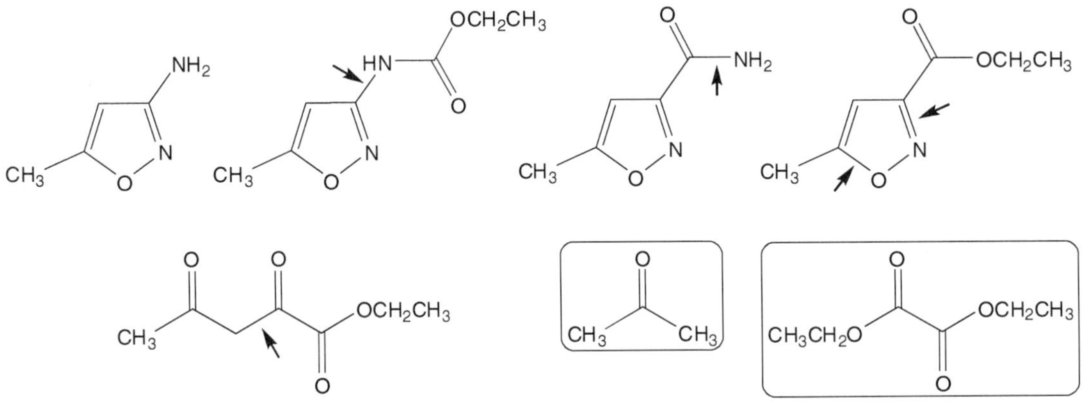

Extended Discussion

Draw the structures of the retrosynthetic analysis of one alternative route to 3-amino-5-methylisoxazole. Include the structures of the retrosynthetic analysis of any organic starting material(s) from petrochemical or biochemical raw materials. List the pros and cons for both routes to 3-amino-5-methylisoxazole and select one route as the preferred route.

Sulfasalazine

Gastrointestinal Medicines/Anti-inflammatory Medicines

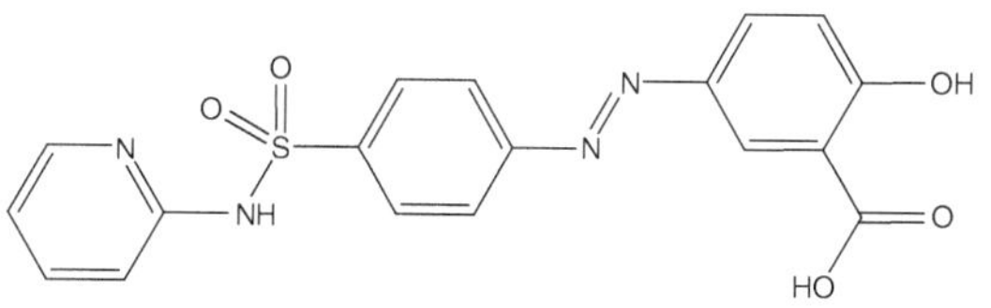

An azo compound is often formed by the coupling of an aromatic diazonium salt with a phenol or aniline.

Discussion. The azo compound is formed in the final step by azo coupling of the diazonium salt with the essential medicine salicylic acid. The diazonium salt is formed from the aniline (sulfapyridine). The amino group of sulfapyridine is released by hydrolysis of the acetamide. The sulfonamide is formed by reaction of 2-aminopyridine with 4-acetylaminobenzenesulfonyl chloride (*N*-acetylsulfanilyl chloride).

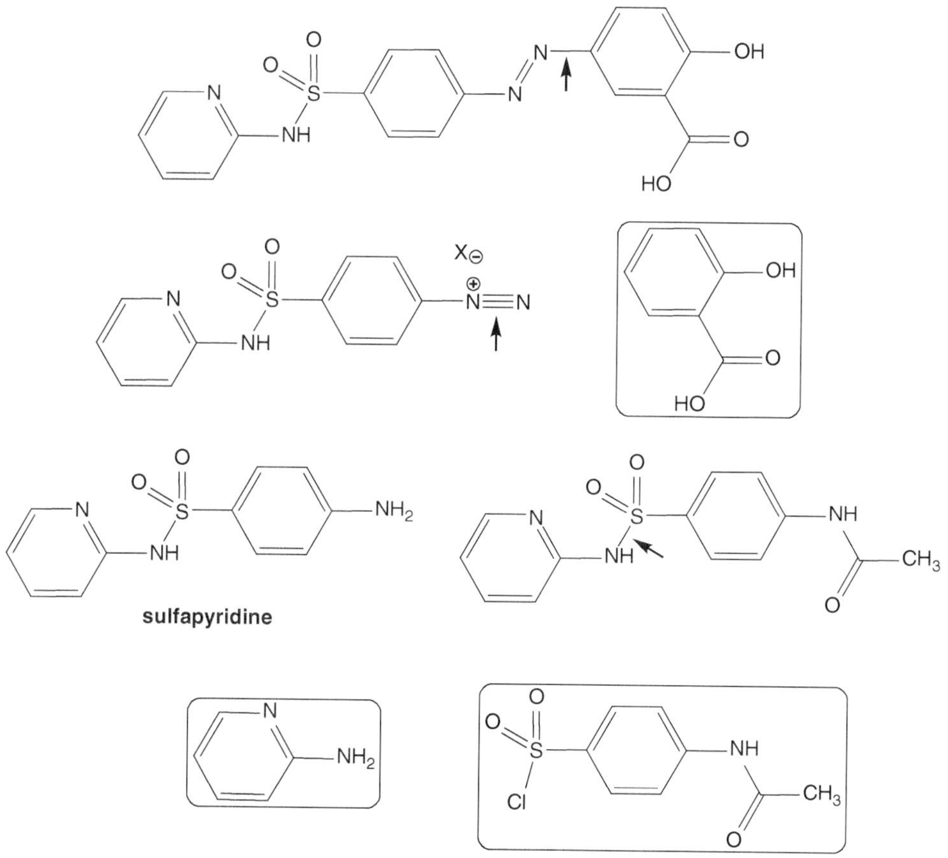

Extended Discussion

Draw the structures of a retrosynthetic analysis of an alternative route to sulfasalazine from 2-bromopyridine. List the pros and cons for both routes and select one route as the preferred route.

Suramin

Anti-infective Medicines/Antiprotozoal Medicines/Antitrypanosomal Medicines/African Trypanosomiasis/ Medicines for the Treatment of First Stage African Trypanosomiasis

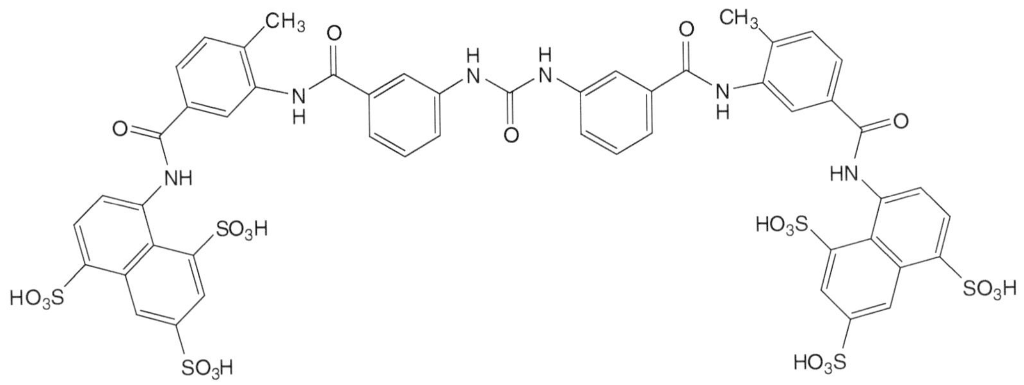

A symmetrical urea is often efficiently formed by the reaction of an amine with phosgene.

Discussion. The final step in the "outside-in" synthesis of suramin is the formation of the symmetrical urea by the reaction of the amine with phosgene.

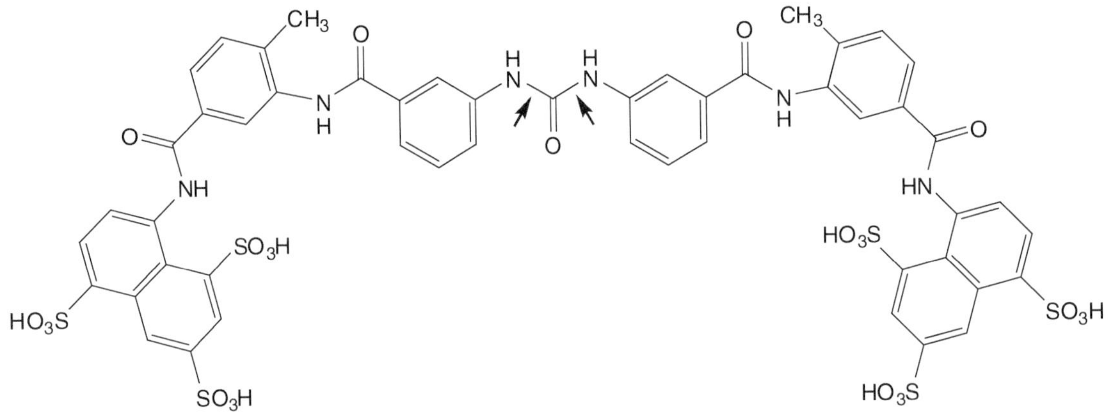

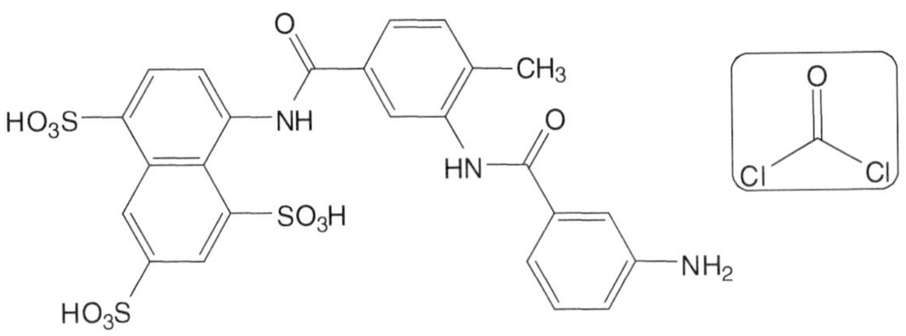

The 3-aminobenzamide is formed by reduction of the 3-nitrobenzamide. The 3-nitrobenzamide is formed from the amine and 3-nitrobenzoyl chloride. The 3-amino-4-methylbenzamide is formed by reduction of the 4-methyl-3-nitrobenzamide. The 4-methyl-3-nitrobenzamide is formed from the amine and 4-methyl-3-nitrobenzoyl chloride. 8-Aminonaphthalene-1,3,5-trisul fonic acid is formed by the reduction of 8-nitronaphthalene-1,3,5-trisulfonic acid. 8-Nitronaphthalene-1,3,5-trisulfonic acid is formed by nitration of naphthalene-1,3,5-trisulfonic acid.

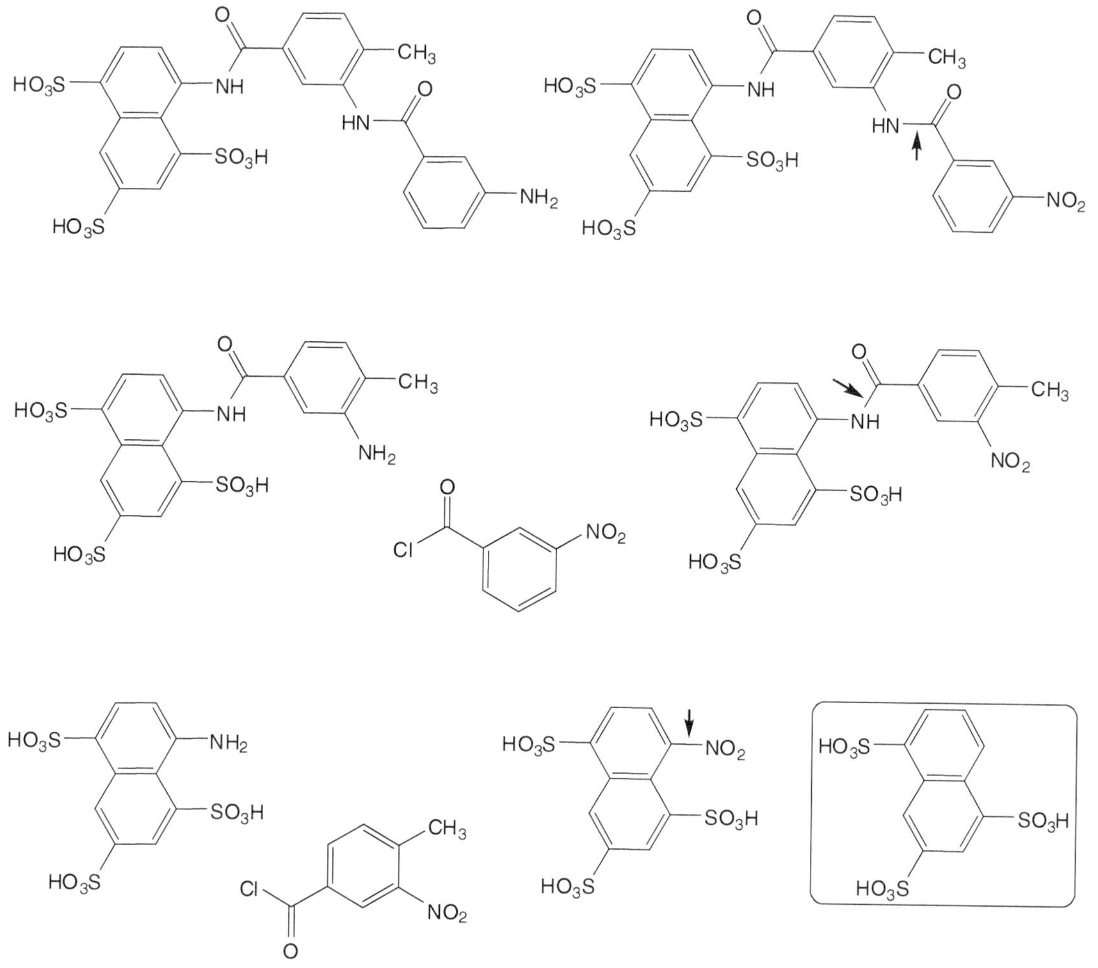

3-Nitrobenzoyl chloride is formed from 3-nitrobenzoic acid. 4-Methyl-3-nitrobenzoyl chloride is formed from 4-methyl-3-nitrobenzoic acid.

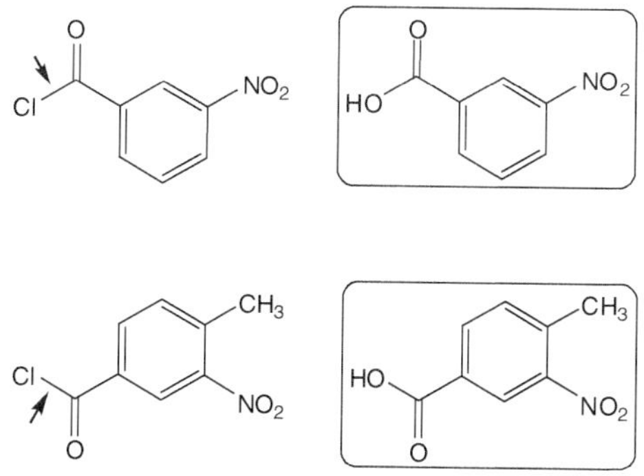

Suxamethonium Chloride

Muscle Relaxants (Peripherally-Acting) and Cholinesterase Inhibitors

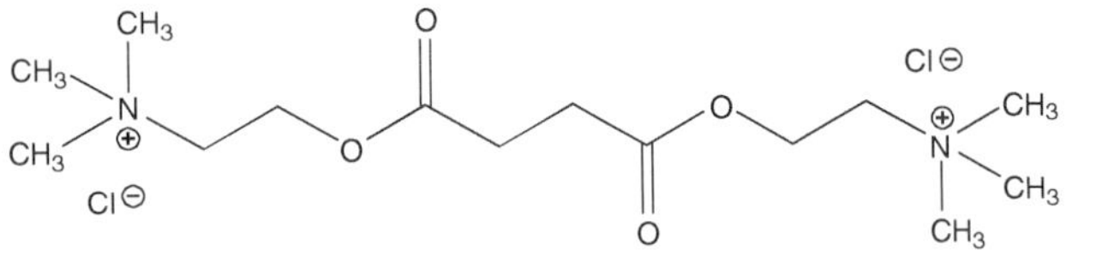

Esters are common in drug structures and intermediates leading to drug structures. An ester is often formed from an alcohol and acid chloride, anhydride, another ester, or a carboxylic acid.

Discussion. The quaternary ammonium salt is formed in the final step by alkylation of the tertiary amine with chloromethane. The succinic acid diester is formed by transesterification from dimethyl succinate and 2-dimethylaminoethanol.

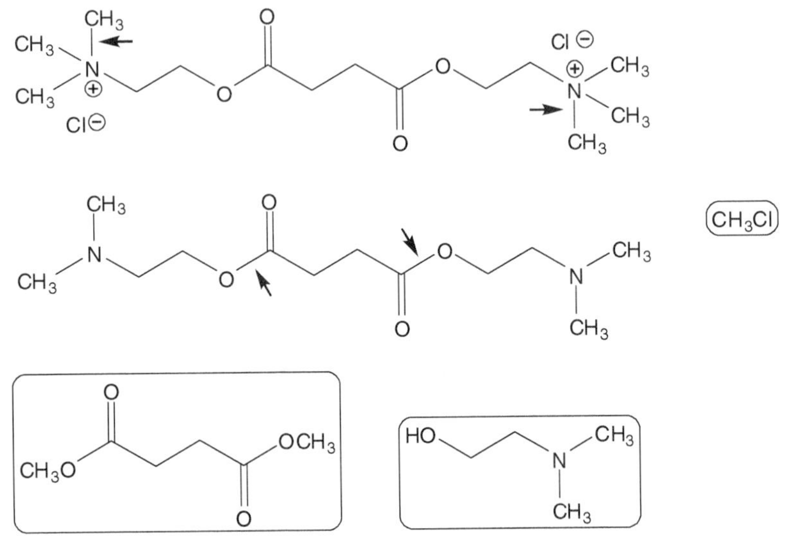

Extended Discussion

Draw the structures of the retrosynthetic analysis of an alternative route to suxamethonium chloride from succinyl chloride. List the pros and cons for both routes and select one route as the preferred route.

T

Tamoxifen

Antineoplastics and Immunosuppressives/Hormones and Antihormones

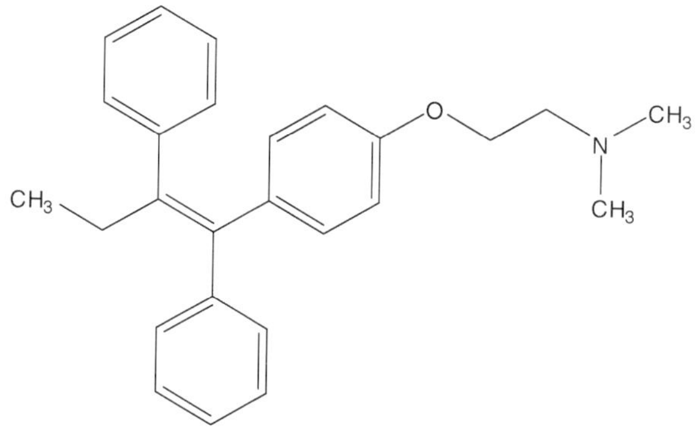

An alkene conjugated to three aromatic rings is often formed by dehydration of an alcohol.

Discussion. Tamoxifen is the (Z)- or *trans*-alkene. Tamoxifen is crystallized from the mixture of (*E*)- and (*Z*)-alkenes. Tamoxifen is then **also** formed by isomerization of the (*E*)-alkene. The mixture of (*E*)- and (*Z*)-alkenes is formed by dehydration of a tertiary alcohol. The tertiary alcohol is formed by the addition of an arylmagnesium bromide to 1,2-Diphenyl-1-butanone (**Grignard Reaction**).

Routes to Essential Medicines: A Workbook for Organic Synthesis, First Edition. Peter J. Harrington.

Companion website: www.wiley.com/go/Harrington/routes_essential_medicine

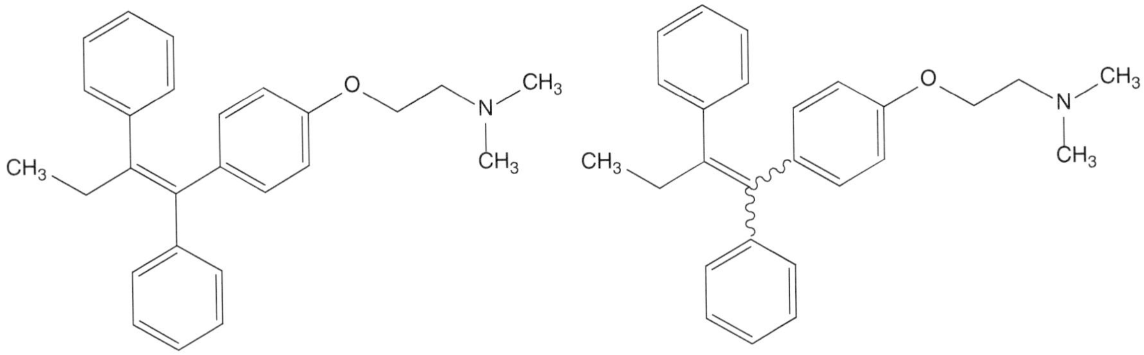

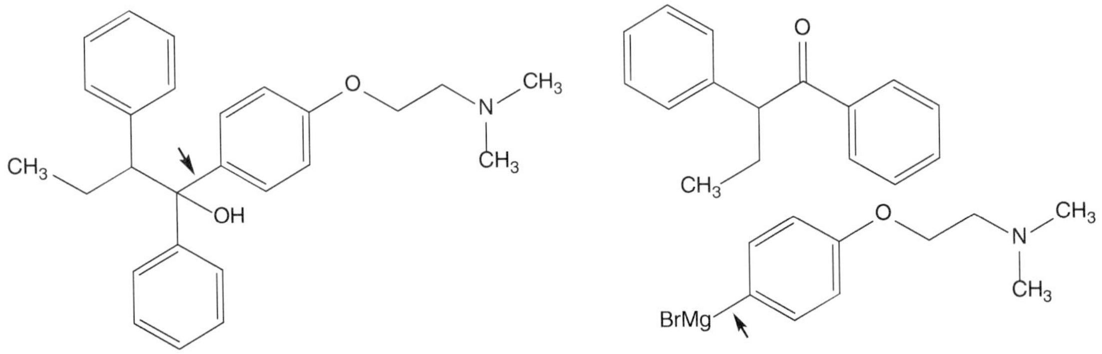

The arylmagnesium bromide is formed from 2-(4-bromophenoxy)-*N*,*N*-dimethylethanamine. The ether C—O bond is formed by chloride displacement from 2-chloro-*N*,*N*-dimethylethanamine by 4-bromophenol (**Williamson Ether Synthesis**).

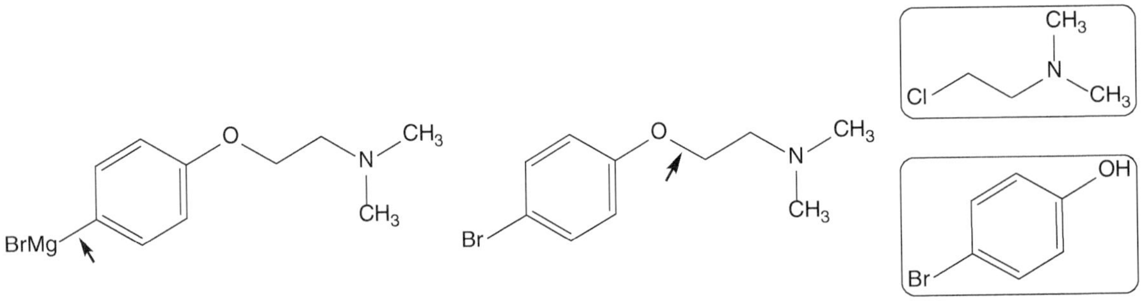

1,2-Diphenyl-1-butanone is formed by α–alkylation of 2-phenylacetophenone (deoxybenzoin) with bromoethane.

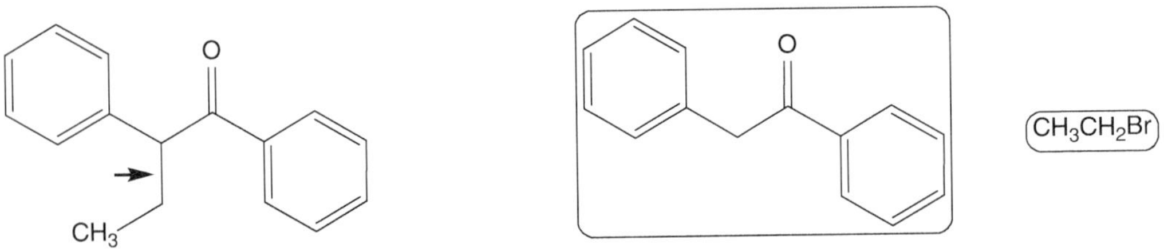

Extended Discussion

Draw the structures of the retrosynthetic analysis of an alternative route to tamoxifen. Include the structures of the retrosynthetic analysis of any organic starting material(s) from petrochemical or biochemical raw materials. List the pros and cons for both routes. Is one route preferred?

Tazobactam

Anti-infective Medicines/Beta-Lactam Medicines

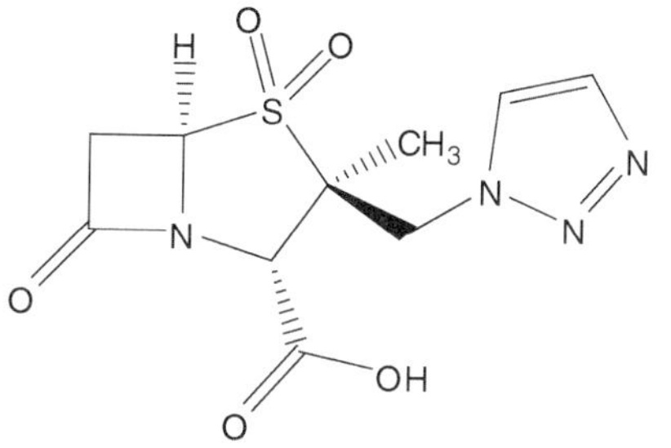

A single-enantiomer molecule with multiple chiral carbons is often formed by the modification of a natural product which has most or all of the chiral carbons already in place.

Discussion. The carboxylic acid is released from a diphenylmethyl (benzhydryl) ester in the final step. The sulfone is formed by oxidation of the sulfide. A C—N bond is formed by chloride displacement by 1,2,3-triazole. (Draw the structure of one side product of this reaction.) The 2β-chloromethylpenam is formed by the reaction of an alkene and disulfide. (Draw the structure of the benzothiazole-containing byproduct of this reaction. How is this byproduct separated from the reaction mixture?) The alkene and disulfide are both formed by the reaction of the 1β-sulfoxide with 2-mercaptobenzothiazole. A 6α-bromine is removed by reduction. The benzhydryl ester is formed by *O*-alkylation of the carboxylic acid. The 1β-sulfoxide is formed by oxidation of the sulfide. The amino group of 6-aminopenicillanic acid (6-APA) is replaced by bromide via the diazonium salt. 6-Aminopenicillanic acid is formed by enzyme-mediated hydrolysis of the side chain amide of penicillin G. Penicillin G (benzylpenicillin) is produced by the fungus *Penicillium chrysogenum*.

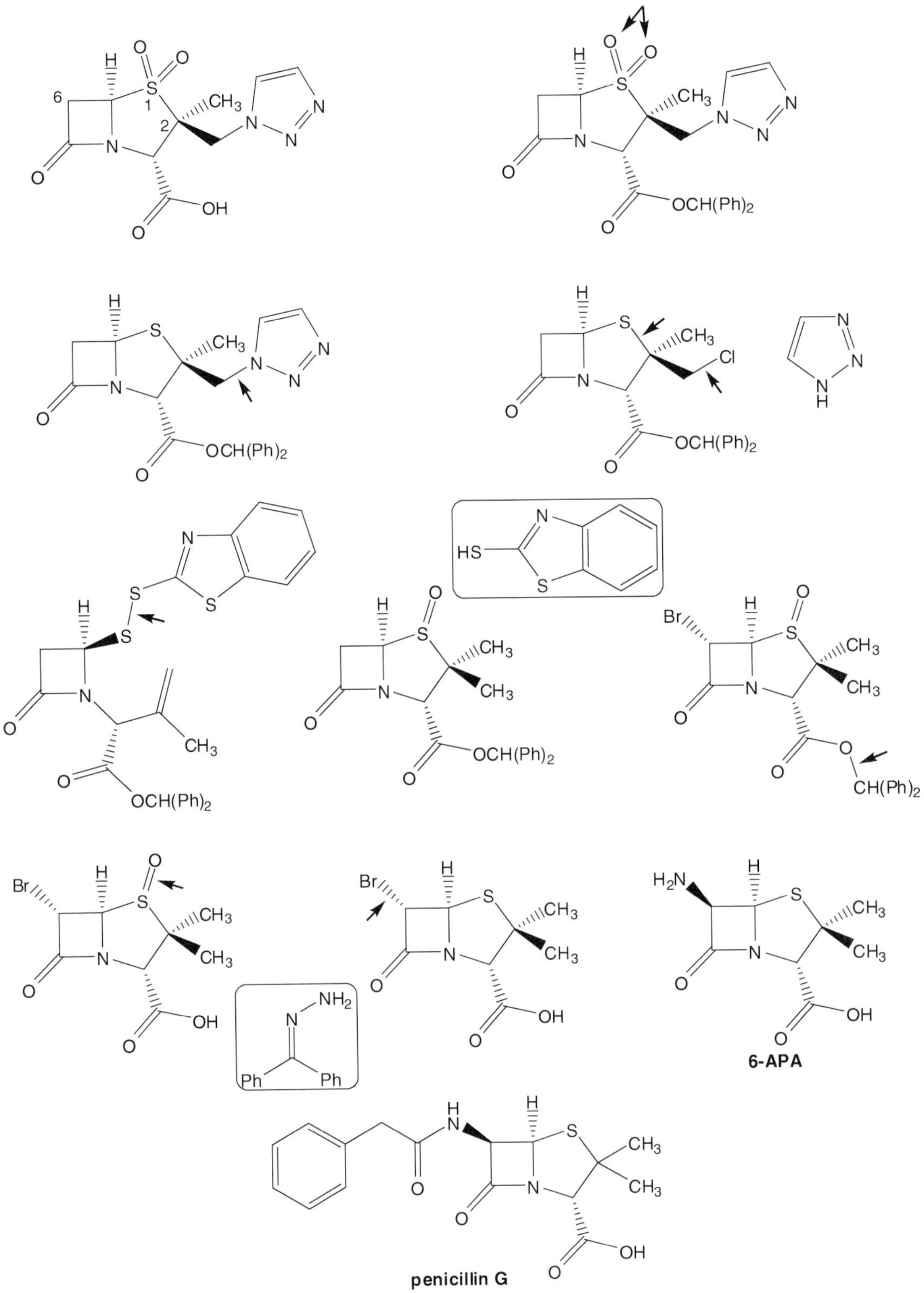

1,2,3-Triazole is released by hydrogenolysis of 1-benzyl-1,2,3-triazole. 1-Benzyl-1,2,3-triazole is formed by [3+2]-cycloaddition of benzyl azide and acetylene. Benzyl azide is formed from benzyl chloride.

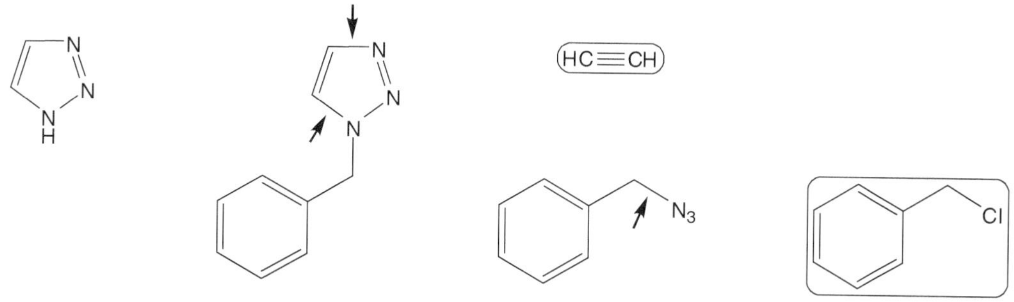

Extended Discussion

Draw the structures of the retrosynthetic analysis of an alternative route to tazobactam that does not use 1,2,3-triazole. List the pros and cons for both routes. Is one route preferred?

Tenofovir Disoproxil

Anti-infective Medicines/Antiviral Medicines/Antiretrovirals/Nucleoside or Nucleotide Reverse Transcriptase Inhibitors
Anti-infective Medicines/Antiviral Medicines/Antihepatitis Medicines/Medicines for Hepatitis B/Nucleoside or Nucleotide Reverse Transcriptase Inhibitors

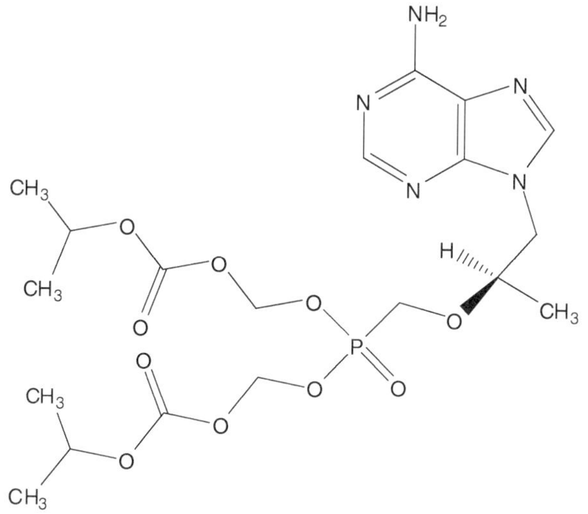

A nucleotide analog is often formed from the nucleoside analog (the phosphate is often introduced last). A nucleoside analog is often formed by displacement of a leaving group on the sugar by N9 of adenine.

Discussion. The phosphonate diester is formed by dialkylation of the phosphonic acid with chloromethyl isopropyl carbonate. The phosphonic acid is formed by dealkylation of the diethyl phosphonate. The ether C—O bond is formed by displacement of *para*-toluenesulfonate (tosylate) by the (*R*)-alcohol (**Williamson Ether Synthesis**). The (*R*)-alcohol is formed by the ring-opening of (*R*)-propylene carbonate with adenine. The tosylate is formed from diethyl hydroxymethylphosphonate and *para*-toluenesulfonyl chloride.

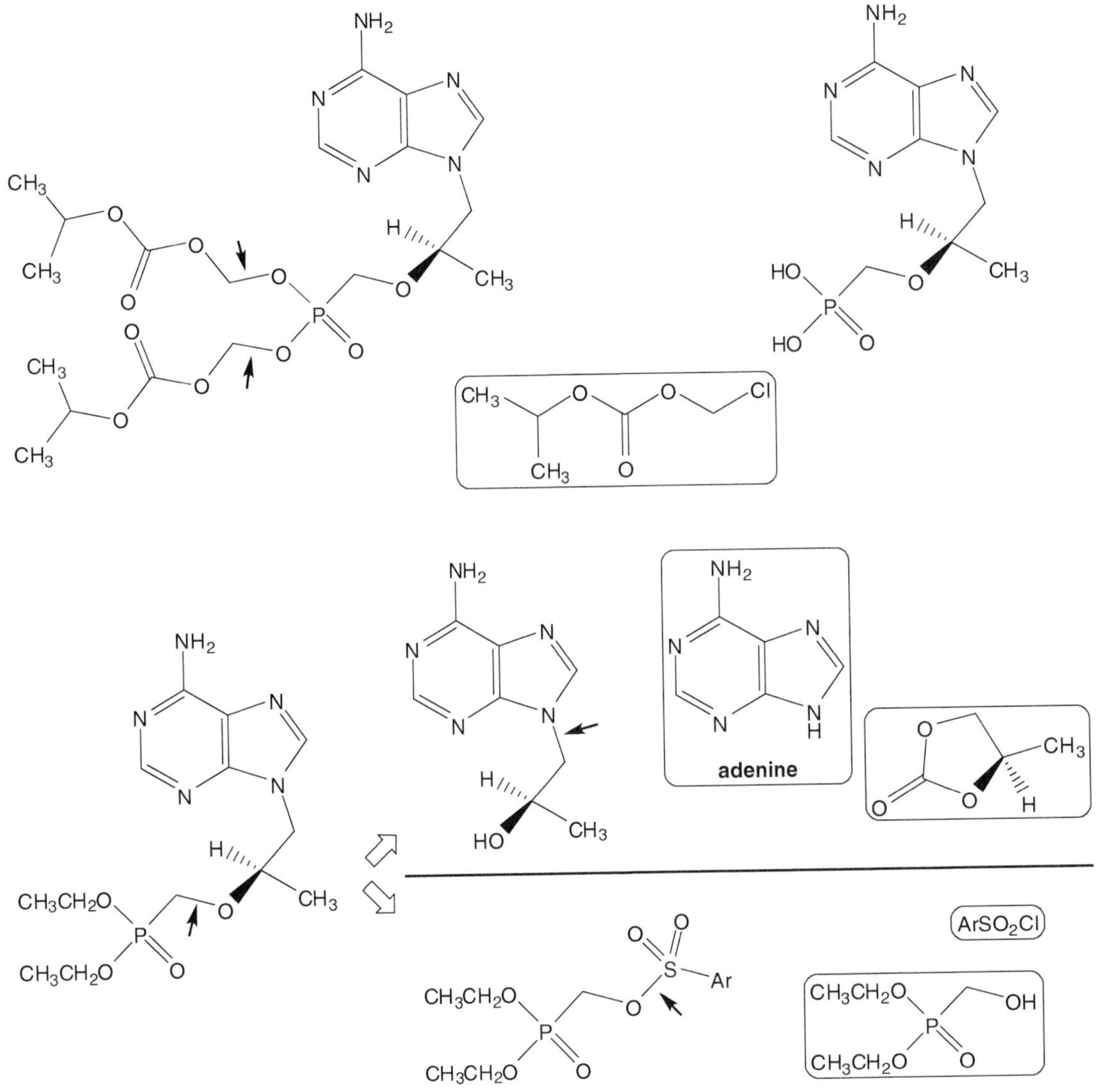

Extended Discussion

Draw the structures of a retrosynthetic analysis of one route to (*R*)-propylene carbonate.

Terbinafine

Dermatological Medicines (Topical)/Antifungal Medicines

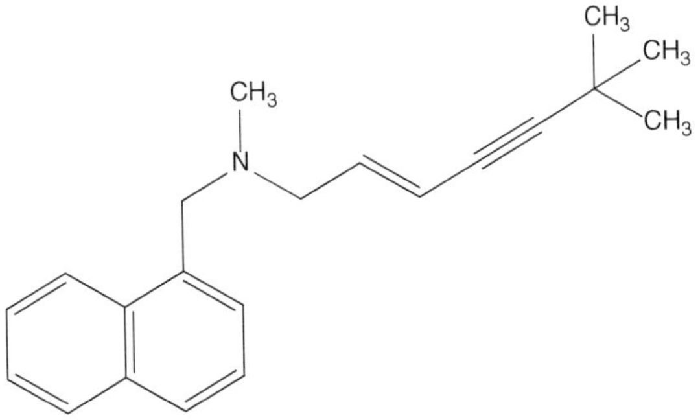

A sp^2C—spC bond is often formed by a transition metal-catalyzed coupling of an aryl or vinyl halide (X = Cl, Br, I) with a terminal alkyne. The (*E*/*Z*)-stereochemistry of the vinyl halide is usually retained in the product.

Discussion. The sp^{2}C—spC bond is formed in the final step by coupling of the vinyl chloride with 3,3-dimethyl-1-butyne (**Sonogashira Coupling**). The tertiary amine is formed by displacement of the allylic chloride of 1,3-dichloropropene by a secondary amine. The secondary amine, *N*-methyl-1-naphthylmethylamine, is formed by chloride displacement from 1-chloromethylnaphthalene by methylamine. (Terbinafine is the *trans*-alkene. 1,3-Dichloropropene is manufactured as a mixture of *cis*- and *trans*-alkenes. At what point(s) in the process are the *cis*- and *trans*-diastereomers separated? Provide details for the separation process(es).)

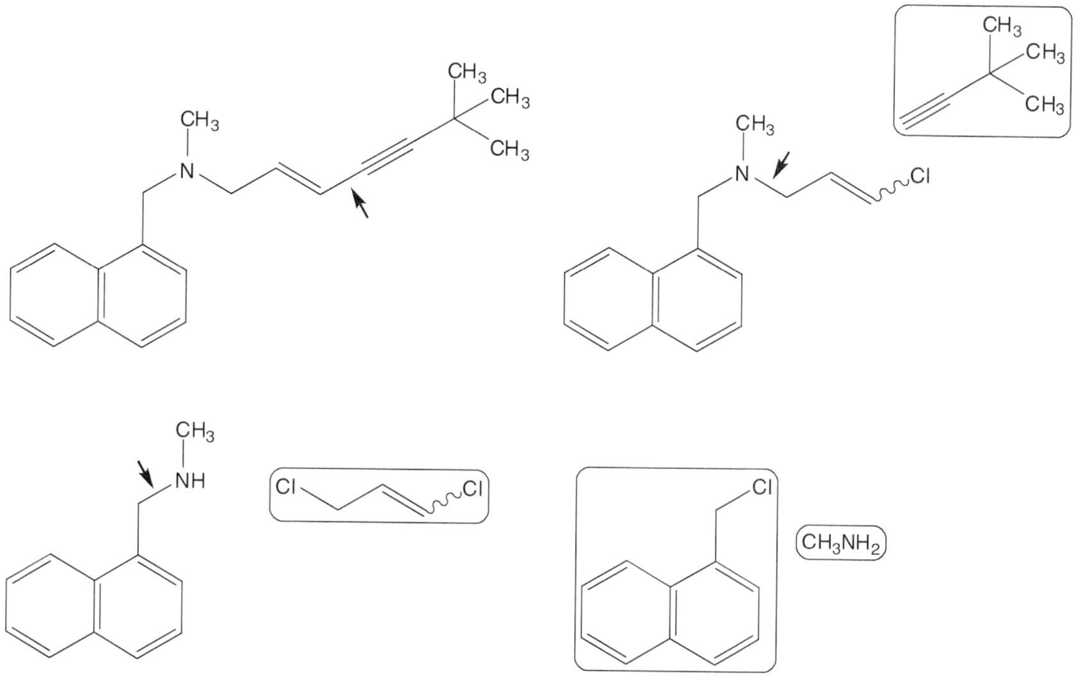

Extended Discussion

Draw the structures of the retrosynthetic analysis of an alternative route to terbinafine where the C-C bond is not formed by a transition metal-catalyzed coupling. Include the structures of the retrosynthetic analysis of any organic starting material(s) from petrochemical or biochemical raw materials. List the pros and cons for both routes and select one route as the preferred route.

Testosterone

Hormones, Other Endocrine Medicines and Contraceptives/Androgens

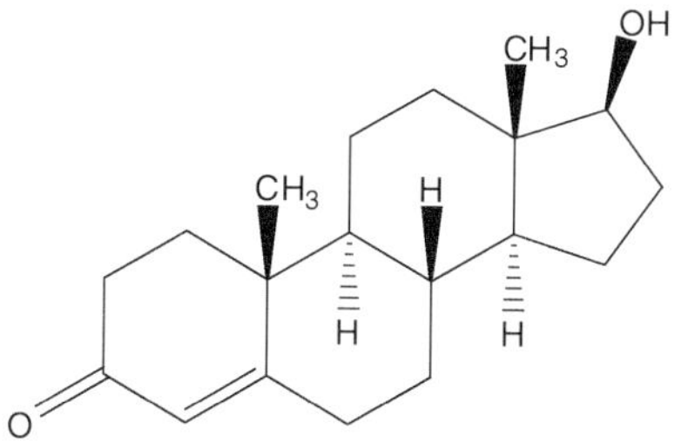

A single-enantiomer molecule with multiple chiral carbons is often formed by the modification of a natural product which has most or all of the chiral carbons already in place. A steroid 4-ene-3-one is often formed from the 5-ene-3-ol by oxidation of the alcohol and double bond migration into conjugation with the ketone.

Discussion. Testosterone is manufactured in three steps from androst-4-ene-3,17-dione and four steps from a phytosterol. In the final step, the 4-ene-3-one is released from the 3-ethoxy-3,5-diene by hydrolysis of the enol ether and double bond migration. The 17β-alcohol is formed by reduction of the C17 ketone. The 3-ethoxy-3,5-diene is formed by reaction of androst-4-ene-3,17-dione with triethyl orthoformate. Androst-4-ene-3,17-dione is formed by microbial oxidation/side chain degradation of phytosterols (plant sterols) including stigmasterol.

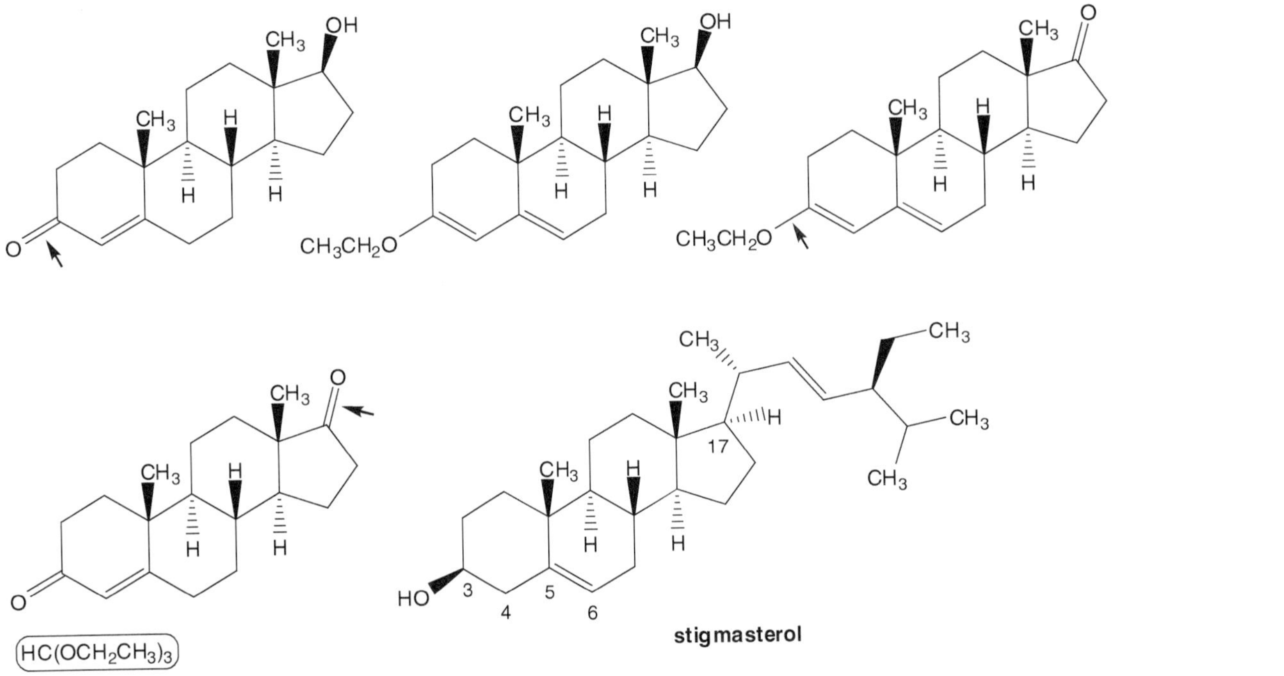

Extended Discussion

Diosgenin is a phytosteroid sapogenin isolated from the tubers of *Discorea* wild yam. Draw structures of the retrosynthetic analysis of an alternative route to testosterone from diosgenin via dehydroepiandrosterone acetate.

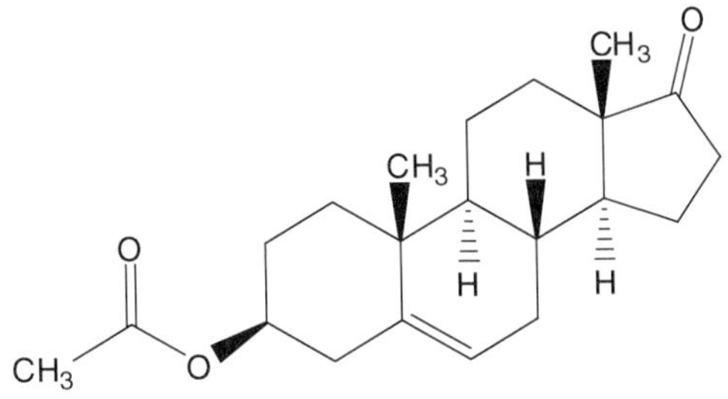

dehydroepiandrosterone acetate

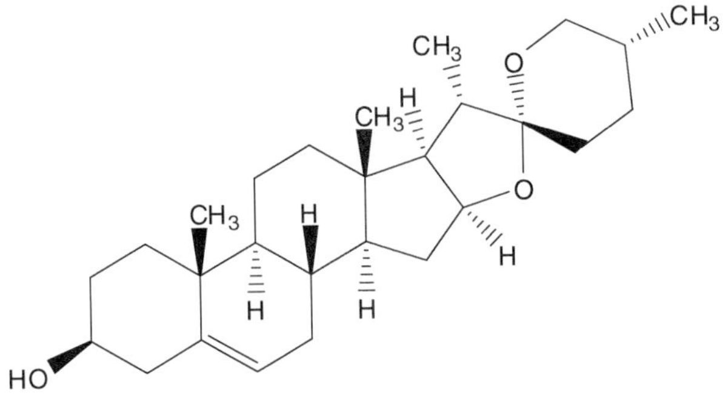

diosgenin

Tetracaine

Ophthalmological Preparations/Local Anesthetics

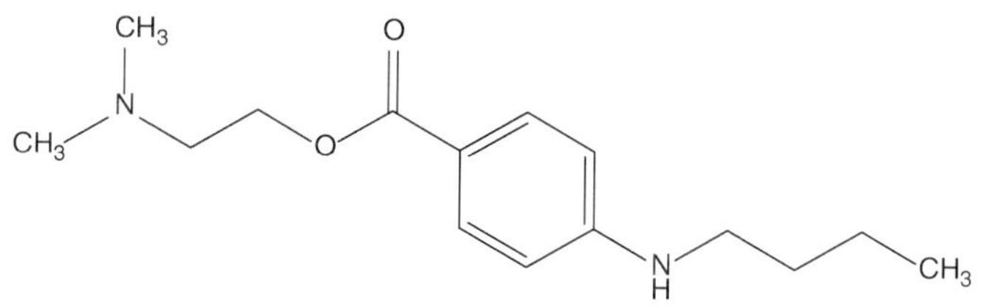

A secondary amine is often formed by reductive amination from a primary amine and an aldehyde or ketone.

Discussion. Tetracaine is formed by transesterification from the ethyl ester and 2-dimethylaminoethanol. The ethyl ester is formed from the carboxylic acid (**Fischer Esterification**). 4-(Butylamino)benzoic acid is formed by reductive amination from 4-aminobenzoic acid and butanal.

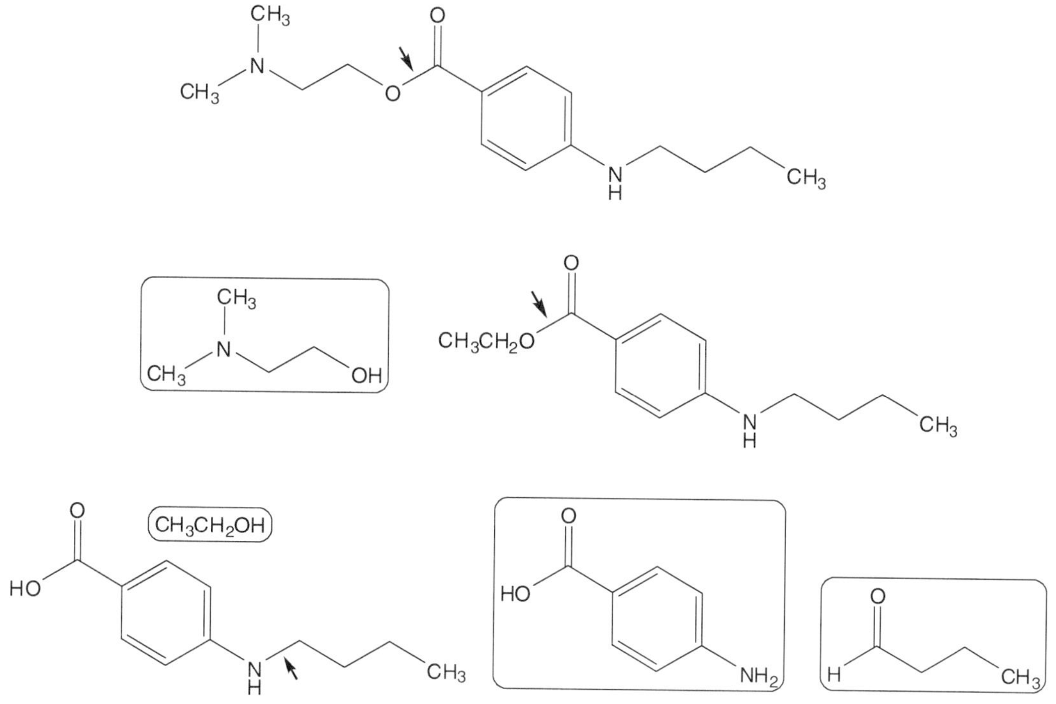

Extended Discussion

Draw the structures of the retrosynthetic analysis of one alternative route to ethyl 4-(butylamino)benzoate. List the pros and cons for both routes to ethyl 4-(butylamino)benzoate. Is one route preferred?

Tetracycline

Opthalmological Preparations/Anti-infective Agents

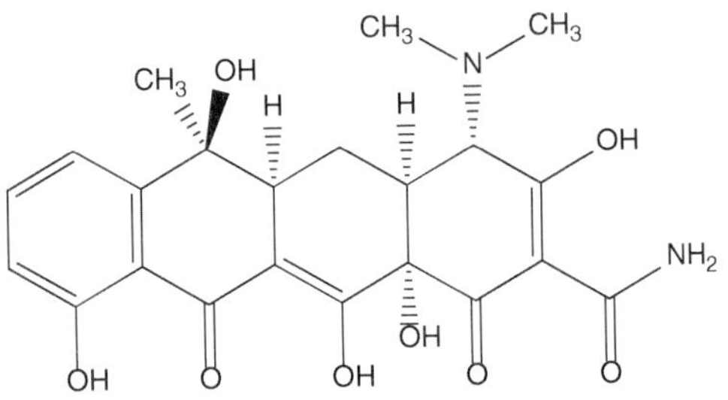

Many tetracyclines are semisynthetic, they are formed by modification of the C and/or D rings of a tetracycline natural product (tetracycline, chlorotetracycline, oxytetracycline) or a tetracycline produced by a mutant strain of *S. aureofaciens* (6-demethyltetracycline, 7-chloro-6-d emethyltetracycline, 7-bromotetracycline).

Discussion. Tetracycline is manufactured by fermentation or by hydrogenolysis of the chlorine of chlorotetracycline.

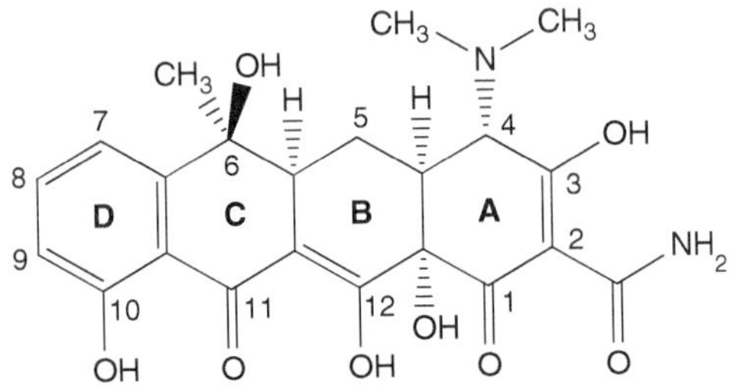

tetracycline

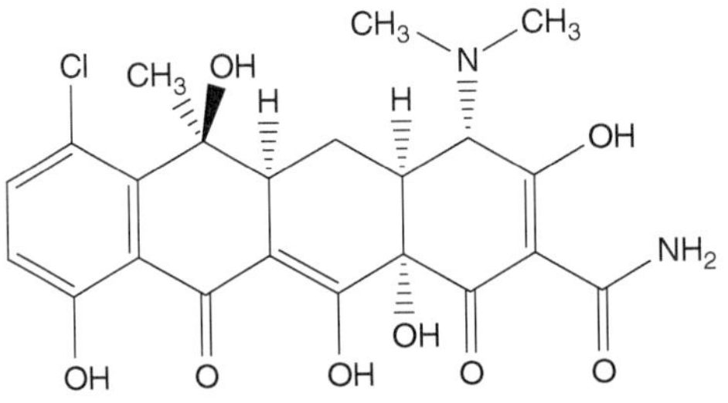

chlorotetracyclin

Thiamine

Vitamins and Minerals

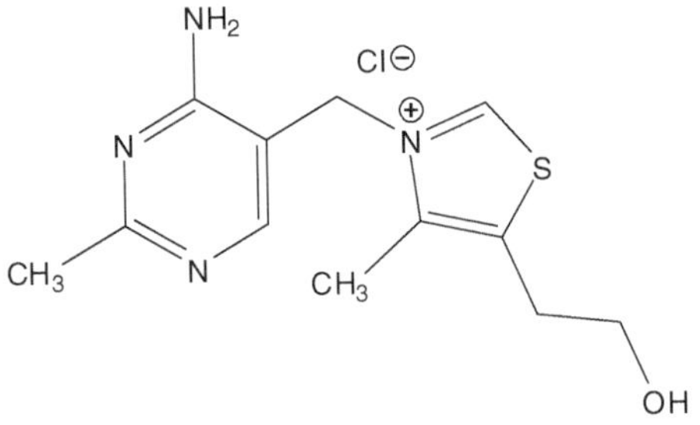

A thiazole is often formed by condensation of a thiourea or thioamide with an α−haloketone.

Discussion. Highly crystalline thiothiamine is oxidized to thiamine in the final step. The reaction of the dithiocarbamate with aqueous acid forms the 3H-thiazole-2-thione and releases the side chain alcohol by acetate ester hydrolysis (**Hantzsch Thiazole Synthesis**). A dithiocarbamate salt formed in situ from 4-amino-5-aminomethyl-2-methylpyrimidine (**Grewe diamine**) and carbon disulfide displaces chloride from an α-chloroketone to form the dithiocarbamate.

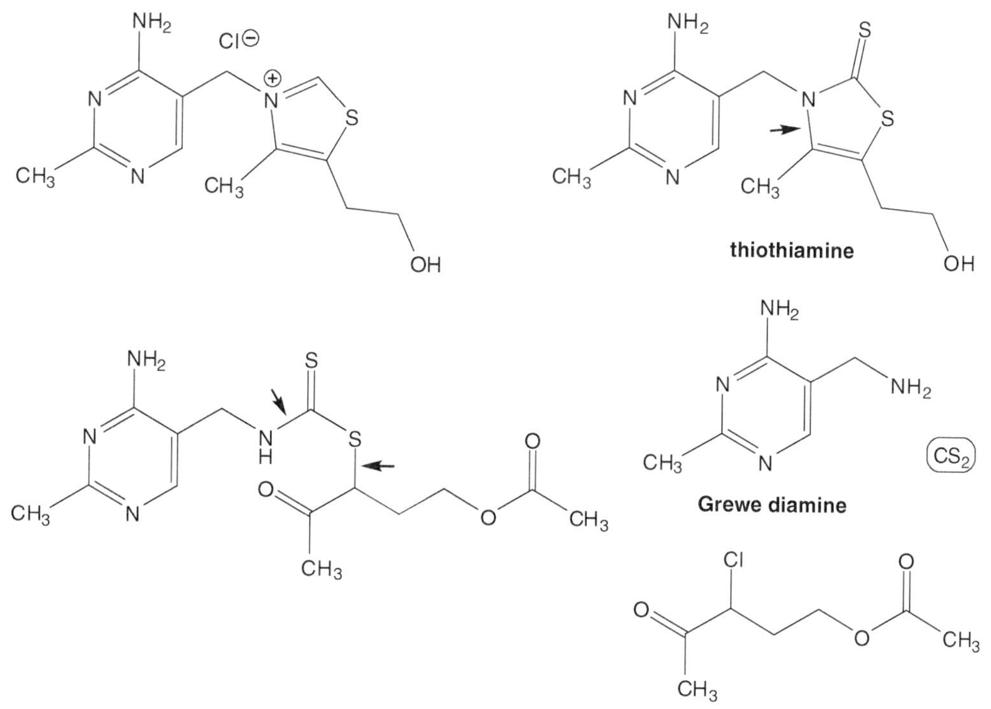

The side chain amine of Grewe diamine is released by hydrolysis of the formamide. The pyrimidine ring is formed from acetamidine and the 3-anilinoacrylonitrile.

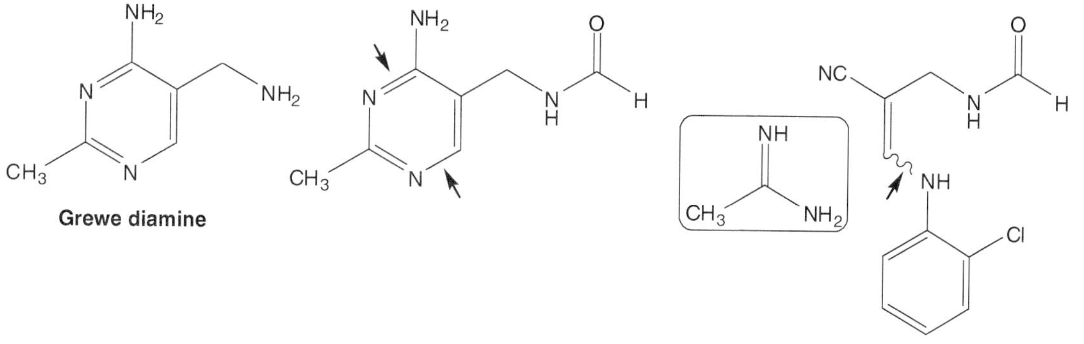

The 3-anilinoacrylonitrile is formed from α–formyl-β-formylaminopropionitrile sodium salt and 2-chloroaniline. α-Formyl-β-formylaminopropionitrile sodium salt is formed from 3-aminopropanenitrile, methyl formate, and carbon monoxide. 3-Aminopropanenitrile is formed from acrylonitrile.

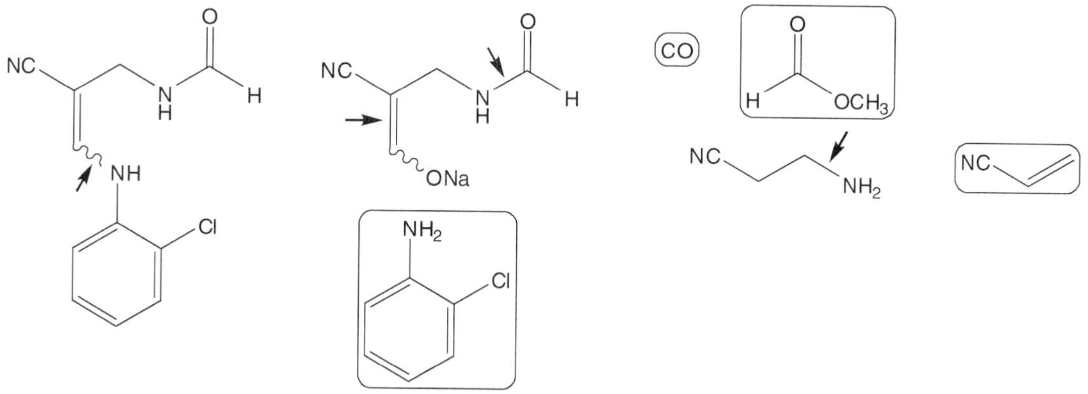

The alpha-chloroketone, 5-acetoxy-3-chloro-2-pentanone, is formed from 2-acetyl-2-chlorobutyrolactone and acetic anhydride by hydrolysis, decarboxylation, and acetylation. The α-chlorolactone is formed by chlorination of α-acetyl-γ-butyrolactone.

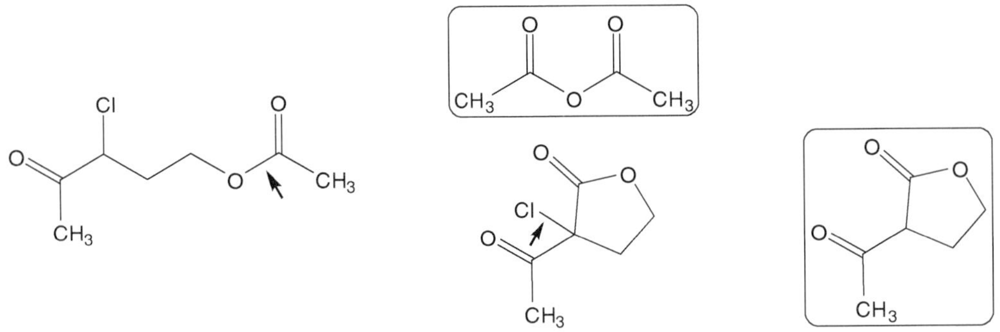

Extended Discussion

Draw the structures of the retrosynthetic analysis of one alternative route to thiamine from Grewe diamine. List the pros and cons for both routes. Is one route preferred?

Tigecycline

Anti-infective Medicines/Antibacterials/Other Antibacterials

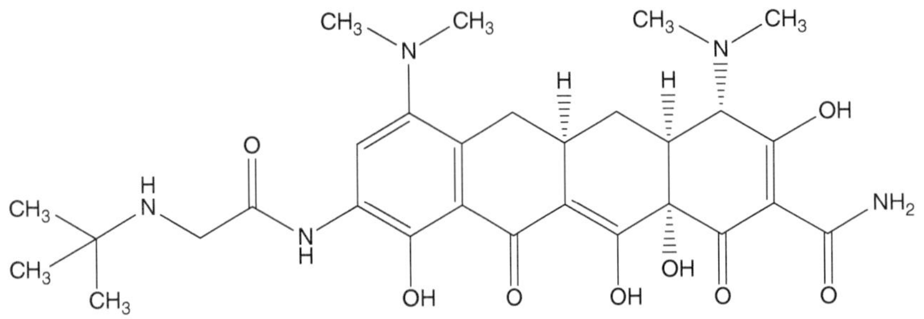

Many tetracyclines are semisynthetic, they are formed by modification of the C and/or D rings of a tetracycline natural product (tetracycline, chlorotetracycline, oxytetracycline) or a tetracycline produced by a mutant strain of *Streptomyces aureofaciens* (6-demethyltetracycline, 7-chloro-6-demethyltetracycline, 7-bromotetracycline).

Discussion. Tigecycline is manufactured from 7-chloro-6-demethyltetracycline. The amide of the side chain at position 9 is formed from the acid chloride and 9-aminominocycline in the final step. 9-Aminominocycline is formed by reduction of the 9-nitrominocycline. 9-Nitrominocycline is formed by nitration of minocycline.

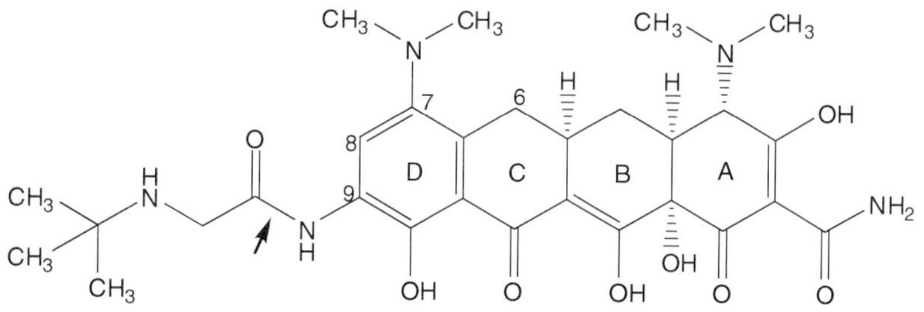

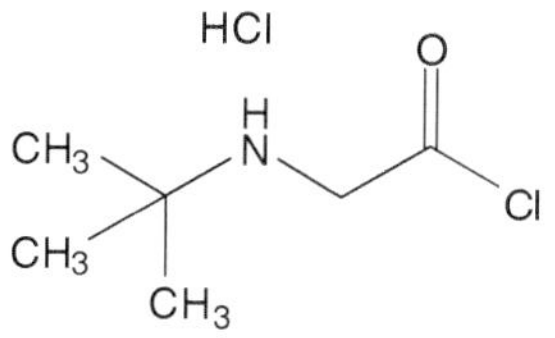

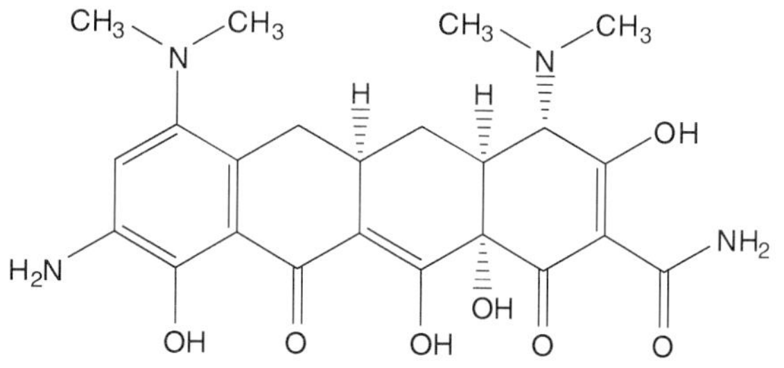

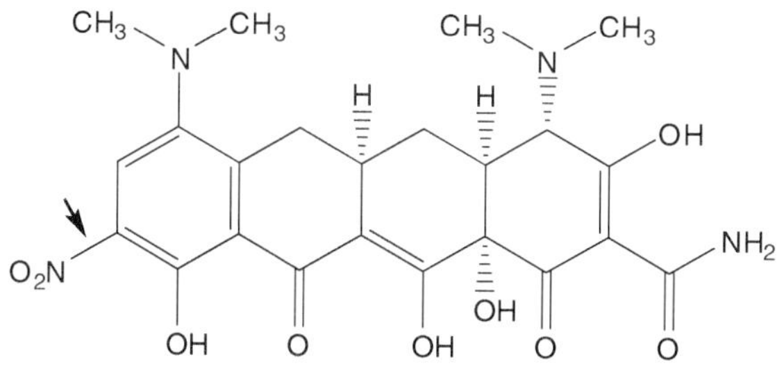

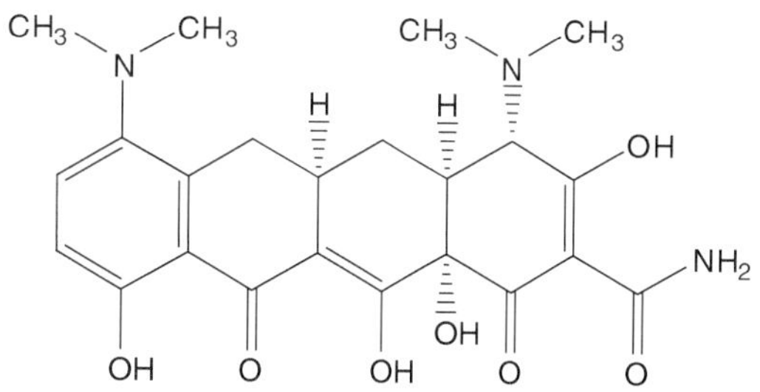

minocycline

The acid chloride is formed from the carboxylic acid, *N-tert*-butylglycine. The carboxylic acid is formed by hydrolysis of the *tert*-butyl ester. The *tert*-butyl ester is formed from *tert*-butyl bromoacetate and *tert*-butylamine.

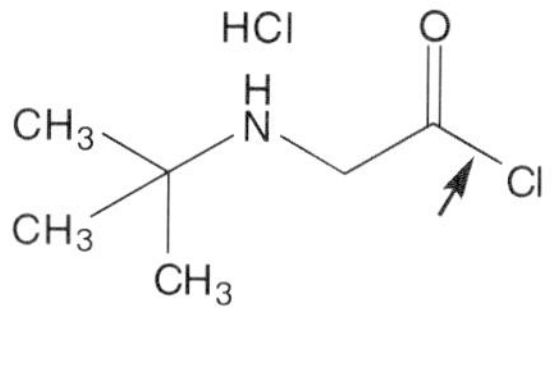

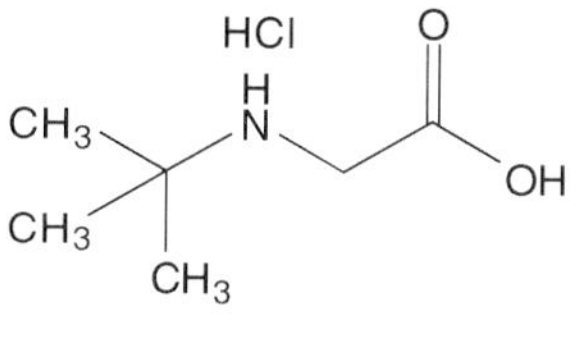

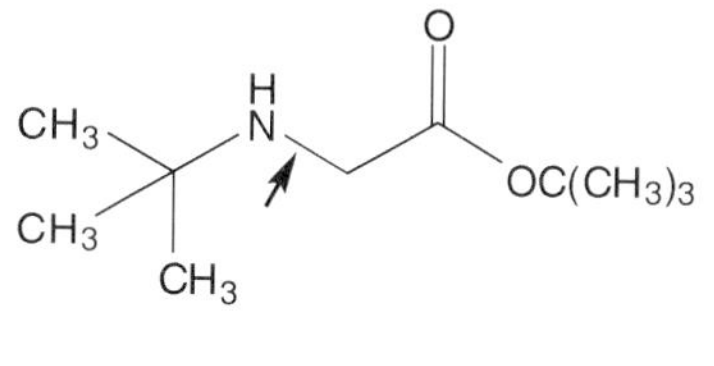

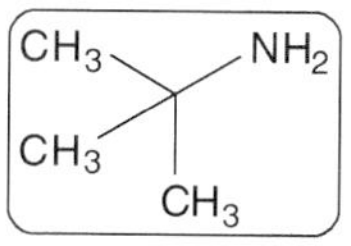

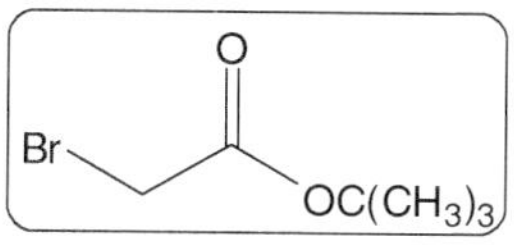

Minocycline is formed by reductive alkylation of 7-nitrosancycline with formaldehyde. 7-Nitrosancyclin is separated from the mixture of 7-nitrosancycline and 9-nitrosancycline formed by nitration of sancycline. (What is the highest ratio of 7-nitrosancycline to 9-nitrosancycline? What reaction conditions are associated with the highest ratio? How is 7-nitrosancycline separated from the 9-nitrosancycline?)

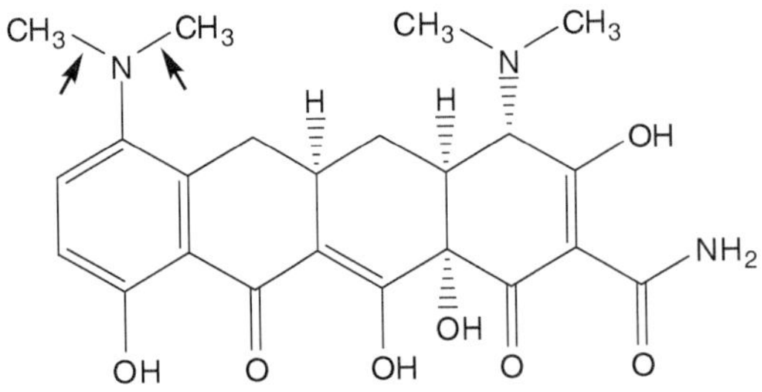

minocycline

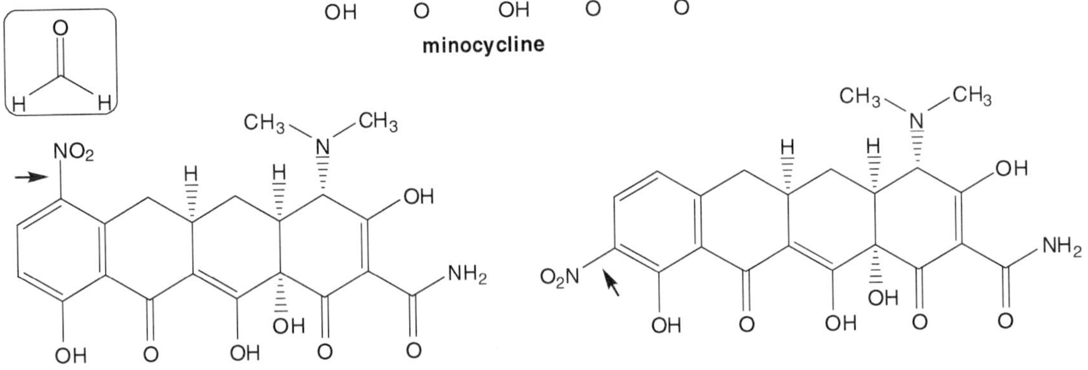

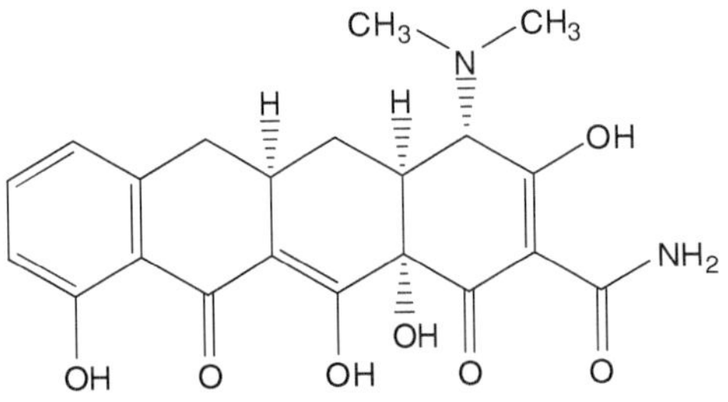

sancycline

Minocycline is also formed in four steps from 9-nitrosancycline. 9-Nitrosancycline is reduced to 9-aminosancycline. 9-Amino-7-nitrosancycline is formed by nitration of 9-aminosancycline. 9-Amino-7-nitrosancycline is converted to the diazonium salt. The diazonium salt is then carried into the reductive alkylation with formaldehyde to form minocycline.

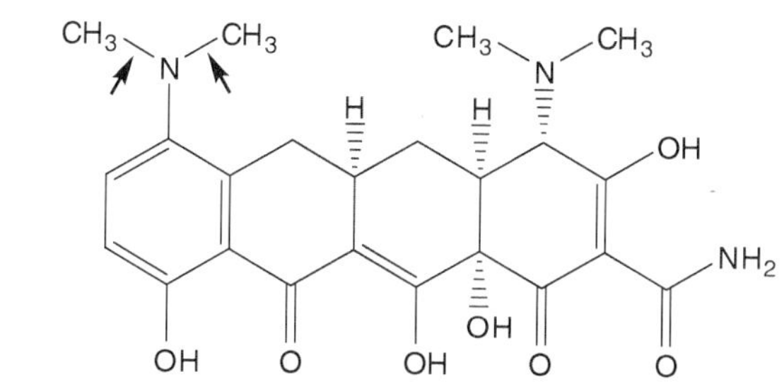

minocycline

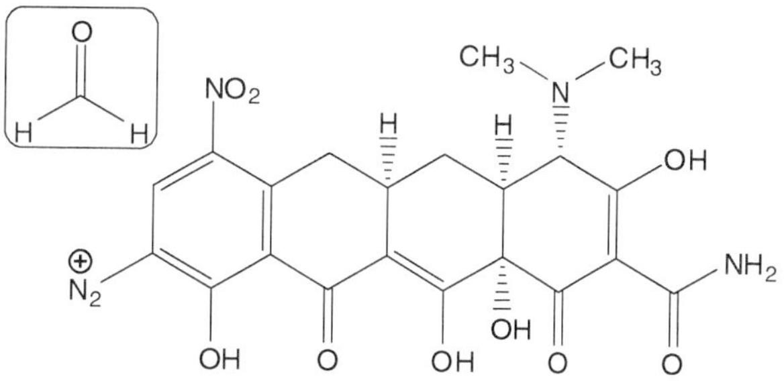

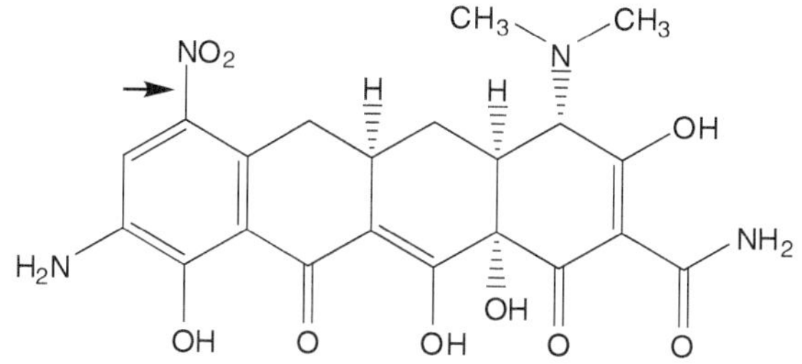

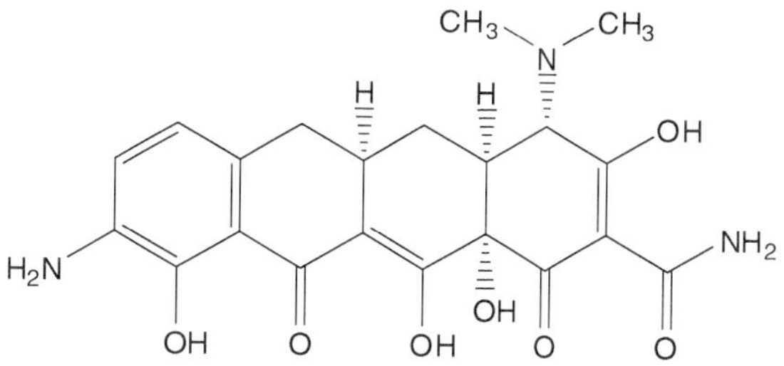

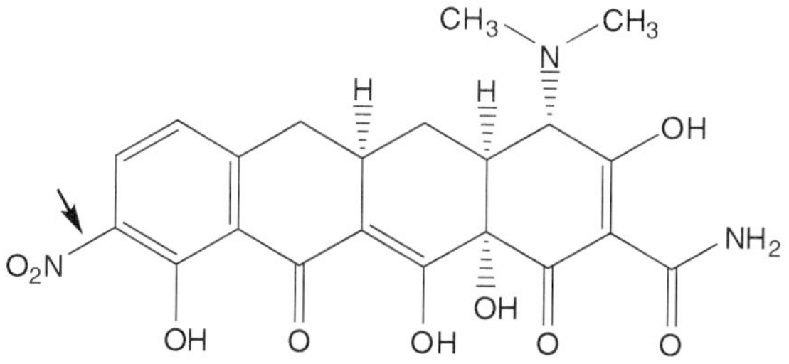

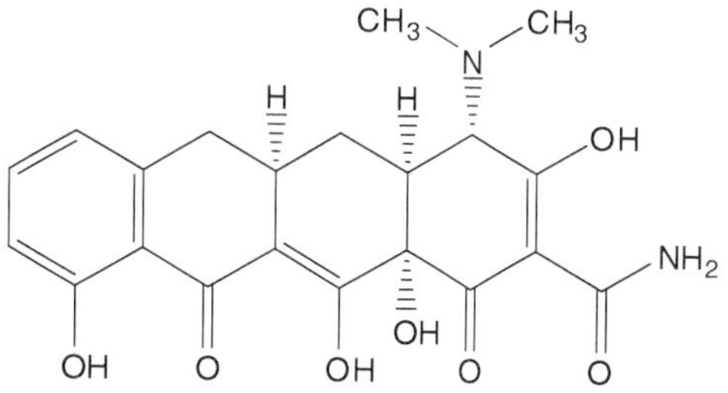

sancycline

Sancycline is formed by hydrogenolysis of the ring chlorine and the benzylic oxygen of 7-chloro-6-demethyltetracycline (demeclocycline).

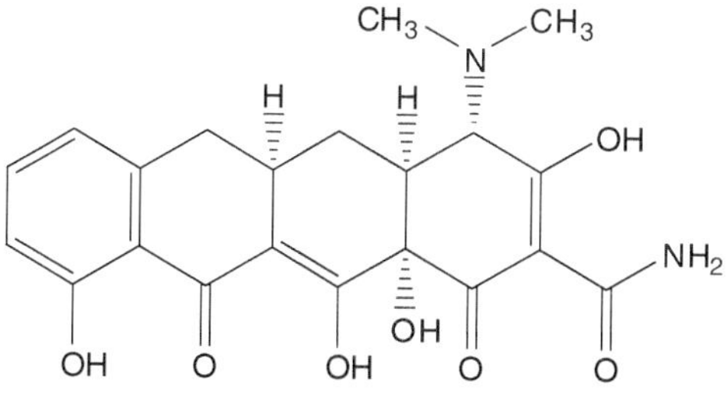

sancycline

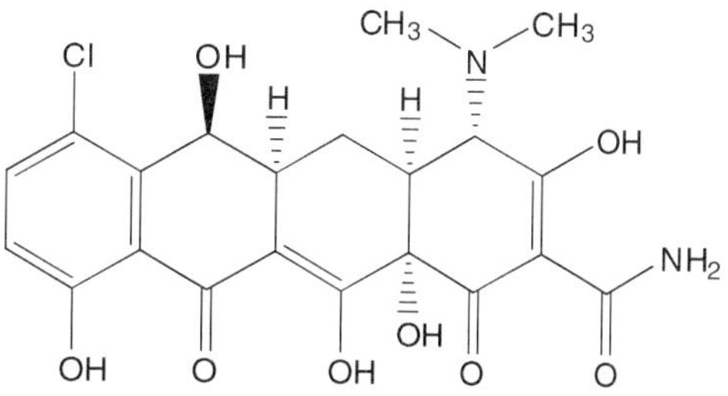

demeclocycline

Extended Discussion

List the pros and cons for manufacturing tigecycline via 7-nitrosancycline, 9-nitrosancycline, or via both 7-nitrosancycline and 9-nitrosancycline.

Timolol

Ophthalmological Preparations/Miotic and Antiglaucoma Medicines

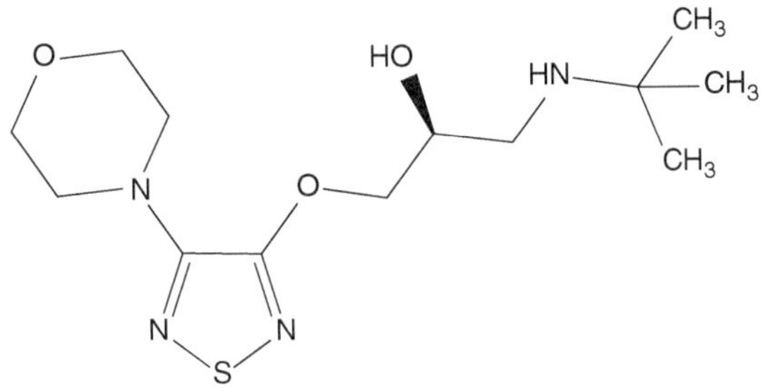

A 3-amino or 3-alkoxy-1,2,5-thiadiazole is often formed by chloride displacement by an amine or alcohol.

Discussion. The alcohol and secondary amine are released by hydrolysis of an oxazolidine in the final step. The thiadiazole ether substituent is introduced by chloride displacement from 3-chloro-4-morpholino-1,2,5-thiadiazole by the oxazolidine-5-methanol. 3-Chloro-4-morpholino-1,2,5-thiadiazole is formed by chloride displacement from 3,4-dichloro-1,2,5-thiadiazole by morpholine. 3,4-Dichloro-1,2,5-thiadiazole is formed from cyanogen and sulfur dichloride.

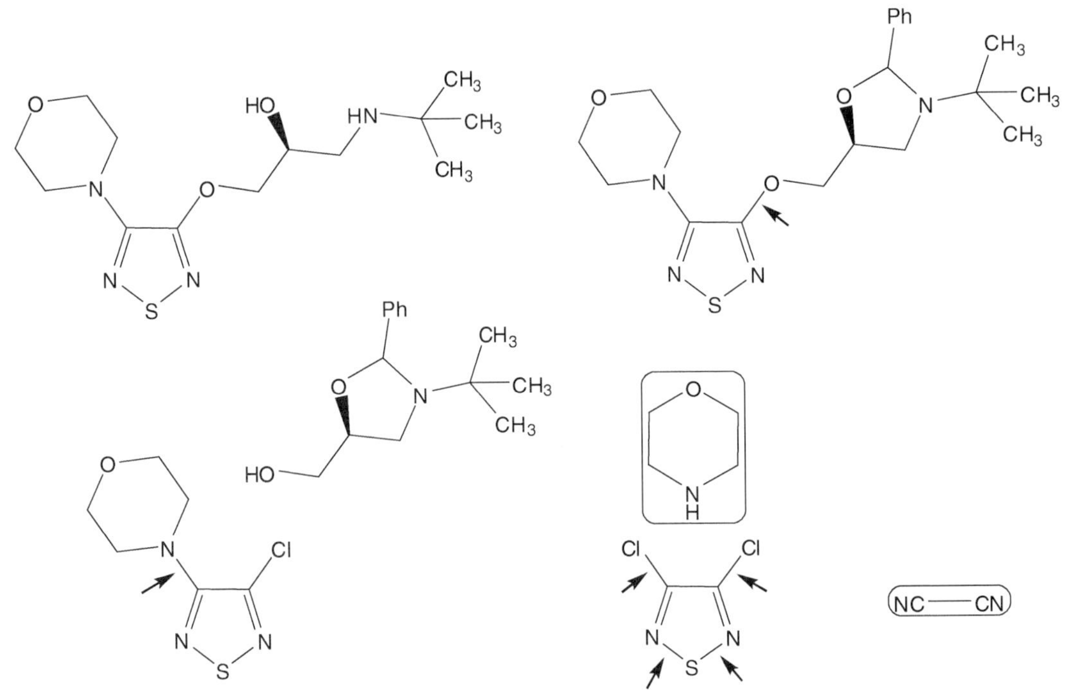

(5*S*)-3-(*tert*-Butyl)-2-phenyloxazolidine-5-methanol is formed by the reaction of (*S*)-3-*tert*-butylamino-1,2-propanediol with benzaldehyde. (*S*)-3-*tert*-Butylamino-1,2-propanediol is separated from the racemic mixture by resolution. 3-*tert*-But ylamino-1,2-propanediol is formed by the ring-opening of glycidol by *tert*-butylamine.

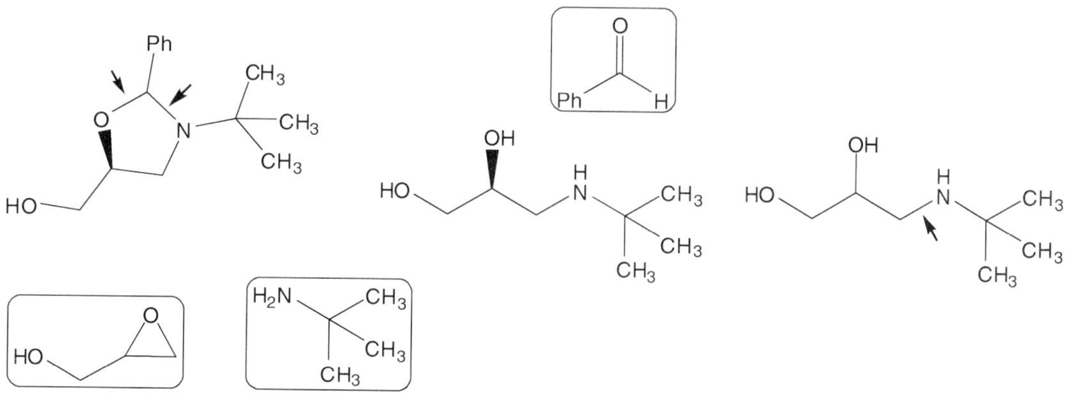

Extended Discussion

Timolol is unstable in the presence of a strong base. Draw structures for two products formed in the reaction of timolol with potassium *tert*-butoxide in *tert*-butanol at 25 °C (**Smiles Rearrangement**).

Tranexamic Acid

Medicines Affecting the Blood/Medicines Affecting Coagulation

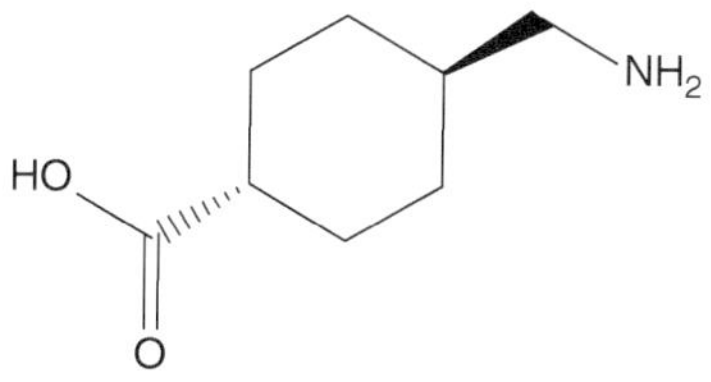

A cyclohexane is often formed by hydrogenation of a carbocyclic aromatic.

Discussion. Hydrogenation of 4-aminomethylbenzoic acid forms a mixture of *cis*-4-aminomethylcyclohexanecarboxylic acid *and trans*-4-aminomethylcyclohexanecarboxylic acid (tranexamic acid). (List the options for hydrogenation catalysts and conditions and the ratio of *cis*-product to *trans*-product associated with each option.) The *cis*-isomer is then converted to the more stable *trans*-isomer and the *trans*-isomer is separated from the mixture by crystallization. 4-Aminomethylbenzoic acid is formed by hydrogenation of 4-cyanobenzoic acid. 4-Cyanobenzoic acid is formed by the oxidation of 4-tolunitrile.

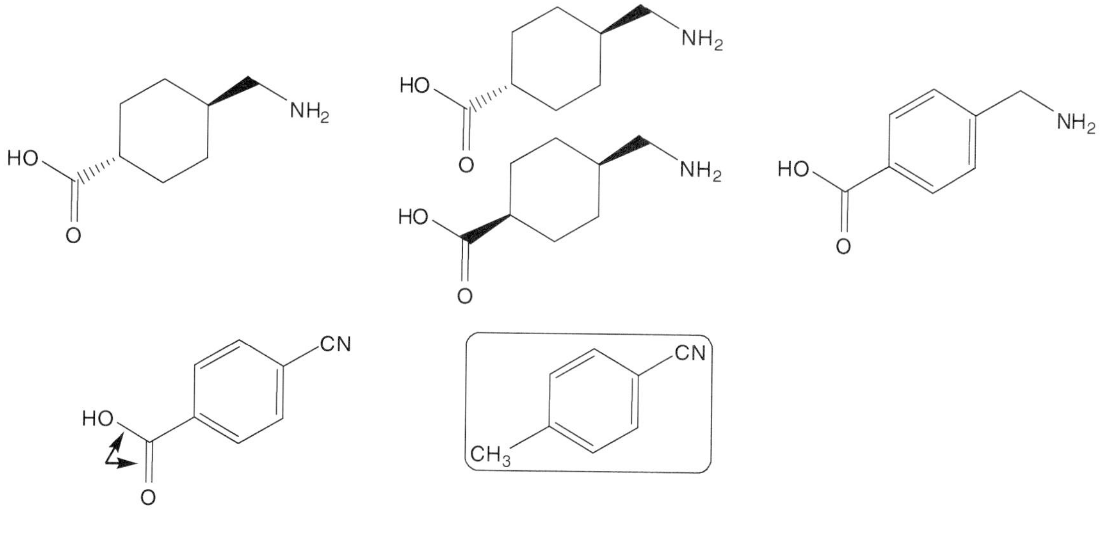

Extended Discussion

Draw the structures of a retrosynthetic analysis of one alternative route to tranexamic acid from ethyl 4-oxocyclohexanecarboxylate. Include the structures of a retrosynthetic analysis of ethyl 4-oxocyclohexanecarboxylate from the organic starting material(s) available from petrochemical or biochemical raw materials.

Triclabendazole

Anti-infective Medicines/Anthelminthics/Antischistosomals and Other Antitrematode Medicines

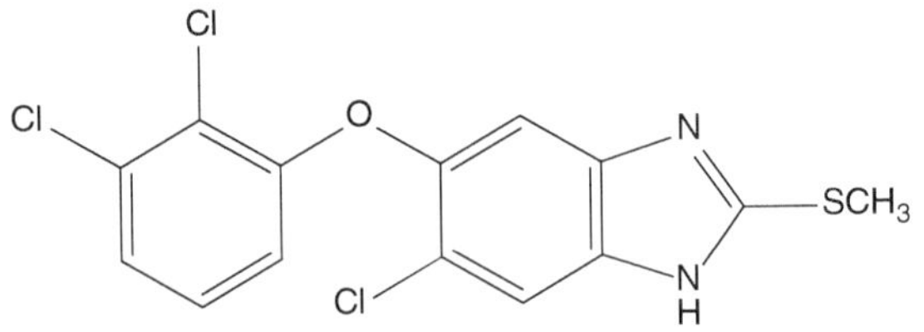

A diaryl ether is often formed by the displacement of an aryl chloride by a phenol.

Discussion. The thioether of triclabenzazole is formed in the final step by *S*-methylation with dimethyl sulfate. The 1,3-dihydro -2H-benzimidazole-2-thione is formed from the benzene-1,2-diamine and carbon disulfide. The benzene-1,2-diamine is formed by the reduction of the 2-nitroaniline. 4-Chloro-5-(2,3-dichlorophenoxy)-2-nitroaniline is formed from 4,5-dichloro-2-nitroaniline by chloride displacement by 2,3-dichlorophenol. 2,3-Dichlorophenol is formed from 1,2,3-trichlorobenzene by chloride displacement by hydroxide. 4,5-Dichloro-2-nitroaniline is formed from 1,2,4-trichloro-5-nitrobenzene by chloride displacement by ammonia. 1,2,4-trichloro-5-nitrobenzene is formed by nitration of 1,2,4-trichlorobenzene.

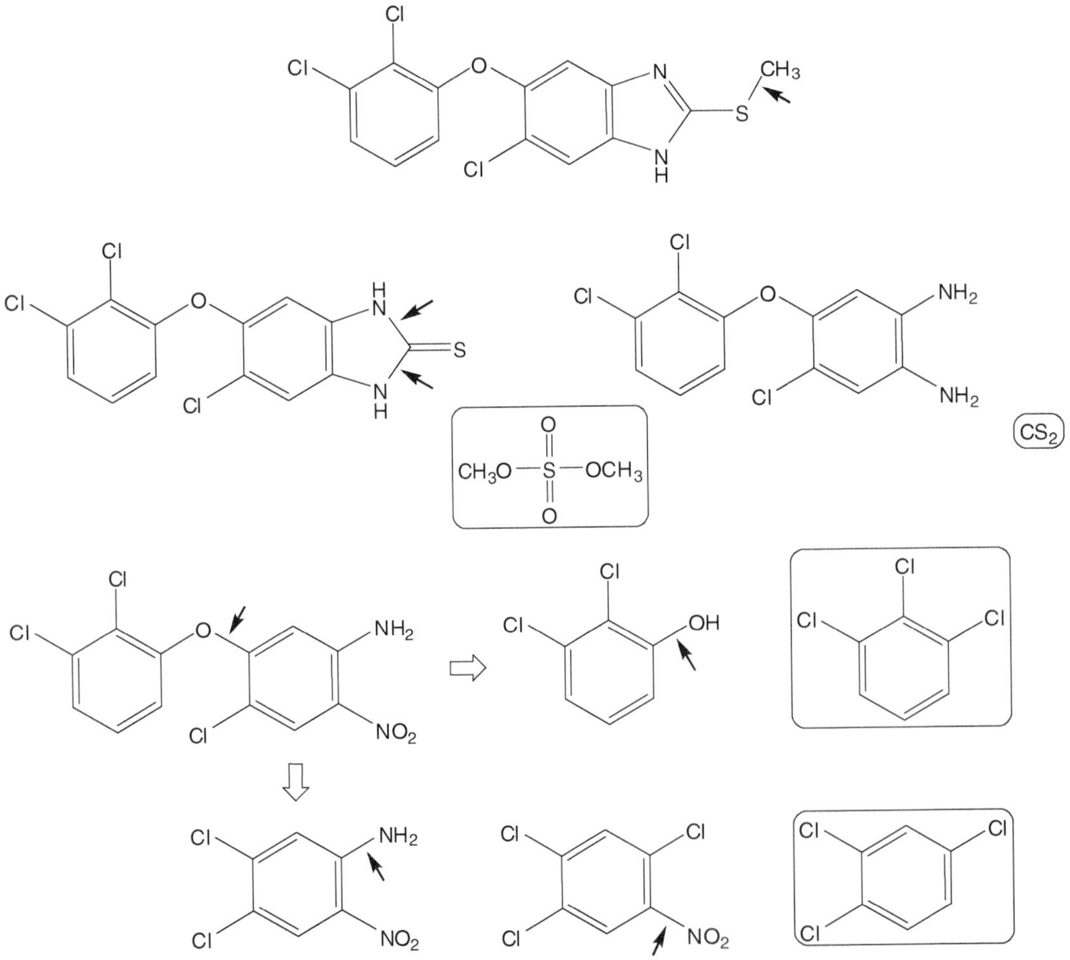

Extended Discussion

Draw the structures of the retrosynthetic analysis of one alternative route to 2,3-dichlorophenol. Include the structures of the retrosynthetic analysis of any organic starting material(s) from petrochemical or biochemical raw materials. List the pros and cons for both routes to 2,3-dichlorophenol and select one route as the preferred route.

Trimethoprim

Anti-infective Medicines/Antibacterials/Other Antibacterials

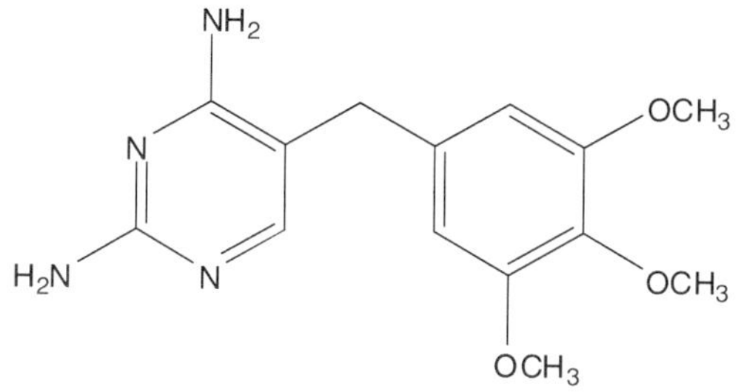

An amidine, thiourea, or urea often provides the carbon at position 2 of a pyrimidine.

Discussion. The pyrimidine ring is formed in the final step by the reaction of guanidine with a 3-anilinoacrylonitrile. 3-Anilino-2-(3,4,5-trimethoxybenzyl)acrylonitrile is formed from 3,4,5-trimethoxybenzaldehyde and 3-anilinopropanenitrile (**Knoevenagel Condensation**).

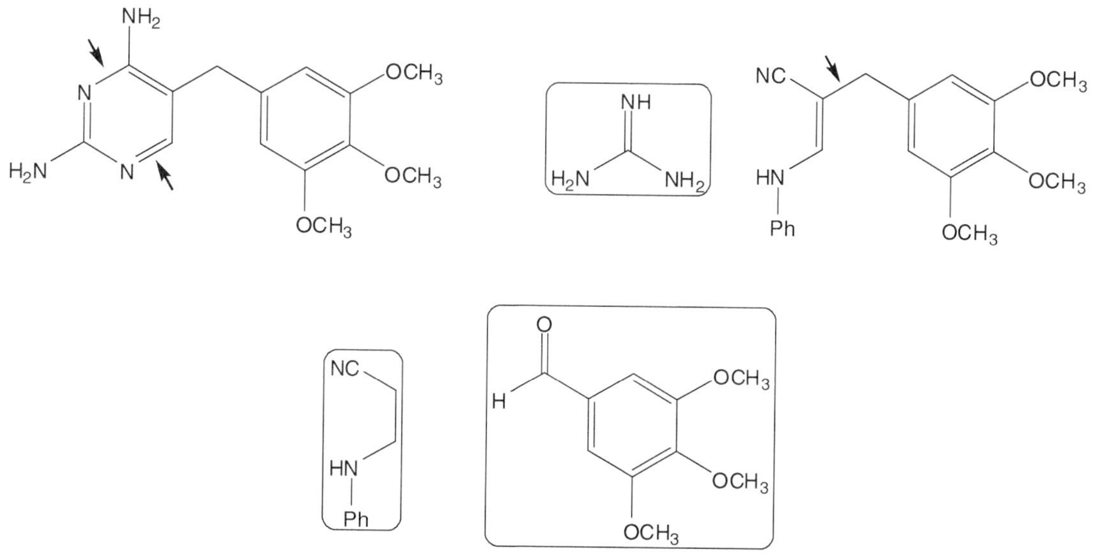

Extended Discussion

Draw the structures of the retrosynthetic analysis of one alternative route to trimethoprim. Include the structures of the retrosynthetic analysis of any organic starting material(s) from petrochemical or biochemical raw materials. List the pros and cons for both routes and select one route as the preferred route.

Tropicamide

Diagnostic Agents/Opthalmic Medicines

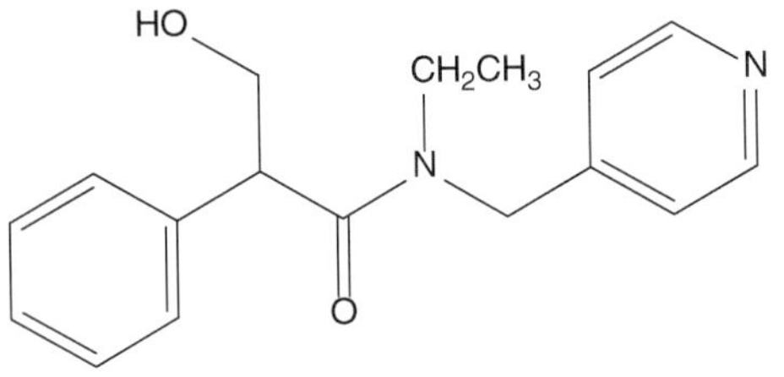

Amides are ubiquitous in drug structures. Amide formation from an amine and acid chloride, anhydride, ester, or carboxylic acid is often very efficient.

Discussion. Tropicamide is a 1 : 1 mixture of the (*R*)- and (*S*)-enantiomers. The primary alcohol is released by hydrolysis of the acetate ester in the final step. The amide is formed from the acid chloride and 4-(ethylaminomethyl)pyridine. The acid chloride is formed from the carboxylic acid. The acetate ester is formed from the primary alcohol and acetyl chloride and then carried directly into the acid chloride formation. Tropic acid is formed by hydrolysis of the ethyl ester. The α–hydroxymethyl ester is formed by the reduction of the α–formyl ester. The α–formyl ester is formed from ethyl phenyl acetate and ethyl formate (mixed **Claisen Condensation**).

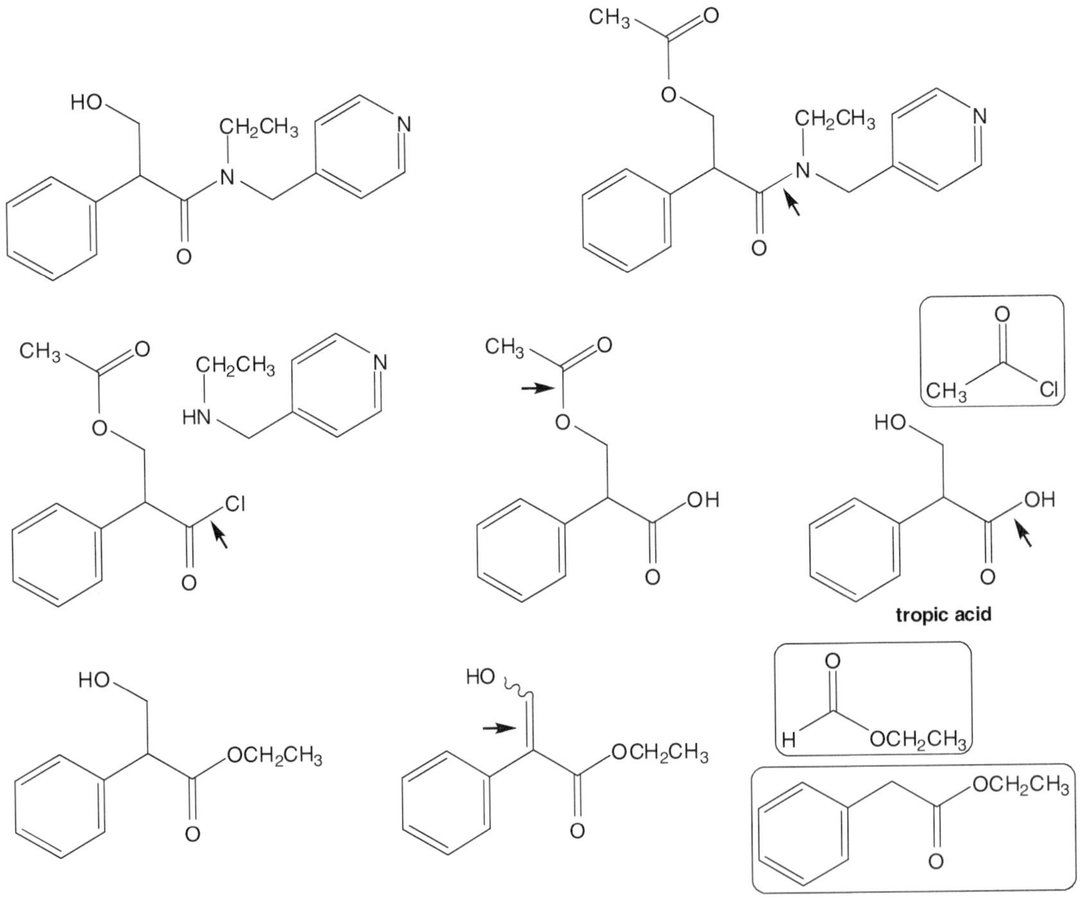

4-(Ethylaminomethyl)pyridine is formed by reduction of the imine. The imine is formed by condensation of 4-pyridinecarboxaldehyde and ethylamine.

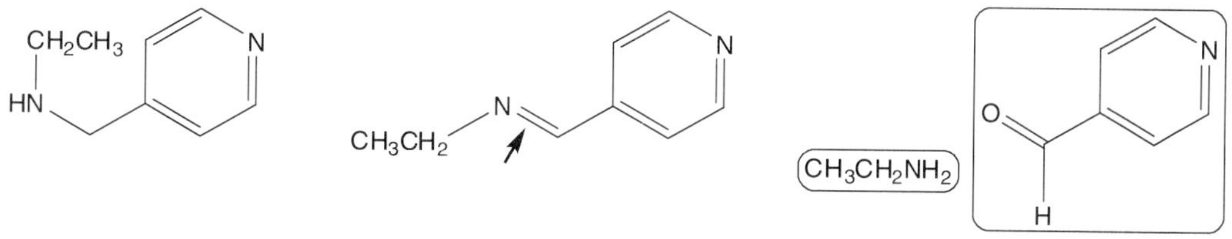

Extended Discussion

Draw the structures of the retrosynthetic analysis of one alternative route to tropicamide. List the pros and cons for both routes and select one route as the preferred route.

U

Ulipristal Acetate

Hormones, Other Endocrine Medicines, and Contraceptives/Contraceptives/Oral Hormonal Contraceptives

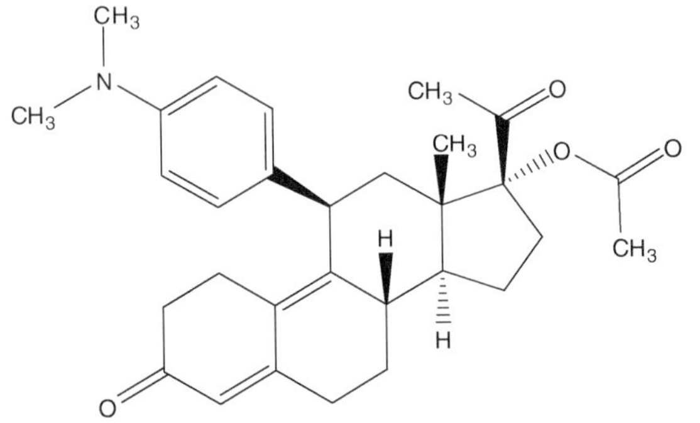

A single-enantiomer molecule with multiple chiral carbons is often formed by the modification of a natural product which has most or all of the chiral carbons already in place. Steroids missing the C19 methyl group (19-nor-steroids) are often formed from sitolactone. Sitolactone is produced by microbial degradation of phytosterols, including β-sitosterol.

Discussion. Ulipristal acetate is formed in four steps from 3,3,20,20-bis(ethylenedioxy)-17α-hydroxy-19-norpregna-5(10),9(11)-diene, in eight steps from 3,3-ethylenedioxyestra-5(10),9(11)-diene-17-one, and in 13 steps from β-sitosterol.

Ulipristal acetate is formed from ulipristal and acetic anhydride. Ulipristal is formed by hydrolysis of the C3 and C20 acetals followed by dehydration. The reaction of the 9(11)-ene-5α(10)-epoxide with an arylmagnesium halide results in C—C bond formation at C11 and epoxide ring opening (**Grignard Reaction**). (An 11α-aryl side product is also formed in this reaction. What is the ratio 11β-aryl product to 11α-aryl side product? How is the 11β-aryl product separated from the 11α-aryl side product?) The Grignard reagent is formed from 4-bromo-*N,N*-dimethylaniline. The 9(11)-ene-5α(10)-epoxide is formed by epoxidation of the 5(10),9(11)-diene. (The 9(11)-ene-5β(10)-epoxide is also formed in the epoxidation. What is the ratio of the 5α(10)-epoxide to the 5β(10)-epoxide? How is the 5α(10)-epoxide separated from the 5β(10)-epoxide?) 3,3,20,20-bis(Ethylenedioxy)-17α-hydroxy-19-norpregna-5(10),9(11)-diene is formed by reaction of the dione with ethylene glycol.

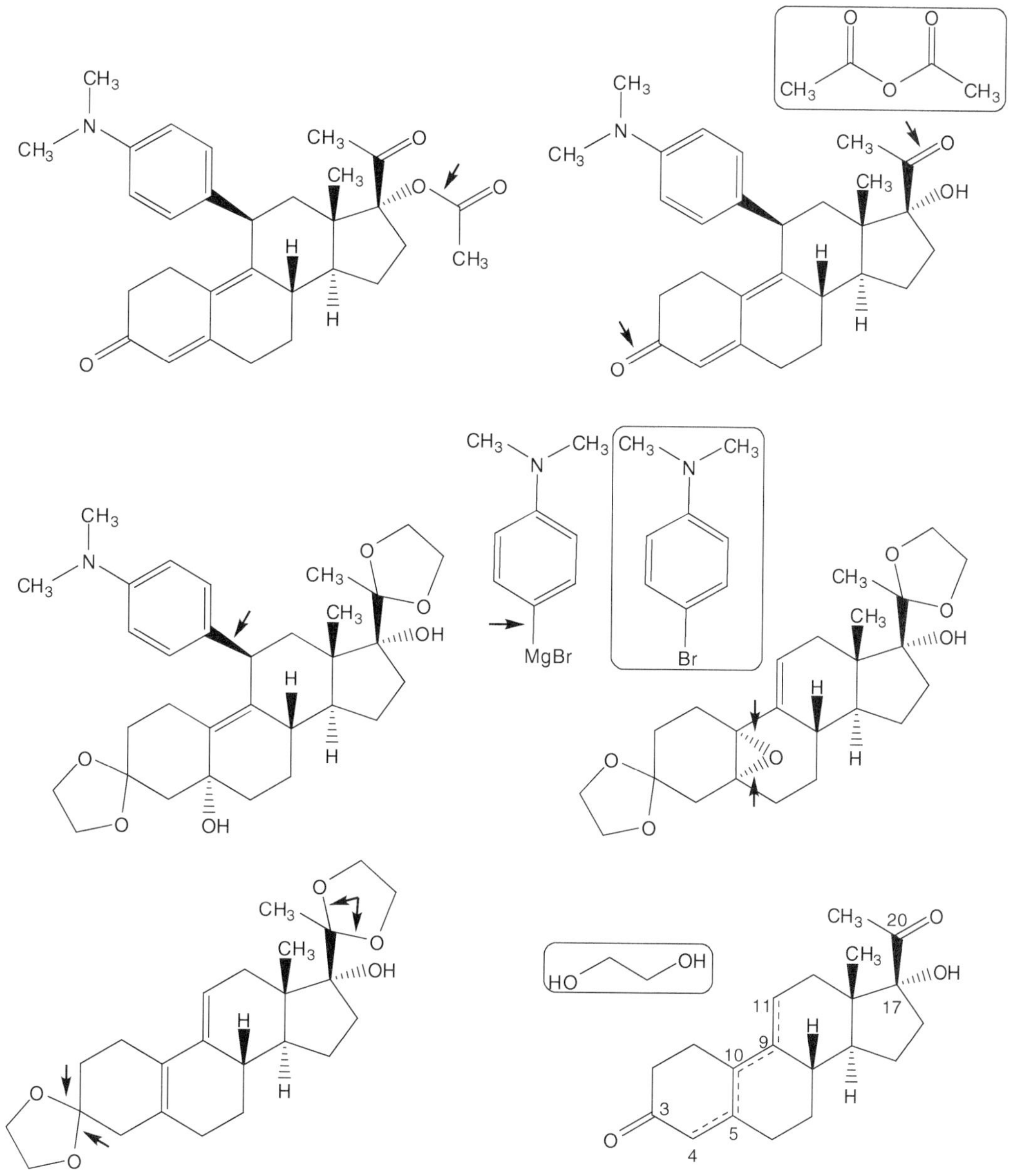

17α-Hydroxy-19-norpregna-4,9(10)-diene-3,20-dione (mixture with the 5(10),9(11)-diene) is formed by addition of methyllithium or a methyl Grignard reagent to a nitrile at C17 followed by acid quench to hydrolyze the imine, the protecting group on the alcohol at C17, and the C3 acetal. The alcohol at C17 is protected. (List the protecting groups used.) The cyanohydrin is formed from the C17 ketone of 3,3-ethylenedioxyestra-5(10),9(11)-diene-17-one.

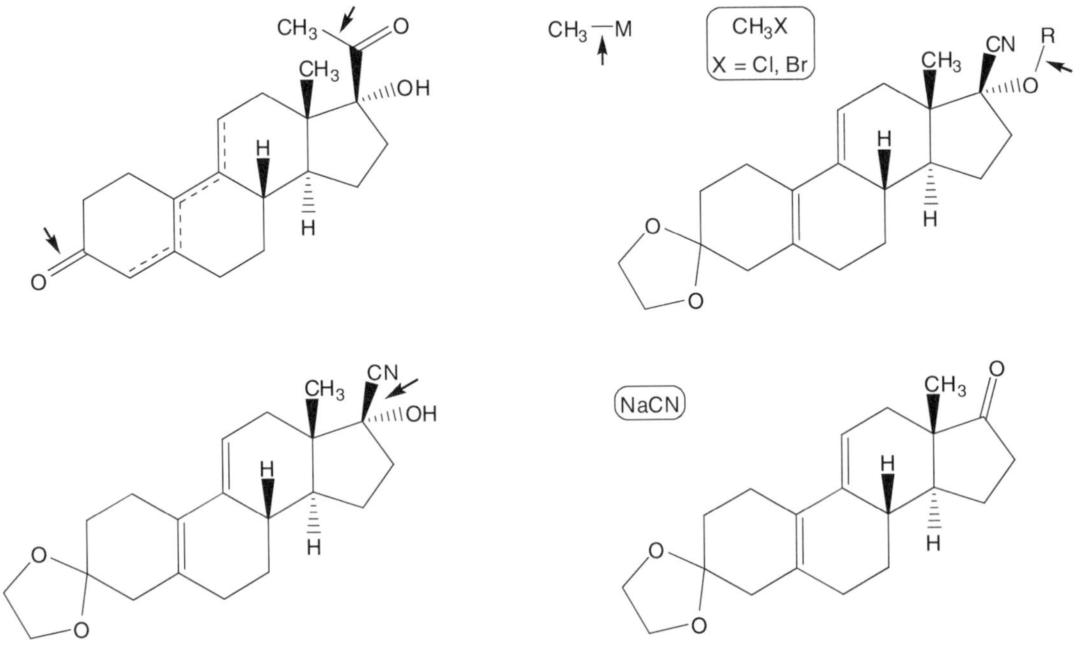

The acetal of 3,3-ethylenedioxyestra-5(10),9(11)-diene-17-one is formed by the reaction of estra-4,9-diene-3,17-dione with ethylene glycol. The 4-ene-3-one and the A ring of estra-4,9-diene-3,17-dione are formed by condensation of the methyl ketone and cyclohexanone of the seco-steroid (**Aldol Condensation**) followed by dehydration. The methyl ketone of the seco-steroid is released by hydrolysis of the acetal and carried directly into the condensation.

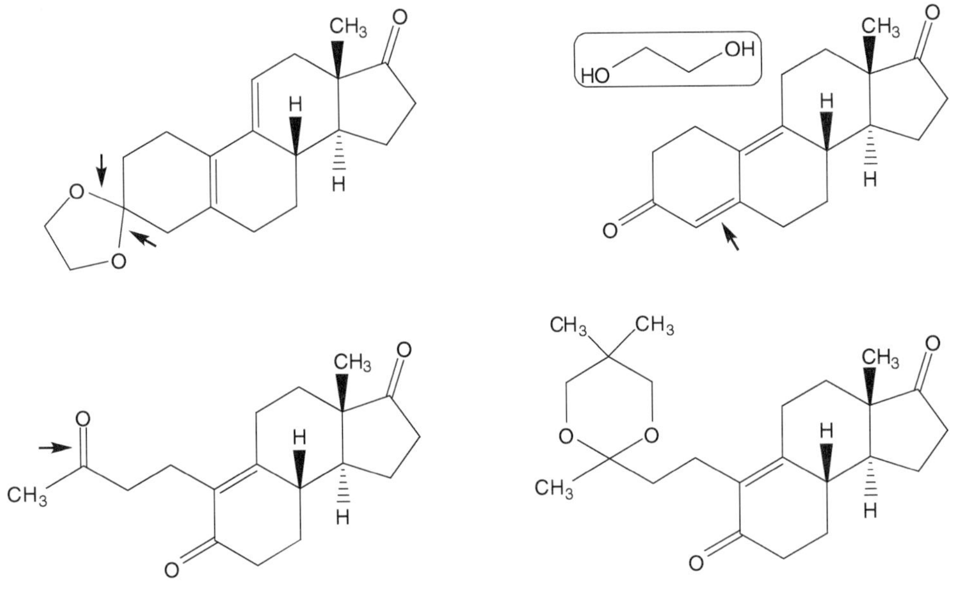

The enone and the B ring are formed by condensation of a ketone on a side chain and a cyclohexanone (**Aldol Condensation**) followed by dehydration. The cyclohexanone is formed by oxidation of the cyclohexanol and carried directly into the condensation. The ketone on the side chain and the cyclohexanol are both formed by reaction of an alkylmagnesium chloride with the lactone carbonyl of sitolactone (**Grignard Reaction**). Sitolactone is formed by microbial degradation of phytosterols, including β–sitosterol.

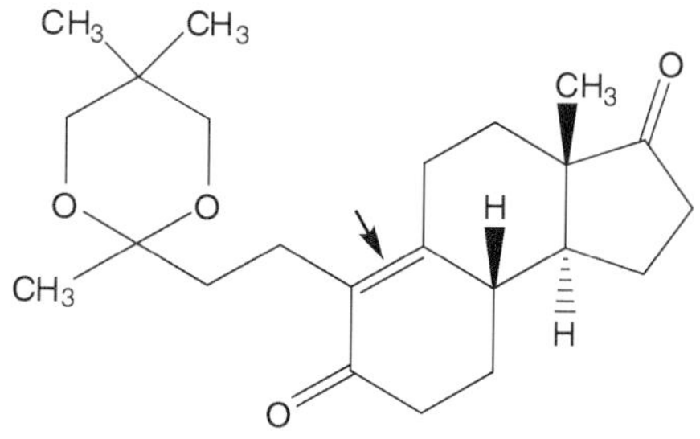

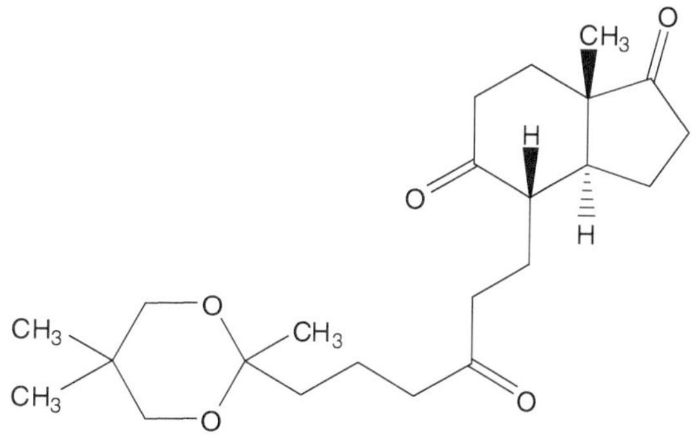

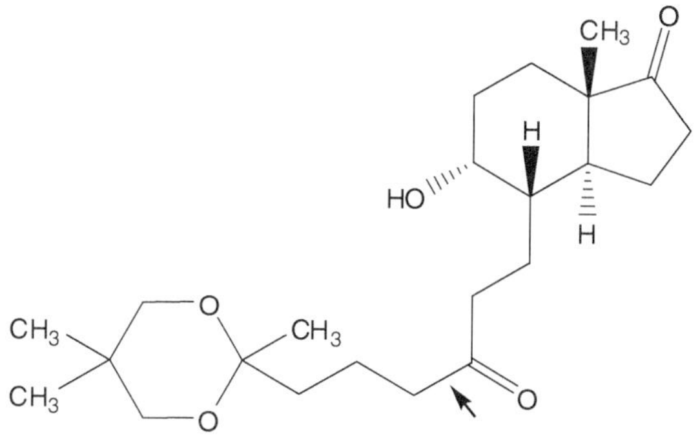

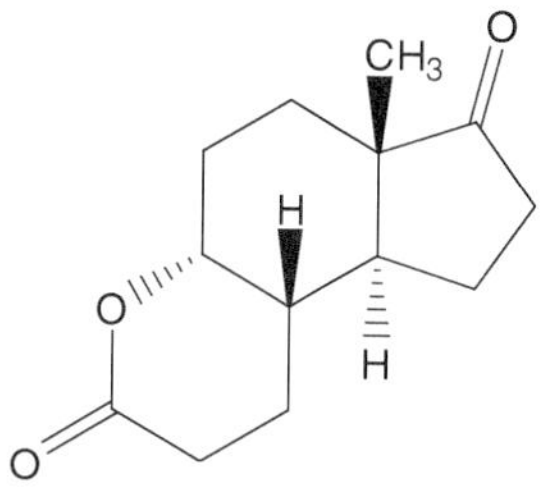

sitolactone

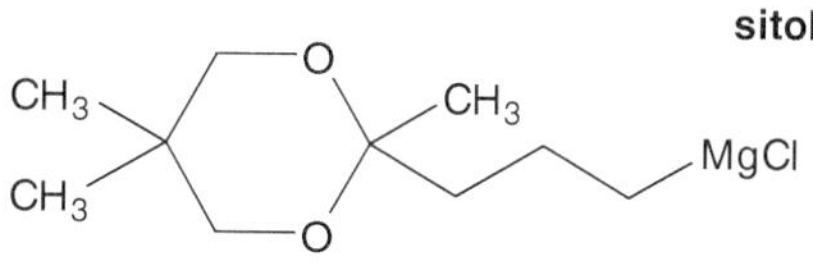

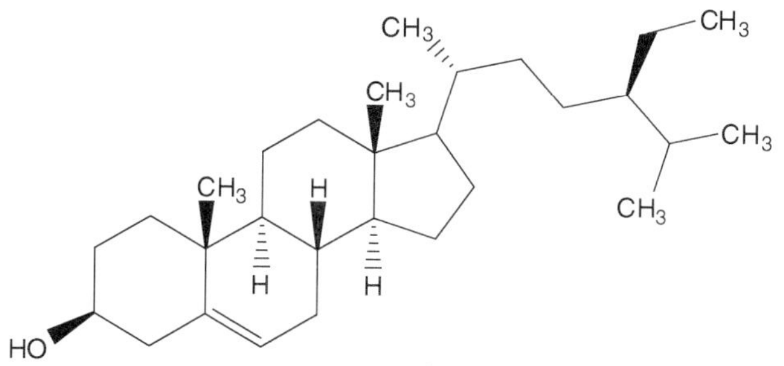

β-sitosterol

The alkylmagnesium chloride is formed from 2-(3-chloropropyl)-2,5,5-trimethyl-1,3-dioxane. The acetal is formed by reaction of 5-chloro-2-pentanone with 2,2-dimethyl-1,3-propanediol (neopentyl glycol).

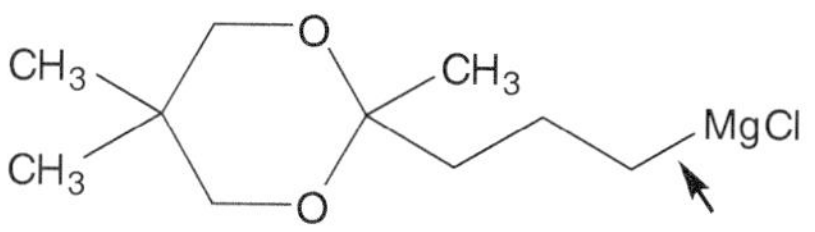

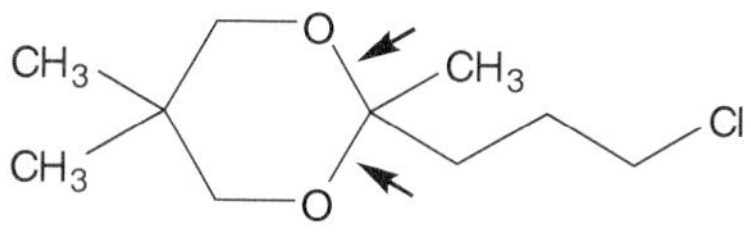

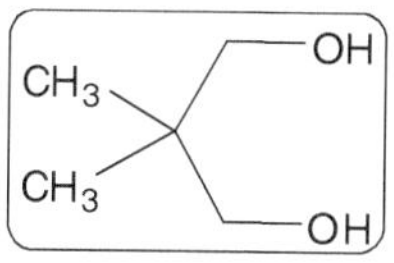

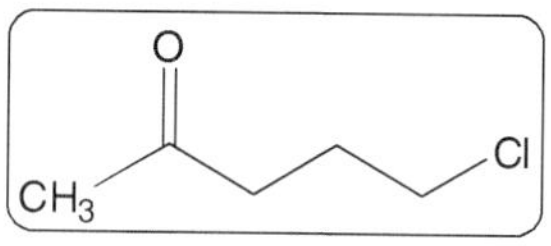

Extended Discussion

Draw the structures of the retrosynthetic analysis of one alternative route to the intermediate 3,3,20,20-bis(ethylenedioxy) -17α-hydroxy-19-norpregna-5(10),9(11)-diene. Include the structures of the retrosynthetic analysis of any organic starting material(s) from petrochemical or biochemical raw materials. List the pros and cons for both routes. Is one route preferred?

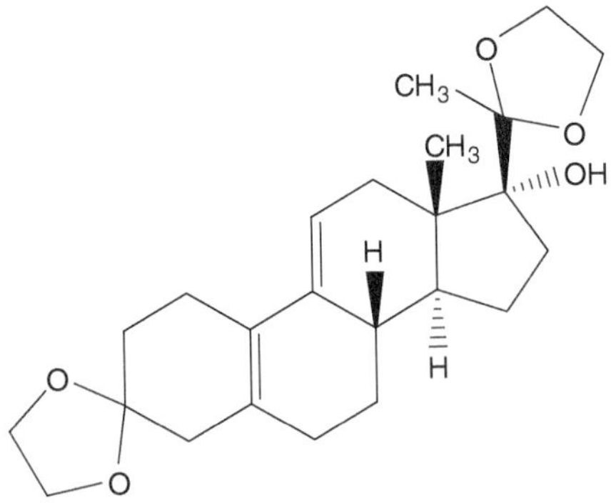

V

Valganciclovir

Anti-infective Medicines/Antiviral Medicines/Other Antivirals

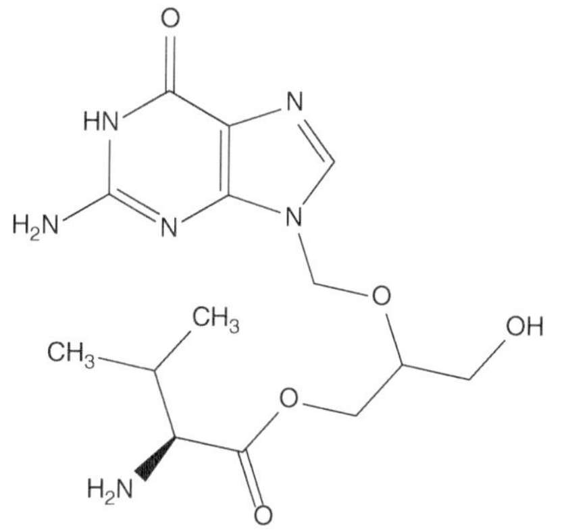

A nucleoside analog is often formed by displacement of a leaving group on the sugar analog by N9 of guanine. Oxygen-based leaving groups include water, alcohols, and carboxylic acids. The oxygen-based leaving group is often germinal to another oxygen atom.

Discussion. Valganciclovir is a nearly 1:1 mixture of the (*S,R*)- and (*S,S*)-diastereomers formed using the amino acid L-valine. Disconnection of the ester C—O bond reveals the two identical hydroxyl groups of ganciclovir and the statistical problem associated with forming the monoester of the diol.

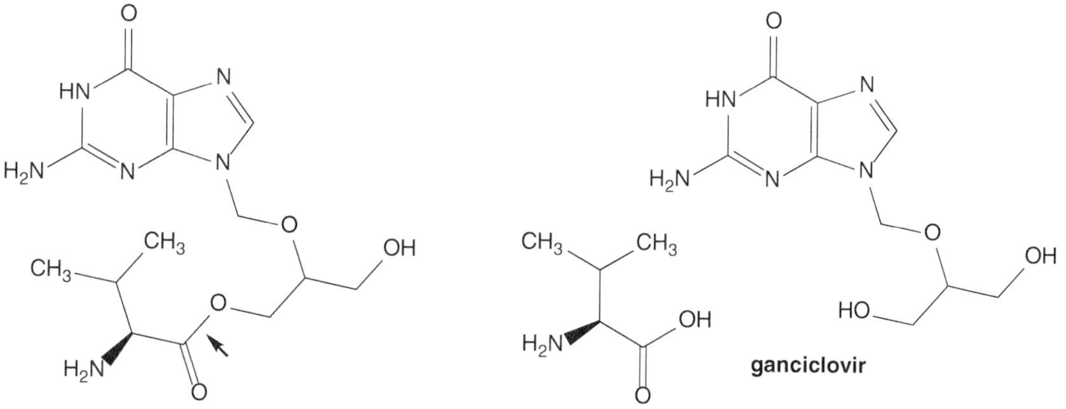

In one preferred route, the final step is hydrogenolysis of a benzyl ether to release the alcohol and hydrogenolysis of a benzyloxycarbamate to release the amine. The ester is formed from the alcohol and *N*-carbobenzyloxy-L-valine *N*-carboxyanhydride (*N*-CBZ-L-valine NCA). The alcohol of monobenzyl ganciclovir is formed by hydrolysis of the

Routes to Essential Medicines: A Workbook for Organic Synthesis, First Edition. Peter J. Harrington.

Companion website: www.wiley.com/go/Harrington/routes_essential_medicine

propanoate ester. The nucleoside C—N bond is formed by displacement of a propanoate leaving group from 1-benzyloxy-3 -propionyloxy-2-(propionyloxy)methoxypropane by guanine N9. (What is the highest N9:N7 ratio for the product mixture? What reaction conditions are associated with the highest ratio?) Acylation of the amino group on guanine with propanoic anhydride and hydrolysis of this group at the same time as the propanoate ester is a processing option. (Does this option result in a higher yield and/or purity of monobenzylganciclovir?)

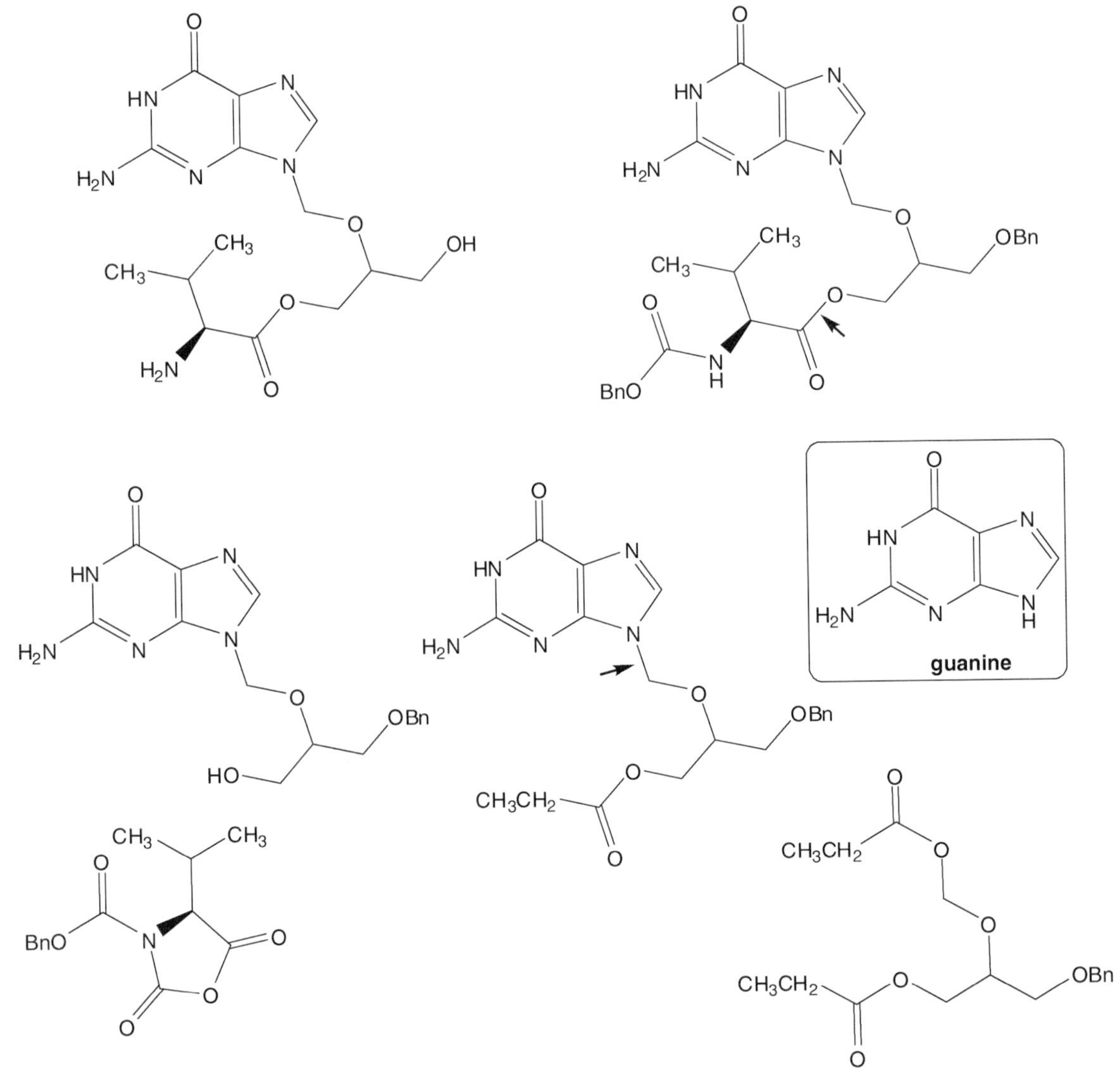

N-Carbobenzyloxy-L-valine *N*-carboxyanhydride is formed from L-valine *N*-carboxyanhydride. L-Valine *N*-carboxyanhydride is formed from L-valine. L-Valine is produced by fermentation.

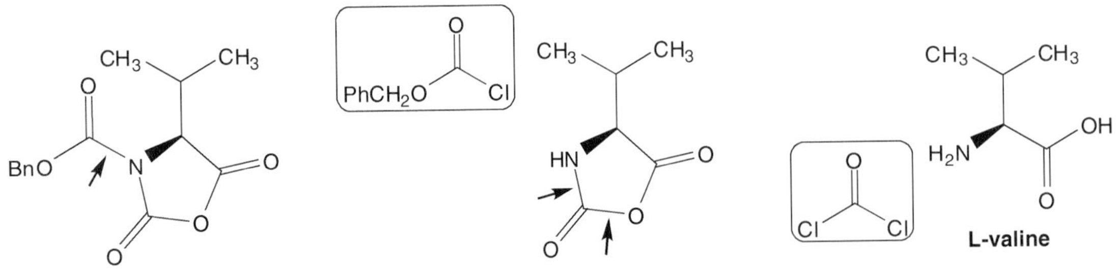

1-Benzyloxy-3-propionyloxy-2-(propionyloxy)methoxypropane is formed in four steps from epichlorohydrin. One propanoate ester is formed by displacement of chloride. The other propanoate ester is formed by the reaction of 1-benzyloxy-3-chloro-2-propanol with methoxymethyl propionate. Methoxymethyl propionate is formed from dimethoxymethane and propanoic anhydride. 1-Benzyloxy-3-chloro-2-propanol is formed by the ring opening of benzyl glycidyl ether with lithium chloride. Benzyl glycidyl ether is formed from epichlorohydrin and benzyl alcohol.

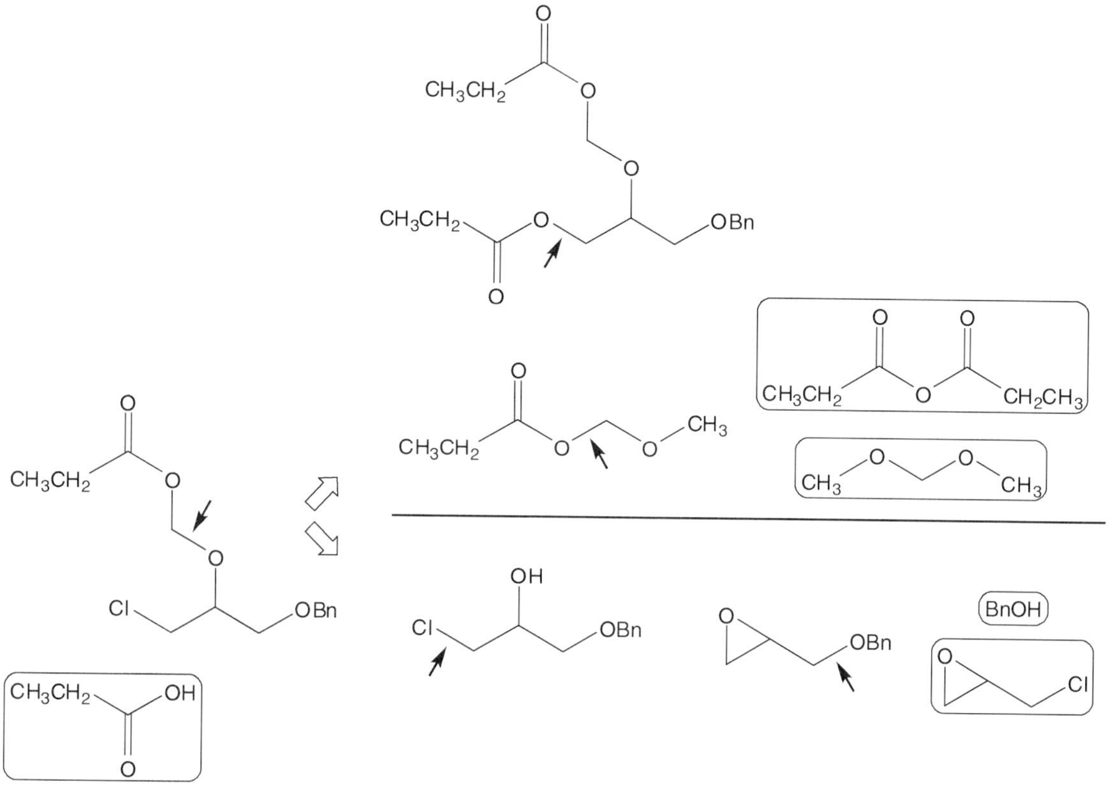

Extended Discussion

Draw the structures of the retrosynthetic analysis of one route to valganciclovir from ganciclovir. Also draw the structures of the retrosynthetic analysis of one route to ganciclovir. Include the structures of the retrosynthetic analysis of any organic starting material(s) from petrochemical or biochemical raw materials. List the pros and cons for both routes to valganciclovir. Is one route preferred?

Valproic Acid

Anticonvulsants/Antiepileptics
Medicines for Mental and Behavioral Disorders/Medicines Used in Mood Disorders/Medicines Used in Bipolar Disorders

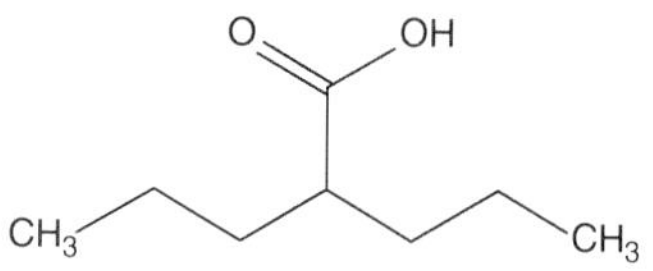

An α-alkyl carboxylic acid is often formed by alkylation of diethyl malonate followed by ester hydrolysis and decarboxylation.

Discussion. Two routes to valproic acid will be presented. In Route A, the carboxylic acid is formed by decarboxylation of α,α-dipropylmalonic acid. The malonic acid is formed by hydrolysis of the malonate diester. Diethyl α,α-dipropylmalonate is formed from diethyl malonate and 1-bromopropane (**Malonic Ester Synthesis**).

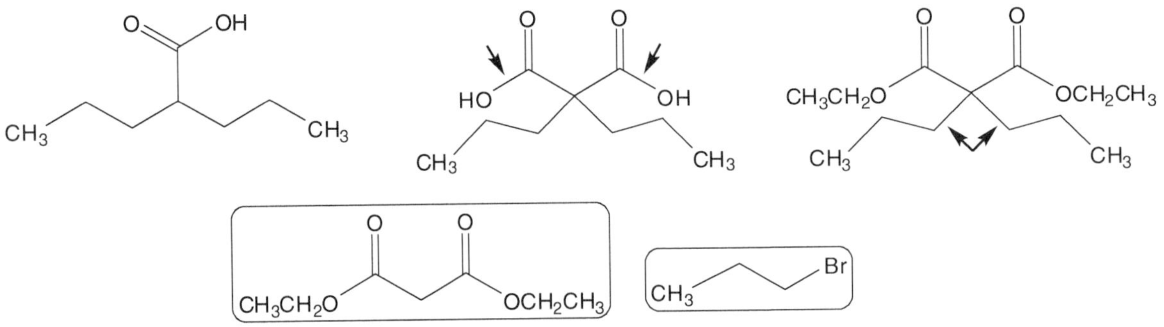

In Route B, the carboxylic acid is formed in the final step by hydrolysis of the methyl ester. The ester is formed from the β-ketoester. (What is the byproduct of this reaction? How is this byproduct separated from the product?) The α,α-dipropyl-β-ketoester is formed by hydrogenation of the α,α-diallyl-β-ketoester. The α,α-diallyl-β-ketoester is formed from methyl acetoacetate and allyl chloride.

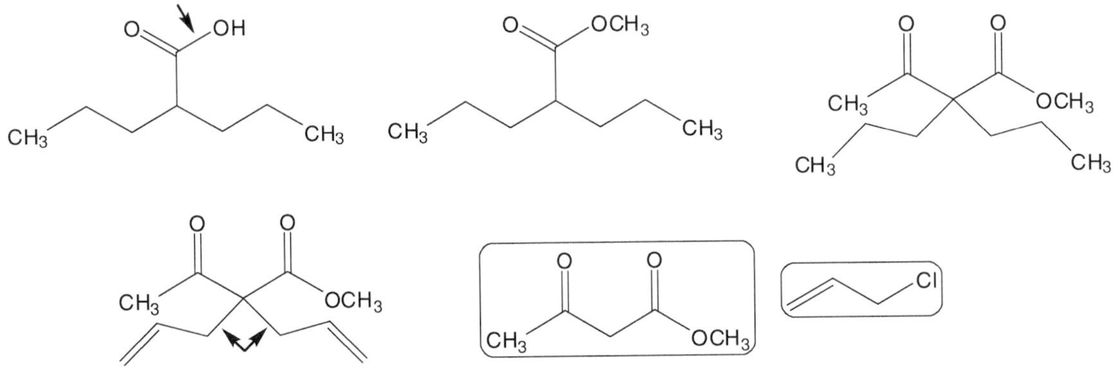

Extended Discussion

Draw the structures of the retrosynthetic analysis of an alternative Route C to valproic acid which does not use 1-bromopropane or allyl chloride. Include the structures of the retrosynthetic analysis of any organic starting material(s) from petrochemical or biochemical raw materials. List the pros and cons for all three routes. Is one route preferred?

Vecuronium Bromide

Muscle Relaxants (Peripherally-Acting) and Cholinesterase Inhibitors

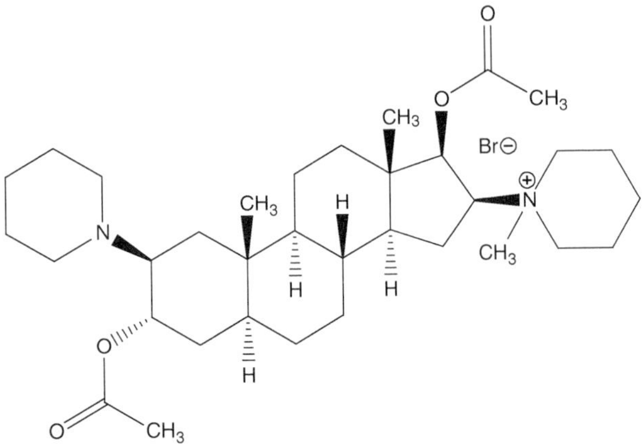

A single-enantiomer molecule with multiple chiral carbons is often formed by the modification of a natural product which has most or all of the chiral carbons already in place. A β-amino alcohol is often formed by ring opening of an epoxide by an amine.

Discussion. Vecuronium bromide is manufactured in 4 steps from the bis-epoxide 2α,3α:16α,17α-diepoxy-17β-acetoxy-5α-androstane, 9 steps from dehydroepiandrosterone, 12 steps from 16-dehydropregnenolone acetate, and 15 steps from diosgenin. Diosgenin is a phytosteroid sapogenin isolated from the tubers of *Discorea* wild yam.

The quaternary ammonium salt is formed in the final step by the reaction of the 16β-amine with bromomethane. (How is quarternization of the 2β-amine avoided?) The diacetate is formed by the reaction of the 3α,17β-diol with acetyl chloride. The 17β-alcohol is formed by the reduction of the C17 ketone. The 3α-alcohol and C17 ketone are formed by ring opening of the epoxides of 2α,3α:16α,17α-diepoxy-17β-acetoxy-5α-androstane with piperidine.

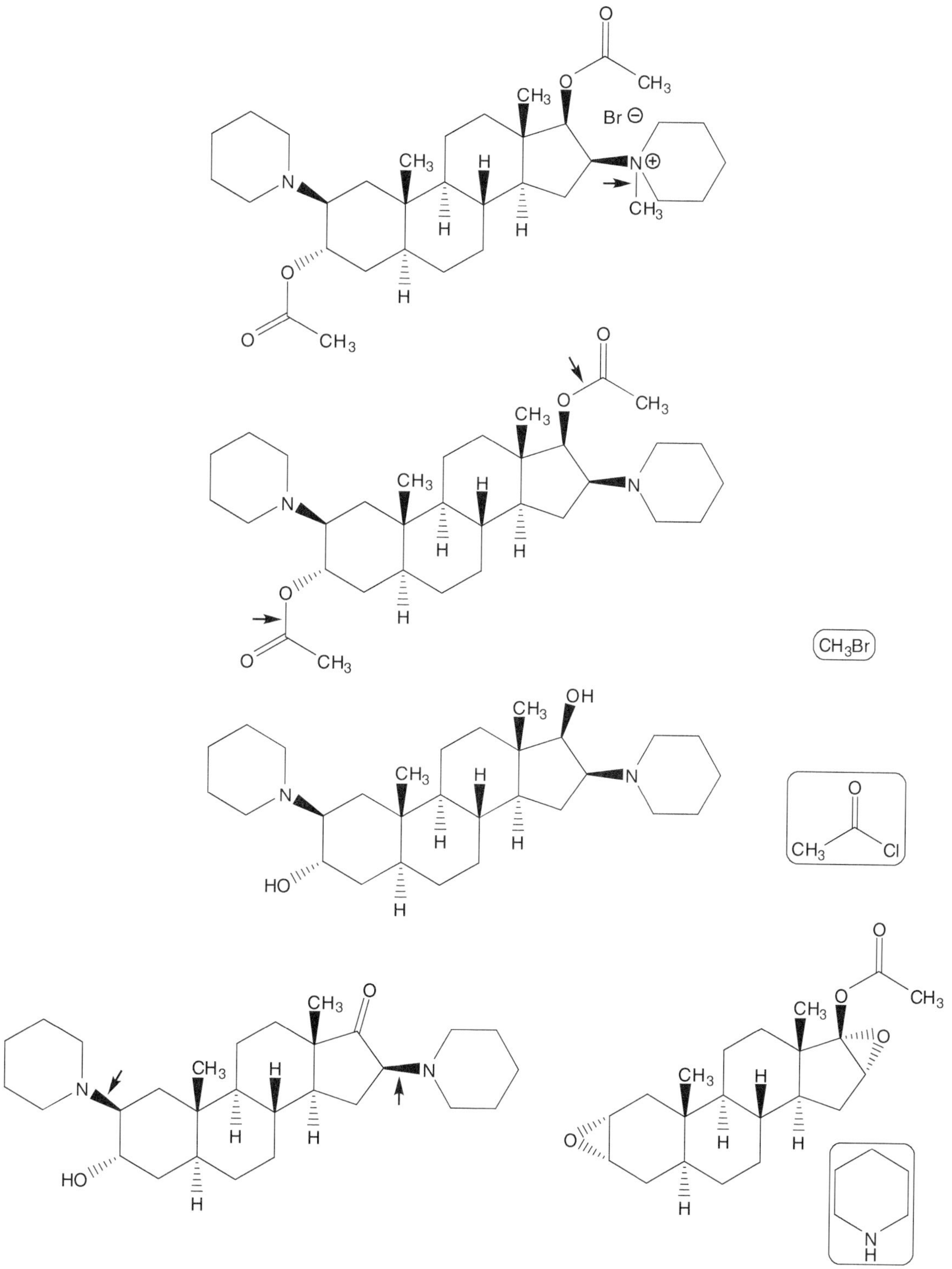

The bis-epoxide is formed by epoxidation of the diene. The enol acetate of 17-acetoxy-5α-androst-2,16-diene is formed by acetylation of 5α-androst-2-en-17-one with isopropenyl acetate. The 2,3-alkene is formed from the 3β-tosylate by elimination. (Draw the structures of other products that are likely present in the mixture after the tosylate elimination. How is 5α-androst-2-en-17-one separated from the reaction mixture?) The 3β-tosylate is formed from the 3β-alcohol and *para*-toluenesulfonyl chloride. The 5α-androstane is formed by catalytic hydrogenation of the 5-ene of dehydroepiandrosterone.

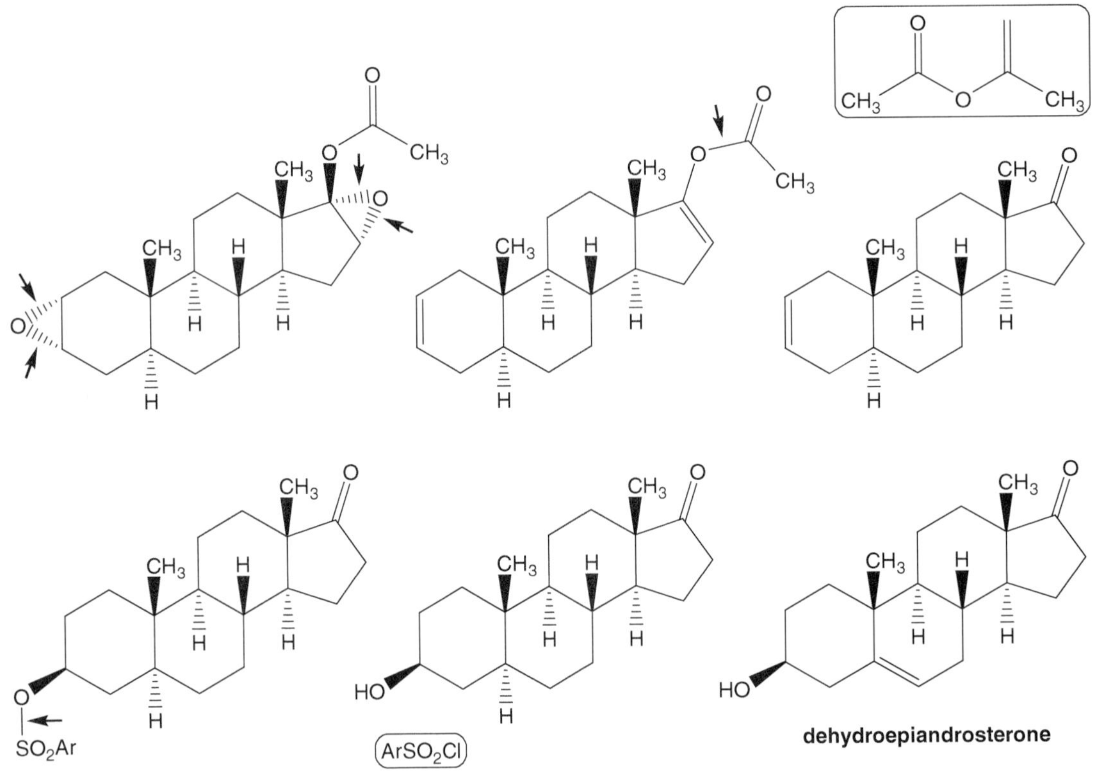

The C17 ketone of dehydroepiandrosterone is formed by hydrolysis of an enamide. The 3β-acetate ester is also hydrolyzed during the enamide hydrolysis. The enamide is formed by rearrangement of the oxime formed from 16-dehydropregnenolone acetate (**Beckmann Rearrangement**).

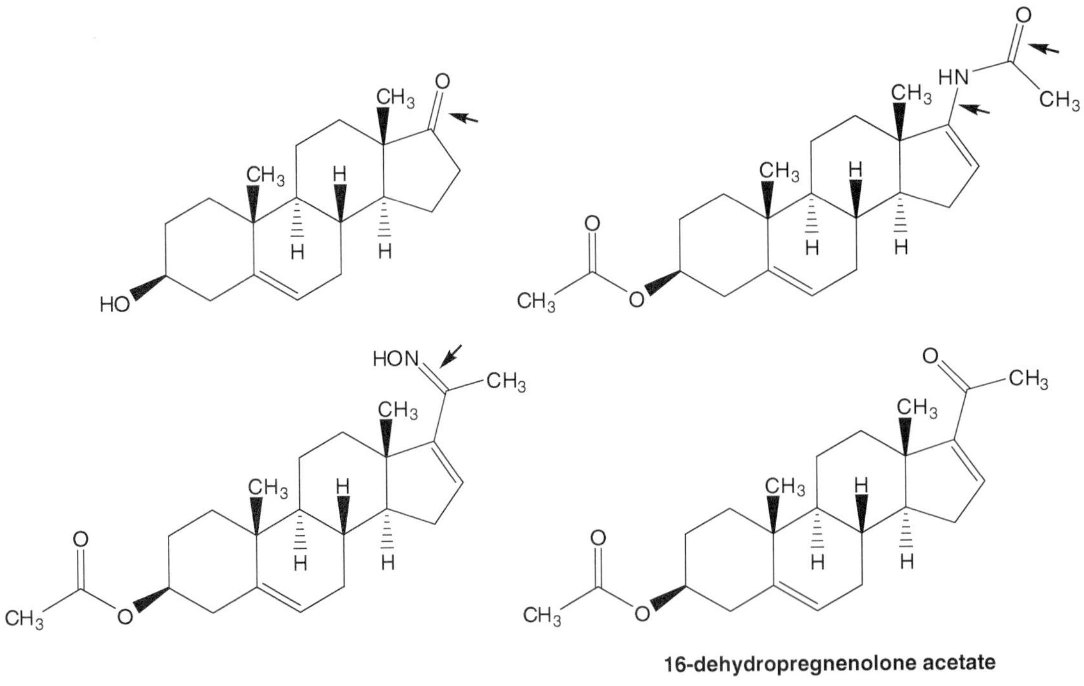

The 16-alkene of 16-dehydropregnenolone acetate is formed from diosone by β-elimination. Diosone is formed by oxidation of the 20(22)-alkene of pseudodiosgenin-3,26-diacetate. Pseudodiosgenin 3,26-diacetate is formed by the reaction of diosgenin with acetic anhydride. The three-step synthesis of 16-dehydropregnenolone acetate from diosgenin by acetylation, oxidation, and elimination is known as the **Marker Degradation**.

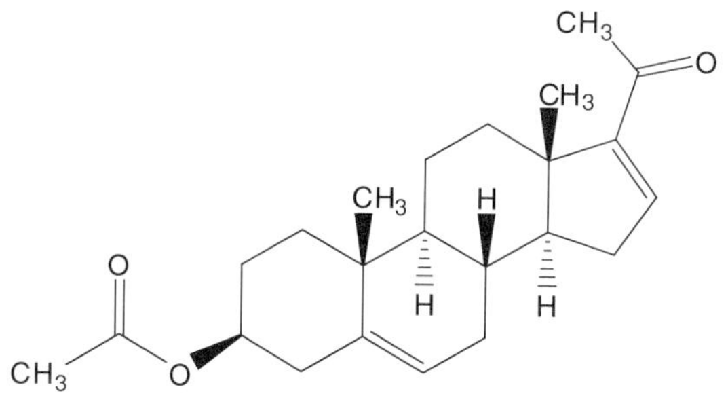

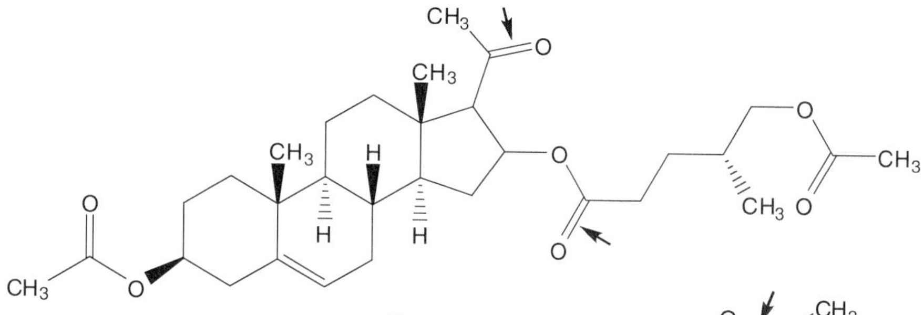

diosone

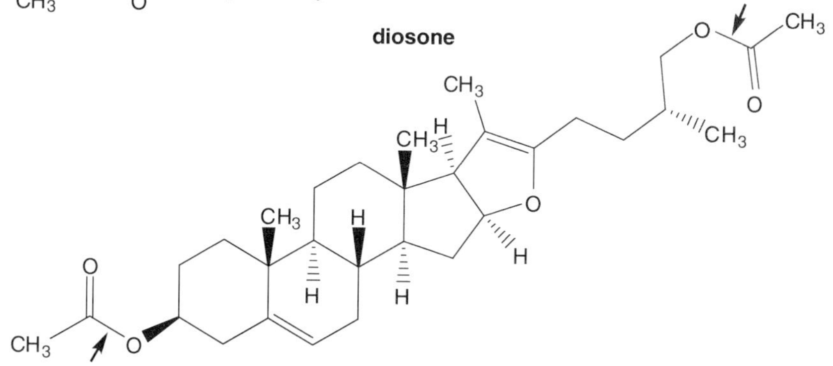

pseudodiosgenin-3,26-diacetate

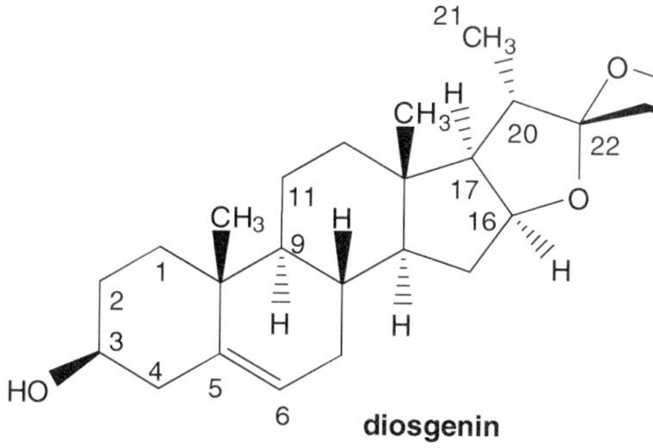

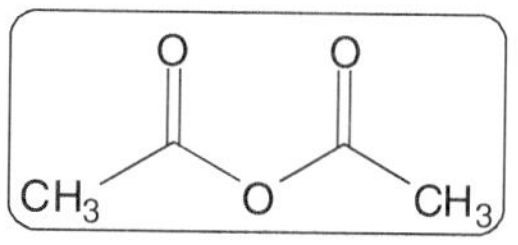

diosgenin

Extended Discussion

Draw the structures of the retrosynthetic analysis of a route to dehydroepiandrosterone that does not use diosgenin as a starting material.

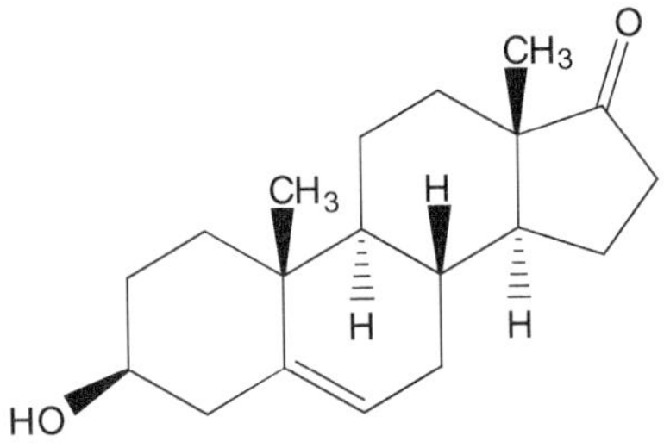

dehydroepiandrosterone

Velpatasvir

Anti-infective Medicines/Antiviral Medicines/Antihepatitis Medicines/Medicines for Hepatitis C/Other Antivirals

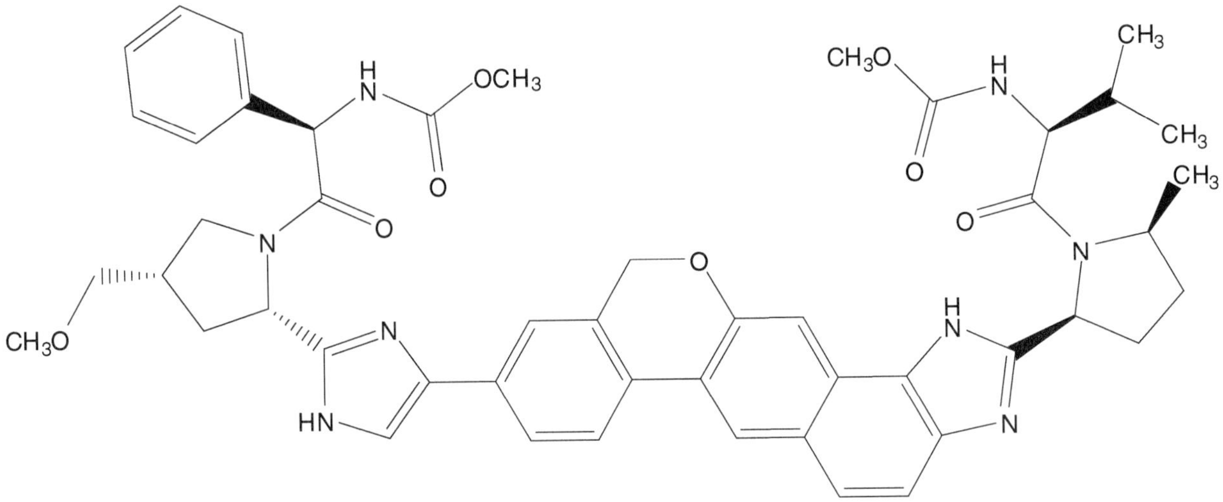

A 2,5-disubstituted imidazole is often formed by the reaction of an α-acyloxyketone with ammonium acetate. Four C—N bonds of the imidazole ring are formed in the reaction.

Discussion. The retrosynthetic analysis will begin with disconnections to reveal five component molecules. Four of the five components are amino acids or amino acid *N*-(methoxycarbonyl) (Moc) derivatives.

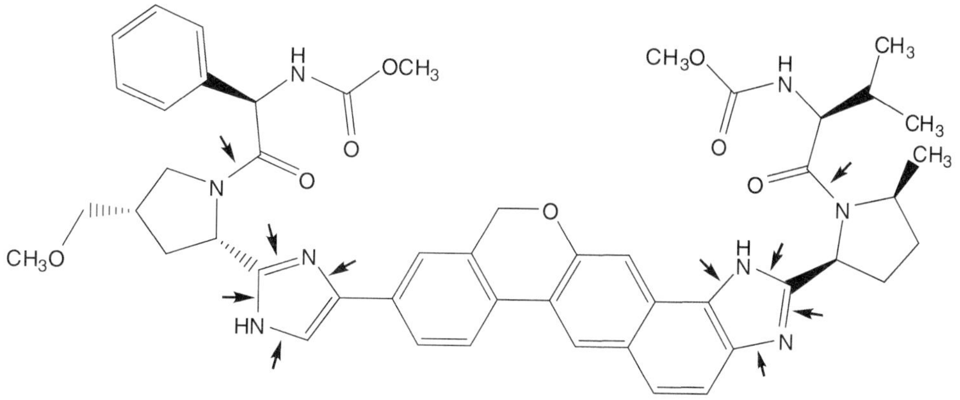

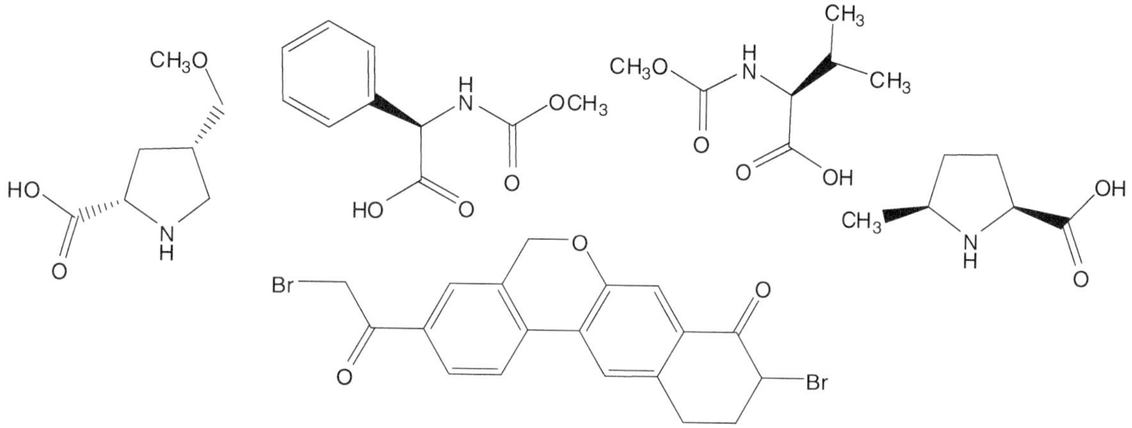

An amide bond is formed by the reaction of an amine with the carboxylic acid of *N*-(methoxycarbonyl)-D-phenylglycine in the final step. The amine is released by hydrolysis of a *tert*-butoxycarbonyl (Boc) protecting group.

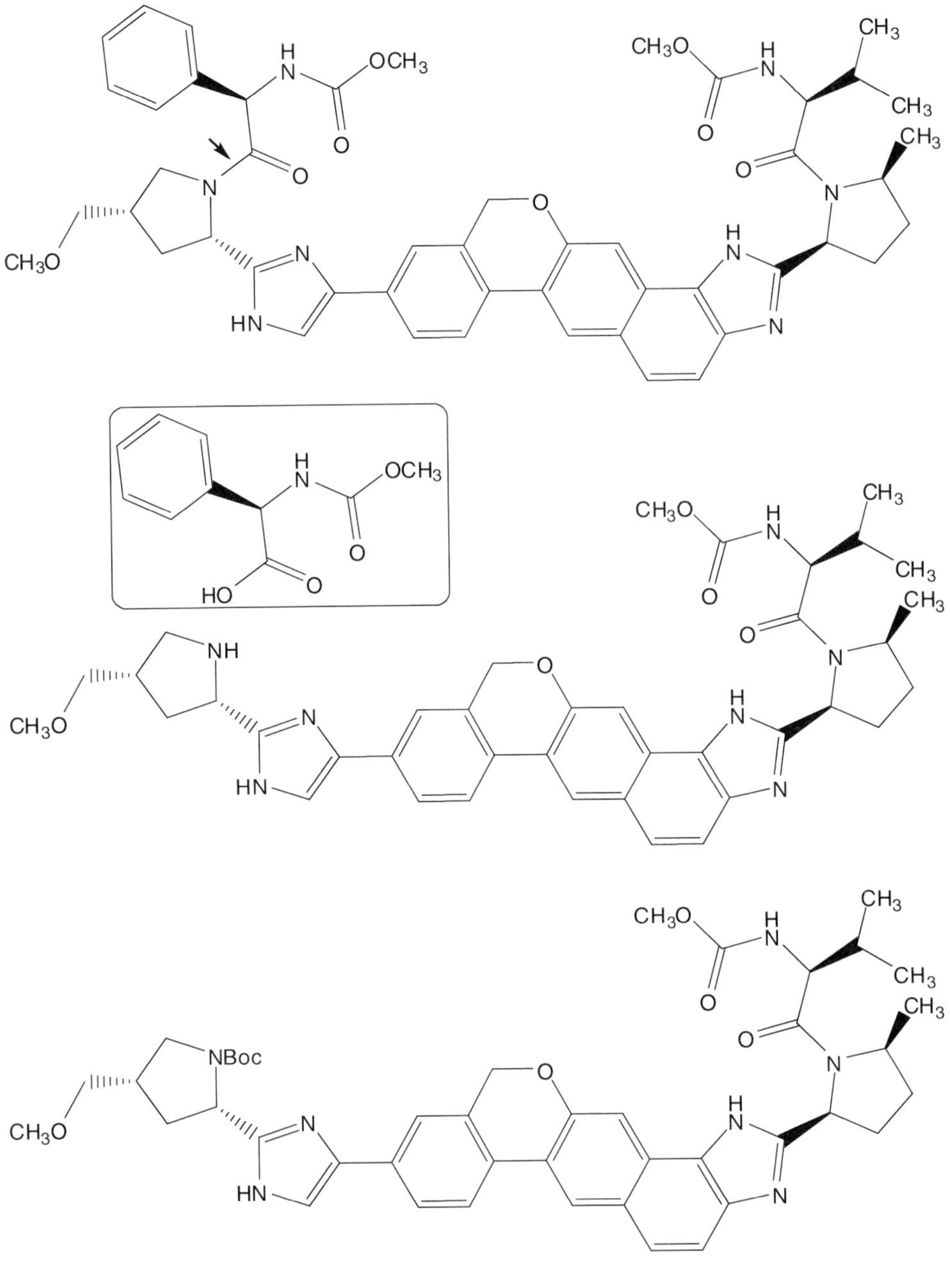

The 10,11-double bond of the 5*H*-dibenzo[c,g]chromene ring system is formed by dehydrogenation. The eight C—N bonds of the two imidazole rings are formed by the reaction of two α-acyloxyketones with ammonium acetate. Both of the α-acyloxyketones are formed by bromide displacement by the carboxylic acid. Since the carboxylic acids are not identical, high selectivity in the displacement of one bromide of the dibromide was necessary for this route to be successful. (Which ester is formed first?)

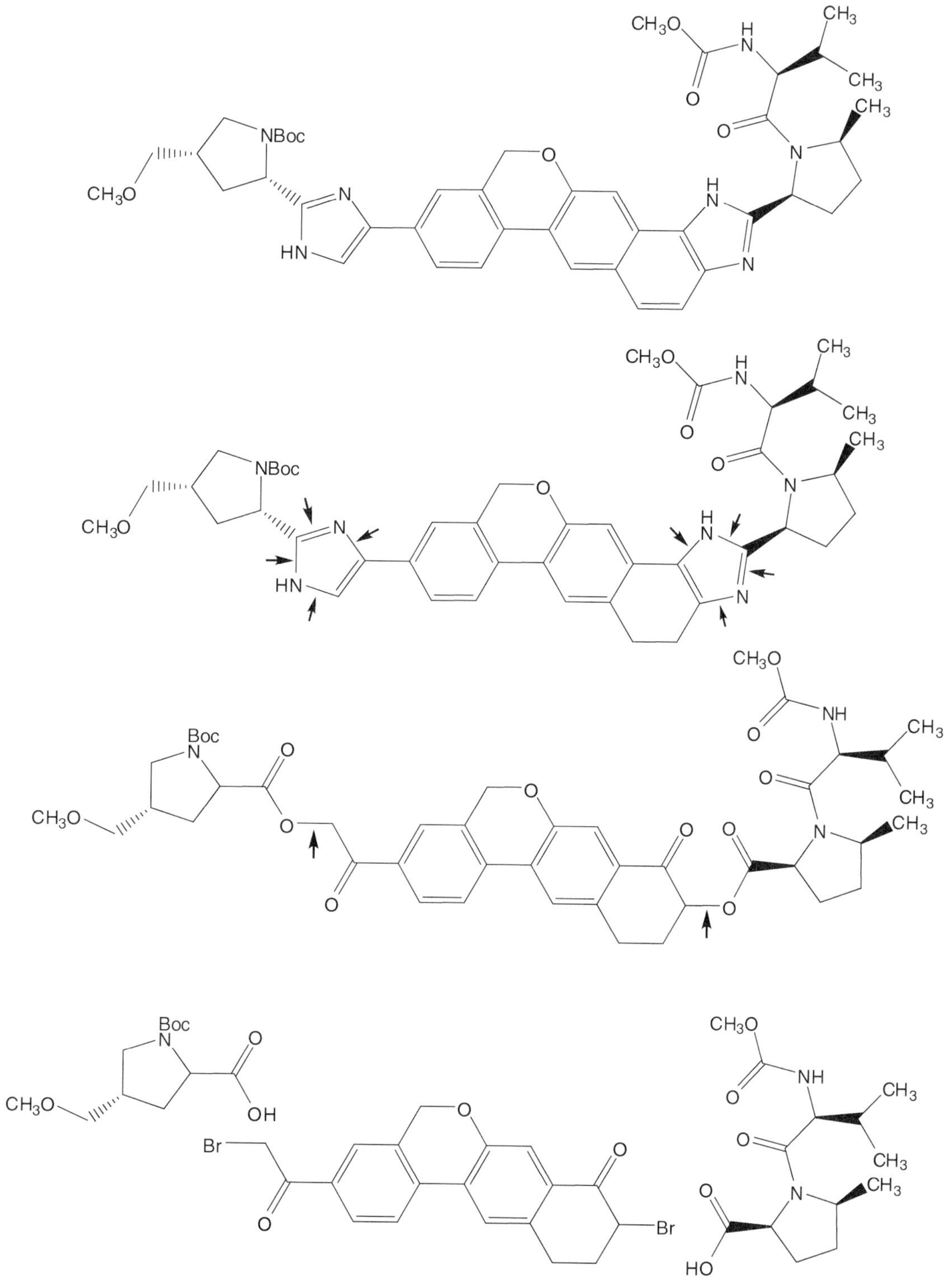

The α-bromoketone at C3 of the 10,11-dihydro-5*H*-dibenzo[c,g]chromen-8(9H)-one is formed by oxidation of the bromohydrin. The bromine at C9 is introduced by α–bromination of the ketone at C8. The bromohydrin is formed from the alkene. The alkene is formed by the coupling of the aryl chloride and potassium vinyltrifluoroborate (**Suzuki-Miyaura Coupling**). A C—C bond is formed by a palladium-catalyzed intramolecular coupling of the aryl bromide with the aromatic ring. The ether is formed by bromide displacement from 1-bromo-2-bromomethyl-4-chlorobenzene by 7-hydroxy-1-tetralone (**Williamson Ether Synthesis**). 1-Bromo-2-bromomethyl-4-chlorobenzene is formed by bromination of 2,5-dichlorotoluene. 7-Hydroxy-1-tetralone is formed from the methyl ether.

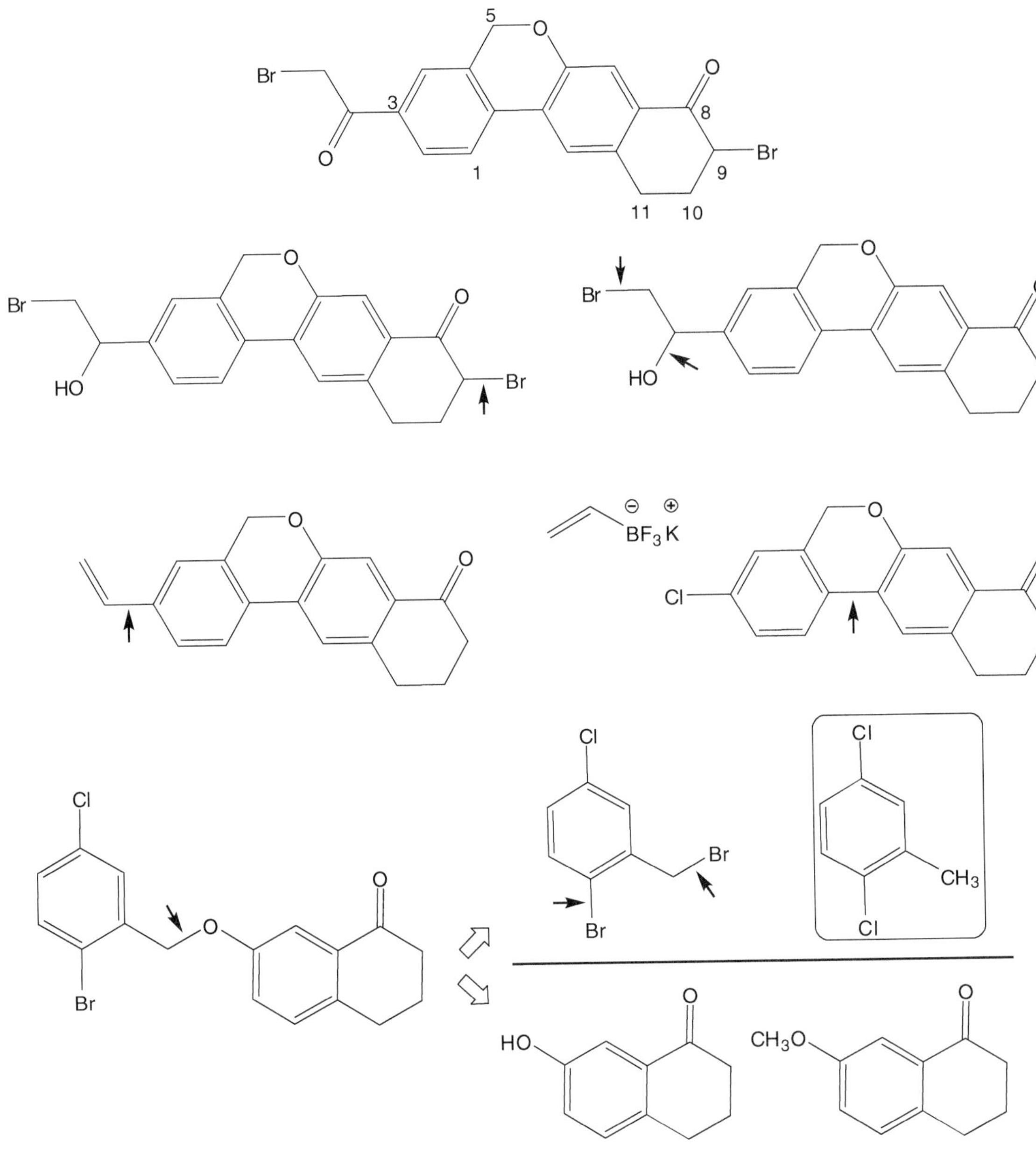

Potassium vinyltrifluoroborate is formed from vinyl bromide.

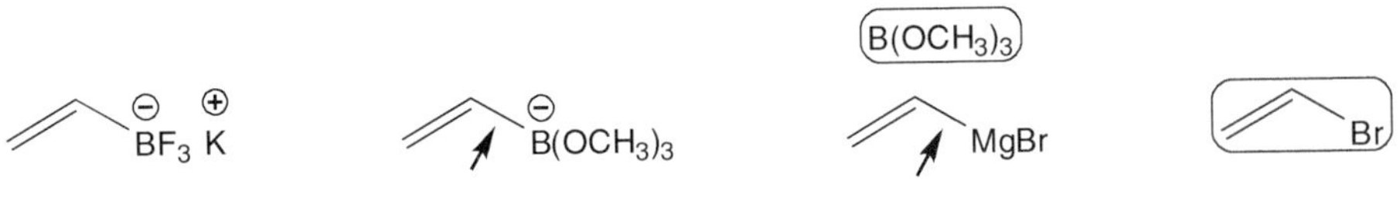

7-Methoxy-1-tetralone is formed from 4-(4-methoxyphenyl)butanoic acid (**Friedel-Crafts Acylation**). 4-(4-Methoxyphenyl) butanoic acid is formed by hydrogenation of 4-(4-methoxyphenyl)-4-oxobutanoic acid. The 4-oxobutanoic acid is formed from anisole and succinic anhydride (**Friedel-Crafts Acylation**).

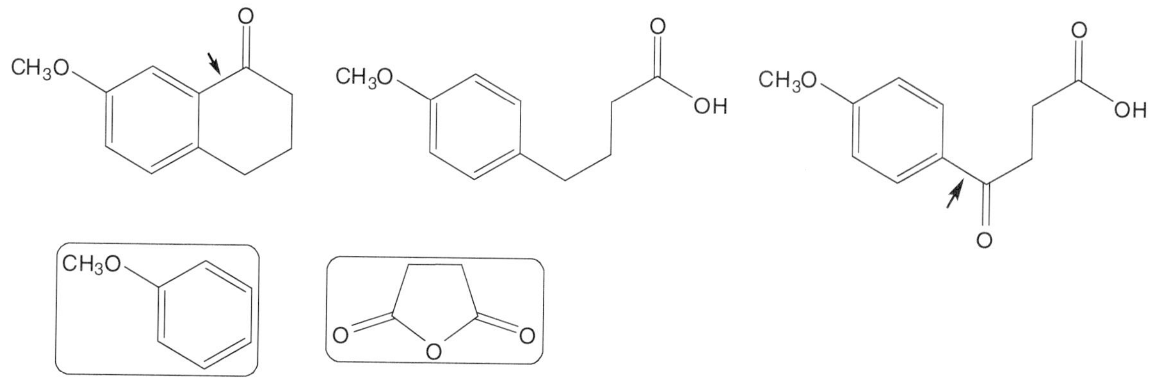

Routes to *N*-(*tert*-butoxycarbonyl)-*cis*-4-(methoxymethyl)-L-proline begin with *N*-(*tert*-butoxycarbonyl)-*trans*-4-hydroxy-L-proline. In one preferred route, *N*-(*tert*-butoxycarbonyl)-*cis*-4-(methoxymethyl)-L-proline is formed by hydrogenation of the enol ether. The enol ether is formed from the ketone *N*-(*tert*-butoxycarbonyl)-4-oxo-L-proline and methoxymethyltriphenylphosphonium chloride (**Wittig Reaction**). The ketone is formed by oxidation of the secondary alcohol of *N*-(*tert*-butoxycarbonyl)-*trans*-4-hydroxy- L-proline.

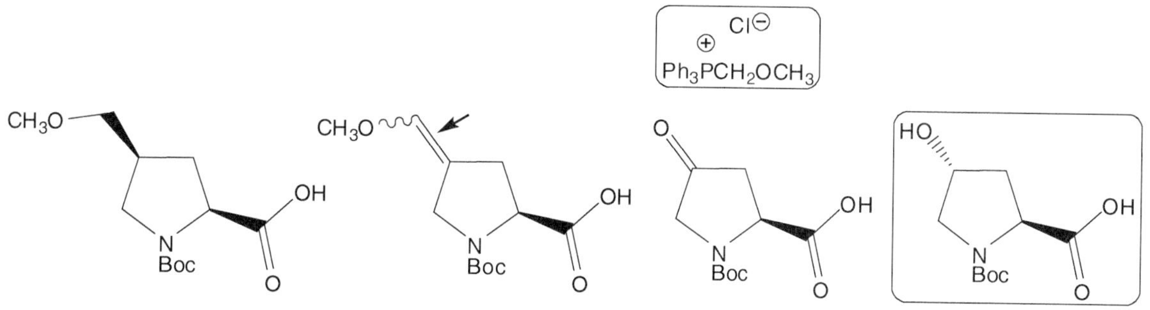

The carboxylic acid used to form the C9 ester is formed by hydrolysis of the methyl ester. The amide is formed between *N*-(methoxycarbonyl)-L-valine and methyl cis-5-methyl-L-prolinate. Methyl cis-5-methyl-L-prolinate is formed from *cis*-5-methyl-L-proline. *Cis*-5-Methyl-L-proline and the (2*R*,5*R*)-enantiomer are separated by resolution. The racemic mixture is formed by reduction of the imine. The imine is formed from the ketone, and the primary amine is formed in situ by amide hydrolysis. Under the amide hydrolysis conditions, a malonate diester is hydrolyzed to the diacid and the diacid decarboxylates. The α–alkylated acetamidomalonate is formed by the reaction of dimethyl acetamidomalonate with methyl vinyl ketone (**Michael Addition**).

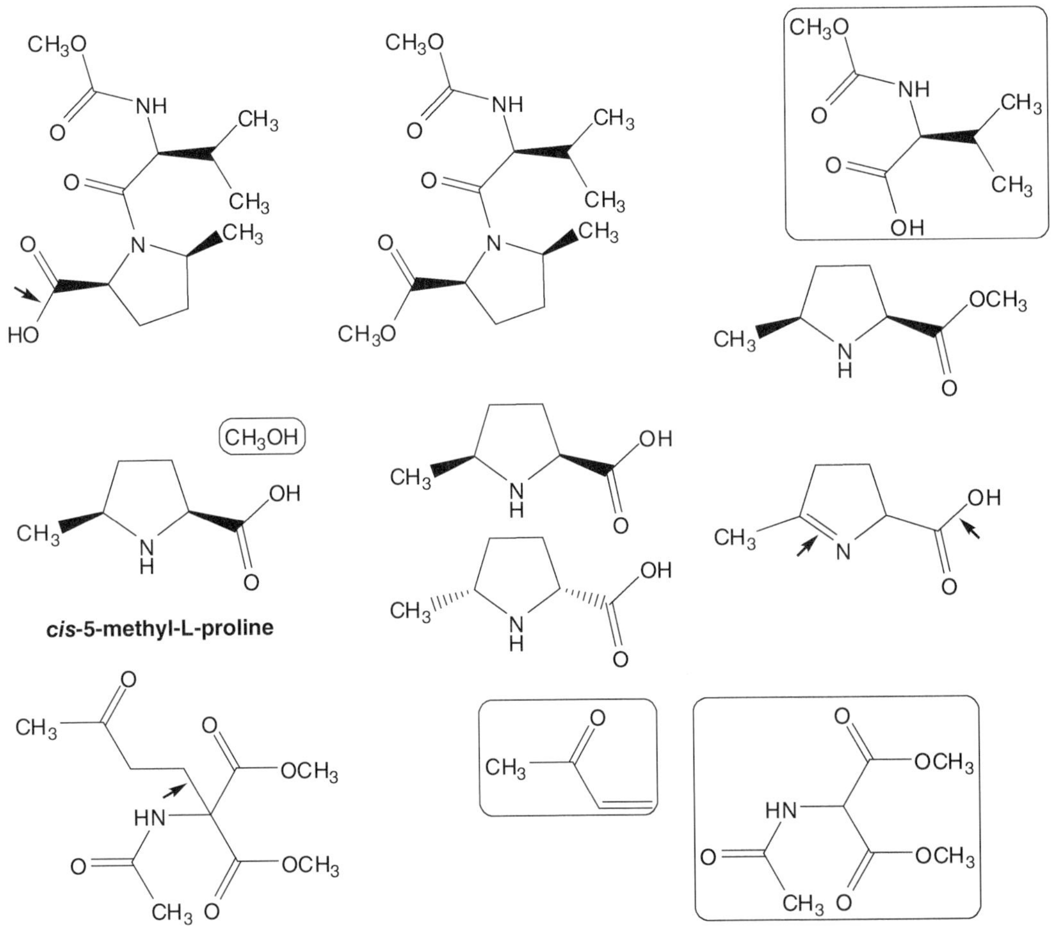

Extended Discussion

Draw the structures of a retrosynthetic analysis of one alternative route to the key dibromide intermediate. Include the structures of the retrosynthetic analysis of any organic starting material(s) from petrochemical or biochemical raw materials. List the pros and cons for both routes and select one route as the preferred route.

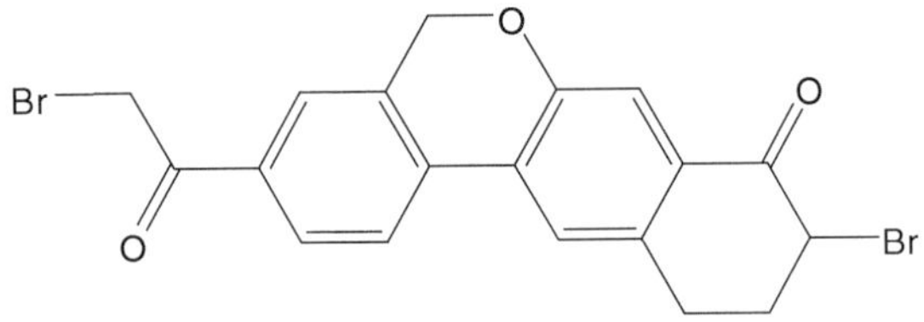

9-bromo-3-(2-bromoacetyl)-10,11-dihydro-5H-dibenzo[c,g]chromen-8(9H)-one

Verapamil

Cardiovascular Medicines/Antiarrhythmic Medicines

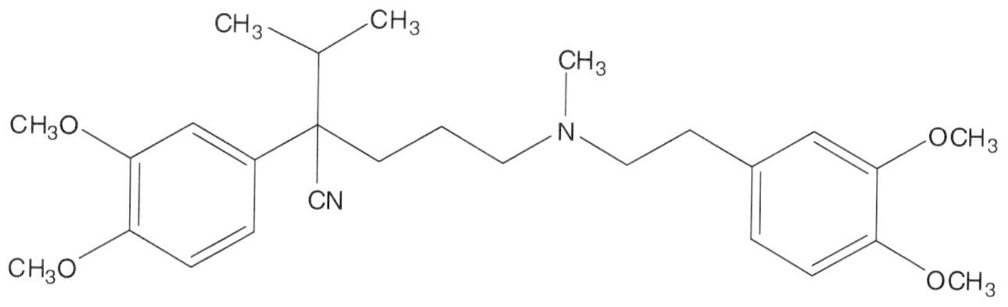

Tertiary amines are ubiquitous in drug structures. A tertiary amine is often formed by alkylation of a secondary amine. The alkylation is most efficient when the amine is used in excess and the carbon with the leaving group (Cl, Br, I, OTs, OMs) is primary or benzylic.

Discussion. Verapamil is a 1 : 1 mixture of the (*R*)- and (*S*)-enantiomers. The quaternary and chiral carbon is formed in the final step by α–alkylation of the nitrile with the primary alkyl chloride.

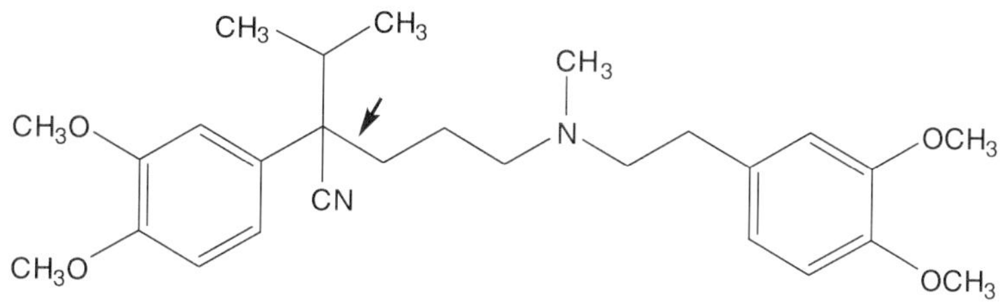

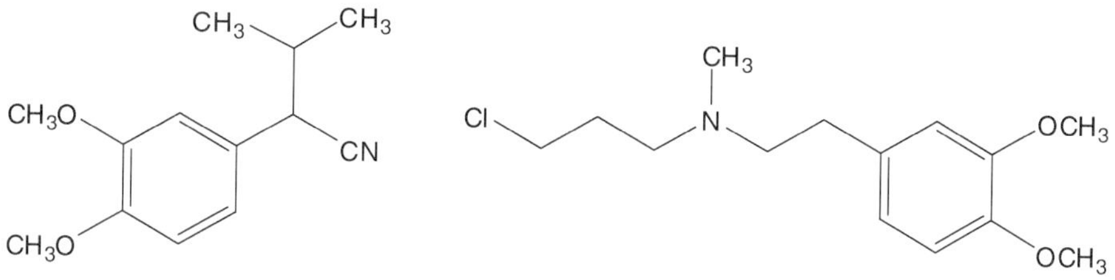

The nitrile, 2-(3,4-dimethoxyphenyl)-3-methylbutanenitrile (α-isopropyl veratryl cyanide), is formed by α–alkylation of (3,4-dimethoxyphenyl)acetonitrile (veratryl cyanide) with 2-bromopropane.

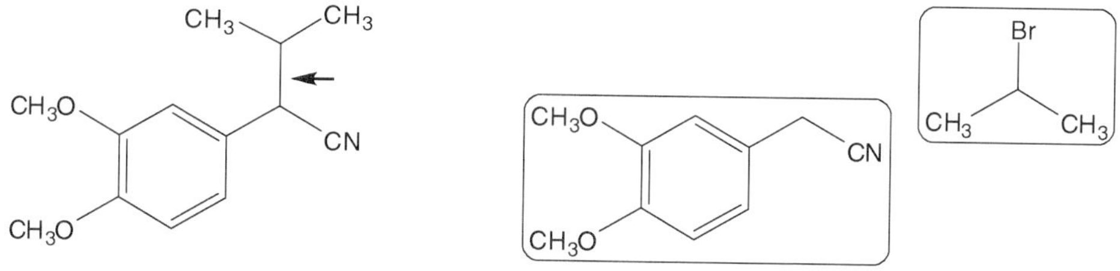

The alkyl chloride, *N*-(3-chloropropyl)-*N*-methylhomoveratrylamine, is formed by alkylation of the secondary amine *N*-methylhomoveratrylamine with 1-bromo-3-chloropropane. (Draw the structures of two side products formed in this reaction.) *N*-Methylhomoveratrylamine is formed from veratryl cyanide and methylamine under hydrogenation conditions.

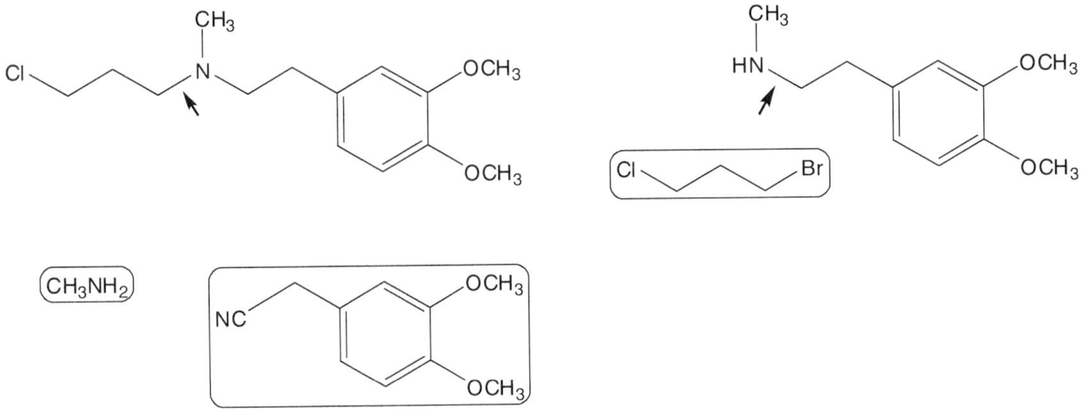

Extended Discussion

Draw the structures of the retrosynthetic analysis of an alternative route to verapamil using the disconnection shown in the final step. Include the structures of the retrosynthetic analysis of any organic starting material(s) from petrochemical or biochemical raw materials. List the pros and cons for both routes. Is one route preferred?

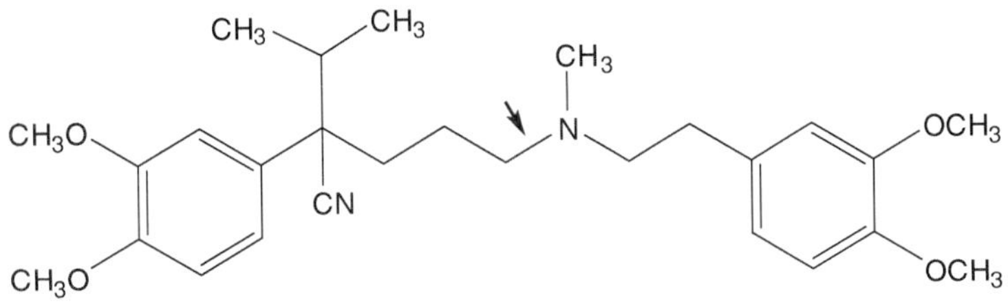

Vincristine

Antineoplastics and Immunosuppressives/Cytotoxic and Adjuvant Medicines

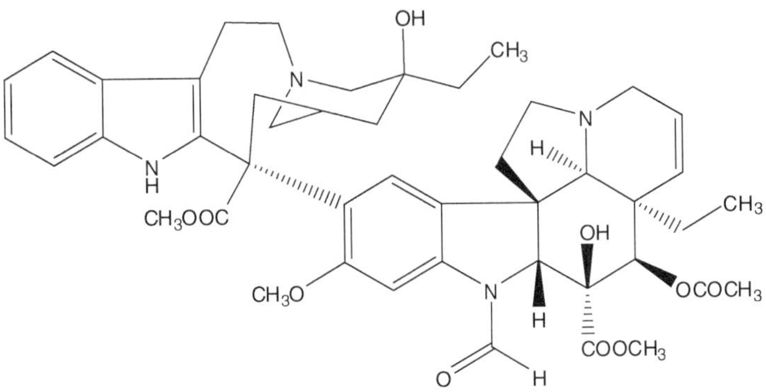

A single-enantiomer molecule with multiple chiral carbons is often formed by the modification of a natural product which has most or all of the chiral carbons already in place.

Discussion. Essential medicines vinblastine and vincristine are alkaloids isolated from the dried leaves of the Madagascar periwinkle *C. roseus*. With isolated yields of 0.01% for vinblastine and just 0.0003% for vincristine, there is an economic incentive to manufacture semisynthetic vincristine by oxidation of vinblastine.

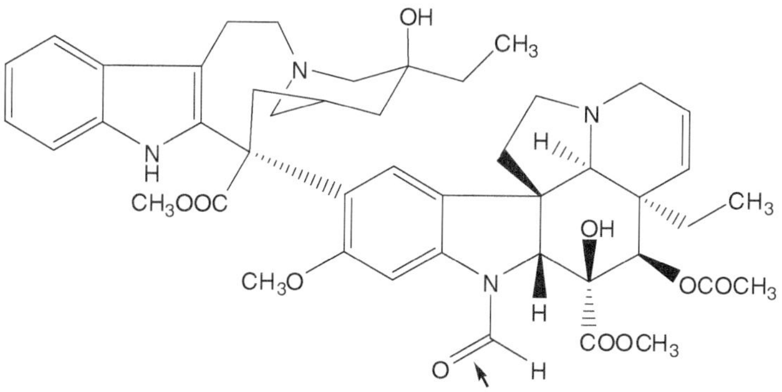

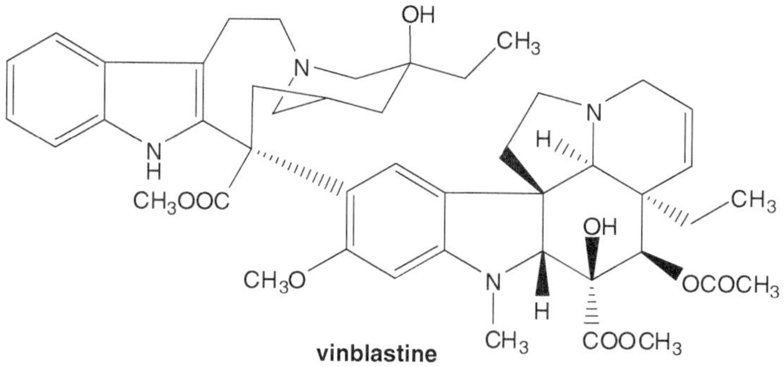

vinblastine

Vinorelbine

Antineoplastics and Immunosuppressives/Cytotoxic and Adjuvant Medicines

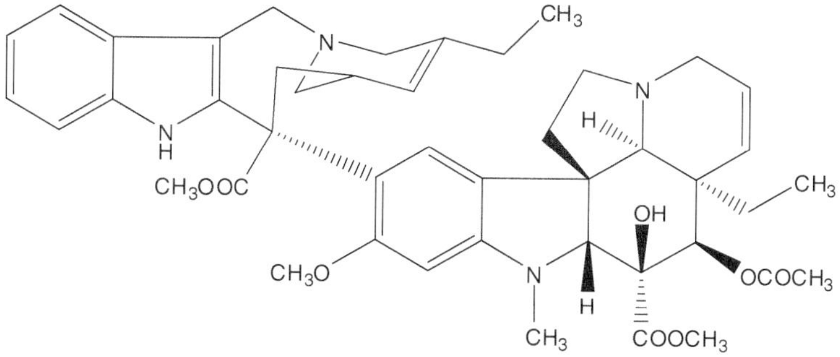

A single-enantiomer molecule with multiple chiral carbons is often formed by modification of a natural product which has most or all of the chiral carbons already in place.

Discussion. Vinorelbine (5′-noranhydrovinblastine) is semisynthetic. In the final step, vinorelbine is formed by loss of C5’ from the 5′,6′-bridge of 7′-bromoanhydrovinblastine. (What is the fate of the lost 5′ carbon?) 7′-Bromoanhydrovinbastine is formed by the bromination of anhydrovinblastine.

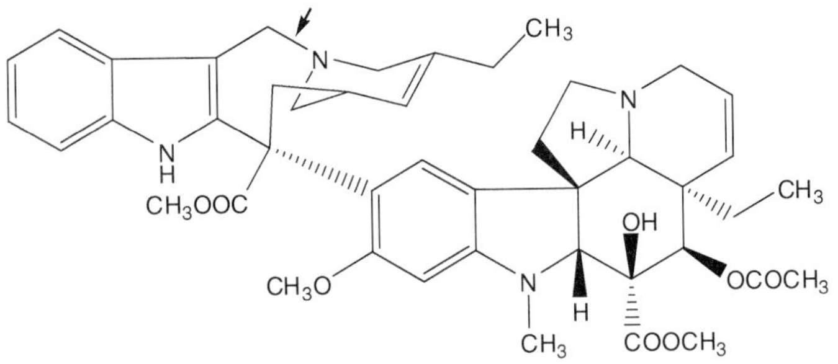

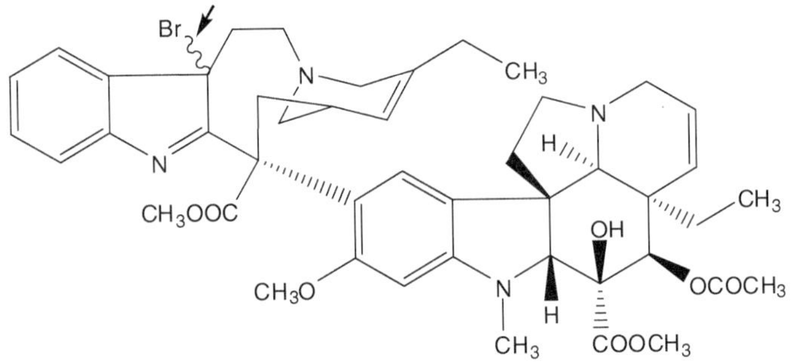

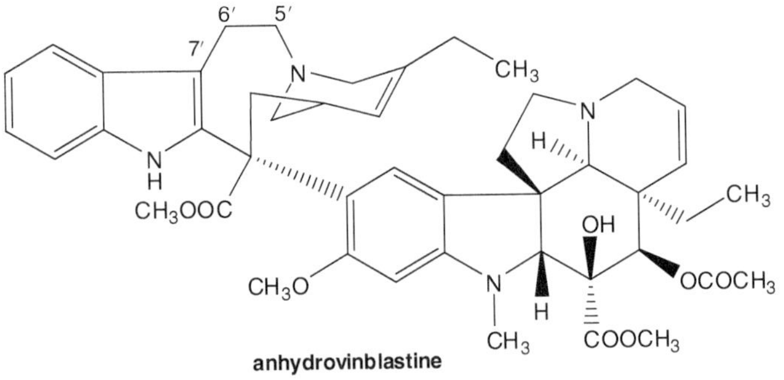

anhydrovinblastine

Anhydrovinblastine is formed by the iron(III) chloride-mediated coupling of vindoline and catharanthine followed by reduction of the resulting iminium ion with borohydride. (Draw the structure of the iminium ion intermediate.) Indole alkaloids catharanthine and vindoline are isolated from the leaves of the Madagascar periwinkle *Catharanthus roseus*.

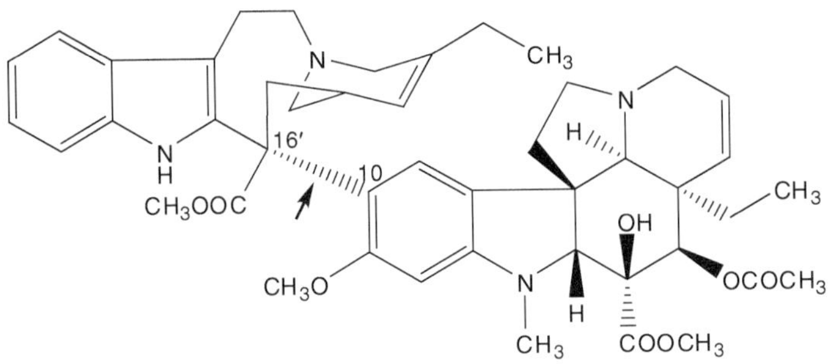

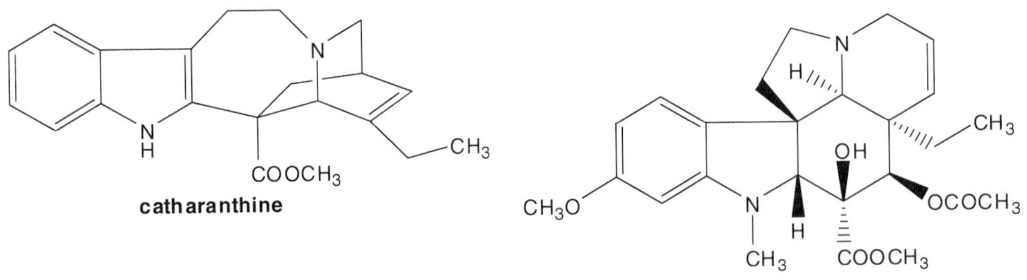

catharanthine

vindoline

Extended Discussion

Draw the structures of the retrosynthetic analysis of one alternative route to vinorelbine utilizing a Polonovski-type fragmentation of an *N*-oxide. List the pros and cons for both routes. Is one route preferred?

Voriconazole

Anti-infective Medicines/Antifungal Medicines

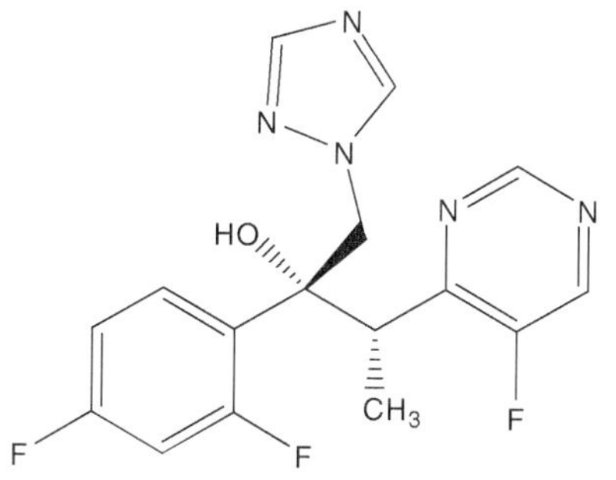

A tertiary alcohol is often formed by the addition of an organometallic reagent RM or RMX (M = Li, Mg, Zn, X = Cl, Br, I) to a ketone.

Discussion. Voriconazole, the (2*R*,3*S*)-enantiomer, is separated from the 1 : 1 mixture of (2*R*,3*S*)- and (2*S*,3*R*)-enantiomers by resolution in the final step. (What chiral acid is used for the resolution?) The 1 : 1 mixture of enantiomers for the final step is formed by hydrogenolysis of the 6-chloropyrimidines. A mixture of four stereoisomers is formed by the addition of the alkylzinc bromide to the ketone. The desired (2*R*,3*S*)- and (2*S*,3*R*)-enantiomer pair is isolated from the mixture by crystallization. (What reagents and reaction conditions are associated with the highest yield of the (2*R*,3*S*)- and (2*S*,3*R*)-enantiomers?) The alkylzinc bromide is formed from 4-(1-bromoethyl)-6-chloro-5-fluoropyrimidine.

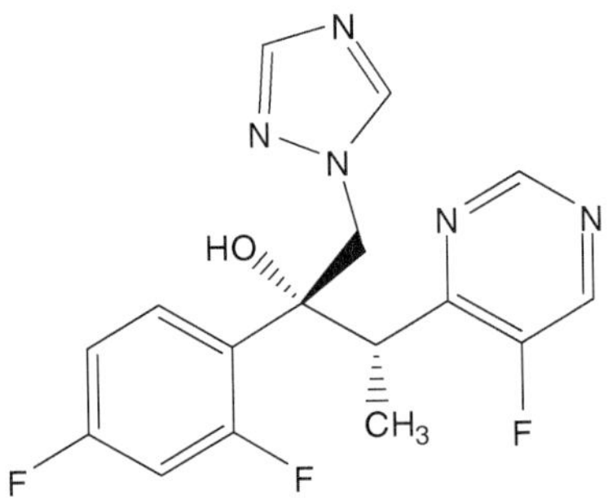

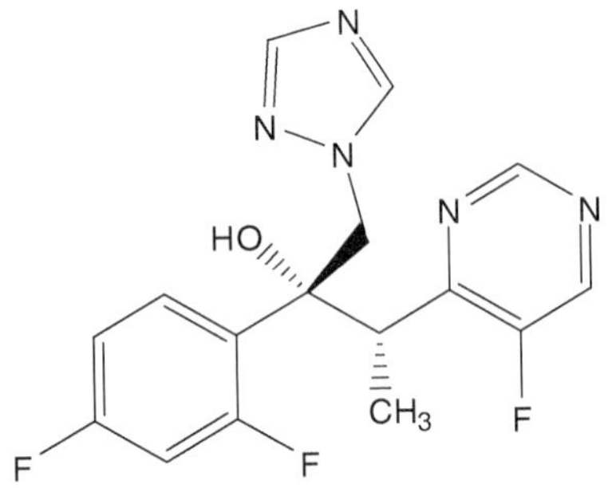

2R,3S

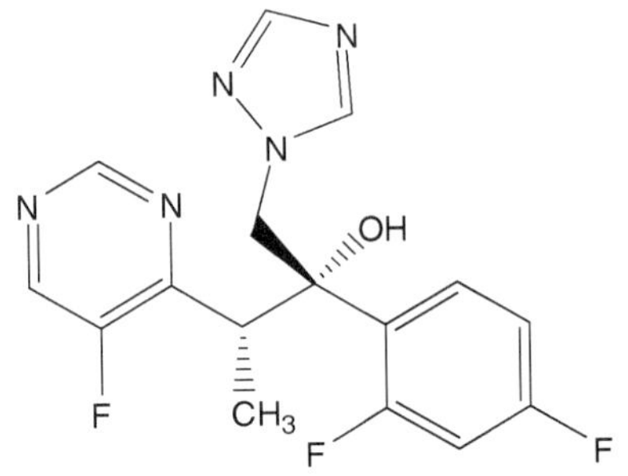

2S,3R

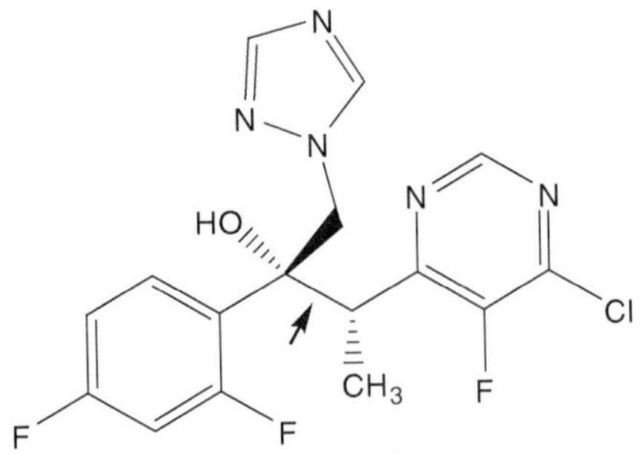

2R,3S

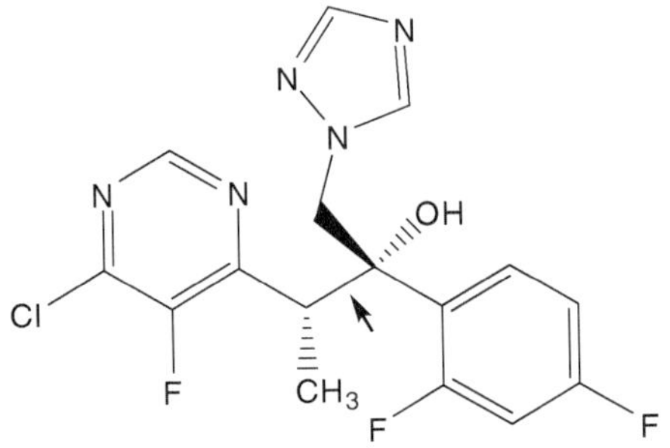

2S,3R

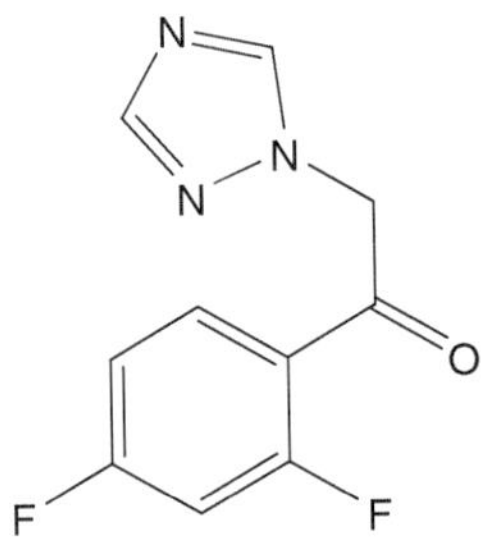

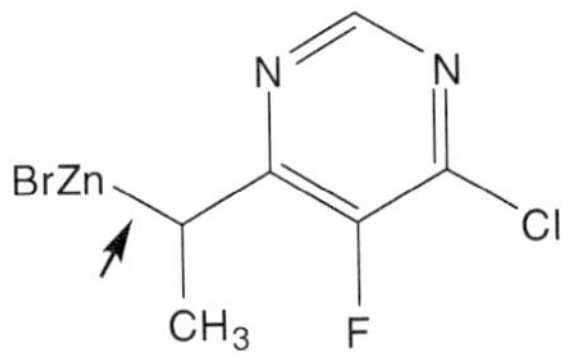

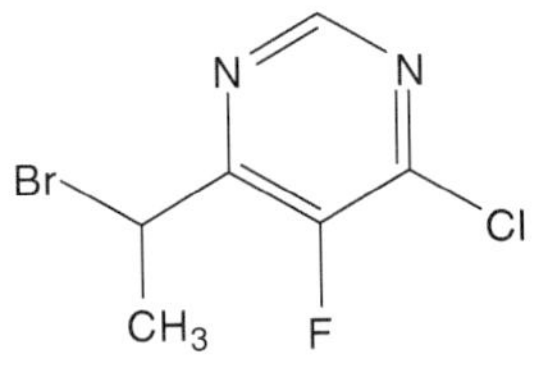

4-(1-Bromoethyl)-6-chloro-5-fluoropyrimidine is formed by bromination of 4-chloro-6-ethyl-5-fluoropyrimidine. The chloropyrimidine is formed from the hydroxypyrimidine/pyrimidinone. The pyrimidine ring is formed by the reaction of the β-ketoester with formamidine (**Pinner Pyrimidine Synthesis**). Methyl 2-fluoro-3-oxopentanoate is formed by fluorination of methyl 3-oxopentanoate. One route to methyl 3-oxopentanoate is from Meldrum's Acid and propanoyl chloride.

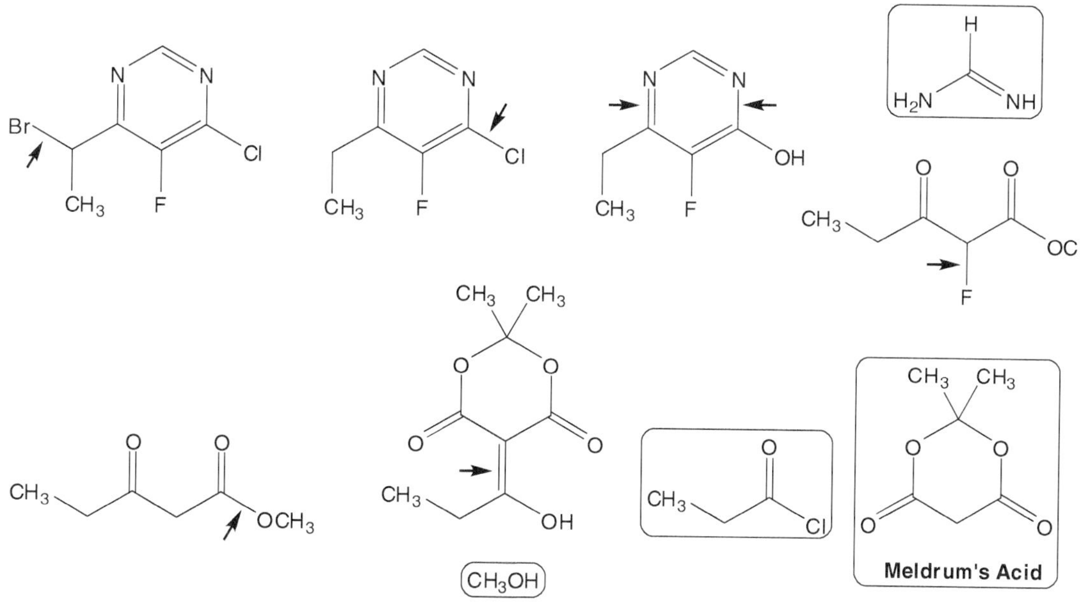

The ketone, 2′,4′-difluoro-2-(1,2,4-triazolyl)acetophenone, is formed by displacement of chloride from the α-chloroketone by 1,2,4-triazole. 2-Chloro-2′,4′-difluoroacetophenone is formed from 1,3-difluorobenzene and chloroacetyl chloride (**Friedel-Crafts Acylation**).

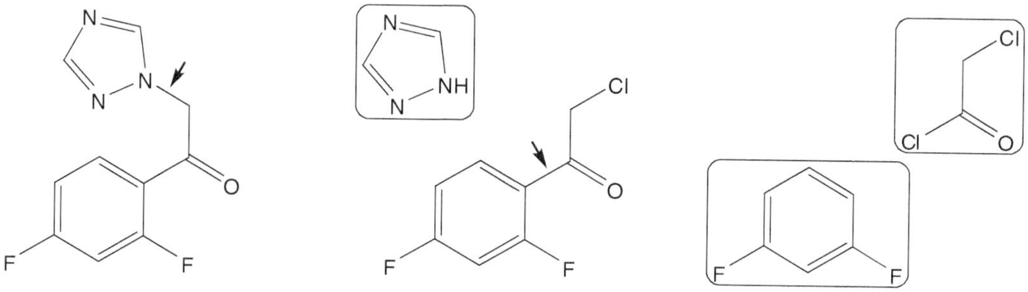

Extended Discussion

Provide details for one alternative route to voriconazole from 4-chloro-6-ethyl-5-fluoropyrimidine and 2′,4′-difluoro-2-(1,2,4-triazolyl)acetophenone. List the pros and cons for both routes.

W

Warfarin

Medicines Affecting the Blood/Medicines Affecting Coagulation

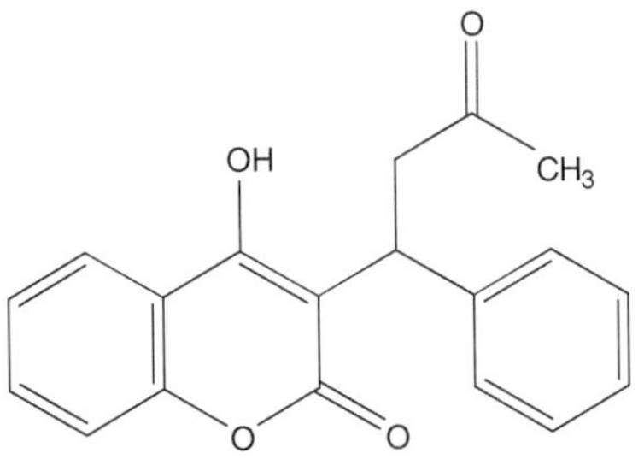

A β-branched ketone is often formed by addition of a carbon nucleophile to an α,β-unsaturated ketone.

Discussion. Warfarin is a 1:1 mixture of the (*R*)- and (*S*)-enantiomers. Warfarin is formed by the reaction of 4-hydroxycoumarin with benzalacetone (**Michael Addition**). 4-Hydroxycoumarin is formed from 2′-hydroxyacetophenone and diethyl carbonate (mixed **Claisen Condensation**).

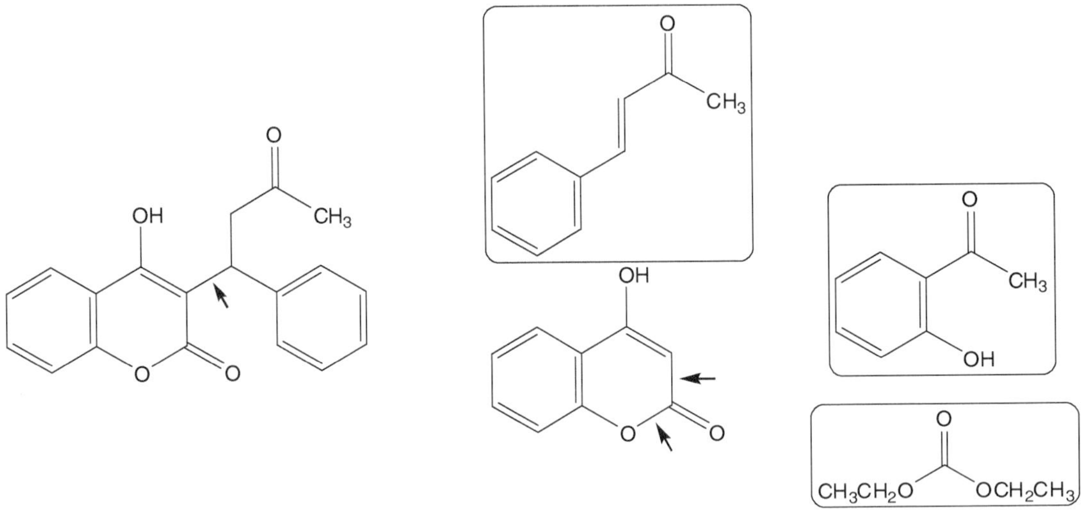

Extended Discussion

Draw the structures of the retrosynthetic analysis of one alternative route to 4-hydroxycoumarin. Include the structures of the retrosynthetic analysis of any organic starting material(s) from petrochemical or biochemical raw materials.

Routes to Essential Medicines: A Workbook for Organic Synthesis, First Edition. Peter J. Harrington.

Companion website: www.wiley.com/go/Harrington/routes_essential_medicine

X

Xylometazoline

Ear, Nose, and Throat Medicines

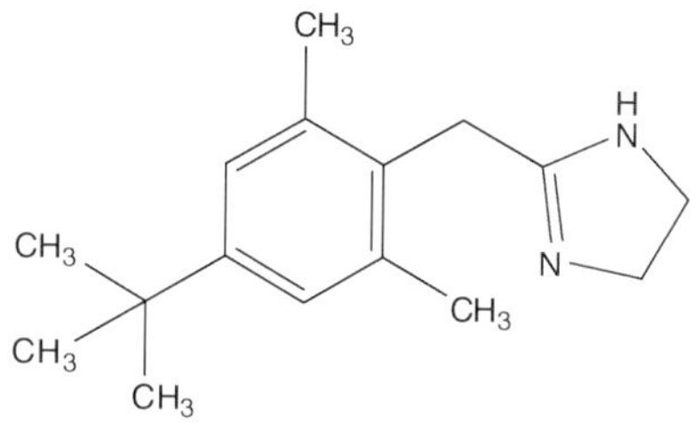

The large *tert*-butyl group of a *tert*-butylbenzene blocks the approach of reagents to the adjacent ring carbons.

Discussion. The imidazoline ring is formed by the reaction of the nitrile with ethylenediamine. 2-(4-*tert*-Butyl-2,6-dimethyl phenyl)acetonitrile is formed by displacement of chloride by cyanide. The benzyl chloride is formed by the reaction of 1-*tert*-butyl-3,5-dimethylbenzene with formaldehyde and hydrochloric acid (**Blanc Reaction**). 1-*tert*-Butyl-3,5-dimethyl benzene is formed by alkylation of *meta*-xylene with *tert*-butyl chloride (**Friedel–Crafts Alkylation**).

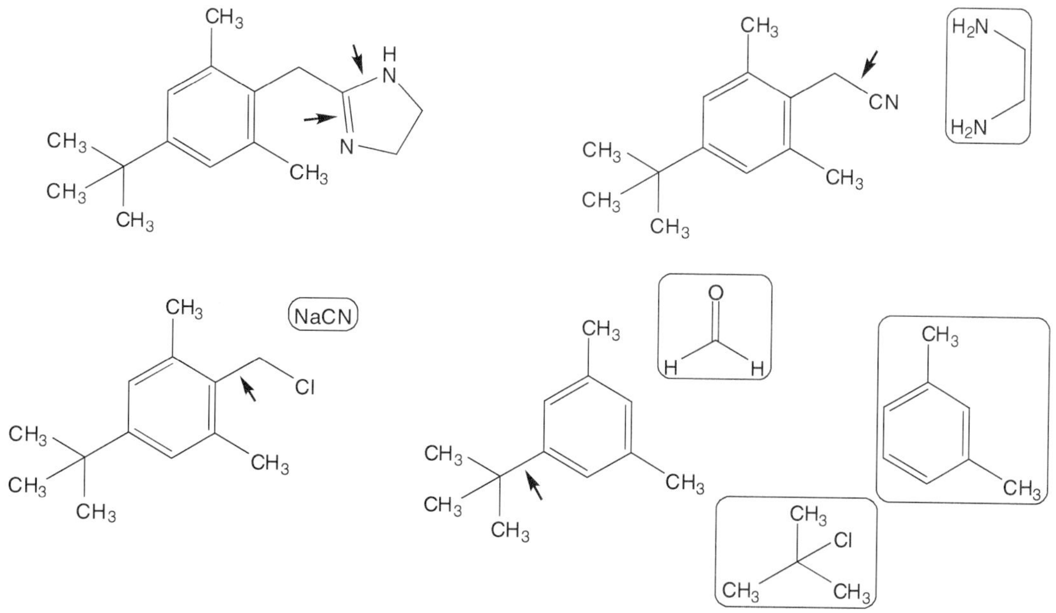

Routes to Essential Medicines: A Workbook for Organic Synthesis, First Edition. Peter J. Harrington.

Companion website: www.wiley.com/go/Harrington/routes_essential_medicine

Extended Discussion

Propose alternative reagent(s) for formation of the benzyl chloride from 1-*tert*-butyl-3,5-dimethylbenzene. Discuss pros and cons for these reagents.

Z

Zidovudine

Anti-infective Medicines/Antiviral Medicines/Antiretrovirals/Nucleoside-Nucleotide Reverse Transcriptase Inhibitors

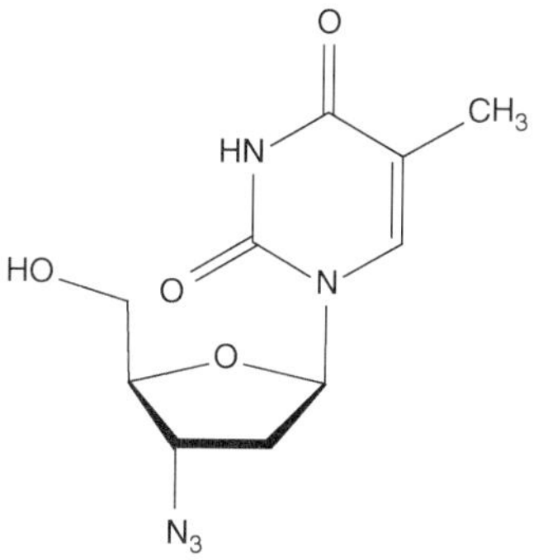

A single-enantiomer molecule with multiple chiral carbons is often formed by modification of a natural product which has most or all of the chiral carbons already in place.

Discussion. The primary alcohol is released by hydrolysis of the 5′-triphenylmethyl (trityl) ether in the final step. The 3′α-azide is formed by nucleophilic displacement of the β-oxygen of 5′-*O*-triphenylmethyl-2,3′-anhydrothymidine by sodium azide. The 2,3′-anhydrothymidine is formed by nucleophilic displacement of the methanesulfonate. The methanesulfonate is formed from the 3′-alcohol of 5′-*O*-tritylthymidine. 5′-*O*-Tritylthymidine is formed from thymidine.

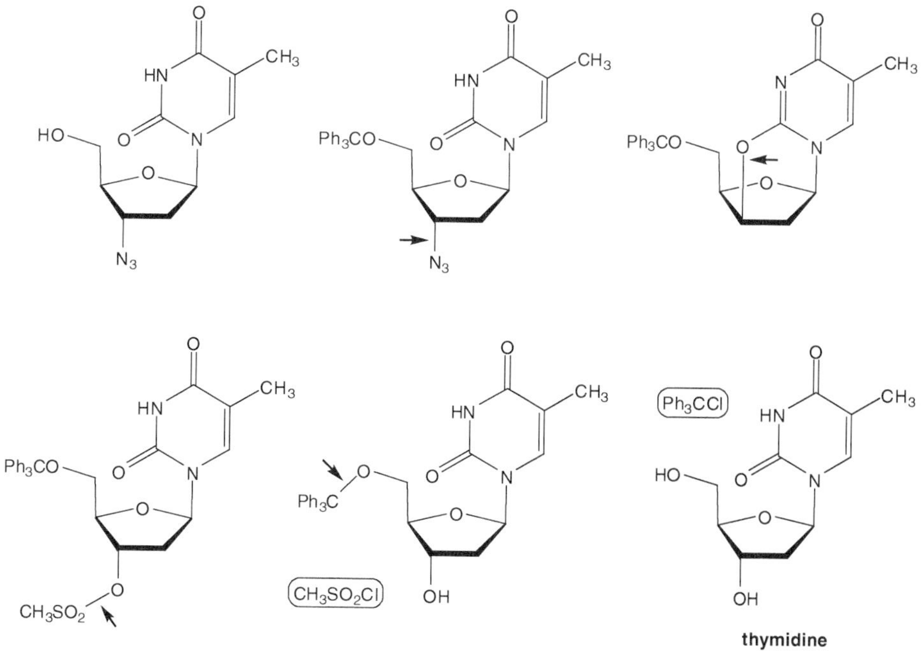

Routes to Essential Medicines: A Workbook for Organic Synthesis, First Edition. Peter J. Harrington.

Companion website: www.wiley.com/go/Harrington/routes_essential_medicine

Thymidine (β-thymidine) is obtained from natural sources or is produced by fermentation or synthesis. In one synthetic route, thymidine is manufactured from thymine via 5-methyluridine. Thymidine is formed when the alcohols are released by cleavage of the acetate esters. The 2′-bromide is cleaved by hydrogenolysis. The 3′ and 5′-acetate esters are formed, and the 2′-alcohol is replaced by bromide in the reaction of 5-methyluridine with acetyl bromide. 5-Methyluridine is formed when the alcohols are released by cleavage of the benzoate (Bz) esters. 2′,3′,5′-Tri-*O*-benzoyl-5-methyluridine is formed by the reaction of 1-*O*-acetyl-2,3,5-tri-*O*-benzoyl-β-D-ribofuranose with thymine.

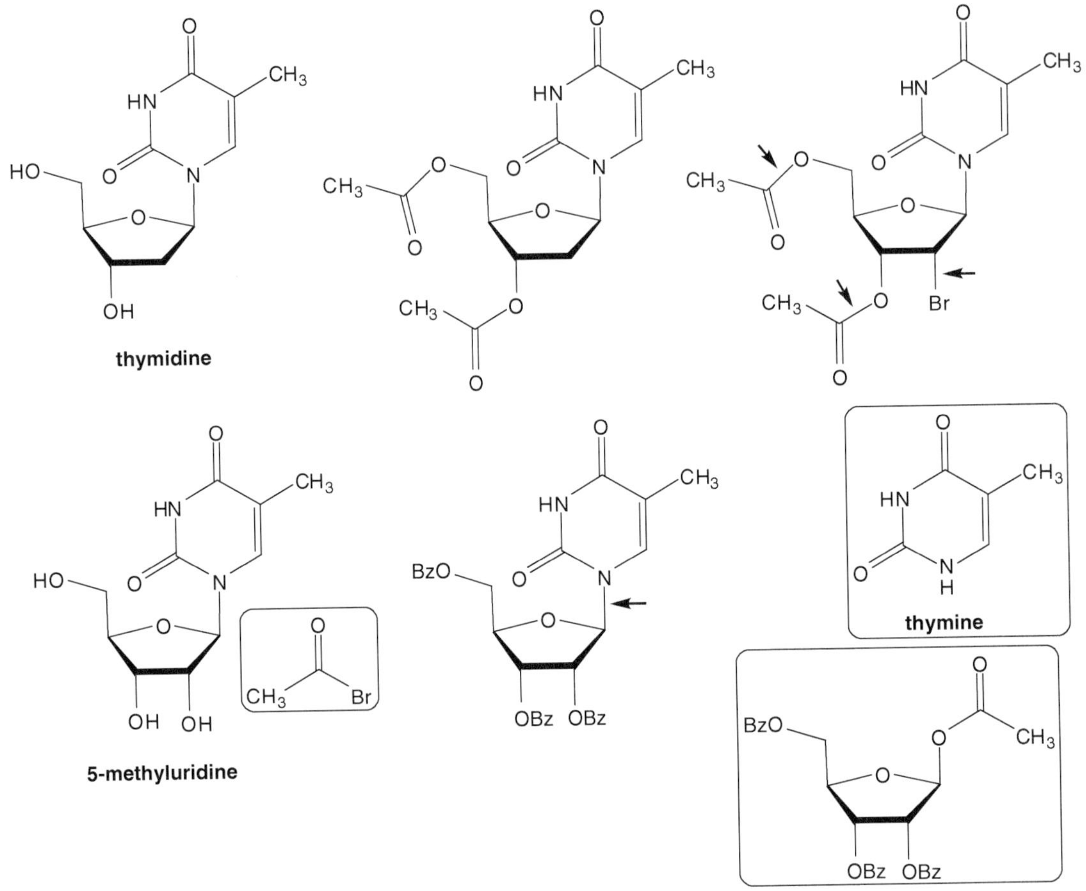

Extended Discussion

How is thymine produced? Draw the structures of a retrosynthetic analysis of one synthetic route to thymine. Include the structures of the retrosynthetic analysis of any organic starting material(s) from petrochemical or biochemical raw materials.

Zoledronic Acid

Antineoplastics and Immunosuppressives/Cytotoxic and Adjuvant Medicines

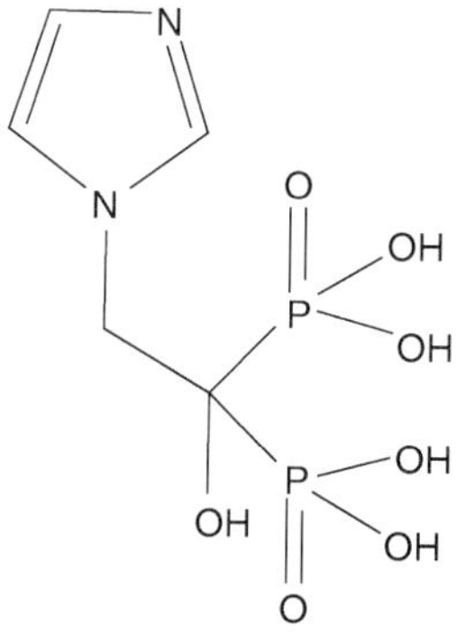

A 1-hydroxy bisphosphonic acid is often formed from the carboxylic acid.

Discussion. Zoledronic acid is formed from the carboxylic acid, 2-(imidazole-1-yl)acetic acid. (There are many options for the phosphorus reagent(s) and the solvent or diluent for this reaction. List the options and the yields associated with each option.) The carboxylic acid is formed by hydrolysis of the methyl ester. Methyl 2-(imidazole-1-yl)acetate is formed by chloride displacement from methyl chloroacetate by imidazole.

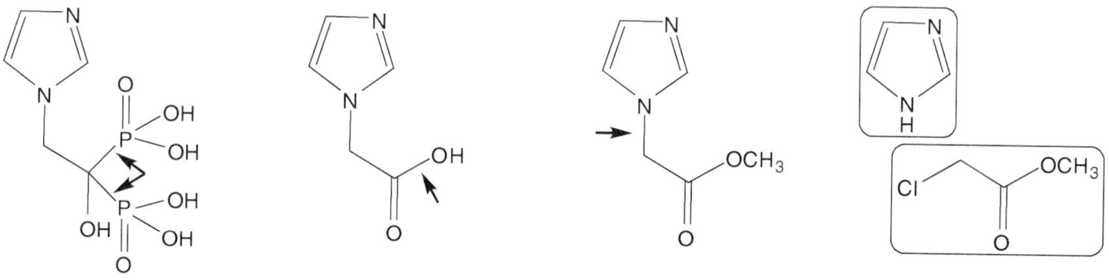

A Demonstration: Amiodarone

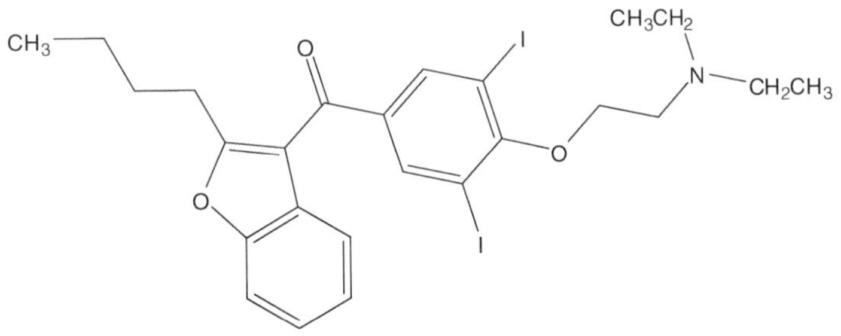

Primary references for the route to amiodarone can be located by many strategies. One strategy using Google search will be presented as a demonstration.

(Each discussion participant could map their search strategy as part of the exercise.)

Search 1: chloroacetonitrile phenol Sugasawa

US4551554

US, EP and WO patents are available for free download at freepatentsonline.com

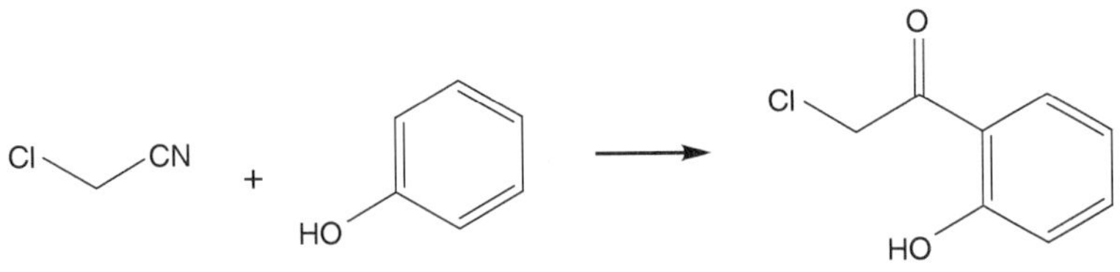

2-Chloro-2′-hydroxyacetophenone (US4551554) [1]:
Phenol, 1.2 equivalents chloroacetonitrile, 1,2 equivalents boron trichloride, 0.5 equivalents aluminum chloride, 1,2-dichloroethane, reflux for 6 hours, then H_2O, 86% yield

Search 2: 2-butylbenzofuran

PubChem from the US National Library of Medicine

patents, literature, and vendors

Molbase site for chemical e-commerce headquartered in Shanghai, China

synthesis routes, precursors and products, reference prices

Routes include the route from 2-chloro-2′-hydroxyacetophenone (*Org. Lett.* **2010**, *12*, 4972) [2].

Routes to Essential Medicines: A Workbook for Organic Synthesis, First Edition. Peter J. Harrington.

Companion website: www.wiley.com/go/Harrington/routes_essential_medicine

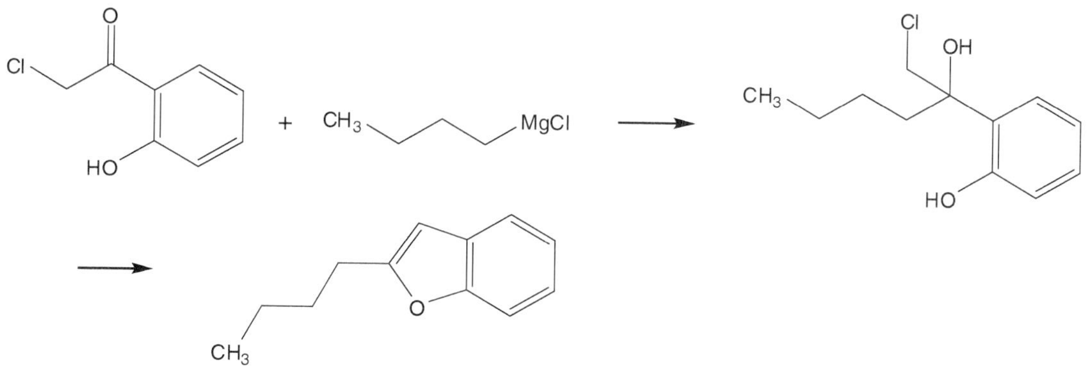

2-Butylbenzofuran (*Org. Lett.* **2010**, *12*, 4972) [2]
2-Chloro-2′-hydroxyacetophenone, 3.0 equivalents butylmagnesium chloride, THF-toluene, −10 to 25 °C for 1 hour, then HCl, isopropanol, 55 °C for 1 hour, 66% yield.

Search 3: *n*-butylmagnesium chloride preparation
Identify one source of a detailed description of this high-yielding procedure.
Handbook of Grignard Reagents, G.S. Silverman and P.E. Rakita, eds., Marcel-Dekker, New York, **1996**, pp. 15–17.

***n*-Butylmagnesium chloride** (*Handbook of Grignard Reagents*) [3]
1-Chlorobutane, 1.1 equivalents magnesium, 0.01 equivalents 1,2-dibromoethane, THF, reflux for 2 hours, settle and decant, 96% yield.

Search 4: 4-methoxybenzoyl chloride
Identify one source of a detailed description of this high-yielding procedure.
PubChem 4-Methoxybenzoyl chloride is formed by the reaction of *para*-anisic acid with thionyl chloride.
Molbase Routes include a procedure from *para*-anisic acid (*J. Med. Chem.* **2012**, *55*, 4189) [4].

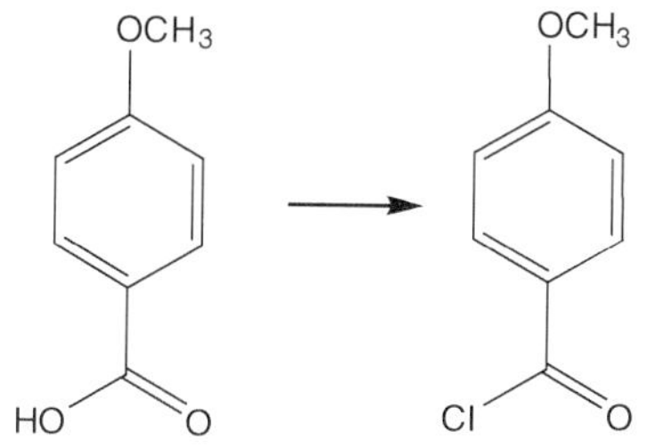

4-Methoxybenzoyl chloride (*J. Med. Chem.* **2012**, *55*, 4189) [4]
para-Anisic acid, excess thionyl chloride, reflux, then distill thionyl chloride, add chlorobenzene, distill thionyl chloride and chlorobenzene, assumed yield 99%.

Search 5: amiodarone synthesis
This search yields many pertinent references including Chinese patent CN104262304 [5]. Two US patents cited in this patent are US4766223 and US5266711 [6, 7]. These two patents have procedures for all the remaining steps in the amiodarone synthesis.

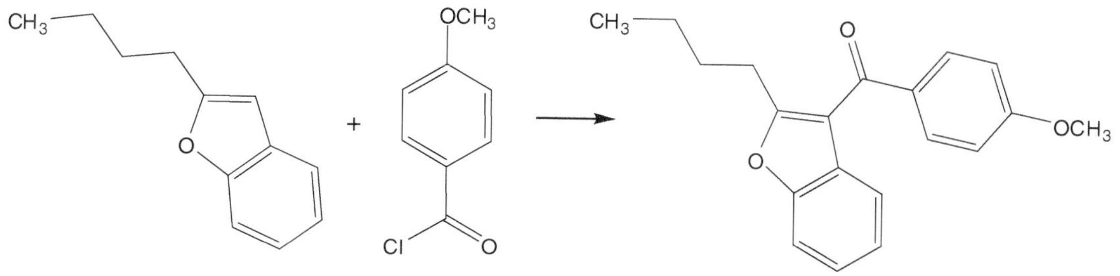

2-Butyl-3-(4-methoxybenzoyl)benzofuran (US4766223) [6]
2-Butylbenzofuran, 1.02 equivalents *para*-anisoyl chloride, 0.8 equivalents ferric chloride, toluene, −10 to 25 °C for 6 hours, then H_2O, 20 °C, 100% yield.

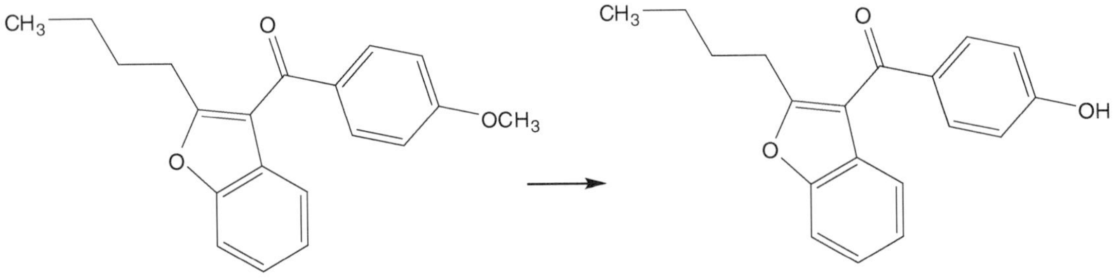

2-Butyl-3-(4-hydroxybenzoyl)benzofuran (US5266711) [7]
2-Butyl-3-(4-methoxybenzoyl)benzofuran, 2.1 equivalents aluminum chloride, 1,2-dichloroethane, reflux for 9 hours, then H_2O, 100% yield.

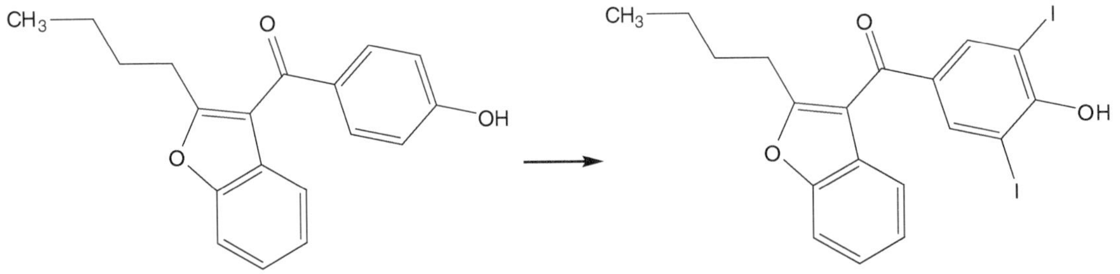

2-Butyl-3-(4-hydroxy-3,5-diiodobenzoyl)benzofuran (US5266711) [7]
2-Butyl-3-(4-hydroxybenzoyl)benzofuran, 2.4 equivalents sodium acetate trihydrate, 2.3 equivalents (small excess) iodine, methanol, 30 °C to reflux, then sodium hydroxide, reflux for 2 hours, sodium bisulfite, aqueous HCl–toluene, carry toluene solution to next step, assume yield 100%.

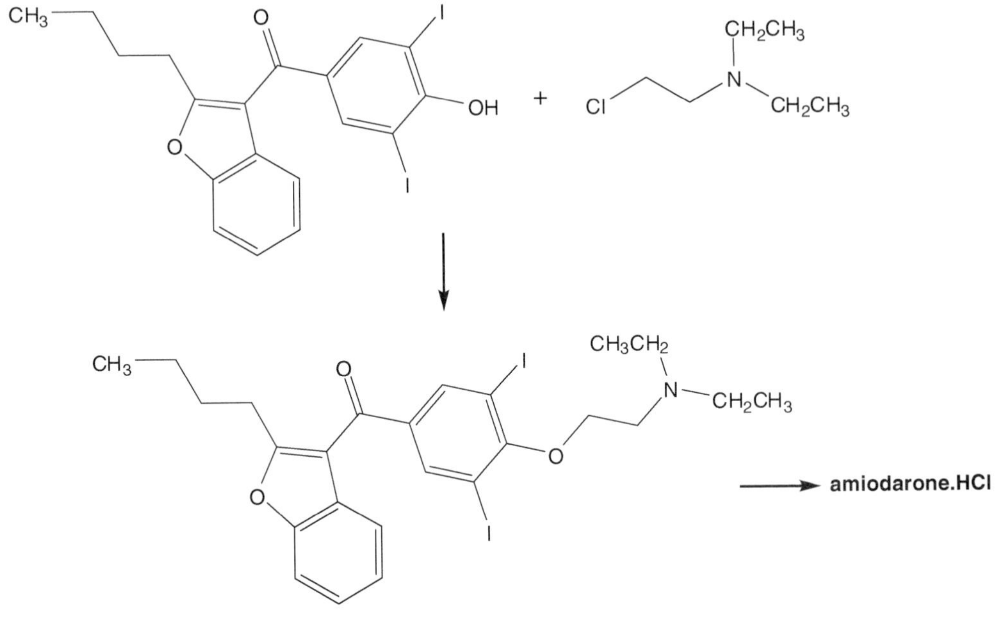

Amiodarone (US5266711) [7]
2-Butyl-3-(4-hydroxy-3,5-diiodobenzoyl)benzofuran in toluene, 1.03 equivalents 2-chloro-N,N-diethylethanamine hydrochloride, H_20, potassium carbonate, 40 °C to reflux, carry toluene solution to next step, assume yield 100%.
Amiodarone hydrochloride (US5266711) [7]
Amiodarone in toluene, hydrogen chloride, 60–75 °C, distill toluene/water to crystallize, 95% yield for three steps from 2-butyl-3-(4-hydroxybenzoyl)benzofuran.

The routes to specialty chemical starting materials (**Appendix A**) can be included in the discussion.
Search 6: chloroacetonitrile
PubChem Use and Manufacturing.
Search 7: phenol
PubChem Use and Manufacturing.
Search 8: *N,N*-diethyl-2-chloroethanamine

Search 9: *para*-anisic acid
guidechem.com

Route Summary

Seven steps from **Appendix A** specialty chemicals
Overall yield 54% from phenol
Lowest yield step is Step 2 (2-butylbenzofuran, 66%)

Proposed Mechanism for 1,2-Aryl Migration in Route to 2-Butylbenzofuran

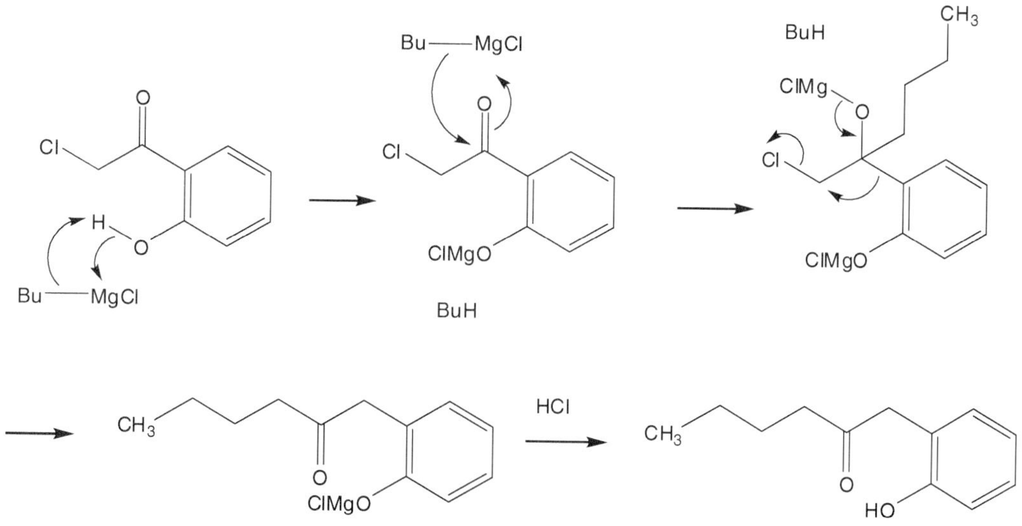

Amiodarone Synthesis Ideas

An aromatic ketone is often formed by Friedel–Crafts Acylation.
A phenyl ether is often formed by alkylation of a phenol.
An *ortho*-iodophenol is often formed by iodination of a phenol.
A phenol is often formed by demethylation of a methyl ether.
A 2-substituted benzofuran is often acylated at C3.

Extended Discussion: Route B to 2-Butylbenzofuran

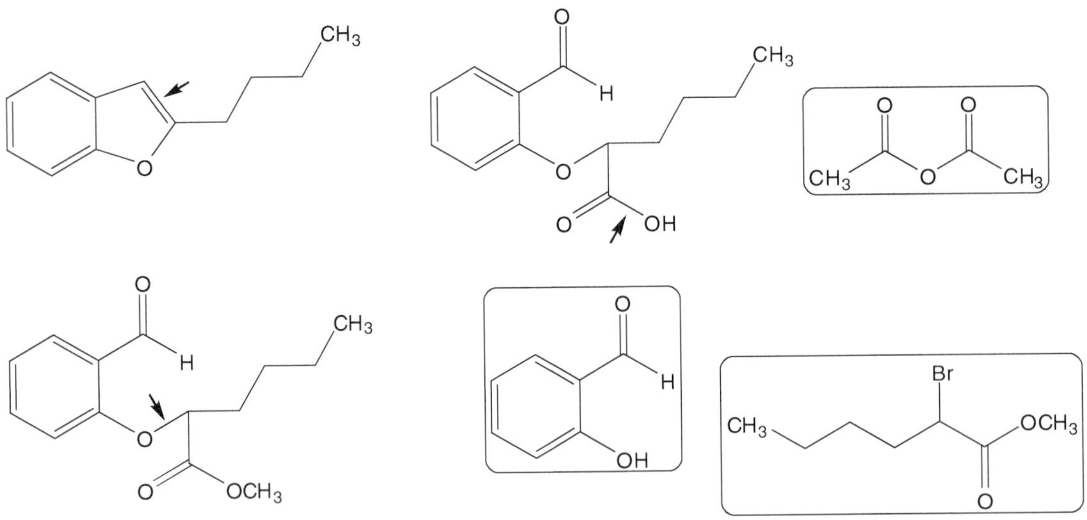

Starting material salicylaldehyde is in **Appendix A**. Starting material methyl 2-bromohexanoate could be added to **Appendix A**. Yields are high for the bromide displacement (99%) and ester hydrolysis (98%) [8]. The reaction of 2-(2-formylphenoxy)hexanoic acid with acetic anhydride likely affords 2-butylbenzofuran in 50–60% yield [9–11].

Methyl 2-bromohexanoate is formed by bromination of hexanoic acid (**Hell–Volhard–Zelinski Reaction**) followed by ester formation. Hexanoic acid is produced by fermentation.

Add to **Appendix A**:

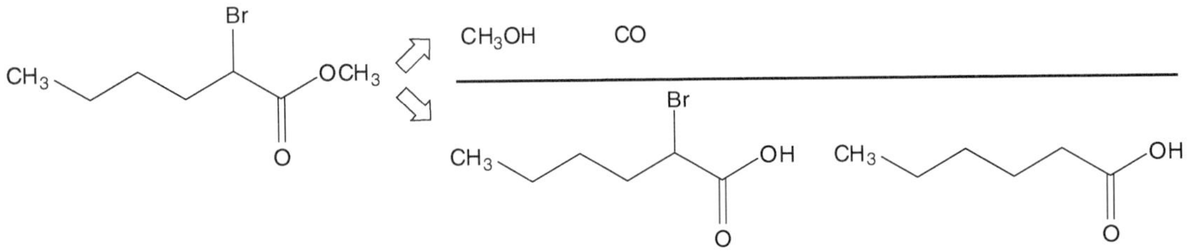

References

1 Sugasawa, T.; Toyoda, T.; Sasakura, K. U. S. Patent 4,551,554 (11/5/1985).
2 Pei, T., Chen, C.-Y., DiMichele, L., and Davies, I.W. (2010). *Org. Lett.* 12: 4972.
3 Silverman, G.S. and Rakita, P.E. (eds.) (1996). *Handbook of Grignard Reagents*, 15–17. New York: Marcel-Dekker.
4 Keenan, M., Abbott, M.J., Alexander, P.W. et al. (2012). *J. Med. Chem.* 55: 4189.
5 CN1042622304 (2014).
6 Grain, C.; Jammot, F. U. S. Patent 4,766,223 (8/23/1988).
7 Boudet, B.; Dormoy, J.R.; Heymes, A. U. S. Patent 5,266,711 (11/30/1993).
8 Schlama, T., Mettling, A.; Karrer, P. U. S. Patent 6,855,842 (2/15/2005).
9 Burgstahler, A.W. and Worden, L.R. (1973). *Org. Syn. Coll.* 5: 251.
10 Miszczyszyn, M. and Kwiecien, H. (2002). *Pol. J. Appl. Chem.* 46: 21.
11 Schlama, T. U. S. Patent 6,555,697 (4/29/2003).

Pros and Cons for Routes A and B to 2-Butylbenzofuran

Early in the course a pros and cons discussion may begin with the overall yield, number of steps, reaction times, and costs for starting materials. As the course progresses, other topics including physical properties of starting materials, reaction and workup solvents, throughput, safety, and environmental impact of byproduct(s) can be added.

Yield of 2-butylbenzofuran from materials found in **Appendix A**:

A) 57%
B) 50–60%

Longest sequence from materials found in **Appendix A** to 2-butylbenzofuran:

A) 3 reactions in 2 steps
B) 3 reactions in 3 steps

Longest reaction times:

A) 6 hours to 2-chloro-2′-hydroxyacetophenone
B) 5–8 hours (5 hours to 2-butyl-5-nitrobenzofuran, 8 hours to benzofuran)

Relative costs for carbon-containing starting materials:

A) phenol, chloroacetonitrile, 1-chlorobutane
B) salicylaldehyde, methyl 2-bromohexanoate

Salicylaldehyde is made from and more expensive than phenol. One reference price for methyl 2-bromohexanoate is $141/kg or $30/mol (Molbase, April 2020). This price, price estimates ($/mol) for chloroacetonitrile and chlorobutane, and a comparison of the routes to produce methyl 2-bromohexanoate, chloroacetonitrile, and 1-chlorobutane suggest that the starting material costs are lower for Route A than Route B. Based on lower costs for carbon-containing starting materials, Route A to 2-butylbenzofuran is the preferred route.

Index

Routes to Essential Medicines: A Workbook for Organic Synthesis, First Edition. Peter J. Harrington.

Companion website: www.wiley.com/go/Harrington/routes_essential_medicine

n

r